67-99

HIGHER
ELECTRICAL
ENGINEERING

Higher Electrical Engineering

J. SHEPHERD

B.Sc., C.Eng., F.I.E.E.
Department of Electrical Engineering
University of Edinburgh

A. H. MORTON

B.Sc., C.Eng., F.I.E.E., F.I.E.R.E.
Department of Electrical Engineering
Paisley College of Technology

L. F. SPENCE

B.Sc., B.Sc.(Econ.), C.Eng., F.I.E.E.
Department of Electrical and Electronic Engineering
Polytechnic of the South Bank

SECOND EDITION

 LONGMAN

Addison Wesley Longman Limited
Edinburgh Gate
Harlow
Essex CM20 2JE, England
and Associated Companies throughout the world

First published in Great Britain by Pitman 1958
Second Edition 1970
Reprinted 1973, 1975, 1977, 1978, 1981, 1982, 1983, 1985
Reprinted by Longman Scientific & Technical 1986, 1988, 1989, 1990, 1993, 1995
Reprinted by Addison Wesley Longman 1997 and 1998

ISBN 0-582-98888-8

Set in 10/11 pt Monotype Times New Roman
Transferred to digital print on demand, 2002
Printed and bound by Antony Rowe Ltd, Eastbourne

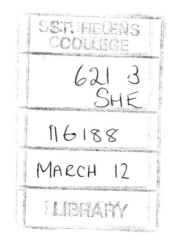

Preface

In preparing this second edition the basic aims of the first edition have not been changed. We have designed this volume to cover the A1 and A2 stages of the Higher National Certificate in Electrical and Electronic Engineering. It does not pretend to be exhaustive, nor specialist, but is designed for a broad background course, with stress laid on the fundamental aspects of each section. The contents will be found to correspond to much of the work in the recent Department of Education and Science outline syllabuses for H.N.C. Courses in England. It accords also with the Scottish H.N.C. scheme. In addition this volume should prove useful in undergraduate C.E.I. Part 1 and H.N.D. courses.

The text has been brought up to date, particularly in the sections on machines and electronics. An introduction to the generalized theory of machines has been given, but conventional treatments (substantially revised) of synchronous and induction machines have been retained. The positive convention for the sign of the voltage induced in a circuit due to a changing flux has been maintained, since further experience has shown that this concept has helped students (and staff) to a clearer understanding of the physical concept of induced e.m.f.

In electronics the shift of emphasis to semiconductor devices is reflected in the omission of valve circuits from large sections of the text, in favour of bipolar transistor circuits. The field-effect transistor is dealt with in a later chapter.

In an age where systems engineering is coming more into prominence it is important that electrical and electronic engineers have some knowledge of reliability. We are grateful to Mr. E. L. Topple of the Polytechnic of the South Bank for undertaking the task of preparing a chapter on this subject for us. A further chapter giving an introduction to logic has also been added.

As in the first edition we have included a large number of worked examples in the text. Problems (with answers) at the end of each chapter give the reader the opportunity of testing his understanding of the text as he proceeds. Thanks are due to the Senate of the University of London, and to the Scottish Association for National Certificates and Diplomas for their willingness to allow us to use examples from their examination papers (designated L.U. and H.N.C. respectively).

The new edition is larger than the authors had hoped and its preparation has not been entirely uncontroversial. Because of this the wife of one author has suggested that the book should be sub-titled "War and Peace"!

We should also like to record our thanks to colleagues who read the manuscript and undertook the corrections at the proof stage. These include in particular Mr. T. Grassie (now of Strathclyde University), Mr. W. R. M. Craig and Mr. A. McKenzie of Paisley College of Technology and Mr. G. Heywood of the Polytechnic of the South Bank. Thanks are also due to those who contributed to the massive typing effort required.

PAISLEY   J.S.
June, 1970   A.H.M.
                                             L.F.S.

Contents

Chapter 1

SYMBOLIC NOTATION

It is assumed that the reader is already familiar with the diagrams which are used to give the relationship between sinusoidal alternating currents and voltages in simple a.c. circuits. These *complexor diagrams* represent the magnitude of any quantity they depict as the length of a line, while the direction of the line gives phase information. Because of similarities in the manipulation of quantities represented in this way with vector methods of addition and subtraction these diagrams used to be called *vector diagrams*. In this book they will be called complexor diagrams, and the lines will be called complexors.*

Problems involving the manipulation of complexors may be solved by representing the complexors as algebraic expressions. The notation used for doing so is called the symbolic notation, and the advantage of the use of this notation is that the processes of manipulation become algebraic processes.

1.1 The Operator j

In complexor diagrams the direction of the X-axis is called the *reference direction*, since it is often used as the reference from which phase angles are measured. The direction of the Y-axis may be called the *quadrate direction*. Fig. 1.1 shows three typical complexors, V_1, V_2 and V_3. The lengths of the lines are proportional to the magnitudes of the quantities they represent. The phase angle of a complexor

* In recent years they have also been called *phasor diagrams* and *phasors*.

is represented as the angle turned through (anticlockwise) from the positive reference direction to the direction of the complexor. Note that the position of a complexor has no bearing on the magnitude

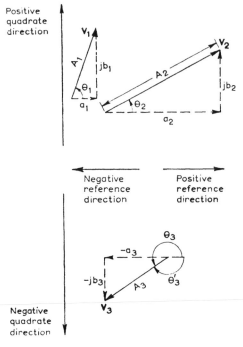

Fig. 1.1   TYPICAL COMPLEXORS

of the quantity it represents, so that in Fig. 1.1, for example, V_1 could be represented by any parallel line of the same length.

If the complexor diagram represents alternating currents or voltages the whole diagram may be assumed to rotate at a constant angular velocity. If the complexor is proportional to the peak value of the alternating quantity, its projection in a particular direction will give the instantaneous value of the quantity.

A complexor may be completely described by

(i) a statement of its magnitude with respect to a given scale unit; and

(ii) a statement of its phase with respect to a reference direction.

Thus in Fig. 1.1, $V_1 = A_1\underline{/\theta_1}$; $V_2 = A_2\underline{/\theta_2}$, etc. The magnitude is sometimes called the *modulus* and is represented by $|V|$ or V.

Note that the symbols for complexors are printed in **bold** type while those for magnitudes are printed in ordinary type.

If the phase angle is greater than 180°, the negative phase angle (i.e. the phase angle measured in a clockwise direction) is often stated for convenience. For example,

$$V_3 = A_3\underline{/\theta_3} \quad \text{or} \quad A_3\underline{/-\theta_3'} \quad \text{or} \quad A_3\overline{/\theta_3'}$$

where $\theta_3' = 360° - \theta_3$

This method of describing a complexor is termed *polar notation*. A and $\underline{/\theta}$ are called *operators*.

A, the magnitude operator or modulus, is the number by which the scale unit should be multiplied to give the magnitude of the complexor. $\underline{/\theta}$, the phase operator, is the anticlockwise angle through which a complexor in the reference direction must be turned in order to take up the direction of the given complexor.

It will be realized that the operations -1, and $\underline{/180°}$ or $\underline{/-180°}$ are identical. Hence $-A$ is a complexor in the negative reference direction.

The operation $\underline{/90°}$ or $\underline{/\pi/2}$ rad is found to occur frequently and is commonly represented by the symbol j:

$$j \equiv \underline{/90°} \tag{1.1}$$

i.e. j represents the operation of turning a complexor through 90° in an anticlockwise direction. Hence

$$jb = b\underline{/90°} \tag{1.2}$$

i.e. jb is a complexor of length b in the quadrate direction. In the same way,

$$-jb = -(jb) = -b\underline{/90°}$$

and is thus a complexor of length b in the negative quadrate direction.

It is very convenient to represent a complexor by the sum of two components, one of which is in either the positive or the negative reference direction, while the other is in either the positive or the negative quadrate direction. Thus, in Fig. 1.1,

$$V_1 = A_1\underline{/\theta_1} = a_1 + jb_1$$

where $a_1 = A_1 \cos \theta_1$ and $b_1 = A_1 \sin \theta_1$, and

$$V_3 = A_3\underline{/\theta_3} = -a_3 - jb_3$$

where

$$a_3 = -A_3 \cos \theta_3 = A_3 \cos (\pi - \theta_3)$$

and

$$b_3 = -A_3 \sin \theta_3 = -A_3 \sin (\pi - \theta_3)$$

The $(a + jb)$ method of describing a complexor is termed *rectangular notation*.

The above complexors may also be expressed in the form

$$V = A (\cos \theta + j \sin \theta) \tag{1.3}$$

this being termed the *trigonometric notation*.

From the geometry of the diagrams,

$$A = \sqrt{(a^2 + b^2)} \tag{1.4}$$

and

$$\theta = \tan^{-1} \frac{b}{a} \tag{1.5}$$

For example,

$$V_A = -3 + j2$$

may be expressed as

$$V_A = \sqrt{(3^2 + 2^2)} \bigg/ \underline{\tan^{-1} \frac{2}{-3}} = 3\cdot61\underline{/180° - 33\cdot6°}$$

Conversely,

$$V_B = 12\underline{/-60°}$$

may be expressed as

$$V_B = 12[\cos(-60°) + j \sin(-60°)] = 6 - j10\cdot39$$

In the preceding paragraphs A, $\underline{/\theta}$, a, b and j are all operators of various types. The combined expressions $A\underline{/\theta}$ and $(a + jb)$ are called complex operators in polar and rectangular forms respectively. A complexor may be expressed as a complex operator when a particular scale unit and reference direction are given.

In the conversion from rectangular to polar form for a complex number, or operator, the square root of the sum of the squares must be calculated. This operation may be conveniently performed on an ordinary slide-rule which has the usual A, B, C, and D scales. For example, suppose it is desired to find $\sqrt{(3^2 + 7^2)}$. Set the smaller of the two numbers (in this case 3) on the C scale against unity on the D scale. Move the cursor to the higher of the two numbers on the C scale, and read off the corresponding figure on the A scale at the top of the rule (5·46). Add one to this figure (giving 6·46). Set the cursor at this new number on the A scale, and read off the desired result on the C scale (7·62). If the two numbers are such that the numerically smaller has a bigger initial figure (e.g. $\sqrt{(7^2 + 12^2)}$) the smaller number is set against 10 on the D scale, the rest of the method being the same.

For numbers which differ by more than a factor of ten it is usually sufficient to take the square root of the sum of the squares as being simply the larger of the two numbers. Thus

$$|10 + j1| = \sqrt{101} \approx 10·05 \approx 10$$

SUCCESSIVE OPERATIONS BY j (Fig. 1.2)

Since j is defined as an operator which turns a complexor through $+90°$ without changing its size, two operations by j will turn a

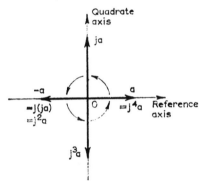

Fig. 1.2 SUCCESSIVE OPERATIONS BY j

complexor through a total of 180° from the original reference direction; i.e. the original direction is reversed. Thus

$$j(ja) = j^2a = -1 \times a \tag{1.6}$$

The operator j^2 is the 180° operator. It is convenient to think of j^2 as being algebraically the same as -1.

If now j^2a is operated on by j (written as $j(j^2a) = j^3a$) the original complexor $+a$ is turned through a total of 270°.

A further operation on j^3a by j brings the complexor back to its original position. Thus

$$j^4a = 1 \times a \tag{1.7}$$

and the operation of j^4 on a complexor leaves it unchanged in size and direction.

THE OPERATION $(-j)$

When a complexor is operated on by $-j$, the operation may be divided into two parts:

(i) operate on the complexor by -1; and
(ii) operate on the resulting complexor by j.

Thus the result is a positive rotation of 270°. Hence

$$-ja = j^3a$$

and the negative sign can be taken to mean that the rotation is clockwise. This conclusion is peculiar to the operator j, and does not apply to any other rotational operator.

1.2   Addition and Subtraction of Complex Operators

ADDITION

The complexors **OP** and **CQ** in Fig. 1.3 may be added graphically by

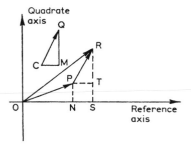

Fig. 1.3   ADDITION OF COMPLEXORS

placing them end to end in order. Let

$$\mathbf{ON} = a \qquad \mathbf{NP} = jb = \mathbf{ST}$$
$$\mathbf{CM} = c = \mathbf{PT} = \mathbf{NS} \qquad \mathbf{MQ} = jd = \mathbf{TR}$$

Then

$$\mathbf{OP} + \mathbf{CQ} = \mathbf{OS} + \mathbf{SR} = \mathbf{ON} + \mathbf{NS} + \mathbf{ST} + \mathbf{TR}$$

Therefore

$$(a + jb) + (c + jd) = a + c + jb + jd = (a + c) + j(b + d)$$

The rule for addition of complex operators is thus seen to be: *Add the reference and quadrate terms separately.* For example,

$$(7 + j9) + (8 - j12) = 15 - j3$$

and

$$(-3 + j7) + (-2 - j10) = -5 - j3$$

SUBTRACTION

If one complexor is to be subtracted from another the graphical method is to reverse the former and then add. Thus in Fig. 1.4,

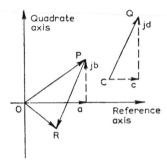

Fig. 1.4   SUBTRACTION OF COMPLEXORS

let $\mathbf{OP} = a + jb$ and $\mathbf{CQ} = c + jd$. Then

$$\mathbf{OP} - \mathbf{CQ} = \mathbf{OP} + \mathbf{PR} = (a + jb) + (-c - jd)$$
$$= (a - c) + j(b - d) = \mathbf{OR}$$

The rule is: *To subtract one complex operator from another, subtract the reference and the quadrate terms separately.*

It should be noted that for both addition and subtraction the normal algebraic rules for signs will operate. For example,

$$(-4 + j7) - (8 + j2) = -12 + j5$$

The polar form of the complex operator is not suitable for addition and subtraction, and complex operators which are expressed in polar form must first be changed to rectangular form if they are to be added or subtracted.

1.3 Multiplication and Division of Complex Operators

MULTIPLICATION

For multiplication in the rectangular form the normal rules of algebra apply, so that

$$(a + jb)(c + jd) = ac + jad + jbc + j^2bd$$
$$= (ac - bd) + j(ad + bc) \tag{1.8}$$

since j^2 may be given the value -1. Particular attention should be paid to the following product:

$$(a + jb)(a - jb) = a^2 + jab - jab - j^2b^2$$
$$= a^2 + b^2 \qquad (1.9)$$

$(a + jb)$ and $(a - jb)$ are termed a pair of *conjugate* complex operators since their product contains no quadrate term. In the same way $(-a + jb)$ and $(-a - jb)$ are also conjugate complex operators.

In the trigonometric form,

$$A(\cos \theta + j \sin \theta) \times B(\cos \phi + j \sin \phi)$$
$$= AB(\cos \theta \cos \phi + j \sin \theta \cos \phi + j \sin \phi \cos \theta + j^2 \sin \theta \sin \phi)$$
$$= AB\{(\cos \theta \cos \phi - \sin \theta \sin \phi) + j(\sin \theta \cos \phi + \sin \phi \cos \theta)\}$$
$$= AB(\cos (\theta + \phi) + j \sin (\theta + \phi)) \qquad (1.10)$$

i.e. in the trigonometric form the product of two complex operators is the product of their moduli taken with the sum of their phase angles.

Since the polar and trigonometric forms of a complex operator are really identical, the same results will hold for both forms. Hence,

$$A\underline{/\theta} \times B\underline{/\phi} = AB\underline{/\theta + \phi} \qquad (1.11)$$

In the same way,

$$A\underline{/\theta} \times \frac{1}{B}\underline{/-\phi} = \frac{A}{B}\underline{/\theta - \phi} \qquad (1.12)$$

It can also be shown that successive multiplication obeys the same rules. Thus

$$A\underline{/\theta} \times B\underline{/\phi} \times C\underline{/\psi} = ABC\underline{/\theta + \phi + \psi} \qquad (1.13)$$

It will be observed that multiplication in the polar form is much less tedious than in the rectangular form, and for this reason it is frequently convenient to convert rectangular operators into the polar form before multiplication.

DIVISION

Division of complex operators in the rectangular form is achieved by *rationalizing* the denominator, i.e. eliminating the quadrate term from the denominator, by multiplying both the numerator and the denominator by the conjugate of the denominator. Thus

$$\frac{a + jb}{c + jd} = \frac{(a + jb)(c - jd)}{(c + jd)(c - jd)} = \frac{ac + bd + j(bc - ad)}{c^2 + d^2}$$

$$= \frac{ac + bd}{c^2 + d^2} + j\frac{bc - ad}{c^2 + d^2}$$

In the same way,

$$\frac{1}{-a + jb} = \frac{-a - jb}{(-a + jb)(-a - jb)}$$

$$= \frac{-a}{a^2 + b^2} - j\frac{b}{a^2 + b^2}$$

In polar form the conjugate of the operator $A\underline{/\theta}$ is $A\underline{/-\theta}$. This follows directly from the expression $A\underline{/\theta} \times A\underline{/-\theta} = A^2$. Similarly $B\underline{/\phi} \times B\underline{/-\phi} = B^2$.

Division of the operator $A\underline{/\theta}$ by the operator $B\underline{/\phi}$ is achieved by multiplying both numerator and denominator by the conjugate of the denominator. Thus

$$\frac{A\underline{/\theta}}{B\underline{/\phi}} = \frac{A\underline{/\theta}\ B\underline{/-\phi}}{B\underline{/\phi}\ B\underline{/-\phi}} = \frac{A}{B}\underline{/\theta - \phi}$$

i.e. the quotient of two complex operators in polar form is the quotient of their moduli taken with the difference of their phase angles.

Also

$$\frac{A\underline{/\theta}}{B\underline{/-\phi}} = \frac{A}{B}\underline{/\theta + \phi}$$

EXAMPLE 1.1 Divide $(10 - j10)$ by $(8 \cdot 66 + j5)$.

Method (i)

$$\frac{10 - j10}{8 \cdot 66 + j5} = \frac{14 \cdot 14\underline{/-45°}}{10\underline{/30°}} = 1 \cdot 414\underline{/-75°} = 0 \cdot 366 - j1 \cdot 37$$

Method (ii)

$$\frac{10 - j10}{8 \cdot 66 + j5} = \frac{(10 - j10)(8 \cdot 66 - j5)}{(8 \cdot 66 + j5)(8 \cdot 66 - j5)}$$

$$= \frac{86 \cdot 6 - 50 - j50 - j86 \cdot 6}{8 \cdot 66^2 + 5^2}$$

$$= \frac{36 \cdot 6}{100} - j\frac{136 \cdot 6}{100} = 0 \cdot 366 - j1 \cdot 37$$

POWERS AND ROOTS

Powers of complex operators simply represent successive multiplication. For numbers in polar form the procedure is to take the power of the modulus and to multiply the phase angle by the index. Thus

$$(A\underline{/\theta})^n = A^n\underline{/n\theta} \tag{1.14}$$

Roots may be dealt with in the same manner by taking the root of the modulus and dividing the phase angle by the root. Thus

$$\sqrt[n]{(B\underline{/\phi})} = \sqrt[n]{B}\ \underline{\bigg/\frac{\phi}{n}} \tag{1.15}$$

It should be noted that in complex operator notation the *n*th root of any number, including pure or reference numbers, has always *n* possible values. For example,

$$\sqrt[3]{(8\underline{/60°})} = 2\underline{/20°}$$

or

$$2\ \underline{\bigg/\frac{60° + 360°}{3}} = 2\underline{/140°}$$

or

$$2\ \underline{\bigg/\frac{60° + 720°}{3}} = 2\underline{/260°}$$

since each of these operators to the power 3 will give $8\underline{/60°}$ or its identicals $8\underline{/60°} + 360°$ and $8\underline{/60°} + 720°$.

In the same way,

$$\sqrt[4]{(16)} = 2\underline{/0°} \quad \text{or} \quad 2\underline{/90°} = j2 \quad \text{or} \quad 2\underline{/180°} = -2$$

or

$$2\underline{/270°} = -j2$$

EQUATIONS

Consider the equation

$$a + jb = c + jd$$

which relates two complex operators V and W, where $V = a + jb$ and $W = c + jd$. Since the operators are equal, their components along the reference axis must be equal, so that $a = c$. Also, for identity, the components along the quadrate axis must be equal, so that $b = d$.

In general, in any complex equation, the sum of the reference components on one side must be equal to the sum of the reference components on the other; and similarly the sum of the quadrate components must be equal on both sides of the equation.

Note that the above identity can be expressed as

$$V = V\underline{/\theta} = W = W\underline{/\phi}$$

where $V = \sqrt{(a^2 + b^2)}$; $W = \sqrt{(c^2 + d^2)}$; $\tan \theta = b/a$;   and $\tan \phi = d/c$. For identity,

$$V = W \quad \text{and} \quad \theta = \phi$$

1.4 Simple Circuits—Impedance

The lines which represent alternating voltages or currents in a complexor diagram can be expressed as complex operators when suitable scale units and a reference direction have been chosen. The complexors may then be summed with or subtracted from other complexors which represent quantities expressed with respect to the same scale unit and reference direction.

Impedance is among those quantities which can be represented by a complex operator.

PURE RESISTANCE

Suppose a sinusoidal current represented by the complexor I, is passed through a pure resistance R. The potential difference across R will be a sinusoidal voltage represented by the complexor V, where V and I are in phase with one another, and where $|V|/|I|$ is equal to R. If I is chosen as the reference complexor, then

$$I = |I|\underline{/0°} = I\underline{/0°}$$

Hence

$$V = |V|\underline{/0°} = V\underline{/0°}$$

and the impedance is given by

$$Z = \frac{V}{I} = \frac{V}{I}\underline{/0°} = R \tag{1.16}$$

Thus the impedance of a pure resistance may be represented by the reference operator R.

PURE INDUCTANCE

Suppose that a sinusoidal current represented by the complexor I is passed through a pure inductance L. The p.d. across L will be a sinusoidal voltage represented by the complexor V, where V leads I by 90° and $|V|/|I|$ is equal to ωL. If I is chosen as the reference complexor then $I = |I| \underline{/0°} = I \underline{/0°}$ and $V = |V| \underline{/90°} = j|V| = V \underline{/90°}$.

Therefore the impedance is given by

$$Z = \frac{V}{I} = \frac{V}{I} \underline{/90°} = j\omega L = jX_L \tag{1.17}$$

Thus the impedance of an inductive reactance may be represented by the quadrate operator $j\omega L$.

PURE CAPACITANCE

Suppose that a sinusoidal current represented by the complexor I is passed through a pure capacitance C. The p.d. across C will be a sinusoidal voltage represented by the complexor V, where V lags I by 90°, and $|V|/|I|$ is equal to $1/\omega C$. If I is chosen as the reference complexor, then $I = I \underline{/0°}$ and $V = V \underline{/-90°} = -jV$.

Therefore the impedance is given by

$$Z = \frac{V}{I} = \frac{V}{I} \underline{/-90°} = \frac{-j}{\omega C} = \frac{-j^2}{j\omega C} = \frac{1}{j\omega C} \tag{1.17}$$

i.e.

$$Z = \frac{1}{j} X_C = -jX_C \tag{1.18}$$

Thus the impedance of a capacitive reactance may be represented by the negative quadrate operator $-j/\omega C$ or $1/j\omega C$.

SERIES CIRCUITS

If the resistances and reactances of a circuit are expressed as reference and quadrate operators then the total impedance of the circuit may be determined by the processes of complex algebra. Thus

$$Z = R + j\omega L$$

represents the impedance of a circuit in which a resistance R is connected in series with an inductive reactance ωL across a sinusoidal supply of frequency $f = \omega/2\pi$ hertz*. For a resistance and capacitance in series the impedance is given by

$$Z = R - j/\omega C$$

* The unit of frequency, the *cycle per second* (c/s) is known as the *hertz* (Hz).

For a circuit with R, L and C in series, the impedance in complex form will be

$$Z = R + j\omega L - \frac{j}{\omega C} \tag{1.19}$$

$$= R + j\left(\omega L - \frac{1}{\omega C}\right)$$

$$= R + j(X_L - X_C) \tag{1.19a}$$

For the series connexion of impedances Z_1, Z_2, ... Z_N, the total effective impedance is

$$Z_{eq} = Z_1 + Z_2 + \ldots + Z_N \tag{1.20}$$

For parallel connexion of N impedances the total effective impedance is given by

$$1/Z_{eq} = 1/Z_1 + 1/Z_2 + \ldots + 1/Z_N \tag{1.21}$$

For two impedances in parallel this reduces to

$$Z_{eq} = \frac{Z_1 Z_2}{Z_1 + Z_2} \tag{1.22}$$

EXAMPLE 1.2 Find the voltage which, when applied to a circuit consisting of a resistance of 120Ω in series with a capacitive reactance of 250Ω, causes a current of $0.9\,A$ to flow. Also find the voltage across each component and the overall power factor of the circuit.

For a series circuit the current is taken as the reference complexor (Fig. 1.5).

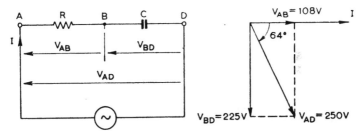

Fig. 1.5

Hence

$$I = 0.9\underline{/0°}\,A$$

The impedance, Z, expressed in complex form, is

$$Z = 120 - j250 = 278\underline{/-64°}\,\Omega$$

Thus

$$V_{AD} = IZ = 0.9\underline{/0°} \times 278\underline{/-64°} = 250\underline{/-64°}\,V$$

The power factor of the circuit is cos 64°, i.e. 0·432 leading.

Also

$$V_{AB} = IR = 0.9\underline{/0°} \times 120 = 108\underline{/0°}\,\text{V}$$

and

$$V_{BD} = I \times jX_C = 0.9\underline{/0°} \times 250\underline{/-90°} = 225\underline{/-90°}\,\text{V}$$

EXAMPLE 1.3 A capacitor of $100\,\Omega$ reactance is connected in parallel with a coil of $70.7\,\Omega$ resistance and $70.7\,\Omega$ reactance to a $250\,\text{V}$ a.c. supply. Find the current in each branch, and the total current taken from the supply. Also find the overall power factor.

For the parallel circuit it is convenient to take the voltage as the reference complexor. Thus

$$V = 250\underline{/0°}\,\text{V}$$

The coil impedance is

$$Z_L = 70.7 + j70.7 = 100\underline{/45°}\,\Omega$$

Therefore the coil current is

$$I_L = \frac{V}{Z_L} = \frac{250\underline{/0°}}{100\underline{/45°}}$$

$$= 2.5\underline{/-45°} \quad \text{or} \quad (1.77 - j1.77)\,\text{A}$$

Impedance of capacitor $= -j100 = 100\underline{/-90°}\,\Omega$

Hence

$$\text{Current in capacitor} = I_C = \frac{250\underline{/0°}}{100\underline{/-90°}}$$

$$= 2.5\underline{/90°} \quad \text{or} \quad j2.5\,\text{A}$$

The total current is

$$I = I_L + I_C$$
$$= 1.77 - j1.77 + j2.5$$
$$= (1.77 + j0.73) \quad \text{or} \quad 1.91\underline{/22.5°}\,\text{A}$$

The power factor is $\cos 22.5°$, i.e. 0.92 leading.

1.5 Parallel Circuits—Admittance

The *admittance*, Y, of a circuit is defined as the r.m.s. current flowing per unit r.m.s. applied voltage. It is thus the reciprocal of the circuit impedance:

$$Y = \frac{I}{V} = \frac{1}{Z} \tag{1.23}$$

The admittance of a circuit may be represented by a complex operator in the same way as the impedance. The reference term of this complex operator is called the *conductance*, and the quadrate term is called the *susceptance*, the symbols for these terms being G and B respectively. Thus an admittance Y may be expressed as

$$Y = G + jB \tag{1.24}$$

If the circuit consists of a pure resistance, R, then

$$Z = R$$

and

$$Y = \frac{1}{Z} = \frac{1}{R} = G \tag{1.25}$$

For a purely inductive reactance, $j\omega L$, the admittance is

$$Y = \frac{1}{j\omega L} = \frac{-j}{\omega L} = -jB \tag{1.26}$$

For a purely capacitive reactance, $1/j\omega C$, the admittance will be

$$Y = \frac{1}{Z} = j\omega C = jB \tag{1.27}$$

From this it is seen that inductive susceptance is negative while capacitive susceptance is positive in the complex form.

If a circuit contains both resistance and reactance in series, the admittance may be derived as follows. Let

$$Z = R + jX$$

Then

$$Y = \frac{1}{Z} = \frac{1}{R + jX} = \frac{R - jX}{R^2 + X^2}$$
$$= \frac{R}{R^2 + X^2} - j\frac{X}{R^2 + X^2}$$

(Note the change of sign.)

The main advantage of the idea of admittance arises when dealing with parallel circuits. In this case the voltage across each element is the same, and the total current is the complexor sum of the branch currents. Thus for three admittances, Y_1, Y_2 and Y_3 in parallel,

$$I = I_1 + I_2 + I_3 = VY_1 + VY_2 + VY_3 = V(Y_1 + Y_2 + Y_3)$$

so that

$$\frac{I}{V} = Y_{eq} = Y_1 + Y_2 + Y_3$$

In general, for N admittances in parallel,

$$Y_{eq} = \sum_{n=1}^{n=N} Y_n \qquad (1.28)$$

The unit of admittance, conductance, and susceptance is the *siemen* (S).

The admittances and impedances of two simple series and of two simple parallel circuits are shown in Fig. 1.6. These show that a series circuit is more easily represented as an impedance, while

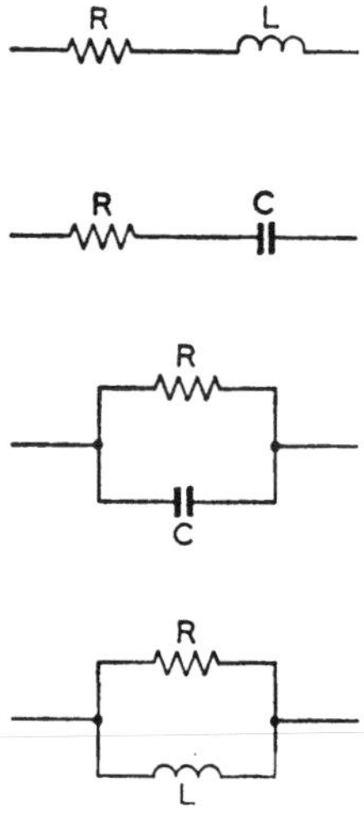

$$Z = R + j\omega L \qquad \qquad Y = \frac{R}{R^2 + \omega^2 L^2} - \frac{j\omega L}{R^2 + \omega^2 L^2}$$

$$Z = R - j/\omega C \qquad \qquad Y = \frac{R}{R^2 + 1/\omega^2 C^2} + \frac{j/\omega C}{R^2 + 1/\omega^2 C^2}$$

$$Y = \frac{1}{R} + j\omega C \qquad \qquad Z = \frac{R}{1 + \omega^2 C^2 R^2} - \frac{j\omega C R^2}{1 + \omega^2 C^2 R^2}$$

$$Y = 1/R - j/\omega L \qquad \qquad Z = \frac{R\omega^2 L^2}{R^2 + \omega^2 L^2} + \frac{j\omega L R^2}{R^2 + \omega^2 L^2}$$

Fig. 1.6 ADMITTANCE AND IMPEDANCE OF FOUR SIMPLE CIRCUITS

a parallel circuit is more easily represented as an admittance. For the case of a resistor R, in parallel with a capacitor C,

$$Y_{eq} = Y_R + Y_C = \frac{1}{R} + j\omega C$$

$$Z_{eq} = \frac{1}{\dfrac{1}{R} + j\omega C}$$

$$= \frac{\dfrac{1}{R} - j\omega C}{\left(\dfrac{1}{R} + j\omega C\right)\left(\dfrac{1}{R} - j\omega C\right)} = \frac{R}{1 + \omega^2 C^2 R^2} - j\frac{\omega C R^2}{1 + \omega^2 C^2 R^2}$$

$$(1.29)$$

It is left for the reader to verify the other results.

EXAMPLE 1.4 Three impedances of $(70 \cdot 7 + j70 \cdot 7)\Omega$, $(120 + j160)\Omega$ and $(120 + j90)\Omega$ are connected in parallel across a 250 V supply. Calculate the admittance of the combination and the total current taken. Also determine the value of the pure reactance which, when connected across the supply, will bring the overall power factor to unity, and find the new value of the total current.

The first step is to express the impedances as admittances in rectangular form, from which the equivalent admittance is easily obtained. Thus

$$Z_1 = 70 \cdot 7 + j70 \cdot 7 = 100\underline{/45°}\,\Omega$$

Therefore

$$Y_1 = \frac{1}{100\underline{/45°}} = 0 \cdot 01\underline{/-45°} = (0 \cdot 00707 - j0 \cdot 00707)\,\mathrm{S}$$

$$Z_2 = 120 + j160 = 200\underline{/53 \cdot 1°}\,\Omega$$

Therefore

$$Y_2 = \frac{1}{200\underline{/53 \cdot 1°}} = 0 \cdot 005\underline{/-53 \cdot 1°} = (0 \cdot 003 - j0 \cdot 004)\,\mathrm{S}$$

$$Z_3 = 120 + j90 = 150\underline{/36 \cdot 9°}\,\Omega$$

Therefore

$$Y_3 = \frac{1}{150\underline{/36 \cdot 9°}} = 0 \cdot 00667\underline{/-36 \cdot 9°} = (0 \cdot 0053 - j0 \cdot 004)\,\mathrm{S}$$

Therefore

$$Y_{eq} = Y_1 + Y_2 + Y_3 = 0 \cdot 0154 - j0 \cdot 015$$
$$= 0 \cdot 0215\underline{/-44 \cdot 3°}\,\mathrm{S}$$

With the voltage as the reference complexor the current will be

$$I = VY_{eq} = 250\underline{/0°} \times 0 \cdot 0215\underline{/-44 \cdot 3°} = 5 \cdot 37\underline{/-44 \cdot 3°}\,\mathrm{A}$$

i.e. the current is 5·37 A at a power factor of 0·71 lagging.

To bring the overall power factor to unity the susceptance required in parallel with the three given impedances must be such that there is no quadrate term in the expression for the resultant admittance. Thus

Susceptance required $= +j0 \cdot 015\,\mathrm{S}$

Therefore

Pure reactance required $= \dfrac{1}{j0 \cdot 015} = -j66 \cdot 6\,\Omega$

i.e. a capacitive reactance of 66·6 Ω.

With this reactance connected across the input, the total admittance will be 0·0154 S (pure conductance); hence new value of current is

$$250\underline{/0°} \times 0 \cdot 0154\underline{/0°} = 3 \cdot 85\,\mathrm{A}$$

EXAMPLE 1.5 Find the parallel combination of resistance and capacitance which takes the same current at the same power factor from a 5 kHz supply as an impedance of $(17 \cdot 3 - j10)\Omega$.

The admittance of the series circuit is

$$Y = \frac{1}{Z} = \frac{1}{17\cdot3 - j10} = 0\cdot05\underline{/30°} = (0\cdot0433 + j0\cdot025)\,S$$

As well as representing the admittance of the series circuit this expression also gives the admittance of a parallel circuit consisting of a pure resistance of

$$R_P = \frac{1}{0\cdot0433} = 23\cdot1\,\Omega$$

in parallel with a purely capacitive susceptance of

$$B_P = 0\cdot025\,S = \omega C_P$$

Hence

$$C_P = \frac{B_P}{\omega} = \frac{0\cdot025}{10{,}000\pi}\,F = \underline{0\cdot796\mu F}$$

Note that these circuits would not be equivalent at any other frequency.

1.6 Impedance and Admittance Diagrams

When an impedance is expressed as a complex operator it may be represented on a diagram which is similar to a complexor diagram, but with the two important differences:

(a) an impedance diagram has two mutually perpendicular axes, and

(b) position on an impedance diagram is important, while position (as distinct from direction) on a complexor diagram is not important. The electrical impedance diagram is equivalent to the mathematician's Argand diagram.

It should be noted that in an *impedance diagram*, such as is shown in Fig. 1.7, both axes must have the same scale, i.e. unit length must represent the same number of ohms on both axes. Pure resistance values are plotted in the reference direction (horizontal), while pure reactance values are plotted above or below the reference axis according to the inductive or capacitive nature of the reactance. For example, point B (Fig. 1.7) represents an impedance which is equivalent to a pure resistance of $40\,\Omega$ in series with an inductive reactance of $10\,\Omega$. The magnitude of the impedance is given by the length OB, and the phase angle by ϕ_B. Also point C on the same diagram represents an impedance which is equivalent to a pure resistance of $20\,\Omega$ in series with a capacitive reactance of $-10\,\Omega$.

If the impedance is not constant then a line on the impedance diagram may be drawn to show all the possible values which the impedance may have. The diagram is then called an *impedance locus diagram*. Fig. 1.7 shows two such impedance loci. The line AB represents the locus for a circuit consisting of a fixed inductance of $10\,\Omega$ in series with a resistance which can be varied between zero and 40Ω. The length OE gives the impedance of the circuit when the resistance is 20Ω, the phase angle then being ϕ_1. In the same way the line CD represents the locus for a circuit consisting of a resistance of 20Ω in series with a capacitive reactance which can be varied from −10Ω to −30Ω. The line OF gives the impedance of this circuit when the capacitive reactance is −20Ω.

In the same way admittance (which has been shown in Section 1.5 to be expressible as a complex operator of the form $Y = G \pm jB$)

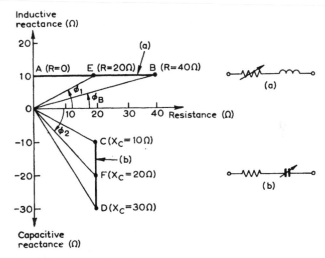

Fig. 1.7   IMPEDANCE LOCI FOR TWO SIMPLE CIRCUITS WITH
ONE VARIABLE

may be represented on an admittance diagram in which the reference axis represents conductance and the quadrate axis represents susceptance. Thus if Fig. 1.7 were an admittance diagram the line AB would represent a fixed capacitive susceptance of 10S in parallel with a resistance whose conductance varied from zero (open-circuit) to 40S. Inductive susceptance (which is negative) is, of course, represented along the negative quadrate axis.

1.7 Current Locus Diagrams for Series Circuits

If a resistance R and a reactance jX are connected in series across a constant voltage supply of V volts, and the voltage V is taken to be in the reference direction, then the current I is given by the expression

$$I = \frac{V}{R + jX}$$

where jX may be either positive (inductive) or negative (capacitive). If the resistance is variable while the reactance is constant, and the above equation is rewritten as

$$I = \frac{\dfrac{V}{jX}}{1 + \dfrac{R}{jX}}$$

i.e.

$$I + \frac{1}{j}\frac{R}{X}I = \frac{V}{jX} = -j\frac{V}{X} \qquad (1.30)$$

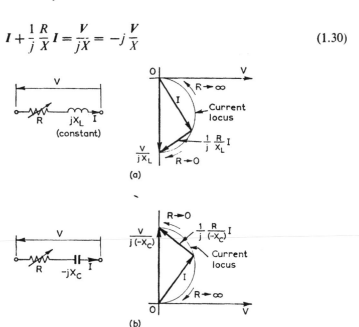

Fig. 1.8 CURRENT LOCUS DIAGRAMS FOR SIMPLE SERIES CIRCUITS
WITH FIXED REACTANCE AND VARIABLE RESISTANCE
(a) Inductive reactance   (b) Capacitance reactance

then the equation may be interpreted as follows. If the current complexor I is drawn from the origin and has the complexor $\dfrac{1}{j}\dfrac{R}{X}I$ added to it, then the sum is the constant complexor $\dfrac{V}{jX}$, and since I and $\dfrac{1}{j}\dfrac{R}{X}I$ are mutually perpendicular, the extremity of I must lie on a circle of diameter $\dfrac{V}{jX}$. The complexor loci for inductive and capacitive circuits with variable resistance are shown in Fig. 1.8.

If the resistance is fixed while the reactance varies, the basic equation is rewritten as

$$I = \frac{\dfrac{V}{R}}{1 + \dfrac{jX}{R}}$$

i.e.

$$I + j\frac{X}{R}I = \frac{V}{R} \qquad\qquad (1.31)$$

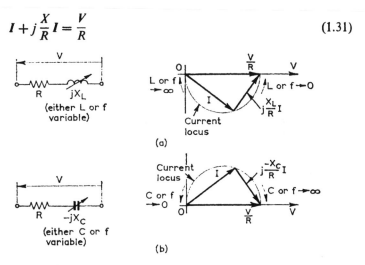

Fig. 1.9 CURRENT LOCI FOR SERIES CIRCUITS WITH VARIABLE REACTANCE
(a) Inductive reactance   (b) Capacitance reactance

This, when interpreted as in the previous case, indicates that the extremity of the current complexor, drawn from the origin, lies on a circle of diameter V/R. The loci for the current in series circuits containing resistance and variable reactance are shown in Fig. 1.9.

EXAMPLE 1.6 A capacitive reactance of $0.5\,\Omega$ is connected in series with a variable resistor R to a 2V, 50kHz supply. Draw the impedance locus, and the locus of current as the resistance varies between zero and $4\,\Omega$. From these loci find the current and its phase angle when the resistance is (a) $0.2\,\Omega$, and (b) $2\,\Omega$. Also determine the maximum power input and the corresponding current, phase angle and resistance.

The impedance locus is shown as the line AB in Fig. 1.10. Since the reactance is constant, this line will be parallel to the reference axis and a distance below it representing the constant capacitive reactance of $0.5\,\Omega$.

(a) OC represents the impedance when $R = 0.2\,\Omega$. By measurement from the diagram OC $= 0.55\,\Omega$ to scale.

The phase angle is $\phi_1 = -69°$

and the current is $\dfrac{2}{0.55} = \underline{\underline{3.64\,A}}$

(b) OD represents the impedance when $R = 2\,\Omega$. From the diagram OD $= 2.09\,\Omega$ to scale.

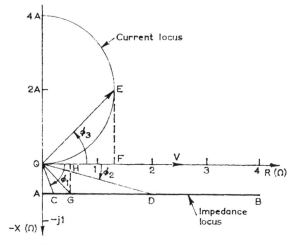

Fig. 1.10

The phase angle is $\phi_2 = -14°$

and the current is $\dfrac{2}{2.09} = \underline{\underline{0.957\,A}}$

From eqn. (1.30), the diameter of the current locus is

$$\frac{V}{-jX} = \frac{2}{-j0.5} = \underline{\underline{j4\,A}}$$

If V is taken in the reference direction then the diameter of the current locus lies along the positive direction of the quadrate axis as shown in the diagram. The maximum power input occurs when the component of the current in the direction of V is a maximum, i.e. when the active component of the current is a maximum. Thus OE gives the current for maximum power and ϕ_3 gives the corresponding phase angle.

$$OE = 2\sqrt{2} = \underline{\underline{2.83\,A}} \qquad \text{and} \qquad \phi_3 = 45°$$

Maximum power $= V \times$ maximum value of active component of current
$$= V \times OF$$
$$= 2 \times 2 = \underline{\underline{4\,W}}$$

The corresponding value of resistance may be obtained by drawing the line OG, making an angle of $-45°$ with the reference axis, to cut the impedance locus at G. Then OG represents the impedance for maximum power, so that the corresponding resistance is OH to scale, i.e. $0.5\,\Omega$.

1.8 Volt-ampere Calculations for Parallel Loads and Generators

If several loads are connected in parallel, the total current flowing is the complexor sum of the individual load currents, i.e.

$$I_{\text{total}} = I_1 + I_2 + I_3 + \ldots$$

The complexor diagram representing the above equation is shown in Fig. 1.11 (a), where the common system voltage is taken as the reference. If each of the current complexors is multiplied by the magnitude of the system voltage V, the above expression becomes

$$VI_{\text{total}} = VI_1 + VI_2 + VI_3 + \ldots$$

The corresponding complexor diagram is shown in Fig. 1.11 (b),

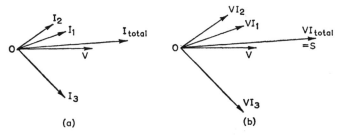

(a)   (b)

Fig. 1.11   CURRENT AND CORRESPONDING VOLT-AMPERE
COMPLEXOR DIAGRAMS FOR PARALLEL CIRCUITS

which is a diagram of complexor volt-amperes. Volt-amperes will be represented by the symbol S.

The reference component of the total volt-amperes is $S \cos \phi$, i.e. $VI_{\text{total}} \cos \phi$, or the total power P absorbed by the load. The quadrature component of the volt-amperes is $S \sin \phi$ or the reactive volt-amperes Q.

$$
\begin{aligned}
S &= S \cos \phi \pm jS \sin \phi = S\underline{/\pm\phi} \\
&= P + jQ \\
&= (P_1 + P_2 + P_3 + \ldots) + j(Q_1 + Q_2 + Q_3 + \ldots)
\end{aligned}
$$
(1.32)
(1.32a)

Note that if the voltage applied to a device is represented by $V = V\underline{/\theta}$ volts, while the corresponding current through the device is represented by $I = I\underline{/\phi}$ amperes, then the product of these complexors will be $VI\,(= VI\underline{/\theta + \phi})$, which is neither the power nor the total volt-amperes absorbed by the device. This product has no practical significance, the power absorbed always being given by

$$P = VI\cos \text{ (phase angle between } V \text{ and } I)$$

EXAMPLE 1.7 An alternator supplies a load of 200kVA at a power factor of 0·8 lagging, and a load of 50kVA at a power factor of 0·6 leading. Find the total kVA, power, and kVAr supplied, and determine the power factor of the alternator.

Load 1 $= 200\underline{/-\cos^{-1} 0{\cdot}8}$ (i.e. 0·8 lag)
       $= 200(0{\cdot}8 - j0{\cdot}6) = 160 - j120\,\text{kVA}$
Load 2 $= 50\underline{/\cos^{-1} 0{\cdot}6}$ (lead)
       $= 50(0{\cdot}6 + j0{\cdot}8) = 30 + j40\,\text{kVA}$

Therefore

Total load $= 190 - j80 = \underline{\underline{206\underline{/-22^\circ\,50'}\,\text{kVA}}}$

Also,

Total power $= \underline{\underline{190\,\text{kW}}}$ and   Total kVAr $= \underline{\underline{80\,\text{kVAr lagging}}}$

Alternator power factor $= \cos 22^\circ\,50'$, i.e. $\underline{\underline{0{\cdot}92\,\text{lagging}}}$

1.9 Mutual Inductance in Networks

When two coils or circuits are linked by a mutual inductance M, then an alternating current I in one, will set up an alternating

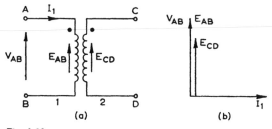

Fig. 1.12 MUTUAL INDUCTANCE

(a) Circuit   (b) Open-circuit complexor diagram

e.m.f. of magnitude $I\omega M$ in the other. A difficulty arises over the direction of the mutually induced e.m.f. in the second coil, since this depends on the relative winding directions and the relative positions of the two coils. The dot notation will be used here to indicate the relative e.m.f. directions. In this notation a dot is placed at an arbitrary end of one coil, and a second dot is placed at the end of the second coil which has the same polarity as the dotted end of the first coil, when the current through either of the coils is changing. In a.c. circuits employing the dot notation for mutual inductance the mutually induced e.m.f. in each coil will be in such a direction as to give the same polarity to the dotted ends of the coils.

In the circuit of Fig. 1.12 a current represented by the complexor I_1 flows in the first coil. Hence, neglecting resistance,

$$V_{AB} = E_{AB} = j\omega L_1 I_1$$

where L_1 is the self-inductance of the first coil.

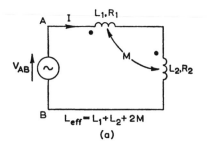

(a)

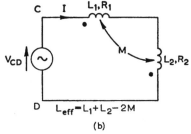

(b)

Fig. 1.13 MUTUAL INDUCTANCE BETWEEN PARTS OF THE
SAME MESH

(a) Coils in series aiding (eqn. (1.35))
(b) Coils in series opposing (eqn. (1.36))

The e.m.f. induced by the mutual inductance M of the circuit is E_{CD}, which leads I_1 by 90°, and is given by

$$E_{CD} = j\omega M I_1 \qquad (1.33)$$

and

$$\text{Mutual inductive reactance} = E_{CD}/I_1 = j\omega M \qquad (1.34)$$

To illustrate the use of the dot notation consider the series circuits of Fig. 1.13. Applying Kirchhoff's laws to the series-aiding circuit at (a),

$$V_{AB} = IR_1 + Ij\omega L_1 + Ij\omega M + IR_2 + Ij\omega L_2 + Ij\omega M$$

since the mutual e.m.f. will have the same polarity as the self-induced e.m.f.

Thus the combined impedance is

$$Z = R_1 + R_2 + j\omega(L_1 + L_2 + 2M) \tag{1.35}$$

Similarly, for the series-opposing circuit at (b),

$$V_{CD} = IR_1 + Ij\omega L_1 - Ij\omega M + IR_2 + Ij\omega L_2 - Ij\omega M$$

since the mutual e.m.f. in each coil will have the opposite polarity to the self-induced e.m.f.

Therefore the combined impedance is

$$Z = R_1 + R_2 + j\omega(L_1 + L_2 - 2M) \tag{1.36}$$

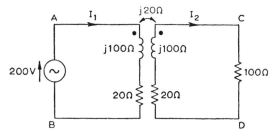

Fig. 1.14

EXAMPLE 1.8 Calculate the currents in each coil of the circuit of Fig. 1.14.

The mesh currents are inserted in arbitrary directions as shown.
For mesh 1,

$$200\underline{/0^\circ} = (20 + j100)I_1 - j20I_2 \tag{i}$$

Since the positive direction of I_2 is shown as entering the second coil at the undotted end, the e.m.f. of self-inductance will make the undotted end of this coil positive. The corresponding mutual e.m.f. in the first coil will make the undotted end of that coil positive; hence, since I_1 leaves the first coil at this end, the voltage drop due to the mutual inductance will be negative (i.e. $-j20I_2$).

Similarly for mesh 2,

$$0 = -j20I_1 + (120 + j100)I_2 \quad \text{i.e.} \quad I_1 = (5 - j6)I_2 \tag{ii}$$

Substituting in eqn. (i),

$$200 = (100 + j500 - j120 + 600 - j20)I_2$$

Hence

$$I_2 = 0.255\underline{/-27.3^\circ}\,\text{A} \quad \text{and} \quad I_1 = 1.99\underline{/-77.2^\circ}\,\text{A}$$

1.10 Equivalent Mutual Inductance Circuits

Consider the inductively coupled circuit shown in Fig. 1.15 (*a*), where Z_{11} is the total self-impedance of the primary loop (including the primary resistance and self-inductance of the mutual inductor M), and Z_{22} is similarly the total self-impedance of the secondary loop

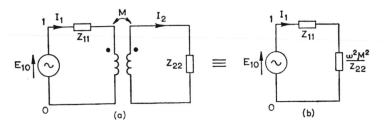

Fig. 1.15 EQUIVALENT CIRCUIT FOR INDUCTIVELY COUPLED
NETWORK

(*a*) Network with pure mutual inductance
(*b*) Equivalent circuit

(including the load impedance). With the winding directions indicated, the mesh equations are

Mesh 1 $E_{10} = Z_{11}I_1 - j\omega M I_2$ (i)
Mesh 2 $0 = -j\omega M I_1 + Z_{22}I_2$ (ii)

Hence from eqn. (i),

$$E_{10}Z_{22} = Z_{11}Z_{22}I_1 - j\omega M Z_{22}I_2$$

and from eqn. (ii),

$$0 = -\omega^2 M^2 I_1 + j\omega M Z_{22}I_2$$

Adding,

$$E_{10}Z_{22} = (Z_{11}Z_{22} + \omega^2 M^2)I_1$$

Therefore

$$\frac{E_{10}}{I_1} = Z_{11} + \frac{\omega^2 M^2}{Z_{22}} \tag{1.37}$$

= Impedance of network looking into the primary circuit

Hence the circuit of Fig. 1.15 (*a*) may be replaced by a simple circuit, consisting of the self-impedance of mesh 1 in series with the term $\{\omega^2 M^2/(\text{self-impedance of mesh 2})\}$ as shown at (*b*). This circuit, when viewed from the generator will always appear to be the same as the original circuit.

It may easily be shown that if the relative winding directions of the mutual inductor are reversed the same result will be obtained.

EXAMPLE 1.9   The coefficient of coupling between the primary and secondary of an air-cored transformer is 0·8. The primary winding has an inductance of 0·6mH and a resistance of 2Ω, while the secondary has an inductance of 5·5mH and a resistance of 20Ω. Calculate the primary current and the secondary terminal voltage when the primary is connected to a 10V, 50kHz supply and the load on the secondary has an impedance of 250Ω at a phase angle of 45° leading.   (*L.U.*)

The circuit diagram is shown in Fig. 1.16.

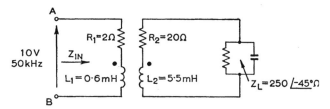

Fig. 1.16

Coupling coefficient $= k = \dfrac{M}{\sqrt{(L_1 L_2)}}$

so that

$$M = 0.8\sqrt{(0.6 \times 5.5)} = 1.45\,\text{mH}$$

Load impedance, $Z_L = 250\underline{/-45^\circ} = (178 - j178)\,\Omega$

Impedance of secondary winding $= 20 + j100\pi \times 5.5$

Therefore

$$Z_{22} = (198 + j1{,}552)\,\Omega$$

Impedance of primary circuit, $\bar{Z}_{11} = (2 + j189)\,\Omega$
Hence from eqn. (1.37),

$$Z_{in} = Z_{11} + \frac{\omega^2 M^2}{Z_{22}}$$

$$= 2 + j189 + \frac{(100\pi)^2 \times 1.45^2}{(198 + j1{,}552)}$$

$$= 18.8 + j56 = 59\underline{/71.4^\circ}\,\Omega$$

Therefore

Primary current $= \dfrac{10}{59}\underline{/-71.4^\circ} = 0.17\underline{/-71.4^\circ}\,\text{A}$

Load voltage $= j\dfrac{\omega M I_1}{Z_{22}} Z_L$

Therefore

$$\text{Magnitude of load voltage} = \frac{100\pi \times 10^3 \times 1\cdot45 \times 10^{-3} \times 0\cdot17 \times 250}{1,565}$$

$$= 12\cdot4\,\text{V}$$

1.11 Series Resonance: Q-factor

The general form of a series *RLC* circuit is shown in Fig. 1.17(*a*).

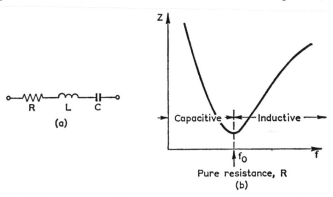

(a)

Capacitive ⟵⟶ Inductive

f_0

Pure resistance, R

(b)

Fig. 1.17 SERIES RESONANCE

The impedance of the circuit is

$$Z = R + j\omega L + \frac{1}{j\omega C} \tag{1.38}$$

$$= R + j\omega L \left(1 - \frac{1}{\omega^2 LC}\right) \tag{1.38a}$$

At the frequency for which $\omega^2 LC = 1$ the quadrate term in this equation will be zero and the impedance will have a minimum value of $Z = R$. The frequency, f_0, at which this occurs is the resonant frequency of the circuit. Thus

$$f_0 = \frac{1}{2\pi\sqrt{(LC)}} \tag{1.39}$$

or

$$\omega_0 = \frac{1}{\sqrt{(LC)}} \tag{1.39a}$$

At resonance $\omega_0 L = 1/\omega_0 C$. At frequencies below the resonant frequency ωL is less than $1/\omega C$, and from eqn. (1.38) the circuit behaves as a capacitive reactance. Above f_0 the circuit behaves as an inductive reactance, since in this case ωL is greater than $1/\omega C$. The variation of impedance of the circuit with frequency is shown in Fig. 1.17(*b*).

It is useful at this point to consider a figure of merit for the coil, known as the coil *Q-factor*, or simply the *Q* of the coil. The *Q*-factor may be defined as the ratio of reactance to resistance of a coil:

$$Q = \frac{\text{Reactance}}{\text{Resistance}} = \frac{\omega L}{R} \tag{1.40}$$

Of particular importance is the *Q*-factor at the resonant frequency of the tuned circuit. This is

$$Q_0 = \frac{\omega_0 L}{R} = \frac{1}{\omega_0 C R} = \frac{1}{R}\sqrt{\frac{L}{C}} \tag{1.41}$$

The expression for impedance given by eqn. (1.38*a*) can be rewritten in terms of Q and ω_0 as

$$Z = R\left\{1 + jQ\left(1 - \frac{\omega_0^2}{\omega^2}\right)\right\}$$

since $\omega_0^2 = 1/LC$.

For frequencies near the resonant frequency this can be written

$$Z = R\left\{1 + jQ_0\left(1 - \frac{\omega_0^2}{\omega^2}\right)\right\}$$

$$= R\left\{1 + jQ_0\left(\frac{\omega^2 - \omega_0^2}{\omega^2}\right)\right\}$$

$$= R\left\{1 + jQ_0\frac{(\omega + \omega_0)(\omega - \omega_0)}{\omega^2}\right\}$$

Again, at frequencies which are near resonance, $\omega \approx \omega_0$ and $\omega + \omega_0 \approx 2\omega$ so that

$$Z \approx R\left\{1 + jQ_0 2\frac{\omega - \omega_0}{\omega_0}\right\}$$

$$\approx R\{1 + j2Q_0\delta\} \tag{1.42}$$

where $\delta = (\omega - \omega_0)/\omega_0 = (f - f_0)/f_0$ is the *per-unit frequency deviation*, or the difference between the circuit frequency, f, and the resonant frequency f_0 expressed as a fraction of f_0. If $f < f_0$, δ is negative, if $f > f_0$, δ is positive.

A convenient way of defining the sharpness of the resonance curve (Fig. 1:17(*b*)) is to find the frequency, f_L, below the resonant frequency and the frequency f_H above it at which the circuit impedance increases to $\sqrt{2}$ of its value at resonance. This figure is chosen since if $Z = \sqrt{2}R$ then from eqn. (1.42),

$$|1 \pm j2Q_0\delta| = \sqrt{2} \quad \text{so that} \quad 2Q_0\delta = \pm 1$$

which is a convenient (but purely arbitrary) criterion.

From this

$$\frac{1}{Q_0} = +2\delta = \frac{2(f_H - f_0)}{f_0} \quad \text{or} \quad \frac{1}{Q_0} = -2\delta = \frac{2(f_0 - f_L)}{f_0}$$

so that, adding the alternative expressions for $1/Q_0$,

$$\frac{1}{Q_0} = \frac{f_H - f_L}{f_0} \quad \text{or} \quad Q_0 = \frac{f_0}{f_H - f_L} \tag{1.43}$$

$f_H - f_L \ (= f_0/Q)$ is often called the *bandwidth* of the tuned circuit. Note that at the frequency f_L the circuit impedance is

$$Z_L = R(1 - j1) = \sqrt{2}R\underline{/-45°}$$

while at the frequency f_H the impedance is

$$Z_H = R(1 + j1) = \sqrt{2}R\underline{/+45°}$$

1.12 Parallel Resonance

The usual form of a parallel resonant circuit is shown in Fig. 1.18(a),

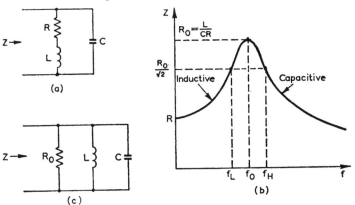

Fig. 1.18 PARALLEL RESONANCE

R being the resistance of the inductor L, and C being a pure capacitance. The impedance of the circuit at an angular frequency ω is

$$Z = \frac{(R + j\omega L)(1/j\omega C)}{R + j\omega L + 1/j\omega C} = \frac{R + j\omega L}{1 - \omega^2 LC + j\omega CR} \tag{1.44}$$

The impedance is R for $\omega = 0$ and zero for $\omega \to \infty$. Between these extremes Z will normally rise to some maximum value as shown at (b). Two cases will be considered, (A) when $R \ll \omega L$, and (B) when R is of the same order as ωL.

CASE (A): $R \ll \omega L$

In this case the impedance can be written

$$Z = \frac{L/C}{R + j\omega L(1 - 1/\omega^2 LC)}$$

The R in the denominator cannot be neglected since the term $\omega L(1 - 1/\omega^2 LC)$ may be small. Rewriting the expression for Z,

$$Z = \frac{L/CR}{1 + \dfrac{j\omega L}{R}\left(1 - \dfrac{1}{\omega^2 LC}\right)} \tag{1.45}$$

At the frequency, f_0, for which $\omega L = 1/\omega C$, the impedance will be resistive and will have a maximum value of L/CR. This is called the *dynamic resistance* of the circuit, R_0. The frequency f_0 is the resonant frequency. Thus

$$R_0 = L/CR \tag{1.46}$$

and

$$f_0 = 1/2\pi\sqrt{(LC)} \quad \text{or} \quad \omega_0 = 1/\sqrt{(LC)} \tag{1.47}$$

Using the Q-factor as defined in Section 1.11, and considering frequencies near resonance (for which $Q = Q_0 = \omega_0 L/R$), the expression for impedance given by eqn. (1.45) can be written

$$Z = \frac{R_0}{1 + jQ_0\left(1 - \dfrac{\omega_0^2}{\omega^2}\right)}$$

Defining $\delta (= (f - f_0)/f_0)$ as in Section 1.11, it is apparent that the expression for impedance can be reduced to

$$Z = \frac{R_0}{1 + j2Q_0\delta} \tag{1.48}$$

in the same manner as for the series circuit. Similarly the bandwidth can be defined in terms of the frequency f_L below f_0 at which the impedance falls to $1/\sqrt{2}$ of its value, R_0, at resonance and the frequency f_H above f_0 at which this same fall in impedance occurs. In both cases therefore

$$|1 \pm j2Q_0\delta| = \sqrt{2}$$

so that, as in the series-resonance case,

$$2Q_0\delta = \pm 1$$

It follows that

$$\frac{1}{Q_0} = \pm 2\delta = \frac{f_H - f_L}{f_0}$$

Note that because Q must always be positive the minus sign in the above expression is required since δ will be negative if $f < f_0$. Hence the bandwidth is

$$f_H - f_L = \frac{f_0}{Q_0} \tag{1.49}$$

as for the series circuit.

The circuit impedances at f_L and f_H are given by

$$Z_L = R_0/(1 - j1) = 0 \cdot 707 R_0 \underline{/45°}$$

$$Z_H = R_0/(1 + j1) = 0 \cdot 707 R_0 \underline{/-45°}$$

Notice that the impedance has an inductive reactive component at frequencies below f_0 while it has a capacitive component at frequencies above f_0.

If the tuned circuit is supplied from a constant-current generator, the voltage across it will fall to $0 \cdot 707$ of its value at resonance at both f_L and f_H. Since the power factor at these frequencies is $0 \cdot 707$, the power dissipated will be $0 \cdot 707^2 (= 0 \cdot 5)$ times the power at resonance. For this reason f_L and f_H are known as the *half-power frequencies*.

A useful equivalent of the parallel tuned circuit is shown in Fig. 1.18(c). From eqn. (1.45), the admittance of the circuit of Fig. 1.18(a) is

$$Y = \frac{1}{Z} = \frac{CR}{L} + j\omega C + \frac{1}{j\omega L} = \frac{1}{R_0} + j\omega C + \frac{1}{j\omega L} \tag{1.50}$$

i.e. the admittance is equivalent to that of a three-element parallel circuit consisting of a pure resistance, R_0, in parallel with a pure capacitance, C, and a pure inductance, L, as shown in Fig. 1.18(c).

EXAMPLE 1.10 The inductance and magnification factor* of a coil are 200μH and 70 respectively. If this coil is connected in parallel with a capacitor of 200pF, calculate the magnitude and phase angle of the impedance of the parallel circuit for a frequency 0·8 per cent below the resonant frequency and the half-power bandwidth. (*H.N.C.*)

---

* Magnification factor is an alternative name for Q-factor.

Resonant angular frequency $= \omega_0 = \dfrac{1}{\sqrt{(LC)}} = \dfrac{10^9}{\sqrt{(200 \times 200)}}$

$$= 5 \times 10^6 \, \text{rad/s}$$

$$Q_0 = 70 = \frac{\omega_0 L}{R}$$

and

$$R_0 = \text{Resonant impedance} = \frac{L}{CR} = \frac{\omega_0 L}{\omega_0 CR} = \frac{Q_0}{\omega_0 C} = Q_0 \omega_0 L$$

Therefore

$$R_0 = Q_0 \omega_0 L = 70 \times 5 \times 10^6 \times 200 \times 10^{-6} = 70,000\,\Omega$$

At 0·8 per cent below resonance

$$\omega = 0 \cdot 992 \omega_0$$

Thus

Impedance, $\mathbf{Z} = \dfrac{70,000}{1 + j70\left(1 - \left(\dfrac{1}{0 \cdot 992}\right)^2\right)}$

$$= \frac{70,000}{1 + j70 \times \dfrac{(0 \cdot 992 - 1)(0 \cdot 992 + 1)}{0 \cdot 992^2}}$$

$$= 46,400 \underline{/48 \cdot 6^\circ}\,\Omega$$

The half-power bandwidth is given by

$$f_H - f_L = f_0/Q_0 = 5 \times 10^6/2\pi \times 70 = \underline{11 \cdot 37\,\text{kHz}}$$

CASE (B): TUNED CIRCUIT WITH APPRECIABLE RESISTANCE

If the tuned circuit of Fig. 1.18(a) has appreciable resistance, the simplifying assumptions made above cannot be used. The impedance, $\mathbf{Z}$, is given by eqn. (1.44):

$$\mathbf{Z} = \frac{(R + j\omega L)(1/j\omega C)}{R + j\omega L + 1/j\omega C} = \frac{R + j\omega L}{1 - \omega^2 LC + j\omega CR}$$

In this case there are three ways of defining the resonant frequency. These are (i) the frequency at which $\omega L = 1/\omega C$ (called the series resonant frequency), (ii) the frequency at which the circuit is purely resistive, (iii) the frequency at which the impedance is a maximum. Rationalizing the denominator of the expression for Z gives

$$\mathbf{Z} = \frac{(R + j\omega L)(1 - \omega^2 LC - j\omega CR)}{(1 - \omega^2 LC)^2 + \omega^2 C^2 R^2}$$

$$= \frac{R + j\omega\{L(1 - \omega^2 LC) - CR^2\}}{(1 - \omega^2 LC)^2 + \omega^2 C^2 R^2}$$

This has a zero quadrate term when

$$L(1 - \omega^2 LC) - CR^2 = 0$$

i.e. when

$$\omega^2 = \frac{1}{LC} - \frac{R^2}{L^2}$$

or

$$\omega = \sqrt{\left(\frac{1}{LC} - \frac{R^2}{L^2}\right)} \tag{1.51}$$

This equation gives the frequency at which the circuit impedance is purely resistive. Notice that, when $R^2/L^2 \geqslant 1/LC$ (i.e. when $R \geqslant \sqrt{(L/C)}$, ω^2 becomes zero, and there will be no frequency at which the circuit is purely resistive.

A special case occurs when there is a *series* resistance of $\sqrt{(L/C)}$ in each branch of the circuit. The impedance is then

$$\begin{aligned}
Z &= \frac{\{\sqrt{(L/C)} + j\omega L\}\{\sqrt{(L/C)} + 1/j\omega C\}}{2\sqrt{(L/C)} + j\omega L + 1/j\omega C} \\
&= \frac{2L/C + j\sqrt{(L/C)}(\omega L - 1/\omega C)}{2\sqrt{(L/C)} + j(\omega L - 1/\omega C)} \\
&= \sqrt{\frac{L}{C}} \tag{1.52}
\end{aligned}$$

i.e. the impedance is a pure resistance, $R = \sqrt{(L/C)}$ at all frequencies.

The impedance is a maximum at a frequency which lies between the series resonant frequency and that for unity power factor. The natural resonant frequency will be dealt with in Chapter 6.

1.13 Reactance/Frequency Graphs

When circuits contain several reactive elements, multiple resonances (both series and parallel) may occur. In many cases the operation of such circuits can be seen by sketching the graph of reactance to a base of frequency. Such graphs (and graphs giving the variation of susceptance with frequency) can be built up from the well-known form of reactance/frequency curves for a pure inductance and a pure capacitance.

For a pure inductance,

$$\text{Impedance} = jX_L = j\omega L$$

so that X_L varies linearly with frequency.

For a pure capacitance,

$$\text{Impedance} = -jX_C = \frac{-j}{\omega C}$$

so that X_C varies inversely with frequency, and may be considered to be negative. The form of these curves is shown in Fig. 1.19.

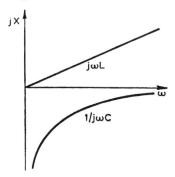

Fig. 1.19   REACTANCE/ANGULAR-FREQUENCY GRAPHS FOR PURE INDUCTANCE AND PURE CAPACITANCE

SIMPLE SERIES CIRCUIT

For a simple series circuit consisting of a pure inductance in series with a pure capacitance, the impedance is $Z = j\omega L - j/\omega C$.

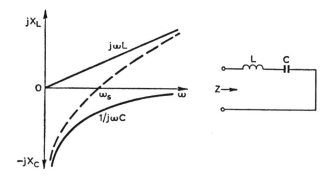

Fig. 1.20   SIMPLE LC SERIES CIRCUIT

The overall reactance graph is obtained by adding the individual graphs of $j\omega L$ and $-j/\omega C$ for each value of ω as shown in Fig. 1.20. The series resonant frequency is then obtained by the intersection of this resultant graph with the axis of ω at ω_s.

SIMPLE PARALLEL CIRCUIT

For a parallel circuit comprising a pure inductance and a pure capacitance, it is simpler to work initially in terms of admittance. Thus for the circuit of Fig. 1.21, the admittance is

$$Y = j\omega C - \frac{j}{\omega L}$$

The capacitive susceptance, $j\omega C$, is positive and varies linearly with ω, the inductive susceptance, $-j/\omega L$, is negative and varies inversely with ω, and the resultant graph of admittance, Y, crosses

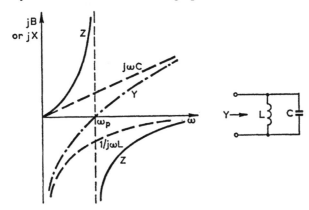

Fig. 1.21   SIMPLE *LC* PARALLEL CIRCUIT

the axis of ω at the parallel-resonant angular frequency, ω_p, as shown. The curve of impedance to a base of ω can now be sketched. Thus when $Y \to -j\infty$ (at the origin), the impedance tends to zero. Between $\omega = 0$ and $\omega = \omega_p$, the admittance is a negative suscep-tance, and hence the impedance must be a positive reactance which rises towards ∞ as ω approaches ω_p (since for $\omega = \omega_p$, $Y = 0$ and hence $Z = \pm j\infty$).

ω_p gives the parallel-resonant angular frequency as stated above.

Above $\omega = \omega_p$ the admittance is positive and rises towards infinity; hence the impedance is negative and rises from $-j\infty$ when $\omega = \omega_p$ (where $Y = 0$) towards zero as $\omega \to \infty$. The curves for Z are shown as the full lines in Fig. 1.21.

1.14   General Form of the Reactance/Frequency Graph

Several general points concerning reactance/frequency graphs are evident from the previous section. Thus these graphs will (i) always have a positive slope, (ii) always start at $X = 0$ or $X = -j\infty$ (since there is no reactance which has any other value for $\omega = 0$), (iii) will always finish either tending towards zero from $-j\infty$ (if the network reactance is capacitive as $\omega \to \infty$) or towards infinity (if the network reactance is inductive as $\omega \to \infty$), and (iv) will have a number of series resonances (where the graph cuts the frequency axis) and a number of parallel resonances (where the graph changes from $+j\infty$ to $-j\infty$) depending on the network configuration. These can easily be obtained by inspection of a given network.

The reactance/frequency graphs for complicated networks of purely inductive and purely capacitive elements can be built up by considering what happens when additional elements are added (*a*) in series or (*b*) in parallel with the original network.

ADDING SERIES REACTANCE

In Fig. 1.22(*a*) the full lines indicate the reactance/frequency graph

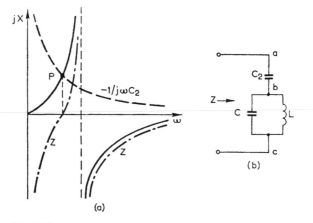

Fig. 1.22

for the parallel *CL* circuit shown at (*b*). The addition of series reactance cannot affect the parallel resonant frequency, but it will cause series resonance whenever the magnitude of the added reactance is equal to that of the original network but is of opposite sign. The angular frequency at which this occurs is readily obtained by plotting the *negative* of the added series reactance on the same graph. In the example shown, the curve of $-1/j\omega C_2$ ($= j/\omega C_2$) is shown dashed at (*a*). The two curves cut at P, which gives the value of the additional series resonance, ω_s. The resultant reactance curve (shown chain dotted) is thus easily constructed. If the added series element is itself a parallel circuit the same rules apply, but in this case there will be a second parallel resonant frequency. Any series resonances in the original network will be modified by the added reactance.

ADDING PARALLEL REACTANCE

Adding reactance in parallel with a given LC network cannot affect the series resonances of the original network, since the network impedance is zero at these frequencies. It will, however, give rise to additional parallel resonances at those frequencies for which the added reactance is equal in magnitude but opposite in sign to that of the original network. The locatio. of these new parallel resonances is readily obtained by superimpos ng the negative of the added reactance on the original reactance/frequency graph. The original parallel resonances will all be modified.

EXAMPLE 1.11 Plot the reactance/angular-frequency graph for the circuit shown in Fig. 1.23(a) over the range 0–3,000 rad/s. Obtain from the graph the angular frequencies at which (i) the reactance is zero, (ii) the reactance is infinite. Over what ranges is the reactance (a) inductive (b) capacitive?

Sketch the reactance/angular-frequency curve if a capacitor C_2 of $2\,\mu$F is now connected in parallel with the whole circuit. (*H.N.C.*)

From the values given the following table can be constructed.

ω	0	500	1,000	2,000	3,000
$\omega C_1(\times 10^{-3})$	0	$j0\cdot5$	$j1$	$j2$	$j3$
$1/\omega L_1(\times 10^{-3})$	$-j\infty$	$-j2$	$-j1$	$-j0\cdot5$	$-j0\cdot33$
$Y_p(\times 10^{-3})$	$-j\infty$	$-j1\cdot5$	0	$j1\cdot5$	$j2\cdot67$
$Z_p = 1/Y_p$	0	$j667$	$\pm j\infty$	$-j667$	$-j375$
$j\omega L_2$	0	$j250$	$j500$	$j1,000$	$j1,500$
$1/j\omega C_2$	0	$-j1,000$	$-j500$	$-j250$	$-j167$

From this the reactance/angular-frequency graph can be plotted as shown at (b).

Following the rules for adding reactances in series, the values of $-j\omega L_2$ are also plotted on the above graph. This curve intersects the curve for Z_p at point X, and hence gives the series-resonant angular frequency as $\omega_s = 1,700$ rad/s. The parallel-resonant angular frequency is not affected by the added series element, and is (from the graph) $\omega_p = 1,000$ rad/s. The total impedance, Z_T, is found by adding $j\omega L_2$ to the parallel circuit impedance (chain dotted curve at (b)). Hence the reactance is zero when $\omega = 0$ or $\omega = \omega_s$ i.e. at 0 and 1,700 rad/s.

Also, the reactance is infinite when $\omega = \omega_p = 1,000$ rad/s.

The circuit is inductive between $\omega = 0$ and $\omega = 1,000$ rad/s and between $\omega = 1,700$ rad/s and ∞, and is capacitive between $\omega = 1,000$ and $\omega = 1,700$ rad/s.

The curves for Z_T are replotted at (c), and the curve of $-1/j\omega C_2$ is superimposed as shown, to cut the Z_T curves at P and Q. These points therefore give the new parallel-resonant frequencies of the circuit when C_2 is connected in parallel with it. For $\omega = 0$, the inductances L_1 and L_2 form a short-circuit, so that the overall impedance is zero. As $\omega \to \infty$ the capacitor C_2 short-circuits the input and hence the impedance tends to zero. Hence the form of the reactance/angular-frequency curve is as shown chain dotted at (c). The series resonance point is, of course, unaffected by the added parallel capacitance, so that the resultant curve must still pass through ω_s.

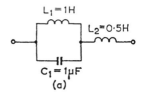

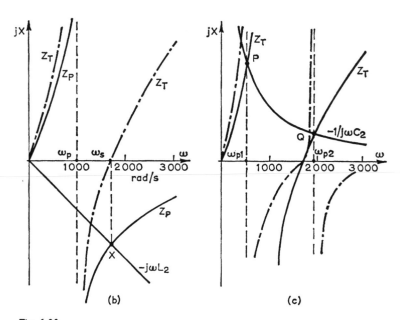

Fig. 1.23

PROBLEMS

The following complex numbers refer to Problems 1.1–1.5:

$A = 1 + j7$, $B = 3 + j3$, $C = -7 - j9$, $D = 5 - j12$, $E = 8 - j4$,
$F = -7 + j2$, $G = -5 + j8$, $H = -10 - j10$, $J = 5 + j13$, $K = 6 + j2$.

1.1 Express these complex numbers in polar form.

Ans. $7{\cdot}07\underline{/81{\cdot}8°}$; $4{\cdot}24\underline{/45°}$; $11{\cdot}4\underline{/-127{\cdot}9°}$; $13\underline{/-67{\cdot}3°}$; $8{\cdot}95\underline{/-26{\cdot}6°}$;
$7{\cdot}28\underline{/164°}$; $9{\cdot}45\underline{/122°}$; $14{\cdot}14\underline{/-135°}$; $13{\cdot}95\underline{/69°}$; $6{\cdot}32\underline{/18{\cdot}4°}$.

1.2 Find $A + B, C + D, E + F, G + H, J + K$, and express the answers in polar form.

Ans. $10.78\underline{/68.2°}$; $21.05\underline{/-95.5°}$; $2.25\underline{/-63.4°}$; $15.1\underline{/-172°}$; $18.6\underline{/53.7°}$.

1.3 Find $A - B, C - D, E - F, G - H, J - K$, and express the answers in polar form.

Ans. $4.47\underline{/116.6°}$; $12.38\underline{/165.9°}$; $16.15\underline{/-21.8°}$; $18.7\underline{/74.4°}$; $11.04\underline{/95.2°}$.

1.4 Find AB, CD, EF, GH, JK, and express the answers in rectangular form.

Ans. $-18 + j24$; $-143 + j39$; $-48 + j44$; $130 - j30$; $4 + j88$.

1.5 Find $A/B, C/D, E/F, G/H, J/K$, and express the answers in rectangular form.

Ans. $1.33 + j1$; $0.432 - j0.763$; $-1.21 + j0.226$; $-0.15 - j0.65$; $1.4 + j1.7$.

1.6 Two impedances $(5 + j7)\Omega$ and $(10 - j5)\Omega$ are connected (*a*) in series, (*b*) in parallel to a 200 V supply. Calculate in each case (*a*) the current drawn from the mains, (*b*) the power supplied, and (*c*) the power factor.

Ans. 13.2 A, 2,620 W, 0.991 lagging; 31.5 A, 5,900 W, 0.937 lagging.

1.7 Explain how alternating quantities can be represented by complex operators. If the potential difference across a circuit is represented by $(40 + j25)$ volts and the circuit consists of a resistance of 20Ω in series with an inductance of 0.06 H and the frequency is 79.5 Hz find the complex operator representing the current in amperes.   (*L.U.*)

Ans. $(1.19 - j0.54)$ A.

1.8 The impedance of a coil at a frequency of 1 MHz can be expressed as $(300 + j400)\Omega$. What do you understand by the symbols of the expression, and what information does this expression convey that is lacking from the simple statement that the impedance of the coil is 500Ω?

This coil is connected in parallel with a capacitor of capacitance 159 pF. Calculate the impedance of the combined circuit, expressing it in the complex form $Z = R + jX$.   (*L.U.*)

Ans. $(667 + j333)\Omega$.

1.9 If the impedance of a circuit is expressed in the form $R + jX$, deduce an expression for the corresponding admittance. Two circuits of impedance $(9 - j12)\Omega$ and $(4 + j3)\Omega$ are connected (*a*) in series, and (*b*) in parallel across a supply of $V = (200 + j150)$ volts. Find the total current, power and power factor in each case.

Ans. 15.8 A, 3,250 W, 0.822 leading; 52.6 A, 12,500 W, 0.948 lagging.

1.10 The series portion of a series-parallel circuit consists of a coil P, the inductance of which is 0.05 H and the resistance 20Ω. The parallel portion consists of two branches A and B. Branch A consists of a coil Q the inductance of which is 0.1 H and the resistance 30Ω, branch B consists of a 100μF capacitor in series with a 15Ω resistor. Calculate the current and power factor from a 230 V a.c. supply when the frequency is 50 Hz.   (*L.U.*)

Ans. 4.4 A; 0.996.

1.11   Two coils of resistance $9\,\Omega$ and $6\,\Omega$ and inductance $0.0159\,H$ and $0.0382\,H$ respectively are connected in parallel across a $200\,V$, $50\,Hz$ supply. Calculate: (a) the conductance, susceptance and admittance of each coil and the entire circuit; (b) the current and power factor for each coil and for the total circuit; and (c) the total power taken from the supply.

Ans. (a) $0.085\,S$, $0.047\,S$, $0.097\,S$; $0.033\,S$, $0.066\,S$, $0.074\,S$; $0.118\,S$, $0.113\,S$. $0.164\,S$. (b) $19.4\,A$, 0.877 lagging; $14.8\,A$, 0.446 lagging; $32.8\,A$, 0.722 lagging. (c) $4,720\,W$.

1.12   Three circuits having impedances of $(10 + j30)$, $(20 + j0)$ and $(1 - j20)\,\Omega$ are connected in parallel across a $200V$ supply. Find the total current flowing and its phase angle.

Ans. $13.1\,A$; $17.6°$ leading.

1.13   Three impedances Z_1, Z_2, and Z_3 are connected in parallel to a $240\,V$, $50\,Hz$ supply. If $Z_1 = (8 + j6)\,\Omega$ and $Z_2 = (12 + j20)\,\Omega$, determine the complex impedance of the third branch if the total current is $35\,A$ at a power factor of 0.9 lagging.

Ans. $(15 - 16.6)\,\Omega$.

1.14   Coils of impedances $(8 + j6)\,\Omega$ and $(15 + j10)\,\Omega$ are connected in parallel. In series with this combination is an impedance of $(20 - j31)\,\Omega$. The supply is $200/\underline{0°}V$ at $50\,Hz$. Calculate in polar form: (a) the total impedance, (b) the current in each branch and the total current, and (c) the power factor.

Ans.   (a) $37.1/\underline{-47.2°}\,\Omega$;   (b) $3.48/\underline{46°}\,A$, $1.93/\underline{49.2°}\,A$, $5.4/\underline{47.2°}\,A$; (c) 0.681 leading.

1.15   Two impedances $Z_1 = (6 + j3)\,\Omega$ and $Z_2 = (5 - j8)\,\Omega$ are connected in parallel and an impedance $Z_3 = (4 + j6)\,\Omega$ is connected in series with them across a $2.5\,V$, $100\,kHz$ supply.

Determine: (a) the complex expressions for the admittance of each section and of the whole circuit, (b) the current and phase angle of the whole circuit, and (c) the total power taken from the supply.

Ans. (a) $0.149/\underline{-26.6°}$, $0.106/\underline{58°}$, $0.139/\underline{-56.3}$, $0.094/\underline{-30.4}$ mho; (b) $0.24\,A$, $30.4°$ lagging; (c) $510\,mW$.

1.16   A series circuit consists of a coil of impedance Z_1 and two other impedances Z_2 and Z_3. A voltmeter V_1 is connected to read the p.d. across Z_1 and a second voltmeter V_2 reads the p.d. across Z_2.

When a $100\,V$, d.c. supply is applied to the circuit, the current is $2\,A$ and V_1 and V_2 read $30\,V$ and $50\,V$ respectively. With a $210\,V$ supply at $50\,Hz$ applied the current in the same circuit is $3\,A$ at a lagging power factor and V_1 and V_2 now read $60\,V$ and $75\,V$ respectively.

Find the complex expressions for the three impedances. If they are now connected in parallel to a supply of $250\,V$, $50\,Hz$, calculate for each branch the current and its phase angle with respect to the supply voltage. The voltmeters read correctly on both a.c. and d.c. supplies. Assume no iron losses and that Z_3 is inductive.

Ans. $Z_1 = 20/\underline{41.4°}\,\Omega$, $Z_2 = 25/\underline{0°}\,\Omega$, $Z_3 = 37.2/\underline{74.4°}\,\Omega$; $12.5/\underline{-41.42}\,A$, $10/\underline{0°}\,A$, $6.72/\underline{-74.4°}\,A$.

1.17 A circuit having a constant resistance of $60\,\Omega$ and a variable inductance of 0 to $0\cdot4\,\text{mH}$ is connected across a $5\,\text{V}$, $50\,\text{kHz}$ supply. Derive from first principles the locus of the extremity of the current complexor.

Find (a) the power, and (b) the inductance when the power factor is $0\cdot8$.

(L.U.)

Ans. (a) $265\,\text{mW}$; (b) $0\cdot143\,\text{mH}$.

1.18 A variable capacitor and a resistance of $300\,\Omega$ are connected in series across $240\,\text{V}$, $50\,\text{Hz}$ mains.

Draw the complexor loci of impedance and current as the capacitance changes from $5\,\mu\text{F}$ to $30\,\mu\text{F}$.

From the diagram find (a) capacitance to give current of $0\cdot7\,\text{A}$, (b) current when capacitance is $10\,\mu\text{F}$.

(L.U.)

Ans. $19\cdot2\,\mu\text{F}$; $0\cdot55\,\text{A}$.

1.19 A circuit consisting of an inductor L in series with a resistance r, which is variable between zero and infinity, is connected to a constant-voltage constant-frequency supply. Prove that the locus of the extremity of the current complexor is a semicircle.

If $V = 400/\underline{0^\circ}\,\text{V}$, $f = 50\,\text{Hz}$, $L = 31\cdot8\,\text{mH}$, draw to scale the current locus.

Calculate the maximum power input, the corresponding current and power factor, and the value of r for this condition.

(H.N.C.)

Ans. $8{,}000\,\text{W}$; $28\cdot28\,\text{A}$; $0\cdot707$; $10\,\Omega$.

1.20 Two coils have inductances of $250\,\mu\text{H}$ and $100\,\mu\text{H}$. They are placed so that their mutual inductance is $50\,\mu\text{H}$. What will be their effective inductance: (a) in series aiding, (b) in series opposing, (c) in parallel aiding, and (d) in parallel opposing? Deduce the formula for the effective inductance in the four cases.

Ans. $450\,\mu\text{H}$; $250\,\mu\text{H}$; $90\,\mu\text{H}$; $50\,\mu\text{H}$.

1.21 The load on an alternator consists of:

    (i) a lighting and heating load of 700 kW at unity p.f.,

    (ii) a motor load of 709 kW which has an average efficiency of $0\cdot9$ and an overall power factor of $0\cdot8$ lagging,

    (iii) a synchronous motor load absorbing $50\,\text{kW}$ at a leading power factor of $0\cdot6$.

Calculate the minimum rating of the alternator, and determine the additional power which it could supply if the load power factors were improved to unity.

(H.N.C.)

Ans. $1{,}625\,\text{kVA}$; $85\,\text{kW}$.

1.22 A single-phase cable has a maximum carrying capacity of $173\,\text{A}$. It supplies the following loads at $440\,\text{V}$:

    (i) a 15 kW lighting load at unity p.f.,

    (ii) a $30\,\text{kVA}$ motor load at a power factor of $0\cdot8$ lagging.

Determine, graphically or otherwise, the maximum kVA rating of an additional load at $0\cdot7$ power factor lagging which could be connected to the cable. If the power factor of the additional load is improved to unity by apparatus which has a constant loss of $2\,\text{kW}$, determine the new maximum kW rating of the additional load which can be installed.

(H.N.C.)

Ans. $35\,\text{kVA}$, $33\,\text{kW}$.

1.23   In the circuit shown in Fig. 1.24 a current of 2A at 50 Hz is fed in at X

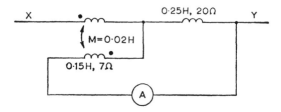

Fig. 1.24

and out at Y. Calculate the current in the ammeter A for the winding directions shown. The ammeter impedance may be neglected.   (*L.U.*)

Ans. 1·17 A.

1.24   Two tuned circuits are coupled inductively, the coupling coefficient being 0·01. The primary consists of a coil of 4 mH and 20 Ω and a capacitor of 200 pF and the secondary has a coil of 1 mH and 10 Ω and is tuned to the resonant frequency of the primary. Calculate the magnitude and phase of the secondary current in terms of primary input voltage. If the primary were driven at half its natural frequency, calculate its input impedance.   (*H.N.C.*)

Ans. $0.0319 V_1 \underline{/90°}$ volts, 6,725 Ω capacitive reactance.

1.25   For a series *LCR* resonant circuit supplied at constant voltage, show that the bandwidth between the half-power (3 dB) points is given by

$$\Delta\omega = \omega_H - \omega_L = \frac{R}{L}$$

The *Q*-factor and resonant frequency of such a circuit are 100 and 51 kHz respectively. Find the bandwidth between the half-power points.   (*H.N.C.*)

Ans. 510 Hz.

1.26   Plot the reactance/angular-frequency curve over the range 0–2.5 × 10⁶ rad/s for a circuit comprising an inductance of 500 μH in parallel with a capacitance of 2,000 pF, all in series with a capacitance of 5,000 pF. If a second inductance of 250 μH is now connected in parallel with the whole circuit, sketch the resultant graph and determine the series- and parallel-resonant angular frequencies.

Ans. $\omega_s = 0.6 \times 10^6$ rad/s; $\omega_p = 0.55 \times 10^6$ and 1.9×10^6 rad/s.

1.27   An inductor having negligible self-capacitance is connected in parallel with a loss-free variable air-capacitor and an electronic voltmeter whose input impedance is equivalent to 6 pF in parallel with 1 MΩ. When the circuit is energized from a constant current source, a maximum indication on the electronic voltmeter is obtained at a frequency of 0·8 MHz, with the capacitor set at 79 pF. Increasing or decreasing this capacitance by 2 pF reduces the indication on the voltmeter to 70·7% of the maximum indication. Calculate the inductance and *Q*-factor of the inductor.

Ans. 466 mH, 47·2.

1.28 A coil, having negligible self-capacitance, has a resistance of $8\,\Omega$ at a frequency of 750 kHz, and a capacitor of 350 pF is required to produce a parallel resonant circuit at this frequency. Calculate the Q-factor of the coil and the resonant impedance of the circuit. What will be the Q-factor and bandwidth of the circuit if a $50\,k\Omega$ resistor is connected in parallel with the capacitor?

Ans. 75·7, 45·7 kΩ, 39·6, 19 kHz.

Chapter 2

CIRCUITS AND CIRCUIT THEOREMS

In any electric circuit the currents and voltages at any point may be found by applying Kirchhoff's laws. As the complexity of the circuit increases, however, the labour involved in the solution becomes multiplied, and several electrical circuit theorems have been developed which reduce the amount of work required for a solution.

The theorems considered here apply to linear circuits, i.e. to circuits with impedances which are independent of the direction and magnitude of the current. They apply to both a.c. and d.c. circuits, provided that in linear a.c. circuits the voltages and currents are expressed as complexors, and the impedances are represented by complex operators. Worked examples will serve to illustrate the method in which the theorems are applied.

In these circuits there are two types of source—the *constant-voltage* source and the *constant-current* source. A constant-voltage source is one which generates a constant predetermined e.m.f., E_E, which may be alternating or direct, and has a series internal impedance Z_i. A constant-current source is one which produces a constant predetermined internal current, I_E, which may be alternating or direct and has a shunt internal impedance Z_i. The graphical symbols for these sources are shown in Fig. 2.1.

In general, a practical source with linear characteristics (having a terminal voltage which falls in proportion to the current) may be

46

represented in either of the above forms. If V_{oc} is the open-circuit terminal voltage and I_{sc} the terminal short-circuit current of a practical source, then the effective internal source impedance is $Z_i = V_{oc}/I_{sc}$. Such a source may be equally well represented by either (*a*) a constant-voltage source with e.m.f. $E_E = V_{oc}$ and series internal impedance Z_i, or (*b*) a constant-current source with $I_E = I_{sc}$

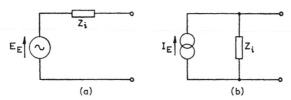

(a)   (b)

Fig. 2.1   REPRESENTATION OF LINEAR SOURCES
(*a*) Constant-voltage generator
(*b*) Constant-current generator

and shunt internal impedance Z_i. It is convenient in calculation to be able to choose either form of source to represent a practical source.

2.1 Multi-mesh Networks

Multi-mesh networks are made up of branches which form closed loops or *meshes*. The junction points of the impedances are known as *nodes*. Kirchhoff's laws in complex form can be used to solve for the currents and voltages in the network. These laws are

1. The complexor sum of the currents at any node in a network is zero.
2. The complexor sum of the e.m.f.s round any closed loop is equal to the complexor sum of the potential drops round the same loop.

Circuits involving multiple meshes may be solved by considering either the meshes (mesh analysis) or the junctions (node analysis). The method chosen will depend on whether a given network gives rise to fewer mesh equations than node equations, or vice versa.

MESH ANALYSIS

Using Maxwell's mesh-current method, each closed loop in the network is assumed to carry a mesh current. The actual current in any branch of the network is then the complexor sum of the mesh currents which flow through that branch. If the internal mesh currents are all assumed to circulate in the same sense, the mesh equations take a standard form. Thus for the circuit of Fig. 2.2,

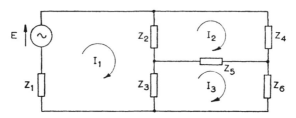

Fig. 2.2

Mesh 1 $E = I_1(Z_1 + Z_2 + Z_3) - I_2Z_2 - I_3Z_3$
Mesh 2 $0 = -I_1Z_2 + I_2(Z_2 + Z_4 + Z_5) - I_3Z_5$ (2.1)
Mesh 3 $0 = -I_1Z_3 - I_2Z_5 + I_3(Z_3 + Z_5 + Z_6)$

These equations are now solved for the unknown currents, remembering that all the quantities are to be expressed in complex form. The same method may be applied to any other network configuration, but it is obvious that the labour of solution increases rapidly with complexity of the network.

The general form of the mesh equations may be seen by writing $Z_{11} = Z_1 + Z_2 + Z_3$ (the self-impedance of mesh 1), $Z_{12} = Z_2$ (the mutual impedance between meshes 1 and 2), etc., so that

$$E_1 = I_1Z_{11} - I_2Z_{12} - I_3Z_{13} - \ldots$$
$$E_2 = -I_1Z_{12} + I_2Z_{22} - I_3Z_{23} - \ldots$$
$$E_3 = -I_1Z_{13} - I_2Z_{23} + I_3Z_{33} - \ldots$$

where $E_1, E_2 \ldots$ are the mesh e.m.f.s.

NODE ANALYSIS

In node analysis, potentials V_1, V_2, etc., are assumed at the circuit nodes. If any sources of e.m.f. are present these are represented by the equivalent constant-current generators. If now the admittances (or, of course, impedances) of each branch are known the node potentials and branch currents can be found. Thus for the circuit of Fig. 2.3:

Node 1 $I = I_1 + I_2 + I_3$
$$= (V_1 - V_4)Y_1 + (V_1 - V_2)Y_2 + (V_1 - V_3)Y_3$$
$$= V_1(Y_1 + Y_2 + Y_3) - V_2Y_2 - V_3Y_3 - V_4Y_1$$
(2.2(i))

Node 2
$$0 = -I_2 - I_4 + I_6$$
$$= -(V_1 - V_2)Y_2 - (V_3 - V_2)Y_4 + (V_2 - V_4)Y_6$$
$$= -V_1Y_2 + V_2(Y_2 + Y_4 + Y_6) - V_3Y_4 - V_4Y_6$$
$$\text{(2.2(ii))}$$

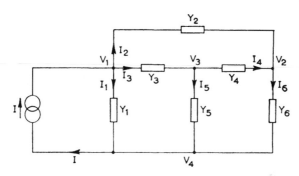

Fig. 2.3

Node 3
$$0 = -I_3 + I_4 + I_5$$
$$= -(V_1 - V_3)Y_3 + (V_3 - V_2)Y_4 + (V_3 - V_4)Y_5$$
$$= -V_1Y_3 - V_2Y_4 + V_3(Y_3 + Y_4 + Y_5) - V_4Y_5$$
$$\text{(2.2(iii))}$$

Node 4
$$-I = -I_1 - I_5 - I_6$$
$$= -(V_1 - V_4)Y_1 - (V_2 - V_4)Y_6 - (V_3 - Y_4)Y_5$$
$$= -V_1Y_1 - V_2Y_6 - V_3Y_5 + V_4(Y_1 + Y_5 + Y_6)$$
$$\text{(2.2(iv))}$$

The general form of the node equations is seen by writing $Y_{11} = Y_1 + Y_2 + Y_3 + \ldots$ as the total admittance at node 1, $Y_{12} = Y_{21}$ as the admittance between nodes 1 and 2, etc., so that

$$\left. \begin{array}{l} I_A = V_1Y_{11} - V_2Y_{12} - V_3Y_{13} - \ldots \\ I_B = -V_1Y_{12} + V_2Y_{22} - V_3Y_{23} - \ldots \end{array} \right\} \tag{2.3}$$

where I_A, I_B, . . . are the currents fed in at nodes 1, 2, . . .

The reader should note that mutual inductance coupling between branches of a network is taken into account in mesh analysis by including appropriately directed e.m.f.s in the meshes. In node analysis the voltage between any two nodes is likewise altered by the e.m.f. induced in any coupled coils between these nodes.

EXAMPLE 2.1 In the circuit of Fig. 2.4, the two sources have the same frequency ($\omega = 5,000$ rad/s). Find the p.d. across resistor R_2 if $E_1 = 10\underline{/0°}$V, $E_2 = 10\underline{/90°}$V; $R_1 = 100\Omega$, $R_2 = 50\Omega$, $R_3 = 50\Omega$; $L_1 = 40$mH, $L_2 = 15$mH, and $M = 10$mH.

Using mesh analysis, the mesh equations are:

Mesh 1 $E_1 - j\omega M I_2 = (R_1 + R_3 + j\omega L_1)I_1 - R_3 I_2$

since the current I_2 entering the dotted end of L_2 gives an e.m.f. $j\omega M I_2$ directed out of the dotted end of L_1.

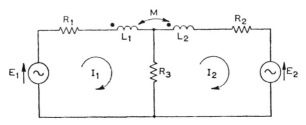

Fig. 2.4

Mesh 2 $-E_2 - j\omega M I_1 = -R_3 I_1 + (R_2 + R_3 + j\omega L_2)I_2$

Hence

$$10\underline{/0^\circ} - j50I_2 = (150 + j200)I_1 - 50I_2 \tag{i}$$

and

$$-j10 - j50I_1 = -50I_1 + (100 + j75)I_2 \tag{ii}$$

so that

$$I_1 = \frac{j10}{50 - j50} + \frac{100 + j75}{50 - j50} I_2$$

Substituting in eqn. (i),

$$10 = \frac{(150 + j200)j10}{50 - j50} + \frac{(150 + j200)(100 + j75)}{50 - j50} I_2 - (50 - j50)I_2$$

so that

$$I_2 = 0{\cdot}089\underline{/-129^\circ} = (0{\cdot}056 - j0{\cdot}07)\,\text{A}$$

Using the current directions of Fig. 2.4, it follows that

$$V_{R2} = I_2 R_2 = -2{\cdot}8 - j3{\cdot}5 = 4{\cdot}5\underline{/231^\circ}\text{V}$$

2.2 Superposition Theorem

"An e.m.f. acting on any linear network produces the same effect whether it acts alone or in conjunction with other e.m.f.s."

Hence a network containing many sources of e.m.f. may be analysed by considering the currents due to each e.m.f. in turn acting alone, the other e.m.f.s being suppressed and represented only by their internal source impedances.

EXAMPLE 2.2 Two a.c. sources, each of internal resistance 20Ω, are connected in parallel across a 10Ω pure resistance load. If the generated e.m.f.s are $50\,\text{V}$

and are 90° out of phase with each other, determine the current which will flow in the 10 Ω resistor and in each generator.

The complete circuit is shown in Fig. 2.5(a). Consider the two e.m.f.s to be $E_1(=50\text{V})$, and $E_2(=j50\text{V})$. Then for E_1 acting alone (Fig. 2.5(b)),

$$I_1 = \frac{E_1}{R_{1eq}} = \frac{50}{20 + \dfrac{10 \times 20}{10 + 20}} = \frac{50}{26\cdot7} = 1\cdot88\,\text{A}$$

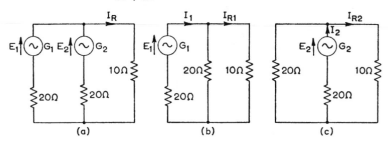

Fig. 2.5

Also,

$$I_{R1} = 1\cdot88\,\frac{20}{20 + 10} = 1\cdot25\,\text{A}$$

For E_2 acting alone (Fig. 2.5(c)),

$$I_2 = \frac{E_2}{R_{2eq}} = \frac{j50}{26\cdot7} = j1\cdot88\,\text{A}$$

and

$$I_{R2} = j1\cdot88\,\frac{20}{30} = j1\cdot25\,\text{A}$$

By the superposition theorem the total current through the 10 Ω resistor with both e.m.f.s acting at once will be

$$I_R = I_{R1} + I_{R2} = 1\cdot25 + j1\cdot25 = 1\cdot77\underline{/45^\circ}\,\text{A}$$

In the same way the current through the first generator G_1 when both e.m.f.s are active will be the algebraic sum of the currents through G_1 when each e.m.f. acts alone.

When E_1 acts alone, the current through G_1 is

$$I_1 = 1\cdot88\,\text{A}$$

and when E_2 acts alone, the current through G_1 is

$$I_2\,\frac{10}{20 + 10} = j0\cdot63\,\text{A}$$

Comparing Figs. 2.5(b) and (c) the above currents through G_1 are found to be oppositely directed. Therefore

$$\text{Total current through } G_1 = (1\cdot88 - j0\cdot63)\,\text{A}$$

Similarly,

Total current through $G_2 = (-0.63 + j1.88)$ A

The superposition theorem may be restated in order to apply to distribution type networks as follows.

"The total current through any branch of a network is equal to the algebraic sum of the currents through the particular branch due to each load current alone, and the no-load current, if any."

EXAMPLE 2.3 A 500 V d.c. generator supplies a load A of 500 A through a 0·02 Ω distributor, and a load B of 200 A through a 0·015 Ω distributor. If A and B are joined by a 0·03 Ω interconnector, determine the interconnector current.

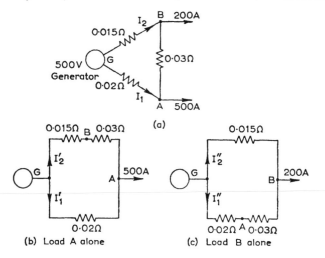

Fig. 2.6

The circuit is shown in Fig. 2.6(*a*). The resistances are assumed to be the total for both "go" and "return" conductors; since the return conductors are then assumed to be perfect conductors which will not give rise to voltage drops, they are omitted.

Applying the superposition theorem, the 500 A load is taken first alone. The circuit then becomes that shown in Fig. 2.6(*b*).

$$I_1' = 500 \times \frac{0.045}{(0.045 + 0.02)} = 346 \,\text{A}$$

and

$$I_2' = 500 - 346 = 154 \,\text{A}$$

Now the 200 A load is taken alone (Fig. 2.6(*c*)). From the diagram,

$$I_2'' = 200 \times \frac{0.05}{0.065} = 152 \,\text{A}$$

Therefore

$$I_1'' = 200 - 152 = 48\,\text{A}$$

By the superposition theorem,

Total current, $I_1 = I_1' + I_1'' = 346 + 48 = 394\,\text{A}$
Total current, $I_2 = I_2' + I_2'' = 154 + 152 = 306\,\text{A}$

The current through the interconnector is made up of 154 A $(=I_2')$ flowing from B to A (Fig. 2.6(b)) and 48 A $(=I_1'')$ flowing from A to B (Fig. 2.6(c)). Therefore

Resultant current through interconnector $= \underline{\underline{106\,\text{A}}}$ flowing from B to A.

2.3 Thévenin's Theorem and Norton's Theorem

The circuit theorems of Thévenin and Norton are extensions of the superposition principle of Section 2.2. Proofs of the theorems will be omitted. It must be understood that the equivalent circuits derived by the use of these theorems are valid only where all elements and actual sources are linear as discussed in the introduction to this chapter. The ideas are simple and of extreme value.

Thévenin: the linear network behind a pair of terminals may be replaced by a constant-voltage generator with an e.m.f. equal to the open-circuit voltage at the terminals and an internal impedance equal to the impedance seen at the actual terminals, with all internal sources removed and replaced by their internal impedances.

Norton: the linear network behind a pair of terminals may be replaced by a constant-current generator with a current equal to the short-circuit current at the terminals and an internal impedance equal to the internal impedance seen at the actual terminals, with all sources removed and replaced by their internal impedances.

Note that Norton's theorem follows directly from Thévenin's theorem and the equivalence of constant-current and constant-voltage sources.

Consider a complicated network of sources and impedances connected to two terminals A and B as in Fig. 2.7(a). Let the voltage across the terminals when they are open-circuited be V_T volts and the impedance measured at the terminals with all the sources suppressed and replaced only by their internal impedances be Z_I, Fig. 2.7(b). Then the circuit, as viewed from the terminals, is exactly equivalent to a generator of V_T volts and internal impedance Z_I ohms, Fig. 2.7(c). The current through the impedance Z connected across the terminals will therefore be

$$I = \frac{V_T}{Z_I + Z} \tag{2.4}$$

The circuit of Fig. 2.7(c) is called the *constant-voltage equivalent circuit*.

Alternatively, if I_{sc} is the short-circuit current from A to B in Fig. 2.7(a), then the circuit, as viewed from the terminals, is exactly equivalent to a generator of I_{sc} amperes and shunt internal impedance

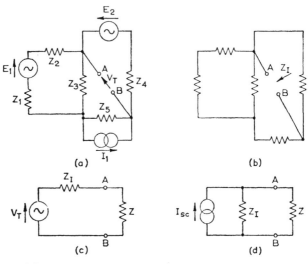

(a)   (b)   (c)   (d)

Fig. 2.7

Z_I ohms (Fig. 2.7(d)). The current through the impedance Z connected across the terminals will therefore be

$$I = I_{sc} \frac{Z_I}{Z_I + Z} \tag{2.5}$$

The circuit of Fig. 2.7(d) is called the *constant-current equivalent circuit*.

EXAMPLE 2.4   Solve Example 2.2 using Thévenin's theorem.

Let the circuit of Fig. 2.5(a) be broken just above the 10 Ω resistor. Then the voltage across the break is

$$V_T = E_1 - \text{Voltage drop across impedance of source } G_1$$

$$\text{Circulating current} = \frac{E_1 - E_2}{40} = \frac{50 - j50}{40} = (1\cdot25 - j1\cdot25)\,\text{A}$$

Thus

$$V_T = 50 - (1\cdot25 - j1\cdot25) \times 20 = (25 + j25)\,\text{V}$$

The impedance looking into the break with the e.m.f.s suppressed is $10 + (20 \times 20)/(20 + 20) = 20\,\Omega$. Therefore

$$\text{Current through } 10\,\Omega \text{ resistor} = \frac{25 + j25}{20} = \underline{\underline{1\cdot25 + j1\cdot25\,\text{A}}}$$

The terminal voltage across the $10\,\Omega$ load resistor is

$$V_R = (12\cdot5 + j12\cdot5)\,\text{V}$$

Therefore

$$\text{Current through } G_1 = \frac{E_1 - V_R}{20} = \frac{37\cdot5 - j12\cdot5}{20} = \underline{\underline{(1\cdot88 - j0\cdot63)\,\text{A}}}$$

and

$$\text{Current through } G_2 = \frac{E_2 - V_R}{20} = \frac{-12\cdot5 + j37\cdot5}{20} = \underline{\underline{(-0\cdot63 + j1\cdot88)\,\text{A}}}$$

For a circuit which has any number of internal sources and impedances, and has two free terminals the short-circuit current between the terminals is

$$I_{sc} = \frac{V_T}{Z_I} \tag{2.6}$$

where V_T is the open-circuit voltage between the terminals and Z_I the internal impedance between them.

If I_{sc} and V_T can be measured, Z_I can be found from this equation.

EXAMPLE 2.5 Find the constant-voltage and constant-current equivalent circuits of the actual circuit of Fig. 2.8(a).

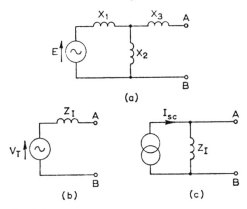

Fig. 2.8

$$\text{Open-circuit voltage, } V_T = E\,\frac{jX_2}{j(X_1 + X_2)} = \frac{EX_2}{X_1 + X_2}$$

Internal impedance, $Z_i = jX_3 + \dfrac{jX_1 \times jX_2}{jX_1 + jX_2}$

$$= j\left(X_3 + \frac{X_1 X_2}{X_1 + X_2}\right)$$

The constant-voltage equivalent circuit is as in Fig. 2.8(b).

Short-circuit current, $I_{sc} = \dfrac{E}{jX_1 + \dfrac{jX_2 jX_3}{jX_2 + jX_3}} \dfrac{jX_2}{jX_2 + jX_3}$

$$= \frac{EX_2}{jX_1 X_2 + jX_2 X_3 + jX_1 X_3}$$

The constant-current equivalent circuit is as in Fig. 2.8(c).

EXAMPLE 2.6 The frequency of the generator shown in Fig. 2.9(a) is the

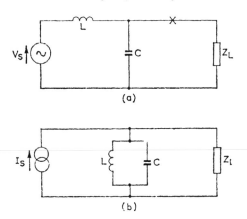

(a)

(b)

Fig. 2.9

series resonant frequency of L and C. Show, by application of Norton's theorem, that the current through the load Z_L is constant independent of the load.

Given that $V_S = 100\,$V, $L = 10\,$mH and $C = 2,000\,$pF, find the value of the voltage across a load consisting of a resistance, R_L, of $1\,$kΩ in parallel with a capacitance, C_L, of $2,000\,$pF.   *(H.N.C.)*

Since L and C form a resonant circuit, $\omega = 1/\sqrt{LC}$. If the load is short-circuited the current through the short-circuit is

$$I_S = V_S/\omega L = V_S \sqrt{\frac{C}{L}}$$

The impedance looking back into the circuit is infinite, since L and C are assumed to be pure circuit elements, and form a parallel tuned circuit when seen from the load terminals. Hence the constant-current equivalent circuit is as shown in Fig. 2.9(b), so that the load current is I_S, independent of the load.

For the values given,

$$I_S = 100 \sqrt{\frac{2{,}000 \times 10^{-12}}{10 \times 10^{-3}}} = 45 \, \text{mA}$$

The load impedance is

$$Z_L = \frac{R_L}{j\omega C_L(R_L + 1/j\omega C_L)} = \frac{R_L}{1 + j\omega C_L R_L} = \frac{R_L}{1 + j\dfrac{C_L R_L}{\sqrt{(LC)}}}$$

$$= \frac{1{,}000}{1 + j0{\cdot}45}$$

Therefore $Z_L = 1{,}000/1{\cdot}1 = 909 \, \Omega$, so that

$$V_L = I_S|Z_L| = 45 \times 10^{-3} \times 909 = \underline{\underline{41 \, \text{V}}}$$

It is left as an exercise for the reader to investigate the effect of a small resistance in series with L, and also to determine whether Thévenin's theorem would result in a suitable simplification of the problem. (The Thévenin impedance would be infinite!)

2.4 Maximum Power Transfer and Matching Theorems

1. A pure resistance load will abstract maximum power from a network when the load resistance is equal to the magnitude of the internal impedance of the network.

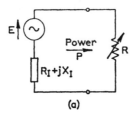

(a)

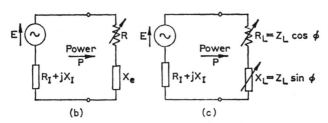

(b)   (c)

Fig. 2.10   RELATING TO MAXIMUM POWER TRANSFER

Assume that the network is represented by the constant-voltage equivalent circuit of Fig. 2.10(a):

$$\text{Load power} = I^2 R = \frac{E^2}{(R + R_I)^2 + X_I^2} R$$

For maximum load power,

$$\frac{d}{dR}\left\{\frac{E^2 R}{(R + R_I)^2 + X_I^2}\right\} = 0$$

whence

$$R = \sqrt{(R_I^2 + X_I^2)} \tag{2.7}$$

i.e. for maximum power the load resistance should equal the magnitude of the internal impedance.

2. A constant-reactance variable-resistance load will abstract maximum power from a network when the resistance of the load is equal to the magnitude of the internal impedance of the network plus the reactance of the load.

The constant-voltage equivalent circuit of Fig. 2.10(b) shows that so far as power transfer is concerned X_e could be grouped with X_I. With this grouping the proof would correspond to the previous one.

3. A variable-impedance load of constant power factor will abstract maximum power from a network when the magnitudes of the load impedance and the internal impedance are equal.

It should be noticed that a constant-power-factor load would be one in which the resistance and reactance were varied in proportion.

Let ϕ be the constant load phase angle, while $Z_L(= \sqrt{(R_L^2 + X_L^2)})$ is the magnitude of the variable load impedance. Fig. 2.10(c) shows the equivalent constant-voltage circuit. As in the previous circuit,

$$\text{Load power} = \frac{E^2 Z_L \cos\phi}{(R_1 + Z_L \cos\phi)^2 + (X_I + Z_L \sin\phi)^2}$$

For maximum power,

$$\frac{d}{dZ_L}\{\text{load power}\} = 0$$

whence

$$R_I^2 + X_I^2 = Z_L^2 \tag{2.8}$$

i.e. for maximum power the magnitude of the load impedance should equal the magnitude of the internal impedance of the generator.

4. If the load resistance and reactance are independently variable, maximum power will be abstracted when the load reactance equals the conjugate of the internal reactance and the load resistance equals the internal resistance.

Clearly, when the two reactances are equal in magnitude but of opposite sign (conjugate), the resultant reactance will be zero and the load resistance will absorb maximum power when it equals the internal resistance according to theorem 1.

Small transformers used in low-power circuits may usually be regarded as ideal: i.e. having a primary-to-secondary voltage ratio V_P/V_S equal to the turns ratio N_P/N_S, and a primary-to-secondary current ratio I_P/I_S equal to the reciprocal of the turns ratio N_S/N_P.

Fig. 2.11 shows a transformer feeding a load impedance Z_L. The

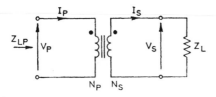

Fig. 2.11   MATCHING WITH AN IDEAL TRANSFORMER

primary input impedance, Z_{LP} (or impedance observed looking into the primary winding), is given by

$$Z_{LP} = \frac{V_P}{I_P} = V_S \frac{N_P}{N_S} \frac{1}{I_S} \frac{N_P}{N_S} = \frac{V_S}{I_S} \left(\frac{N_P}{N_S}\right)^2$$

$$= Z_L \left(\frac{N_P}{N_S}\right)^2 \tag{2.9}$$

Thus a transformer may, for circuit work, be regarded as a device which transforms impedance by the square of the turns ratio of the transformer.

A transformer is often used to obtain a maximum power transfer condition—the transformer so used being termed a *matching transformer*.

EXAMPLE 2.7   A variable-frequency generator is represented by an e.m.f., V_S, a resistance R_1 and an inductance L in series. It is to be connected, by an ideal matching transformer, to a load consisting of a resistor R_2 and a capacitor C in series. If $R_1 = 100\,\Omega$, $L = 0.1\,H$, $R_2 = 1\,k\Omega$, $C = 1\,\mu F$ and $V_S = 10\,V$, calculate (*a*) the turns ratio of the transformer to give maximum power in the load, (*b*) the frequency at which this maximum power is obtained, and (*c*) the value of the maximum power.

(*a*) The turns ratio for maximum power transfer must be such that eqn. (2.9) is satisfied for the resistive terms. The frequency at which maximum power is obtained is such that with the above turns ratio the capacitive reactance reflected into the primary side is equal to the inductive reactance of the source. Thus

$$R_1 = R_2 \left(\frac{N_P}{N_S}\right)^2$$

or

$$\frac{N_P}{N_S} = \sqrt{\frac{R_1}{R_2}} = \sqrt{\frac{100}{1,000}} = \frac{1}{3\cdot16}$$

(*b*) Also the reflected capacitive reactance is

$$X_C = \frac{1}{\omega C}\frac{N_P{}^2}{N_S{}^2} = \frac{1}{10\omega C}$$

so that, for maximum power,

$$\frac{1}{10\omega C} = \omega L \quad \text{or} \quad \omega = \sqrt{\frac{1}{10LC}} = 1,000$$

and

$$f = \omega/2\pi = \underline{\underline{159\,\text{Hz}}}$$

(*c*) For the matched condition, the power delivered from the source is

$$P_{max} = \frac{V^2}{2R_1} = \frac{100}{200} = 0\cdot5\,\text{W}$$

Hence the maximum load power is $\underline{\underline{0\cdot25\,\text{W}}}$

2.5 Millman's Theorem

Problems in which interest is centred on one particular node of a circuit (such as in an unbalanced 3-phase star-connected load or an

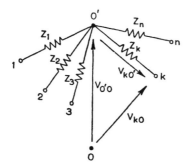

Fig. 2.12   RELATING TO MILLMAN'S THEOREM

electronic amplifier) may often be simplified by a circuit theorem due to J. E. Millman. This theorem (also called the parallel-generator theorem) states:

If any number of linear impedances $Z_1, Z_2, Z_3 \ldots$, etc., meet at a common point $0'$, and the voltages from another point 0 to the free ends of these impedances are known, the voltage $V_{0'0}$ is given by

$$V_{0'0} = \frac{\sum_{k=1}^{n} V_{k0} Y_k}{\sum_{k=1}^{n} Y_k} \tag{2.10}$$

where V_{k0} = Voltage of point k with respect to point 0

$\qquad Y_k$ = Admittance of Z_k

Proof. In Fig. 2.12, $0'$ is the common point of the impedances $Z_1, Z_2, \ldots Z_n$, and the potential differences between the point 0 and the ends $(1, 2, 3, \ldots n)$ of the impedances are known. Then round the closed loop $00'k$, the sum of the p.d.s is zero. Thus

$$V_{0'0} + V_{k0'} + V_{0k} = 0$$

or

$$V_{0'0} + V_{k0'} - V_{k0} = 0$$

whence

$$V_{k0'} = V_{k0} - V_{0'0}$$

The current through Z_k is

$$I_{k0'} = \frac{V_{k0'}}{Z_k} = V_{k0'} Y_k = (V_{k0} - V_{0'0}) Y_k$$

By Kirchhoff's first law, the sum of the currents at $0'$ is zero:

$$I_{10'} + I_{20'} + \ldots + I_{k0'} + \ldots + I_{n0'} = 0$$
$$(V_{10} - V_{0'0}) Y_1 + (V_{20} - V_{0'0}) Y_2 + \ldots$$
$$+ (V_{k0} - V_{0'0}) Y_k + \ldots = 0$$
$$V_{10} Y_1 + V_{20} Y_2 + \ldots + V_{k0} Y_k + \ldots$$
$$= V_{0'0}(Y_1 + Y_2 + \ldots + Y_k + \ldots)$$

Therefore

$$V_{0'0} = \frac{V_{10} Y_1 + V_{20} Y_2 + \ldots}{Y_1 + Y_2 + \ldots} = \frac{\sum_{k=1}^{n} V_{k0} Y_k}{\sum_{k=1}^{n} Y_k}$$

It should be noted that the impedances between the point 0 and points 1, 2, 3, etc., need not be known.

2.6 General Star–Mesh Transformation

It is always possible to transform a network of n admittances which are connected to a star point from n terminals into a corresponding mesh of admittances connecting each pair of terminals. It is, however, possible to find a unique transform from a mesh to a star only in the case of three elements (the delta–star transformation).

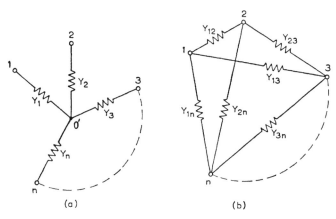

(a)   (b)

Fig. 2.13   GENERAL STAR–MESH TRANSFORMATION

Consider the star of admittances $Y_1, Y_2, \ldots Y_n$ shown in Fig. 2.13(a). If terminal 1 is taken as the reference point and $0'$ as the star point then (noting that $V_{11} = 0$) Millman's theorem gives

$$V_{0'1} = \frac{\sum_{k=1}^{n} V_{k1} Y_k}{\sum_{k=1}^{n} Y_k} = \frac{V_{21} Y_2 + V_{31} Y_3 + \ldots}{Y_1 + Y_2 + Y_3 + \ldots}$$

Hence

$$I_1 = V_{0'1} Y_1 = \frac{V_{21} Y_1 Y_2 + V_{31} Y_1 Y_3 + \ldots}{\Sigma Y_k}$$

This is the same current as would flow into terminal 1 if it were connected to terminal 2 by $Y_{12} = Y_1 Y_2/\Sigma Y_k$), to terminal 3 by $Y_{13} = (Y_1 Y_3/\Sigma Y_k)$, etc., as shown in Fig. 2.13(b). In the same way it may be shown that the current into terminal 2 in the star connection is equivalent to the current which would flow if admittances of

$Y_{21} = (Y_2 Y_1 / \Sigma Y_k)$, $Y_{23} = (Y_2 Y_3 / \Sigma Y_k)$, etc., were connected between terminal 2 and terminals 1, 3, etc., respectively. Hence the element of the equivalent mesh between any two terminals p and q is

$$Y_{pq} = \frac{Y_p Y_q}{\Sigma Y_k} \qquad (2.11)$$

This is known as *Rosen's theorem*.

2.7 Star–Delta Transformations

The particular case of the transformation from a 3-element star to a delta is important and is shown in Fig. 2.14. For this circuit, and using eqn. (2.11),

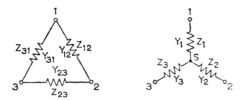

Fig. 2.14 DELTA–STAR OR STAR–DELTA TRANSFORMATION

$$Y_{12} = \frac{Y_1 Y_2}{Y_1 + Y_2 + Y_3} \qquad (2.12(\text{i}))$$

$$Y_{23} = \frac{Y_2 Y_3}{Y_1 + Y_2 + Y_3} \qquad (2.12(\text{ii}))$$

$$Y_{31} = \frac{Y_3 Y_1}{Y_1 + Y_2 + Y_3} \qquad (2.12(\text{iii}))$$

Inverting these gives the impedance forms,

$$Z_{12} = \frac{1}{Y_{12}} = Z_1 + Z_2 + \frac{Z_1 Z_2}{Z_3} \qquad (2.13)$$

with similar expressions for Z_{23} and Z_{31}.

2.8 Delta–Star Transformation

From the general star–mesh transformation it will be seen that a 3-element star becomes a 3-branch mesh or delta, whereas a 4-element star forms a 6-branch mesh. In general there are more branches in the mesh than there are elements in the corresponding star. Thus any arbitrary mesh cannot be replaced by a star since there are a greater number of variables in a mesh than in a star. The 3-branch mesh or delta is exceptional, and here the inverse transform exists.

From eqn. (2.12(i)), writing $Y = Y_1 + Y_2 + Y_3$,

$$Y_{12}Y = Y_1Y_2 \tag{i}$$

Similarly,

$$Y_{31}Y = Y_1Y_3 \tag{ii}$$

and

$$Y_{23}Y = Y_2Y_3 \tag{iii}$$

where $Y = Y_1 + Y_2 + Y_3$.

(iii)/(i) yields $Y_3 = Y_1Y_{23}/Y_{12}$
(iii)/(ii) yields $Y_2 = Y_1Y_{23}/Y_{31}$

Substituting for Y_3 and Y_2 in eqn. (2.12(i)) yields

$$Y_{12} = \frac{Y_1{}^2 Y_{23}/Y_{31}}{Y_1 + Y_1Y_{23}/Y_{31} + Y_1Y_{23}/Y_{12}}$$

so that

$$Y_1 = Y_{12} + Y_{31} + \frac{Y_{12}Y_{31}}{Y_{23}} \tag{2.14}$$

Inverting and simplifying this equation,

$$Z_1 = \frac{Z_{12}Z_{31}}{Z_{12} + Z_{23} + Z_{31}} \tag{2.15(i)}$$

Similarly,

$$Z_2 = \frac{Z_{23}Z_{12}}{Z_{12} + Z_{23} + Z_{31}} \tag{2.15(ii)}$$

and

$$Z_3 = \frac{Z_{23}Z_{31}}{Z_{12} + Z_{23} + Z_{31}} \tag{2.15(iii)}$$

EXAMPLE 2.8 Find the input impedance of the circuit shown in Fig. 2.15(a).

The circuit is first simplified by applying the Δ–Y transformation to the inductive reactances between terminals 1, 2 and 3, when the circuit becomes that shown in Fig. 2.14(b). Thus

$$Z_{10} = X_{L1} = \frac{j5 \times j10}{j30} = j1·67\,\Omega$$

$$Z_{20} = X_{L2} = \frac{j5 \times j15}{j30} = j2{\cdot}5\,\Omega$$

$$Z_{30} = X_{L3} = \frac{j10 \times j15}{j30} = j5\,\Omega$$

Hence

$$Z_{024} = j2{\cdot}5 - j22{\cdot}5 = -j20 = 20\underline{/-90^\circ}\,\Omega$$

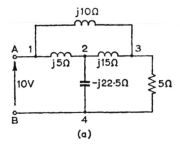
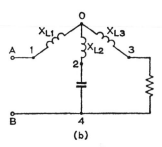

Fig. 2.15

Therefore

$$Y_{024} = \frac{1}{20\underline{/-90^\circ}} = 0{\cdot}05\underline{/90^\circ} = j0{\cdot}05\ \text{S}$$

and

$$Z_{034} = 5 + j5 = 7{\cdot}07\underline{/45^\circ}\,\Omega$$

Therefore

$$Y_{034} = \frac{1}{7{\cdot}07\underline{/45}} = 0{\cdot}141\underline{/-45^\circ} = (0{\cdot}1 - j0{\cdot}1)\text{S}$$

so that

$$Y_{0B} = Y_{024} + Y_{034} = j0{\cdot}05 + 0{\cdot}1 - j0{\cdot}1 = 0{\cdot}112\underline{/-26^\circ\ 34'}\text{S}$$

Therefore

$$Z_{0B} = 8{\cdot}95\underline{/26^\circ\ 34'} = (8 + j4)\,\Omega$$

and

$$Z_{AB} = j1{\cdot}67 + (8 + j4) = 8 + j5{\cdot}67 = 9{\cdot}8\underline{/35^\circ\ 45'}\,\Omega$$

2.9 Reciprocity Theorem

If an e.m.f. acting in one branch of a network causes a current I to flow in a second branch, the same e.m.f. acting in the second branch would produce the same current in the first branch.

An obvious conclusion from the theorem is that in a Wheatstone

bridge network the source and the galvanometer may be interchanged. The principal application of the reciprocity theorem is to
four-terminal and transmission-line networks.

2.10   Compensation Theorem

If a change, ΔZ say, is made to the impedance of any branch of a
network where the current was originally I, then the change of current
at any other point in the network may be calculated by assuming
that an e.m.f. of $-I\Delta Z$ has been introduced into the changed branch,
while all other sources have their e.m.f.s suppressed and are represented by their internal impedances only.

EXAMPLE 2.9   A 100 V battery supplies the current shown in Fig. 2.16(*a*).

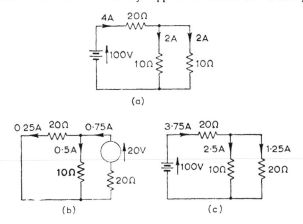

Fig. 2.16   ILLUSTRATING THE COMPENSATION THEOREM

Calculate the new currents if one of the 10 Ω resistors were increased to 20 Ω.

The circuit for the calculation of the current change by the compensation
theorem is shown in Fig. 2.16(*b*). Since the change of resistance was an increase,
ΔZ is positive, and the generator voltage opposes the original current, the equivalent e.m.f. being $-I\Delta Z$. The compensating currents resulting from the application of this e.m.f. are shown in Fig. 2.16(*b*). If these currents are added to the
original currents in the corresponding limbs, the final current distribution will
be as indicated at (*c*). In this simple case the result may readily be verified by a
series-parallel method.

PROBLEMS

2.1   Three batteries A, B and C have their negative terminals connected together.
The positive terminal of A is connected to the positive terminal of B by a resistance

of $0 \cdot 3\,\Omega$, and the positive terminal of B is connected to the positive terminal of C by a resistance of $0 \cdot 45\,\Omega$. The respective e.m.f.s and resistances of the batteries are: battery A, 100 V, $0 \cdot 25\,\Omega$; battery B, 105 V, $0 \cdot 2\,\Omega$; and battery C, 95 V, $0 \cdot 15\,\Omega$. Calculate the current in each of the external resistors and the p.d. across the battery B.   (*C. & G. Inter.*)

Ans. 3·6 A, 11·65 A, 102 V.

2.2 In the circuit shown in Fig. 2.17 transform the star, ABC, to a delta and then apply Thévenin's theorem to find the voltage across the $30\,\Omega$ resistor.

(*H.N.C. part question*)

Ans. 3·33 V.

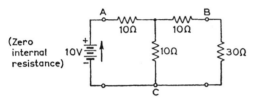

Fig. 2.17

2.3 A network is composed of the following resistances: $AB = 0 \cdot 1\,\Omega$; $BC = 0 \cdot 2\,\Omega$; $CD = 0 \cdot 1\,\Omega$; $DA = 0 \cdot 1\,\Omega$; $AC = 0 \cdot 2\,\Omega$.

A current of 80 A is fed into the network at A and currents of 25 A, 35 A, and 20 A leave at the points B, C and D respectively. Calculate the current in AC.

(*H.N.C.*)

Ans. 20 A.

2.4 A generator A supplies a load B of 50 A through a $0 \cdot 1\,\Omega$ distributor, and a load C of 30 A through a $0 \cdot 15\,\Omega$ distributor. B and C are joined by a $0 \cdot 2\,\Omega$ interconnector. Find the magnitude and direction of the current in the interconnector by (*a*) Thévenin's theorem, (*b*) superposition theorem.   (*H.N.C.*)

Ans. 1·11 A from C to B.

2.5 A constant-voltage generator has an internal resistance of $5{,}000\,\Omega$, and the generated e.m.f. is 200 V at a frequency of 1 kHz. Deduce the equivalent constant-current generator. If the load on the generator consists of a resistance of $4{,}000\,\Omega$ in parallel with a capacitance of $0 \cdot 1\,\mu\mathrm{F}$ determine, using Norton's theorem, the voltage across the capacitor.

Ans. 51·8 V.

2.6 A generator has an output impedance of $(600 + j50)\,\Omega$. Calculate the turns ratio of an ideal transformer necessary to match the generator to a load of $(65 + j30)\,\Omega$ for maximum transfer of power. Prove any formula used.

(*L.U.*)

Ans. 2·9.

2.7 A Wheatstone bridge network has the following components: $AB = 1\,\Omega$; $BC = 1 \cdot 7\,\Omega$; $AD = 4\,\Omega$; $DC = 6\,\Omega$.

A 10 V d.c. supply of internal impedance $2\,\Omega$ is connected across terminals A and C. Determine, using Thévenin's theorem or the compensation theorem. the current in a $100\,\Omega$ resistor connected between terminals B and D.

Ans. 1·49 mA.

2.8 Transform the star-connected impedances C, C, and R_2 to a delta in the bridged-T circuit shown in Fig. 2.18 and hence show that the voltage between D and E is zero when

$$R_1 = \frac{1}{R_2\omega^2C^2} \quad \text{and} \quad \omega L = \frac{2}{\omega C}$$

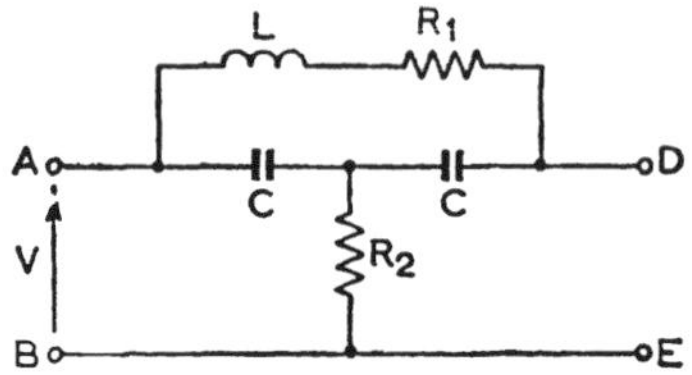

Fig. 2.18

2.9 A small transformer has primary and secondary inductances of 450μH and 35μH respectively and negligible resistance. The load on the secondary is a resistance of 15Ω, and the primary forms part of a series resonant circuit to which an e.m.f. of frequency $5/2\pi$ MHz is applied by a generator of internal resistance 20Ω. Find the mutual inductance between the primary and secondary windings, and the setting of the tuned circuit capacitor to make the power developed in the load a maximum.

(*H.N.C.*)

Ans. $40\cdot6\mu$H; $99\cdot4$pF.

HINT. Convert the inductive coupling to an equivalent circuit as in Section 1.10.

2.10 A generating station A with a line voltage of 11kV supplies two substations B and C through two independent feeders, the substations also being interconnected by another feeder.

The impedances of the feeders are: A to B, $(2 + j4)\Omega$; A to C, $(2 + j3)\Omega$; B to C, $(3 + j5)\Omega$. The load at B is 100A at $0\cdot8$ power factor lagging, and at C, 70A at $0\cdot9$ lagging.

Calculate the current flowing in each feeder, and also the voltage between B and C if the feeder BC is removed.

(*L.U.*)

Ans. 86A, 84A, $14\cdot1$A, $196\cdot4$V.

2.11 Show that the voltage $V_{0'0}$ in Fig. 2.19 is given by

$$V_{0'0} = \frac{E_1Y_1 + E_2Y_2 + E_3Y_3}{Y_1 + Y_2 + Y_3}$$

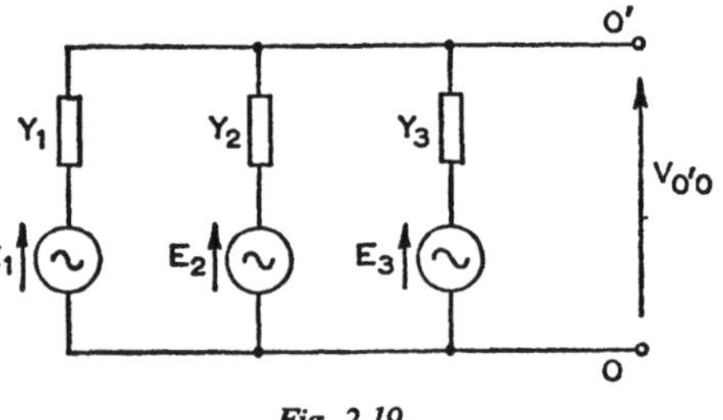

Fig. 2.19

where the generators marked E_1, E_2 and E_3 are all of the same frequency. If the instantaneous e.m.f.s are $e_1 = -10\sqrt{2} \sin 1{,}000t$, $e_2 = 20\sqrt{2} \cos 1{,}000t$ and $e_3 = 15\sqrt{2} \sin 1{,}000t$, and if $Y_1 = -j0\cdot1$, $Y_2 = j0\cdot05$ and $Y_3 = 0\cdot067$ S find the r.m.s. value of the magnitude of $V_{0'0}$ and its phase relationship to E_1.

(*H.N.C.*)

Ans. 12V, 53·3 lagging.

Chapter 3

MEASUREMENT CIRCUITS

The Wheatstone bridge network may be adapted for a.c. measurements by making the supply an alternating one of the frequency desired, and using a detector which is sensitive to alternating currents. The a.c. bridge network is used for comparison measurements of resistance, inductance and capacitance to a high degree of precision.

3.1 Standards

Primary electrical standards have values which may be determined with reference to the units of mass, length, time, and one arbitrarily chosen electrical constant, which is the unit of electric current, namely the ampere. A determination of any electrical quantity in the above terms is called an absolute measurement. Electrical units and standards form part of the internationally adopted Système Internationale d'Unités, or SI units, in which there are six fundamental units, the kilogramme, metre, second, ampere, kelvin and candela.*

Primary electrical standards of mutual inductance (the Campbell mutual inductance at the National Physical Laboratory) and self-inductance have been constructed so that the inductances are accurately known in terms of their physical dimensions and the ampere. The standard mutual inductor is used to calibrate variable

* The kelvin (formerly the degree kelvin) is the absolute unit of thermodynamic temperature. The candela is the absolute unit of luminous intensity.

laboratory standard mutual inductors which may then be used for further measurements. For instance, a resistor may be calibrated in terms of the product of a mutual inductance and a speed of rotation of a disc, as in the Lorenz method.

No primary reference standard of current is possible, but current can be measured in absolute terms by measuring the force exerted between two circuits carrying the same current. If the current is passed through a standard resistor, the resulting known p.d. may be used to calibrate a standard of e.m.f. (i.e. the standard cell).

The three laboratory reference standards are thus

(*a*) A standard variable mutual inductor (mutual inductometer) which usually has a range of up to 11·1 mH from zero.

(*b*) Standard resistors, calibrated as has already been indicated.

(*c*) A standard e.m.f. derived from a Weston cadmium cell.

Laboratory measurements can be made in terms of these reference standards.

It is customary also to have standards of capacitance available. Primary standards, in which the capacitance is measured in terms of the physical dimensions, are theoretically possible, but the difficulties involved are great and normally only capacitors whose values are determined by comparison measurements are used. These are called secondary standards, and their values are usually known to a very high degree of accuracy.

The following a.c. bridges will illustrate methods by which inductance and capacitance may be compared with mutual inductance and resistance, and further circuits will illustrate the use of the secondary standards of capacitance. Bridges employing self-inductance standards are uncommon.

3.2 Balance Conditions

In the same way as for the d.c. Wheatstone bridge network, an alternating-current bridge is said to be balanced when the current through the detector is zero. Fig. 3.1 shows a generalized bridge circuit, in which Z_1, Z_2, Z_3 and Z_4 are the impedances (in complex form) of the bridge arms. If the current through the detector is zero, then the current I_1 in Z_1 must also flow through Z_2, and the current I_4 in Z_4 must also flow through Z_3. Equating the voltage drops between A and B, and A and D, gives

$$I_1 Z_1 = I_4 Z_4$$
$$I_1 Z_2 = I_4 Z_3$$

whence

$$Z_1 Z_3 = Z_2 Z_4 \tag{3.1}$$

This equality represents the balance conditions of the bridge.
Normally one of the bridge arms contains the unknown impedance
while the other arms contain known fixed or variable comparison

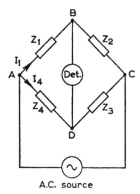

Fig. 3.1   GENERAL A.C. BRIDGE CIRCUIT

standards. The bridge is set up, and the current through the detector
is reduced to zero by successive adjustments of the variable circuit-
elements (usually only two elements in the bridge are variable).
When this is achieved the unknown impedance may be expressed
in terms of the comparison standards.

It should be noted that eqn. (3.1) will be in complex form, so
that for balance the reference terms on each side must be equated,
and also the quadrate terms. Balance will obviously be achieved
quickly if there is one variable in the reference terms, and the other
variable in the quadrate terms. If both variables appear in either
of the resulting balance equations, balance will not be achieved
quickly; the bridge is then said to be slow to converge.

3.3 Detectors

The detectors used to determine the balance point in an a.c. bridge
vary with the type of bridge and with the frequency at which it is
operated.

The cathode-ray tuning indicator can be used over a very wide
range of frequencies.

For mains-operated bridges, a suitable detector is the vibration
galvanometer. This consists essentially of a narrow moving coil
which is suspended on a fine phosphor-bronze wire between the

poles of a magnet. The mechanical resonant frequency of the suspension is made equal to the electrical frequency of the coil current, so that, when a current of the correct frequency flows through the coil, it is set into vibration. A small mirror attached to the coil reflects a spot of light on to a scale. When the coil is vibrating this spot appears as an extended band of light. Balance of the bridge is indicated when the band reduces to a spot.

The vibration galvanometer is insensitive to frequencies other than the one to which it is tuned. This tuning may be adjusted by altering the tension of the suspension. Vibration galvanometers are constructed for frequencies ranging from 10 to about 300 Hz.

For audio frequencies a telephone headset is often used as a bridge detector. Since the human ear is very sensitive in the 1,000 Hz frequency region, this is a common a.c. bridge frequency. The disadvantage of telephones is the fact that if other frequencies are present in the a.c. supply (harmonics) these will also be heard, and zero current may not be obtained when the bridge is balanced.

Frequency-sensitive heterodyne detectors are used where extreme sensitivity in detection is desired or when frequencies are above the audio range.

3.4 Owen Bridge for Inductance

The derivation of the balance conditions for an a.c. bridge are illustrated by the Owen bridge (Fig. 3.2), which measures the resistance, R_x, and inductance, L_x, of a coil in terms of fixed and variable

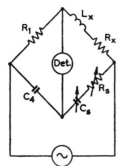

Fig. 3.2 OWEN BRIDGE

resistors and capacitors. By suitable choice of bridge elements a wide range of inductance can be measured. The bridge may be simply modified to measure the inductance and losses in iron-cored coils which are subject to both a.c. and d.c. magnetization.

The balance conditions are

$$(R_x + j\omega L_x)\frac{1}{j\omega C_4} = R_1 \left(R_s + \frac{1}{j\omega C_s} \right)$$

Equating the quadrate terms,

$$R_x = R_1 \frac{C_4}{C_s} \tag{3.2}$$

and, from the reference terms,

$$L_x = C_4 R_1 R_s \tag{3.3}$$

Note that the variable standards affect each balance condition independently, i.e. varying C_s will not affect the balance for L_x, and varying R_s will not affect the balance for R_x. This means that a very quick and accurate balance can be obtained.

3.5 Anderson Bridge for Inductance

This bridge may be used to measure inductances ranging from a few microhenrys up to the order of one henry. The schematic

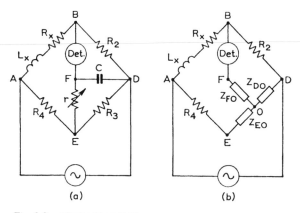

Fig. 3.3 ANDERSON BRIDGE

circuit is shown in Fig. 3.3(a). The arm AB contains the coil whose inductance, L_x, and resistance, R_x, it is desired to measure. The components of the other arms of the bridge are known standards. Balance may be obtained by variation of r and either R_2, R_3 or R_4.

To obtain the balance conditions the delta of impedances FDE Fig. 3.3(a)) is first transformed to an equivalent star. The bridge) then reduces to the circuit shown at (*b*). If O is the star point, then from eqn. (2.15),

$$Z_{EO} = \frac{rR_3}{r + R_3 + 1/j\omega C}$$

Also,

$$Z_{DO} = \frac{R_3/j\omega C}{r + R_3 + 1/j\omega C}$$

The impedance Z_{FO} does not affect the balance of the bridge since it is in series with the detector. Thus at balance,

$$(R_x + j\omega L_x) \frac{R_3}{j\omega C(r + R_3 + 1/j\omega C)} = R_2 \left(R_4 + \frac{rR_3}{r + R_3 + 1/j\omega C} \right)$$

so that

$$R_x + j\omega L_x = \frac{R_2}{R_3} j\omega C\{R_4(r + R_3 + 1/j\omega C) + rR_3\}$$

Equating the reference terms,

$$R_x = \frac{R_2 R_4}{R_3} \tag{3.4}$$

and equating the quadrate terms,

$$L_x = R_2 C \left(\frac{R_4 r}{R_3} + R_4 + r \right) \tag{3.5}$$

3.6 Loss in Capacitors

In capacitors with solid dielectrics there is a power loss due to leakage currents, and also a dielectric heat loss (analogous to the hysteresis loss in magnetic circuits) when an alternating voltage is applied to the capacitor. The total loss may be represented as the loss in an additional resistance connected between the plates. The dielectric loss will normally exceed the leakage loss except in air dielectrics. The total alternating current passing through the capacitor will be made up of (*a*) the capacitive current plus (*b*) the loss current, and the equivalent circuit of such a capacitor will consist of a pure capacitance, C_p, in parallel with a high resistance R_p as shown in Fig. 3.4(*a*). The complexor diagram for the arrangement is shown at (*b*), where I_a is the loss current, and I_c is the capacitive current, the voltage across the capacitor being V.

Normally the loss current will be very much smaller than the capacitive current, so that the resultant current will lead the voltage by an angle which is nearly 90°. The difference between 90° and the actual phase angle of the capacitor is the angle δ in Fig. 3.4(*b*), and this is termed the *loss angle* of the capacitor.

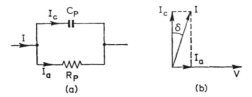

Fig. 3.4   IMPERFECT CAPACITOR
(*a*) Equivalent circuit   (*b*) Complexor diagram

It is also possible to represent an imperfect capacitor by an equivalent series circuit, in which a capacitance C_s is connected in series with a low resistance R_s.

For the equivalent parallel circuit,

$$\tan \delta = I_a/I_c = 1/R_p\omega C_p \tag{3.6}$$

and the impedance is approximately $1/\omega C_p$, since the parallel loss resistance will be very much greater than the capacitive reactance. For the equivalent series circuit, $\tan \delta$ is $R_s\omega C_s$, and the impedance is approximately $1/\omega C_s$ since the series loss resistance will be very much smaller than the capacitive reactance. For equivalence between the series and the parallel circuits we have

$$C_s \approx C_p = C \tag{3.7}$$

and

$$R_s \approx 1/R_p\omega^2C^2 \tag{3.8}$$

3.7 Modified Carey Foster Bridge for Capacitance

The circuit of the bridge is shown in Fig. 3.5. The unknown capacitor is represented by its equivalent series circuit, and a known variable resistor R_3 is connected in series with it to form the arm BD. The arm EA is a short-circuit. The mutual inductometer M must be connected with the winding directions indicated by the dot notation, or balance will be impossible.

Let L be the self-inductance of the secondary of the mutual inductometer, and let the resistance R_4 include the resistance of this winding. Then the total current I taken from the bridge supply will divide at point E, into I_A and I_B.

If the detector current is zero, the voltage drop between E and B must be zero giving

$$I_B(R_4 + j\omega L) - j\omega M(I_A + I_B) = 0$$

Therefore

$$j\omega M I_A = (R_4 + j\omega L - j\omega M)I_B \qquad (3.9)$$

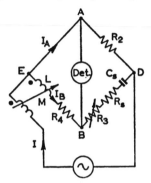

Fig. 3.5 MODIFIED CAREY FOSTER BRIDGE

In the same way the voltage drop from A to D must be equal to that from B to D, and further, the current I_A must flow through arm AD and the current I_B through BD. Hence

$$I_A R_2 = \left(R_3 + R_s - j\frac{1}{\omega C_s}\right) I_B \qquad (3.10)$$

Dividing eqn. (3.9) by eqn. (3.10) and cross-multiplying,

$$j\omega M\{(R_3 + R_s) + 1/j\omega C_s\} = R_2\{R_4 + j\omega(L - M)\}$$

Equating the reference terms,

$$C_s = \frac{M}{R_2 R_4} \qquad (3.11)$$

and equating the quadrate terms,

$$R_s = \frac{R_2}{M}(L - M) - R_3 \qquad (3.12)$$

3.8 Schering Bridge for Capacitance

The bridges so far considered have operated with supply voltages of the order of 10 V. The Schering bridge was developed to measure the loss resistance of dielectrics, line insulators, cables and high-voltage capacitors under high-voltage conditions (up to 100 kV). The bridge is shown in Fig. 3.6. The unknown capacitance is represented by the equivalent series circuit (C_s, R_s). C_1 is a fixed high-voltage air capacitor, whose value is of the order of 50 pF. R_2 is a fixed non-inductive resistor, while the resistor R_3 and the capacitor C_2 (mica type) are the variable elements. R_2 and R_3 are normally of the order of a few hundred ohms, so that their impedance will be negligible compared with that of C_s or C_1. This means

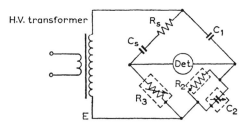

Fig. 3.6   HIGH-VOLTAGE SCHERING BRIDGE

that the voltage across the variable arms of the bridge (and the detector) will be only a very small fraction of the supply voltage, so that the bridge will be safe to operate despite the high voltage. Protective earthed screening is always employed to reduce the danger to the operator, and to stabilize stray leakage currents.

Substituting the appropriate values in the balance equation (3.1),

$$\frac{R_3}{j\omega C_1} = \left(R_s + \frac{1}{j\omega C_s}\right)\frac{1}{\dfrac{1}{R_2} + j\omega C_2}$$

Therefore

$$C_s = C_1 \frac{R_2}{R_3} \tag{3.13}$$

and

$$R_s = R_3 \frac{C_2}{C_1} \tag{3.14}$$

The loss angle of the test capacitor is

$$\delta = \tan^{-1} R_s \omega C_s = \tan^{-1} \omega C_2 R_2 \tag{3.15}$$

3.9 Campbell–Heaviside Equal-ratio Bridge

This bridge (Fig. 3.7) is used for the determination of small inductances in terms of a laboratory standard mutual inductance. R_1 and R_4 are two equal low-inductance standard resistors, while R_3 is a variable resistor whose value is of the same order as the resistance of the unknown coil (L_x and r_x). The mutual inductometer has mutual inductances M_x and M_y between the primary and the two halves of the secondary, and M_s is the mutual inductance between these halves themselves. The self-inductances of the secondary windings

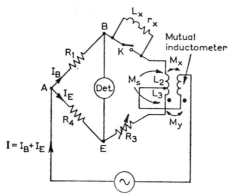

Fig. 3.7 CAMPBELL–HEAVISIDE EQUAL-RATIO BRIDGE

are L_2 and L_3. At balance there is no current through the detector, so that the total input current I will divide equally between paths AB and AE (since R_1 and R_4 are equal). Thus

$$I_B = I_E = \tfrac{1}{2}I \tag{3.16}$$

With key K open,

$$I_B(r_x + j\omega L_x) + I_B j\omega L_2 - I j\omega M_x - I_E j\omega M_s$$
$$= I_E(R_3 + j\omega L_3) + I j\omega M_y - I_B j\omega M_s$$

Substituting from eqn. (3.16),

$$\tfrac{1}{2}I\{r_x + j\omega(L_x + L_2)\} - I j\omega(M_x + \tfrac{1}{2}M_s)$$
$$= \tfrac{1}{2}I(R_3 + j\omega L_3) + I j\omega(M_y - \tfrac{1}{2}M_s)$$

Equating the reference terms,

$$r_x = R_3 \tag{3.17}$$

and equating the quadrate terms,

$$L_x + L_2 - L_3 = 2(M_x + M_y) \tag{3.18}$$

$(M_x + M_y)$ is the dial reading on the mutual inductometer. The method for small inductances is to balance the bridge first with key K open, and then with the key closed. The difference between the results gives the values of the unknown inductance and resistance, all residual errors (which would be important in this case) being cancelled out.

3.10 Frequency-dependent Bridges

In many forms of a.c. bridge the balance conditions are dependent on the frequency of the source. A bridge of this nature may therefore

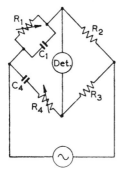

Fig. 3.8 WIEN PARALLEL BRIDGE

be used to measure the frequency of an a.c. supply. A simple bridge of this type has a series resonant circuit (with variable capacitance) as one arm, the other arms being pure resistors of suitable value. At resonance the resonant circuit has no reactance, and an adjustment of one of the resistors will give balance.

Another frequency-dependent bridge which does not rely on inductance is the Wien parallel bridge illustrated in Fig. 3.8. R_3 is a standard resistor of twice the value of R_2, and C_1 and C_4 are equal standard capacitors. R_1 and R_4 are resistors which can be varied together so that their values remain equal.

At balance,

$$R_3 \frac{1}{\dfrac{1}{R_1} + j\omega C_1} = R_2 \left(R_4 + \frac{1}{j\omega C_4} \right)$$

Hence

$$\frac{R_3}{R_2} = \frac{R_4}{R_1} + \frac{C_1}{C_4} + j \left(\omega C_1 R_4 - \frac{1}{\omega C_4 R_1} \right)$$

and

$$\omega^2 = \frac{1}{C_1 C_4 R_1 R_4}$$

Therefore

$$f = \frac{1}{2\pi CR} \tag{3.19}$$

where $C = C_1 = C_4$, and $R = R_1 = R_4$.

Note that with the values chosen for R_1, R_2, R_3, R_4, C_1 and C_4,

$$\frac{R_3}{R_2} = 2 = \frac{R_4}{R_1} + \frac{C_1}{C_4}$$

If f is known the bridge may be used to measure capacitance.

3.11 Accuracy of Bridges

In bridge measurements it is desirable to be able to calculate the accuracy of the measurement in terms of the accuracy of the known bridge elements, and the accuracy with which balance is achieved. The comparison standards used in bridges normally have an error of about ± 0.02 per cent or less. The accuracy with which balance conditions are achieved varies from bridge to bridge, but the error should be less than ± 0.5 per cent.

Suppose that the balance equation for a bridge is in the form

$$X = \frac{AB}{C}$$

Then

$$\log_e X = \log_e A + \log_e B - \log_e C$$

Therefore

$$\frac{dX}{X} = \frac{dA}{A} + \frac{dB}{B} - \frac{dC}{C} \tag{3.20}$$

If the percentage errors of A, B and C are known, then eqn. (3.20) shows that the percentage error with which X is determined is simply the *sum* of the percentage errors of A, B, and C. The negative sign in front of C does not count since errors are always given as plus or minus.

If the balance conditions result in the subtraction of two quantities, then the actual limits of each quantity must be determined (not the percentage error). These limits are then added to give the limits of the final result. If the two subtracted quantities are almost equal, the resultant error can be large.

EXAMPLE 3.1   The Schering bridge shown in Fig. 3.6 is used to measure the capacitance and loss resistance of a length of concentric cable. The supply voltage is 100 kV at 50 Hz. C_1 is an air capacitor of 40 pF, R_2 is fixed at $1,000/\pi$ ohms, R_3 is $122 \pm 0.5\Omega$, and C_2 is $0.921 \pm 0.001 \mu F$.

Determine (a) the cable capacitance, (b) the parallel loss resistance, (c) the loss angle of the cable, and (d) the power loss in the cable.

$$\text{Percentage error of } R_3 = \pm \frac{0.5}{122} \times 100 = \pm 0.4\%$$

$$\text{Percentage error of } C_2 = \frac{0.001}{0.921} \times 100 = 0.108\%$$

(a) From the balance condition of eqn. (3.13),

$$C_s = 40 \times 10^{-12} \times \frac{1,000}{\pi(122 \pm 0.4\%)} = \underline{104.3\,\text{pF} \pm 0.4\%}$$

(b) From the balance condition of eqn. (3.14),

$$R_s = \frac{(122 \pm 0.4\%)(0.921 \pm 0.108\%) \times 10^{-6}}{40 \times 10^{-12}} = 2.81\,\text{M}\Omega \pm 0.508\%$$

The parallel loss resistance of the cable is the equivalent parallel resistance corresponding to R_s, namely

$$R_p = \frac{1}{R_s \omega^2 C_s^2} = \frac{1}{\omega^2}\frac{C_1}{R_3 C_2}\left(\frac{R_3}{C_1 R_2}\right)^2 = \frac{1}{\omega^2}\frac{R_3}{C_1 C_2 R_2^2}$$

$$= \frac{1}{314^2}\frac{122 \pm 0.4\%}{(40 \times 10^{-12})(0.921 \pm 0.108\%)10^{-6}(1,000/\pi)^2}$$

$$= \underline{332\,\text{M}\Omega \pm 0.508\%}$$

(c) From eqn. (3.15),

$$\delta = \tan^{-1}\left(100\pi \times 0.921 \times \frac{1,000}{\pi} \times 10^{-6} \pm 0.108\%\right)$$

$$= 0.0921\,\text{rad} \pm 0.108\%$$

since for small angles $\tan \delta = \delta$.

(d) Since R_3 is so small compared with R_s, the whole supply voltage may be considered to be across C_s and R_s. Therefore

$$\text{Power loss} = \frac{V^2}{R_p} = \frac{10^{10}}{332 \times 10^6} = \underline{30.1\,\text{W}}$$

3.12 Stray Effects and Residuals

Where extreme accuracy is desired from a.c. bridges the effects of residual inductance and capacitance in the standard resistors, and losses in capacitors, should be allowed for by including appropriate terms in the balance equations. In addition to those effects there will be stray effects which are generally not calculable.

Stray electromagnetic coupling between components may be minimized by (a) using non-inductive resistors, (b) having all inductance elements in toroidal form so that there is little external

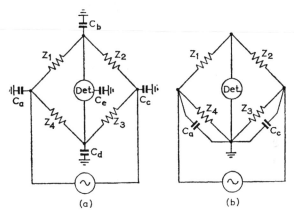

Fig. 3.9 STRAY CAPACITANCES IN A.C. BRIDGES

magnetic field, and (c) twisting the leads to the various components where possible.

Stray electrostatic coupling between components is minimized by ensuring adequate spacing between them.

More serious is the capacitance between components and earth. Where possible screened components are used, the capacitance between the components and the screen then being measurable and fixed. The stray earth capacitance of the detector will also cause a current through it which will make true balance impossible. This is termed "head effect" and can only be eliminated if the detector is at earth potential at balance.

Fig. 3.9(a) shows diagrammatically the stray earth capacitances in an a.c. bridge. The effect of connecting one end of the detector to earth is shown at (b). C_b at (a) simply shunts the detector, and so may be neglected. So also may C_e since the detector is at earth potential. C_a and C_c, however, now appear across the arms Z_4 and Z_3 so introducing an unknown factor into the balance conditions.

For this reason direct earthing of the detector is not employed.

A device which overcomes this difficulty is the *Wagner earth*, shown in Fig. 3.10. Z_1 and Z_2 are preferably the fixed arms of the bridge, while Z_5 and Z_6 are additional elements which must be made to balance with Z_1 and Z_2. The junction point of Z_5 and Z_6 is solidly earthed so that the stray capacitances C_a and C_c are in parallel with them. An approximate balance is obtained with the switch in position 1. The switch is then moved to position 2, and a further balance is obtained by varying Z_5 and/or Z_6. When this is achieved the point A must be at earth potential. The switch is then moved back to position 1 and a further balance is obtained. When this is achieved the detector must still be at earth potential

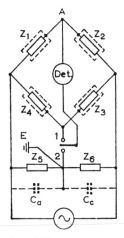

Fig. 3.10   WAGNER EARTHING DEVICE

and so there will be no stray capacitance effects. The values of Z_5 and Z_6 need not, of course, be accurately known.

3.13 Transformer Ratio-arm Bridges

Unlike the bridge circuits considered so far, the transformer ratio-arm bridge depends on an ampere-turn balance in a transformer. Admittance measurements over a wide range of frequencies up to some 250 MHz are possible. The basic circuit is shown in Fig. 3.11(a), where Y_u is the unknown admittance and Y_s is a standard variable. Assuming ideal transformers the detector will indicate a null when there are no net ampere-turns in the output transformer T, i.e. When

$$I_u N_1 = I_s N_2$$

Neglecting leakage reactance and winding resistance there is

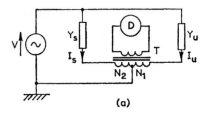

(a)

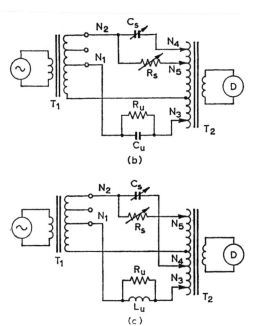

(b)

(c)

Fig. 3.11 TRANSFORMER RATIO-ARM BRIDGE

then no voltage drop across either input section of the transformer. Hence

$$I_u = VY_u \quad \text{and} \quad I_s = VY_s$$

so that

$$\frac{Y_u}{Y_s} = \frac{I_u}{V}\frac{V}{I_s} = \frac{N_2}{N_1}$$

or

$$Y_u = \frac{N_2}{N_1} Y_s \qquad (3.21)$$

In practice the leakage reactances of the transformers can be made to balance out and resistances can be made negligible. An important feature of the bridge is that one terminal of both source and detector can be earthed. Further, if the centre tap of transformer T is earthed, then at balance one end of both the standard and the unknown will be earthed. Strays may readily be taken into account by setting the standard admittance to zero, and balancing without the unknown connected, by means of auxiliary variables in the standard arm which need not be calibrated. In this way the effect of long leads to the unknown may be eliminated.

A simplified set-up for the measurement of an unknown capacitance is shown in Fig. 3.11(*b*). Balancing components are used to eliminate stray effects, and the bridge is rearranged to increase its flexibility.

Taking resistive and reactive balance conditions separately,

$$R_u = \frac{N_1 N_3}{N_2 N_5} R_s \quad \text{and} \quad C_u = \frac{N_2 N_4}{N_1 N_3} C_s \tag{3.22}$$

To measure an unknown inductance, the standard capacitor is connected to the same side of transformer T_2 as the unknown, as shown in Fig. 3.11(*c*). At balance,

$$R_u = \frac{N_1 N_3}{N_2 N_5} R_s \quad \text{and} \quad \frac{1}{j\omega L_u} = -\frac{N_2 N_4}{N_1 N_3} j\omega C_s$$

or

$$L_u = \frac{N_1 N_3}{N_2 N_4} \frac{1}{\omega^2 C_s} \tag{3.23}$$

For this measurement the frequency must be accurately known.

In commercial bridges the turns-ratio terms are directly read as multipliers (usually decades) on dials.

Errors can be reduced to around 0·1 per cent in audio-frequency bridges, while from 50 to 250 MHz, accuracy within 1 or 2 per cent is still possible.

3.14 Bridged-T and Parallel-T Networks

These networks have the advantage over 4-arm bridges that one end of both source and detector may be solidly earthed. Such networks can be used for measurements at frequencies above the audio-frequency range.

Consider the bridged-T network shown in Fig. 3.12(a). At balance there is no current through the detector, and hence

$$I_2 = -I_4$$

Also point A must then be at earth potential, so that

$$I_4 = \frac{V}{Z_4} \quad \text{and} \quad I_2 = \frac{V}{\{Z_1 + Z_2 Z_3/(Z_2 + Z_3)\}} \frac{Z_3}{Z_2 + Z_3}$$

$$= \frac{V Z_3}{Z_1 Z_2 + Z_1 Z_3 + Z_2 Z_3}$$

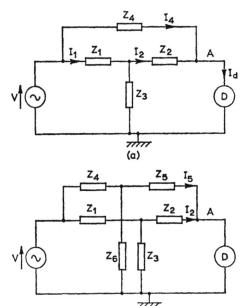

(a)

(b)

Fig. 3.12   BRIDGED-T AND PARALLEL-T NETWORKS

Hence

$$-Z_4 = Z_1 + Z_2 + Z_1 Z_2/Z_3$$

or

$$Z_1 + Z_2 + Z_4 + Z_1 Z_2/Z_3 = 0 \qquad (3.24)$$

For the parallel-T circuit shown in Fig. 3.12(b), if the detector current is zero it follows that

$$I_2 = -I_5$$

Since point A is then at earth potential,

$$I_5 = \frac{VZ_6}{Z_4Z_5 + Z_4Z_6 + Z_5Z_6}$$

and

$$I_2 = \frac{VZ_3}{Z_1Z_2 + Z_1Z_3 + Z_2Z_3}$$

Hence

$$Z_4 + Z_5 + Z_4Z_5/Z_6 = -(Z_1 + Z_2 + Z_1Z_2/Z_3) \tag{3.25}$$

represents the balance conditions.

EXAMPLE 3.2 The bridged-T circuit of Fig. 3.13 is used to measure the inductance L and series resistance R of a coil at 3·18 MHz. If $C = 45·9\,\text{pF}$ and $R_3 = 10·4\,\text{k}\Omega$ find L and R.

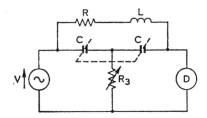

Fig. 3.13 TYPICAL BRIDGED-T CIRCUIT

In this circuit the two capacitors are ganged together to form one variable, while R_3 forms the other.

From eqn. (3.24), at balance

$$\frac{1}{j\omega C} + \frac{1}{j\omega C} - \frac{1}{\omega^2 C^2 R_3} + R + j\omega L = 0$$

Equating reference terms,

$$R = \frac{1}{\omega^2 C^2 R_3} = \frac{1}{4 \times 10^{14} \times 21 \times 10^{-22} \times 10·4 \times 10^3} = \underline{\underline{115\,\Omega}}$$

and equating quadrate terms

$$L = \frac{2}{\omega^2 C} = \frac{2}{4 \times 10_{14} \times 45·9 \times 10^{-12}} = \underline{\underline{109\,\mu\text{H}}}$$

3.15 The Q-meter

At radio frequencies it is often convenient to be able to measure the Q-factor of a coil directly. The Q-meter is designed specifically to do this, and can also be used to measure inductance and capacitance. The method is not a null but a resonance method, and since it depends on the calibration of a meter, may involve errors of 1 or 2 per cent. However, since the measurement is at maximum current the effect of strays is minimized.

The basic circuit of a Q-meter is shown in Fig. 3.14. A variable-frequency oscillator is loosely coupled to a very low resistance, r. The unknown coil is inserted in series with a calibrated standard variable air capacitor C_s to form a series resonant circuit. If r is much smaller than R, the voltage, V, applied to the tuned circuit will be constant and equal to Ir. The current I is measured by a thermocouple ammeter, A, and can be set at some standard known value by altering the coupling between the oscillator and the load

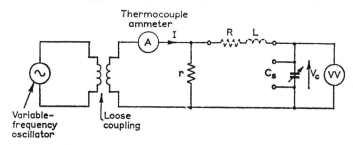

Fig. 3.14 THE Q-METER

circuit. With the oscillator set at the desired frequency $f_0\ (=\ \omega_0/2\pi)$, the capacitor C_s is tuned to give a maximum indication on the electronic voltmeter VV. In this condition,

$$V_c = \frac{I_c}{j\omega_0 C_s} = \frac{V}{Rj\omega_0 C_s} = \frac{Ir}{j\omega_0 C_s R}$$

Also, at resonance and assuming a reasonably large value of Q,

$$\omega L = 1/\omega C_s \qquad\qquad (3.26)$$

so that the magnitude of V_c is given by

$$V_c = Ir\,\frac{\omega_0 L}{R} = \text{constant} \times Q_0 \qquad\qquad (3.27)$$

The electronic voltmeter can be calibrated to read directly in units of Q. The range of Q measured can be altered by changing the current I to a new standard on the ammeter scale.

MEASUREMENT OF CAPACITANCE

The value of an unknown capacitance within the range of the standard C_s can be readily determined by the method of substitution.

A standard coil is used, and the circuit is tuned to resonance at a suitable frequency. The unknown capacitor is then connected in parallel with C_s and the circuit retuned to resonance by varying C_s. The decrease in C_s is then equal to the value of the unknown capacitor. By noting the change in Q for the two conditions the parallel loss resistance of the unknown capacitor can also be determined.

MEASUREMENT OF INDUCTANCE

The unknown inductance is connected in series with the standard capacitor, C_s, and the circuit is tuned to resonance. The unknown inductance, L_u, is then given by

$$\omega L_u = \frac{1}{\omega C_s}$$

SELF-CAPACITANCE OF A COIL

An equivalent circuit of a coil is shown in Fig. 3.15(*a*). At high frequencies the self-capacitance C_0, may materially affect the coil

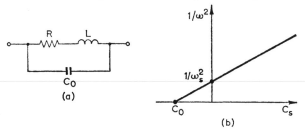

(a)

(b)

Fig. 3.15   REACTANCE VARIATION METHOD OF FINDING THE
SELF-CAPACITANCE OF A COIL

impedance, and generally coils are not operated at more than one-third of their self-resonant frequency, f_0 ($= 1/2\pi\sqrt{(LC_0)}$). C_0 can be measured on the Q-meter by the method of reactance variation. The values of standard capacitor setting, C_s, to achieve resonance are noted for a range of frequencies. Since the resistance r (Fig. 3.14) is so low, the self-capacitance may be taken as being in parallel with C_s, and resonance occurs when

$$\omega L = \frac{1}{\omega(C_s + C_0)}$$

Hence

$$\frac{1}{\omega^2 L} = C_s + C_0 \tag{3.28}$$

If $1/\omega^2$ is plotted to a base of C_s, a straight-line graph is obtained, which may be extrapolated as shown in Fig. 3.15(b) to give C_0 and the self-resonant angular frequency $\omega_s = 1/\sqrt{(LC_0)}$.

PROBLEMS

3.1 An a.c. bridge network for the measurement of inductance consists of four impedances arranged as a closed loop ABCD, where AB is the unknown inductance coil; BC is a variable resistance Q, in series with a capacitor C_1; CD is a capacitor C_2; and DA is a resistance R.

State the conditions for balance and obtain expressions for the inductance and resistance of the coil.

Hence calculate the values of the inductance and resistance of a coil if balance is obtained when $Q = 250\,\Omega$, $C_1 = 5\mu\text{F}$, $C_2 = 2\mu\text{F}$, and $R = 1,000\,\Omega$.

(H.N.C.)

Ans. 0·5 H, 400 Ω.

3.2 A coil having an inductance of the order of 1 H is measured by an a.c. bridge method at a frequency of 1 kHz. In order to bring the impedance to be measured within the range of the bridge, the coil is shunted by a non-inductive resistance of 500 Ω and the equivalent series impedance of the combination is measured, the values obtained being 6·35 mH and 487 Ω.

Determine the inductance and resistance of the coil. Indicate the general effect of this procedure on the possible accuracy of the measurement. *(L.U.)*

Ans. 0·92 H, 1,420 Ω.

3.3 A modified Carey Foster bridge is arranged as follows.

Arm AB is a non-inductive resistance of 10 Ω.
Arm BC is a non-inductive resistance of 500 Ω.
Arm CD is a variable resistor, R, in series with a 1·0μF capacitor.
Arm DA is the secondary of a variable mutual inductor the primary of which is connected between A and the source of supply, the other lead of the supply being taken to C. A detector is across BD. The secondary of the mutual inductor has a resistance of 15 Ω and $\omega = 5{,}000$ rad/s. At balance $R = 185\,\Omega$.

Find the corresponding mutual inductance and the self-inductance of the secondary. Also calculate the current in each arm of the bridge assuming $V_{AC} = 3\,\text{V}$.

Draw a complexor diagram representing the currents and p.d.s across the arms. *(L.U.)*

Ans. 5·65 mH, 6·94 mH, 5·88 mA, 10·8 mA.

3.4 The conditions at balance of a Schering bridge set up to measure the capacitance and loss angle of a paper-dielectric capacitor are as follows:

$f = 500\,\text{Hz}$
Z_1 = a pure capacitance of 0·1 μF
Z_2 = a resistance of 500 Ω shunted by a capacitance of 0·0033 μF
Z_3 = pure resistance of 163 Ω
Z_4 = capacitor under test

Calculate the approximate values of the loss resistance of the capacitor assuming: (*a*) series loss resistance, and (*b*) shunt loss resistance.   (L.U.)

Ans. $5·37\,\Omega$, $197\,k\Omega$.

3.5   An a.c. bridge network consists of the following four arms: AB—a fixed resistor R_1; BC—a variable resistor R_2 in series with a variable capacitor C; CD—a fixed resistor R_3; DA—a coil of unknown inductance L, and loss resistance R.

Derive expressions for L and R when the bridge is balanced at a frequency f. (This is Hay's bridge.) Evaluate L and R, with their limits of possible error if the values of the components when the bridge is balanced are

$R_1 = 1{,}000\,\Omega \pm 1$ part in 10,000
$R_2 = 2{,}370\,\Omega \pm 0·1\,\Omega$
$C = 4{,}210\,pF \pm 1\,pF$
$R_3 = 1{,}000\,\Omega \pm 1$ part in 10,000

The frequency of the bridge supply is 1,595 Hz to an accuracy of $\pm 1\,Hz$.
    (L.U.)

Ans. $4{,}170 \pm 1·88\,\mu H$, $4·26 \pm 0·009\,\Omega$.

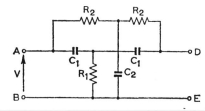

Fig. 3.16

3.6   Using the star-delta transformation show that the voltage between D and E in the parallel-T circuit shown in Fig. 3.16 is zero when

$$\frac{2}{\omega C_1} = R_2{}^2 \omega C_2 \quad \text{and} \quad \frac{1}{R_1(\omega C_1)^2} = 2R_2$$

3.7   Solve Problem 3.6 by any other method.

Chapter **4**

ADVANCED THREE-PHASE THEORY

Symbolic notation may be simply extended to cover 3-phase systems.
It allows solutions to be obtained more easily for problems involving
unbalanced loads, these problems being extremely awkward without
symbolic methods.

4.1 The 120° Operator

It is important to maintain a conventional positive direction in
which to measure voltages in a 3-phase system. To facilitate this
the following double-subscript notation will be used. V_{RY} denotes
the voltage of the red line with respect to the yellow line, V_{YB}
denotes the voltage of the yellow line with respect to the blue line,
and V_{BR} denotes the voltage of the blue line with respect to the red
line. These directions are illustrated in Fig. 4.1(a), from which it is
clear that $V_{YB} = -V_{BY}$, etc.

In any 3-phase system there are two possible sequences in which
the voltages may pass through their maximum positive values,
namely red → yellow → blue, or red → blue → yellow. By convention the first of these sequences is called the *positive sequence*,
and the second is called the *negative sequence*. The conventional
positive sequence is the one which is most common for electricity
supply and will be assumed in the following sections unless specifically stated otherwise. In Fig. 4.1(b) the line voltage complexor
diagrams are drawn for both positive and negative sequences.

In 3-phase systems, the voltage complexors are displaced from one another by 120°, so that it is convenient to have an operator which rotates a complexor through this angle. This operator is a.*

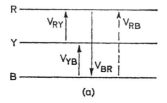

(a)

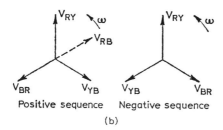

Positive sequence   Negative sequence

(b)

Fig. 4.1   VOLTAGE COMPLEXORS FOR SYMMETRICAL 3-PHASE SYSTEM

Any complexor when multiplied by a, remains unchanged in magnitude, and has 120° added to its phase angle. Thus

$$a = 1\underline{/120°} \tag{4.1}$$

$$= -\tfrac{1}{2} + j\frac{\sqrt{3}}{2} \tag{4.2}$$

Also

$$a^2 = 1\underline{/120°} \times 1\underline{/120°} = 1\underline{/240°} = -\tfrac{1}{2} - j\sqrt{3}/2$$

The operator a^2 will turn a complexor through 240° in an anticlockwise direction. This is the same as turning it through 120° in a clockwise direction. Thus

$$a^2 = 1\underline{/-120°}$$

In the same way,

$$a^3 = 1\underline{/360°} = 1$$

It will be recalled that the operation $-j$ results in the complexor concerned being turned through an angle of $-90°$. It should be

---

* The symbol h may also be used to represent the 120° operator.

noted that the operation $-a$, however, does not turn a complexor through $-120°$. This can be seen as follows:

$$-a = a \times (-1) = a \times 1\underline{/180°} = 1\underline{/120°} \times 1\underline{/180°}$$
$$= 1\underline{/300°} = 1\underline{/-60°} \tag{4.3}$$

Thus $-a$ turns a complexor through $60°$ in a clockwise direction. From the rectangular forms for the operators, the following important identity may be verified:

$$a^2 + a + 1 = 0 \tag{4.4}$$

4.2 Four-wire Balanced Star

In the balanced 4-wire star-connected system shown in Fig. 4.2(a), the voltages are assumed to be symmetrical. The positive direction

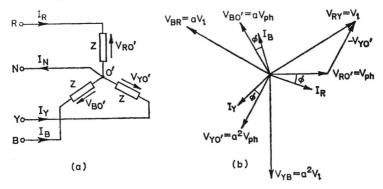

(a)   (b)

Fig. 4.2 BALANCED 4-WIRE STAR

of phase voltage at the load is assumed to be the potential of the line terminal with respect to the star point $0'$. $0'$ is in this case the neutral point of the system. Let the phase voltages have magnitude V_{ph}, and take the voltage between the neutral and the red line as the reference complexor, i.e.

$$V_{RO'} = V_{ph}\underline{/0°}$$

Then

$$V_{YO'} = V_{ph}\underline{/-120°} = a^2 V_{ph}$$

and

$$V_{BO'} = V_{ph}\underline{/+120°} = a V_{ph}$$

The line voltage is the difference between the phase voltages concerned. Thus

$$V_{RY} = V_{RO'} - V_{YO'} \text{ (see Fig. 4.2}(b))$$

$$= V_{ph} - a^2 V_{ph}$$

$$= V_{ph} \left\{ 1 - \left(-\frac{1}{2} - j\frac{\sqrt{3}}{2} \right) \right\}$$

$$= V_{ph} \left\{ \frac{3}{2} + j\frac{\sqrt{3}}{2} \right\} = \sqrt{3} V_{ph} \underline{/30°} \qquad (4.5)$$

In the same way,

$$V_{YB} = V_{YO'} - V_{BO'} = a^2 V_{ph} - a V_{ph} = \sqrt{3} V_{ph} \underline{/-90°}$$

and

$$V_{BR} = V_{BO'} - V_{RO'} = a V_{ph} - V_{ph} = \sqrt{3} V_{ph} \underline{/150°}$$

In each case the line voltage is $\sqrt{3}$ times the phase voltage in magnitude, and leads the corresponding phase voltage by 30°.

In the same way, with I_R as the reference quantity,

$$I_R = I_{ph} = I_l = V_{ph}/Z_{ph}$$

$$I_Y = a^2 I_l$$

$$I_B = a I_l$$

The current through the neutral line, by Kirchhoff's law, is

$$I_N = I_R + I_Y + I_B$$

$$= I_l + a^2 I_l + a I_l$$

$$= (1 + a^2 + a) I_l$$

$$= 0 \text{ (from eqn. (4.4))}$$

4.3 Three-wire Balanced Star

Since there will be no neutral wire current in a 4-wire star with symmetrical supply voltages and balanced loads, the neutral wire may be removed, and the familiar 3-wire system is obtained.

4.4 Balanced Delta-connected Load

If the current I_1 through the load connected between the red and yellow lines of the delta-connected system shown in Fig. 4.3(a) is taken as the reference complexor at (b) then, since all the load impedances are equal, each load current will lag by the same angle behind its respective line voltage. Since the line voltages are assumed to be symmetrical, they will be 120° displaced from one another, and the load currents will thus be displaced by the same amount.

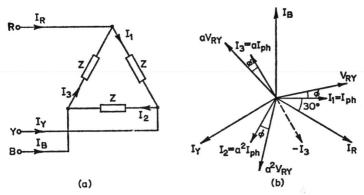

(a)　　　　　　　　　　　　**(b)**

Fig. 4.3 BALANCED DELTA-CONNECTED LOAD

Let

$$I_1 = I_{ph} = \text{Reference complexor}$$

Then

$$I_2 = I_1 \underline{/-120^\circ} = a^2 I_{ph}$$

and

$$I_3 = I_1 \underline{/+120^\circ} = a I_{ph}$$

The line currents, by Kirchhoff's law, will be the difference between the phase currents at the corresponding terminals, i.e.

$$I_R = I_1 - I_3 = I_{ph} - a I_{ph}$$

$$= I_{ph} \left(\frac{3}{2} - j \frac{\sqrt{3}}{2} \right) = \sqrt{3} I_{ph} \underline{/-30^\circ}$$

In the same way,

$$I_Y = I_2 - I_1 = a^2 I_{ph} - I_{ph} = \sqrt{3} I_{ph} \underline{/-150^\circ}$$

and

$$I_B = I_3 - I_2 = a I_{ph} - a^2 I_{ph} = \sqrt{3} I_{ph} \underline{/90^\circ}$$

The sum of the three line currents is

$$I_R + I_Y + I_B = I_{ph}(1 - a) + I_{ph}(a^2 - 1) + I_{ph}(a - a^2) = 0$$

EXAMPLE 4.1 A symmetrical 3-phase 450 V system supplies a balanced delta-connected load of 12 kW at 0·8 p.f. lagging. Calculate (a) the phase currents, (b) the line currents, and (c) the effective impedance per phase.

The complexor diagram corresponds to that of Fig. 4.3(b).

Power per phase $= \frac{1}{3} \times$ total power $= 4 \text{kW}$

Therefore

$$\text{kVA per phase} = \frac{4}{0·8} = 5 \text{kVA}$$

and

$$\text{Current per phase} = \frac{5 \times 10^3}{450} = 11·1 \text{A at } 0·8 \text{ power factor lagging}$$

(a) Take V_{RY} as the reference complexor; then the phase currents will be

$$I_1 = 11·1\underline{/-\cos^{-1}0·8} = 11·1\underline{/-36° \ 52'} \text{A} = (8·88 - j6·66) \text{A}$$

$$I_2 = a^2 I_1 = 11·1\underline{/-36° \ 52'} \times 1\underline{/-120°} = 11·1\underline{/-156° \ 52'}$$
$$= (-10·2 - j4·36) \text{A}$$

and

$$I_3 = a I_1 = 11·1\underline{/-36° \ 52'} \times 1\underline{/120°} = 11·1\underline{/83° \ 8'}$$
$$= (1·33 + j11·0) \text{A}$$

(b) The line currents are found by subtraction:

$$I_R = I_1 - I_3 = 8·88 - j6·66 - 1·33 - j11·0 = 7·55 - j17·6$$
$$= 19·2\underline{/-66° \ 52'} \text{A}$$

$$I_Y = I_2 - I_1 = -19·1 + j2·3 = 19·2\underline{/-186° \ 52'} \text{A}$$

$$I_B = I_3 - I_2 = 11·5 + j15·4 = 19·2\underline{/53° \ 8'} \text{A}$$

Note that, once I_R is found, I_Y and I_B follow for a balanced load by subtracting and adding 120° to the phase angle of I_R.

(c) The impedance per phase is given by

$$Z_{ph} = \frac{V_{RY}}{I_1} = \frac{450\underline{/0°}}{11·1\underline{/-36° \ 52'}} = 40·5\underline{/36° \ 52'} = (32·4 + j24·3) \Omega$$

EXAMPLE 4.2 A short 3-phase transmission line has an effective resistance per conductor of 0·6 Ω and an effective inductive reactance per conductor of 0·8 Ω. Find the sending-end line voltage and power factor when the line supplies a balanced load of 1,800 kVA at 5·2 kV and 0·8 power factor lagging.

Since the load and transmission system are balanced, this problem may be treated in the same way as a single-phase problem.

Receiving-end phase voltage, $V_{RN} = (5,200/\sqrt{3})\underline{/0°} = 3,000\underline{/0°} \text{V}$

(This voltage is taken as the reference complexor.)

Line current $= 1,800/(\sqrt{3} \times 5·2) = 200 \text{A}$

With respect to the reference phase voltage this current may be expressed as $200/{-36.9°}$ A. Thus

Line voltage drop $= IZ = 200/{-36.9°}(0.6 + j0.8) = (192 + j56)$ V

Sending-end phase voltage $= 3,000 + 192 + j53 \approx 3,190/\underline{1°}$ V

and

Sending-end line voltage $= \underline{\underline{5.51 \text{ kV}}}$

Also,

Sending-end power factor $= \underline{\underline{\cos 37.9°}}$

4.5 Unbalanced Four-wire Star-connected Load on System of Negligible Line Impedance (Fig. 4.4)

This is the simplest case of an unbalanced load, and may be treated as three separate single-phase systems with a common return lead.

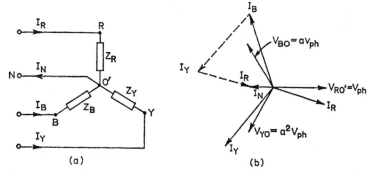

Fig. 4.4 UNBALANCED 4-WIRE STAR

The phase voltages will be equal in magnitude and displaced by 120° from each other because the voltage of the load star point is the same as that of the generator star point. The neutral wire current is the complexor sum of the three line currents.

If the voltage across the red phase is taken to be in the reference direction, then

$$V_{RO'} = V_{ph}; \quad V_{YO'} = a^2 V_{ph} = V_{ph}/{-120°}$$

and

$$V_{BO'} = aV_{ph} = V_{ph}/\underline{120°}$$

The corresponding phase currents are

$$I_R = \frac{V_{ph}}{Z_R} \qquad I_Y = \frac{V_{ph}/{-120°}}{Z_Y} \qquad I_B = \frac{V_{ph}/\underline{120°}}{Z_B}$$

These currents will also be the line currents in the system. The current through the neutral wire will then be given by

$$I_N = I_R + I_Y + I_B = \frac{V_{ph}}{Z_R} + \frac{V_{ph}\underline{/-120°}}{Z_Y} + \frac{V_{ph}\underline{/120°}}{Z_B} \qquad (4.6)$$

The impedances are, of course, in complex form. The currents are shown in Fig. 4.4(*b*).

This result may be obtained from the complexor diagram as follows.

1. Draw the three phase voltages, equal in magnitude and 120° apart.
2. Calculate, by single-phase theory, the current in each phase and the phase angle relative to the corresponding phase voltage, taking each voltage in turn as a reference quantity $(I = V_{ph}/Z)$.
3. Draw these currents in the complexor diagram in the correct phase relationship to the corresponding phase voltages.
4. Find, by complexor addition, the sum of the three phase currents. This will give the neutral wire current in magnitude and phase.

The overall power factor of an unbalanced load is taken to be the ratio of total kW to total kVA.

EXAMPLE 4.3   The 440V, 50Hz, 3-phase 4-wire main to a workshop provides power for the following loads.

(*a*) Three 3 kW induction motors each 3-phase, 85 per cent efficient, and operating at a lagging power factor of 0·9.

(*b*) Two single-phase electric furnaces of 250 V rating each consuming 6kW at unity power factor.

(*c*) A general lighting load of 3kW, 250V at unity power factor.

If the lighting load is connected between one phase and neutral, while the furnaces are connected one between each of the other phases and neutral, calculate the current in each line and the neutral current at full load.   (*H.N.C.*)

$$\text{Total motor power input} = \frac{3 \times 3}{0 \cdot 85} = 10 \cdot 6 \text{kW}$$

Therefore

$$\text{Motor kVA input} = \frac{\text{kW}}{\text{power factor}} = \frac{10 \cdot 6}{0 \cdot 9} = 11 \cdot 8 \text{kVA}$$

Also

$$\text{Each motor kVAr} = \text{kVA} \times \sin(\cos^{-1} 0 \cdot 9) = 5 \cdot 1 \text{kVAr}$$

$$\text{Current in each line due to motor load} = \frac{11 \cdot 8 \times 1,000}{\sqrt{3} \times 440} = 15 \cdot 4 \text{A}$$

This current will lag behind the corresponding voltage by $\cos^{-1} 0 \cdot 9 = 25° \, 50'$. Thus the current through each line due to the motor, and with reference to each phase voltage, will be $15 \cdot 4\underline{/25° \, 50'} = (14 - j6 \cdot 7)\text{A}$. The line current for each furnace will be $6,000/250 = 24$A at $0°$ phase angle (i.e. $24\underline{/0°}$, or $24 + j0$), with respect to the corresponding phase voltage. The line current for the lighting load will be $3,000/250 = 12$A at unity p.f. (i.e. $12\underline{/0°}$ or $12 + j0$) with respect to the corresponding phase voltage. The total current in each furnace line will then be

$$14 - j6·7 + 24 = 38 - j6·7 = 38·5\underline{/-10°}\,\text{A}$$

with respect to the corresponding phase voltage.
The current in the third line will be

$$14 - j6·7 + 12 = 26 - j6·7 = 26·8\underline{/-14°}\,\text{A}$$

The complexor diagram corresponds to that of Fig. 4.4(b).

To find the neutral current, the three line current complexors may be added graphically or by the following method using symbolic notation.

To simplify the calculation, it should be noted that, since the motors form balanced loads, the motor currents will not give rise to a neutral current and may be neglected in the calculation of the neutral current.

Let the lighting load phase voltage be taken as the reference complexor. Then

Lighting load current $= (12 + j0)$A
First heating load current $= (24 + j0)a^2 = (-12 - j20·8)$A
Second heating load current $= (24 + j0)a = (-12 + j20·8)$A

Therefore

$$\begin{aligned}
\text{Neutral current} &= I_R + I_Y + I_B \\
&= (12 + j0) + (-12 - j20·8) + (-12 + j20·8) \\
&= -12 + j0 = 12\underline{/180°}\,\text{A}
\end{aligned}$$

4.6 Unbalanced Delta-connected Load (Fig. 4.5)

In the case of the delta-connected unbalanced load with symmetrical line voltages, full line voltage will be across each load phase.

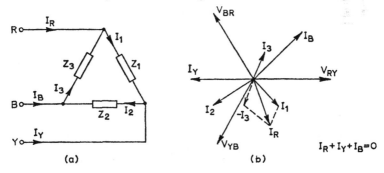

Fig. 4.5 UNBALANCED DELTA-CONNECTED LOAD

The problem thus resolves itself into three independent single-phase systems supplied with voltages which are $120°$ apart in phase.

In the analytical solution the line voltage V_{RY} will be taken as the reference complexor (Fig. 4.5(b)), so that

$$V_{RY} = V_l \qquad V_{YB} = V_l\underline{/-120°} \qquad V_{BR} = V_l\underline{/120°}$$

The complex impedances of the load are Z_1, Z_2 and Z_3, connected as shown in Fig. 4.5(a).
Then the phase currents are easily obtained from the equations

$$I_1 = \frac{V_l}{Z_1} \qquad I_2 = \frac{V_l\underline{/-120°}}{Z_2} \qquad I_3 = \frac{V_l\underline{/120°}}{Z_3}$$

By Kirchhoff's first law, the line currents will be the differences between the corresponding phase currents. Since there is no neutral wire, the complexor sum of these three line currents must be zero.

TRIANGULAR COMPLEXOR DIAGRAMS

In some cases it is convenient to represent the line voltage and line current complexor diagrams of 3-phase circuits in a triangular

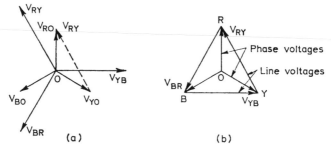

Fig. 4.6   PHASE- AND LINE-VOLTAGE COMPLEXOR DIAGRAMS IN TRIANGULAR FORM

form. In Fig. 4.6(a) the conventional complexor diagram for line and phase voltages is shown, while Fig. 4.6(b) shows the triangular diagram. The line voltage complexors form a closed triangle. These complexors correspond exactly in magnitude and direction to the line voltage complexors shown at (a).

For symmetrical line voltages, the line voltage triangle will be equilateral and 0 will be the centroid of the triangle.

In a 3-wire system, there is no resultant current so that the three line current complexors must form a closed triangle. Fig. 4.7 shows the diagram for a delta-connected system.

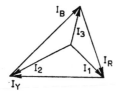

Fig. 4.7 LINE CURRENTS IN A 3-WIRE SYSTEM

EXAMPLE 4.4 Three impedances, $Z_1 = (10 + j10)\Omega$, $Z_2 = (8\cdot66 + j5)\Omega$, and $Z_3 = (12 + j16)\Omega$ are delta-connected to a 380V, 3-phase system. Determine the line currents and draw the complexor diagram.

The circuit is the same as that of Fig. 4.5(a). Take V_{RY} as the reference. Then

$$V_{YB} = 380\underline{/240°}\,\text{V} \quad \text{and} \quad V_{BR} = 380\underline{/120°}\,\text{V}$$

so that

$$I_1 = \frac{V_{RY}}{Z_1} = \frac{380\underline{/0°}}{10 + j10} = \frac{380\underline{/0°}}{14\cdot14\underline{/45°}} = 26\cdot8\underline{/-45°} = (19 - j19)\,\text{A}$$

$$I_2 = \frac{V_{YB}}{Z_2} = \frac{380\underline{/240°}}{8\cdot66 + j5} = \frac{380\underline{/240°}}{10\underline{/30°}} = 38\underline{/210°} = (-32\cdot9 - j19)\,\text{A}$$

$$I_3 = \frac{V_{BR}}{Z_3} = \frac{380\underline{/120°}}{12 + j16} = \frac{380\underline{/120°}}{20\underline{/53\cdot1°}} = 19\underline{/66\cdot9°} = (7\cdot45 + j17\cdot5)\,\text{A}$$

Therefore

$$I_R = I_1 - I_3 = 11\cdot5 - j36\cdot5 = 38\cdot2\underline{/-72\cdot5°}\,\text{A}$$

$$I_Y = I_2 - I_1 = -51\cdot9 - j0 = 51\cdot9\underline{/180°}\,\text{A}$$

$$I_B = I_3 - I_2 = 40\cdot4 + j36\cdot5 = 54\cdot3\underline{/42\cdot1°}\,\text{A}$$

The complexor diagram is that shown in Fig. 4.5(b).

4.7 Unbalanced Three-wire Star-connected Load

The unbalanced 3-wire star load is the most difficult unbalanced 3-phase load to deal with, but several methods are available. One method is to apply the star–mesh transformation to the load. The problem is then solved as a delta-connected system, and the line currents are obtained. A second method is to apply Maxwell's mesh equations to the system, using the complex notation for impedances, voltages and currents. Both of these methods involve a fairly large amount of arithmetical work, which, while not eliminated, is at least simplified by the use of Millman's theorem.

The circuit and complexor diagrams are shown in Fig. 4.8. In this case 0 is the star point of the generator or the neutral of the supply (normally zero potential). The voltages between 0 and the end points of Z_R, Z_Y and Z_B are the phase voltages of the supply. Hence, by Millman's theorem, the voltage of 0' with respect to 0 is given by

$$V_{0'0} = \frac{V_{R0}Y_R + V_{Y0}Y_Y + V_{B0}Y_B}{Y_R + Y_Y + Y_B} \tag{4.7}$$

The voltage across each phase of the load is derived by considering that, for example, the voltage $V_{R0'}$ is the voltage of line R with respect to 0'. This is then the voltage of line R with respect to 0 less the voltage of 0' with respect to 0. In symbols this gives

$$\left.\begin{aligned} V_{R0'} &= (V_{R0} - V_{0'0}) \\ V_{Y0'} &= (V_{Y0} - V_{0'0}) \\ V_{B0'} &= (V_{B0} - V_{0'0}) \end{aligned}\right\} \tag{4.8}$$

The line currents are then given by

$$\left.\begin{aligned} I_{R0'} &= (V_{R0} - V_{0'0})Y_R \\ I_{Y0'} &= (V_{Y0} - V_{0'0})Y_Y \\ I_{B0'} &= (V_{B0} - V_{0'0})Y_B \end{aligned}\right\} \tag{4.9}$$

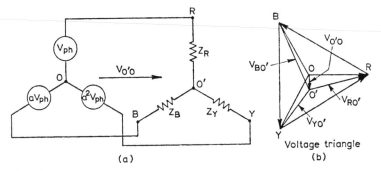

Fig. 4.8 UNBALANCED 3-WIRE STAR

EXAMPLE 4.5 Loads of 2, $2 + j2$ and $-j5$-ohm elements are connected in star to a 433 V 50 Hz 3-phase 3-wire symmetrical system. Find (a) the potential of the load star point with respect to the supply neutral, (b) the load phase voltages, and (c) the line currents. Draw the complete complexor diagram.

The circuit diagram is shown in Fig. 4.8. V_{R0} is chosen as the reference complexor. The corresponding complexor diagram is shown at (b). Then

$$V_{R0} = V_{ph}\underline{/0°} = \frac{433}{\sqrt{3}}\underline{/0°} = 250\underline{/0°}\,\text{V}$$

The admittances of the load are

$$Y_R = \frac{1}{2} = 0.5\,S$$

$$Y_Y = \frac{1}{2 + j2} = \frac{1}{2.83\underline{/45°}} = 0.353\underline{/-45°} = (0.25 - j0.25)\,S$$

$$Y_B = \frac{1}{-j5} = j0.2 = 0.2\underline{/90°}\,S$$

(a) Applying Millman's theorem, eqn. (4.7),

$$V_{0'0} = \frac{(250\underline{/0°} \times 0.5) + (250\underline{/240°} \times 0.353\underline{/-45°}) + (250\underline{/120°} \times 0.2\underline{/90°})}{0.5 + 0.25 - j0.25 + j0.2}$$

This reduces to

$$V_{0'0} \approx \underline{-j64\,V}$$

(b) The load phase voltages are then, by eqns (4.8),

$$V_{RO'} = V_{RO} - V_{0'0} = 250 + j64 = \underline{258\underline{/14° 21'}\,V}$$

$$V_{YO'} = V_{YO} - V_{0'0} = a^2 250 + j64 = -125 - j152$$
$$= \underline{197\underline{/230° 34'}\,V}$$

$$V_{BO'} = V_{BO} - V_{0'0} = a250 + j64 = -125 + j280$$
$$= \underline{307\underline{/114° 4'}\,V}$$

(c) Having obtained the load phase voltages, the load currents follow simply from the expressions

$$I_{RO'} = V_{RO'}Y_1 = 258\underline{/14° 21'} \times 0.5 = \underline{129\underline{/14° 21'}\,A}$$

$$I_{YO'} = V_{YO'}Y_2 = 197\underline{/230° 34'} \times 0.353\underline{/-45°} = \underline{69.4\underline{/185°34'}\,A}$$

$$I_{BO'} = V_{BO'}Y_3 = 307\underline{/114° 4'} \times 0.2\underline{/90°} = \underline{61.4\underline{/204°4'}\,A}$$

Note that these three currents must form a closed complexor triangle.

4.8 Effect of Line Impedance

If the impedances of the lines connecting the generator to the load are appreciable, then for the 3-wire system with a star-connected load, the line impedances may be lumped with the load impedances to obtain the line currents. In the case of a 4-wire system, a solution may be readily effected by Millman's theorem. Fig. 4.9 shows a star-connected load, supplied through lines of impedance Z_L, and having a fourth wire of impedance Z_N. Then the total impedances in the lines are $(Z_L + Z_R)$, $(Z_L + Z_Y)$ and $(Z_L + Z_B)$. Let N be the end of the fourth wire so that 0 coincides with N; then

$$
V_{0'0} = \frac{V_{R0}\dfrac{1}{Z_L + Z_R} + V_{Y0}\dfrac{1}{Z_L + Z_Y} + V_{B0}\dfrac{1}{Z_L + Z_B} + V_{N0}\dfrac{1}{Z_N}}{\dfrac{1}{Z_L + Z_R} + \dfrac{1}{Z_L + Z_Y} + \dfrac{1}{Z_L + Z_B} + \dfrac{1}{Z_N}}
$$

$$
= \frac{V_{R0}Y_1 + V_{Y0}Y_2 + V_{B0}Y_3}{Y_1 + Y_2 + Y_3 + Y_N} \tag{4.10}
$$

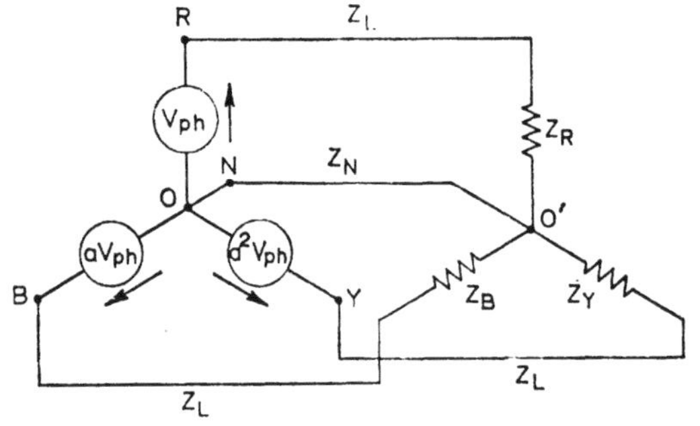

Fig. 4.9   UNBALANCED 4-WIRE STAR WITH APPRECIABLE LINE IMPEDANCES

since $V_{N0} = 0$.

Then

$$I_N = (V_{0'0})Y_N$$

and

$$I_R = (V_{R0} - V_{0'0})Y_1, \text{ etc.}$$

4.9 Power Measurement in General Three-phase Systems

The theoretically simplest method of measuring unbalanced 3-phase power in a 3-wire system is to insert a wattmeter in each line, with the voltage coils connected together to a common point (Fig. 4.10). The total power is then the sum of the three wattmeter readings. It can be shown that this method gives the correct result whether the wattmeters are identical or not.

There is a rather unexpected theorem related to polyphase power measurement, called *Blondel's theorem*. This states that the minimum number of wattmeters required to measure the power in a polyphase system is one less than the number of wires carrying current in the system.

Thus for a 3-phase 4-wire system three wattmeters are required, but for a 3-phase 3-wire system only two wattmeters are required. This two-wattmeter method will now be considered in detail.

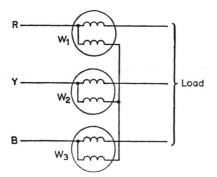

Fig. 4.10 THREE-WATTMETER POWER MEASUREMENT IN A 3-WIRE SYSTEM

The connexions for the two-wattmeter method are shown in Fig. 4.11. The wattmeters have their current coils connected in any two

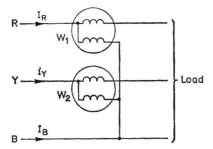

Fig. 4.11 TWO-WATTMETER METHOD

lines, while their voltage coils are connected between the corresponding lines and the third line. Then the sum of the two readings gives the total power irrespective of balance of load or waveform of supply.

Proof. Let P_1 and P_2 be the average value of the two wattmeter readings. Then

$$P_1 = \frac{1}{T} \int_0^T v_{RB} i_R \, dt \quad \text{and} \quad P_2 = \frac{1}{T} \int_0^T v_{YB} i_Y \, dt$$

Note that with both wattmeter voltage coils connected in the same way (i.e. corresponding ends joined to the blue line) the voltage across the voltage coil of W_2 is v_{YB} and not v_{BY}. Now,

$$v_{RB} = v_{RO'} - v_{BO'} \quad \text{and} \quad v_{YB} = v_{YO'} - v_{BO'}$$

Also, for a 3-wire system,

$$i_B = -(i_R + i_Y)$$

Therefore

$$P_1 + P_2 = \frac{1}{T} \int_0^T (v_{RO'}i_R - v_{BO'}i_R + v_{YO'}i_Y - v_{BO'}i_Y)\,dt$$

$$= \frac{1}{T} \int_0^T (v_{RO'}i_R + v_{YO'}i_Y + v_{BO'}i_B)\,dt$$

$$= \frac{1}{T} \int_0^T (\text{total instantaneous power})\,dt$$

$$= \text{Average total power}$$

In this analysis no assumptions have been made with regard to phase sequence, balance of load or waveform.

4.10 Special Case of Balanced Loads and Sine Waveforms

If the currents and voltages are sinusoidal they may be represented on a complexor diagram, this being done in Fig. 4.12 for the case of the balanced load and a phase sequence RYB. It is assumed that W_1 is the wattmeter in the leading line (e.g. if the wattmeters were in the R and B lines, then the wattmeter in the B line would be called W_1).

If the circuit current leads the phase voltage by an angle ϕ, and the wattmeters are connected as in Fig. 4.11, then

Power indicated by W_1 = (voltage across voltage coil)

$\times$ (current in current coil) $\times$ cos (angle between them)

i.e.

$$P_1 = V_{RB}I_R \cos(\phi + 30°) \quad \text{(from the complexor diagram)}$$

$$= V_l I_l \cos(\phi + 30°) \tag{4.11}$$

In the same way the power indicated by W_2 is

$$P_2 = V_{YB}I_Y \cos(\phi - 30°) = V_l I_l \cos(\phi - 30°) \tag{4.12}$$

For the case of a lagging phase angle, ϕ will be negative in the above equations.

For phase angles greater than 60° (either leading or lagging) one wattmeter will have a negative deflexion, in which case the connexions to the voltage coil of that wattmeter should be reversed to

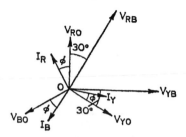

Fig. 4.12 COMPLEXOR DIAGRAM FOR TWO-WATTMETER METHOD WITH BALANCED LOADS

give an upscale reading. Such a reversed reading must always be subtracted to give the total power.

For balanced loads the power factor may be obtained from the wattmeter readings. Thus, if W_1 is in the leading line, then from eqn. (4.11) the reading is

$$P_1 = V_l I_l \cos(\phi + 30°)$$
$$= V_l I_l (\cos\phi \cos 30° - \sin\phi \sin 30°)$$
$$= V_l I_l \left(\frac{\sqrt{3}}{2}\cos\phi - \tfrac{1}{2}\sin\phi\right) \tag{4.13}$$

In the same way, from eqn. (4.12),

$$P_2 = V_l I_l \left(\frac{\sqrt{3}}{2}\cos\phi + \tfrac{1}{2}\sin\phi\right) \tag{4.14}$$

Therefore

$$P_1 + P_2 = \sqrt{3}V_l I_l \cos\phi$$

(i.e. the sum of readings gives the total circuit power)

Also

$$P_2 - P_1 = V_l I_l \sin\phi$$

Dividing this equation by the previous one,

$$\frac{P_2 - P_1}{P_1 + P_2} = \frac{1}{\sqrt{3}}\tan\phi \tag{4.15}$$

But

Power factor, $\cos \phi = \dfrac{1}{\sec \phi} = \dfrac{1}{\sqrt{(1 + \tan^2 \phi)}}$

Therefore

$$\cos \phi = \frac{1}{\sqrt{\left\{1 + 3\left(\dfrac{P_2 - P_1}{P_1 + P_2}\right)^2\right\}}} \qquad (4.16)$$

It should be noted that the total reactive volt-amperes, Q, may be obtained as

$$Q = \sqrt{3}V_l I_l \sin \phi = \sqrt{3}(P_2 - P_1) \qquad (4.17)$$

The two-wattmeter method may be used with a single wattmeter if suitable switching is provided. The switches consist of make-before-break ammeter switches and a reversing switch.

In the polyphase wattmeter, two wattmeter elements are mounted in the same housing. They are screened and insulated from each other, but their moving coils are fixed to the same spindle and rotate against the same control spring. The meter indication is then the sum of the separate indications. The connexions are the same as for two separate meters.

EXAMPLE 4.6 A 3-phase induction motor develops 11·2 kW when running at 85 per cent efficiency and at a power factor of 0·45 lagging. Calculate the readings on each of two wattmeters connected to read the input power.

Let P_1 and P_2 be the readings on the "leading" and "lagging" phase wattmeters respectively.

Total input power $= \dfrac{11,200}{0\cdot85} = 13,100\,\text{W}$

i.e.

$P_1 + P_2 = 13,100\,\text{W}$ \hfill (i)

Total input VAr $= 13,100 \times \tan \phi$

$\qquad\qquad = -13,100 \times \dfrac{\sqrt{(1 - 0\cdot45^2)}}{0\cdot45} = -26,000\,\text{VAr}$

Therefore

$P_2 - P_1 = \dfrac{-26,000}{\sqrt{3}} = -15,000\,\text{W}$ \hfill (ii)

Adding eqns. (i) and (ii),

$2P_2 = -1,900\,\text{W}$

Therefore

$\underline{\underline{P_2 = -950\,\text{W}}}$ \qquad and \qquad $\underline{\underline{P_1 = 14,050\,\text{W}}}$

EXAMPLE 4.7 Two wattmeters W_1 and W_2 connected to read the input to a 3-phase induction motor running unloaded, indicate 3 kW and 1 kW respectively. On increasing the load, the reading on W_1 increases while that on W_2 decreases and eventually reverses.

Explain the above phenomenon and find the unloaded power and power factor of the motor.   (*H.N.C.*)

' When the load increases the power factor may be assumed to improve. If the unloaded power factor were less than 0·5 lagging (i.e. 60° lag), then reference to eqns. (4.11) and (4.12) shows that, if W_1 were in the leading line and W_2 in the lagging line, an increase in power would have the results described. This also assumes that the connexions to the voltage coil of W_2 have been reversed. As the phase angle decreases P_2 falls to zero and P_1 increases; any further decrease in phase angle causes P_2 to increase in the opposite sense.

Hence, assuming that in the unloaded condition W_2 reads a negative power,

$$\text{Input power} = 3 - 1 = \underline{\underline{2\,\text{kW}}}$$

$$\text{Input power factor} = \frac{1}{\sqrt{\{1 + 3(\tfrac{1}{2})^2\}}} = \frac{1}{\sqrt{13}} = \underline{\underline{0\cdot278\ \text{lagging}}}$$

4.11 Power Measurement in Balanced Three-phase Systems

If it is known that the load is balanced, then one wattmeter is sufficient to measure the power in a 3-phase system. There are two methods in which one wattmeter may be applied.

ARTIFICIAL-STAR METHOD (Fig. 4.13)

The wattmeter will read the phase power, and the total power will be

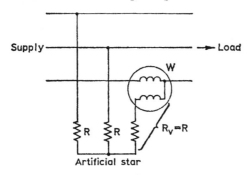

Fig. 4.13   ARTIFICIAL-STAR METHOD FOR BALANCED 3-PHASE POWER MEASUREMENT

three times the wattmeter reading. No indication is given of whether the power factor is leading or lagging.

DOUBLE-READING METHOD (Fig. 4.14)

Two readings are taken, with the wattmeter current coil in the same line for both readings and with the voltage coil connected across the current coil line and each of the other lines in turn. Suppose the current coil is inserted in the R line and the phase sequence is RYB. Let P_1 be the wattmeter reading when the voltage coil is

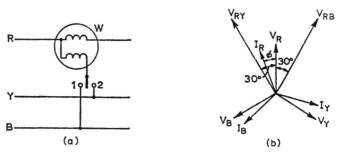

Fig. 4.14   DOUBLE-READING METHOD FOR BALANCED 3-PHASE
POWER MEASUREMENT

connected to the leading line (B in this case) and P_2 be the wattmeter reading when the voltage coil is connected to the lagging line (Y in this case). Now,

Wattmeter reading = (voltage across voltage coil)
× (current through current coil)
× cos (phase angle between above current and voltage)

i.e.

$$P_1 = V_l I_l \cos (\phi + 30°) \qquad \text{(see Fig. 4.14(b))}$$

and

$$P_2 = V_l I_l \cos (\phi - 30°)$$

Thus

$$P_1 = V_l I_l (\cos \phi \cos 30° - \sin \phi \sin 30°)$$

and

$$P_2 = V_l I_l (\cos \phi \cos 30° + \sin \phi \sin 30°)$$

Therefore

$$P_1 + P_2 = \sqrt{3} V_l I_l \cos \phi = \text{total power}$$

$$P_2 - P_1 = V_l I_l \sin \phi = (1/\sqrt{3}) \times \text{total reactive volt-amperes}$$

and

$$\tan \phi = \frac{\sqrt{3}(P_2 - P_1)}{P_1 + P_2}$$

4.12 Phase Sequence Determination

Suppose that a 3-phase supply, with balanced line voltages, is brought out to three terminals marked A, B, C, but the sequence of the supply is unknown, e.g. the sequence might be $A \to B \to C$ or $A \to C \to B$ as illustrated by the complexor diagrams of Fig. 4.15(*a*) and (*b*). For many purposes it is essential to determine

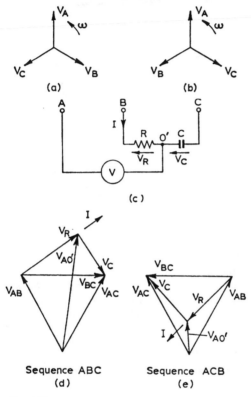

Fig. 4.15 PHASE SEQUENCE DETERMINATION

which of the two possible sequences the phase voltages follow. The determination may be carried out by various circuits, one of which is illustrated in Fig. 4.15(*c*).

The complexor diagram of Fig. 4.15(*d*) is drawn for the circuit when the phase sequence is $A \to B \to C$, and Fig. 4.15(*e*) applies when the phase sequence is $A \to C \to B$. It should be particularly noted that V_{BC} appears in anti-phase in the second case with respect to its direction in the first case. Since the circuit between terminals

B and C is capacitive, the current I through this circuit must lead the voltage V_{BC} across it.

Hence the current complexor may be drawn leading V_{BC} in both cases. V_R may then be drawn in the same direction as the current complexor. The voltage $V_{A0'}$ will then be represented by the complexor $V_{A0'}$ in each diagram. It is apparent that

(a) $V_{A0'}$ will exceed the line voltage for sequence A → B → C.

(b) $V_{A0'}$ will be less than the line voltage for sequence A → C → B.

The following points should be noted:

1. The voltmeter current has been neglected, so that a high-impedance voltmeter is necessary.
2. The clearest differentiation between the readings will be obtained when the resistance and the capacitive reactance are equal.

4.13 Symmetrical Components

Any unbalanced system of 3-phase currents may be represented by the superposition of a balanced system of 3-phase currents having *positive phase sequence*, a balanced system of 3-phase currents having the opposite or *negative phase sequence*, and a system of three currents equal in phase and magnitude and called the *zero phase sequence*. For example, Fig. 4.16(a) shows an unbalanced system of 3-phase currents, and Figs. 4.16(b), (c) and (d) show the positive, negative and zero phase-sequence components.

Adopting the nomenclature indicated in Fig. 4.16, where I_R, I_Y and I_B represent any unbalanced system of 3-phase currents,

$$I_R = I_{R+} + I_{R-} + I_{R0} \tag{4.18}$$

$$I_Y = I_{Y+} + I_{Y-} + I_{Y0} \tag{4.19}$$

$$I_B = I_{B+} + I_{B-} + I_{B0} \tag{4.20}$$

Evidently,

$$I_{Y+} = a^2 I_{R+} \tag{4.21}$$

$$I_{B+} = a I_{R+} \tag{4.22}$$

$$I_{Y-} = a I_{R-} \tag{4.23}$$

$$I_{B-} = a^2 I_{R-} \tag{4.24}$$

$$I_{R0} = I_{Y0} = I_{B0} \tag{4.25}$$

The original unbalanced currents may now be expressed in terms of red-phase symmetrical components only by substitution in eqns. (4.19) and (4.20). Repeating eqn. (4.18) for convenience, this gives

$$I_R = I_{R+} + I_{R-} + I_{R0} \tag{4.18}$$
$$I_Y = a^2 I_{R+} + a I_{R-} + I_{R0} \tag{4.26}$$
$$I_B = a I_{R+} + a^2 I_{R-} + I_{R0} \tag{4.27}$$

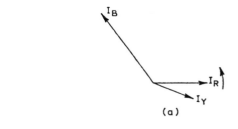

(a)

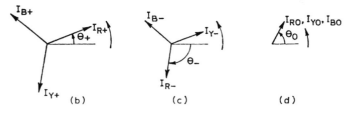

(b)   (c)   (d)

Fig. 4.16   SYMMETRICAL COMPONENTS

(a) Unbalanced 3-phase currents
(b) Positive phase-sequence currents
(c) Negative phase-sequence currents
(d) Zero phase-sequence currents

The symmetrical components I_{R+}, I_{R-} and I_{R0} may now be expressed in terms of the original unbalanced system I_R, I_Y, I_B.
Addition of eqns. (4.18), (4.26) and (4.27) gives

$$I_R + I_Y + I_B = (1 + a^2 + a)I_{R+} + (1 + a + a^2)I_{R-} + 3I_{R0}$$
$$I_{R0} = \tfrac{1}{3}(I_R + I_Y + I_B) \tag{4.28}$$

bearing in mind that

$$1 + a + a^2 = 0 \tag{4.4}$$

If eqn. (4.26) is multiplied by a and eqn. (4.27) by a^2, this gives

$$I_R = I_{R+} + I_{R-} + I_{R0} \tag{4.18}$$
$$a I_Y = a^3 I_{R+} + a^2 I_{R-} + a I_{R0} \tag{4.29}$$
$$a^2 I_B = a^3 I_{R+} + a^4 I_{R-} + a^2 I_{R0} \tag{4.30}$$

Addition of eqns. (4.18), (4.29) and (4.30) gives

$$I_{R+} = \tfrac{1}{3}(I_R + aI_Y + a^2I_B) \tag{4.31}$$

bearing in mind that $a^3 = 1$ and $a^4 = a$.

If eqn. (4.26) is multiplied by a^2 and eqn. (4.27) by a it can be shown that

$$I_{R-} = \tfrac{1}{3}(I_R + a^2I_Y + aI_B) \tag{4.32}$$

Eqns. (4.28), (4.31) and (4.32) show that it is possible to find zero, positive and negative phase-sequence components of current in terms of the original unbalanced system.

An unbalanced system of 3-phase voltages may similarly be represented by symmetrical components and the value of these components found by equations of the same form as (4.28), (4.31) and (4.32).

EXAMPLE 4.8 The currents flowing in an unbalanced 4-wire system are

$$I_R = 100\underline{/0^\circ}\,\text{A} \qquad I_Y = 200\underline{/-90^\circ}\,\text{A} \qquad I_B = 100\underline{/120^\circ}\,\text{A}$$

Find the positive, negative and zero phase-sequence components of these currents

$$I_R = 100\underline{/0^\circ} = (100 + j0)\,\text{A}$$

$$I_Y = 200\underline{/-90^\circ} = (0 - j200)\,\text{A}$$

$$aI_Y = 200\underline{/-90^\circ} \times 1\underline{/120^\circ} = 200\underline{/30^\circ} = (173\cdot2 + j100)\,\text{A}$$

$$a^2I_Y = 200\underline{/-90^\circ} \times 1\underline{/-120^\circ} = 200\underline{/150^\circ} = (-173\cdot2 + j100)\,\text{A}$$

$$I_B = 100\underline{/120^\circ} = (-50 + j86\cdot6)\,\text{A}$$

$$aI_B = 100\underline{/120^\circ} \times 1\underline{/120^\circ} = 100\underline{/-120} = (-50 - j86\cdot6)\,\text{A}$$

$$a^2I_B = 100\underline{/120^\circ} \times 1\underline{/-120^\circ} = 100\underline{/0^\circ} = (100 + j0)\,\text{A}$$

$$I_{R+} = \tfrac{1}{3}(I_R + aI_Y + a^2I_B)$$

$$3I_{R+} = 373\cdot2 + j100$$

$$\underline{\underline{I_{R+} = 129\underline{/15^\circ}\,\text{A}}}$$

$$I_{R-} = \tfrac{1}{3}(I_R + a^2I_Y + aI_B)$$

$$3I_{R-} = -123\cdot2 + j13\cdot4$$

$$\underline{\underline{I_{R-} = 41\cdot3\underline{/174^\circ}\,\text{A}}}$$

$$I_{R0} = \tfrac{1}{3}(I_R + I_Y + I_B)$$

$$3I_{R0} = 50 - j113\cdot4$$

$$\underline{\underline{I_{R0} = 41\cdot3\underline{/-66\cdot2^\circ}\,\text{A}}}$$

A graphical solution is shown in Fig. 4.17. The complexors I_R, I_Y and I_B are first drawn using a suitable scale. The complexors aI_Y, a^2I_Y, aI_B and a^2I_B are then drawn 120° leading and lagging on I_Y and I_B respectively. The required

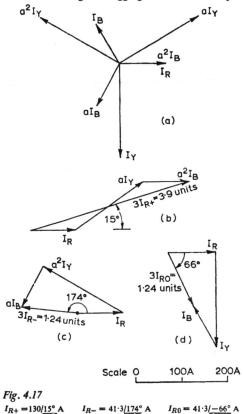

(a)

(b)

(c)

(d)

Scale 0   100A   200A

Fig. 4.17

$I_{R+} = 130\underline{/15°}$ A $I_{R-} = 41\cdot3\underline{/174°}$ A $I_{R0} = 41\cdot3\underline{/-66°}$ A

complexor additions are then performed, and these are shown in Figs. 4.17(*b*), (*c*) and (*d*).

4.14 Applications of Symmetrical Components

The method of symmetrical components may be applied to the solution of all kinds of unbalanced 3-phase network problems such as those discussed in Sections 4.5, 4.6 and 4.7. However, when the source impedance is assumed to be negligible, the methods of solution adopted there are easier than the application of the method of symmetrical components. Where source impedance is taken into account the method of symmetrical components *must* be applied if the source is a synchronous machine or a transformer

supplied by a synchronous machine. This is because the impedance
of a synchronous machine to positive phase-sequence currents is
different from its impedance to negative phase-sequence currents.
The source impedance is likely to be of significance under fault
conditions in 3-phase networks. Indeed for a fault at the generator
terminals the generator impedance is the only impedance present.
The main field of application of symmetrical components, therefore,
is the analysis of 3-phase networks under asymmetrical fault
conditions.

4.15 Synchronous Generator supplying an Unbalanced Load

Three-phase generators are specifically designed so that the phase

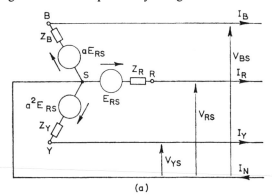

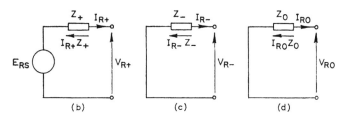

Fig. 4.18   SYNCHRONOUS GENERATOR SUPPLYING AN UNBALANCED LOAD

    (a) Actual network
    (b) Positive phase-sequence network
    (c) Negative phase-sequence network
    (d) Zero phase-sequence network

e.m.f.s form a balanced 3-phase system. It is assumed, therefore,
that the e.m.f. of the synchronous generator shown in Fig. 4.18 is
of positive phase sequence only. Then the generator phase e.m.f.s
are

$$E_{RS}, \quad E_{YS} \,(= a^2 E_{RS}), \quad E_{BS} \,(= a E_{RS})$$

If the generator is connected by four wires to an unbalanced star-connected load (the most general case), the generator phase currents will be unbalanced, and as shown in Section 4.13, may be represented by the superposition of positive, negative and zero phase-sequence components of current. These components are:

$$I_{R+} = \tfrac{1}{3}(I_R + aI_Y + a^2I_B) \tag{4.33}$$

$$I_{Y+} = a^2I_R \tag{4.34}$$

$$I_{B+} = aI_R \tag{4.35}$$

$$I_{R-} = \tfrac{1}{3}(I_R + a^2I_Y + aI_B) \tag{4.36}$$

$$I_{Y-} = aI_{R-} \tag{4.37}$$

$$I_{B-} = a^2I_{R-} \tag{4.38}$$

$$I_{R0} = \tfrac{1}{3}(I_R + I_Y + I_B) \tag{4.39}$$

$$I_{Y0} = I_{R0} \tag{4.40}$$

$$I_{BO} = I_{R0} \tag{4.41}$$

Assuming, as is usual, that the impedances of the generator phases to positive phase-sequence current are equal, i.e.

$$Z_{R+} = Z_{Y+} = Z_{B+} = Z_+$$

then

  Voltage drop in R-phase due to positive phase-sequence current
    $= I_{R+}Z_+$

  Voltage drop in Y-phase due to positive phase-sequence current
    $= I_{Y+}Z_+ = a^2I_{R+}Z_+$

  Voltage drop in B-phase due to positive phase-sequence current
    $= I_{B+}Z_+ = aI_{R+}Z_+$

That is, the voltage drops due to the positive phase-sequence currents are positive phase sequence only. This would not be the case if Z_{R+}, Z_{Y+} and Z_{B+} were different.

Similarly, if the impedances of the generator phases to negative phase-sequence current are equal, i.e.

$$Z_{R-} = Z_{Y-} = Z_{B-} = Z_-$$

then

  Voltage drop in R-phase due to negative phase sequence current
    $= I_{R-}Z_-$

  Voltage drop in Y-phase due to negative phase sequence current
    $= I_{Y-}Z_- = aI_{R-}Z_-$

  Voltage drop in B-phase due to negative phase sequence current
    $I_{B-}Z_- = a^2I_{R-}Z_-$

That is, the voltage drops due to the negative phase-sequence currents are of negative phase sequence only. Again this result depends on Z_{R-}, Z_{Y-} and Z_{B-} being equal.

Also, if the impedances of the generator phases to zero phase-sequence current are equal, i.e.

$$Z_{R0} = Z_{Y0} = Z_{B0} = Z_0$$

then

Voltage drop in R-phase due to negative phase-sequence current
$$= I_{R0}Z_0$$

Voltage drop in Y-phase due to negative phase-sequence current
$$= I_{Y0}Z_0 = I_{R0}Z_0$$

Voltage drop in B-phase due to negative phase-sequence current
$$= I_{B0}Z_0 = I_{R0}Z_0$$

E.M.F. induced in R-phase
$$= \text{(Terminal voltage of R-phase)}$$
$$+ \text{(Internal voltage drop in R-phase)}$$

$$E_{RS} = V_{RS} + I_{R+}Z_+ + I_{R-}Z_- + I_{R0}Z_0 \tag{4.42}$$

Similarly,

$$E_{YS} = a^2 E_{RS} = V_{YS} + a^2 I_{R+}Z_+ + a I_{R-}Z_- + I_{R0}Z_0 \tag{4.43}$$

and

$$E_{BS} = a E_{RS} = V_{BS} + a I_{R+}Z_+ + a^2 I_{R-}Z_- + I_{R0}Z_0 \tag{4.44}$$

Adding these three equations,

$$0 = V_{RS} + V_{YS} + V_{BS} + 3 I_{R0}Z_0$$
$$\text{or} \quad 0 = V_{R0} + I_{R0}Z_0 \tag{4.45}$$

where V_{R0} is the zero phase-sequence component of the generator terminal voltage V_{RS}, i.e.

$$V_{R0} = \tfrac{1}{3}(V_{RS} + V_{YS} + V_{BS})$$

If eqn. (4.43) is multiplied by a^2 and eqn. (4.44) by a,

$$E_{RS} = V_{RS} + I_{R+}Z_+ + I_{R-}Z_- + I_{R0}Z_0 \tag{4.42}$$
$$a^4 E_{RS} = a^2 V_{YS} + a^4 I_{R+}Z_+ + a^3 I_{R-}Z_- + a^2 I_{R0}Z_0 \tag{4.46}$$
$$a^2 E_{RS} = a V_{BS} + a^3 I_{R+}Z_+ + a^3 I_E-Z_- + a I_{R0}Z_0 \tag{4.47}$$

Adding these three equations,

$$0 = V_{RS} + a^2 V_{YS} + a V_{BS} + 3 I_{R-}Z_-$$
$$\text{or} \quad 0 = V_{R-} + I_{R-}Z_- \tag{4.48}$$

where V_{R-} is the negative phase-sequence component of the generator terminal voltage V_{RS}.

If eqn. (4.43) is multiplied by a and eqn. (4.44) by a^2,

$$E_{RS} = V_{RS} + I_{R+}Z_+ + I_{R-}Z_- + I_{R0}Z_0 \tag{4.42}$$

$$a^3 E_{RS} = aV_{YS} + a^3 I_{R+}Z_+ + a^2 I_{R-}Z_- + aI_{R0}Z_0 \tag{4.49}$$

$$a^3 E_{RS} = a^2 V_{BS} + a^3 I_{R+}Z_+ + a^4 I_{R-}Z_- + a^2 I_{R0}Z_0 \tag{4.50}$$

Adding the above equations,

$$3E_{RS} = V_{RS} + aV_{YS} + a^2 V_{BS} + 3I_{R+}Z_+$$

$$E_{RS} = \tfrac{1}{3}(V_{RS} + aV_{YS} + a^2 V_{BS}) + I_{R+}Z_+$$

$$E_{RS} = V_{R+} + I_{R+}Z_+ \tag{4.51}$$

Examination of eqns. (4.45), (4.48) and (4.51) reveals that, when an unbalanced load is imposed on a system, the positive, negative and zero phase-sequence components may be considered separately if the impedance of each phase of the network is balanced. Under these conditions the positive, negative and zero phase-sequence networks, which are interpretations of eqns. (4.45), (4.48) and (4.51), may be drawn, as in Fig. 4.18. It should be noted particularly that negative and zero phase-sequence currents may flow even when there is no negative or zero phase-sequence e.m.f., since such current components may arise due to unbalanced loading.

4.16 Analysis of Asymmetrical Faults

In the following analysis it will be assumed that

1. The positive, negative and zero phase-sequence impedances of the generator and any interconnected plant are known.
2. The generator e.m.f. system is of positive phase sequence only.
3. No current flows in the network other than that due to the fault.
4. The network impedances in each phase are balanced.
5. The impedance of the fault is zero.

ONE-LINE-TO-EARTH FAULT

The circuit diagram of Fig. 4.19 shows the fault conditions. Evidently, $V_{RS} = 0$ since the R-phase terminal and the star point are earthed. Also

$$I_Y = 0 \quad \text{and} \quad I_B = 0$$

since only fault current is assumed to flow.

The phase sequence components of current are, from eqns. (4.31). (4.32) and (4.28),

$$I_{R+} = \tfrac{1}{3}(I_R + aI_Y + a^2I_B) = \tfrac{1}{3}I_R$$
$$I_{R-} = \tfrac{1}{3}(I_R + a^2I_Y + aI_B) = \tfrac{1}{3}I_R$$
$$I_{R0} = \tfrac{1}{3}(I_R + I_Y + I_B) = \tfrac{1}{3}I_R$$

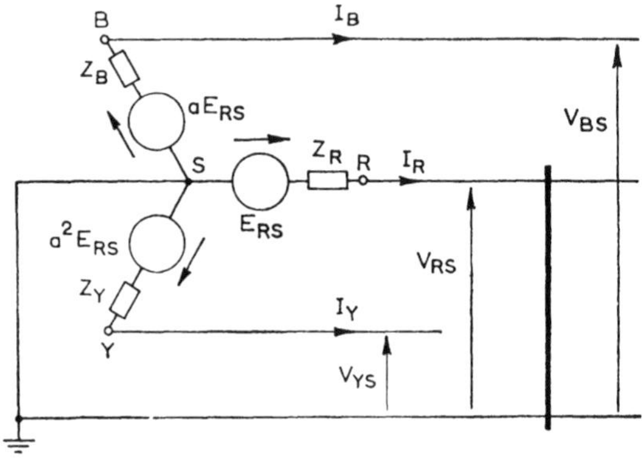

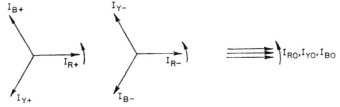

Fig. 4.19   ONE-LINE-TO-EARTH FAULT

Therefore

$$I_{R+} = I_{R-} = I_{R0} = \tfrac{1}{3}I_R \qquad (4.52)$$

A complexor diagram of the phase-sequence components of current is given in Fig. 4.20.

Fig. 4.20   SYMMETRICAL COMPONENTS OF ONE-LINE-TO-EARTH CURRENT

Since the generator e.m.f. system is of positive phase sequence only, and since the network impedances in the three phases are balanced so that each phase sequence may be considered separately as proved in Section 4.15, then

$$E_{RS} = V_{R+} + I_{R+}Z_+ \qquad (4.51)$$
$$0 = V_{R-} + I_{R-}Z_- \qquad (4.48)$$
$$0 = V_{R0} + I_{R0}Z_0 \qquad (4.45)$$

Adding these three equations,

$$E_{RS} = V_{R+} + V_{R-} + V_{R0} + I_{R+}Z_+ + I_{R-}Z_- + I_{R0}Z_0$$

But $V_{R+} + V_{R-} + V_{R0} = \frac{1}{3}V_{RS} = 0$, and $I_{R+} + I_{R-} + I_{R0} = \frac{1}{3}I_R$, so that

$$E_{RS} = \frac{1}{3}I_R(Z_+ + Z_- + Z_0) \tag{4.53}$$

The fault current is

$$I_F = I_R = \frac{3E_{RS}}{Z_+ + Z_- + Z_0} \tag{4.54}$$

Examination of this equation shows that an equivalent circuit from which the fault current may be calculated is as given in Fig. 4.21(a). Since the positive, negative and zero phase-sequence

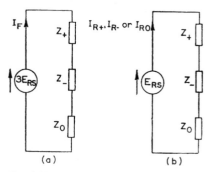

Fig. 4.21 EQUIVALENT CIRCUITS FOR ONE-LINE-TO-EARTH FAULT

components of current in the faulted phase are all equal and in phase, the equivalent circuit from which each phase-sequence component of current may be calculated is given in Fig. 4.21(b).

SHORT-CIRCUIT BETWEEN TWO LINES

The circuit diagram of Fig. 4.22 shows the fault conditions. Evidently

$$I_Y + I_B = 0 \qquad I_R = 0 \qquad \text{and} \qquad V_{YS} = V_{BS}$$

The phase-sequence components of current are, from eqns. (4.31), (4.32) and (4.28),

$$I_{R+} = \frac{1}{3}(I_R + aI_Y + a^2I_B) = \frac{1}{3}(a - a^2)I_Y$$
$$I_{R-} = \frac{1}{3}(I_R + a^2I_Y + aI_B) = \frac{1}{3}(a^2 - a)I_Y$$
$$I_{R0} = \frac{1}{3}(I_R + I_Y + I_B) = 0$$

The above equations show that

$$I_{R0} = 0 \quad \text{and} \quad I_{R+} = -I_{R-}$$

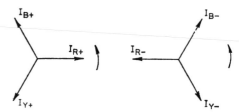

Fig. 4.22   SHORT-CIRCUIT BETWEEN TWO LINES

A complexor diagram of the phase-sequence components of current is given in Fig. 4.23.

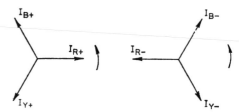

Fig. 4.23   SYMMETRICAL COMPONENTS OF FAULT CURRENT FOR A SHORT-CIRCUIT BETWEEN TWO LINES

Since the generator e.m.f. system is of positive phase sequence only and the network impedances in each phase are balanced,

$$E_{RS} = V_{R+} + I_{R+}Z_+ \tag{4.51}$$

$$0 = V_{R-} + I_{R-}Z_- \tag{4.48}$$

$$0 = V_{R0} + I_{R0}Z_0 \tag{4.45}$$

Also,

$$V_{R+} = \tfrac{1}{3}(V_{RS} + aV_{YS} + a^2V_{BS}) = \tfrac{1}{3}(V_{RS} + (a + a^2)V_{YS}) \tag{4.55}$$

$$V_{R-} = \tfrac{1}{3}(V_{RS} + a^2 V_{YS} + a V_{BS}) = \tfrac{1}{3}(V_{RS} + (a^2 + a)V_{YS})$$
$$(4.56)$$

since $V_{YS} = V_{BS}$.

From eqns. (4.55) and (4.56),

$$V_{R+} = V_{R-}$$

Therefore, subtracting eqn. (4.48) from eqn. (4.51),

$$E_{RS} = V_{R+} - V_{R-} + I_{R+}Z_+ - I_{R-}Z_- = I_{R+}(Z_+ + Z_-) \quad (4.57)$$

since $V_{R+} = V_{R-}$ and $I_{R+} = -I_{R-}$.

From eqn. (4.57),

$$I_{R+} = \frac{E_{RS}}{Z_+ + Z_-} \qquad (4.58)$$

Also,

$$I_{R-} = \frac{- E_{RS}}{Z_+ + Z_-} \qquad (4.59)$$

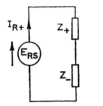

Fig. 4.24 EQUIVALENT CIRCUITS FOR A SHORT-CIRCUIT BETWEEN TWO LINES

Examination of these two equations shows that the equivalent circuits from which the positive and negative phase-sequence components of current may be calculated are as shown in Figs. 4.24(*a*) and (*b*).

The fault current is

$$\begin{aligned}
I_F = I_Y &= I_{Y+} + I_{Y-} \\
&= a^2 I_{R+} + a I_{R-} \\
&= a^2 I_{R+} - a I_{R+} \\
&= \frac{E_{RS}}{Z_+ + Z_-}(a^2 - a) \\
&= \frac{-j\sqrt{3}E_{RS}}{Z_+ + Z_-} \qquad (4.60)
\end{aligned}$$

TWO-LINES-TO-EARTH FAULT

The circuit diagram of Fig. 4.25 shows the fault conditions. Evidently,

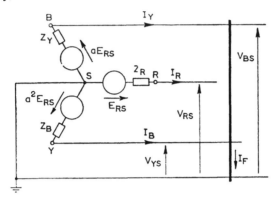

Fig. 4.25   TWO-LINES-TO-EARTH FAULT

$$I_R = 0$$
$$V_{YS} = V_{BS} = 0$$
$$I_F = I_Y + I_B$$
$$I_R = I_{R+} + I_{R-} + I_{R0} = 0 \tag{4.61}$$

The phase sequence components of the generator terminal voltage are

$$V_{R+} = \tfrac{1}{3}(V_{RS} + aV_{YS} + a^2V_{BS}) = \tfrac{1}{3}V_{RS}$$
$$V_{R-} = \tfrac{1}{3}(V_{RS} + a^2V_{YS} + aV_{BS}) = \tfrac{1}{3}V_{RS}$$
$$V_{R0} = \tfrac{1}{3}(V_{RS} + V_{YS} + V_{BS}) = \tfrac{1}{3}V_{RS}$$

Therefore

$$V_{R+} = V_{R-} = V_{R0} = \tfrac{1}{3}V_{RS} \tag{4.62}$$

Since the generator e.m.f. system is of positive sequence only, and since the network impedances in each phase are balanced,

$$E_{RS} = V_{R+} + I_{R+}Z_+ \tag{4.51}$$
$$0 = V_{R-} + I_{R-}Z_- \tag{4.48}$$
$$0 = V_{R0} + I_{R0}Z_0 \tag{4.45}$$

The phase-sequence components of the generator terminal voltage may be eliminated from these equations by subtraction:

$$(4.51) - (4.48) \qquad E_{RS} = I_{R+}Z_+ - I_{R-}Z_- \qquad (4.63)$$

$$(4.51) - (4.45) \qquad E_{RS} = I_{R+}Z_+ - I_{R0}Z_0 \qquad (4.64)$$

$$(4.48) - (4.45) \qquad 0 = I_{R-}Z_- - I_{R0}Z_0 \qquad (4.65)$$

Substituting (from eqn. (4.61)) $-I_{R0} = I_{R+} + I_{R-}$ in eqn. (4.64),

$$E_{RS} = I_{R+}(Z_+ + Z_0) + I_{R-}Z_0 \qquad (4.66)$$

I_{R-} can now be eliminated between eqns. (4.63) and (4.66) by multiplying eqn. (4.63) by Z_0 and eqn. (4.66) by Z_-; thus

$$E_{RS}Z_0 = I_{R+}Z_+Z_0 - I_{R-}Z_-Z_0$$

$$E_{RS}Z_- = I_{R+}(Z_+ + Z_0)Z_- + I_{R-}Z_-Z_0$$

Adding,

$$E_{RS}(Z_0 + Z_-) = I_{R+}(Z_+Z_0 + Z_+Z_- + Z_-Z_0)$$

Therefore,

$$I_{R+} = \frac{E_{RS}(Z_0 + Z_-)}{Z_+Z_0 + Z_+Z_- + Z_-Z_0} = \frac{E_{RS}}{Z_+ + \dfrac{Z_-Z_0}{Z_0 + Z_-}} \qquad (4.67)$$

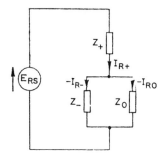

Fig. 4.26 EQUIVALENT CIRCUIT FOR TWO-LINES-TO-EARTH FAULT

Examination of this equation shows that the equivalent circuit from which the positive phase-sequence component of current may be determined is as shown in Fig. 4.26.

Further, from eqn. (4.63),

$$-I_{R-} = \frac{E_{RS} - I_{R+}Z_+}{Z_-} \qquad (4.68)$$

$E_{RS} - I_{R+}Z_+$ is the voltage drop across the parallel branch of the equivalent circuit in Fig. 4.26. Therefore the current flowing in Z_- of this equivalent circuit is $-I_{R-}$.

Also, from eqn. (4.64),

$$-I_{R0} = \frac{E_{RS} - I_{R+}Z_+}{Z_0} \qquad (4.69)$$

Therefore the current flowing in Z_0 of the equivalent circuit of Fig. 4.26 is $-I_{R0}$.

EXAMPLE 4.9 A 3-phase 75 MVA 0·8 p.f. (lagging) 11·8 kV star-connected alternator having its star point solidly earthed supplies a feeder. The relevant per-unit (p.u.)* impedances, based on the rated phase voltage and phase current of the alternator are as follows:

	Generator Z_G	Feeder Z_F
Positive sequence impedance (p.u.)	$j1·70$	$j0·10$
Negative sequence impedance (p.u.)	$j0·18$	$j0·10$
Zero sequence impedance (p.u.)	$j0·12$	$j0·30$

Determine the fault current and the line-to-neutral voltages at the generator terminals for a one-line-to-earth fault occurring at the distant end of the feeder. The generator e.m.f. per phase is of positive sequence only and is equal to the rated voltage per phase.

$$\text{Rated voltage per phase} = \frac{11·8 \times 10^3}{\sqrt{3}} = 6,820 \, \text{V}$$

$$\text{Rated current per phase} = \frac{75 \times 10^6}{\sqrt{3} \times 11·8 \times 10^3} = 3,670 \, \text{A}$$

The circuit diagram is shown in Fig. 4.27. The fault is assumed to occur on the red phase.

Let $Z_{T+} = Z_{G+} + Z_{F+} = j1·80 \, \text{p.u.}$
$Z_{T-} = Z_{G-} + Z_{F-} = j0·28 \, \text{p.u.}$
$Z_{T0} = Z_{G0} + Z_{F0} = j0·42 \, \text{p.u.}$

Take E_{RS} as the reference complexor, i.e.

$E_{RS} = 1\underline{/0°} \, \text{p.u.}$

Then

$$I_{R+} = I_{R-} = I_{R0} = \frac{E_{RS}}{Z_{T+} + Z_{T-} + Z_{T0}} = \frac{1\underline{/0°}}{2·50\underline{/90°}} = 0·4\underline{/-90°} \, \text{p.u.}$$

and

$$\text{Fault current, } I_F = 3I_{R+} = 3 \times 0·4\underline{/-90°} = 1·2\underline{/-90°} \, \text{p.u.}$$

i.e.

$$I_F = 1·2 \times 3,670 = \underline{\underline{4,400 \, \text{A}}}$$

* See Section 9.16.

The positive, negative and zero sequence networks are shown in Figs. 4.27(*c*), (*d*) and (*e*).

$$E_{RS} = (V_{rs})_+ + I_{R+}Z_{G+}$$
$$0 = (V_{rs})_- + I_{R-}Z_{G-}$$
$$0 = (V_{rs})_0 + I_{R0}Z_{G0}$$

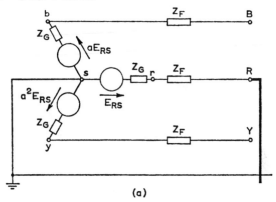

(a)

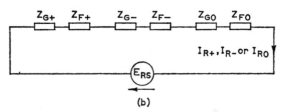

(b)

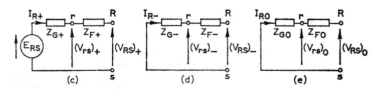

(c)   (d)   (e)

Fig. 4.27   CIRCUIT DIAGRAMS FOR EXAMPLE 4.9

(a) Actual network
(b) Equivalent circuit for phase-sequence components of current
(c) Positive phase-sequence network
(d) Negative phase-sequence network
(e) Zero phase-sequence network

where $(V_{rs})_+$, $(V_{rs})_-$ and $(V_{rs})_0$ are the positive, negative and zero phase-sequence components of the generator red-phase terminal voltage.

$$(V_{rs})_+ = E_{RS} - I_R Z_G = 1\underline{/0°} - (0{\cdot}4\underline{/-90°} \times 1{\cdot}70\underline{/90°}) = 0{\cdot}32\underline{/0°}\,\text{p.u.}$$
$$(V_{rs})_- = -I_R{\cdot}Z_{G-} = -0{\cdot}4\underline{/-90°} \times 0{\cdot}18\underline{/90°} = 0{\cdot}072\underline{/180°}\,\text{p.u.}$$
$$(V_{rs})_0 = -I_{R0}Z_{G0} = -0{\cdot}4\underline{/-90°} \times 0{\cdot}12\underline{/90°} = 0{\cdot}048\underline{/180°}\,\text{p.u.}$$

$$V_{rS} = (V_{rS})_+ + (V_{rS})_- + (V_{rS})_0 = 0 \cdot 32 - 0 \cdot 072 - 0 \cdot 048 = 0 \cdot 20\underline{/0^\circ}\,\text{p.u.}$$

$$V_{yS} = a^2(V_{rS})_+ + a(V_{rS})_- + (V_{rS})_0$$
$$= (1\underline{/-120^\circ} \times 0 \cdot 32\underline{/0^\circ}) + (1\underline{/120^\circ} \times 0 \cdot 072\underline{/180^\circ}) + 0 \cdot 048\underline{/180^\circ}$$
$$= -0 \cdot 16 - j0 \cdot 277 + 0 \cdot 036 - j0 \cdot 0624 - 0 \cdot 048 = 0 \cdot 38\underline{/-116 \cdot 9^\circ}\,\text{p.u.}$$

$$V_{bS} = a(V_{rS})_+ + a^2(V_{rS})_- + (V_{rS})_0$$
$$= (1\underline{/120^\circ} \times 0 \cdot 32\underline{/0^\circ}) + (1\underline{/-120^\circ} \times 0 \cdot 072\underline{/180^\circ}) + 0 \cdot 048\underline{/180^\circ}$$
$$= -0 \cdot 16 + j0 \cdot 277 + 0 \cdot 036 + j0 \cdot 0624 - 0 \cdot 048 = 0 \cdot 38\underline{/-116 \cdot 9^\circ}\,\text{p.u.}$$

$$V_{rS} = 0 \cdot 20 \times 6{,}820 = \underline{\underline{1{,}364\,\text{V}}}$$

$$V_{yS} = V_{bS} = 0 \cdot 38 \times 6{,}820 = \underline{\underline{2{,}550\,\text{V}}}$$

PROBLEMS

4.1 A symmetrical 3-phase 400 V system supplies a balanced mesh-connected load. The R Y branch current is $20\underline{/40^\circ}$ A. Find (a) the line currents, and (b) the total power.

Draw a complexor diagram showing the voltages and currents in the lines and phases. Assume V_{RY} is the reference complexor.

Ans. $34 \cdot 6\underline{/10^\circ}$ A, $34 \cdot 6\underline{/-110^\circ}$ A, $34 \cdot 6\underline{/130^\circ}$ A, $18 \cdot 24$ kW.

4.2 An unbalanced 4-wire star-connected load has balanced line voltages of 440 V. The loads are

$$Z_R = 10\,\Omega \qquad Z_Y = (5 + j10)\,\Omega \qquad Z_B = (15 - j5)\,\Omega$$

Calculate the value of the current in the neutral wire and its phase relationship to the voltage across the red phase. The phase sequence is RYB.

Hence sketch the complexor diagram.

Ans. $15 \cdot 2$ A, $127 \cdot 3^\circ$ lagging.

4.3 A 3-phase, 4-wire, 440 V system is loaded as follows:

(i) An induction motor load of 350 kW at power factor 0·71 lagging.

(ii) Resistance loads of 150 kW, 250 kW and 400 kW connected between neutral and the R, Y and B lines respectively. Calculate (a) the line currents, (b) the current in the neutral, and (c) the power factor of the system. Phase sequence, RYB.   (*H.N.C.*)

Ans. 1,145 A, 1,512 A, 2,080 A, 855 A, 0·957 lagging.

4.4 Three impedances Z_1, Z_2 and Z_3 are mesh connected to a symmetrical 3-phase 400 V 50 Hz supply of phase sequence R Y B.

$$Z_1 = (10 + j0)\,\Omega \text{ and is connected between lines R and Y}$$
$$Z_2 = (8 + j6)\,\Omega \text{ connected between lines Y and B}$$
$$Z_3 = (5 - j5)\,\Omega \text{ connected between lines B and R}$$

Calculate the phase and line currents and the total power consumed.

(*H.N.C.*)

Ans. 40 A, 40 A, 56·6 A; 95·7 A, 78·4 A, 35·2 A; 44·8 kW.

4.5 A symmetrical 3-phase supply, of which the line voltage is 380 V, feeds a mesh-connected load as follows:

Load A: 19 kVA at p.f. 0·5 lagging connected between lines R and Y
Load B: 30 kVA at p.f. 0·8 lagging connected between lines Y and B
Load C: 10 kVA at p.f. 0·9 leading connected between lines B and R

Determine the line currents and their phase angles. Phase sequence, RYB.
(H.N.C.)

Ans. 74·6$\underline{/-51°}$ A, 98·6$\underline{/173°}$ A, 68·3$\underline{/41·8}$ A.

4.6 Determine the line currents in an unbalanced star-connected load supplied from a symmetrical 3-phase 440 V 3-wire system. The branch impedances of the load are $Z_1 = 5\underline{/30°}\Omega$, $Z_2 = 10\underline{/45°}\Omega$ and $Z_3 = 10\underline{/60°}\Omega$. The phase sequence is RYB. Draw the complexor diagram. *(H.N.C.)*

Ans. 35·7 A, 32·8 A, 27·7 A.

4.7 A 440 V symmetrical 3-phase supply feeds a star-connected load consisting of three non-inductive resistances of 10, 5 and 12 Ω connected to the R, Y and B phases respectively.

Calculate the line currents and the voltage across each resistor. Phase sequence, RYB. *(H.N.C.)*

Ans. 28·9 A, 36·5 A, 25·4 A; 290 V, 182 V, 304 V.

4.8 The power input to a 2,000 V 50 Hz 3-phase motor running on full load at an efficiency of 90 per cent is measured by two wattmeters which indicate 300 kW and 100 kW. Calculate (*a*) the input, (*b*) the power factor, (*c*) the line current, (*d*) the power output.

Ans. 400 kW, 0·756, 152 A, 360 kW.

4.9 The wattmeter readings in an induction motor circuit are 34·7 W and 4·7 W respectively, the latter reading being obtained after reversal of the connexions of the instrument voltage coil. Calculate the power factor at which the motor is working assuming normal two-wattmeter connexion of the wattmeters.

Ans. 0·4.

4.10 A 500 V 3-phase motor has an output of 37·8 kW and operates at a p.f. of 0·85 with an efficiency of 90 per cent. Calculate the reading on each of two watt-meters connected to measure the power input.

Ans. 28·2 kW, 13·3 kW.

4.11 Give the circuit arrangement and the theory of the two-wattmeter method of measuring power in a 3-phase 3-wire system.

A balanced star-connected load, each phase having a resistance of 10 Ω and an inductive reactance of 30 Ω is connected to a 400 V 50 Hz supply. The phase rotation is RYB. Wattmeters connected to read the total power have their current coils in the red and blue lines respectively. Calculate the reading of each wattmeter. *(L.U.)*

Ans. 2,190 W; −583 W.

4.12 Three impedances are mesh connected to a symmetrical 3-phase 440 V 50 Hz supply of phase sequence RYB.

$Z_1 = (5 + j10)\Omega$ is connected between lines R and Y
$Z_2 = (5 + j5)\Omega$ is connected between lines Y and B
$Z_3 = (6 - j4)\Omega$ is connected between lines B and R

Two wattmeters connected to measure the power input have their current coils in lines R and Y respectively. Calculate the line currents and the wattmeter readings.   *(H.N.C.)*

Ans. 95·5 A, 79·4 A, 43·3 A; 39·8 kW, 9·6 kW.

4.13 A 3-phase transmission line delivers a current at 33 kV to a balanced load having an equivalent impedance to neutral of $(240 + j320)\Omega$. The line has a resistance per conductor of $20\,\Omega$ and a reactance per conductor of $30\,\Omega$.
   Calculate the voltage at the generator end.
   If the load is made $(280 + j370)\Omega$, calculate the receiving-end voltage if the voltage at the generator end is unchanged.   *(L.U.)*

Ans. 36 kV; 33·5 kV.

4.14 A 3-phase, 50 Hz, transmission line is 25 km long and delivers, 2,500 kW at 30 kV, 0·8 power factor lagging. Calculate the voltage at the sending end if each conductor has $R = 0.8\,\Omega$ and $X_L = 1.0\,\Omega$ per km. If an extra load consisting of capacitors having $C = 1.5\,\mu\mathrm{F}$ to neutral is connected at the middle of the line, calculate the voltage at the sending end.   *(H.N.C.)*

Ans. 34·1 kV, 33·1 kV.

4.15 What are the advantages of overhead lines as compared with underground cables for transmission at very high voltages?
   A 3-phase, 50 Hz transmission line is 100 km long and has the following constants:

   Resistance/phase/km = $0.2\,\Omega$
   Inductance/phase/km = 2 mH
   Capacitance (line-to-neutral) per km = $0.015\,\mu\mathrm{F}$

   If the line supplies a load of 50 MW at 0·8 p.f. lagging and 132 kV, determine, using the nominal-T method, the sending-end voltage, current and power factor and the line efficiency.   *(L.U.)*

Ans. 156 kV; 248 A; 0·8 lagging; 93·5 per cent.

4.16 Determine the voltage at the sending end, and the efficiency of a 3-phase transmission line given:

   Output of line—250 A, 132 kV at 0·8 p.f. lagging
   Line reactance—42 Ω in each wire
   Line resistance—12 Ω in each wire
   Line susceptance—3.75×10^{-4} mho line-to-neutral
   Line leakance—negligible

   The capacitance may be assumed to be located wholly at the centre of the line.   *(L.U.)*

Ans. 146 kV; 96 per cent.

4.17 The resistance, reactance and line-to-neutral susceptance of a 3-phase transmission line are 15 Ω, 45 Ω and 4×10^{-4} S respectively, leakance being negligible. The load at the receiving end of the line is 50 MVA at 130 kV, 0·7 p.f. lagging.
   Assuming the susceptance to be lumped half at each end of the line, find the sending-end voltage and current, and the efficiency of transmission.
                                                                      (H.N.C.)

Ans. 145 kV; 200 A; 94·6 per cent.

4.18 The three currents in a 3-phase system are

$$I_A = (120 + j60)\text{A}, \ I_B = (120 - j120)\text{A and } I_C = (-150 + j100)\text{A}$$

Find the symmetrical components of these currents.

Ans. $I_{A+} = (109 + j100)\text{A}; \ I_{A-} = (-18\cdot5 - j54\cdot7)\text{A}; \ I_{A0} = (30 + j13\cdot3)\text{A}.$

4.19 A 3-phase 75 MVA 11·8 kV star-connected alternator with a solidly earthed star point has the following p.u. impedances based on rated phase voltage and rated phase current: positive phase sequence impedance, $j2\cdot0$ p.u.; negative phase sequence impedance, $j0\cdot16$ p.u.; zero phase sequence impedance, $j0\cdot08$ p.u. Determine the steady-state fault current for the following: (a) a 3-phase symmetrical short-circuit, (b) a one-line-to-earth fault, and (c) a two-line-to-earth fault. The generator e.m.f. per phase is equal to the rated voltage.

Ans. 1,840 A; 4,920 A; 3,580 A.

4.20 Two similar 3-phase 50 MVA 11 kV star-connected alternators, A and B, are connected in parallel to 3-phase busbars to which an 11 kV 3-phase feeder is also connected. The star point of generator A is solidly earthed, but that of generator B is insulated from earth. The p.u. impedances of each generator and the feeder based on rated phase current and rated phase voltage are:

	Generator	Feeder
Positive phase-sequence impedance	$j2\cdot00$ p.u.	$j0\cdot20$ p.u.
Negative phase-sequence impedance	$j0\cdot16$ p.u.	$j0\cdot20$ p.u.
Zero phase-sequence impedance	$j0\cdot08$ p.u.	$j0\cdot60$ p.u.

Determine the total fault current in each line, the fault current to earth and the fault current supplied by each generator, for a two-line-to-earth fault at the distant end of the feeder. Each generator has an e.m.f. per phase equal to rated voltage. Assume the blue and yellow lines to be the faulted lines.

Notes

1. Since only zero phase sequence current flows in the earth, generator B, the star point of which is not earthed, cannot supply any zero phase sequence current.
2. In the equivalent circuit the generator positive and negative phase sequence impedances are in parallel and in series with that of the feeder. The zero phase sequence impedance of generator B does not appear in the equivalent circuits since it supplies no zero sequence current.

Ans. $I_R = 0; \ I_Y = 1,680\underline{/162\cdot8°}\text{A}; \ I_B = 1,680\underline{/17\cdot2°}\text{A}; \ I_E = 1,000\underline{/90°}\text{A}$

Generator A: $I_R = 0; \ I_Y = 900\underline{/152\cdot9°}\text{A}; \ I_B = 900\underline{/27\cdot1°}\text{A}$

Generator B: $I_R = 0; \ I_Y = 807\underline{/174\cdot2°}\text{A}; \ I_B = 807\underline{/5\cdot8°}\text{A}$

Chapter 5

HARMONICS

Up to this stage it has been assumed that all alternating currents and voltages have been sinusoidal in waveform, this being by far

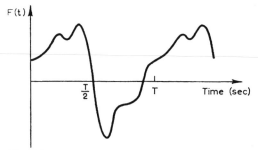

Fig. 5.1 A COMPLEX WAVEFORM

the most important form of alternating quantity which is met in electrical engineering. An alternating waveform which is not sinusoidal is said to be *complex*, and the complex wave (subject to certain mathematical conditions which are always complied with in electrical engineering applications) may be shown to be built up of a series of sinusoidal waves whose frequencies are integral multiples of the frequency of the *fundamental*, or basic, wave. The sinusoidal components of a complex wave are called the *harmonics*, the second harmonic having a frequency of twice the fundamental, the third harmonic three times the fundamental and so on.

Fig. 5.1 shows the graph to a base of time of a complex wave,

134

whose period is T second, i.e. the fundamental frequency is $f = 1/T$ hertz (cycles per second).

Although it is desirable to have sinusoidal currents and voltages in a.c. systems, this is not always possible, and currents and voltages with complex waveforms do occur in practice. From the above, these complex voltages and currents may be represented by a series of sinusoidal waves naving frequencies which are integral multiples of the fundamental. The resultant effect of complex currents and/or voltages in a linear electric circuit is simply the sum of the individual effects of each harmonic by itself. This applies to instantaneous values and not to r.m.s. values, as will be seen later.

5.1 General Equation for a Complex Wave

Consider a complex voltage wave to be built up of a fundamental plus harmonic terms, each of which will have its own phase angle with respect to zero time. The instantaneous value of the resultant voltage will be given by the expression

$$v = V_{1m} \sin(\omega t + \psi_1) + V_{2m} \sin(2\omega t + \psi_2) + \dots$$
$$+ V_{nm} \sin(n\omega t + \psi_n) \qquad (5.1)$$

where $f (= \omega/2\pi)$ is the frequency of the fundamental of the complex wave, and V_{nm} is the peak value of the nth harmonic.

In the same way the instantaneous value of a complex alternating current is given by

$$i = I_{1m} \sin(\omega t + \phi_1) + I_{2m} \sin(2\omega t + \phi_2) + \dots$$
$$+ I_{nm} \sin(n\omega t + \phi_n) \qquad (5.2)$$

If eqns. (5.1) and (5.2) refer to the voltage across and the current through a given circuit, then the phase angles between harmonic currents and voltages are $(\psi_1 - \phi_1)$ for the fundamental, $(\psi_2 - \phi_2)$ for the second harmonic, and so on.

It should be noted that eqn. (5.2) may be rewritten in the form

$$i = I_{1m} \sin(\omega t + \phi_1') + I_{2m} \sin 2(\omega t + \phi_2') + \dots$$
$$+ I_{nm} \sin n(\omega t + \phi_n') \qquad (5.2a)$$

In this equation the phase angle ϕ_1', ϕ_2', ϕ_n', etc., refer to the scale of the fundamental wave. A similar expression for the complex voltage wave of eqn. (5.1) may also be used.

5.2 Harmonic Synthesis

Figs. 5.2–5.5 show some complex waves which have been built up from simple harmonics. This synthesis is carried out by adding the instantaneous values of the fundamental and the harmonics for given instants in time.

In Fig. 5.2 a second harmonic, $E_{2m} \sin 2\omega t$, has been added to a fundamental $E_{1m} \sin \omega t$. Since E_{2m} is about 35 per cent of E_{1m}

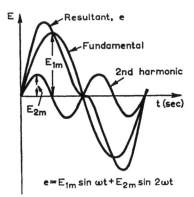

$$e = E_{1m} \sin \omega t + E_{2m} \sin 2\omega t$$

Fig. 5.2 SYNTHESIS OF FUNDAMENTAL AND SECOND HARMONIC

the resultant complex wave is said to contain 35 per cent second harmonic. The effect of a phase change of the harmonic with respect to the fundamental is shown in Fig. 5.3, where the added second

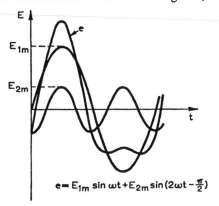

$$e = E_{1m} \sin \omega t + E_{2m} \sin \left(2\omega t - \frac{\pi}{2}\right)$$

Fig. 5.3 EFFECT OF PHASE SHIFT OF SECOND HARMONIC ON RESULTANT WAVEFORM

harmonic is $E_{2m} \sin \left(2\omega t - \frac{\pi}{2}\right)$. It can be seen that the shape of the resultant wave has been completely changed, although the percentage of second harmonic remains the same.

Fig. 5.4 shows a fundamental with about 30 per cent third harmonic added, while Figs. 5.5(*a*) and (*b*) illustrate the effect of a phase shift of the harmonic with respect to the fundamental.

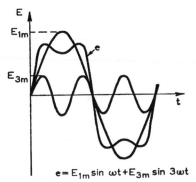

$$e = E_{1m}\sin \omega t + E_{3m}\sin 3\omega t$$

Fig. 5.4 SYNTHESIS OF FUNDAMENTAL AND THIRD HARMONIC

From these diagrams it can be seen that the complex wave produced by a fundamental and third harmonic has identical positive and negative half-cycles, while the wave produced by adding a

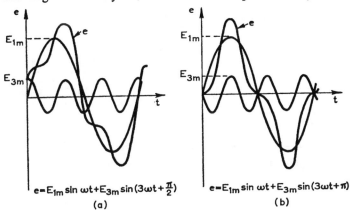

$$e = E_{1m}\sin \omega t + E_{3m}\sin\left(3\omega t + \frac{\pi}{2}\right)$$

(a)

$$e = E_{1m}\sin \omega t + E_{3m}\sin(3\omega t + \pi)$$

(b)

Fig. 5.5 EFFECT OF PHASE SHIFT OF THIRD HARMONIC ON RESULTANT WAVEFORM

second harmonic to the fundamental has dissimilar positive and negative half-cycles. In general a complex wave with identical positive and negative half-cycles can have no even-harmonic components. Since most rotating machines produce similar positive and negative half-cycles there will be only odd harmonics in their output waves. Non-linear circuit elements (i.e. circuit elements whose impedance varies with the direction and/or magnitude of the applied voltage, such as thermionic valves and rectifiers) will produce non-symmetrical current waves which must therefore contain even harmonics.

5.3 R.M.S. Value of a Complex Wave

Consider a current given by eqn. (5.2). The effective or r.m.s. value of this current is

$$I = \sqrt{(\text{Average value of } i^2)}$$

Now,

$$i^2 = I_{1m}^2 \sin^2 (\omega t + \phi_1) + \ldots + I_{nm}^2 \sin^2 (n\omega t + \phi_n)$$
$$+ 2I_{1m}I_{2m} \sin (\omega t + \phi_1) \sin (2\omega t + \phi_2) + \ldots \quad (5.3)$$

The right-hand side of this equation is seen to consist of two types of term: (a) harmonic self-products of the form $I_{pm}^2 \sin^2 (p\omega t + \phi_p)$ for the pth harmonic, and (b) products of different harmonics of the form $I_{pm}I_{qm} \sin (p\omega t + \phi_p) \sin (q\omega t + \phi_q)$ for the pth and qth harmonics. The average value of i^2 is the sum of the average values of each term in eqn. (5.3). For the general self-product,

Average value of $I_{pm}^2 \sin^2 (p\omega t + \phi_p)$ over one cycle of the fundamental

$$= \frac{1}{2\pi} \int_0^{2\pi} I_{pm}^2 \sin^2 (p\omega t + \phi_p) \, d(\omega t)$$

$$= \frac{I_{pm}^2}{2\pi} \int_0^{2\pi} \tfrac{1}{2}\{1 - \cos 2(p\omega t + \phi_p)\} \, d(\omega t)$$

$$= \frac{I_{pm}^2}{2} - 0 - \frac{I_{pm}^2}{4\pi} \left\{ \frac{1}{2p} \sin 2(p \cdot 2\pi + \phi_p) - \frac{1}{2p} \sin (2\phi_p) \right\}$$

$$= \frac{I_{pm}^2}{2} \quad (5.4)$$

For the products of different harmonics,

Average value of $I_{pm}I_{qm} \sin (p\omega t + \phi_p) \sin (q\omega t + \phi_q)$

$$= \frac{1}{2\pi} \int_0^{2\pi} I_{pm}I_{qm} \sin (p\omega t + \phi_p) \sin (q\omega t + \phi_q) \, d(\omega t)$$

$$= \frac{I_{pm}I_{qm}}{2\pi} \int_0^{2\pi} \tfrac{1}{2}\{\cos ((p - q)\omega t + \phi_p - \phi_q)$$
$$- \cos ((p + q)\omega t + \phi_p + \phi_q)\} \, d(\omega t)$$

$$= \frac{I_{pm}I_{qm}}{4\pi} \left[\frac{1}{p - q} \sin \{(p - q)\omega t + \phi_p - \phi_q\} \right.$$
$$\left. - \frac{1}{p + q} \sin \{(p + q)\omega t + \phi_p + \phi_q\} \right]_0^{2\pi}$$

$$= 0 \quad (5.5)$$

i.e. the average value of the product of two different harmonic sinusoidal waves is zero. Therefore

$$\text{Average value of } i^2 = \frac{I_{1m}^2}{2} + \frac{I_{2m}^2}{2} + \ldots \frac{I_{nm}^2}{2}$$

so that

$$I = \sqrt{\left(\frac{I_{1m}^2}{2} + \frac{I_{2m}^2}{2} + \ldots \frac{I_{nm}^2}{2}\right)} \tag{5.6}$$

The r.m.s. values of each harmonic can be inserted in eqn. (5.6), which will then become

$$I = \sqrt{(I_1^2 + I_2^2 \ldots + I_n^2)} \tag{5.7}$$

where $I_1 = I_{1m}/\sqrt{2}$, etc.

Exactly similar expressions may be derived for the r.m.s. value of a complex voltage wave.

EXAMPLE 5.1   A complex voltage of r.m.s. value 240 V has 22 per cent third-harmonic content, and 5 per cent fifth-harmonic content. Find the r.m.s. values of the fundamental and of each harmonic.

The expression for voltage corresponding to eqn. (5.7) is

$$V = \sqrt{(V_1^2 + V_3^2 + V_5^2)}$$

But $V_3 = 0.22V_1$, and $V_5 = 0.05V_1$, so that

$$V = \sqrt{(V_1^2 + 0.0484V_1^2 + 0.0025V_1^2)} = 1.03V_1$$

Therefore

$$V_1 = \frac{240}{1.03} = 233\,\text{V}$$

Hence

$$V_3 = 0.22 \times 233 = 51.3\,\text{V}$$

and

$$V_5 = 0.05 \times 233 = 11.7\,\text{V}$$

5.4   Power Associated with Complex Waves

Consider a voltage wave given by

$$v = V_{1m} \sin \omega t + V_{2m} \sin 2\omega t + V_{3m} \sin 3\omega t + \ldots$$

applied to a circuit and causing a current given by

$$i = I_{1m} \sin(\omega t - \phi_1) + I_{2m} \sin(2\omega t - \phi_2) + I_{3m} \sin(3\omega t - \phi_3) + \ldots$$

The power supplied to this circuit at any instant is the product vi. This product will involve multiplying every term in the voltage wave in turn by every term in the current wave. The average power supplied over a cycle will be the sum of the average values over one cycle of each individual product term. From eqn. (5.5) it can be seen that the average value of all product terms involving harmonics of different frequencies will be zero, so that only the products of current and voltage harmonics of the same frequency need be considered. For the general term of this nature, the average value of the product over one cycle of the fundamental is given by

$$P_n = \frac{1}{2\pi} \int_0^{2\pi} V_{nm}I_{nm} \sin n\omega t \sin (n\omega t - \phi_n) d(\omega t)$$

$$= \frac{V_{nm}I_{nm}}{2\pi} \int_0^{2\pi} \tfrac{1}{2}\{\cos \phi_n - \cos (2n\omega t - \phi_n)\} d(\omega t)$$

$$= V_n I_n \cos \phi_n \qquad (5.8)$$

where V_n and I_n are r.m.s. values.

The total power supplied by complex voltages and currents is thus the sum of the powers supplied by each harmonic component acting independently. The average power supplied per cycle of the fundamental is

$$P = V_1 I_1 \cos \phi_1 + V_2 I_2 \cos \phi_2 + V_3 I_3 \cos \phi_3 + \ldots \qquad (5.9)$$

When harmonics are present the overall power factor is defined as

$$\text{Overall power factor} = \frac{\text{Total power supplied}}{\text{Total r.m.s. voltage} \times \text{Total r.m.s. current}}$$

$$= \frac{V_1 I_1 \cos \phi_1 + V_2 I_2 \cos \phi_2 + \ldots}{VI} \qquad (5.10)$$

Alternatively, if R_s is the equivalent series resistance of the circuit the total power is given by

$$P = I_1^2 R_s + I_2^2 R_s + I_3^2 R_s + \ldots \qquad (5.11)$$

$$= I^2 R_s \qquad (5.12)$$

If the effective parallel resistance of the whole circuit (R_p) is known, then the power supplied will be

$$P = \frac{V^2}{R_p} \qquad (5.13)$$

where V is the r.m.s. value of the complex voltage wave.

EXAMPLE 5.2 A voltage given by

$$v = 50 \sin \omega t + 20 \sin (3\omega t + 30°) + 10 \sin (5\omega t - 90°) \text{ volts}$$

is applied to a circuit and the resulting current is found to be given by

$$i = 0.5 \sin (\omega t - 37°) + 0.1 \sin (3\omega t - 15°)$$
$$+ 0.09 \sin (5\omega t - 150°) \text{ amperes}$$

Find the total power supplied and the overall power factor.

The best method of tackling problems involving harmonics is to deal with each harmonic separately. Thus,

$$\text{Power at fundamental frequency} = \frac{V_{1m}I_{1m}}{2} \times \cos 37° = \frac{50 \times 0.5}{2} \times 0.8$$

$$= 10 \, \text{W}$$

$$\text{Power at third harmonic} = \frac{20 \times 0.1}{2} \times \cos 45° = 0.707 \, \text{W}$$

$$\text{Power at fifth harmonic} = \frac{10 \times 0.09}{2} \times \cos 60° = 0.23 \, \text{W}$$

Therefore

$$\text{Total power} = 10 + 0.707 + 0.23 = \underline{10.9 \, \text{W}}$$

Also,

$$I = \sqrt{\left(\frac{0.5^2}{2} + \frac{0.1^2}{2} + \frac{0.09^2}{2}\right)} = 0.365 \, \text{A}$$

and

$$V = \sqrt{\left(\frac{50^2}{2} + \frac{20^2}{2} + \frac{10^2}{2}\right)} = 38.8 \, \text{V}$$

Therefore

$$\text{Overall power factor} = \frac{10.9}{38.8 \times 0.365} = \underline{0.77}$$

It should be noted that the overall power factor of a circuit when harmonics are present cannot be stated as leading or lagging; it is simply the ratio of the power to the product of r.m.s. voltage and current.

5.5 Harmonics in Single-phase Circuits

If an alternating voltage containing harmonics is applied to a single-phase circuit containing linear circuit elements, then the current which will result will also contain harmonics. By the superposition principle the effect of each voltage harmonic term may be considered separately. In the following paragraphs a voltage given by

$$v = V_{1m} \sin \omega t + V_{2m} \sin 2\omega t + V_{3m} \sin 3\omega t + \ldots \quad (5.14)$$

will be applied to various circuits, and the resulting current will be determined.

PURE RESISTANCE

The impedance of a pure resistance is independent of frequency and the current and voltage will be in phase for each harmonic, so that the expression for the current will be

$$i = \frac{V_{1m}}{R} \sin \omega t + \frac{V_{2m}}{R} \sin 2\omega t + \frac{V_{3m}}{R} \sin 3\omega t + \dots \quad (5.15)$$

This equation shows that the percentage harmonic content in the current wave will be exactly the same as that in the voltage wave. The current and voltage waves will therefore be identical in shape.

PURE INDUCTANCE

If the voltage of eqn. (5.14) is applied to an inductance of L henrys, the inductive reactance $(2\pi f L)$ will vary with the harmonic frequency. For every harmonic term, however, the current will lag behind the voltage by $90°$. Hence the general expression for the current is

$$i = \frac{V_{1m}}{\omega L} \sin (\omega t - 90°) + \frac{V_{2m}}{2\omega L} \sin (2\omega t - 90°)$$

$$+ \frac{V_{3m}}{3\omega L} \sin (3\omega t - 90°) + \dots \quad (5.16)$$

This shows that for the nth harmonic the percentage harmonic content in the current wave is only $1/n$ of the corresponding harmonic content in the voltage wave.

PURE CAPACITANCE

The capacitive reactance $1/2\pi f C$ of a capacitor C will vary with the harmonic frequency, but for every harmonic the current will lead the voltage by $90°$. For the nth harmonic the reactance will be $1/n\omega C$, so that the peak current at this frequency will be

$$I_{nm} = V_{nm} n\omega C$$

Hence if the voltage of eqn. (5.14) is applied to a capacitor C, the current will be

$$i = V_{1m}\omega C \sin (\omega t + 90°) + V_{2m} \times 2\omega C \sin (2\omega t + 90°)$$
$$+ V_{3m} \times 3\omega C \sin (3\omega t + 90°)$$
$$(5.17)$$

This shows that the percentage harmonic content of the current wave is larger than that of the voltage wave—for the nth harmonic, for instance, it will be n times larger.

EXAMPLE 5.3 A resistance of $10\,\Omega$ is connected in series with a coil of inductance $6\cdot36\,\text{mH}$. The supply voltage is given by

$$v = 300 \sin 314t + 50 \sin 942t + 40 \sin 1{,}570t \text{ volts}$$

Find (a) an expression for the instantaneous value of the current, and (b) the power dissipated.

For the fundamental, ω is $314\,\text{rad/s}$. The inductive reactance is therefore

$$X_1 = \omega L = 314 \times 6\cdot36 \times 10^{-3} = 2\,\Omega$$

The impedance at fundamental frequency is

$$Z_1 = \sqrt{(R^2 + X_1{}^2)} = \sqrt{(10^2 + 2^2)} = 10\cdot2\,\Omega$$

Also the phase angle is

$$\tan^{-1}\frac{X_1}{R} = \tan^{-1} 0\cdot2 = 11\cdot3° \text{ lagging}$$

At the third-harmonic frequency, the inductive reactance, the impedance and the phase angle are

$$X_3 = 3\omega L = 3X_1 = 6\,\Omega$$
$$Z_3 = \sqrt{(10^2 + 6^2)} = 11\cdot7\,\Omega$$
$$\tan^{-1}\frac{X_3}{R} = \tan^{-1} 0\cdot6 = 31° \text{ lagging}$$

At the fifth-harmonic frequency the reactance will have increased to $5X_1$, so that the impedance will now be

$$Z_5 = \sqrt{(10^2 + 10^2)} = 14\cdot1\,\Omega$$

and the phase angle will be

$$\tan^{-1}\frac{10}{10} = 45° \text{ lagging}$$

(a) The resultant expression for the total current (amperes) will be

$$i = \frac{300}{10\cdot2} \sin(\omega t - 11\cdot3°) + \frac{50}{11\cdot7} \sin(3\omega t - 31°) + \frac{40}{14\cdot1} \sin(5\omega t - 45°)$$

$$= 29\cdot4 \sin(\omega t - 11\cdot3°) + 4\cdot28 \sin(3\omega t - 31°) + 2\cdot83 \sin(5\omega t - 45°)$$

(b) In this case the power dissipated will be the product of the resistance and the square of the r.m.s. current.

$$I^2 = \frac{29\cdot4^2}{2} + \frac{4\cdot28^2}{2} + \frac{2\cdot83^2}{2} = 445\cdot4$$

Therefore

$$\text{Power dissipated} = I^2 R_s = 4{,}454\,\text{W}$$

EXAMPLE 5.4 A capacitor of $3 \cdot 18 \mu F$ capacitance is connected in parallel with a resistance of $1,000 \Omega$, the combination being connected in series with a $1,000 \Omega$ resistor to a voltage given by

$$v = 350 \sin \omega t + 150 \sin (3\omega t + 30°) \text{ volts}$$

(a) Determine the power dissipated in the circuit if $\omega = 314 \text{rad/sec}$.
(b) Obtain an expression for the voltage across the series resistor.
(c) Determine the percentage harmonic content of the resultant current.

$$\text{Reactance of capacitor at fundamental frequency} = \frac{10^6}{314 \times 3 \cdot 18}$$

$$= 1,000 \Omega$$

The complex impedance of the circuit at this frequency is

$$Z_1 = 1,000 + \frac{1,000(-j1,000)}{1,000 - j1,000} = 1,500 - j500 = 1,580 \underline{/-18 \cdot 5°} \, \Omega$$

The reactance of the capacitor at the third-harmonic frequency will be one-third of that at the fundamental frequency. Therefore the complex impedance of the circuit at the third-harmonic frequency is

$$Z_3 = 1,000 + \frac{1,000(-j333)}{1,000 - j333} = 1,000 - j300 = 1,140 \underline{/-15 \cdot 3°} \, \Omega$$

Thus

$$i = \frac{350}{1,580} \sin (\omega t + 18 \cdot 5°) + \frac{150}{1,140} \sin (3\omega t + 45 \cdot 3)$$

$$= 0 \cdot 222 \sin (\omega t + 18 \cdot 5°) + 0 \cdot 131 \sin (3\omega t + 45 \cdot 3°)$$

(a) From eqn. (5.9), the total power is

$$P = \frac{350}{\sqrt 2} \times \frac{0 \cdot 222}{\sqrt 2} \cos 18 \cdot 5° + \frac{150}{\sqrt 2} \times \frac{0 \cdot 131}{\sqrt 2} \cos 15 \cdot 3° = \underline{\underline{46 \cdot 3 \text{ W}}}$$

(b) The voltage across the series resistor is

$$v_r = iR$$

$$= \underline{\underline{222 \sin (\omega t + 18 \cdot 5°) + 131 \sin (3\omega t + 45 \cdot 3) \text{ volts}}}$$

(c) The percentage harmonic content of the current wave is

$$\frac{131}{222} \times 100 = \underline{\underline{59 \text{ per cent}}}$$

5.6 Selective Resonance

If a voltage which is represented by a complex wave is applied to a circuit containing both inductance and capacitance, the resulting current can be found by the methods previously described. It may happen that the circuit resonates at one of the harmonic frequencies of the applied voltage, and this effect is termed *selective resonance*. If series selective resonance occurs, then large currents at the resonant frequency may be produced, and in addition large harmonic voltages may appear across both the inductance and the capacitance. If parallel selective resonance occurs, on the other hand, the resultant current from the supply at the resonant frequency will be a minimum, but the current at this frequency through both the inductance and the capacitance will be large (i.e. current magnification).

The possibility of selective resonance is one reason why it is undesirable to have harmonics in a supply voltage. Selective resonance is used, however, in some wave analysers, which are instruments for determining the harmonic content of alternating waveforms. A simple form of analyser may consist of a series resonant circuit, which can be tuned over the range of harmonic frequencies to be measured. The voltage across the inductance or capacitance in the circuit will then give a measure of the size of the harmonic to which the circuit is tuned.

5.7 Effect of Harmonics on Single-phase Measurements

If measurements of impedance are made in circuits containing reactive elements, the presence of harmonics in the current and voltage waveforms may cause considerable errors unless they are allowed for.

Consider a capacitor C, across which the voltage is

$$V = V_{1m} \sin \omega t + V_{2m} \sin 2\omega t + V_{3m} \sin 3\omega t + \ldots$$

Then from eqn. (5.17) the current flowing through the capacitor will be

$$i = V_{1m}\omega C \sin (\omega t + 90°) + V_{2m} 2\omega C \sin (2\omega t + 90°)$$
$$+ V_{3m} 3\omega C \sin (3\omega t + 90°) + \ldots$$

If the r.m.s. value of the voltage is V, then

$$V = \frac{1}{\sqrt{2}} \sqrt{(V_{1m}^2 + V_{2m}^2 + V_{3m}^2 \ldots)}$$

and if the r.m.s. value of the current is I, then

$$I = \frac{\omega C}{\sqrt{2}} \sqrt{(V_{1m}^2 + 4V_{2m}^2 + 9V_{3m}^2 + \ldots)}$$

V and I will be the indications on r.m.s. measuring instruments if these are connected in circuit. From the above equations, the value of C will be given by

$$C = \frac{I}{\omega V} \sqrt{\left(\frac{V_{1m}^2 + V_{2m}^2 + V_{3m}^2 + \ldots}{V_{1m}^2 + 4V_{2m}^2 + 9V_{3m}^2 + \ldots} \right)} \qquad (5.18)$$

If the effect of the harmonics were neglected, the value of the capacitance would appear to be $C' = I/\omega V$, from normal circuit theory. The true capacitance will be less than this.

In a similar manner it can be shown that the true inductance of a coil will be less than its apparent value in a circuit where harmonics are actually present but have been neglected.

If a wattmeter is included in a circuit when harmonics are present, the power indicated will be the true total power including the harmonic power. This follows since only sinusoidal currents of the same frequency in the fixed and moving coils will produce a resultant torque in the instrument.

This fact may be used to allow the wattmeter to be employed as a wave analyser. The current which is to be analysed is passed through the current coils of the wattmeter, while a variable-frequency sinusoidal generator supplies the voltage coil. The frequency of this generator is set successively at the fundamental and the harmonic frequencies, the voltage being maintained constant. The meter indication at each frequency setting will then be proportional to the magnitude of the component of the same frequency in the complex current wave. Hence the relative sizes of the fundamental and of each harmonic component may be obtained.

5.8 Superimposed Alternating and Direct Current

A special case of a complex wave is obtained when both an alternating and a direct current flow through the same circuit. The

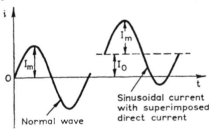

Fig. 5.6 ALTERNATING CURRENT WITH SUPERIMPOSED
DIRECT CURRENT

effect of the d.c. component is to cause a shift of the whole a.c. wave by an amount equal to the magnitude of the direct current. In other words, the effective base line of the wave is moved relatively by an amount depending on the direct current (Fig. 5.6). This form of complex wave is found in electronic circuits and in saturable reactors. As far as calculations are concerned the d.c. component may be treated in the same way as any other harmonic term, provided that the following points are remembered.

(i) The r.m.s. value of a d.c. component is the actual value of the component, hence the r.m.s. value of a complex wave containing a d.c. term, I_0, is

$$I = \sqrt{\left(I_0^2 + \frac{I_{1m}^2}{2} + \frac{I_{2m}^2}{2} + \frac{I_{3m}^2}{2} + \ldots\right)} \qquad (5.19)$$

$$= \sqrt{(I_0^2 + I_1^2 + I_2^2 + I_3^2 + \ldots)} \qquad (5.20)$$

where I_{1m}, etc., are the peak values of the a.c. components, and I_1, etc., are the r.m.s. values of these components.

(ii) The steady voltage drop across a pure inductance due to a direct current is zero.

(iii) No direct current will flow through a capacitor.

EXAMPLE 5.5   A voltage given by

$$v = 30 + 25 \sin \omega t - 20 \sin 2\omega t \text{ volts}$$

is applied to the circuit shown in Fig. 5.7. Find the reading on each instrument

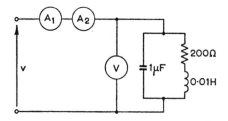

Fig. 5.7

if $\omega = 10,000$ rad/sec. A_1 is a thermocouple ammeter, A_2 a moving-coil ammeter, and V is an electrostatic voltmeter.

The thermocouple and electrostatic instruments will read r.m.s. values, while the moving-coil instrument will read average values, i.e. the moving-coil meter will record the d.c. component of current only. This d.c. component is given by

$$I_0 = \frac{V_{dc}}{200} = \frac{30}{200} = 0\cdot15 \text{A}$$

This follows since no direct current will flow through the capacitive arm of the network.

The equivalent impedance at the fundamental frequency is

$$Z_1 = \frac{(R + jX_{L1})(-jX_{C1})}{R + jX_{L1} - jX_{C1}} = \frac{(200 + j100)(-j100)}{200}$$

$$= 112\underline{/-63° \, 26'} \, \Omega$$

The r.m.s. fundamental current is therefore

$$I_1 = \frac{25}{\sqrt{2} \times 112} = 0.158\,\text{A}$$

The equivalent impedance at second harmonic is

$$Z_2 = \frac{(R + jX_{L2})(-jX_{C2})}{R + jX_{L2} - jX_{C2}} = \frac{(200 + j200)(-j50)}{200 + j150}$$

$$= 56.6\underline{/-81°\,51'}\,\Omega$$

Therefore the r.m.s. second-harmonic current is

$$I_2 = \frac{20}{\sqrt{2} \times 56.6} = 0.25\,\text{A}$$

The reading on the thermocouple ammeter is, from eqn. (5.20),

$$I = \sqrt{(I_0^2 + I_1^2 + I_2^2)} = \sqrt{(0.0225 + 0.025 + 0.0625)} = \underline{\underline{0.332\,\text{A}}}$$

The voltmeter reading is, from eqn. (5.19),

$$V = \sqrt{\left(900 + \frac{625}{2} + \frac{400}{2}\right)} = \underline{\underline{37.6\,\text{V}}}$$

5.9 Production of Harmonics

Harmonics may be produced in the output waveform of an a.c. generator, due to a non-sinusoidal air-gap flux distribution, or to *tooth ripple* which is caused by the effect of the slots which house the windings. In large supply systems, the greatest care is taken to ensure a sinusoidal output from the generators, but even in this case any non-linearity in the circuit will give rise to harmonics in the current waveform. Some non-linear circuit elements will be considered in the following sections.

RECTIFIERS

A rectifier is a circuit-element which has a low impedance to the flow of current in one direction, and a nearly infinite impedance to the flow of current in the opposite direction. Thus, when an alternating voltage is applied to the rectifier circuit, current will flow through it during the positive half-cycles only, being zero during the negative half-cycles. The current waveform is shown in Fig. 5.8. This waveform is seen to correspond roughly in shape to that shown in Fig. 5.3, but in this case the presence of a d.c. component brings the negative half-cycle up to the zero current position. Since the average value of a sine wave over one half-cycle is $(2/\pi) \times$ (maximum value), the average value taken over one cycle, with the negative half-cycle zero, will be $(1/\pi) \times$ (maximum value), and this then represents the d.c. component of the wave shown in Fig. 5.8. Also the wave must contain a large proportion of second harmonic.

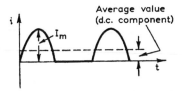

Fig. 5.8 CURRENT WAVEFORM FOR AN IDEAL RECTIFIER SUPPLIED WITH A SINUSOIDAL VOLTAGE

EXAMPLE 5.6 A battery charger is connected to a sinusoidal 220 V supply through a 20 Ω resistor. The equipment takes 5·5 A (r.m.s.) with 30 per cent second harmonic. Calculate the total circuit power, the overall power factor, and the power factor of the charging equipment alone. *(L.U.)*

The circuit is shown in Fig. 5.9(*a*). Since there is no second harmonic in the supply voltage, the battery charger may be regarded as a second-harmonic generator whose output is dissipated in the 20 Ω resistor. The circuit then becomes

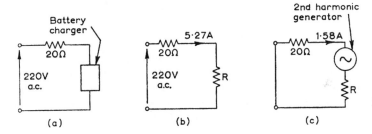

Fig. 5.9

that shown at (*b*) for the fundamental, and that shown at (*c*) for the second harmonic. *R* is the equivalent resistance of the charging unit.

Let I_1 = r.m.s. value of the fundamental current

I_2 = r.m.s. value of second-harmonic current

V_1 = r.m.s. value of fundamental supply voltage (=220 V)

V_2 = r.m.s. value of second-harmonic voltage across the 20 Ω resistor

It may be assumed that the charger will be connected to the supply circuit by a transformer so that there will be no direct current in the supply system.

Then $I_2 = 0 \cdot 3I_1$, and from eqn. (5.7),

$$5 \cdot 5 = \surd(I_1{}^2 + (0 \cdot 3I_1)^2) = 1 \cdot 044I_1$$

Therefore

$$I_1 = 5 \cdot 27 \text{A} \quad \text{and} \quad I_2 = 1 \cdot 58 \text{A}$$

The total power supplied must all be at the fundamental frequency since only a fundamental frequency voltage is applied. The value will be $V_1 I_1 \cos \phi_1$ watts. Since the circuit contains resistance only, $\cos \phi_1$ will be unity.

Power supplied $= V_1 I_1 = 220 \times 5 \cdot 27 = \underline{1,160\,\text{W}}$

Overall power factor (from eqn. (5.10)) $= \dfrac{1,160}{220 \times 5 \cdot 5} = \underline{0 \cdot 96}$

The power supplied to the charger may be found by subtracting the power dissipated in the $20\,\Omega$ resistor from the total power; thus

Power dissipated in $20\,\Omega$ resistor $= I^2 R = 5 \cdot 5^2 \times 20 = 605\,\text{W}$

Power supplied to charger $= 1,160 - 605 = 555\,\text{W}$

Also, the fundamental-frequency voltage drop (r.m.s.) across the charger is

$V_{C1} = 220 - (5 \cdot 27 \times 20) = 115\,\text{V}$ (Fig. 5.9(b))

and the second-harmonic voltage is

$V_2 = 1 \cdot 58 \times 20 = 31 \cdot 6\,\text{V}$ (Fig. 5.9(c))

Thus

R.M.S. value of charger voltage $= \sqrt{(V_{C1}^2 + V_2^2)} = 119\,\text{V}$

Therefore

Charger power factor $= \dfrac{555}{119 \times 5 \cdot 5} = \underline{0 \cdot 85}$

IRON-CORED COILS WITH SINUSOIDAL APPLIED VOLTAGE

Iron-cored coils are a source of harmonic generation in a.c. circuits owing to the non-linear character of the B/H curve and hysteresis loop, especially if saturation occurs. Consider a sinusoidal voltage, applied to an iron-cored coil of N turns and of cross-section A square metres. The instantaneous voltage is

$$v = N \frac{d\Phi}{dt}$$

where Φ is the flux produced in the iron core. If B is the core flux density,

$$v = NA \frac{dB}{dt}$$

Therefore

$$\int dB = \frac{1}{NA} \int v\,dt = \frac{1}{NA} \int V_m \sin \omega t\,dt \tag{5.21}$$

and

$$B = -\frac{V_m}{\omega NA} \cos \omega t = \frac{V_m}{\omega NA} \sin (\omega t - 90°) \tag{5.22}$$

Hence, if the applied voltage is sinusoidal, the flux density in the iron core must also vary sinusoidally. Note that eqn. (5.21) leads to the familiar relation

$$B_m = \frac{V_m}{2\pi f N A} = \frac{V}{4 \cdot 44 f N A} \tag{5.23}$$

where V is the r.m.s. value of the applied voltage.

The hysteresis loop of a specimen of iron, for a given applied voltage, is shown in Fig. 5.10, the peak value of the flux density

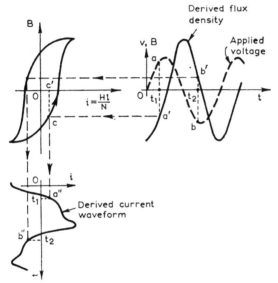

Fig. 5.10   CURRENT THROUGH AN IRON-CORED COIL WITH A SINUSOIDAL APPLIED VOLTAGE

being found from eqn. (5.23). The base is taken to a current scale $i = Hl/N$, where l is the length of the core. To derive the current waveform, sinusoidal voltage and flux density curves to a base of time are first drawn as shown. The current waveform is then derived by the point-by-point method indicated, care being taken to move round the hysteresis loop in the correct direction. Thus point a on the voltage curve corresponds to point a' on the flux density curve and to point c on the hysteresis loop, so that the current at this instant is $0c'$. This current is then plotted from the vertical time scale to give the derived point a'' on the current curve. By continuing this process for other points on the voltage curve, the current curve will be obtained.

It is seen that the current curve has identical positive and negative half-cycles, so that it contains no even harmonics. Comparison with Fig. 5.5(a) shows that there is a pronounced third-harmonic content.

FREE AND FORCED MAGNETIZATION

If the resistance of the circuit containing an iron-cored coil is high compared with the inductive reactance, then the current flowing from a sinusoidal supply will tend to be sinusoidal. This means

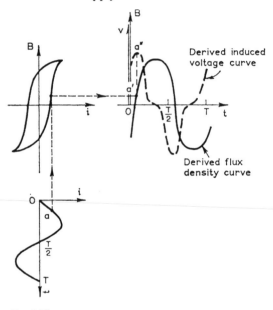

Fig. 5.11   WAVEFORMS OF CURRENT AND VOLTAGE UNDER FORCED MAGNETIZATION

that the core flux density cannot be sinusoidal, since it is related to the current by the hysteresis loop. In turn, this means that the induced voltage due to the alternating flux will not be sinusoidal. This condition is termed *forced magnetization*.

The condition of low circuit resistance relative to reactance gives sinusoidal flux from a sinusoidal supply voltage, and is called *free magnetization*.

To determine the shape of the induced voltage wave under forced magnetization, the hysteresis loop corresponding to the peak value of circuit current is drawn, and the flux density curve is derived from a sinusoidal current wave as shown in Fig. 5.11. The induced voltage is related to the flux density by the equation

$$v = NA \frac{dB}{dt}$$

so that the curve of flux density must be graphically differentiated and multiplied by the number of turns on the coil and the cross-sectional area of the iron to obtain the voltage curve. The result is shown by the dotted curve in the diagram. Comparing this curve with that of Fig. 5.5(*b*) it is seen that it contains a prominent third harmonic.

5.10 Harmonics in Three-phase Systems

Harmonics may be produced in 3-phase systems in the same way as in the single-phase systems already considered, and calculations are carried out similarly by considering each harmonic separately. As will be seen, however, care must be exercised when dealing with the third, and all multiples of the third harmonic (called the *triple-n* harmonics). It is unusual for even harmonics to be present.

PHASE E.M.F.S

Consider a 3-phase alternator with identical phase windings, in which harmonics are generated. The phase e.m.f.s will then be

$$e_R = E_{1m} \sin (\omega t + \psi_1) + E_{3m} \sin (3\omega t + \psi_3)$$
$$+ E_{5m} \sin (5\omega t + \psi_5) + \ldots \quad (5.24)$$

$$e_Y = E_{1m} \sin \left(\omega t - \frac{2\pi}{3} + \psi_1 \right) + E_{3m} \sin \left\{ 3 \left(\omega t - \frac{2\pi}{3} \right) + \psi_3 \right\}$$

$$+ E_{5m} \sin \left\{ 5 \left(\omega t - \frac{2\pi}{3} \right) + \psi_5 \right\} + \ldots$$

$$= E_{1m} \sin \left(\omega t - \frac{2\pi}{3} + \psi_1 \right) + E_{3m} \sin (3\omega t - 2\pi + \psi_3)$$

$$+ E_{5m} \sin \left(5\omega t - \frac{10\pi}{3} + \psi_5 \right) + \ldots$$

$$= E_{1m} \sin \left(\omega t - \frac{2\pi}{3} + \psi_1 \right) + E_{3m} \sin (3\omega t + \psi_3)$$

$$+ E_{5m} \sin \left(5\omega t - \frac{4\pi}{3} + \psi_5 \right) + \ldots \quad (5.25)$$

$$e_B = E_{1m} \sin\left(\omega t - \frac{4\pi}{3} + \psi_1\right) + E_{3m} \sin\left\{3\left(\omega t - \frac{4\pi}{3}\right) + \psi_3\right\}$$

$$+ E_{5m} \sin\left\{5\left(\omega t - \frac{4\pi}{3}\right) + \psi_5\right\} + \ldots$$

$$= E_{1m} \sin\left(\omega t - \frac{4\pi}{3} + \psi_1\right) + E_{3m} \sin\left(3\omega t + \psi_3\right)$$

$$+ E_{5m} \sin\left(5\omega t - \frac{2\pi}{3} + \psi_5\right) + \ldots \quad (5.26)$$

It can be seen from these expressions that all the third harmonics are in time phase, and that the fifth harmonics have a negative phase sequence, the fifth harmonic in the blue phase reaching its maximum value before that in the yellow phase. In the same way it can be shown that

(a) All triple-n harmonics are in phase (3rd, 9th, 15th, etc.).
(b) The 7th, 13th, 19th, etc., harmonics have a positive phase sequence.
(c) The 5th, 11th, 17th, etc., harmonics have a negative phase sequence.

LINE VOLTAGES FOR STAR CONNEXION

If the windings are star connected, the line voltages will be the difference between successive phase voltages, and hence will contain no third-harmonic terms, since these are identical in each phase. The fundamental will have a line value of $\sqrt{3}$ times the phase value and so too will the fifth harmonic.

It should be noted that the r.m.s. value of the line voltage in this case will be less than $\sqrt{3}$ times the r.m.s. value of the phase voltage, owing to the absence of the third-harmonic term from the expression for the line voltage.

LINE VOLTAGES FOR DELTA CONNEXION

If the alternator windings are delta connected, the resultant e.m.f. acting round the closed loop will be the sum of the phase e.m.f.s. This sum is zero for the fundamental, and for the 5th, 7th, 11th, etc., harmonics. All the third harmonics, however, are in phase, and hence there will be a resultant third-harmonic e.m.f. of three times the phase value acting round the mesh. This will cause a circulating current whose value will depend on the impedance of the windings at the third-harmonic frequency. Hence the third-harmonic e.m.f. is effectively short-circuited by the windings, so that there can be no third-harmonic voltage across the lines. The same applies to all triple-n harmonic terms. The line voltage will then be the phase voltage without the triple-n terms.

EXAMPLE 5.7 A 3-phase alternator has a generated e.m.f. per phase of 250V, with 10 per cent third and 6 per cent fifth harmonic content. Calculate (*a*) the r.m.s. line voltage for star connexion, and (*b*) the r.m.s. line voltage for mesh connexion.

Let V_1, V_3, V_5 be the r.m.s. values of the phase e.m.f.s. Then

$$V_3 = 0.1V_1 \quad \text{and} \quad V_5 = 0.06V_1$$

Hence, from eqn. (5.7),

$$250 = \sqrt{(V_1^2 + 0.01V_1^2 + 0.0036V_1^2)}$$

Therefore

$$V_1 = \frac{250}{\sqrt{1.0136}} = 248\,\text{V} \qquad V_3 = 24.8\,\text{V} \quad \text{and} \quad V_5 = 14.9\,\text{V}$$

(*a*) R.M.S. value of fundamental line voltage $= \sqrt{3} \times 248 = 430\,\text{V}$

     R.M.S. value of third-harmonic line voltage $= 0$

     R.M.S. value of fifth-harmonic line voltage $= \sqrt{3} \times 14.9 = 25.8\,\text{V}$

Therefore

     R.M.S. value of line voltage $= \sqrt{(430^2 + 25.8^2)} = \underline{\underline{431\,\text{V}}}$

(*b*) In delta there is again no third harmonic component in the line voltage.

Thus

     R.M.S. value of line voltage $= \sqrt{(248^2 + 14.9^2)} = \underline{\underline{249\,\text{V}}}$

FOUR-WIRE SYSTEMS

In a 3-phase system there cannot be any third-harmonic term in the line voltage, as has already been seen. In a 4-wire system, however, each line-to-neutral voltage may contain a third-harmonic component, and if one is actually present a third-harmonic current will then flow in the star-connected load. If the load is balanced, the resulting third-harmonic currents will all be in phase, so that the neutral wire must carry three times the third-harmonic line current. There will be no neutral wire current at the fundamental frequency or at any harmonic frequency other than the triple-n frequencies.

5.11 Harmonics in Transformers

The flux density in transformer cores is usually fairly high to keep the volume of iron required to a minimum. It therefore follows that, owing to the non-linearity of the magnetization curve, there will be some third-harmonic distortion produced (Section 5.9(*b*)). There is usually a small percentage of fifth harmonic also. For single-phase transformers the conditions are the same as those already described for iron-cored coils with sinusoidal applied voltage, namely the magnetizing current will contain a proportion of mainly third harmonic depending on the size of the applied voltage, and the flux will be sinusoidal.

In 3-phase transformers the method of connexion and the type of construction will affect the production of harmonics, as the following cases will show.

PRIMARY WINDING IN DELTA

Each phase of the winding may be considered as separately connected across a sinusoidal supply. The flux will be sinusoidal, so that the magnetizing current will contain a third-harmonic component (in addition to other harmonics of higher order but of relatively small magnitude). These third-harmonic currents will be in phase in each winding, and will constitute a circulating current, so that there will be no third-harmonic component in the line current.

PRIMARY WINDING IN FOUR-WIRE STAR

Again in this case each primary phase may be considered as separately connected to a sinusoidal supply. The core flux will be sinusoidal, and hence the output voltage will also be sinusoidal. The magnetizing current will contain a third-harmonic component which is in phase in each winding and will therefore return through the neutral.

PRIMARY WINDING IN THREE-WIRE STAR

Since, in the absence of a neutral, there is no return path for the third-harmonic components of the magnetizing current, no such currents can flow, and a condition of forced magnetization must therefore exist. The core flux must then contain a third-harmonic component which is in phase in each limb (Section 5.9(*c*)). In the shell type of 3-phase transformer, or in the case of three single-phase units, there will be a magnetic path for these fluxes, but in the 3-limb core type of transformer the third-harmonic component of flux must return via the air (or through the steel tank in an oil-cooled transformer). The high reluctance of the magnetic path for the 3-limb core type of construction reduces the third-harmonic flux in this case to a fairly small value.

If the secondary is delta connected, then the third-harmonic flux component will give rise to a third-harmonic circulating current in the secondary winding. This current, by Lenz's law, tends to oppose the original effect which causes it, and hence minimizes the third-harmonic flux component. There is, of course, no circulating current at any but the triple-*n* frequencies.

If the secondary is star-connected, it is usual to have an additional delta-connected tertiary winding in which the third-harmonic currents can flow. This winding also preserves the magnetic equilibrium of the transformer on unbalanced loads. In this way the output voltage is kept reasonably sinusoidal.

5.12 Harmonic Analysis

It has been seen that a complex wave may be represented by an equation of the form

$$y = Y_{1m} \sin (\omega t + \psi_1) + Y_{2m} \sin (2\omega t + \psi_2) + \dots \quad (5.27)$$

If the shape of the complex wave is known, the process of harmonic analysis is one of finding the coefficients Y_{1m}, Y_{2m}, . . . etc., and the phase angles ψ_1, ψ_2, . . . etc., in this equation. If a mathematical expression for y is possible, then the analysis is the standard Fourier analysis found in appropriate advanced mathematics textbooks. Generally, however, the shape of a complex wave is readily obtainable, but the wave has no simple mathematical expression, and the following methods illustrate how complex waves may be analysed under such conditions.

SUPERPOSITION METHOD (WEDMORE'S METHOD)

This method is used mainly for the analysis of third-harmonic content, but may be extended to fifth and higher harmonics. It depends on the fact that, if a sine wave is divided into any number of equal parts and the parts are then superimposed, the sum of the ordinates at any point will be zero. Thus consider the sine wave given by

$$y = Y_{1m} \sin \omega t$$

The following table may then be drawn up:

ωt	$0°$	$30°$	$60°$	$90°$	$120°$	$150°$	
y	0	$0.5 Y_{1m}$	$\dfrac{\sqrt{3}}{2} Y_{1m}$	Y_{1m}	$\dfrac{\sqrt{3}}{2} Y_{1m}$	$0.5 Y_{1m}$	
ωt	$180°$	$210°$	$240°$	$270°$	$300°$	$330°$	$360°$
y	0	$-0.5 Y_{1m}$	$-\dfrac{\sqrt{3}}{2} Y_{1m}$	$-Y_{1m}$	$-\dfrac{\sqrt{3}}{2} Y_{1m}$	$-0.5 Y_{1m}$	0

Dividing this cycle into, say, three parts, and adding gives the following:

$y(0–120°)$	0	$0.5 Y_{1m}$	$\dfrac{\sqrt{3}}{2} Y_{1m}$	Y_{1m}	$\dfrac{\sqrt{3}}{2} Y_{1m}$
$y(120°–240°)$	$\dfrac{\sqrt{3}}{2} Y_{1m}$	$0.5 Y_{1m}$	0	$-0.5 Y_{1m}$	$-\dfrac{\sqrt{3}}{2} Y_{1m}$
$y(240°–360°)$	$-\dfrac{\sqrt{3}}{2} Y_{1m}$	$-Y_{1m}$	$-\dfrac{\sqrt{3}}{2} Y_{1m}$	$-0.5 Y_{1m}$	0
Sum	0	0	0	0	0

This is shown graphically in Fig. 5.12(*a*). If a third harmonic of this sine wave is treated in the same way (as shown at (*b*)), the resultant is three times the third harmonic, in the correct phase. Hence if a complex wave containing a third-harmonic component is divided into three equal parts and the ordinates are added, the resultant will be three times the third harmonic only. It may readily be verified that all harmonics other than the third (and multiples of the third) are absent from this resultant.

If the complex wave is divided into five equal sections and the ordinates are added, then it can be shown that the resultant will be five times the fifth harmonic, and will contain no fundamental or third-harmonic components. The method is not very suitable if the complex wave contains large percentages of harmonics higher than the fifth.

EXAMPLE 5.8 A complex current wave has the following shape over one half-cycle of the fundamental, the negative half-cycle being similar:

ωt	0	15°	30°	45°	60°	75°	90°	105°	120°	135°	150°	165°	180°
i	3	5.1	5.3	5.6	6	7.2	10	11.5	12	11	8	3	−3

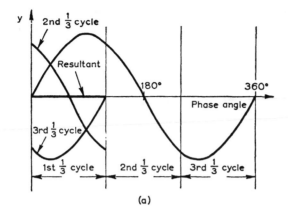

(a)

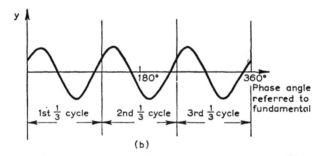

(b)

Fig. 5.12 SUPERPOSITION METHOD

(a) Sine wave   (b) Third harmonic

Find the magnitude and phase angle of the third-harmonic component, and express this as a percentage of the fundamental.

One cycle of the wave is divided into three equal sections and the ordinates are added. The resultant wave is plotted in Fig. 5.13 from the figures obtained from the following table:

Abscissae	0°	15°	30°	45°	60°	75°	90°	105°	120°
$i(0–120°)$	3	5·1	5·3	5·6	6	7·2	10	11·5	12
$i(120–240°)$	12	11	8	3	−3	−5·1	−5·3	−5·6	−6
$i(240–360°)$	−6	−7·2	−10	−11·5	−12	−11	−8	−3	3
Sum	9	8·9	3·3	−2·9	−9	−8·9	−3·3	2·9	9

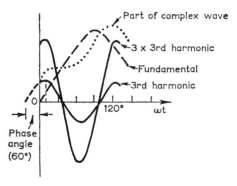

Fig. 5.13

This will represent three times the third harmonic so that the third harmonic itself must have the co-ordinates

Abscissae	0°	15°	30°	45°	60°	75°	90°	105°	120°
Ordinates	3	2·97	1·1	−0·97	−3	−2·97	−1·1	0·97	3

This wave is also plotted in Fig. 5.13, and from it the phase angle of the third harmonic is estimated to be 20° on the fundamental scale.

To obtain the fundamental, the third harmonic is subtracted from the complex wave, assuming there is no higher harmonic. Thus:

Angle ωt	0°	15°	30°	45°	60°	75°	90°	105°	120°	135°	150°	165°	180°
Complex wave	3	5·1	5·3	5·6	6	7·2	10	11·5	12	11	8	3	−3
Third harmonic	3	2·97	1·1	−0·97	−3	−2·97	−1·1	0·97	3	2·97	1·1	−0·97	3
Difference	0	2·13	4·2	6·57	9	10·17	11·1	10·53	9	9·03	6·9	3·97	0

When this wave is plotted it is seen to be almost sinusoidal in shape, indicating that the complex wave contained mainly fundamental and third-harmonic frequencies. From the diagram, the peak value of the fundamental (neglecting higher harmonics) is 11·1, and that of the third harmonic is 3·3. Therefore

$$\text{Percentage third harmonic} = \frac{3·3}{11·1} \times 100 = \underline{\underline{29·7 \text{ per cent}}}$$

The equation of the complex wave is

$$i = 11·1 \sin \omega t + 3·3 \sin 3(\omega t + 20°) = 11·1 \sin \omega t + 3·3 \sin (3\omega t + 60°)$$

TWENTY-FOUR ORDINATE METHOD

The complex wave given by eqn. (5.27) may be expressed in a slightly different form by expanding the sine terms. Thus, for the fundamental term,

$$y_1 = Y_{1m} \sin (\omega t + \psi_1)$$
$$= Y_{1m} \cos \psi_1 \sin \omega t + Y_{1m} \sin \psi_1 \cos \omega t$$
$$= A_1 \sin \omega t + B_1 \cos \omega t$$

where $A_1 = Y_{1m} \cos \psi_1$ and $B_1 = Y_{1m} \sin \psi_1$, and hence

$$\psi_1 = \tan^{-1} \frac{B}{A} \qquad Y_{1m} = \sqrt{(A_1{}^2 + B_1{}^2)}$$

Thus

$$y = A_1 \sin \omega t + A_2 \sin 2\omega t + \ldots$$
$$+ B_1 \cos \omega t + B_2 \cos 2\omega t + \ldots \quad (5.28)$$

In the evaluation of A_1, A_2, . . . B_1, B_2, . . . etc., the following results of integral calculus will be used:

(a) $$\int_0^{2\pi} \sin^2 m\omega t \, d(\omega t) = \pi = \int_0^{2\pi} \cos^2 m\omega t \, d(\omega t) \quad (5.29)$$

(b) $$\int_0^{2\pi} \sin m\omega t \times \sin n\omega t \, d(\omega t) = 0 \qquad \text{for } n \neq m \quad (5.30)$$

(c) $$\int_0^{2\pi} \cos m\omega t \times \cos n\omega t \, d(\omega t) = 0 \qquad \text{for } n \neq m \quad (5.31)$$

(d) $$\int_0^{2\pi} \sin m\omega t \times \cos n\omega t \, d(\omega t) = 0 \quad (5.32)$$

To evaluate A_n, each term in eqn. (5.28) is multiplied by $\sin n\omega t$ and integrated between 0 and 2π, giving

$$\int_0^{2\pi} y \sin n\omega t \, d(\omega t) = \int_0^{2\pi} \{A_1 \sin \omega t \sin n\omega t + A_2 \sin 2\omega t \sin n\omega t$$
$$+ \ldots + A_n \sin^2 n\omega t + \ldots$$
$$+ B_1 \cos \omega t \sin n\omega t + B_2 \cos 2\omega t \sin n\omega t$$
$$+ \ldots + B_n \cos n\omega t \sin n\omega t + \ldots \} d(\omega t)$$
$$= \pi A_n$$

Therefore

$$A_n = 2 \times \frac{1}{2\pi} \int_0^{2\pi} y \sin n\omega t \, d(\omega t)$$

Thus A_n is twice the average value of $y \sin n\omega t$ over one cycle of the fundamental, and in the same way it follows that

$A_1 = 2 \times$ average value of $y \sin \omega t$ over one cycle

$A_2 = 2 \times$ average value of $y \sin 2\omega t$ over one cycle

$B_1 = 2 \times$ average value of $y \cos \omega t$ over one cycle . . . etc.

If the wave is known to contain odd harmonics only, the analysis need only be carried out over one half-cycle, since the positive and negative half-cycles are identical. The integrals (b)–(d) are true over one half-cycle, and the integral (a) has a value of $\pi/2$ for a half-cycle. The coefficients are then $2 \times$ average value of $y \sin n\omega t$ or $y \cos n\omega t$ over one half-cycle.

To obtain the average value of $y \sin n\omega t$ or $y \cos n\omega t$ the complex wave is divided into 24 equal parts (or more if increased accuracy is desired) and 24 ordinates are erected, one at each division starting from zero. Each ordinate is then multiplied by $\sin n\omega t$ (or $\cos n\omega t$) and the sum of these terms is divided by 24. Thus,

$$A_1 = \tfrac{2}{24}(y_0 \sin 0° + y_{15} \sin 15° + y_{30} \sin 30°$$
$$+ \ldots y_{345} \sin 345°) \qquad (5.33)$$

where y_0, y_{15}, y_{30} . . . etc., are the ordinates erected at $0°$, $15°$, $30°$, etc. (to give 24 intervals), and

$$A_3 = \tfrac{2}{24}\{y_0 \sin (3 \times 0°) + y_{15} \sin (3 \times 15°)$$
$$+ y_{30} \sin (3 \times 30°)\} \quad (5.34)$$

For symmetrical waves, twelve ordinates are erected over one half-cycle, and the coefficients are then of the form

$$A_n = \tfrac{2}{12}\{y_0 \sin (n \times 0°) + y_{15} \sin (n \times 15°)$$
$$+ \ldots y_{165} \sin (n \times 165°)\} \quad (5.35)$$

EXAMPLE 5.9 Analyse the wave given in Example 5.8 by the 24-ordinate method, assuming only third and fifth harmonics to be present.

The best approach is to use a tabular method. In this case, since the positive and negative half-cycles are identical, the average over one half-cycle only need be considered.

SINE TERMS

ωt	i	$\sin \omega t$	$i \sin \omega t$	$\sin 3\omega t$	$i \sin 3\omega t$	$\sin 5\omega t$	$i \sin \omega t$
0	3·0	0·0	0·0	0·0	0·0	0·0	0·0
15°	5·1	0·26	1·32	0·707	3·6	0·97	4·95
30°	5·3	0·5	2·65	1·0	5·3	0·5	2·65
45°	5·6	0·707	3·9	0·707	3·96	−0·707	−3·96
60°	6·0	0·87	5·2	0·0	0·0	−0·87	−5·2
75°	7·2	0·97	7·0	−0·707	−5·1	0·26	1·87
90°	10·0	1·0	10·0	−1·0	−10·0	1·0	10·0
105°	11·5	0·97	11·15	−0·707	−8·15	0·26	3·0
120°	12·0	0·87	10·45	0·0	0·0	−0·87	10·45
135°	11·0	0·707	7·8	0·707	7·8	−0·707	−7·8
150°	8·0	0·5	4·0	1·0	8·0	0·5	4·0
165°	3·0	0·26	0·78	0·707	2·12	0·97	2·9
Sum			64·25		7·53		1·96
$\frac{2}{12} \times$ Sum		$A_1 = 10\cdot74$		$A_3 = 1\cdot25$		$A_5 = 0\cdot327$	

COSINE TERMS

ωt	i	$\cos \omega t$	$i \cos \omega t$	$\cos 3\omega t$	$i \cos 3\omega t$	$\cos 5\omega t$	$i \cos 5\omega t$
0	3·0	1·0	3·0	1·0	3·0	1·0	3·0
15°	5·1	0·97	4·95	0·707	3·6	0·26	1·32
30°	5·3	0·87	4·6	0·0	0·0	−0·87	−4·6
45°	5·6	0·707	3·96	−0·707	−3·96	−0·707	−3·96
60°	6·0	0·5	3·0	−1·0	−6·0	0·5	3·0
75°	7·2	0·26	1·87	−0·707	−5·1	0·97	7·0
90°	10·0	0·0	0·0	0·0	0·0	0·0	0·0
105°	11·5	−0·26	−3·0	0·707	8·15	−0·97	−11·15
120°	12·0	−0·5	−6·0	1·0	12·0	−0·5	−6·0
135°	11·0	−0·707	−7·8	0·707	7·8	0·707	7·8
150°	8·0	−0·87	−6·95	0·0	0·0	0·87	6·95
165°	3·0	−0·97	−2·9	−0·707	−2·12	−0·26	−0·78
Sum			−5·27		17·37		2·58
$\frac{2}{12} \times$ Sum		$B_1 = -0\cdot88$		$B_3 = 2\cdot90$		$B_5 = 0\cdot43$	

From these results

$$I_{1m} = \sqrt{(A_1{}^2 + B_1{}^2)} = 10\cdot8$$
$$I_{3m} = \sqrt{(A_3{}^2 + B_3{}^2)} = 3\cdot18$$
$$I_{5m} = \sqrt{(A_5{}^2 + B_5{}^2)} = 0\cdot54$$

Also

$$\Psi_1 = \tan^{-1} - \frac{0\cdot88}{10\cdot74} = -5°$$

$$\Psi_3 = \tan^{-1} \frac{2\cdot90}{1\cdot25} = 66\cdot7°$$

$$\Psi_5 = \tan^{-1} \frac{0\cdot43}{0\cdot327} = 53\cdot7°$$

Thus

$$i = 10\cdot8 \sin(\omega t - 5°) + 3\cdot18 \sin(3\omega t + 66\cdot7°)$$
$$+ 0\cdot43 \sin(5\omega t + 53\cdot7°)$$

This is a more accurate result than that obtained by the superposition method, and clearly shows the presence of a small fifth harmonic. The results for the fundamental and third harmonic compare favourably in each case.

5.13 Form Factor

The form factor, k_f, of any alternating waveform may be defined as

$$k_f = \frac{\text{R.M.S. value}}{\text{Full-wave-rectified mean value}} \tag{5.36}$$

The full-wave-rectified waveform has its negative-going portions inverted, as shown in Fig. 5.14. Its mean value is found by integrat-

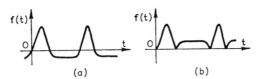

(a)   (b)

Fig. 5.14

ing over the period of the wave. If the wave has identical positive and negative half-cycles, the integration may be performed over one half-cycle, between zero values.

For a full sine wave, $i = I_m \sin \omega t$, and since positive and negative half-cycles are identical, and zeros occur at times $t = 0$ and $t = T/2$, the r.m.s. value is

$$I = \sqrt{\left\{\frac{1}{T/2} \int_0^{T/2} I_m{}^2 \sin^2 \omega t \, dt\right\}} = \frac{I_m}{\sqrt{2}}$$

and the full-wave-rectified mean is

$$I_{av} = \frac{1}{T/2} \int_0^{T/2} I_m \sin \omega t \, dt = \frac{2}{\pi} I_m$$

so that the form factor is

$$k_f = \frac{I_m/\sqrt{2}}{I_m 2/\pi} = 1 \cdot 11 \tag{5.37}$$

For a square wave with no d.c. component the r.m.s. and mean values are both equal to the peak value and hence the form factor is unity.

The form factors of complex waves whose positive and negative half-cycles are identical (and hence contain odd harmonics only), and which have only two zero values per cycle can be readily evaluated. The r.m.s. value is obtained as in Section 5.3 as the square root of the sum of the squares of the r.m.s. values of each harmonic. The mean value is obtained by integrating between two zeros (i.e. the half-cycle mean). Thus consider the sine series

$$i = I_{1m} \sin \omega t + I_{3m} \sin 3\omega t + I_{5m} \sin 5\omega t + \ldots$$

The r.m.s. value is (from eqn. (5.6))

$$I = \sqrt{\left(\frac{I_{1m}^2}{2} + \frac{I_{3m}^2}{2} + \frac{I_{5m}^2}{2} + \ldots\right)}$$

while the mean value is

$$I_{av} = \frac{1}{T/2} \int_0^{T/2} i \, dt$$

since zeros occur when $t = 0$ and when $t = T/2$. Hence

$$I_{av} = \frac{2}{T} \left(\frac{2}{\omega} I_{1m} + \frac{2}{3\omega} I_{3m} + \frac{2}{5\omega} I_{5m} + \ldots\right)$$

$$= \frac{2}{\pi} \left(I_{1m} + \frac{I_{3m}}{3} + \frac{I_{5m}}{5} + \ldots\right) \tag{5.38}$$

where $\omega = 2\pi/T$. It follows that the form factor is

$$k_f = \frac{\pi}{2\sqrt{2}} \frac{\sqrt{(I_{1m}^2 + I_{3m}^2 + I_{5m}^2 + \ldots)}}{(I_{1m} + I_{3m}/3 + I_{5m}/5 + \ldots)} \tag{5.39}$$

In a similar way a cosine series of odd harmonics yields an easily evaluated form factor. Let

$$i = I_{1m} \cos \omega t + I_{3m} \cos 3\omega t + I_{5m} \cos 5\omega t + \ldots$$

The r.m.s. value is the same as for the sine series which has just been evaluated. Zeros occur at $t = -T/4$ and $t = +T/4$, so that the mean value is

$$
\begin{aligned}
I_{av} &= \frac{1}{T/2} \int_{-T/4}^{T/4} i\, dt \\
&= \frac{2}{T}\left(\frac{2}{\omega} I_{1m} - \frac{2}{3\omega} I_{3m} + \frac{2}{5\omega} I_{5m} - \dots \right) \\
&= \frac{2}{\pi}\left(I_{1m} - \frac{I_{3m}}{3} + \frac{I_{5m}}{5} - \dots \right)
\end{aligned}
\tag{5.40}
$$

and the form factor is

$$
k_f = \frac{\pi}{2\sqrt{2}} \frac{\sqrt{(I_{1m}^2 + I_{3m}^2 + I_{5m}^2 + \dots)}}{(I_{1m} - I_{3m}/3 + I_{5m}/5 - \dots)}
\tag{5.41}
$$

In the general case it is necessary to determine the instants in the cycle at which the wave has zero values, and integrate separately over each period of time between zeros, adding the integrals so obtained arithmetically, and dividing the result by the period of the wave. This can be a cumbersome process.

The form factor of a complex wave can sometimes cause errors in instrument readings. Moving-iron, electrodynamic and electrostatic instruments will always read true r.m.s. values independent of form factor. Moving-coil rectifier instruments, on the other hand, have a deflexion which is proportional to the full-wave-rectified mean current which flows in the moving coil. These instruments are normally calibrated in r.m.s. values assuming sine wave inputs (i.e. form factors of $1 \cdot 11$). The readings on such instruments are really the mean values multiplied by $1 \cdot 11$. Hence if the form factor of a wave measured on a rectifier instrument is k_f, the mean value is the instrument reading divided by $1 \cdot 11$, and

$$
\text{True r.m.s. value} = \frac{\text{Instrument reading}}{1 \cdot 11} k_f
\tag{5.42}
$$

It follows that the form factor of a complex wave can be found experimentally by measuring its value on a true r.m.s. instrument and on a rectifier instrument. The form factor is then

$$
k_f = \frac{\text{R.M.S. instrument reading}}{\text{Rectifier instrument reading}} \times 1 \cdot 11
\tag{5.43}
$$

One further important application of form factor in complex waves occurs when iron-cored coils are excited into saturation by an alternating current. It has already been seen in Section 5.9(*c*) that the flux waveform in this case is flat-topped, and analysis shows it to be represented by a sine series of odd harmonics

$$\Phi = \Phi_{1m} \sin \omega t + \Phi_{3m} \sin 3\omega t + \ldots$$

The induced e.m.f. in a coil of N turns linking this flux is given by

$$e = N\frac{d\Phi}{dt} = N\omega\Phi_{1m} \cos \omega t + 3N\omega\Phi_{3m} \cos 3\omega t + \ldots$$

From eqn. (5.40) the average e.m.f. is therefore

$$E_{av} = \frac{2}{\pi}(N\omega\Phi_{1m} - N\omega\Phi_{3m} + \ldots)$$
$$= 4fN(\Phi_{1m} - \Phi_{3m} + \ldots)$$
$$= 4fN\Phi_{max}$$

since in this instance the peak flux (Φ_{max}) is given by

$$\Phi_{max} = \Phi_{1m} - \Phi_{3m} + \Phi_{5m} - \ldots$$

If the form factor of the e.m.f. wave if k_f then the r.m.s. value is

$$E = 4k_f fN\Phi_{max} \tag{5.44}$$

PROBLEMS

5.1 Show that the current wave through a capacitor contains a larger percentage of harmonics than the voltage wave across it.

A voltage of 200V (r.m.s.), containing 20 per cent third harmonic is applied to a circuit containing a resistor and a capacitor in series. The current is 3A (r.m.s.) with 30 per cent harmonic. Determine the resistance and capacitance of the circuit, and the overall power factor. (The fundamental frequency is 50 Hz.)

(*H.N.C.*)

Ans. 41·8 Ω, 59 μF, 0·626.

5.2 The magnetization curve for a ferromagnetic material is given in the following table:

B (T)*	0	0·42	0·8	0·97	1·08	1·15
H (At/m)	0	100	200	300	400	500

The material is used for a transformer working from a 250V 50Hz supply. The supply waveform may be assumed sinusoidal and the resistance of the transformer primary negligible. The net iron cross-sectional area of the transformer is 0·004 m^2 and there are 250 turns on the primary. Deduce the waveform of the magnetizing current, neglecting hysteresis.

5.3 Using Wedmore's method find the third-harmonic component of the magnetizing current in the above example.

Ans. 15 per cent.

* See page 221.

5.4 A voltage wave is given by the following expression:

$$v = 30(\sin 314t + \tfrac{1}{3}\sin 942t + \tfrac{1}{5}\sin 1{,}570t) \text{ volts}$$

It is applied to a circuit consisting of a $32 \cdot 7\,\Omega$ resistor in series with a parallel combination of a 100mH pure inductance and a $4 \cdot 06\,\mu$F capacitor.

Calculate the total power delivered to the circuit and also the total r.m.s. current through the capacitance.

Ans. $6 \cdot 95$ W; $0 \cdot 032$ A.

5.5 A 3-phase 50 Hz alternator has a phase voltage

$$v = 100\sin \omega t + 10\sin 3\omega t + 5\sin 5\omega t \text{ volts}$$

What are the line voltages if the alternator is (*a*) star connected, (*b*) mesh connected?

Three similar star-connected coils of $50\,\Omega$ resistance and $0 \cdot 1$ H inductance are supplied from the alternator, which is star connected. Calculate the line current and the current through the neutral when the neutral is connected.

Ans. $122 \cdot 6$ V; $70 \cdot 8$ V; $1 \cdot 2$ A; $0 \cdot 196$ A.

5.6 A series circuit consists of a coil of inductance $0 \cdot 1$ H and resistance $25\,\Omega$ and a variable capacitor. Across this circuit is applied a voltage whose instantaneous value is given by

$$v = 100\sin \omega t + 20\sin (3\omega t + 45°) + 5\sin (5\omega t - 30°) \text{ volts}$$

where $\omega = 314$ rad/s.

Determine the value of C which will produce resonance at the third-harmonic frequency, and with this value of C find (*a*) an expression for the current in the circuit, (*b*) the r.m.s. value of this current, (*c*) the total power absorbed.

(*H.N.C.*)

Ans. $11 \cdot 25\,\mu$F: (*a*) $i = 0 \cdot 398\sin (\omega t + 84 \cdot 3°) + 0 \cdot 8\sin (3\omega t + 45°) + 0 \cdot 0485 \sin (5\omega t + 106°)$, (*b*) $0 \cdot 633$ A, (*c*) 10 W.

5.7 With the aid of clear diagrams, explain the anticipated waveform of:

(*a*) The current in a reactor with negligible resistance when a sinusoidal voltage, sufficient to saturate the reactor, is applied.

(*b*) The e.m.f. in a reactor with negligible resistance when a sinusoidal current, sufficient to saturate the reactor, flows through the reactor.

(*c*) The current when a sinusoidal voltage is applied to a rectifier with a forward resistance of $1{,}000\,\Omega$ and a back resistance of $100{,}000\,\Omega$.

What general type of harmonic would you expect to be present or absent in each case?

In sections (*a*) and (*b*) a B/H curve, neglecting hysteresis, may be considered.

(*H.N.C.*)

5.8 Derive an expression for the r.m.s. value of the complex voltage wave represented by the equation

$$v = V_0 + V_{1m}\sin (\omega t + \phi_1) + V_{3m}\sin (3\omega t + \phi_3)$$

A voltage $v = 200\sin 314t + 50\sin (942t + 45°)$ volts is applied to a circuit consisting of a resistance of $20\,\Omega$, an inductance of 20mH and a capacitance of $56 \cdot 3\,\mu$F all connected in series.

Calculate the r.m.s. values of the applied voltage and the current.

Find also the total power absorbed by the circuit.

(*H.N.C.*)

Ans. 146 V; $3 \cdot 16$ A; 200 W.

5.9 The e.m.f. of one phase of a 3-phase mesh-connected alternator is represented by the following expression:

$$e = 500 \sin \theta + 60 \sin 3\theta - 40 \sin 5\theta \text{ volts}$$

The fundamental frequency is 50 Hz and each phase of the windings has a resistance of 3Ω and an inductance of 0.01 H. Calculate the r.m.s. value of (a) the current circulating in the windings, and (b) the current through a 100μF capacitor connected across a pair of line wires. *(L.U.)*

Ans. 4·28 A; 14·7 A.

(Hint. Do not neglect internal impedance.)

5.10 If the voltage applied to a circuit be represented by

$$V_1 \sin \omega t + V_n \sin n\omega t$$

and if the current is

$$I_1 \sin (\omega t - \phi_1) + I_n \sin (n\omega t - \phi_n)$$

derive an expression for the average power in the circuit.

A voltage represented by $250 \sin \omega t$ volts is applied to a circuit consisting of a non-inductive resistance of 30Ω in series with an iron-cored inductance. The corresponding current is represented approximately by

$$3 \sin \left(\omega t - \frac{\pi}{3} \right) + 1.2 \sin \left(3\omega t - \frac{\pi}{2} \right) \text{ amperes}$$

Calculate (a) the power absorbed by the resistance, (b) the effective value of the voltage across the inductance, and (c) the power factor of the whole circuit.

Draw to scale the waveform of the fundamental and third-harmonic currents showing their phase relation.

Explain why accurately calibrated rectifier and dynamometer [electrodynamic] ammeters would read differently when placed in the above circuit. *(L.U.)*

Ans. 156 W; 157 V; 0·47.

5.11 A p.d. of the form $v = 400 \sin \omega t + 30 \sin 3\omega t$ volts is applied to a rectifier having a resistance of 50Ω in one direction and 200Ω in the reverse direction. Find the average and effective values of the current and the p.f. of the circuit.

(L.U.)

Ans. 1·96 A; 4·1 A; 0·51.

5.12 The following table gives the characteristics of each of the four elements of a copper-oxide bridge rectifier.

Voltage (V)	0·1	0·15	0·2	0·24	0·28	0·34	0·38
Current (mA)	0·2	0·4	1	2	4	8	12

This bridge is connected directly across a supply voltage represented by $(1.0 \sin \theta + 0.1 \sin 3\theta)$ volts, and a milliammeter having a resistance of 20Ω is connected across appropriate points of the bridge. Determine the reading on the milliammeter, assuming the reverse current to be negligible. *(L.U.)*

Ans. 5·4 mA.

Chapter 6

TRANSIENTS

The steady direct current which flows in a circuit connected to a battery or a d.c. generator may easily be calculated. Similarly the alternating current, which flows in a circuit connected to an alternator, may be calculated by the methods previously discussed. These are called *steady-state* currents, for it is assumed that the components in the circuits are unvarying and have been previously connected to the generator for so long that any peculiar disturbance, associated with the initial connexion or switching on of the apparatus, has had time to resolve itself.

In most cases the connexion and disconnexion of apparatus causes a disturbance which dies out in a short time, i.e. a *transient* disturbance. In this chapter the effect of suddenly switching on and off various circuits will be considered. In each case the resultant current is assumed to be the steady-state or normal current, with a transient or disturbance current superimposed.

The transient currents are found to be associated with the changes in stored energy in inductors and capacitors. Since there is no stored energy in a resistor there will be no transient currents in a pure-resistance circuit, i.e. the steady-state direct or alternating current will be attained immediately when the supply is connected.

6.1 Inductive Circuits

At any instant in a series circuit containing resistance R and inductance L, the applied voltage v is equal to the sum of the voltage drops across the resistance and the inductance, i.e.

$$v = iR + L\frac{di}{dt} \tag{6.1}$$

where i is the instantaneous value of the current.

The circuit current i is composed, as has already been noted, of two parts, i_s, the steady-state current, and i_t the transient current, i.e.

$$i = i_s + i_t$$

When the transient current has ceased, the steady-state current must still satisfy eqn. (6.1); therefore

$$v = i_s R + L\frac{di_s}{dt} \tag{6.2}$$

During the period in which the transient exists,

$$v = (i_s + i_t)R + L\frac{d}{dt}(i_s + i_t)$$

$$= i_s R + L\frac{di_s}{dt} + i_t R + L\frac{di_t}{dt} \tag{6.3}$$

whence

$$i_t R + L\frac{di_t}{dt} = 0 \tag{6.4}$$

The current i_s is obtained mathematically by solving eqn. (6.2). Since this is done implicitly when normal circuit theory is applied, the formal mathematical solution need not be given here. It therefore remains only to solve eqn. (6.4) for i_t. Thus, rearranging the terms,

$$\frac{di_t}{i_t} = -\frac{R}{L}dt$$

so that

$$\int \frac{di_t}{i_t} = -\frac{R}{L}\int dt$$

and

$$\log_e i_t + \log_e A = -\frac{R}{L}t$$

where $\log_e A$ is the constant of integration. Continuing,

$$\log_e Ai_t = -\frac{R}{L}t$$

so that

$$Ai_t = e^{-(R/L)t}$$

and

$$i_t = Be^{-(R/L)t} \tag{6.5}$$

where $B = 1/A$ = constant.

The complete solution is then

$$i = i_s + Be^{-(R/L)t} \tag{6.6}$$

The current i_s is found from simple circuit theory, and the constant B is then determined by substituting a known set of values for i and t in eqn. (6.6). These known values are normally the initial conditions in the circuit.

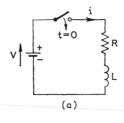

(a)

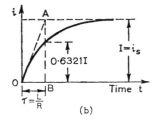

(b)

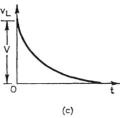

(c)

Fig. 6.1 GROWTH OF CURRENT IN AN INDUCTIVE D.C. CIRCUIT

.2 Growth of Current in an Inductive Circuit (D.C.)

uppose the switch in the circuit shown in Fig. 6.1(a) is closed t a datum time, taken as $t = 0$. Before the switch is closed the urrent is obviously zero. At the moment when the switch is closed 1e current will remain at an instantaneous value of zero since the urrent through an inductor cannot immediately change (Lenz's 1w). Thus in the present problem at $t = 0$, $i = 0$.

Also, from d.c. theory, $i_s = V/R = I$. Substituting in eqn. (6.6) at the instant $t = 0$,

$$0 = \frac{V}{R} + Be^0 \quad \text{whence} \quad B = -\frac{V}{R}$$

Again from eqn. (6.6),

$$i = \frac{V}{R} - \frac{V}{R}e^{-(R/L)t}$$

$$= \frac{V}{R}(1 - e^{-(R/L)t})$$

$$= I(1 - e^{-(R/L)t}) \tag{6.7}$$

The curve of i plotted to a base of time is shown in Fig. 6.1(b). It is called an *exponential-growth curve*.

The rate of change of current is found by differentiating eqn. (6.7). Thus

$$\frac{di}{dt} = \frac{V}{R}\left(\frac{R}{L}e^{-(R/L)t}\right) = \frac{V}{L}e^{-(R/L)t}$$

The initial rate of change of current is then

$$\frac{di}{dt}\bigg|_{t=0} = \frac{V}{L} \quad \text{amperes/second} \tag{6.8}$$

Consider now the value of the current when $t = L/R$ seconds. Then from eqn. (6.7),

$$i = I(1 - e^{-1}) = 0.6321I$$

L/R is called the *time constant*, τ (tau), of the RL circuit. It may easily be verified that this would be the time required for the current to reach its final value if the initial rate of increase were continued (i.e. point A on Fig. 6.1(b)). The time constant is defined as the time required for the current to reach 63.21 per cent (i.e. approximately $\frac{2}{3}$) of its final value.

The voltage across the resistor R is easily obtained as

$$v_R = iR = I(1 - e^{-(R/L)t})R = V(1 - e^{-(R/L)t}) \tag{6.9}$$

The voltage across the inductor at any instant is then,

$$v_L = V - v_R = Ve^{-(R/L)t} \tag{6.10}$$

This is an exponential decay curve, and is shown in Fig. 6.1(c).

It should be noted that the general form for the curve of exponential growth is $y = Y(1 - e^{-t/\tau})$ where τ is the time constant. The corresponding expression for an exponential decay curve is $y = Ye^{-t/\tau}$.

The energy relations in the circuit may be derived as follows. The energy supplied by the battery in time dt is $Vi\,dt$ joules. Hence the total energy supplied in t seconds is $\int_0^t Vi\,dt$. This energy is partly dissipated as heat in the resistor R, and is partly stored in the magnetic field of the coil. If the coil current at a given instant is i amperes, then the stored energy at the same instant is $\frac{1}{2}Li^2$ joules. The energy dissipated in the resistor R up to an instant t is $\int_0^t i^2R\,dt$, this being easily calculated as the total energy supplied minus the energy stored in the magnetic field.

6.3 Decay of Current in an Inductive Circuit

Consider the circuit shown in Fig. 6.2(*a*). At the datum time $t = 0$, the switch is opened, disconnecting the inductor L and

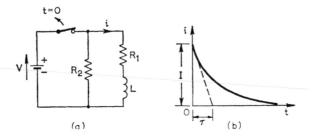

Fig. 6.2   DECAY OF CURRENT IN AN INDUCTIVE D.C. CIRCUIT

resistor R_1 from the supply. Since there is no continuous source of e.m.f. in the circuit formed by L, R_1 and R_2, the steady-state current will be zero. Hence eqn. (6.6) becomes,

$$i = Be^{-(R/L)t} \tag{6.11}$$

where R = total circuit resistance = $R_1 + R_2$.

As before, at the instant of switching the current through the inductor remains momentarily unchanged, i.e. it has the same value (V/R_1 amperes) as it had before the switch was operated. Hence at $t = 0$, $i = V/R_1 = I$. Therefore from eqn. (6.11),

$$\frac{V}{R_1} = B$$

Substituting in eqn. (6.11),

$$i = \frac{V}{R_1} e^{-(R/L)t} = Ie^{(-R/L)t} \tag{6.12}$$

This is the exponential decay curve shown in Fig. 6.2(*b*). The time constant for this circuit is L/R, i.e. $L/(R_1 + R_2)$ seconds. The total energy available is $\frac{1}{2}LI^2$ joules; all of this energy is eventually dissipated as heat in the resistances of the circuit.

It should be noted that if R_2 is omitted the energy stored in the magnetic field will cause a spark at the switch contacts, or will destroy the insulation of the coil (owing to the large induced e.m.f.).

EXAMPLE 6.1 A coil of 10 H inductance, and 5 Ω resistance is connected in parallel with a 20 Ω resistor across a 100 V d.c. supply which is suddenly disconnected. Find:

    (*a*) The initial rate of change of current after switching.
    (*b*) The voltage across the 20 Ω resistor initially, and after 0·3 s.
    (*c*) The voltage across the switch contacts at the instant of separation.
    (*d*) The rate at which the coil is losing stored energy 0·3 s after switching.
                                            (*H.N.C.*)

    (*a*) The steady-state current is zero; hence

$$i = Be^{-(R/L)t}$$

where R is the total circuit resistance after switching (25 Ω).

At $t = 0$, the current is $100/5 = 20$ A, i.e. the current through the coil immediately prior to the opening of the switch is 20 A. Thus

$$20 = Be^0 = B$$

whence

$$i = 20e^{-2·5t}$$

$$\text{Initial rate of change of current} = \frac{di}{dt}\bigg|_{t=0} = -20 \times 2·5e^{-2·5t}\bigg|_{t=0}$$

$$= -50 \text{ A/s}$$

The negative sign indicates that the current is decreasing.

    (*b*) The current through the 20 Ω resistor after the supply has been disconnected is i ampere.

$$\text{Initial voltage across 20 Ω resistor} = (\text{current at } t = 0) \times 20$$
$$= 20 \times 20 = 400 \text{ V}$$

Current after 0·3 sec $= 20e^{-0·75} = 9·44$ A

Therefore

Voltage across 20 Ω resistor after 0·3 s $= 9·44 \times 20 = 188$ V

    (*c*) Since the e.m.f. induced in the inductor tends to maintain the current through it in the original direction, the direction of the current through R_2 (20 Ω) will be upwards. The voltage across the switch contacts will therefore be the supply voltage plus the voltage across R_2. Therefore

Initial voltage across contacts $= 100 + 400 = \underline{\underline{500\,V}}$

(*d*) Rate at which coil loses stored energy = Power
$$= \text{Coil e.m.f.} \times \text{Current}$$

At $0.3\,s$, $i = 9.44\,A$.
The rate of change of current at $t = 0.3$ is

$$\left.\frac{di}{dt}\right|_{t=0.3} = -20 \times 2.5 \times e^{-0.75} = -23.6\,A/s$$

The rate at which the coil loses stored energy is

$$L\frac{di}{dt} \times i = -10 \times 23.6 \times 9.44 = \underline{\underline{-2,230\,J/s}}$$

The negative sign indicates a decrease in stored energy.

6.4 Growth of Current in an Inductive Circuit (A.C.) (Fig. 6.3)

Let a voltage given by $v = V_m \sin(\omega t + \psi)$ be suddenly applied to an RL series inductive circuit at the instant $t = 0$, i.e. the voltage

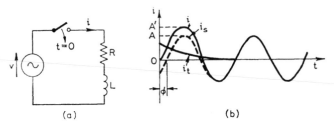

Fig. 6.3 A.C. SWITCHING TRANSIENTS

is suddenly applied when it is passing through the value $V_m \sin \psi$. Since the contacts may close at any instant in the cycle, ψ may have any value from zero to 2π radians. The voltage equation is then

$$V_m \sin(\omega t + \psi) = iR + L\frac{di}{dt} \qquad (6.13)$$

As has already been explained, the steady state-current i_s is easily found by normal circuit theory. If the circuit impedance is $\sqrt{(R^2 + \omega^2 L^2)}$, the peak steady-state current is

$$I_m = \frac{V_m}{\sqrt{(R^2 + \omega^2 L^2)}}$$

This current lags behind the applied voltage by ϕ radians, where

$\phi = \tan^{-1} \omega L/R$. The expression for the instantaneous value of the steady-state current is therefore

$$i_s = \frac{V_m}{\sqrt{(R^2 + \omega^2 L^2)}} \sin\left(\omega t + \psi - \tan^{-1}\frac{\omega L}{R}\right)$$
$$= I_m \sin(\omega t + \psi - \phi) \qquad (6.14)$$

The transient current has already been derived from eqn. (6.4) as

$$i_t = Be^{-(R/L)t}$$

Hence

$$i = i_s + i_t = I_m \sin(\omega t + \psi - \phi) + Be^{-(R/L)t} \qquad (6.15)$$

As before, the condition that the current though an inductor is instantaneously the same before and after switching is used to evaluate the constant B. In this case at $t = 0$, $i = 0$. Substituting in eqn. (6.15),

$$0 = I_m \sin(\psi - \phi) + B$$

whence

$$B = -I_m \sin(\psi - \phi)$$

Therefore

$$i = I_m \sin(\omega t + \psi - \phi) - I_m \sin(\psi - \phi)e^{-(R/L)t} \qquad (6.15a)$$

From this it will be seen that the value of B, and hence the size of the switching transient, depends on the value of ψ, i.e. on the instant in the cycle at which the contacts close. Three cases will be investigated.

Case 1. At $t = 0$, the voltage is passing through zero and is positive going, i.e. $\psi = 0$.

$$i = I_m \sin(\omega t - \phi) - I_m \sin(-\phi)e^{-(R/L)t}$$
$$= I_m (\sin(\omega t - \phi) + \sin\phi e^{-(R/L)t})$$

The curve of i to a base of ωt is shown in Fig. 6.3(b). This shows that the maximum instantaneous peak current (OA′) may be larger than the normal peak current (OA).

Case 2. At $t = 0$, the voltage is passing through $V_m \sin\phi$, i.e. $\psi = \phi$, and $(\psi - \phi) = 0$.
In this case, $B = 0$, and there is no switching transient ($i_t = 0$). This corresponds to the contacts closing at the instant when the steady-state current will itself be zero.

Case 3. At $t = 0$, the voltage is passing through $V_m \sin\left(\phi \pm \frac{\pi}{2}\right)$, i.e.

$$\psi = \phi \pm \frac{\pi}{2} \quad \text{and} \quad (\psi - \phi) = \pm \frac{\pi}{2} \qquad (6.16)$$

The transient term in this case is

$$i_t = -I_m \sin\left(\pm \frac{\pi}{2}\right) e^{-(R/L)t}$$

$$= \pm I_m e^{-(R/L)t}$$

i.e. the transient now has its maximum possible initial value.

EXAMPLE 6.2   A 50 Hz alternating voltage of peak value 300 V is suddenly applied to a circuit which has a resistance of $0.1\,\Omega$ and an inductance of 3.18 mH. Determine the first peak value of the resultant current when the transient term has a maximum value.

Inductive reactance, $X_L = 2\pi f L = 2\pi \times 50 \times 0.00318 = 1\,\Omega$

Therefore

Circuit impedance $= 0.1 + j1 \approx 1\underline{/84.3°}$

whence

Peak steady-state current $= 300/1 = 300\,\text{A}$

If $v = 300 \sin(\omega t + \psi)$, then the maximum transient will occur when $\psi = \phi \pm \pi/2$, where ϕ is the phase angle of the current with respect to the voltage (i.e. 84.3°). Therefore

$$\psi = 84.3 \pm 90 = -5.7° \qquad \text{(choosing the negative value)}$$

Hence

$$i = 300 \sin(\omega t - 90°) + Be^{-31.4t} \qquad (6.15)$$

At $t = 0$, $i = 0$. Therefore

$$0 = 300 \sin(-90°) + B$$

so that $B = 300$, and

$$i = 300 \sin(\omega t - 90°) + 300e^{-31.4t}$$

To obtain an exact solution for the first peak of the current, the above expression must be differentiated, equated to zero, and the resulting expression solved graphically for t. It is usually sufficiently accurate to determine the instant at which the steady-state term reaches its first maximum positive value, and to add the value of the transient term at this instant to the peak value of the steady-state term. Thus the first maximum positive value of the steady-state term occurs when

$$(\omega t - 90°) = 90° = \pi/2\,\text{rad}$$

i.e. when $t = 0.01$ sec. At this time $i_t = 300e^{-0.314} = 219\,\text{A}$. Therefore

Resultant current at this instant $= 300 + 219 = \underline{\underline{519\,\text{A}}}$

6.5 Capacitive Circuits

The voltage equation for a circuit consisting of a capacitor C in series with a resistor R is

$$v = iR + \frac{q}{C} \tag{6.17}$$

where q is the instantaneous charge on the capacitor.

As for the inductive circuit the current i is expressed as the sum of the steady-state current i_s, and the transient current i_t. The transient current is the solution of the equation

$$i_t R + \frac{q_t}{C} = 0$$

Differentiating,

$$R \frac{di_t}{dt} + \frac{1}{C} \frac{dq_t}{dt} = 0$$

i.e.

$$\frac{1}{C} i_t + R \frac{di_t}{dt} = 0$$

since $i = dq/dt$.

This equation has the same form as eqn. (6.4) for the case of the inductive circuit, and the solution for i_t will follow exactly that previously derived, if $1/C$ is substituted for R in eqn. (6.4) and R is substituted for L. Hence

$$i_t = Be^{-(1/CR)t} \tag{6.18}$$

The complete solution for i is therefore

$$i = i_s + Be^{-(1/CR)t} \tag{6.19}$$

where i_s is found from normal circuit theory, and B is a constant obtained by substituting known values in eqn. (6.19).

The initial condition, which is used to determine the constant B in eqn. (6.19), is that the charge on a capacitor cannot instantaneously change since an instantaneous change of charge would require an infinite current and hence an infinite rate of change of voltage. In effect, since the capacitance is constant, this means that the voltage across a capacitor is momentarily the same before and after any sudden change in the circuit conditions.

6.6 Charging of a Capacitor through a Resistor

Consider a circuit consisting of a resistor R in series with a capacitor C, connected to a battery of voltage V, at time $t = 0$ (Fig. 6.4(a)).

The steady-state current in this case is obviously zero, since no current will flow from a d.c. supply through a capacitor. Hence $i_s = 0$, and eqn. (6.19) becomes

$$i = Be^{-(1/CR)t} \tag{6.20}$$

When the switch is closed there will be momentarily no voltage across the capacitor, so that the battery voltage V must all appear across the resistor R. Hence the initial current from the battery must be $i = V/R$ amperes, i.e. at $t = 0$, $i = V/R = I$, say. Therefore

$$i = \frac{V}{R}e^{-(1/CR)t} = Ie^{(1/CR)t} \tag{6.21}$$

This equation represents the exponential decay curve drawn in Fig. 6.4(*b*).

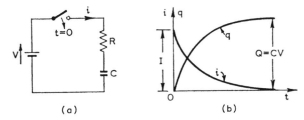

Fig. 6.4   CHARGING A CAPACITOR THROUGH A RESISTOR

The voltage v_R across the resistor at any instant is

$$v_R = iR = IRe^{-(1/CR)t} = Ve^{-(1/CR)t} \tag{6.22}$$

From this, the voltage v_c across the capacitor is

$$v_c = V - v_R = V(1 - e^{-(1/CR)t}) \tag{6.23}$$

This equation represents an exponential growth to a final value of V volts. From it, the charge q on the capacitor may be found at any instant. Thus

$$q = v_cC = VC(1 - e^{-(1/CR)t}) = Q(1 - e^{-(1/CR)t}) \tag{6.24}$$

where $Q = VC$ = final charge on the capacitor.

The time constant τ of the CR circuit is defined as the time required for the charge on the capacitor to attain 63·21 per cent of its final value, i.e. the index $(1/CR)t$ must be unity when $t = \tau$. Therefore

$$\tau = CR \quad \text{seconds} \tag{6.25}$$

Energy will only be supplied from the battery during the time required to charge the capacitor. Some of this energy will be dissipated as heat in the resistor, and some will be stored in the electric field of the capacitor.

$$\text{Total energy from the battery in } t \text{ seconds} = \int_0^t Vi\,dt \text{ joules}$$

$$\text{Energy stored in electric field in } t \text{ seconds} = \tfrac{1}{2}Cv_c^2$$

where v_c is the voltage across the capacitor after t seconds. The energy dissipated in the resistor will be the difference between the total energy taken from the battery and that stored in the capacitor.

6.7 Discharge of a Capacitor through a Resistor

Suppose that a capacitor which is originally charged to V_c volts is discharged through a resistor of R ohms (Fig. 6.5(a)). Since there is

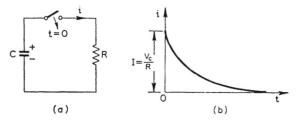

Fig. 6.5   DISCHARGING A CAPACITOR THROUGH A RESISTOR

no generator in the circuit the steady-state current must be zero, so that, from eqn. (6.19), the general equation for the circuit current is

$$i = Be^{-(1/CR)t}$$

The initial condition is that the voltage across the capacitor must be the same after the switch is closed as it was before, i.e. V_c volts. Hence at $t = 0$ the voltage across R is V_c volts, and the current through it is $i = V_c/R = I$, say. The general equation for the current must therefore be

$$i = \frac{V_c}{R}\,e^{-(1/CR)t} = Ie^{-(1/CR)t} \tag{6.26}$$

This is the exponential decay curve shown in Fig. 6.5(b).

The voltage at any instant is the same across both the capacitor and the resistor. Let this voltage be v. Then

$$v = iR = V_c e^{-(1/CR)t} \tag{6.27}$$

The charge on the capacitor at any instant is

$$q = Cv$$
$$= CV_c e^{-(1/CR)t}$$
$$= Q e^{-(1/CR)t} \text{ coulombs} \tag{6.28}$$

where $Q \, (= CV_c)$ is the initial charge on the capacitor.

EXAMPLE 6.3 A simple sawtooth voltage generator consists of a $10{,}000\,\Omega$ resistor in series with a $0.25\,\mu\text{F}$ capacitor across a $200\,\text{V}$ d.c. supply. A gas discharge tube with a striking voltage of $120\,\text{V}$ is connected across the capacitor. Determine (a) the frequency of the oscillation, and (b) the average power required from the d.c. supply.
Also sketch three cycles of the output voltage.

The circuit is shown in Fig. 6.6(a). The operation of the circuit is briefly as follows. The capacitor charges up exponentially through the resistor until the

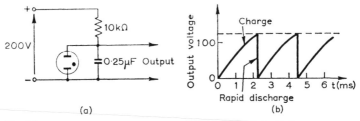

(a)   (b)

Fig. 6.6

voltage across it is $120\,\text{V}$. Up to this point the gas discharge tube is inoperative, but at this voltage the gas becomes conducting, and the capacitor discharges quickly through it. The voltage across the capacitor falls rapidly to zero, and the discharge tube again becomes non-conducting, so allowing the capacitor to charge up once more, and so on. The voltage across the capacitor is thus the saw-tooth wave shown in Fig. 6.6(b), the discharge time being so small that it may be taken as instantaneous.

(a) In order to determine the repetition frequency, the time taken to charge the capacitor to $120\,\text{V}$ must be found.

$$V_c = 120 = 200(1 - e^{-(100/0.25)t})$$

$$200\,e^{-400t} = 80$$

$$e^{400t} = \frac{200}{80} = 2.5$$

$$400t = \log_e 2.5 = 0.916$$

Therefore $t = 0.00229\,\text{s}$, so that the repetition frequency f is

$$f = \frac{1}{t} = 436\,\text{Hz}$$

(b) The energy taken from the d.c. supply per cycle is $\int_0^{0.00229} Vi\,dt$, where i is the instantaneous current in the circuit. This current is given by

$$i = \frac{V}{R}e^{-(1/CR)t} = \frac{200}{10,000}e^{-400t} = 0.02e^{-400t}$$

Thus

$$\text{Energy per cycle} = \int_0^{0.00229} 200 \times 0.02e^{-400t}\,dt$$

$$= 4\left[-\frac{1}{400}e^{-400t}\right]_0^{0.00229}$$

$$= 0.006\,\text{J}$$

Hence the average power taken from the supply is $0.006/0.00229$ J/s, or $2.62\,\text{W}$.

6.8 Transients in Capacitive A.C. Circuits

When an alternating voltage is applied to a capacitive circuit the resultant current may be determined by a method similar to that employed in Section 6.4 for inductive a.c. transients. An expression for the instantaneous value of the steady-state term is found from normal circuit theory. This is substituted in eqn. (6.19), and the constant of integration is then found from the known initial conditions.

6.9 Thermal Transients

In electrical apparatus the losses cause a rise in temperature with a final value determined by the magnitude of the losses and the rate of cooling. The rate of cooling approximately obeys *Newton's law of cooling*, which states that the rate of loss of heat from a body is proportional to the temperature rise, θ, of the body above the ambient temperature (i.e. the temperature of the surroundings). If the loss power is P watts, then in dt seconds the energy supplied as heat is $P\,dt$ joules. Suppose that in this time the temperature of the body rises from a value θ above ambient to $(\theta + d\theta)$, and that the heat stored in the body per deg C rise in temperature is H joules. Then

$$\begin{bmatrix}\text{Energy supplied as heat} \\ \text{in } dt \text{ seconds}\end{bmatrix} = \begin{bmatrix}\text{Heat stored in} \\ dt \text{ seconds}\end{bmatrix} + \begin{bmatrix}\text{Heat lost in} \\ dt \text{ seconds}\end{bmatrix}$$

or

$$P\,dt = H\,d\theta + K\theta\,dt$$

since $K\theta$ is, by Newton's law, the rate of heat loss, where K is a constant which depends on the convection, conduction and radiation heat loss. Hence

$$P = H\frac{d\theta}{dt} + K\theta \tag{6.29}$$

Comparing this with eqn. (6.1), it will be seen that the solution consists of a steady-state term, θ_s, and a transient term, θ_t, so that, solving for θ as in section 6.1,

$$\theta = \theta_s + Be^{-(K/H)t} = \theta_s + Be^{-t/\tau} \tag{6.30}$$

where B is a constant which is determined from known conditions, and τ ($= H/K$) is the thermal time-constant.

HEATING CURVE

Consider a constant loss power, P, existing in an electrical apparatus. When a steady temperature has been reached (i.e. when the rate of heat loss is just equal to the rate of heat supplied), the value of $d\theta/dt$ will be zero, so that, from eqn. (6.29), the steady temperature rise, θ_s, attained will be

$$\theta_s = \frac{P}{K} \text{ kelvins or degrees Celsius above ambient}$$

At the instant when the apparatus is switched on its temperature will be simply the ambient temperature. This can be expressed mathematically by writing

at $t = 0$, $\theta = 0$

Substituting this in eqn. (6.30) gives a particular solution from which B can be evaluated. Thus

$$0 = \theta_s + Be^0 \quad \text{or} \quad B = -\theta_s$$

The complete solution for θ is obtained by substituting this value of B back in eqn. (6.30) to give

$$\theta = \theta_s(1 - e^{-t/\tau}) \tag{6.31}$$

or

$$\theta = \frac{P}{K}(1 - e^{-(K/H)t}) \tag{6.31a}$$

This is an exponential growth curve, as shown in Fig. 6.7(*a*).

It should be noted that, in order to reduce θ_s, the loss power P must be reduced or the cooling constant K must be increased.

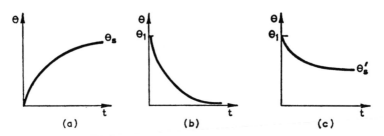

Fig. 6.7 THERMAL TRANSIENTS

COOLING CURVE

If a piece of apparatus which has attained a temperature θ_1 is allowed to cool, by switching off the supply, then $P = 0$ and the final temperature rise is zero (i.e. $\theta_s = 0$). Also at $t = 0$, $\theta = \theta_1$ and for this value of t, eqn. (6.30) becomes

$$\theta_1 = Be^0 \quad \text{or} \quad B = \theta_1$$

Hence the general expression for θ is

$$\theta = \theta_1 e^{-t/\tau} \tag{6.32}$$

This is the equation to the cooling curve of Fig. 6.7(*b*).

Suppose, however, that when the apparatus has attained some temperature θ_1, the loss power is reduced to some lower value P'. As before, the steady-state temperature rise is obtained from eqn. (6.29) by putting $d\theta/dt = 0$, so that now

$$\theta'_s = \frac{P'}{K}$$

Also at $t = 0$ the temperature rise is θ_1, and substituting this condition in eqn. (6.30) gives

$$\theta_1 = \theta'_s + Be^0 \quad \text{or} \quad B = (\theta_1 - \theta'_s)$$

so that the general solution is

$$\theta = \theta'_s + (\theta_1 - \theta'_s)e^{-t/\tau} \tag{6.33}$$

as is shown in Fig. 6.7(*c*)

6.10 Double-energy Transients

Circuits containing both inductance and capacitance involve both electromagnetic and electrostatic stored energies, and hence any

sudden change in the circuit conditions will involve the redistribution of two forms of stored energy. The transient currents resulting from this redistribution are called double-energy transients.

Consider the general series circuit of resistance R, inductance L
and capacitance C. If v_R, v_L, v_C are the instantaneous voltages
across R, L and C respectively, then the supply voltage, v, is given by

$$v = v_R + v_L + v_C$$

$$= iR + L\frac{di}{dt} + \frac{q}{C} \tag{6.34}$$

where i is the instantaneous circuit current and q is the instantaneous
charge on the capacitor. As before, the complete solution of this
equation will have two parts—the steady-state current i_s, and the
transient current i_t. The steady-state current may readily be obtained
from normal circuit theory, while the transient current is the solution
of the equation

$$i_t R + L\frac{di_t}{dt} + \frac{q_t}{C} = 0 \tag{6.35}$$

Differentiating,

$$L\frac{d^2 i_t}{dt^2} + R\frac{di_t}{dt} + \frac{i_t}{C} = 0$$

since $dq_t/dt = i_t$. On rearranging this equation it becomes

$$\frac{d^2 i_t}{dt^2} + \frac{R}{L}\frac{di_t}{dt} + \frac{i_t}{LC} = 0 \tag{6.36}$$

The solution of this equation for i_t may be obtained in several
ways. An operational method will be used here, the operator p
standing for d/dt and the operator p^2 standing for d^2/dt^2, i.e.

$$pi_t \equiv \frac{di_t}{dt} \quad \text{and} \quad p^2 i_t \equiv \frac{d^2 i_t}{dt^2}$$

Use will be made of the following operational relationships which
may be verified by differentiation.

(a) $(p - m)Ce^{mt} = \dfrac{d}{dt}(Ce^{mt}) - mCe^{mt} = 0$ \hfill (6.37)

(b) $(p - m)^2(B + Ct)e^{mt} = \dfrac{d^2}{dt^2}(B + Ct)e^{mt}$

$$- 2m\frac{d}{dt}(B + Ct)e^{mt} + m^2(B + Ct)e^{mt} = 0 \tag{6.38}$$

(c) $(p - m_1)(p - m_2)(Be^{m_1 t} + Ce^{m_2 t})$

$$= \frac{d^2}{dt^2}(Be^{m_1 t} + Ce^{m_2 t}) - (m_1 + m_2)\frac{d}{dt}(Be^{m_1 t} + Ce^{m_2 t})$$

$$+ m_1 m_2(Be^{m_1 t} + Ce^{m_2 t}) = 0 \quad (6.39)$$

where B and C are constants.

In the operational notation eqn. (6.36) becomes,

$$p^2 i_t + \frac{R}{L} p i_t + \frac{1}{LC} i_t = 0$$

i.e.

$$\left(p^2 + \frac{R}{L} p + \frac{1}{LC}\right) i_t = 0$$

The expression in brackets may be factorized to $(p - m_1)(p - m_2)$, where

$$m_1 = \frac{-R/L + \sqrt{(R^2/L_2 - 4/LC)}}{2} = \frac{-R}{2L} + \sqrt{\left(\frac{R^2}{4L^2} - \frac{1}{LC}\right)}$$

$$(6.40)$$

and

$$m_2 = \frac{-R/L - \sqrt{(R^2/L^2 - 4/LC)}}{2} = \frac{-R}{2L} - \sqrt{\left(\frac{R^2}{4L^2} - \frac{1}{LC}\right)}$$

$$(6.41)$$

Hence

$$(p - m_1)(p - m_2) i_t = 0 \quad (6.42)$$

Comparing eqns. (6.42) and (6.39),

$$i_t = Be^{m_1 t} + Ce^{m_2 t} \quad (6.43)$$

excepting where $m_1 = m_2 = m$, say, when eqn. (6.42) becomes

$$(p - m)^2 i_t = 0 \quad (6.44)$$

and hence, by comparison with eqn. (6.38),

$$i_t = (B + Ct)e^{mt} \quad (6.45)$$

According to the values of m_1 and m_2 four different conditions of the circuit are distinguishable.

CASE 1: LOSSLESS CIRCUIT: $R = 0$, i.e. UNDAMPED

In this case,

$$m_1 = \sqrt{\left(-\frac{1}{LC}\right)} = j\frac{1}{\sqrt{(LC)}} = j\omega' \quad \text{from eqn. (6.40)}$$

and

$$m_2 = -\sqrt{\left(-\frac{1}{LC}\right)} = -j\frac{1}{\sqrt{(LC)}} = -j\omega' \quad \text{from eqn. (6.41)}$$

Eqn. (6.43) gives the solution for i_t as

$$i_t = Be^{j\omega't} + Ce^{-j\omega't} \tag{6.46}$$

But

$$e^{j\omega't} = \cos \omega't + j \sin \omega't$$

and

$$e^{-j\omega't} = \cos \omega't - j \sin \omega't$$

Therefore

$$i_t = D \cos \omega't + E \sin \omega't \tag{6.47}$$

where

$$D = (B + C) \quad \text{and} \quad E = j(B - C)$$

Eqn. (6.47) may be still further reduced to

$$i_t = I_m \sin (\omega't + \psi) \tag{6.47a}$$

where $I_m = \sqrt{(D^2 + E^2)}$ and $\psi = \tan^{-1} D/E$. Hence the transient current in this case is a sine wave of constant peak value, and of frequency $f' = \dfrac{1}{2\pi\sqrt{(LC)}}$ as shown in Fig. 6.8(a). It will be observed that the solution contains two constant terms, namely I_m and ψ, which must be determined in any particular case from a knowledge of *two* initial circuit conditions. These conditions are

(a) The initial current in the inductance.
(b) The initial voltage across the capacitance.

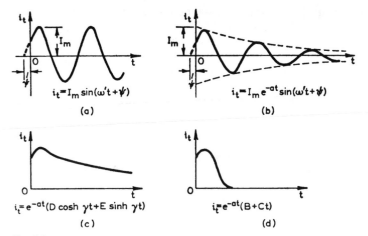

Fig. 6.8 TYPICAL DOUBLE-ENERGY TRANSIENT WAVEFORMS

 (a) Undamped
 (b) Underdamped
 (c) Overdamped
 (d) Critically damped

CASE 2: LOW-LOSS CIRCUIT: $R^2/4L^2 < 1/LC$, i.e. UNDERDAMPED

In this case, as in the theoretical Case 1, the term under the square root sign in eqns. (6.40) and (6.41) is negative, so that m_1 and m_2 will be conjugate complex numbers. Let

$$m_1 = -a + j\omega'$$

where

$$a = \frac{R}{2L} \quad \text{and} \quad \omega' = \sqrt{\left(\frac{1}{LC} - \frac{R^2}{4L^2}\right)}$$

Then

$$m_2 = -a - j\omega'$$

Substituting these values in eqn. (6.43),

$$i_t = Be^{(-a+j\omega')t} + Ce^{(-a-j\omega')t} = e^{-at}(Be^{j\omega't} + Ce^{-j\omega't})$$

This may be reduced to

$$i_t = I_m e^{-at} \sin(\omega't + \psi) \tag{6.48}$$

where I_m and ψ are constants. This is the equation of a damped oscillation as shown in Fig. 6.8(b).

The factor e^{-at}, which accounts for the decay of the oscillation, is called the *damping factor*. The ratio between successive positive (or negative) peak values of the oscillation is $1:e^{-a\tau}$, where τ is the period of the oscillation.

The frequency of the damped oscillation is

$$f' = \frac{1}{2\pi}\sqrt{\left(\frac{1}{LC} - \frac{R^2}{4L^2}\right)}$$

and is called the *natural frequency* of the circuit. In a great many cases $\dfrac{R^2}{4L^2} \ll \dfrac{1}{LC}$, and with quite sufficient accuracy $f' = \dfrac{1}{2\pi\sqrt{(LC)}}$

CASE 3. HIGH-LOSS CIRCUIT: $R^2/4L^2 > 1/LC$: i.e. OVERDAMPED

If $R^2/4L^2$ is greater than $1/LC$, then the term under the square root sign in eqns. (6.40) and (6.41) will be positive, so that m_1 and m_2 will be pure numbers. Let

$$m_1 = -a + \gamma'$$

where

$$a = \frac{R}{2L} \quad \text{and} \quad \gamma' = \sqrt{\left(\frac{R^2}{4L^2} - \frac{1}{LC}\right)}$$

Then

$$m_2 = -a - \gamma'$$

Therefore

$$i_t = Be^{(-a+\gamma')t} + Ce^{(-a-\gamma')t} \quad \text{(from eqn. (6.43))}$$
$$= e^{-at}(Be^{\gamma't} + Ce^{-\gamma't})$$

But

$$e^{\gamma't} = \sinh\gamma't + \cosh\gamma't$$

and

$$e^{-\gamma't} = \cosh\gamma't - \sinh\gamma't$$

Therefore

$$i_t = e^{-at}\{(B + C)\cosh\gamma't + (B - C)\sinh\gamma't\}$$
$$= e^{-at}\{D\cosh\gamma't + E\sinh\gamma't\} \tag{6.49}$$

A typical curve of this equation is shown in Fig. 6.8(c).

CASE 4. CRITICAL DAMPING: $R^2/4L^2 = 1/LC$

When $R^2/4L^2$ is equal to $1/LC$, m_1 and m_2 become equal, each having a value of $-R/2L$. Hence, from eqn. (6.45),

$$i_t = e^{-(R/2L)t}(B + Ct) \tag{6.50}$$

In this case i_t reduces to almost zero in the shortest possible time. A typical curve for i_t is shown in Fig. 6.8(*d*).

To summarize,

$$\text{Transient term is oscillatory if } R < 2\sqrt{\frac{L}{C}} \qquad (6.51)$$

$$\text{Transient term is non-oscillatory if } R \geqslant 2\sqrt{\frac{L}{C}} \qquad (6.52)$$

$$\text{Critical damping occurs when } R = 2\sqrt{\frac{L}{C}} \qquad (6.53)$$

EXAMPLE 6.4 A $4\mu F$ capacitor is discharged suddenly through a coil of inductance 1 H and resistance $100\,\Omega$. If the initial voltage on the capacitor is 10 V, derive an expression for the resulting current, and find the additional resistance required to give critical damping.

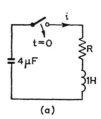 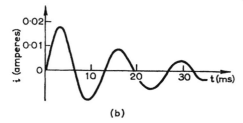

(a)   (b)

Fig. 6.9

The circuit is shown in Fig. 6.9(*a*). Since there is no generator in the circuit the steady-state current must be zero, so that the resultant current is simply the transient current.

The value of $2\sqrt{(L/C)}$ is 1,000, and hence from the inequality (6.51) the circuit is originally oscillatory. The transient current is therefore

$$i_t = I_m e^{-at} \sin(\omega' t + \psi)$$

where

$$a = \frac{R}{2L} = \frac{100}{2} = 50$$

and

$$\omega' = \sqrt{\left(\frac{1}{LC} - \frac{R^2}{4L^2}\right)} = \sqrt{(250{,}000 - 2{,}500)} = 497 \, \text{rad/s}$$

Therefore

$$i_t = I_m e^{-50t} \sin(497t + \psi) = i \qquad (6.54)$$

The two known initial conditions are (*a*) at $t = 0$, $i = 0$, and (*b*) at $t = 0$, capacitor voltage, $v_C = 10 \, \text{V}$. Applying condition (*a*) to equation (6.54) gives

$$0 = I_m \sin \psi$$

whence

$$\psi = 0$$
$$i = I_m e^{-50t} \sin 497t \qquad (6.55)$$

At time $t = 0$, the voltage across the inductor must be 10 V (from condition (b)), since the current through the resistor is zero, i.e.

$$L\frac{di}{dt}\bigg|_{t=0} = 10\,\text{V}$$

Therefore

$$\frac{di}{dt}\bigg|_{t=0} = \frac{10}{L} = 10\,\text{A/s} \qquad (6.56)$$

But from eqn. (6.55),

$$\frac{di}{dt} = -50I_m e^{-50t} \sin 497t + 497I_m e^{-50t} \cos 497t$$

At $t = 0$, this becomes

$$\frac{d}{dt}\bigg|_{t=0} = 497I_m = 10\,\text{A/s (from eqn. (6.56))}$$

Therefore

$$I_m = \frac{10}{497} = 0{\cdot}0201\,\text{A}$$

Hence the general expression for the current is

$$i = 0{\cdot}0201 e^{-50t} \sin 497t \text{ amperes}$$

The first few cycles of this current are shown in Fig. 6.9(b).
From eqn. (6.53) the total resistance required for critical damping is

$$R = 2\sqrt{\frac{L}{C}} = 1{,}000\,\Omega$$

Therefore the additional resistance required is $900\,\Omega$.

EXAMPLE 6.5 A damped oscillation is given by the equation

$$i = 100e^{-10t} \sin (500t) \text{ amperes}$$

Determine the number of oscillations which will occur before the amplitude decays to $\frac{1}{10}$th of its undamped value.

The decay of the peak of the oscillations is given by the term $100e^{-10t}$. Thus

$$\tfrac{1}{10} \times 100 = 100e^{-10t_1}$$

where t_1 is the time required for the oscillation to die to $\frac{1}{10}$th of its undamped value.

$$e^{10t_1} = 10$$
$$10t_1 = \log_e 10 = 2{\cdot}303$$
$$t_1 = 0{\cdot}2303\,\text{s}$$
$$\text{Frequency of oscillation} = \frac{500}{2\pi}\,\text{Hz}$$

Therefore the number of oscillations before decay to $\frac{1}{10}$th amplitude is

$$n = 0.2303 \times \frac{500}{2\pi} = \underline{\underline{18.4}}$$

6.11 Energy Transformations

With the underdamped or oscillatory transients discussed in Section 6.10 it will be noticed that at some instants the current is zero. At these instants there will be no energy stored in the magnetic field. However, at succeeding instants current does flow and there must be an associated magnetic energy. This energy must be stored as electrostatic energy during the instants when the current is zero, so that there is a continuous transformation of energy from an electrostatic to an electromagnetic form, and vice versa. Consider the circuit illustrated in Fig. 6.10. It is assumed for simplicity that

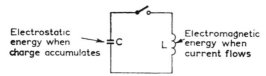

Fig. 6.10 ENERGY TRANSFORMATION

the circuit is lossless ($R = 0$). Initially the capacitor is charged to a potential V volts and the switch is closed at the instant $t = 0$. The steady-state current must be zero, and

$$i_t = I_m \sin (\omega' t + \psi) \quad \text{(from eqn. (6.47))}$$

At $t = 0$, $i = 0$ due to the inductor action; therefore $\psi = 0$. Hence

$$i = I_m \sin \omega' t \qquad (6.57)$$

At $t = 0$, $i = 0$, magnetic energy $= 0$, electrostatic energy $= \frac{1}{2}CV^2$.

At $t = \pi/2\omega'$, $i = I_m$, magnetic energy $= \frac{1}{2}LI_m^2$, electrostatic energy $= 0$.

At $t = \pi/\omega'$, $i = 0$, magnetic energy $= 0$, electrostatic energy $= \frac{1}{2}CV^2$.

Since there is assumed to be no energy loss from this circuit and no energy supplied after $t = 0$, the peak stored magnetic energy must equal the peak stored electrostatic energy:

$$\tfrac{1}{2}CV^2 = \tfrac{1}{2}LI_m^2 \qquad (6.58)$$

If the circuit is of the low-loss type rather than the imaginary lossless type, then at each interchange of energy there is a small loss of energy from the system. Many low-loss circuits may be regarded as obeying eqn. (6.58) for one energy cycle.

Example 6.6 If a break occurs at the point X in the circuit of Fig. 6.11, determine the voltage across the break. It may be assumed that, prior to the break, the circuit current had reached a steady-state value.

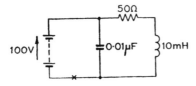

Fig. 6.11

Before the break, inductor current $= 2\,A$

Energy initially stored in inductor $= \tfrac{1}{2}LI^2$

$$= \tfrac{1}{2} \times 10^{-2} \times 4 = 2 \times 10^{-2}\,J$$

Energy initially stored in capacitor $= \tfrac{1}{2}CV^2$

$$= \tfrac{1}{2} \times 10^{-8} \times 10^4\,J \quad \text{(negligible)}$$

After the break occurs the energy initially stored in the magnetic field of the inductor will be transferred to the capacitor. Neglecting energy loss in the first transfer,

Maximum energy stored in capacitor $= 2 \times 10^{-2}\,J$

Therefore

$$\tfrac{1}{2}CV_m^2 = 2 \times 10^{-2}$$

and

Peak voltage across capacitor $= V_m = \sqrt{\left(\dfrac{2 \times 10^{-2} \times 2}{10^{-8}}\right)} = 2{,}000\,V$

Therefore

Maximum voltage across break $= 2{,}000 + 100 = 2{,}100\,V$

The voltage will be oscillatory as the energy alternates between the inductor and the capacitor.

Frequency of voltage oscillations $= f = \dfrac{1}{2\pi\sqrt{(LC)}}$

$$= \dfrac{1}{2\pi\sqrt{(10^{-2} \times 10^{-8})}} = 15{,}900\,Hz$$

Damping factor $= e^{-(R/2L)t} = e^{-(50/2 \times 10^{-2})t} = e^{-2{,}500t}$

Therefore

Voltage across break $= (2{,}000\,e^{-2{,}500t}\sin 10^5 t + 100)\,volts$

PROBLEMS

6.1 The field circuit of an alternator has an effective inductance of 100 H and a resistance of 10 Ω. Calculate the time required to increase the excitation current from I to 99 per cent of $2I$ amperes if the supply voltage is doubled.

Ans. 39·1 s.

6.2 Deduce from first principles an expression for the current growth in an inductive circuit.

A 15 H inductance coil of 10 Ω resistance is suddenly connected to a 20 V d.c. supply. Calculate

(a) the initial rate of change of current,
(b) the current after 2 s,
(c) the rate of change of current after 2 s,
(d) the energy stored in the magnetic field in this time,
(e) the energy lost as heat in this time, and
(f) the time constant. (*H.N.C.*)

Ans. 1·33 A/s; 1·47 A; 0·352 A/s; 16·3 J; 19·5 J; 1·5 s.

6.3 Derive an expression for the value of the current in a circuit of resistance R ohms and inductance L henrys, t seconds after the sudden application of a constant voltage V to the circuit.

A constant voltage of 100 V is suddenly applied to a circuit of resistance 2 Ω and inductance 10 H. After 7·5 s the voltage is suddenly increased to 200 V. What will be the value of the current after a further 2 s?

Sketch the approximate shape of the current/time graph. (*H.N.C.*)

Ans. 59·0 A.

6.4 A coil of resistance r ohms and inductance L henrys is connected in parallel with an R-ohm resistor to a d.c. supply of V volts. After a "long period" the supply is suddenly disconnected; derive an expression for the current t seconds later.

6.5 Sketch the shape of the graph showing the growth of current in an inductive circuit when a steady voltage is applied.

Explain the desirability of a discharge resistance when such a circuit is switched off.

What must be the greatest permissible value of the suppressor resistance used in conjunction with a 500 V field circuit having a resistance of 50 Ω in order that the voltage across the terminals of the field winding shall not exceed 750 V when the circuit is opened? (*C. & G. Inter.*)

Ans. 75 Ω.

6.6 A 12 μF capacitor is allowed to discharge through its own leakage resistance, and a fall of p.d. from 120 V to 100 V is recorded in 300 s by an electrostatic voltmeter. Calculate the leakage resistance of the capacitor. Prove any formula used. (*L.U. Part I*)

Ans. 137 MΩ.

6.7 A 2 μF capacitor is charged to 100 V. It is then discharged through a 1 MΩ resistor in parallel with a 1 μF capacitor. Find the voltage across the resistor 2 s after connexion, and also determine the energy dissipated up to this time.

Ans. 34·2 V, 0·00825 J.

6.8 A $1\,\mu F$ capacitor is charged from a 2 V, d.c. supply and is then discharged through a $10\,M\Omega$ resistor. After 5 s of discharge the capacitor is connected across a ballistic galvanometer of negligible resistance, and causes a deflection of 1·2 divisions. Calculate

    (a) the voltage across the capacitor after the 5 s of slow discharge,
    (b) the sensitivity of the galvanometer in microcoulombs per division, and
    (c) the energy expended in heating the $10\,M\Omega$ resistor during the 5 s of discharge. *(H.N.C.)*

Ans. (a) 1·21 V; (b) $1\,\mu C$/div; (c) $1\cdot27 \times 10^{-6}$ J.

6.9 A single-phase 50 Hz transformer fed from an "infinite" supply has an equivalent impedance of $(1 + j10)$ ohms referred to the secondary. The open-circuit secondary voltage is 200 V. Find

    (a) the steady-state secondary short-circuit current,
    (b) the transient secondary short-circuit current assuming that the short-circuit occurs at the instant when the voltage is passing through zero going positive.
    (c) the total short-circuit current under the same conditions.

Plot these curves to a base of time for the period of three cycles from the instant when the short-circuit occurs. Neglect saturation. *(H.N.C.)*

Ans. $28\cdot3 \sin (314t - 84\cdot3°)$; $28\cdot3\,e^{-31\cdot4t} \sin 84\cdot3°$; $28\cdot3(e^{-31\cdot4t} + \sin (314t - 84\cdot3°))$.

6.10 A single-phase 11,000/1,100 V 50 Hz transformer is supplied from 11 kV "infinite" busbars. The transformer leakage impedance referred to the low-voltage side is $0\cdot08\,\Omega$ resistance and $0\cdot8\,\Omega$ reactance. Calculate

    (a) the r.m.s. steady-state short-circuit current which might develop on the secondary side for a short-circuit at the secondary terminals,
    (b) the corresponding initial transient short-circuit current assuming the short-circuit to occur at the "worst" instant in the voltage cycle,
    (c) the instantaneous total current magnitude at the instant 0·04 s after the short-circuit has occurred.

What are the "worst" and "best" instants in the voltage cycle? *(H.N.C.)*

Ans. (a) 1,370 A; (b) ±1,930 A, (c) ±1,380 A.

6.11 A coil having a resistance R ohms and an inductance L henrys is suddenly connected to a voltage of constant r.m.s. value V and varying in time according to the law $v = V_m \sin 2\pi ft$. Deduce an expression for the current at any instant.

In the case of an alternator suddenly short-circuited explain how the expression for the current would differ from the above and enumerate the factors causing the difference. *(L.U.)*

6.12 A $4\,\mu F$ capacitor is initially charged to 300 V. It is discharged through a 10 mH inductance and a resistor in series. Find

    (a) the frequency of the discharge if the resistance is zero,
    (b) how many cycles at the above frequency would occur before the discharge oscillation decays to $\frac{1}{10}$th of its initial value if the resistance is $1\,\Omega$,
    (c) the value of the resistance which would just prevent oscillation. *(H.N.C.)*

Ans. 796 Hz; 36·6; 100 Ω.

6.13 Derive an expression for the instantaneous current in a circuit consisting of a resistance of R ohms in series with a capacitance of C farads at a time t

seconds after applying a sinusoidal voltage, the switch being closed t_1 seconds after the voltage had passed through its zero value.

If the values of R and C are 1,000 Ω and 10 μF respectively and if the voltage is 200 V at 50 Hz, calculate the value of the voltage at the instant of closing the switch such that no transient current is set up. (*L.U.*)

Ans. 269 V.

6.14 A capacitor C, initially charged to 350 V, is discharged through a coil of inductance 8 mH and resistance R ohms. The amplitude of the resulting oscillation dies away to 0·1 of its initial value in 2·3 ms. If the value of resistance for critical damping is 113·2 Ω, calculate the natural frequency and the actual value of R. (*H.N.C.*)

Ans. 1,130 Hz; 1·6 Ω.

6.15 An *RLC* series circuit has $R = 5$ Ω, $L = 10$ mH and $C = 400$ μF. Show that the current immediately after switch closure on to a direct-voltage source is oscillatory and of gradually decreasing amplitude. Calculate the frequency of oscillation and the damping factor. (Part question, *H.N.C.*)

Ans. 69 Hz; e^{-250t}.

6.16 A coil having a resistance of R ohms and an inductance of L henrys is connected in series with a resistance of R_1 ohms to a d.c. supply of V volts. After the current has reached a steady value, R_1 is short-circuited by a switch. Deduce, from first principles, an expression for the current in the coil at a time t seconds after the switch is closed.

If $L = 3$ H, $R = 50$ Ω and $R_1 = 30$ Ω, calculate the time taken for the current to increase by 10 per cent after the switch is closed. (*H.N.C.*)

Ans. 10·9 ms.

6.17 Discuss the factors which determine the kVA rating of a transformer.

A transformer has a heating time constant of 4 hours and its temperature rises by 18 °C from ambient after 1 hour on full load. The winding loss (proportional to I^2) is twice the core loss (constant) at full load. Estimate the final temperature rise in service after a consecutive loading of 1·5 hours at full load, 0·5 hour at one-half full load, and 1 hour at 25 per cent overload. (*H.N.C.*)

Ans. 46 °C.

Chapter **7**

ELECTRIC AND MAGNETIC FIELD THEORY

Electrostatic fields, magnetic fields and conduction fields exhibit similar characteristics, and all may be analysed by similar processes. In this chapter some linear fields of each type will be dealt with, linear fields being those which exist in materials which have constant electrical properties.

7.1 Streamlines and Current Tubes in Conduction Fields

A *conduction field* is the region in which an electric current flows,

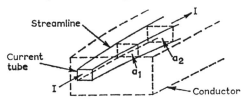

Fig. 7.1 STREAMLINES FORMING A CURRENT TUBE IN A CONDUCTION FIELD

and a *streamline* is a line drawn in such a field with a direction which is everywhere parallel to the direction of current flow; current will never flow across a streamline. If a series of streamlines is taken to enclose a tube of current as in Fig. 7.1, then the total

198

current across any cross-section of the tube will be the same. The most important cross-sections are those normal to the direction of current flow (such as a_1 and a_2 in the diagram), and these will be assumed to be taken unless otherwise stated.

Suppose the total current enclosed by the tube in Fig. 7.1 is I amperes. Then the current density at any point inside the tube is

$$J = \frac{I}{a} \quad \text{amperes/metre}^2$$

where a is the cross-sectional area of the current tube at the particular point. It is assumed that the current density within the tube is sufficiently uniform to be taken as constant at each cross-section. This will be the case if the tube is relatively small, or, in the limit, where $a \to 0$.

The current density at a point is a vector quantity having magnitude J (usually measured in amperes/metre^2), and having the same direction as the current or streamline at the point.

7.2 Equipotential Surfaces

The *potential* of a point in a conduction field is the work done in moving a unit charge from a specified point of zero potential (usually

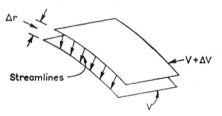

Fig. 7.2 RELATING TO ELECTRIC FIELD STRENGTH

the "earthed" point) to the point in question. There will in general be a large number of points with the same potential and the surface which contains these points is called an *equipotential surface*.

Since all points on an equipotential surface have, by definition, the same potential it follows that there will be no current flow between any points on the surface. Thus a streamline (which is in the direction of current flow) must intersect an equipotential surface at right angles. If it did not, there would then be at least a component of current flow along the equipotential surface which contradicts the previous statements. Thus the surfaces a_1 and a_2 in Fig. 7.1 are equipotential surfaces.

7.3 Electric Field Strength

Consider two equipotential surfaces at potentials V and $V + \Delta V$, where ΔV is a small potential step. The two surfaces lie a small distance Δr apart (Fig. 7.2). Since streamlines cross equipotentials normally, and since here the equipotentials are very close together, the streamlines may be taken to cross the intervening space normally.

Work done in moving a unit charge, or coulomb, across inter-space

$$= \Delta V \text{ volts} = -E \times \Delta r \quad \text{newton-metres/coulomb}$$

where E is the force per coulomb in the direction of a streamline. This is called the *electric field strength*.

$$\text{Electric field strength} = E \quad \text{newtons/coulomb}$$
$$= -\frac{\Delta V}{\Delta r} \quad \text{volts/metre}$$

As $\Delta r \to 0$,

$$E = -\frac{dV}{dr} \quad \text{volts/metre or newtons/coulomb} \tag{7.1}$$

The negative sign is included, since if ΔV is a positive increase in potential, the direction of the force on the charge will be towards the lower potential surface.

7.4 Relationship between Field Strength and Current Density

At a point in a conduction field both the field strength E and the current density J are vector quantities with the same direction. Then the conductivity σ at the point is given by the equation,

$$\sigma = \frac{J}{E} \quad \frac{\text{amperes/metre}^2}{\text{volts/metre}} \quad \text{or siemens/metre}$$

Also

$$\sigma = \frac{1}{\rho}$$

where ρ is the resistivity of the conductor. By Ohm's law, σ is a constant which is independent of J and E for most materials.

Consider a uniform conductor of overall length l metres and cross-sectional area a metres². If a voltage V across the conductor gives rise to a current I through it, then

$$\text{Current density in conductor, } J = \frac{I}{a} \quad \text{amperes/metre}^2$$

and

Potential gradient throughout the conductor, $E = \dfrac{V}{l}$ volts/metre

Therefore

$$\sigma = \frac{J}{E} = \frac{I}{V}\frac{l}{a} = \frac{l}{Ra} \tag{7.2}$$

where $R\left(= \rho\dfrac{l}{a}\right)$ is the total resistance of the uniform conductor.

7.5 Boundary Conditions

In almost every case a conductor has a well-defined edge. Outside this edge, or boundary, there is an insulating material whose conductivity is negligible compared with that of the conductor. In effect, then, no current will pass across the edge of the conductor, and hence the edge must be a streamline. Also, since equipotential surfaces intersect streamlines at right angles, it follows that these surfaces must cross the conductor edges normally.

7.6 Field Plotting Methods

If the boundaries of the field cannot be simply expressed mathematically, an approximate estimate of the conductance may be obtained by a "mapping" method. This applies to plane fields, i.e. fields whose variations may be represented on a flat plane (e.g. the field of two long parallel conductors). The field between two spheres could not be tackled simply by this method since there are variations in all planes.

The basis of the method is the division of the plane of the field into a number of squares formed between adjacent streamlines and adjacent equipotentials. Since the streamlines and equipotentials will in general be curved lines rather than straight lines, true squares will not be formed. However, since the streamlines and equipotentials intersect normally, "square-like" figures are formed—these are usually called *near*, or *curvilinear*, squares. The test for a given figure being a near square is that it should be capable of subdivision into smaller squares which tend to be true squares with equal numbers of the true squares along each side of the original square.

Fig. 7.3 shows a pair of adjacent equipotentials and a pair of adjacent streamlines forming a large square-like figure (90° corners).

The square-like figure is a near square since on successive sub-division by equal numbers of intermediate streamlines and equipotentials the smaller figures are seen to approach the true square form.

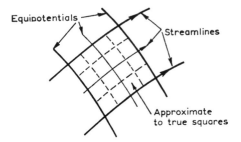

Fig. 7.3   FORMATION OF CURVILINEAR SQUARES

Consider a true square formed between an adjacent pair of streamlines and an adjacent pair of equipotentials (Fig. 7.4). Let this square be the end of d metres depth of the field so that the portion of the field behind the square forms a current tube between the

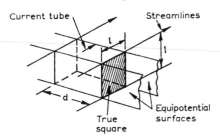

Fig. 7.4   CONDUCTANCE OF A CURRENT TUBE WITH SQUARE
CROSS-SECTION

adjacent equipotential surfaces. Let l be the length of a side of the square, and σ be the conductivity of the medium in which the current tube is situated.

$$\text{Conductance of the current tube} = \sigma \times \frac{\text{area}}{\text{length}} = \sigma \times \frac{d \times l}{l}$$

$$= \sigma d \text{ siemens}$$

Therefore the conductance of a tube whose end is a true square is σd mhos, independent of the size of the end square.

In Fig. 7.5 a rectangular block carrying a uniform current has been mapped into a number of tubes whose ends are true squares. The conductance of each tube is σd siemens, and

$$\text{Total conductance} = \sigma d \times \frac{m}{n} \tag{7.3}$$

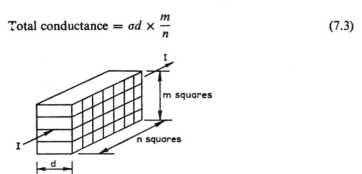

Fig. 7.5   CONDUCTANCE OF A RECTANGULAR BLOCK

where m is the number of parallel squares (across the direction of current flow), and n is the number of series squares (i.e. along the direction of current flow). Neither n nor m need be an integer.

In other cases the field will not be reducible to true squares but to near squares—the conductance per near-square-ended tube still being σd, and the total conductance still being given by eqn. (7.3). This follows since the definition of a near square is that it should be capable of subdivision to form small figures which approach to being true squares each corresponding to a conductance σd mhos (independent of size). Thus the conductance of any plane field may be estimated by a field plot which divides the field into a number of curvilinear squares. The plotting may be performed by eye as in succeeding examples. Alternatively, the electrolytic trough or the rubber membrane method may be used.

EXAMPLE 7.1   A 6 × 3 cm sheet of high-resistance conducting material of uniform depth is soldered to massive copper blocks at either end. Find the fractional increase in resistance when a thin slot is cut halfway across the sheet, as shown in Fig. 7.6.

Since the end pieces are of copper and of large section, voltage drops in them should be negligible compared with voltage drops in the material of the sheet, so that the ends of the sheet may be taken as equipotentials.

The vertical centre-line is, by symmetry in this particular case, an equipotential, and the conduction field on each side should be symmetrical. The edges are, of course, streamlines.

To solve the problem the area must be mapped by a series of equipotentials and streamlines forming good near squares. This may be performed by successive graphical approximation, as follows.

1. Draw in the middle streamline which divides the total current in half; this will start a little lower than mid-point at one edge (say at A), and cut the central equipotential a little above mid-point (say at B). C is similarly placed to A.

2. Draw in the two quarter streamlines DEF and GHI; note that XH < HB < BE < EY and RF < FC < CI < IT. The increases are judged by eye.

3. Commence drawing equipotentials cutting the above streamlines orthogonally, and as far as possible completing near squares. In this case start drawing the equipotentials either at one of the ends or at the central equipotential. The

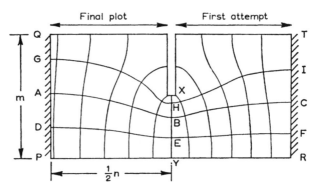

Fig. 7.6

right-hand side of the figure shows the plot at this stage. Manifestly all the areas are not near squares.

4. Either rub out the previously drawn streamlines, leaving the equipotentials, or better, take a tracing of the figure with equipotentials but not streamlines. Now redraw the streamlines cutting the equipotentials orthogonally and forming near squares. It will be unlikely that an integral number of near squares will fit into the total area. The resultant trace at this stage is probably fairly accurate and sufficiently good for most purposes. It is shown on the left-hand side of the plot. If, however, greater accuracy is required, the figure should be retraced excluding the equipotentials but including the new streamlines and so on. It must be emphasized that, while one retracing provides fair accuracy after some practice, the result without the retracing operation has a very poor accuracy and is unsatisfactory except for the roughest estimations. Less than 5 per cent error may often be attained with one retracing only.

Suppose there are n squares in the current direction and m squares normal to the current direction where m and n are not necessarily integral numbers.

The conductance per square element is σd siemens, where σ is the conductivity of the material and d is the uniform depth. Then

$$\text{Total conductance} = \sigma d \times \frac{m}{n} \text{ siemens} \tag{7.3}$$

From the field plot, $m = 4$ and $n = 2 \times 5 \cdot 1 = 10 \cdot 2$. Had the sheet been without the slot, the number of squares in the current direction would have been proportional to the length l and the number normal to the current direction proportional to the breadth b. Then

$$\text{Total conductance without slot} = \sigma d \times \frac{b}{l} \quad \text{siemens}$$

$$\frac{\text{Conductance with slot}}{\text{Conductance without slot}} = \frac{ml}{nb} = \frac{4}{10 \cdot 2} \times \frac{6}{3} = 0 \cdot 78$$

Therefore

$$\frac{\text{Resistance with slot}}{\text{Resistance without slot}} = \frac{1}{0 \cdot 78} = \underline{\underline{1 \cdot 28}}$$

The symmetry of the figure determines in each particular case whether it is better to start the field plot by drawing the streamlines or the equipotentials. The number of such lines must also be judged for each case on its own. If too few lines are chosen, then the result will be inaccurate owing to the difficulty of estimating near squares. If too many lines are chosen, the plot becomes too confused to be useful.

It will be noticed in the above example that a five-sided near square appears at the corners of the slot. This is admissible, since on further subdivision the figure will (in the limit) give true squares except for one square.

7.7 Streamlines and Tubes of Electric Flux in Electrostatic Fields

A streamline in an electrostatic field is a line so drawn that its direction is everywhere parallel to the direction of the electrostatic flux. It is also a line of force and has the same properties as a streamline in a conduction field. Several streamlines may be taken as enclosing a *tube of electric flux*. Let the total flux through a tube be Ψ coulombs; then the *electric flux density* at a point in the tube is

$$D = \frac{\Psi}{a} \quad \text{coulombs/metre}^2 \qquad (7.4)$$

where a square metres is the cross-section of the tube at the particular point and it is assumed that the flux density within the tube is sufficiently uniform to be taken as constant over that area.

7.8 Equipotential Surfaces and Electric Field Strength

Equipotential surfaces have the same definition and properties in electrostatic as in conduction fields. Hence the electric field strength in an electrostatic field is given by

$$E = -\frac{dV}{dr} \quad \text{volts/metre} \qquad (7.1)$$

where V is the potential at the point, and r is in the direction in which E is measured.*

* Electric field strength is also often called *electric intensity, electric stress* or *potential gradient*.

7.9 Relationship between Field Strength and Flux Density

At all points in an electrostatic field both the field strength, E, and the electric flux density, D, are vector quantities in the direction of the streamline through the point. From electrostatic theory,

$$D = \epsilon E$$

where ϵ is the permittivity of the dielectric material, and

$$\epsilon = \epsilon_0 \epsilon_r = \frac{1}{36\pi \times 10^9} \epsilon_r$$

ϵ_r being the relative permittivity of the material. D is measured in coulombs per square metre, and E is measured in volts per metre (or newtons per coulomb).

7.10 Boundary Conditions

Usually an electrostatic field is set up in the insulating medium between two good conductors. The conductors have such a high conductivity that any voltage drop within the conductors (except at very high frequencies) is negligible compared with the potential differences across the insulator. All points in the conductors are therefore at the same potential so that the conductors form the boundary equipotentials for the electrostatic field.

Streamlines, which must cut all equipotentials at right angles, will leave one boundary at right angles, pass across the field, and enter the other boundary at right angles. Curvilinear squares may again be formed between the streamlines and the equipotentials.

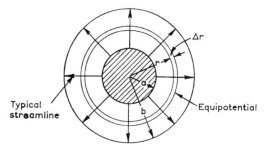

Fig. 7.7 CAPACITANCE BETWEEN CONCENTRIC CYLINDERS

7.11 Capacitance between Concentric Cylinders

The electrostatic field between two concentric conducting cylinders is illustrated in Fig. 7.7. The boundary equipotentials are concentric cylinders of radii a and b, and the streamlines are radial lines cutting the equipotentials normally.

Let Q' be the charge per unit length (coulombs per metre run) of the inner conductor. Then the total electric flux, Ψ, across the dielectric per unit length is Q' coulombs per metre. This total flux will pass through the elemental cylinder of thickness Δr at radius r and a distance of 1 metre into the plane of the paper. Thus

$$\text{Flux density at radius } r = D = \frac{\Psi}{A} = \frac{Q'}{2\pi r \times 1}$$

Therefore

$$\text{Electric stress at radius } r = E = \frac{D}{\epsilon} = \frac{Q'}{2\pi\epsilon r} \qquad (7.5)$$

Also

$$\text{Potential difference across element} = \Delta V$$

Therefore

$$\Delta V = -E\Delta r = -\frac{Q'}{2\pi\epsilon r}\,\Delta r$$

$$\text{Total potential difference between boundaries, } V = -\frac{Q'}{2\pi\epsilon}\int_b^a \frac{dr}{r}$$

Therefore

$$V = \frac{Q'}{2\pi\epsilon}\log_e\frac{b}{a} \quad \text{volts} \qquad (7.6)$$

But

$$\text{Capacitance per unit length} = C' = \frac{\text{Charge per unit length}}{\text{Potential difference}}$$

Therefore

$$C' = \frac{Q'}{V} = \frac{2\pi\epsilon}{\log_e\dfrac{b}{a}} \quad \text{farads/metre} \qquad (7.7)$$

7.12 Electric Stress in a Single-core Cable

A single-core cable with a metal sheath has the same electrostatic field as a pair of concentric cylinders. From eqn. (7.6),

$$\frac{Q'}{2\pi\epsilon} = \frac{V}{\log_e\dfrac{b}{a}}$$

Substituting this expression in eqn. (7.5),

$$E = \frac{V}{r \log_e \dfrac{b}{a}} \tag{7.8}$$

Thus the stress at any point in the dielectric varies inversely as r, and will have a maximum value at the minimum radius, i.e. when $r = a$.
  Thus

$$E_{max} = \frac{V}{a \log_e \dfrac{b}{a}} \tag{7.9}$$

  When designing a cable it is important to obtain the most economical dimensions. The greater the value of the permissible maximum stress, E_{max}, the smaller the cable may be for a given voltage V. The maximum permissible stress, however, is limited to the safe working stress for the dielectric material. With V and E_{max} both fixed, the relationship between b and a will be given by

$$a \log_e \frac{b}{a} = \frac{V}{E_{max}} = R$$

where R is a constant. Therefore

$$\log_e \frac{b}{a} = \frac{R}{a}$$

i.e.

$$b = a\mathrm{e}^{R/a} \tag{i}$$

  For the most economical cable, b will be a minimum, and hence

$$\frac{db}{da} = 0 = \mathrm{e}^{R/a} + a\left(-\frac{R}{a^2}\right)\mathrm{e}^{R/a}$$

i.e.

$$a = R = V/E_{max} \tag{7.10}$$

and from (i)

$$b = a\mathrm{e} = 2{\cdot}718a \tag{7.11}$$

  The core diameter given by eqn. (7.10) is usually found to give a larger conductor cross-sectional area than is necessary for the economical transmission of current through a high-voltage cable. The high cost of the unnecessary copper may be reduced by (*a*) making the core of hollow construction, or (*b*) having a hemp-cord centre for the core, or (*c*) constructing the centre of the core of a cheaper metal.

EXAMPLE 7.2 A single-core concentric cable is to be manufactured for a 100 kV 50 Hz transmission system. The paper used has a maximum permissible safe stress of 10^7 V/m (r.m.s.) and a dielectric constant (relative permittivity) of 4. Calculate the dimensions for the most economical cable, and the charging current per kilometre run with this cable.

By eqn. (7.10),

$$\text{Core radius} = a = \frac{V}{E_{max}} = \frac{10^5}{10^7} = 10^{-2}\,\text{m} = 1\,\text{cm}$$

By eqn. (7.11),

Internal sheath radius $= b = \mathrm{e}a = \underline{\underline{2\cdot718\,\text{cm}}}$

By eqn. (7.7),

$$\text{Capacitance per metre} = C' = \frac{2\pi \times 4}{36\pi \times 10^9}\,\text{F}$$

since $\log_e (b/a) = \log_e \mathrm{e} = 1$. Thus

$$C' = 0\cdot222 \times 10^{-3}\,\mu\text{F/m}$$

Therefore

Capacitance for 1 km $= C = 0\cdot222\,\mu\text{F}$

and

Charging current per km $= V\omega C = 10^5 \times 314 \times 0\cdot222 \times 10^{-6} = \underline{\underline{7\,\text{A}}}$

7.13 Capacitance of an Isolated Twin Line

Fig. 7.8(*a*) shows the actual field distribution round a pair of oppositely charged long conductors each of radius *a*. The conductors are spaced so that the distance between their centres is *D*, where *a* and *D* are measured in the same units and $D \gg a$. Fig. 7.8(*b*) shows the field of each conductor separately, this being the field assumed for calculation purposes.

Let conductor A carry a charge of $+q'$ coulombs per metre length while conductor B is uncharged. Consider a cylindrical element of radius *r* about conductor A, of depth 1 metre and of thickness Δr.

Total electric flux through element, $\Psi = q'$

whence

$$\text{Electric flux density at element, } D' = \frac{\Psi}{\text{Area of element}}$$

$$= \frac{q'}{2\pi r \times 1}$$

Thus

Field strength at element, $E = \dfrac{D'}{\epsilon} = \dfrac{q'}{2\pi\epsilon r}$

and

Voltage drop across element $= -E\Delta r = -\dfrac{q'\Delta r}{2\pi\epsilon r}$

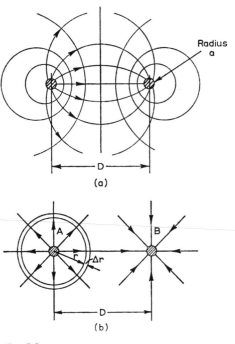

Fig. 7.8 FIELD BETWEEN PARALLEL CONDUCTORS
(*a*) Actual field
(*b*) Field of each conductor alone

Suppose that at a large radius R, the potential may be considered zero. Then

Potential of conductor A above zero $= -\dfrac{q'}{2\pi\epsilon}\displaystyle\int_R^a \dfrac{dr}{r}$

$$= \dfrac{q'}{2\pi\epsilon}\log_e \dfrac{R}{a} \text{ volts} \quad (7.12)$$

and

Potential at conductor B above zero $= -\dfrac{q'}{2\pi\epsilon}\displaystyle\int_R^D \dfrac{dr}{r}$

$$= \dfrac{q'}{2\pi\epsilon}\log_e \dfrac{R}{D}$$

since conductor B lies in the electrostatic field of conductor A.

Let conductor B carry a charge of $-q'$ coulombs per metre length while conductor A is uncharged. Then

Potential of conductor B below zero $= -\dfrac{q'}{2\pi\epsilon}\log_e \dfrac{R}{a}$

and

$\left.\begin{array}{l}\text{Potential at conductor A below zero}\\ \quad\text{due to the charge on conductor B}\end{array}\right\} = -\dfrac{q'}{2\pi\epsilon}\log_e \dfrac{R}{D}$

When both conductors are charged simultaneously,

Total potential of A above zero $= \dfrac{q'}{2\pi\epsilon}\left(\log_e \dfrac{R}{a} - \log_e \dfrac{R}{D}\right)$

$$= \dfrac{q'}{2\pi\epsilon}\log_e \dfrac{D}{a} \tag{7.13}$$

and

Total potential of B below zero $= -\dfrac{q'}{2\pi\epsilon}\log_e \dfrac{D}{a}$.

Therefore

Potential difference between A and B $= 2\dfrac{q'}{2\pi\epsilon}\log_e \dfrac{D}{a}$

Capacitance between A and B per metre length

$$= C' = \dfrac{q'}{\text{P.D. between A and B}}$$

Therefore

$$C' = \dfrac{1}{2}\dfrac{2\pi\epsilon}{\log_e \dfrac{D}{a}} \quad \text{farads/metre} \tag{7.14}$$

7.14 Potential and Electric Field Strength for an Isolated Twin Line

Consider a twin line where the potential difference between conductors is V volts and the system is balanced to earth so that the potential of conductor A with respect to earth is $\frac{1}{2}V$ and the potential of conductor B with respect to earth is $-\frac{1}{2}V$ (Fig. 7.9).

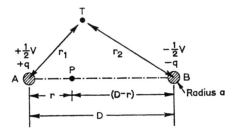

Fig. 7.9 ISOLATED TWIN LINE

By eqn. (7.13),

$$\text{Potential of conductor A above earth} = \frac{1}{2}V = \frac{q'}{2\pi\epsilon} \log_e \frac{D}{a}$$

Therefore

$$\frac{q'}{2\pi\epsilon} = \frac{V}{2 \log_e \dfrac{D}{a}}$$

Also, the potential at a point T, distant r_1 from conductor A, is

$$V_{r1} = -\frac{q'}{2\pi\epsilon} \int_R^{r_1} \frac{dr_1}{r_1} = \frac{q'}{2\pi\epsilon} \log_e \frac{R}{r_1} = \frac{V}{2 \log_e \dfrac{D}{a}} \log_e \frac{R}{r_1}$$

due to the change on conductor A. The point T will also have a potential due to the charge on conductor B. Suppose T is distant r_2 from conductor B; then

$$\text{Potential at T due to charge on B} = \frac{-V}{2 \log_e \dfrac{D}{a}} \log_e \frac{R}{r_2}$$

The total potential at T will be the sum of the potentials due to A and B separately. Hence

$$\text{Potential at T} = \frac{V}{2 \log_e \dfrac{D}{a}} \log_e \frac{r_2}{r_1} \tag{7.15}$$

The electric field strength has its highest value along the line joining the conductor centres, since on this line the forces due to each line charge will be acting in the same direction. At the point P, which is distant r from A, the potential will be

$$V_p = \frac{V}{2 \log_e \dfrac{D}{a}} \log_e \left(\frac{D-r}{r}\right)$$

Electric field strength at P, $E_p = -\dfrac{dV}{dr}$ (7.1)

$$= \frac{-V}{2 \log_e D/a} \frac{r}{D-r} \frac{\{-r-(D-r)\}}{r^2}$$

$$= \frac{V}{2 \log_e D/a} \frac{D}{r(D-r)} \quad \text{volts/metre} \quad (7.16)$$

From eqn. (7.16) the field strength E_p will have a maximum value when either r or $(D-r)$ has a minimum value. This occurs when $r = a$ or when $(D-r) = a$, i.e. at each conductor surface. The maximum field strength is almost independent of the conductor spacing, and is almost inversely proportional to the conductor radius. High-voltage conductors should not be of a small radius (even if the current is small) or the stress at the conductor surface will exceed the breakdown strength of the surrounding air.

7.15 Lines Above a Conducting Earth

In Section 7.10 it was seen that a good conductor would be an equipotential boundary for an electrostatic field. If a conductor is introduced into an electrostatic field in a random manner and with a random potential, then the field will be greatly altered, but if a conductor of negligible thickness and at the correct potential is introduced into a field so that it lies entirely in the corresponding equipotential surface then the field will be unaltered.

Fig. 7.10(a) shows the field between two long isolated parallel conductors, which have potentials of $+1,000$ V and $-1,000$V respectively. The edges of several equipotential surfaces between the conductors are drawn in. If a cylindrical conductor at a potential of 500 V were inserted in the field to coincide with the 500 V equipotential surface, then the resultant field would not be changed. It would, however, now be divided into two separate parts, screened from each other by the conducting cylinder, so that either the enclosed or the outer conductor could be removed without affecting the field in the opposite part.

The zero equipotential surface is a plane, and if a plane conducting sheet at zero potential is inserted in the field to coincide with this equipotential surface the field will be divided, as above, into two separate parts, each independent of the other. The negative conductor could therefore be removed and replaced by any other charged system, or by a solid conductor, without affecting the shape

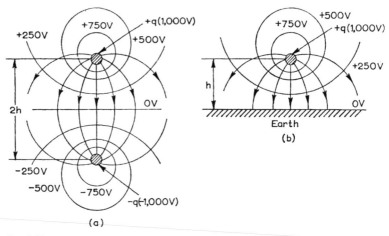

Fig. 7.10   FIELD BETWEEN A CONDUCTOR AND A PARALLEL PLANE

of the field between the positive conductor and the zero equipotential surface.

Fig. 7.10(*b*) shows one conductor at a height *h* metres above an infinite conducting earth. The shape of the field between the conductor and the earth must, by the preceding argument, be the same as the shape of the field which would exist above the zero-potential surface in an isolated twin-line system where the conductor spacing is 2*h*.

Thus, to analyse the field of a single charged conductor (or any other system) above a conducting earth, the earth is replaced by an image system, carrying the opposite charge to the real system and placed the same distance below earth as the real system is above earth. Of course, only the field above earth actually exists.

Consider the single wire of radius *a*, suspended at height *h* above a uniform conducting earth, and having a line charge of *q′* coulombs per metre run. The earth effect is represented by the image conductor with a line charge −*q′* as shown in Fig. 7.11. Then the potential of the actual wire to earth due to its own charge and to that on the image conductor is given by eqn. (7.13) as

$$V = \frac{q'}{2\pi\epsilon} \log_e \frac{2h}{a}$$

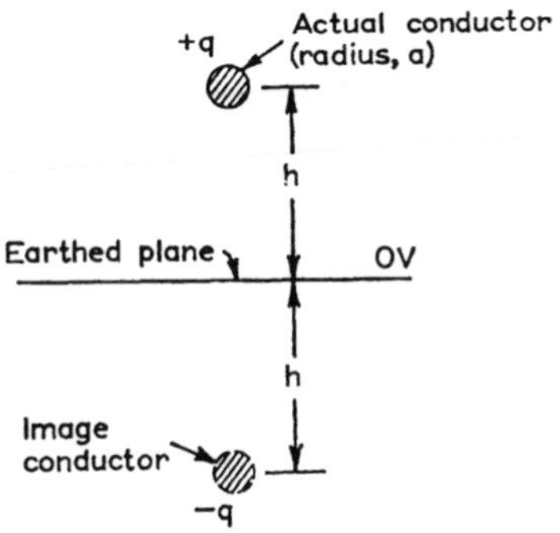

Actual conductor (radius, a)

+q

h

Earthed plane   0V

h

Image conductor   −q

Fig. 7.11   RELATING TO CAPACITANCE BETWEEN A CONDUCTOR AND A PARALLEL PLANE

where $2h$ is the distance between the wire and its image. Thus

$$\text{Capacitance to earth, } C' = \frac{q'}{V} = \frac{2\pi\epsilon}{\log_e \dfrac{2h}{a}} \quad \text{farads/metre}$$

(7.17)

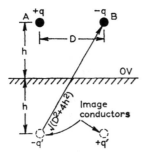

A +q   −q B

D

h

0V

h

$\sqrt{(D^2 + 4h^2)}$ Image conductors

−q'   +q'

Fig. 7.12   RELATING TO CAPACITANCE OF A TWIN LINE ABOVE A CONDUCTING PLANE

To find the capacitance for a twin line with each conductor of radius a at a height h above earth and with a line spacing of D, the earth effect is represented by the image pair as shown in Fig. 7.12. Let R be the large radius at which the potential is zero. Let a potential above zero be represented as positive and a potential below zero as negative. Then the total potential of conductor A, due to its own charge and to the charges on B and the image pair is

$$V_A = \frac{q'}{2\pi\epsilon} \log_e \frac{R}{a} - \frac{q'}{2\pi\epsilon} \log_e \frac{R}{D} - \frac{q'}{2\pi\epsilon} \log_e \frac{R}{2h}$$

$$+ \frac{q'}{2\pi\epsilon} \log_e \frac{R}{\sqrt{(D^2 + 4h^2)}}$$

$$= \frac{q'}{2\pi\epsilon} \log_e \left\{ \frac{D}{a} \frac{2h}{\sqrt{(D^2 + 4h^2)}} \right\}$$

Conductor B will be at a similar potential below zero, so that the p.d. between A and B will be twice V_A.

Capacitance per metre, $C' = \dfrac{q'}{2V_A}$

$$= \frac{1}{2} \frac{2\pi\epsilon}{\log_e \left\{ \dfrac{D}{a} \dfrac{2h}{\sqrt{(D^2 + 4h^2)}} \right\}} \quad \text{farads/metre} \quad (7.18)$$

7.16 Equivalent Phase Capacitance of an Isolated Three-phase Line

Fig. 7.13(a) shows the cross-section of a typical 3-phase line in which the conductors all have equal radii a, but different spacings D_{12}, D_{23} and D_{31}. There will actually be capacitances C_{12}', C_{23}' and C_{31}' per metre length between each pair of lines, but it is more convenient to represent these as equivalent phase capacitances C_1', C_2' and C_3' per metre length as shown in Fig. 7.13(b). With an irregular spacing C_1', C_2' and C_3' will be unequal.

Assuming that the line is part of a 3-wire system, then

$$i_1 + i_2 + i_3 = 0$$

where i_1, i_2 and i_3 are the instantaneous charging currents.

Then $\int(i_1 + i_2 + i_3)dt = \text{constant} = 0$, since in an a.c. system there are no constant charges. Thus

$$q_1 + q_2 + q_3 = 0 \text{ (total charge)}$$

or

$$q_1' + q_2' + q_3' = 0 \text{ (charge per metre)}$$

Therefore

$$q_2' + q_3' = -q_1 \tag{7.19}$$

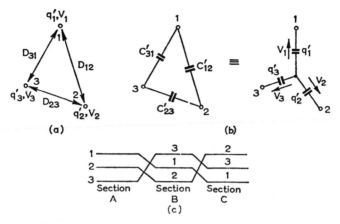

Fig. 7.13 CAPACITANCE OF A 3-PHASE LINE: UNIFORM TRANSPOSITION

The total potential of line 1 due to its own charge and to the charges on the other lines is

$$V_1 = \frac{q_1'}{2\pi\epsilon} \log_e \frac{R}{a} + \frac{q_2'}{2\pi\epsilon} \log_e \frac{R}{D_{12}} + \frac{q_3'}{2\pi\epsilon} \log_e \frac{R}{D_{31}}$$

$$= \frac{q_1'}{2\pi\epsilon} \log_e \frac{R}{a} + \frac{q_2'}{2\pi\epsilon} \log_e \frac{R}{D_{12}} - \frac{q_1' + q_2'}{2\pi\epsilon} \log_e \frac{R}{D_{31}}$$

i.e.

$$V_1 = \frac{q_1'}{2\pi\epsilon} \log_e \frac{D_{31}}{a} + \frac{q_2'}{2\pi\epsilon} \log_e \frac{D_{31}}{D_{12}}$$

In the same way, the total potential of line 2 due to its own charge and to the charges on the other lines is

$$V_2 = \frac{q_2'}{2\pi\epsilon} \log_e \frac{D_{23}}{a} + \frac{q_1'}{2\pi\epsilon} \log_e \frac{D_{23}}{D_{12}}$$

Therefore

$$V_1 - V_2 = \frac{q_1'}{2\pi\epsilon} \log_e \frac{D_{12}D_{31}}{aD_{23}} - \frac{q_2'}{2\pi\epsilon} \log_e \frac{D_{12}D_{23}}{aD_{31}} \qquad (7.20)$$

Also, from the equivalent circuit of star-connected capacitances of Fig. 7.13(b),

$$V_1 - V_2 = \frac{q_1'}{C_1'} - \frac{q_2'}{C_2'} \qquad (7.21)$$

From eqns. (7.20) and (7.21),

$$C_1' = \frac{2\pi\epsilon}{\log_e \dfrac{D_{12}D_{31}}{aD_{23}}} \quad \text{farads/metre} \tag{7.22}$$

$$C_2' = \frac{2\pi\epsilon}{\log_e \dfrac{D_{23}D_{12}}{aD_{31}}} \quad \text{farads/metre} \tag{7.23}$$

Similarly it can be shown that

$$C_3' = \frac{2\pi\epsilon}{\log_e \dfrac{D_{31}D_{23}}{aD_{12}}} \quad \text{farads/metre} \tag{7.24}$$

Fig. 7.13(c) illustrates a uniformly transposed line where line 1 runs for one-third of its length in position 1, one-third of its length in position 2 and one-third of its length in position 3. Conductors 2 and 3 are similarly transposed. Transposition of this nature is often adopted for practical lines since there is the obvious advantage of equalizing phase capacitances of the line, and also the advantage of minimizing stray potentials induced in parallel telephone lines or other conductors.

An approximate solution for the effective phase capacitance of a uniformly transposed line may be obtained by assuming that the charge per unit length of line is uniform despite the transpositions. The potential of a line must then change at each transposition. The approximate value of capacitance is based on finding the average value for the potential of a line. The method is approximate since each line is, in fact, equipotential. Successive transposed sections could have different potentials only if they were insulated from each other.

For section A and Fig. 7.13(c), the total potential of line 1 is

$$V_{1A} = \frac{q_1'}{2\pi\epsilon} \log_e \frac{R}{a} + \frac{q_2'}{2\pi\epsilon} \log_e \frac{R}{D_{12}} + \frac{q_3'}{2\pi\epsilon} \log_e \frac{R}{D_{31}}$$

For section B the total potential of line 1 is

$$V_{1B} = \frac{q_1'}{2\pi\epsilon} \log_e \frac{R}{a} + \frac{q_2'}{2\pi\epsilon} \log_e \frac{R}{D_{23}} + \frac{q_3'}{2\pi\epsilon} \log_e \frac{R}{D_{12}}$$

For section C the total potential of line 1 is

$$V_{1C} = \frac{q_1'}{2\pi\epsilon} \log_e \frac{R}{a} + \frac{q_2'}{2\pi\epsilon} \log_e \frac{R}{D_{31}} + \frac{q_3'}{2\pi\epsilon} \log_e \frac{R}{D_{23}}$$

The average potential of line 1 is

$$V_1 = \tfrac{1}{3}(V_{1A} + V_{1B} + V_{1C})$$

$$= \frac{1}{3}\left(\frac{3q_1'}{2\pi\epsilon}\log_e\frac{R}{a} + \frac{q_2'}{2\pi\epsilon}\log_e\frac{R^3}{D_{12}D_{23}D_{31}}\right.$$

$$\left. + \frac{q_3'}{2\pi\epsilon}\log_e\frac{R^3}{D_{12}D_{23}D_{31}}\right)$$

$$= \frac{q_1'}{2\pi\epsilon}\log_e\frac{R}{a} + \frac{q_2' + q_3'}{2\pi\epsilon}\log_e\frac{R}{\sqrt[3]{(D_{12}D_{23}D_{31})}}$$

Since $q_2' + q_3' = -q_1'$,

$$V_1 = \frac{q_1'}{2\pi\epsilon}\log_e\frac{\sqrt[3]{(D_{12}D_{23}D_{31})}}{a}$$

The average capacitance of each line is

$$C' = C_1' = \frac{q_1'}{V_1} = \frac{2\pi\epsilon}{\log_e\dfrac{\sqrt[3]{(D_{12}D_{23}D_{31})}}{a}} \quad \text{farads/metre} \qquad (7.25)$$

7.17 Electrostatic Field Plotting Methods

The mapping or field plotting methods which were developed for conduction fields in Section 7.6 are equally applicable to plane electrostatic fields. The electrostatic field should be divided by streamlines and equipotentials into a number of curvilinear squares.

For unit depth of field behind each curvilinear square the capacitance of the flux tube formed will be ϵ farads, where ϵ is the permittivity of the dielectric material. The capacitance of the flux tube is independent of the size of the curvilinear square. The total capacitance of the field will be given by

$$C = \epsilon d \times \frac{m}{n} \quad \text{farads} \qquad (7.26)$$

where $d =$ Depth of field in metres

$\quad m =$ Number of "parallel" squares measured along each equipotential

$\quad n =$ Number of "series" squares measured along each streamline

EXAMPLE 7.3 A cross-section of a parallel-strip transmission line is shown in Fig. 7.14(a); the conductors have each a breadth of $2c$ metres and are spaced *c* metres apart. Compare the values of capacitance per metre length found by (a) neglecting fringing at the edges, and (b) estimating the capacitance by a field plotting method.

Comment on the electric stress distribution.

(a) *Neglecting fringing*

$$\text{Capacitance/metre length} = \frac{\epsilon \times \text{Area of plates}}{\text{Separation}}$$

$$= \epsilon \times \frac{2c \times 1}{c} = \underline{\underline{2\epsilon \text{ farads/metre}}}$$

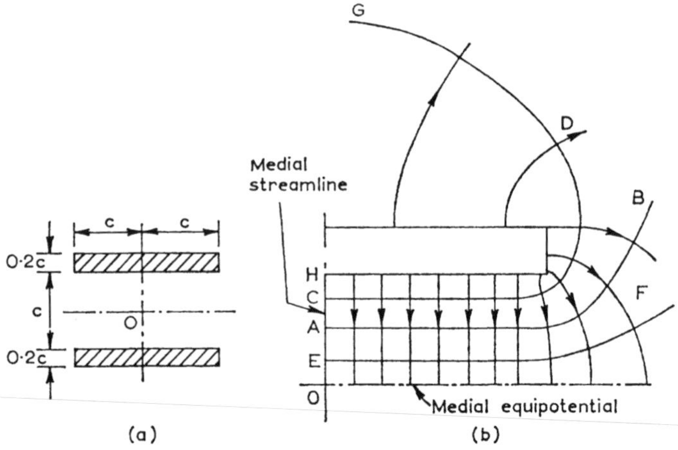

Fig. 7.14

(b) *Mapping*. The centre-line joining the two plates will be the medial streamline, by the symmetry of the arrangement; also the centre-line between the two plates will be the medial equipotential. Between the medial equipotential and the medial streamline the whole field may be divided into four separate symmetrical parts. One such part is enlarged for mapping in Fig. 7.14(b).

The mapping may be as follows.

1. Estimate the position of the equipotential AB which has the mean potential between that of the plate and that of the medial equipotential. The end B is not taken too far, since if it were, the exact position of the line would be very difficult to estimate. Point A will lie slightly closer to H than to O.

2. Estimate the positions of the intermediate equipotentials CD and EF, stopping these curves when the position which they should occupy becomes difficult to estimate.

3. Draw in a series of streamlines to cut the equipotentials normally, and to form, as far as possible, curvilinear squares.

4. Continue the equipotential CD to the point G, making the curve normal to the vertical at G, and forming squares with the streamlines.

5. Erase the equipotentials and redraw to fit the streamlines.

6. Repeat this procedure as necessary.

In this case it should be noted that between the plates the field is almost uniform, giving a field plot consisting of true squares in this region. At the corner of the plate the squares are smaller—hence there is a greater electric stress here. On the top of the plate the squares become extremely large, indicating that the main field exists between the plates.

From Fig. 7.14(*b*),

$$\text{Total capacitance per unit depth} = \epsilon \times \frac{\text{Number of parallel squares}}{\text{Number of series squares}}$$

$$= \frac{\epsilon \times 2 \times 13 \cdot 2}{2 \times 4}$$

$$= 3 \cdot 3\epsilon \text{ farads/metre}$$

From the field plot it is possible to estimate the electric stress at any point in the dielectric except just at a sharp edge where the field is greatly affected by the "sharpness" of the edge. Since there is the same potential difference across each curvilinear square and since the electric stress over one square is approximately uniform, the electric stress at any point is given approximately by

$$\frac{\text{Potential drop across a square}}{\text{Length of one side of an adjacent square}}$$

Thus the electric stress is inversely proportional to the length of the sides of a curvilinear square, and hence the stress is highest where the squares are smallest.

7.18 Streamlines and Tubes of Magnetic Flux

A streamline in a magnetic field is a line so drawn that its direction is everywhere parallel to the direction of the magnetic flux. It is also a line of *magnetic field strength* and has the same properties as a streamline in a conduction field. Several streamlines may be taken as enclosing a *tube of magnetic flux*.

Let the total flux through a tube be Φ webers. Then the *magnetic flux density* at a point in the tube is

$$B = \frac{\Phi}{a} \text{ webers/metre}^2 \text{ or teslas*}$$

where a square metres is the cross-section of the tube at the particular point and it is assumed that the flux density within the tube is sufficiently uniform to be taken as constant over the area a.

7.19 Equipotential Surfaces and Magnetic Field Strength

The term "potential" is less frequently used with respect to magnetic fields than with respect to conduction and electrostatic fields, probably owing to the difficulty in fixing a zero or reference potential. Fig. 7.15 shows a typical magnetic field system where the magnetic

* The *tesla* (T) is the SI unit of magnetic flux density. Some writers prefer to retain the *weber per square metre* (Wb/m²) since this conveys the idea of surface density.

flux is partly in air and partly in iron. An equipotential surface in this field will be a surface over which a *magnetic pole* could be moved without the expenditure of work or energy. Various lines are drawn across the flux path of Fig. 7.15 to show the edges of typical equipotential surfaces. Any one of the equipotential surfaces can be taken as the zero or reference potential surface. The work done in moving a *unit magnetic pole* from the zero-potential surface

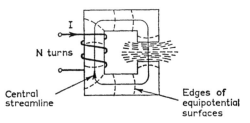

Fig. 7.15   MAGNETIC EQUIPOTENTIALS IN A TYPICAL
MAGNETIC FIELD SYSTEM

to a second equipotential will give the *magnetic potential* of the second surface.

If the unit pole is moved completely round the magnetic circuit so that its completed path links the total magnetizing current, *I* amperes, in the *N* turns of the magnetizing coil, then the total work done is *IN* joules per pole, or *IN* ampere-turns. Thus the zero magnetic potential surface may also be considered to have a magnetic potential of *IN* ampere-turns; in the same way all points in a magnetic field may be considered to have many potential values, but only the basic value denoting the movement of a unit pole round a fraction of the magnetic circuit from the reference surface need be considered. Thus the magnetic potential of a point (*F* ampere-turns) is equal to the work done in moving a unit pole from the arbitrary zero potential surface to the particular point.

Total magnetic p.d. round a complete loop = *IN* ampere-turns

Let *H* ampere-turns per metre (or newtons per unit pole) be the field strength in any given direction *r*, at a point in a magnetic field where the magnetic potential is *F* ampere-turns. Then,

$$H = -\frac{dF}{dr} \quad \text{ampere-turns/metre} \tag{7.27}$$

In the same way as for the conduction and electrostatic fields, the minus sign is included since *H* acts in the opposite direction with respect to *r* to that in which *F* increases.

At all points in a magnetic field both the field strength H and the flux density B are vector quantities in the direction of the streamline through the point. They are related by the equation

$$B = \mu H$$

where μ is the permeability of the material at the point $(= \mu_0 \mu_r)$; μ_0 being the permeability of space $(= 4\pi \times 10^{-7})$, and μ_r the permeability of the material relative to the permeability of space.

7.20 Boundary Conditions

The magnetic field in the air gap between two iron surfaces may be considered as bounded by two equipotentials which follow the

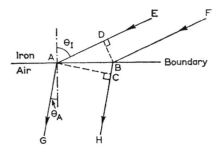

Fig. 7.16   REFRACTION OF MAGNETIC FLUX AT AN IRON-AIR
BOUNDARY

iron surfaces. This follows from the following proof, which shows that at an iron-air boundary a line of force will emerge from the iron in a direction almost normal to the iron-air boundary.

Consider unit depth of an iron-air boundary as depicted in Fig. 7.16. Let AE and BF be two adjacent streamlines (or lines of force) in the iron which strike the boundary at an angle θ_I to the normal at the boundary. The streamlines continue in the air as AG and BH at an angle θ_A to the boundary normal.

For unit depth at the boundary a tube of magnetic flux is formed between the streamlines EAG and FBH. Let Φ webers be the flux in this tube.

$$\text{Flux density in iron, } B_I = \frac{\Phi}{DB \times 1} = \frac{\Phi}{AB \cos \theta_I}$$

$$\text{Flux density in air, } B_A = \frac{\Phi}{AC \times 1} = \frac{\Phi}{AB \cos \theta_A}$$

Therefore

$$\text{Field strength in iron, } H_I = \frac{B_I}{\mu_I} = \frac{\Phi}{\mu_I \text{AB cos } \theta_I}$$

and

$$\text{Field strength in air, } H_A = \frac{B_A}{\mu_A} = \frac{\Phi}{\mu_A \text{AB cos } \theta_A}$$

Consider that a unit pole is moved round the closed path ACBDA —the total work done must be zero since no current is linked. No work will be done in the movements along AC and DB since these paths are normal to the field direction, i.e. equipotentials. Thus

Work done in movement CB = Work done in movement DA

i.e.

$$H_A \times \text{CB} = H_I \times \text{AD}$$

or

$$\frac{\Phi}{\mu_A \text{AB cos } \theta_A} \text{ AB sin } \theta_A = \frac{\Phi}{\mu_I \text{AB cos } \theta_I} \text{ AB sin } \theta_I$$

Therefore

$$\frac{\tan \theta_A}{\mu_A} = \frac{\tan \theta_I}{\mu_I}$$

and

$$\theta_A = \tan^{-1} \left(\frac{\mu_A}{\mu_I} \tan \theta_I \right)$$

Since $\mu_A \ll \mu_I$, $(\mu_A/\mu_I) \tan \theta_I$ is very small, and hence θ_A must be nearly zero. Thus a streamline will cross an iron-air boundary almost normally.

A non-varying magnetic field (i.e. one that is set up by a permanent-magnet or a direct current) will be unaffected by the presence of a conductor provided that the conductor is non-magnetic. A permanent field, e.g. the earth's magnetic field, will, for instance, penetrate a block of brass. If the field is varying (i.e. set up by an alternating current), then in general it will not penetrate a conducting material since the eddy currents which would be set up within the material would oppose the magnetic field. The actual depth to which a magnetic field will effectively penetrate a conductor decreases as the frequency increases. The depth of penetration also depends on the resistivity and permeability of the conductor.

If the depth of penetration of flux is small, then the eddy currents in the conductor must be effectively neutralizing any flux which is tending to enter the conductor. Thus all the flux must be parallel to the conductor surface, i.e. the streamlines will be parallel to the conducting surface, and the magnetic equipotentials will intersect the conducting surface normally.

7.21 Shielding for Static Magnetic Fields

The object of shielding (or screening) is to prevent a magnetic field from existing at some particular point. For steady (or static) fields, the only method of achieving shielding is to provide a low-reluctance magnetic path for the stray flux, in such a way that this flux bypasses the shielded point. Since no material has infinite

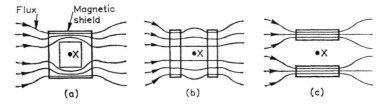

Fig. 7.17   PRINCIPLE OF SCREENING FROM STATIC FIELDS
(a) Very good screening
(b) Negligible screening
(c) Good screening

permeability, perfect shielding is not possible, but shields made of materials which have high permeabilities at low flux densities give satisfactory results.

The best type of shield is indicated in Fig. 7.17(a), where the shielded point (X) is entirely surrounded by a magnetic container.

It should be noted that for weak fields the thickness of the shield is relatively unimportant, since the flux tends to follow the outside edge of the shield. Indeed better shielding is usually provided by two thin shields with an air gap between them, than by one thick shield.

In Fig. 7.17(b) the effect of removing the sides of the shield which are parallel to the magnetic field is shown. The field at the shielded point X is actually stronger than if no shielding materials were present, owing to the concentration of flux in the magnetic material. If, however, the other sides of the shield are removed (as in Fig. 7.17(c)), it will be seen that the shielding is still fairly effective, since the interfering field bypasses the point X through the magnetic material.

7.22 Shielding for Alternating Magnetic Fields

Conducting sheets are used to restrict the extent of alternating magnetic fields. The thickness of the shielding plate should be greater than the depth of penetration of the field at the operating frequency, though considerable shielding may be obtained with plates of smaller thickness.

When shielding an alternating magnetic field (e.g. that of a coil) it is important to arrange that the eddy currents in the shield do not give a high power loss. This usually means that the shield should not be too small.

Sometimes sufficient shielding can be obtained by a few short-circuited copper turns, placed round the object to be shielded in

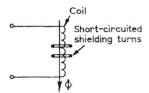

Fig. 7.18   USE OF SHORT-CIRCUITED TURNS FOR SHIELDING FROM ALTERNATING FIELDS

such a direction that the axis of the turns is in the direction of the magnetic field. Fig. 7.18 shows such an arrangement, which may be used either to prevent an external field from affecting the coil, or to restrict the field of the coil itself. An external field linking the coil would also link the short-circuited turns and induce eddy currents in them. These eddy currents would set up an m.m.f. opposing the flux. Similarly the field set up by a current in the coil would be restricted to the region within the short-circuited turns, since any flux beyond this limit would link the short-circuited turns and be opposed by the m.m.f. of the eddy-currents in the turns.

7.23 Skin Effect

A direct current flowing in a uniform conductor will distribute itself uniformly over the cross-section of the conductor. An alternating current on the other hand always tends to flow at the surface of a conductor. This is called *skin effect*. The effect is more pronounced at high frequencies and with conductors of large cross-section.

Consider the round conductor of Fig. 7.19 and suppose that a direct current is flowing into the plane of the paper. There will be a magnetic field outside the conductor and there will also be a magnetic field inside the conductor, as shown by the dotted line. This inner field is produced by the current at the centre of the conductor. The portion of the conductor inside the dotted line may be regarded as a separate conductor in parallel with the portion of the conductor outside the dotted line. The inner conductor is linked by the magnetic field from the dotted line outwards, while the outer conductor is only linked by the magnetic field from the conductor surface outwards. Thus the inner conductor may be regarded as having a larger inductance than the outer conductor, since the

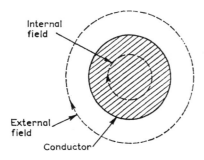

Internal field

External field

Conductor

Fig. 7.19   PERTAINING TO SKIN EFFECT

former links a larger magnetic field than the latter. The larger inductance does not affect direct currents, but obviously the inner conductor will have a larger impedance to alternating currents than the outer, and hence the current and current density will tend to be greater in the outer conductor. The effect increases with frequency, until at high frequencies the current is almost entirely in the "outer skin" of the conductor.

Exact analysis shows that the depth of penetration of the current is given by the same equation as is the depth of penetration of magnetic flux.

7.24 Inductance due to Low-frequency Internal Linkages

For a conductor used at high frequencies (i.e. where the depth of penetration is small compared with the conductor cross-section), the internal linkages are negligible and the circuit inductance is simply the inductance due to the fields in the surrounding space. At very low frequencies (i.e. where the current distribution may be considered uniform over the section of the conductor), the inductance due to internal linkages has a maximum value. At other frequencies this inductance will have a value between this maximum value and zero.

Consider a conductor of radius a carrying a total current I amperes uniformly distributed over the conductor cross-section (Fig. 7.20).

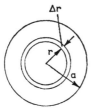

Fig. 7.20   PERTAINING TO LOW-FREQUENCY INTERNAL LINKAGES
IN A CONDUCTOR

Current density at all points on the cross-section $= \dfrac{I}{\pi a^2}$

Consider a unit magnetic pole moved round a path of radius r within the conductor.

Current enclosed by path of radius r

$$= \text{Current density} \times \text{Area enclosed}$$

$$= \frac{I}{\pi a^2} \times \pi r^2 = \frac{Ir^2}{a^2}$$

By the work law,

Work done in moving unit pole round closed path

$$= \text{Ampere-turns linked}$$

Thus

$$H_r \times 2\pi r = \frac{Ir^2}{a^2} \times 1 \quad \text{and} \quad H_r = \frac{Ir}{2\pi a^2}$$

where H_r is the field strength at radius r and there is only one turn.

Flux density at radius $r = B_r = \mu H_r = \dfrac{\mu Ir}{2\pi a^2}$

For 1 metre depth of the conductor,

Flux within element of thickness $\Delta r = \dfrac{\mu Ir \Delta r}{2\pi a^2}$

The flux in the element links the portion r^2/a^2 of the total conductor. Therefore

Linkages due to flux in element $= \dfrac{\mu I r^3}{2\pi a^4}\,\Delta r$

and

Total linkages per metre due to flux in conductor

$$= \frac{\mu I}{2\pi a^4}\int_0^a r^3\,dr = \frac{1}{4}\frac{\mu I}{2\pi}$$

Inductance per metre due to internal flux

$=$ Internal flux linkages per ampere

$$= \frac{1}{4}\frac{\mu}{2\pi}\ \text{henrys/metre} \tag{7.28}$$

It is notable that this inductance is independent of the conductor radius.

7.25 Inductance of a Concentric Cable

Assume that the core of a concentric cable carries a current of I amperes in one direction while the sheath carries the same current in the opposite direction.

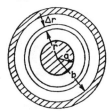

Fig. 7.21 INDUCTANCE OF A CONCENTRIC CABLE

Consider a unit pole moved once round a path of radius r in the interspace between core and sheath, i.e. so that it links the core current (Fig. 7.21). By the work law,

$H_r \times 2\pi r = I$

or

$$H_r = \frac{I}{2\pi r}$$

where H_r is the field strength at radius r.

Flux density at radius $r = B_r = \mu H_r$

Consider unit depth of the element of thickness Δr at radius r.

Flux within unit depth of the element $= B_r \Delta r \times 1$

$$= \frac{\mu I}{2\pi r} \Delta r$$

This flux links the core of the cable, i.e. links the loop of the cable formed by the core and the sheath.

Flux linkages due to element flux per metre of cable $= \dfrac{\mu I \Delta r}{2\pi r}$

Total flux linkages per metre $= \dfrac{\mu I}{2\pi} \displaystyle\int_a^b \dfrac{dr}{r} = \dfrac{\mu I}{2\pi} \log_e \dfrac{b}{a}$

Therefore

$$\text{Inductance per metre} = \mu \, \frac{\log_e \dfrac{b}{a}}{2\pi} \text{ henrys/metre} \qquad (7.29)$$

To this should be added terms representing the inductances due to internal core linkages and internal sheath linkages. Both of these are negligible at high frequencies, and the sheath linkages are also negligible (to a good accuracy) at low frequencies since the sheath is usually relatively thin and has a weak magnetic field over most of its cross-section.

Total inductance per metre at low frequencies

$$= \frac{1}{4} \frac{\mu}{2\pi} + \mu \, \frac{\log_e \dfrac{b}{a}}{2\pi} \text{ henrys/metre} \qquad (7.30)$$

7.26 Inductance of a Twin Line

Consider two isolated, long, straight, parallel conductors of radius a metres, spaced D metres apart, and each carrying a current of I amperes in opposite directions. D is assumed large compared with a. The magnetic field surrounding the conductors is shown in Fig. 7.22(a). The field is most easily analysed by considering each conductor alone in turn.

Consider conductor A only carrying current (Fig. 7.22(b)).

Magnetic field strength at radius $r = H_r = \dfrac{I}{2\pi r}$

Flux density at radius $r = B_r = \dfrac{\mu I}{2\pi r}$

Therefore

Total flux in 1 metre depth of element $= B_r\Delta r \times 1 = \dfrac{\mu I}{2\pi r}\,\Delta r$

This flux links conductor A once.

Linkage with conductor A due to flux in element $= \dfrac{\mu I}{2\pi}\dfrac{\Delta r}{r}$

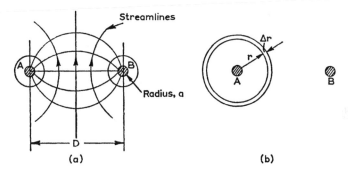

Fig. 7.22 INDUCTANCE OF A TWIN LINE

Therefore

Total linkages with conductor A due to current in conductor A

$$= \frac{\mu I}{2\pi}\int_a^R \frac{dr}{r} = I\frac{\mu}{2\pi}\log_e\frac{R}{a}$$

where R is a very large radius at which the magnetic field strength may be regarded as zero. Similarly,

Total linkages with conductor B due to current in A

$$= \frac{\mu I}{2\pi}\int_D^R \frac{dr}{r} = I\frac{\mu}{2\pi}\log_e\frac{R}{D}$$

since magnetic flux at a radius greater than D about A will link B.

Similarly when only B carries current of $-I$ amperes,

Total linkages with B due to current in B $= -I\dfrac{\mu}{2\pi}\log_e\dfrac{R}{a}$

and

Total linkages with A due to current in B $= -I\dfrac{\mu}{2\pi}\log_e\dfrac{R}{D}$

Therefore

$$\text{Total linkages with A} = I\frac{\mu}{2\pi}\log_e\frac{R}{a} - I\frac{\mu}{2\pi}\log_e\frac{R}{D}$$

$$= I\frac{\mu}{2\pi}\log_e\frac{D}{a} \text{ weber-turns/metre}$$

and similarly,

$$\text{Total linkages with B} = I\frac{\mu}{2\pi}\log_e\frac{D}{a}$$

For 1 metre length of twin lines,

Total inductance = Flux linkages/ampere

$$= 2\mu\frac{\log_e\dfrac{D}{a}}{2\pi} \text{ henrys/metre} \qquad (7.31)$$

This does not include linkages internal to each line. If these are to be included, the inductance at low frequency is

Total inductance per loop metre

$$= \frac{1}{2}\frac{\mu}{2\pi} + 2\mu\frac{\log_e\dfrac{D}{a}}{2\pi} \text{ henrys/metre} \qquad (7.32)$$

7.27 Inductance of a Single Line above a Conducting Plane

It has been shown (Section 7.20) that an alternating flux cannot penetrate (beyond a certain depth) a conducting sheet; thus, if a "thick" conducting sheet is to be introduced into an alternating magnetic field without affecting the field, the sheet must be so introduced that its surface is everywhere along streamlines. In Fig. 7.22(*a*) an infinite flat sheet could be introduced along the medial streamline between the two conducting wires without affecting the field. If the frequency were such that the sheet thickness was considerably greater than the penetration depth, then the sheet would completely divide the field into two independent parts. Either line could then be removed without affecting the field around the other line.

Applying these deductions in reverse to the actual system of Fig. 7.23, the field between the actual wire and the conducting sheet will be the same as it would be if the conducting sheet were replaced by an *image conductor* carrying the same current as the actual conductor (but in the opposite direction), at a depth h below the surface of the conducting sheet. The calculation of the inductance of the actual conductor system may be made on this basis.

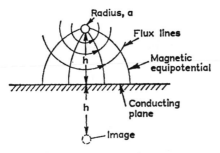

Fig. 7.23 INDUCTANCE OF A WIRE ABOVE A PARALLEL CONDUCTING PLANE

Linkages with actual conductor due to its own current

$$= \frac{\mu I}{2\pi} \log_e \frac{R}{a}$$

Linkages with actual conductor due to image current

$$= -\frac{\mu I}{2\pi} \log_e \frac{R}{2h}$$

where R is a very large distance, according to the method of the previous section. Therefore

Total linkages with actual conductor per metre $= \dfrac{\mu I}{2\pi} \log_e \dfrac{2h}{a}$

Inductance of actual conductor per metre

$$= \frac{\mu}{2\pi} \log_e \frac{2h}{a} \quad \text{henrys/metre} \quad (7.33a)$$

To this must be added the inductance due to the internal flux linkages within the conductor, giving, at low frequency only,

$$L_{eff} = \frac{1}{4} \frac{\mu}{2\pi} + \mu \frac{\log_e \dfrac{2h}{a}}{2\pi} \quad \text{henrys/metre} \quad (7.33b)$$

7.28 Equivalent Phase Inductance of a Three-phase Line

A simple expression may be deduced for the equivalent phase inductance of an isolated 3-phase 3-wire line if the line is uniformly transposed and the line spacings are considerably greater than the line diameters. By *equivalent phase inductance* is meant the total linkages with a given wire under 3-phase conditions—this inductance may be considered as a series inductance in each line.

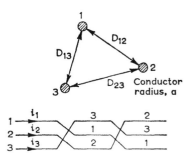

Fig. 7.24 SPACING AND TRANSPOSITION OF A 3-PHASE LINE

The system cross-section and the transposition scheme are shown in Fig. 7.24. Since the currents are part of a three-wire system,

$$i_1 + i_2 + i_3 = 0$$

and $(i_2 + i_3) = -i_1$ at all instants.

Following the method of Section 7.26,

Total linkages with conductor 1 in first position

$$= \frac{\mu i_1}{2\pi} \log_e \frac{R}{a} + \frac{\mu i_2}{2\pi} \log_e \frac{R}{D_{12}} + \frac{\mu i_3}{2\pi} \log_e \frac{R}{D_{13}}$$

Total linkages with conductor 1 in second position

$$= \frac{\mu i_1}{2\pi} \log_e \frac{R}{a} + \frac{\mu i_2}{2\pi} \log_e \frac{R}{D_{23}} + \frac{\mu i_3}{2\pi} \log_e \frac{R}{D_{12}}$$

and

Total linkages with conductor 1 in third position

$$= \frac{\mu i_1}{2\pi} \log_e \frac{R}{a} + \frac{\mu i_2}{2\pi} \log_e \frac{R}{D_{13}} + \frac{\mu i_3}{2\pi} \log_e \frac{R}{D_{23}}$$

Therefore

Average linkage with conductor 1

$$= \frac{1}{3}\left(\frac{3\mu i_1}{2\pi}\log_e\frac{R}{a} + \frac{\mu i_2}{2\pi}\log_e\frac{R^3}{D_{12}D_{23}D_{31}} + \frac{\mu i_3}{2\pi}\log_e\frac{R^3}{D_{12}D_{23}D_{31}}\right)$$

$$= \frac{\mu i_1}{2\pi}\log_e\frac{R}{a} + \frac{\mu(i_2 + i_3)}{2\pi}\log_e\frac{R}{\sqrt[3]{(D_{12}D_{23}D_{31})}}$$

$$= \frac{\mu i_1}{2\pi}\log_e\frac{R}{a} - \frac{\mu i_1}{2\pi}\log_e\frac{R}{\sqrt[3]{(D_{12}D_{23}D_{31})}}$$

$$= \frac{\mu i_1}{2\pi}\log_e\frac{\sqrt[3]{(D_{12}D_{23}D_{31})}}{a}$$

Thus

Average equivalent inductance/phase/metre

$$= \mu\frac{\log_e\dfrac{\sqrt[3]{(D_{12}D_{23}D_{31})}}{a}}{2\pi} \quad \text{henrys/metre} \tag{7.34}$$

This does not include the inductance due to internal linkages.

Total low-frequency equivalent inductance/phase/metre

$$= \frac{1}{4}\frac{\mu}{2\pi} + \mu\frac{\log_e\dfrac{\sqrt[3]{(D_{12}D_{23}D_{31})}}{a}}{2\pi} \quad \text{henrys/metre} \tag{7.35}$$

7.29 Determination of Air-gap Permeance by a Mapping Method

The *reluctance* of an air gap is commonly used in elementary treatments, but to treat magnetic fields in the same manner as electrostatic and conduction fields it is preferable to deal with its reciprocal, namely *permeance*:

$$\text{Permeance} = \frac{1}{\text{Reluctance}}$$

or, in symbols,

$$\Lambda = \frac{1}{S}$$

For a uniform field of area a square metres and length l metres,

$$\Lambda = \frac{1}{S} = \frac{\mu a}{l}$$

Supposing a magnetic field is mapped into a number of curvilinear squares, and a flux tube is considered of d metres depth and with curvilinear squares for its ends, as in Fig. 7.4. Then

$$\text{Permeance of tube} = \mu \times \frac{\text{area}}{\text{length}} = \mu \frac{dl}{l} = \mu d$$

where l is the length of one side of the curvilinear square. Therefore

$$\text{Total permeance} = \mu d \frac{m}{n} \quad \text{webers/ampere-turn} \qquad (7.36)$$

where m is the number of parallel squares, i.e. the number of squares across the flux direction, and n is the number of series squares, i.e. the number of squares along the flux direction.

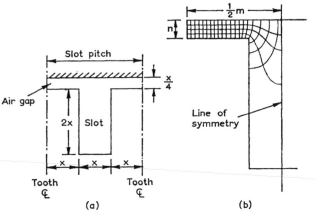

Fig. 7.25

EXAMPLE 7.4 The arrangement of Fig. 7.25(a) resembles a slot pitch in an electrical machine. Estimate the reluctance of the slot pitch if the rotor length is 20 cm. The field may be considered uniform over this length.

Assuming that there are similar slots on either side of the one depicted, the teeth centre-lines will be streamlines. Also, by symmetry, the slot centre-line will be a streamline dividing the slot pitch into two symmetrical parts. One part of the slot pitch is enlarged for mapping in Fig. 7.25(b).

The iron surfaces are boundary equipotentials which the streamlines intersect normally (by Section 7.20). The division into curvilinear squares is carried out by the same method as previously (Examples 7.1 and 7.3). By eqn. (7.36),

$$\text{Total permance per slot pitch} = \mu_0 d \times \frac{m}{n} = \mu_0 \times \frac{20}{100} \times \frac{2 \times 20 \cdot 1}{4}$$

$$= 2 \cdot 52 \times 10^{-6} \, \text{Wb/At}$$

Therefore

$$\text{Reluctance per slot pitch} = \frac{1}{2 \cdot 52} \times 10^6 = \underline{\underline{397,000 \, \text{At/Wb}}}$$

7.30 Determination of Inductance by a Mapping Method

In simple magnetic fields (e.g. fields without iron magnetic circuits) it is generally the inductance of the arrangement rather than the reluctance or permeance which is required. The inductance may be simply calculated from a field plot, provided that all the magnetic flux may be assumed to link all the turns of the conductor arrangement producing the field.

$$\text{Flux, } \Phi = \frac{\text{M.M.F.}}{\text{Reluctance}} = \frac{IN}{S} = IN\Lambda$$

But

$$\text{Inductance, } L = \text{Flux linkages/ampere} = \frac{\Phi N}{I} = N^2\Lambda$$

i.e.

$$L = N^2\mu d\,\frac{m}{n} \quad \text{henrys} \tag{7.37}$$

EXAMPLE 7.5 For an isolated twin-wire line with a spacing-to-diameter ratio of 10, calculate the inductance per metre of the line by (a) direct calculation, and (b) a field plot. Neglect linkages within the conductors.

$$\text{Inductance per metre} \approx 2\mu_0 \frac{\log_e \dfrac{D}{a}}{2\pi}$$

$$= \frac{2 \times 4\pi \times 10^{-7} \times \log_e 20}{2\pi} = 1\cdot2 \times 10^{-6}\,\text{H/m}$$

Fig. 7.26(a) on the next page shows the conductor system. The centre-line through the conductors is a symmetrical equipotential and the centre-line between the conductors is a symmetrical streamline. The field is thus divided into four symmetrical parts, and one of these is enlarged in Fig. 7.26(b) for field plotting. From the field plot,

$$\text{Inductance per metre} = \mu_0 \times 1 \times \frac{m}{n} = 4\pi \times 10^{-7} \times \frac{2 \times 4\cdot5}{2 \times 5\cdot1}$$

$$= 1\cdot11 \times 10^{-6}\,\text{H/m}$$

PROBLEMS

7.1 A concentric cable has a core diameter of 1cm and a sheath diameter of 5cm. If the core is displaced from the true centre of the cable by 0·75cm, calculate the capacitance per metre run by field plotting. ϵ_r for the dielectric is 3. From the same field plot write down the inductance per metre run of cable.

(*H.N.C.*)

Ans. 0·00012μF, 0·28μH.

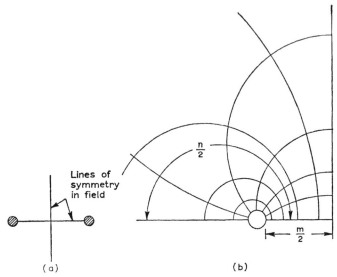

Fig. 7.26
(a) Actual spacing   (b) Enlarged spacing for field plot

7.2 The plan view of a poor electrolytic plating bath is shown in Fig. 7.27. A is the anode plate and the central uniform cylinder is the object to be plated,

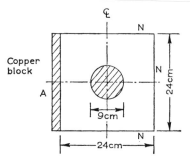

Fig. 7.27

while N is the non-conducting wall of the container. Determine the ohmic resistance of the bath if the depth of the electrolyte is 6 cm and its resistivity is $100\,\Omega$-cm. Draw also a polar diagram showing the variation in thickness of plating round the circumference of the cylinder.

Ans. $7 \cdot 2\,\Omega$.

7.3 Fig. 7.28 represents very approximately a conductor lying in an open slot. Plot the electrostatic field for the system and estimate the capacitance per metre length of conductor. Explain the dependence of the electric stress on the shape of the corners.

Ans. 85 pF.

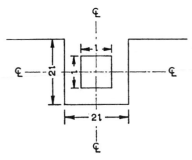

Fig. 7.28

7.4 Explain why magnetic flux may be assumed to emerge from an iron boundary into air at right angles to the surface.

Use field plotting to determine the reluctance per unit depth of the air gap, shown in Fig. 7.29, between two long flat iron plates. Determine the percentage

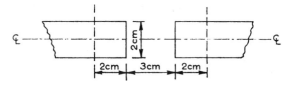

Fig. 7.29

error which is caused by neglecting fringing. Neglect the field beyond the dashed lines. (*H.N.C.*)

Ans. 450,000 At/Wb, 160 per cent.

7.5 Two parallel infinitely long straight conductors of diameter 0·2 cm are spaced 10 cm apart and relatively far from conducting objects. Draw the shape of the field surrounding these conductors and from it estimate the inductance and capacitance per unit length of the conductors. Check the result by calculation on the usual basis.

Ans. 1·84 μH/m, 6·03 pF/m.

7.6 From Ohm's law in its usual circuit form derive an expression for Ohm's law applicable to a point in an electrical conductor (i.e. the conduction field form).

Explain the meaning of the terms (*a*) flowline, (*b*) equipotential, with reference to a conductor. Show that these must necessarily intersect at right angles.

By a field-plotting method estimate the percentage increase in the resistance of the busbar length shown in Fig. 7.30 due to the four holes. Assume the busbar thickness uniform and the current distribution uniform at each end.

(*H.N.C.*)

Ans. 20 per cent.

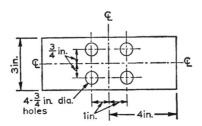

Fig. 7.30

7.7 If the electromagnetic field in the interspace between two long parallel concentric conductors is "mapped" by a number of "curvilinear squares" derive an expression for the inductance per metre run of the conductors in terms of numbers of squares and the permeability of the interspace material. (Neglect inductance due to flux linkages within the conductors.)

A conductor of cross-section 1 in. × ½ in. is enclosed in a tube of square cross-section whose internal side is 2 in. (Fig. 7.31). Draw (several times full

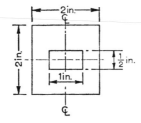

Fig. 7.31

size) a symmetrical part of this cross-section and hence estimate the inductance of the arrangement per metre run. (Assume a non-magnetic interspace and neglect flux linkages within the conductors.)

(*H.N.C.*)

Ans. 0·157 μH/m.

7.8 A liquid rheostat is to be constructed with its outer electrode a hollow steel cylinder of 1·5 m internal diameter; this cylinder is also the liquid container. The inner electrode is a rod of 0·5 m diameter. Calculate the resistance between electrodes per metre length of the rheostat when the liquid has a resistivity 100 Ω-m and the centre-line of the inner electrode is coincident with the centre-line of the outer cylinder.

By means of a field plot find the resistance when the centre-line of the rod is parallel to the centre-line of the cylinder but spaced 0·25 m from it.   (*H.N.C.*)

Ans. 17·5 Ω, 15 Ω.

7.9 Calculate the loop inductance and capacitance of a 1 km length of single-phase line having conductors of diameter 1 cm and spaced 72 cm apart. If this line is to be converted to a 3-phase line by the addition of a third conductor of the same cross-section as the originals, calculate the phase-to-neutral inductance and capacitance when the third conductor (*a*) is in line with the first two and 72 cm from the nearest, (*b*) forms an equilateral triangle with the first two. Assume regular transposition and neglect earth effects.

Ans. 0·00212 H/loop km;   0·00111 H/km;   0·00106 H/km;   0·0056 μF/km; 0·0104 μF/km; 0·0112 μF/km.

7.10 Derive an expression for the inductance per metre of a coaxial cable of core diameter *d* metres and internal sheath diameter *D* metres. (The sheath thickness may be neglected.)

A coaxial cable 8·05 km long has a core 1 cm diameter and a sheath 3 cm diameter of negligible thickness. Calculate the inductance and capacitance of the cable, assuming non-magnetic materials and a dielectric of relative permittivity 4.

For this cable calculate also (*a*) the charging current when connected to a 66 kV, 50 Hz supply (neglect inductance), (*b*) the surge impedance of the cable,* (*c*) the surge velocity for the cable.*

Ans. 21·7 × 10^{-4} H, 1·63 μF, (*a*) 33·8 A, (*b*) 33 Ω, (*c*) 1·5 × 10^{8} m/s.

7.11 A single-core lead-sheathed cable has a conductor of 10 mm diameter and two layers of insulating material each 10 mm thick. The permittivities are 4 and 2·5, and the layers are placed to allow the greater applied voltage. Calculate the potential gradient at the surface of the conductor when the potential difference bet·ween the conductor and the lead sheathing is 60 kV.   (*H.N.C.*)

Ans. 6·29 kV/mm.

7.12 A single-core cable has a conductor diameter *d* and an inside sheath diameter *D*. Show that the maximum voltage which can be applied so as not to exceed the permissible electric stress *E* is given by $\dfrac{Ed}{2} \times \log_e \dfrac{D}{d}$.

Find this voltage for a cable in which $D = 8$ cm, $d = 1$ cm and $E = 50$ kV/cm (r.m.s.).

If *D* is fixed at 8 cm, find the most suitable value for *d* so that the greatest voltage can be applied to the cable.   (*H.N.C.*)

Ans. 52 kV, 2·95 cm.

---

* *See* Chapter 16.

Chapter 8

TWO-PORT NETWORKS

Networks in which electrical energy is fed in at one pair of terminals and taken out at a second pair of terminals are called *two-port*

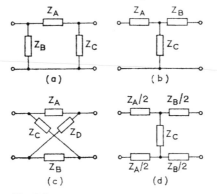

Fig. 8.1 BASIC PASSIVE TWO-PORT NETWORKS

(a) π-network
(b) T-network
(c) Lattice network
(d) Balanced T-network

networks. Thus a transmission line (whether for power or communications) is a form of two-port network. So too are attenuators, electric wave filters, electronic amplifiers, transformers, etc. The network between the input port and the output port is a transmission network for which a known relationship exists between the input and output voltages and currents.

242

Two-port networks are said to be *passive* if they contain only passive circuit-elements, and *active* if they contain sources of e.m.f. Thus an electronic amplifier is an active two-port. Passive two-port networks are symmetrical if they look identical from both the input and output ports. Fig. 8.1 shows some typical passive two-port networks. The π-section shown at (*a*) is symmetrical if $Z_B = Z_C$. At (*b*) is shown a T-section, which is symmetrical if $Z_A = Z_B$. The lattice section at (*c*) is symmetrical and balanced if $Z_A = Z_B$ and $Z_C = Z_D$. This network is simply a rearranged Wheatstone bridge. The π- and T-networks shown at (*a*) and (*b*) have one common terminal, which may be earthed, and are therefore said to be *unbalanced*. The *balanced* form of the T-circuit is shown at (*d*).

In this simple introduction to two-port theory, only passive symmetrical circuits will be considered.

8.1 Characteristic Impedance

The input impedance of a network is the complex ratio of voltage to current at the input terminals. For a two-port network the input

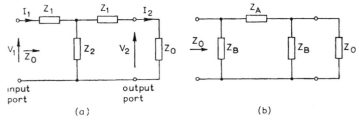

Fig. 8.2 ILLUSTRATING THE CONCEPT OF CHARACTERISTIC IMPEDANCE

impedance will normally vary according to the load impedance across the output terminals. It is found that for any passive two-port network a load impedance can always be found which will give rise to an input impedance which is the same as the load impedance. For an unsymmetrical network this is called the *iterative impedance*, and its value will depend upon which port (i.e. which pair of terminals) is taken to be the input and which the output port. For a symmetrical network there is only one value for iterative impedance, and this is called the *characteristic impedance*, Z_0, of the symmetrical two-port network.

Consider the symmetrical T-circuit of Fig. 8.2(*a*) terminated in an impedance Z_0 such that

$$\frac{V_1}{I_1} = Z_0 = \frac{V_2}{I_2}$$

From simple circuit theory the impedance seen looking into the input port is

$$\frac{V_1}{I_1} = Z_1 + \frac{Z_2(Z_1 + Z_0)}{Z_1 + Z_2 + Z_0}$$

$$= \frac{Z_1{}^2 + Z_1 Z_2 + Z_1 Z_0 + Z_2 Z_1 + Z_2 Z_0}{Z_1 + Z_2 + Z_0}$$

If this has to be equal to Z_0, then

$$Z_1{}^2 + 2Z_1 Z_2 + Z_0(Z_1 + Z_2) = Z_0(Z_1 + Z_2 + Z_0)$$
$$= Z_0(Z_1 + Z_2) + Z_0{}^2$$

or

$$Z_0 = \sqrt{\{Z_1(Z_1 + 2Z_2)\}} \tag{8.1}$$

A useful general expression for Z_0 can be deduced by investigating the input impedance of the network when the output terminals are (*a*) short-circuited, and (*b*) open-circuited. Thus the short-circuit input impedance is

$$Z_{sc} = Z_1 + \frac{Z_1 Z_2}{Z_1 + Z_2} = \frac{Z_1(Z_1 + 2Z_2)}{Z_1 + Z_2}$$

and the open-circuit input impedance is

$$Z_{oc} = Z_1 + Z_2$$

Hence

$$Z_{sc}Z_{oc} = Z_1(Z_1 + 2Z_2)$$

Comparison of this with eqn. (8.1) yields the important general relation

$$Z_0 = \sqrt{(Z_{sc}Z_{oc})} \tag{8.2}$$

In the same way, for the symmetrical-π circuit of Fig. 8.2 (*b*), if the input impedance when the network is terminated by Z_0 is equal to Z_0, then

$$Z_0 = \frac{Z_B(Z_A + Z_B Z_0/(Z_B + Z_0))}{Z_B + Z_A + Z_B Z_0/(Z_B + Z_0)}$$

whence, after some manipulation,

$$Z_0 = \sqrt{\frac{Z_A Z_B{}^2}{Z_A + 2Z_B}} \tag{8.3}$$

It is not difficult to show that eqn. (8.2) applies in this case also. Hence for any passive symmetrical network,

$$Z_0 = \sqrt{(Z_{sc}Z_{oc})} \tag{8.2a}$$

since by suitable manipulation all such circuits can be represented by an equivalent π- or T-section (note, however, that the equivalent sections may contain unrealizable circuit-elements such as negative resistance).

The concept of characteristic impedance is important since it facilitates the design of networks with specific transmission properties. It is important to realize that terminating a network in its characteristic impedance does not mean that the input and output voltages (or currents) are equal—only that $V_1/I_1 = V_2/I_2$.

8.2 Logarithmic Ratios: the Decibel

It is often convenient to express the ratio of two powers P_1 and P_2 in logarithmic form. If natural (Napierian) logarithms are used, the ratio is said to be in *nepers*. Thus

$$\text{Power ratio in nepers (Np)} = \tfrac{1}{2} \log_e \frac{P_1}{P_2} \tag{8.4}$$

If logarithms to base 10 are used, then the ratio is said to be in *bels*. The bel is rather a large unit, and one-tenth of a bel, or *decibel* (dB) is more commonly used. Thus

$$\text{Power ratio in decibels} = 10 \log_{10} \frac{P_1}{P_2} \tag{8.5}$$

The decibel is roughly the smallest difference in power level between two sound waves which is detectable as a change in volume by the human ear.

If the powers P_1 and P_2 refer to power developed in *two equal resistors*, R, then

$$P_1 = \frac{V_1^2}{R} \quad \text{and} \quad P_2 = \frac{V_2^2}{R}$$

so that the ratio can be expressed as

$$\text{Ratio in dB} = 10 \log_{10} \frac{V_1^2/R}{V_2^2/R} = 20 \log_{10} \frac{V_1}{V_2} \tag{8.6}$$

This is called the *logarithmic voltage ratio*, but it is really a power ratio. Strictly it should not be applied to the ratio of two voltages across different resistances, but this is sometimes done.

Similarly, if currents I_1 and I_2 in two equal resistors, R, give powers P_1 and P_2, then

$$\text{Ratio in dB} = 10 \log_{10} \frac{I_1^2 R}{I_2^2 R} = 20 \log_{10} \frac{I_1}{I_2} \tag{8.6a}$$

It should be particularly noted that the decibel notation applies to the *sizes* of voltages and currents—it gives no phase information.

For example, if a two-port network is terminated in its characteristic impedance and if the input voltage is $\sqrt{2}$ times the output voltage, then since the input and the output voltages refer to the equal impedances, the voltage ratio is given by

$$\text{Input/output ratio in dB} = 20 \log_{10} \frac{V_1}{V_2} = 20 \log_{10} \sqrt{2} \approx 3\,\text{dB}$$

Note that the output/input voltage ratio can be expressed in logarithmic units in the same manner. Thus in the above case,

$$\text{Output/input ratio in dB} = 20 \log_{10} \frac{V_2}{V_1} = 20 \log_{10} \frac{1}{\sqrt{2}}$$

$$= 20(\log_{10} 1 - \log_{10} \sqrt{2}) \approx -3\,\text{dB}$$

i.e. the minus sign indicates that the ratio denotes a power loss, while the positive sign indicates a power gain.

8.3 Insertion Loss

When a two-port network is inserted between a generator and a load, the voltage and current at the load will generally be less than

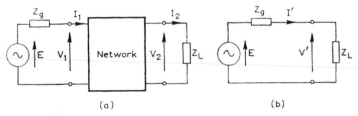

(a)   (b)

Fig. 8.3   PERTAINING TO INSERTION LOSS

those which would arise if the load were connected directly to the generator. The *insertion loss ratio* of a two-port network is defined as the ratio of the voltage across the load when it is connected directly to the generator to the voltage across the load when the two-port network is inserted. Referring to Fig. 8.3,

$$\text{Insertion loss ratio, } A_L = \frac{V'}{V_2} = \frac{I'}{I_2} \tag{8.7}$$

Since V' and V_2 both refer to voltages across the same impedance, Z_L, it is permissible to express the insertion loss ratio as

$$\text{Insertion loss} = 20 \log_{10} \frac{V'}{V_2} \quad \text{decibels} \tag{8.8}$$

$$= 20 \log_{10} \frac{I'}{I_2} \quad \text{decibels} \tag{8.9}$$

Of particular importance is the case when the network is terminated in its characteristic impedance, Z_0. The network is said to be matched, and in this case the input impedance is also Z_0 so that the insertion loss is simply the ratio of input to output voltage (or current), i.e.

$$\text{Insertion loss} = 20 \log_{10} \frac{V_1}{V_2} \text{ (for network terminated in } Z_0)$$

(8.10)

$$= 20 \log_{10} \frac{I_1}{I_2}$$

(8.11)

It will be seen that in the matched condition, the impedance of the generator does not affect the insertion loss.

EXAMPLE 8.1 For the attenuator pad shown in Fig. 8.4 determine the characteristic impedance and the insertion loss when the attenuator feeds a matched load. $R_1 = 312\,\Omega$ and $R_2 = 423\,\Omega$.

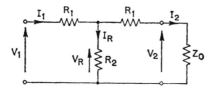

Fig. 8.4 SYMMETRICAL T-NETWORK

From eqn. (8.2) the characteristic impedance is

$$Z_0 = \sqrt{(Z_{sc}Z_{oc})} = \sqrt{\left\{\left(R_1 + \frac{R_1 R_2}{R_1 + R_2}\right)(R_1 + R_2)\right\}}$$

$$= \sqrt{(R_1^2 + 2R_1 R_2)} = \sqrt{\{312^2 + (2 \times 312 \times 423)\}}$$

$$= 600\,\Omega$$

The ratio of V_1/V_2 is obtained as follows. Since $V_2 = I_2 Z_0$, it follows that

$$A_L = \frac{V_1}{V_2} = \frac{I_1 Z_0}{I_2 Z_0} = \frac{I_1}{I_1 \dfrac{R_2}{R_1 + Z_0 + R_2}} = \frac{R_1 + R_2 + Z_0}{R_2}$$

$$= \frac{1,335}{423} = 3 \cdot 16$$

or

$$\text{Insertion loss} = 20 \log_{10} 3 \cdot 16 = 10\,\text{dB}$$

EXAMPLE 8.2 For the simple π-section shown in Fig. 8.5 determine the range of frequencies over which the insertion loss is zero, when the section is terminated in its characteristic impedance.

It is apparent that when the frequency is zero, there is no insertion loss (since the inductor behaves as a short-circuit), and that when the frequency is infinite the capacitors act as a short-circuit and the insertion loss is infinite. The network is a *low-pass filter* section, i.e. it permits signals up to a certain *cut-off frequency*

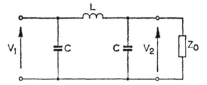

Fig. 8.5   SIMPLE LOW-PASS FILTER

to pass through unattenuated, while attenuating all signals above the cut-off frequency.

The insertion loss ratio is

$$A_L = \frac{V_1}{V_2} = \frac{j\omega L + \dfrac{Z_0/j\omega C}{Z_0 + 1/j\omega C}}{\dfrac{Z_0/j\omega C}{Z_0 + 1/j\omega C}}$$

$$= \frac{\left(j\omega L Z_0 + \dfrac{L}{C} + \dfrac{Z_0}{j\omega C}\right) j\omega C}{Z_0} = 1 - \omega^2 LC + \frac{j\omega L}{Z_0}$$

From eqn. (8.3),

$$Z_0 = \sqrt{\frac{j\omega L}{2j\omega C - j\omega^3 LC^2}} = \sqrt{\frac{L}{2C - \omega^2 LC^2}}$$

Note that Z_0 will vary with frequency, and will be infinite when $\omega^2 LC = 2$. At low frequencies Z_0 will be approximately constant and equal to $\sqrt{(L/2C)}$. It is, however, a reference quantity and will therefore represent a pure resistance provided that

$$2C > \omega^2 LC^2 \quad \text{or} \quad \omega < \sqrt{\frac{2}{LC}}$$

For this condition,

$$A_L = 1 - \omega^2 LC + \frac{j\omega L \sqrt{(2C - \omega^2 LC^2)}}{\sqrt{L}}$$

so that

$$A_L = \sqrt{\{(1 - \omega^2 LC)^2 + \omega^2 L(2C - \omega^2 LC^2)\}} = 1$$

This means that for all frequencies below that for which $\omega = \sqrt{(2/LC)}$ the insertion loss ratio is unity, i.e. there is no attenuation. This gives the cut-off frequency of the filter.

EXAMPLE 8.3   A resistance voltage divider with a total resistance of 5,000 Ω s connected across the output of a generator of internal resistance 600 Ω.

Determine the insertion loss ratio if a load of 1,000Ω is connected across the divider at a tapping of (a) 2,500Ω, (b) 1,250Ω. The circuit is shown in Fig. 8.6(a).

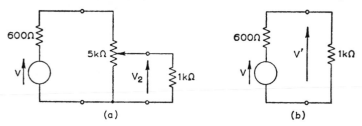

(a)   (b)

Fig. 8.6 VOLTAGE DIVIDER AS A TWO-PORT NETWORK

Without the voltage divider the voltage across the 1 kΩ load resistor is

$$V' = \frac{V}{1,600} \times 1,000 = 0.625 \text{ V}$$

where V is the generator e.m.f.

(a) With the 2·5 kΩ tapping,

$$V_2 = \frac{V}{600 + 2,500 + \dfrac{2,500 \times 1,000}{3,500}} \times \frac{2,500 \times 1,000}{3,500} = 0.187 \text{ V}$$

Hence the insertion loss ratio is

$$A_L = \frac{V'}{V_2} = \frac{0.625}{0.187} = \underline{3.34}$$

(b) With the 1·25 kΩ tapping the voltage across the load is

$$V_2 = \frac{V}{600 + 3,750 + \dfrac{1,250 \times 1,000}{2,250}} \times \frac{1,250 \times 1,000}{2,250} = 0.113 \text{ V}$$

Hence the insertion loss ratio is

$$A_L = \frac{0.625}{0.113} = \underline{5.5}$$

Note that the insertion loss is not doubled by halving the tapping.

8.4 Equivalent Circuit of a Short Transmission Line

A single-phase a.c. transmission line consists of two conductors, conventionally called the "go" and "return" conductors. These have series resistance due to the finite resistivity of both conductors. They also have

(a) a distributed leakage resistance between the conductors which depends on the conductivity of the insulation,

(b) a distributed inductance since they form a current-carrying loop which sets up and links a magnetic field, and

(c) a distributed shunt capacitance, since the conductors form the electrodes of an electric field.

The line is a two-port network.

In an overhead line not exceeding 100 km, or a cable not exceeding 20 km, at a frequency of 50 Hz (and proportionately shorter at higher frequencies), the shunt capacitance and leakage may normally be neglected, and the equivalent circuit of the line will then

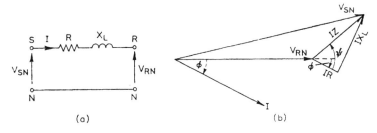

Fig. 8.7   SHORT TRANSMISSION LINE SUPPLYING A LOAD AT A
LAGGING PHASE ANGLE ϕ
(a) Equivalent circuit   *(b)* Complexor diagram

consist of a resistance in series with an inductive reactance. For convenience the impedance of both "go" and "return" conductors are lumped together in one line as shown in Fig. 8.7.

The sending-end voltage V_{SN} is the complexor sum of the receiving-end voltage V_{RN} and the line voltage drop IZ, where $Z = R + jX_L$, i.e.

$$V_{SN} = V_{RN} + IZ$$

The *regulation* of the transmission line is defined as the rise in voltage at the receiving end of the line when the load is removed, with the sending-end voltage remaining constant. Hence

$$\text{Regulation} = \frac{V_{SN} - V_{RN}}{V_{RN}} \quad \text{per unit} \tag{8.12}$$

Since V_{RN} will depend on the load, the regulation will also depend on the load current and power factor.

8.5 Medium-length Lines

When 50 Hz lines exceed 100 km in length (and about 20 km in the case of cables), the capacitance can no longer be neglected. For the purposes of calculation the line may then be approximately represented by a nominal-π or nominal-T network, as shown in

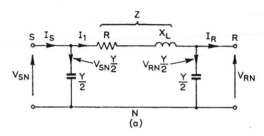

(a)

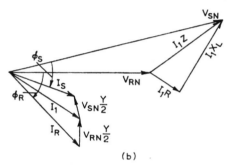

(b)

Fig. 8.8 NOMINAL-π EQUIVALENT CIRCUIT AND COMPLEXOR DIAGRAM (LAGGING LOAD)

Figs. 8.8 and 8.9 respectively. These approximate representations do not hold for 50 Hz lines of over 350 km, or for lines operating at frequencies higher than normal power frequencies.

Consider the nominal-π circuit of Fig. 8.8. Let the "go" and "return" resistance of the line be R ohms, the loop inductance L henrys, and the capacitance between the two lines C farads. The series impedance of the line is

$$Z = R + j\omega L = R + jX_L$$

The total admittance between conductors is

$$Y = j\omega C$$

Each "leg" of the π-circuit will have half this admittance.

The complexor diagram shown at (*b*) is constructed taking the receiving-end voltage, V_{RN}, as the reference complexor, and is self-explanatory.

In the nominal-T circuit the resistance and inductance are divided into two, and the capacitance is considered to be concentrated at the centre of the line as shown in Fig. 8.9(*a*).

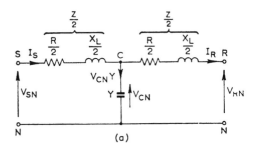

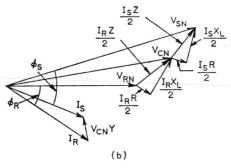

(b)

Fig. 8.9   NOMINAL-T EQUIVALENT CIRCUIT AND COMPLEXOR
DIAGRAM (LAGGING LOAD)

EXAMPLE 8.4   A single-phase 50 Hz transmission line has the following line constants—resistance 25 Ω; inductance 200 mH; capacitance 1·4 μF. Calculate the sending-end voltage, current and power factor when the load at the receiving-end is 273 A at 76·2 kV with a power factor of 0·8 lagging, using (*a*) a nominal-π circuit, and (*b*) a nominal-T circuit, to represent the line.

Series impedance, $Z = R + j\omega L = 25 + j62\cdot8 = 67\cdot6\underline{/68\cdot3°}\,\Omega$

Shunt admittance, $Y = j\omega C = 0\cdot44 \times 10^{-3}\underline{/+90°}$ S

Receiving-end voltage, $V_{RN} = 76\cdot2 \times 10^3\underline{/0°}$ V (reference complexor)

and

Receiving-end current, $I_R = 273\underline{/-36\cdot9°}$ A

(a) Using the nominal-π circuit (Fig. 8.8):
Mid-section current $= I_1 =$ receiving-end current
$\qquad\qquad\qquad\qquad$ + current through receiving-end half of Y
$$= I_R + V_{RN}Y/2$$
$$= 273\underline{/-36\cdot9^\circ} + (72\cdot6 \times 10^3\underline{/0^\circ} \times 0\cdot22 \times 10^{-3}\underline{/90^\circ})$$
$$= 264\underline{/-33\cdot9^\circ}\,\text{A}$$

Sending-end voltage $= V_{SN} = V_{RN} + I_1Z$
$$= (76\cdot2 \times 10^3\underline{/0^\circ}) + (264\underline{/-33\cdot9^\circ} \times 67\cdot6\underline{/68\cdot3^\circ})$$
$$= (91 + j10\cdot1) \times 10^3 = 91{,}500\underline{/6\cdot6^\circ}\,\text{V}$$

Sending-end current $= I_S = I_1 + V_{SN}Y/2$
$$= (219 - j147) + (91\cdot5 \times 10^3\underline{/6\cdot6^\circ} \times 0\cdot22 \times 10^{-3}\underline{/90^\circ})$$
$$= 251\underline{/-30\cdot3^\circ}\,\text{A}$$

Sending-end power factor $= \cos(6\cdot6 - (-30\cdot3))^\circ = 0\cdot8$ lagging

(b) Using the nominal-T circuit (Fig. 8.9):
Mid-section voltage $= V_{CN} = V_{RN} + I_RZ/2$
$$= (76\cdot2 \times 10^3\underline{/0^\circ}) + (273\underline{/-36\cdot9^\circ} \times 33\cdot8\underline{/68\cdot3^\circ})$$
$$= (84\cdot1 + j4\cdot8) \times 10^3 = 84\cdot1 \times 10^3\underline{/3^\circ}\,\text{V}$$

Sending-end current $= I_S =$ Receiving-end current + current in mid-section
$$= (219 - j164) + (84\cdot1 \times 10^3\underline{/3^\circ} \times 0\cdot44 \times 10^{-3}\underline{/(90^\circ)})$$
$$= 252\underline{/-30\cdot3^\circ}\,\text{A}$$

Sending-end voltage
$$= V_{SN} = V_{CN} + I_SZ/2$$
$$= (84\cdot1 + j4\cdot8) \times (10^3 + 252\underline{/30\cdot3^\circ} \times 33\cdot8\underline{/68\cdot3^\circ})$$
$$= 91\cdot3 \times 10^3\underline{/6\cdot5^\circ}\,\text{V}$$

Sending-end power factor $= \cos(6\cdot5 - (-30\cdot3))^\circ = 0\cdot8$ lagging

8.6 ABCD Constants

For any linear passive two-port network there will be a linear relationship between the input voltage and current (V_1, I_1) and the output voltage and current (V_2, I_2). This relationship can be expressed in the form

$$\left.\begin{array}{l} V_1 = AV_2 + BI_2 \\ I_1 = CV_2 + DI_2 \end{array}\right\} \qquad\qquad (8.13)$$

where **A, B, C** and **D** are constants. It is obvious from the form of the equations that **A** and **D** are dimensionless, **B** has the dimensions of an impedance, and **C** has the dimensions of an admittance.

For passive networks there is a fixed relationship between the ABCD constants. Consider any passive network represented by the "black box" in Fig. 8.10(a) with its output terminals short-circuited (and hence with $V_2 = 0$). Then,

$$\left.\begin{array}{l} V_1 = \mathbf{B}I_{2sc} \\ I_{1sc} = \mathbf{D}I_{2sc} \end{array}\right\} \tag{8.14}$$

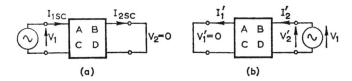

(a)   *(b)*

Fig. 8.10

If now the generator is connected across the output terminals, and the input terminals are short-circuited as at (b), then, taking into account the current directions, the fact that V_1' is zero, and choosing $V_2' = V_1$,

$$0 = \mathbf{A}V_1 - \mathbf{B}I_2' \quad \text{or} \quad I_2' = \frac{\mathbf{A}V_1}{\mathbf{B}} \tag{8.15}$$

and

$$-I_1' = \mathbf{C}V_1 - \mathbf{D}I_2' \tag{8.16}$$

Since the network is passive, the reciprocity theorem (Section 2.9) applies, so that $I_1' = I_{2sc}$ and hence from eqns. (8.15) and (8.16),

$$-I_{2sc} = \mathbf{C}V_1 - \frac{\mathbf{D}\mathbf{A}V_1}{\mathbf{B}}$$

But $I_{2sc} = V_1/\mathbf{B}$ (from eqn. (8.14)), so that

$$-\frac{V_1}{\mathbf{B}} = \mathbf{C}V_1 - \frac{\mathbf{D}\mathbf{A}V_1}{\mathbf{B}}$$

or

$$\mathbf{AD} - \mathbf{BC} = 1 \tag{8.17}$$

This important relationship can be simplified for symmetrical networks, since in this case the ratio of V/I at any pair of terminals with the other pair short-circuited must result in the same value (the network looks the same from either port). For the left-hand port,

$$\frac{V_1}{I_{1sc}} = \frac{\mathbf{B}}{\mathbf{D}}$$

For the right hand port,

$$\frac{V_2'}{I_2'} = \frac{V_1}{I_2'} = \frac{\mathbf{B}}{\mathbf{A}} \quad \text{(from eqn. (8.15))}$$

Since these two expressions must be equal it follows that $\mathbf{A} = \mathbf{D}$ for a symmetrical network, and hence eqn. (8.17) becomes

$$\mathbf{A}^2 - \mathbf{BC} = 1 \tag{8.18}$$

8.7 Evaluation of ABCD Constants

The ABCD constants of a passive network can be found from measurements of input and output currents and voltages under open- and short-circuit conditions. If the actual configuration of a network is known, the ABCD constants can be found from the circuit by inspection (or short calculation).

Thus if the output port is open-circuited it follows that $I_2 = 0$, and from eqn. (8.13),

$$V_{1oc} = \mathbf{A}V_{2oc}$$

or

$$\mathbf{A} = \frac{V_1}{V_2}\bigg|_{oc} \tag{8.19}$$

and

$$I_{1oc} = \mathbf{C}V_{2oc}$$

or

$$\mathbf{C} = \frac{I_1}{V_2}\bigg|_{oc} \tag{8.20}$$

If the output is short-circuited and the input current and voltage and the output current are measured, then, since $V_2 = 0$,

$$V_{1sc} = \mathbf{B}I_{2sc}$$

or

$$\mathbf{B} = \frac{V_1}{I_2}\bigg|_{sc} \tag{8.21}$$

and

$$I_{1\,sc} = DI_{2\,sc}$$

or

$$D = \left.\frac{I_1}{I_2}\right|_{sc} \tag{8.22}$$

EXAMPLE 8.5 Determine the A BCD constants for the symmetrical-T circuit shown in Fig. 8.11.

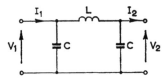

Fig. 8.11 SYMMETRICAL-T TWO-PORT NETWORK

First imagine the output to be open-circuited so that $I_2 = 0$. Then for an input voltage V_1, eqn. (8.19) gives

$$A = \left.\frac{V_1}{V_2}\right|_{oc} = \frac{V_1}{\dfrac{V_1}{j\omega L + 1/j\omega C}\dfrac{1}{j\omega C}} \quad \text{by inspection}$$

$$= j\omega C(j\omega L + 1/j\omega C) = \underline{1 - \omega^2 LC}$$

Since the network is symmetrical $A = D = 1 - \omega^2 LC$.

If the output is now short-circuited, then from eqn. (8.21),

$$B = \left.\frac{V_1}{I_2}\right|_{sc} = \frac{V_1}{V_1/j\omega L} = \underline{j\omega L}$$

The constant C can be found either from eqn. (8.20) or by using the relation for a passive symmetrical two-port network, $A^2 - BC = 1$. Using the latter method,

$$C = \frac{A^2 - 1}{B} = \frac{\omega^4 L^2 C^2 - 2\omega^2 LC}{j\omega L} = \underline{2\omega C - j\omega^3 LC^2}$$

8.8 Characteristic Impedance in terms of ABCD Constants

It has been seen that the characteristic impedance, Z_0, of a symmetrical network is obtained from the relation

$$Z_0 = \sqrt{(Z_{sc}Z_{oc})}$$

The input impedances under short- and open-circuit *output* conditions can be obtained from eqns. (8.13). Thus for short-circuited output, $V_2 = 0$, and

$$Z_{in\,sc} = \left.\frac{V_1}{I_1}\right|_{sc} = \frac{B}{D}\left(= \frac{B}{A} \quad \text{for a symmetrical network}\right)$$

Similarly for an open-circuited output, $I_2 = 0$, and

$$Z_{in\,oc} = \left.\frac{V_1}{I_1}\right|_{oc} = \frac{A}{C}$$

It follows directly that, for a symmetrical network,

$$Z_0 = \sqrt{\frac{B}{A}\frac{A}{C}} = \sqrt{\frac{B}{C}} \qquad (8.23)$$

EXAMPLE 8.6 A symmetrical two-port network has $A = D = 0\cdot8\underline{/30°}$; $B = 100\underline{/60°}\,\Omega$; $C = 5 \times 10^{-3}\underline{/90°}$ S. Determine (*a*) the characteristic impedance, and (*b*) the output voltage and current when the input voltage is $120\underline{/0°}$ V and the output load is a pure resistance of $141\,\Omega$. The circuit is shown in Fig. 8.12.

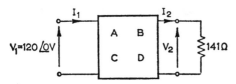

Fig. 8.12

The characteristic impedance, Z_0, is readily obtained from eqn. (8.23) as

$$Z_0 = \sqrt{\frac{B}{C}} = \sqrt{\frac{100\underline{/60°}}{5 \times 10^{-3}\underline{/90°}}} = \underline{\underline{141\underline{/-15°}\,\Omega}}$$

From Fig. 8.12,

$$\frac{V_2}{I_2} = 141\underline{/0°} \qquad \text{or} \qquad V_2 = 141I_2 \qquad (i)$$

Hence, from eqn. (8.13),

$$V_1 = A \times 141I_2 + BI_2$$

or

$$I_2 = \frac{120\underline{/0°}}{(0\cdot8\underline{/30°} \times 141) + 100\underline{/60°}} = \underline{\underline{0\cdot58\underline{/-44°}\,A}}$$

and from (i),

$$V_2 = 141 \times 0\cdot58\underline{/-44°} = \underline{\underline{82\underline{/-44°}\,V}}$$

8.9 Two-port Networks in Cascade

Very frequently two-port networks are connected in cascade as shown in Fig. 8.13. Thus an attenuator or a filter may consist of several cascaded sections in order to achieve a given desired overall performance. If the chain of networks is designed on an iterative basis, then each section will have the same characteristic impedance (assuming symmetrical sections), and the last network will be terminated in Z_0. It follows that each network will have a matched

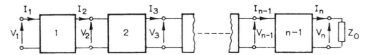

Fig. 8.13   TWO-PORT NETWORKS IN CASCADE

termination, and hence the insertion loss in decibels of section 1 is $m_1 = 20 \log_{10}(V_1/V_2)$; of section 2 is $m_2 = 20 \log_{10}(V_2/V_3)$; etc.
The overall insertion loss will be

$$m = 20 \log_{10}(V_1/V_n)$$

$$= 20 \log_{10} \left(\frac{V_1}{V_2} \times \frac{V_2}{V_3} \times \frac{V_3}{V_4} \cdots \frac{V_{n-1}}{V_n} \right)$$

$$= 20 \log_{10} \frac{V_1}{V_2} + 20 \log_{10} \frac{V_2}{V_3} + \ldots + 20 \log_{10} \frac{V_{n-1}}{V_n}$$

$$= m_1 + m_2 + \ldots + m_{n-1} \tag{8.24}$$

This illustrates the great convenience of the use of matched sections in cascade—the overall insertion loss is simply the sum of the insertion losses of the sections. If the sections do not have the same characteristic impedance or are not terminated in Z_0, eqn. (8.24) is no longer valid, and the calculation of insertion loss may be very tedious.

PROBLEMS

8.1   In a symmetrical-T attenuator pad each series arm has a resistance of $30\,\Omega$ and the shunt arm has a resistance of $100\,\Omega$. Determine the characteristic impedance, and the insertion loss when feeding a matched load.
   Ans. $83\,\Omega$; $6 \cdot 6\,\mathrm{dB}$.

8.2   A symmetrical-π attenuator pad has a series arm of $2\,\mathrm{k}\Omega$ resistance, and each shunt arm of $1\,\mathrm{k}\Omega$ resistance. Determine the characteristic impedance, and the insertion loss when feeding a matched load.
   Ans. $707\,\Omega$; $15 \cdot 4\,\mathrm{dB}$.

8.3   A 10:1 voltage divider has a total resistance of $1{,}000\,\Omega$. Calculate from first principles the insertion loss when the divider is connected between a generator of internal resistance 600 Ω and a load of resistance 200 Ω. Repeat for a generator internal resistance of 5 Ω.

Ans. 5·9 (15·4 dB); 14·2 (23 dB).

8.4 A symmetrical attenuator pad has a characteristic impedance of 75 Ω and an insertion loss of 20 dB. It feeds a load of 75 Ω resistance. Calculate the load voltage when the attenuator is connected to a generator of e.m.f. 20 V and internal resistance (a) 75 Ω, (b) 600 Ω.

Ans. 1 V; 0·22 V.

8.5 A symmetrical-T section has each series arm of 100 mH pure inductance and a shunt arm of 0·1 μF capacitance. Determine (a) the frequency at which $Z_0 = 0$; (b) the insertion loss of the network below this frequency, when correctly matched. Plot the variation of Z_0 with frequency from zero up to the frequency found for (a).

Ans. 2·25 kHz; 1.

8.6 A single-phase 50 Hz line, 5 km long, supplies 5,000 kW at a p.f. of 0·71 lagging. The line has a resistance of 0·0345 Ω per km for each wire, and a loop inductance per km of 1·5 mH. The receiving-end voltage is 10 kV. A capacitor is connected across the load to raise its p.f. to 0·9 lagging. Calculate (a) the capacitance of the capacitor, (b) the sending-end voltage with the capacitor in use and out of use, and (c) the efficiency of transmission in each case.[*]

Ans. 82·4 μF; 10·8 kV, 11·4 kV; 98 per cent, 96 per cent.

8.7 A single-phase transmission line 50 miles long, delivers 4,000 kW at a voltage of 38 kV and a power factor of 0·85 lagging. Find the sending-end voltage, current, and power factor by the nominal-T method. The resistance, reactance and susceptance per mile are 0·3 Ω, 0·7 Ω and 12 × 10⁻⁶ S.

Ans. 41·5 kV; 112·5 A, 0·9 lagging.

8.8 Repeat Problem 8.7 using the nominal-π equivalent circuit.

8.9 Calculate the ABCD constants of the line in Problem 8.6.
Ans. 1; 0·17 + j2·36; 0; 1.

8.10 Calculate the ABCD constants of the line in Problem 8.7.
Ans. 1; 15 + j35 Ω; j6 × 10⁻⁴ S; 1.

8.11 A 3-phase transmission line has the following constants:

$$\mathbf{A} = \mathbf{D} = 1\underline{/0°} \qquad \mathbf{B} = 50\underline{/60°} \qquad \mathbf{C} = 10^{-3}\underline{/90°}$$

Determine the input voltage, current and power factor when the output current is 100 A lagging behind the output phase voltage of 20 kV by 37°.

Ans. 39 kV; 90 A; 0·79.

---

[*] Efficiency of transmission = $\dfrac{\text{Output power}}{\text{Input power}}$.

Chapter 9

TRANSFORMERS

Two circuits are said to be *mutually inductive* when some of the magnetic flux caused by the excitation of one circuit links with some or all of the turns of the second circuit. Transformers are mutual inductors designed so that the magnetic coupling between the two circuits is *tight*, i.e. the greater part of the flux linking each circuit is mutual. This is achieved by providing a low-reluctance path for the mutual flux.

Equivalent circuits consisting of resistive and self-inductive (not mutually inductive) elements can be derived for mutual inductors or transformers. The self-inductive elements may be expressed either in terms of the total circuit inductances and the mutual inductance or in terms of leakage inductances and magnetizing inductances.

9.1 Winding Inductances

Fig. 9.1 shows a pair of magnetically coupled coils. At (a) is shown the flux distribution when only the winding denoted by subscript 1 carries current; at (b) is shown the flux distribution when only the winding denoted by subscript 2 carries current, while the fluxes when both windings carry current are shown at (c). The dot notation is used to indicate the relative coil polarities; thus currents entering the dotted ends of the coils produce aiding mutual fluxes. At (c), current is shown entering the dotted end of one winding and leaving the dotted end of the other. When both windings carry

260

current simultaneously these current directions are the more common as they represent power transmission through the mutual coupling.

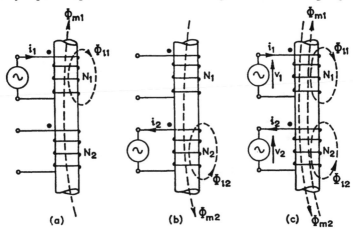

Fig. 9.1 MAGNETICALLY COUPLED COILS

(a) Primary winding carrying current
(b) Secondary winding carrying current
(c) Both windings carrying current

The diagrams show the total flux linkage of the coils to be due to four component fluxes, namely

Φ_{m1}, a component, due to the current in winding 1, which links all the turns of both windings.

Φ_{l1}, a component, due to the current in winding 1, which links all the turns of winding 1 but none of those of winding 2 (this is the primary leakage flux).

Φ_{m2}, a component, due to the current in winding 2, which links all the turns of both windings.

Φ_{l2}, a component, due to the current in winding 2, which links only the turns of winding 2 (this is the secondary leakage flux).

These component fluxes may be used to define various inductances. It is assumed that the magnetic circuit is unsaturated so that the defined inductances are constant.

$$\text{Primary self-inductance, } L_{11} = \frac{(\Phi_{m1} + \Phi_{l1})N_1}{i_1} \tag{9.1}$$

$$\text{Secondary self-inductance, } L_{22} = \frac{(\Phi_{m2} + \Phi_{l2})N_2}{i_2} \tag{9.2}$$

$$\text{Mutual inductance, } L_{12} = \frac{\Phi_{m1}N_2}{i_1} \tag{9.3}$$

Alternatively the mutual inductance is

$$L_{21} = \frac{\Phi_{m2}N_1}{i_2} \tag{9.4}$$

These mutual inductances are equal,

i.e. $\quad L_{12} = L_{21}$ \hfill (9.5)

The coupling coefficient, k, of magnetically coupled coils is defined as

$$k = \frac{L_{12}}{\sqrt{(L_{11}L_{22})}} \tag{9.6}$$

k has a maximum value of 1, corresponding to the case when no leakage flux links either coil.

Equivalent circuits for magnetically coupled coils may be obtained by representing the flux linkage in terms of L_{11}, L_{22} and L_{12}, and this is undertaken in the following section. Equivalent circuits may also be obtained by representing the flux linkage in terms of leakage and magnetizing inductances, and these inductances are now defined.

Primary leakage inductance, $L_{l1} = \dfrac{\Phi_{l1}N_1}{i_1}$ \hfill (9.7)

Secondary leakage inductance, $L_{l2} = \dfrac{\Phi_{l2}N_2}{i_2}$ \hfill (9.8)

Primary magnetising inductance, $L_{m1} = \dfrac{\Phi_{m1}N_1}{i_1}$ \hfill (9.9)

Secondary magnetising inductance, $L_{m2} = \dfrac{\Phi_{m2}N_2}{i_2}$ \hfill (9.10)

9.2 Equivalent Circuits for Magnetically Coupled Coils

Applying Kirchhoff's second law to the magnetically coupled coils shown in Fig. 9.1(c) gives the equations

$$v_1 = R_1i_1 + N_1 \frac{d}{dt}(\Phi_{l1} + \Phi_{m1} - \Phi_{m2}) \tag{9.11}$$

$$v_2 = -R_2i_2 + N_2 \frac{d}{dt}(\Phi_{m1} - \Phi_{m2} - \Phi_{l2}) \tag{9.12}$$

where R_1 and R_2 are the resistances of the primary and secondary coils respectively. The primary and secondary flux linkages may now be expressed in terms of the self- and mutual inductances according to eqns. (9.1)–(9.3). Rearranging eqns. (9.11) and (9.12),

$$v_1 = R_1 i_1 + N_1 \frac{d}{dt}(\Phi_{l1} + \Phi_{m1}) - N_1 \frac{d}{dt}\Phi_{m2}$$

$$v_2 = -R_2 i_2 + N_2 \frac{d}{dt}\Phi_{m1} - N_2 \frac{d}{dt}(\Phi_{m2} + \Phi_{l2})$$

Therefore,

$$v_1 = R_1 i_1 + L_{11}\frac{di_1}{dt} - L_{12}\frac{di_2}{dt} \tag{9.13}$$

$$v_2 = -R_2 i_2 - L_{22}\frac{di_2}{dt} + L_{12}\frac{di_1}{dt} \tag{9.14}$$

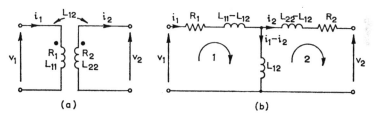

Fig. 9.2   EQUIVALENT T-CIRCUIT FOR MAGNETICALLY COUPLED COILS
(a) Mutually inductive coils
(b) Equivalent T-circuit

If the direction of i_2 had been chosen as entering the dotted end of the secondary, then all fluxes would have been additive and no minus signs would have occurred in eqns. (9.13) and (9.14).

Fig. 9.2(a) shows the circuit representation for mutually inductive coils. Eqns. (9.13) and (9.14) may be rewritten

$$v_1 = R_1 i_1 + L_{11}\frac{di_1}{dt} - L_{12}\frac{di_1}{dt} + L_{12}\frac{di_1}{dt} - L_{12}\frac{di_2}{dt} \tag{9.15}$$

adding and subtracting $L_{12}(di_1/dt)$ in eqn. (9.13); and

$$v_2 = -R_2 i_2 - L_{22}\frac{di_2}{dt} + L_{12}\frac{di_2}{dt} + L_{12}\frac{di_1}{dt} - L_{12}\frac{di_2}{dt} \tag{9.16}$$

adding and subtracting $L_{12}(di_2/dt)$ in eqn. (9.14).

When Kirchhoff's second law is applied to meshes 1 and 2 of Fig. 9.2(b), eqns. (9.15) and (9.16) respectively are obtained. Fig. 9.2(b) is therefore the equivalent T-circuit of the mutually inductive coils. This equivalent circuit is convenient for the consideration of networks containing mutual inductance since it consists of resistive and self-inductive elements only.

9.3 Mutual Inductance in Networks

Eqns. (9.13) and (9.14) may be made to refer to steady-state sinusoidal operation by

(a) Replacing all the time-varying quantities by complexor quantities, putting i.e. V_1 for v_1, I_1 for i_1, etc.

(b) Replacing d/dt operators by $j\omega$.

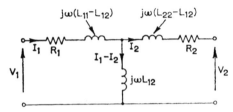

Fig. 9.3   EQUIVALENT T-CIRCUIT FOR STEADY-STATE A.C. OPERATION

Eqns. (9.13) and (9.14) then become

$$V_1 = R_1 I_1 + j\omega L_{11} I_1 - j\omega L_{12} I_2 \tag{9.17}$$

$$V_2 = -R_2 I_2 - j\omega L_{22} I_2 + j\omega L_{12} I_1 \tag{9.18}$$

The corresponding equivalent circuit is shown in Fig. 9.3.

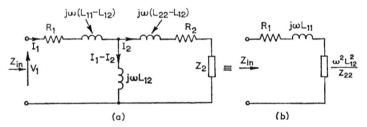

Fig. 9.4   CALCULATION OF INPUT IMPEDANCE

(a) Equivalent T-circuit
(b) Equivalent series circuit: $Z_{22} = R_2 + j\omega L_{22} + Z_2$

Fig. 9.4(a) shows the equivalent T-circuit with a secondary load of impedance Z_2. Fig. 9.4(b) shows the series equivalent circuit. Since the equivalent T-circuit does not contain any mutual coupling (but only self-inductances whose values are expressed in terms of the mutual inductance), the input impedance may be calculated using series-parallel circuit theory. Thus the input impedance measured at the primary terminals is

$$Z_{in} = R_1 + j\omega(L_{11} - L_{12})$$

$$+ \frac{j\omega L_{12}\{R_2 + j\omega(L_{22} - L_{12}) + Z_2\}}{j\omega L_{12} + R_2 + j\omega(L_{22} - L_{12}) + Z_2} \quad (9.19)$$

Putting $R_2 + j\omega L_{22} + Z_2 = Z_{22}$ gives, after simplification,

$$Z_{in} = R_1 + j\omega L_{11} + \frac{\omega^2 L_{12}{}^2}{Z_{22}} \quad (9.20)$$

This confirms the result found in Section 1.10.

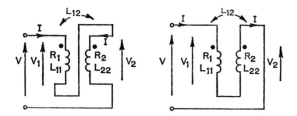

Fig. 9.5   IMPEDANCE OF INDUCTIVELY COUPLED COILS

   (a) Series aiding: $Z = R_1 + R_2 + jw(L_{11} + L_{22} + 2L_{12})$
   (b) Series opposing: $Z = R_1 + R_2 + jw(L_{11} + L_{22} - 2L_{12})$

In Fig. 9.5(a) two mutually coupled coils are connected in series. The connexion is called *series aiding*, since current enters the dotted ends of the coils, which thus produce aiding fluxes. Eqns. (9.17) and (9.18) will apply with $I_1 = -I_2 (= I$, say), since the current in the second coil is reversed compared with the direction assumed when these equations were established. This gives

$$V_1 = R_1 I + j\omega L_{11} I + j\omega L_{12} I$$

$$V_2 = R_2 I + j\omega L_{22} I + j\omega L_{12} I$$

$$V = V_1 + V_2 = (R_1 + R_2)I + j\omega(L_{11} + L_{22} + 2L_{12})I$$

i.e.

$$Z = \frac{V}{I} = R_1 + R_2 + j\omega(L_{11} + L_{22} + 2L_{12}) \quad (9.21)$$

Fig. 9.5(b) shows a *series-opposing* connexion of mutually coupled coils. In this case the current enters the dotted end of coil 1 and leaves the dotted end of coil 2. Eqns. (9.17) and (9.18) again apply with $I_1 = I$ and $I_2 = I$. This gives

$$V_1 = R_1I + j\omega L_{11}I - j\omega L_{12}I$$
$$V_2 = -R_2I - j\omega L_{22}I + j\omega L_{12}I$$
$$V = V_1 - V_2 = (R_1 + R_2)I + j\omega(L_{11} + L_{22} - 2L_{12})I$$

i.e.

$$Z = \frac{V}{I} = R_1 + R_2 + j\omega(L_{11} + L_{22} - 2L_{12}) \tag{9.22}$$

Eqns. (9.21) and (9.22) confirm the results found in Section 1.10.

9.4 The Ideal Transformer

The terminal voltages of a pair of magnetically coupled coils as shown in Fig. 9.1(c) are, from eqns. (9.11) and (9.12),

$$v_1 = R_1i_1 + N_1 \frac{d}{dt}(\Phi_{l1} + \Phi_{m1} - \Phi_{m2}) \tag{9.11}$$

$$v_2 = -R_2i_2 + N_2 \frac{d}{dt}(\Phi_{m1} - \Phi_{m2} - \Phi_{l2}) \tag{9.12}$$

Let e_1, e_2 be the voltages induced in the primary and secondary coils respectively due to the mutual flux only; then

$$e_1 = N_1 \frac{d}{dt}(\Phi_{m1} - \Phi_{m2}) \tag{9.23}$$

$$e_2 = N_2 \frac{d}{dt}(\Phi_{m1} - \Phi_{m2}) \tag{9.24}$$

Dividing eqn. (9.23) by eqn. (9.24),

$$\frac{e_1}{e_2} = \frac{N_1}{N_2} \tag{9.25}$$

This equation defines an ideal voltage transformation. Examination of eqns. (9.11) and (9.12) shows that an ideal voltage transformation would be obtained from a pair of magnetically coupled coils in which the winding resistances and leakage fluxes were zero. In practice there must always be some leakage flux, but a close approximation to the ideal is obtained if the two windings are placed physically close to each other on a common magnetic core.

The mutual flux linking an ideal voltage transformation is

$$\Phi = \Phi_{m1} - \Phi_{m2} = \Lambda(i_1N_1 - i_2N_2) \tag{9.26}$$

where Λ represents the permeance of the magnetic circuit carrying the mutual flux Φ, and $i_1N_1 - i_2N_2$ is the net m.m.f. For the circuit of Fig. 9.1(c) the net m.m.f. is as stated since i_1 is shown entering the dotted end of the primary winding and i_2 is shown leaving the dotted end of the secondary winding.

For a magnetic circuit of finite permeance, i_1N_1 must be greater than i_2N_2 for there to be resultant magnetization. For a core of infinite permeance, however,

$$i_1N_1 - i_2N_2 = 0 \quad \text{or} \quad i_1N_1 = i_2N_2 \tag{9.27}$$

This equation defines the ideal current transformation that would be obtained from a pair of magnetically coupled coils linked by a magnetic circuit which had infinite permeance. In practice the magnetic circuit must have a finite permeance, but a close approximation to the ideal is obtained by making the magnetic circuit of a high-permeability material.

A pair of magnetically coupled coils which provides both an ideal voltage transformation and an ideal current transformation is called an *ideal transformer*. Eqns. (9.25) and (9.27) both apply to such a device:

$$\frac{e_1}{e_2} = \frac{N_1}{N_2} \tag{9.25}$$

$$i_1N_1 = i_2N_2 \tag{9.27}$$

Substituting for N_1/N_2 in eqn. (9.25) in terms of the winding currents,

$$\frac{e_1}{e_2} = \frac{i_2}{i_1} \quad \text{or} \quad e_1i_1 = e_2i_2 \tag{9.28}$$

Therefore, for an ideal transformer,

$$\left.\begin{matrix}\text{Instantaneous power absorbed} \\ \text{by primary winding}\end{matrix}\right\} = \left\{\begin{matrix}\text{Instantaneous power deliver-} \\ \text{ed by secondary winding}\end{matrix}\right.$$

9.5 Transformer Equivalent Circuit

In Section 9.2 an equivalent circuit for magnetically coupled coils is obtained by representing the flux linkage in terms of the coil self-inductances L_{11} and L_{22} and the mutual inductance L_{12}. An alternative approach is to represent the flux linkage in terms of the coil leakage and magnetizing inductances. This approach leads to an equivalent circuit which will be called the transformer equivalent circuit.

The terminal voltages of a pair of magnetically coupled coils as shown in Fig. 9.1(c) are, from eqns. (9.11) and (9.12),

$$v_1 = R_1 i_1 + N_1 \frac{d}{dt}(\Phi_{l1} + \Phi_{m1} - \Phi_{m2}) \tag{9.11}$$

$$v_2 = -R_2 i_2 + N_2 \frac{d}{dt}(\Phi_{m1} - \Phi_{m2} - \Phi_{l2}) \tag{9.12}$$

From eqn. (9.7),

Primary leakage flux linkage, $N_1 \Phi_{l1} = L_{l1} i_1$

From eqn. (9.8),

Secondary leakage flux linkage, $N_2 \Phi_{l2} = L_{l2} i_2$

Substituting for these flux linkages in eqns. (9.11) and (9.12),

$$v_1 = R_1 i_1 + L_{l1} \frac{di_1}{dt} + N_1 \frac{d}{dt}(\Phi_{m1} - \Phi_{m2}) \tag{9.29}$$

$$v_2 = -R_2 i_2 - L_{l2} \frac{di_2}{dt} + N_2 \frac{d}{dt}(\Phi_{m1} - \Phi_{m2}) \tag{9.30}$$

Substituting e_1 and e_2 in terms of eqns. (9.23) and (9.24) in the above equations gives

$$v_1 = R_1 i_1 + L_{l1} \frac{di_1}{dt} + e_1 \tag{9.31}$$

$$v_2 = -R_2 i_2 - L_{l2} \frac{di_2}{dt} + e_2 \tag{9.32}$$

Fig. 9.6 shows an equivalent circuit which is consistent with eqns. (9.31) and (9.32), in which the resistance and leakage inductance of the two windings are shown as series elements external to the windings. The windings are represented as wound on a common core in close physical proximity to give an ideal voltage transformation element.

In an actual transformer the core does not have infinite permeance and it is necessary to represent the effects of finite core permeance in the equivalent circuit. The core permeance is

$$\Lambda = \frac{\text{Flux}}{\text{M.M.F.}} = \frac{\Phi_{m1}}{i_1 N_1} \tag{9.33}$$

Also, from eqn. (9.9),

$$\frac{\Phi_{m1}}{i_1} = \frac{L_{m1}}{N_1}$$

Substituting for Φ_{m1}/i_1 in eqn. (9.33),

$$\Lambda = \frac{L_{m1}}{N_1{}^2} \tag{9.34}$$

or

$$L_{m1} = \Lambda N_1{}^2 \tag{9.35}$$

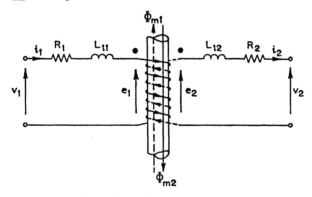

Fig. 9.6 TRANSFORMER EQUIVALENT CIRCUIT INCORPORATING
IDEAL VOLTAGE TRANSFORMATION

In a similar manner it can be shown that

$$L_{m2} = \Lambda N_2{}^2 \tag{9.36}$$

Dividing eqn. (9.35) by eqn. (9.36),

$$\frac{L_{m1}}{L_{m2}} = \frac{N_1{}^2}{N_2{}^2} \tag{9.37}$$

Substituting for Λ from eqn. (9.34) in eqn. (9.26),

$$\Phi_{m1} - \Phi_{m2} = \frac{L_{m1}}{N_1{}^2}(i_1 N_1 - i_2 N_2)$$

i.e.

$$(\Phi_{m1} - \Phi_{m2})N_1 = L_{m1}\left(i_1 - i_2 \frac{N_2}{N_1}\right)$$

or

$$(\Phi_{m1} - \Phi_{m2})N_1 = L_{m1}(i_1 - i_2') \tag{9.38}$$

where

$$i_2' = i_2 \frac{N_2}{N_1}$$

i_2' is called the *secondary current referred to the primary*. The effect of finite core permeance may therefore be represented in the equivalent circuit by the inclusion of an element of inductance L_{m1} carrying a current $i_1 - i_2'$. The voltage v across the element is, from eqn. (9.38),

$$v = L_{m1} \frac{d}{dt}(i_1 - i_2') = N_1 \frac{d}{dt}(\Phi_{m1} - \Phi_{m2}) = e_1$$

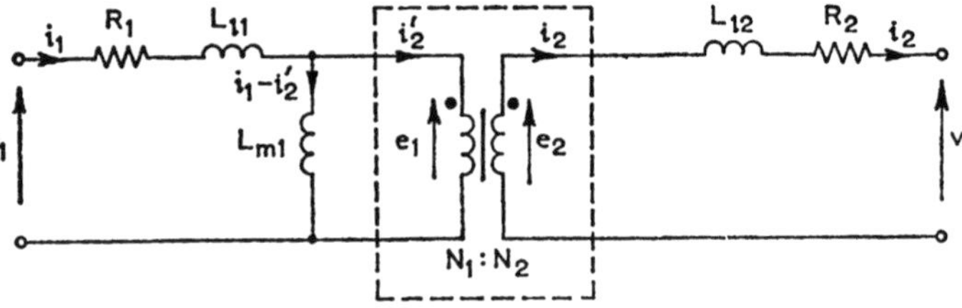

Fig. 9.7   TRANSFORMER EQUIVALENT CIRCUIT INCORPORATING AN IDEAL TRANSFORMER

Fig. 9.7 shows the transformer equivalent circuit with the circuit-element so connected that the voltage across its ends is e_1. The current through L_{m1} will be $i_1 - i_2'$ provided that the element in the "box" is an ideal transformer ($e_1/e_2 = N_1/N_2$ and $i_2'N_1 = i_2N_2$).

Fig. 9.2(b), which is repeated in Fig. 9.8(a), shows the T-equivalent circuit for a pair of magnetically coupled coils the inductive circuit elements of which are expressed in terms of L_{11}, L_{22} and L_{12}. Fig. 9.7 shows the transformer equivalent circuit in which the inductive elements are expressed in terms of L_{l1}, L_{l2} and L_{m1}. This latter equivalent circuit differs from that of Fig. 9.8(a) in that it includes an ideal transformer of turns ratio N_1/N_2.

When the equivalent circuit of Fig. 9.8(a) is rearranged by including in it at section XX the ideal transformer shown in Fig. 9.8(b), the equivalent circuit shown in Fig. 9.8(c) is obtained. This is identical with the transformer equivalent circuit of Fig. 9.7. Using eqns. (9.1)–(9.9), it is easy to show that

$$L_{l1} = L_{11} - \frac{N_1}{N_2}L_{12}$$

$$L_{l2} = L_{22} - \frac{N_2}{N_1}L_{12} \quad \text{and} \quad L_{m1} = \frac{N_1}{N_2}L_{12}$$

For identity between the equivalent circuits Figs. 9.7 and 9.8(*c*), $v_1' = e_1$, so that the shunt element of Fig. 9.8(*b*) is $(N_1/N_2)L_{12}$. Comparison of the mesh equations of Figs. 9.8(*a*) and (*c*) shows that these circuits are equivalent.

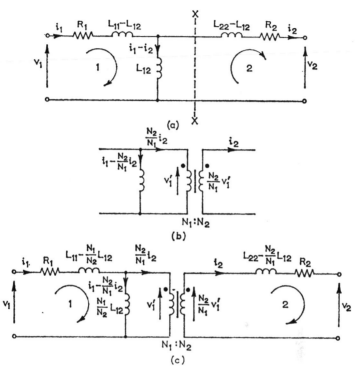

Fig. 9.8   TRANSFORMER EQUIVALENT T-CIRCUIT WITH IDEAL N_1/N_2 TRANSFORMATION

EXAMPLE 9.1   Measurements on an air-cored mutual inductor gave the following results:

Input impedance of primary (secondary open-circuited) . $(34.0 + j413)\Omega$
Input impedance of secondary (primary open-circuited) . $(40{\cdot}8 + j334)\Omega$
Input impedance when windings are joined in series
aiding . . . . $(74{\cdot}8 + j1,1190)\Omega$

The measurements were carried out at a frequency of 1,592 Hz ($\omega = 10,000$ rad/s). The primary winding has 1,200 turns and the secondary 1,000 turns.
   Determine the self-inductance of each winding, the mutual inductance and the coupling coefficient. Draw the equivalent T-circuit.
   Determine also the primary and secondary leakage and magnetizing inductances and draw the transformer equivalent circuit.

Determine the input impedance to the primary for a secondary load of $(200 + j0)\Omega$, and the input impedance to the secondary when the same load is connected to the primary terminals. The angular frequency in both cases is 10,000 rad/s.

$$L_{11} = \frac{X_1}{\omega} = \frac{413 \times 10^3}{10^4} = 41\cdot3\,\text{mH}$$

$$L_{22} = \frac{X_2}{\omega} = \frac{334 \times 10^3}{10^4} = 33\cdot4\,\text{mH}$$

From eqn. (9.21), the total inductance for a series-aiding connection is

$$L_T = L_{11} + L_{22} + 2L_{12}$$

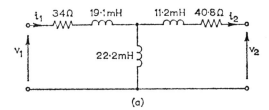

(a)

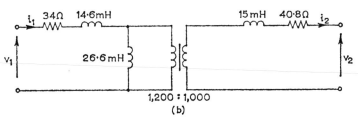

1,200 : 1,000
(b)

Fig. 9.9

Thus

$$L_T = \frac{1,190}{\omega} = \frac{1,190 \times 10^3}{10^4} = 119\,\text{mH}$$

and the mutual inductance is

$$L_{12} = \frac{L_T - L_{11} - L_{22}}{2} = \frac{119 - 41\cdot3 - 33\cdot4}{2} = 22\cdot2\,\text{mH}$$

Coupling coefficient, $k = \dfrac{L_{12}}{\sqrt{(L_{11}L_{22})}} = \dfrac{22\cdot2}{\sqrt{41\cdot3 \times 33\cdot4}} = 0\cdot537$

$$L_{11} - L_{12} = 41\cdot3 - 22\cdot2 = 19\cdot1\,\text{mH}$$

$$L_{22} - L_{12} = 33\cdot4 - 22\cdot2 = 11\cdot2\,\text{mH}$$

The equivalent-T circuit is shown in Fig. 9.9(a).

Primary leakage inductance, $L_{l1} = L_{11} - \dfrac{N_1 L_{12}}{N_2}$

$$= 41 \cdot 3 - \frac{1,200}{1,000} \times 22 \cdot 2 = \underline{\underline{14 \cdot 6 \mathrm{mH}}}$$

Secondary leakage inductance, $L_{l2} = L_{22} - \dfrac{N_2 L_{12}}{N_1}$

$$= 33 \cdot 4 - \frac{1,000}{1,200} \times 22 \cdot 2 = \underline{\underline{15 \cdot 0 \mathrm{mH}}}$$

Primary magnetizing inductance, $L_{m1} = \dfrac{N_1 L_{12}}{N_2} = \dfrac{1,200}{1,000} \times 22 \cdot 2 = \underline{\underline{26 \cdot 6 \mathrm{mH}}}$

Secondary magnetizing inductance, $L'_{m2} = \dfrac{N_2 L_{12}}{N_1} = \dfrac{1,000}{1,200} \times 22 \cdot 2$

$$= \underline{\underline{18 \cdot 5 \mathrm{mH}}}$$

The transformer equivalent circuit is shown in Fig. 9.9(b). From eqn. (9.20) the input impedance for a secondary load impedance Z_2 is

$$Z_{in} = R_1 + j\omega L_{11} + \frac{\omega^2 L_{12}{}^2}{Z_{22}}$$

where

$$Z_{22} = R_2 + j\omega L_{22} + Z_2$$

Thus

$$Z_{in} = 34 + j413 + \frac{10^8 \times 22 \cdot 2^2 \times 10^{-6}}{40 \cdot 8 + j334 + 200 + j0}$$

$$= 34 + j413 + \frac{22 \cdot 2^2 \times 10^2}{412^2}(241 - j334)$$

$$= 104 + j306 = \underline{\underline{321 \underline{/71 \cdot 2^\circ} \, \Omega}}$$

By analogy with eqn. (9.20), the secondary input impedance for a primary load Z_1 is

$$Z_{in} = R_2 + j\omega L_{22} + \frac{\omega^2 L_{12}{}^2}{Z_{11}}$$

where

$$Z_{11} = R_1 + j\omega L_{11} + Z_1$$

Thus

$$Z_{in} = 40 \cdot 8 + j334 + \frac{10^8 \times 22 \cdot 2^2 \times 10^{-6}}{34 + j413 + 200 + j0}$$

$$= 40 \cdot 8 + j334 + \frac{22 \cdot 2^2 \times 10^2}{475^2}(234 - j413)$$

$$= 91 \cdot 8 + j244 = \underline{\underline{260 \underline{/69 \cdot 4^\circ} \, \Omega}}$$

9.6 Power Transformers

Power transformers are normally operated in a.c. circuits under approximately constant voltage and constant frequency conditions. The remainder of this chapter will refer mainly to power transformers used in this way.

The transformation element enclosed by the broken lines in Fig. 9.7 is ideal and satisfies not only the conditions both for ideal voltage transformation and ideal current transformation but also, from eqn. (9.28),

$$e_1 i_1 = e_2 i_2 \tag{9.39}$$

i.e. the instantaneous power input is equal to the instantaneous power output.

An actual power transformer will tend towards this ideal if the terminal voltages and currents as defined in Fig. 9.7 tend to conform to eqns. (9.25), (9.26) and (9.28) so that

$$\frac{v_1}{v_2} = \frac{N_1}{N_2} \tag{9.40}$$

$$i_1 N_1 = i_2 N_2 \tag{9.41}$$

$$v_1 i_1 = v_2 i_2 \tag{9.42}$$

To make a transformer for which eqn. (9.40) is approximately true the voltage drops in the series elements must be small compared with the respective terminal voltages. This can be realized practically by making the winding resistances, R_1 and R_2, and the leakage inductances, L_{l1} and L_{l2} small. To obtain low values of leakage inductance the coupling between the windings must be tight. In practice this is achieved by using sandwich or concentric coils (see page 277).

To make a transformer for which eqn. (9.41) is approximately true the current flowing in the magnetizing inductance L_{m1} must be small. This is achieved by winding the transformer coils on a core of high permeance as explained in Section 9.4. A large magnetizing flux per ampere is thus produced in the core, the magnetizing current is kept low and the magnetizing inductance is high.

In short, an actual power transformer tends towards the ideal when the series resistances and inductances in its equivalent circuit have low values and the shunt resistances and inductances have high values.

The power transformer is normally connected to a supply voltage which varies sinusoidally with time. As a result the magnetizing flux in the core varies sinusoidally with time (see page 288) and eddy-current and hysteresis losses occur in the core.

The core losses are primarily dependent on the maximum flux density attained and on the frequency of the alternating flux. In a power transformer operating in a constant-voltage constant-frequency system, the core loss will be substantially constant, and may be represented approximately by introducing into the equivalent

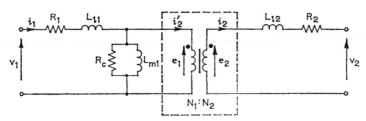

Fig. 9.10 TRANSFORMER EQUIVALENT CIRCUIT TAKING ACCOUNT OF CORE LOSS

circuit of Fig. 9.7 a resistance R_c the power dissipation in which, e_1^2/R_c, is equal to the core loss. Such an equivalent circuit is shown in Fig. 9.10.

For an efficient transformer the internal losses must be small compared with the input or output power at any instant. The internal losses consist of winding I^2R losses and core losses. Winding losses may be made small by making the winding resistances R_1 and R_2 small. The core hysteresis loss can be made small by using a core material which has a relatively narrow hysteresis loop. The core eddy-current loss is made small by laminating the core (see page 277).

9.7 Construction

MAGNETIC CIRCUIT

There are three distinct types of construction: core type, shell type and Berry type. In the core type (Fig. 9.11(a)), half of the primary winding and half of the secondary winding are placed round each limb. This reduces the effect of flux leakage which would seriously affect the operation if the primary and secondary were each wound separately on different limbs. The limbs are joined together by an iron yoke.

In the shell type (Fig. 9.11(b)), both windings are placed round a central limb, the two outer limbs acting simply as a low-reluctance flux path.

In the Berry type of construction (Fig. 9.11(c)), the core may be considered to be placed round the windings. It is essentially a shell type of construction with the magnetic paths distributed evenly

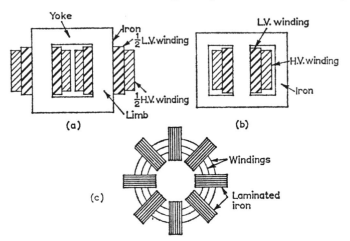

Fig. 9.11   CONSTRUCTION OF SINGLE-PHASE TRANSFORMERS
(*a*) Core type   (*b*) Shell type   (*c*) Berry type (plan view)

round the windings. Owing to constructional difficulties it is not so common as the first two.

Three-phase transformers are usually of the unsymmetrical core type (Fig. 9.12(*a*)). It will be seen that the flux path for the phase

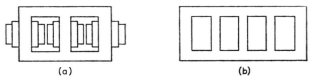

Fig. 9.12   THREE- PHASE TRANSFORMERS
(*a*) Core type   (*b*) Shell type

which is wound on the central limb has a lower reluctance, or higher permeance, than that of a phase which is wound on an outer limb, but the effect is normally small. The shell type (Fig. 9.12(*b*)) has five limbs, the central three of which carry the windings. The cross-section of the yoke in this case can be a little less than that required for the core type.

Solid magnetic cores are inadmissible, since the core material is an electrical conductor, and the core would, in effect, form a single short-circuited turn, which would carry large induced eddy currents (Fig. 9.13(*a*)). Eddy currents are reduced by using high-resistivity silicon steel, made up in insulated laminations which are usually 0·355 mm. (0·014 in) thick. The length and the resistance of

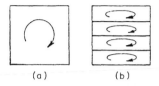

(a)   (b)

Fig. 9.13   USE OF LAMINATIONS TO REDUCE EDDY CURRENTS

the eddy-current paths are increased by this means, so that the heat loss due to the eddy currents is reduced to small proportions (Fig. 9.13(*b*)).

WINDINGS

There are two main forms of winding: the concentric cylinder (Fig. 9.14(*a*)) and the sandwich (Fig. 9.14(*b*)). The coils are made

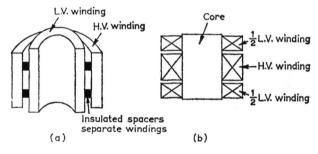

Fig. 9.14   TYPES OF TRANSFORMER WINDING
(*a*) Concentric   (*b*) Sandwich

of varnished-cotton or paper-covered wire or strip and are circular in shape to prevent high mechanical stresses at corners. Due to the problems of insulation, the low-voltage windings are normally placed nearest the core in the concentric cylinder winding. For small transformers each layer is separated from the others by a thin paper, and the high-voltage winding is kept well insulated by special insulating sheets. In the larger transformers each winding is on a separate former, the two windings being well separated by insulating distance pieces.

In the sandwich winding, which has the advantage of reducing the leakage flux, the windings are arranged in layers, with a half-layer of the low-voltage winding at the top and bottom.

In all forms of winding for transformers used in power supply systems the end turns are specially insulated. This is necessary to prevent the destruction of the transformer when voltage surges arise in the system. Surges are described in a later chapter.

COOLING

Heat is produced in a transformer by the eddy-current and hysteresis losses in the core, and by the I^2R loss in the windings. To prevent undue temperature rise this heat is removed by cooling. In small transformers natural air cooling is usually employed. For larger transformers oil cooling is needed, especially where high voltages are in use. Oil-cooled transformers must be enclosed in steel tanks. Oil has the following advantages over air as a cooling medium:

(i) It has a larger specific heat than air, so that it will absorb larger quantities of heat for the same temperature rise.

(ii) It has a greater heat conductivity than air, and so enables the heat to be transferred to the oil more quickly.

(iii) It has about six times the breakdown strength of air—ensuring increased reliability at high voltages.

9.8 E.M.F. Equation

The e.m.f.s induced in the primary and secondary winding due to the mutual flux in a transformer are given by eqns. (9.23) and (9.24). The e.m.f.s induced due to the leakage fluxes are accounted for in the equivalent circuit by the inclusion of leakage-inductance elements.

Suppose the voltage applied to the primary varies sinusoidally with time, i.e.

$$v_1 = V_{1m} \sin \omega t$$

If the voltage drop in the primary resistance and leakage inductance is small enough to be neglected compared with the applied voltage, the induced primary e.m.f. at any instant is equal to the applied voltage, i.e.

$$e_1 = v_1 = V_{1m} \sin \omega t \tag{9.43}$$

Substituting this expression for e_1 in eqn. (9.23),

$$V_{1m} \sin \omega t = N_1 \frac{d}{dt} (\Phi_{m1} - \Phi_{m2})$$

Therefore the mutual flux at any instant is

$$\Phi = \Phi_{m1} - \Phi_{m2} = \frac{1}{N_1} \int V_{1m} \sin \omega t \, dt$$

$$= -\frac{V_{1m}}{N_1 \omega} \cos \omega t + \text{constant}$$

Since there is no d.c. flux component the constant of integration is zero.

The peak value of the mutual flux is

$$\Phi_m = \frac{V_{1m}}{N_1 \omega}$$

so that the peak value of the applied voltage is

$$V_{1m} = \omega \Phi_m N_1$$

Substituting this expression for V_{1m} in eqn. (9.43),

$$e_1 = \omega \Phi_m N_1 \sin \omega t$$

The r.m.s. value of the induced e.m.f. in the primary is

$$E_1 = \frac{\omega \Phi_m N_1}{\sqrt{2}} \tag{9.44}$$

Similarly the induced e.m.f. in the secondary is

$$E_2 = \frac{\omega \Phi_m N_2}{\sqrt{2}} \tag{9.45}$$

The above equations are often written in the following form:

$$E_1 = \frac{2\pi f}{\sqrt{2}} \Phi_m N_1 = 4\cdot44 f \Phi_m N_1 \tag{9.44a}$$

Similarly,

$$E_2 = 4\cdot44 f \Phi_m N_2 \tag{9.45a}$$

Substituting $\Phi_m = B_m A$ (where A is the cross-sectional area of the core) in eqn. (9.44a),

$$E_1 = 4\cdot44 f B_m A N_1 \tag{9.46}$$

9.9 Primary Current Waveform

The equivalent circuits derived for the transformer have assumed that magnetic saturation is absent and that as a result the various inductances are constant. In practice, transformers are nearly always designed so that their magnetic circuits are driven into saturation to reduce the magnetic cross-section required and so produce the most economical design. In these circumstances the inductances are not constant but vary with the degree of magnetic saturation present. Since power transformers are operated at approximately constant voltage and frequency the degree of saturation at maximum flux density does not vary greatly, and appropriate average values may be found for the inductances to give reasonably accurate predictions.

Another result of saturation of the magnetic core is that the magnetizing current becomes non-sinusoidal. The method of deriving the current waveform is given in Section 5.9.

The no-load current is considerably distorted in transformers of normal design. It may be represented by its *sine-wave equivalent*— the current having the same r.m.s. value as the actual no-load current and producing the same mean power. It is this current which is displayed on complexor diagrams and used in calculations based on the indicated value of no-load current.

9.10 Approximate Equivalent Circuit

In a power transformer the current taken by the shunt arm of the equivalent circuit, as given in Fig. 9.10, is small compared with the

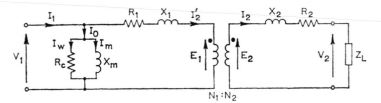

Fig. 9.15 APPROXIMATE TRANSFORMER EQUIVALENT CIRCUIT

rated primary current. Further, the rated primary current will cause only a small voltage drop in the series elements so that the drop in these elements due to the current in the shunt elements will be negligible. The shunt elements may therefore be placed at the input terminals without serious error. This results in a simplification of the equivalent circuit which is shown in Fig. 9.15, where the instantaneous currents and voltages are replaced by the corresponding complexor quantities, and the inductances by their corresponding reactances. The new circuit is convenient for predicting the steady-state response of the transformer to an input voltage varying sinusoidally with time.

9.11 Equivalent Circuit referred to Primary

For steady-state a.c. operation the instantaneous values for the voltage and current relations in eqns. (9.25) and (9.27) may be replaced by the corresponding r.m.s. values:

$$\frac{E_1}{E_2} = \frac{N_1}{N_2} \tag{9.47}$$

and

$$I_2'N_1 = I_2N_2 \tag{9.48}$$

Applying Kirchhoff's law to the primary and secondary meshes of the equivalent circuit of Fig. 9.15,

$$V_1 = (R_1 + jX_1)I_2' + E_1 \tag{9.49}$$
$$E_2 = (R_2 + jX_2)I_2 + V_2 \tag{9.50}$$

If all the terms of eqn. (9.50) are multiplied by N_1/N_2 this gives

$$\frac{N_1}{N_2} E_2 = (R_2 + jX_2) \frac{N_1}{N_2} I_2 + \frac{N_1}{N_2} V_2$$

Substituting $I_2 = \dfrac{N_1}{N_2} I_2'$ gives

$$\frac{N_1}{N_2} E_2 = (R_2 + jX_2) \left(\frac{N_1}{N_2}\right)^2 I_2' + \frac{N_1}{N_2} V_2$$

or

$$E_1 = (R_2' + jX_2')I_2' + V_2' \tag{9.51}$$

where

$$R_2' = \left(\frac{N_1}{N_2}\right)^2 R_2 \tag{9.52}$$

$$X_2' = \left(\frac{N_1}{N_2}\right)^2 X_2 \tag{9.53}$$

$$V_2' = \frac{N_1}{N_2} V_2 \tag{9.54}$$

R_2' is called the *secondary resistance referred to the primary*; X_2' is the *secondary reactance referred to the primary*; and V_2' is the *secondary load voltage referred to the primary*.

Eqns. (9.49) and (9.50) are consistent with the equivalent circuit of Fig. 9.16(*a*), which is the transformer equivalent circuit referred to the primary. The load impedance, Z_L, may also be referred to the primary if desired to give an equivalent primary impedance, Z_L', where

$$Z_L' = \frac{V_2'}{I_2'} = \frac{\dfrac{N_1}{N_2} V_2}{\dfrac{N_2}{N_1} I_2} = \left(\frac{N_1}{N_2}\right)^2 \frac{V_2}{I_2} = \left(\frac{N_1}{N_2}\right)^2 Z_L \qquad (9.55)$$

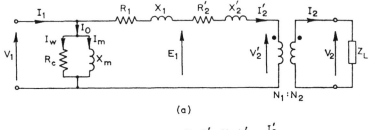

(a)

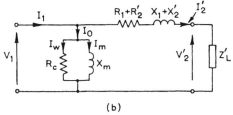

(b)

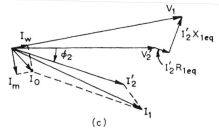

(c)

Fig. 9.16   TRANSFORMER EQUIVALENT CIRCUIT REFERRED TO PRIMARY

The output quantities now become V_2' and I_2', and the ideal transformer element may be omitted. This equivalent circuit is shown in Fig. 9.16(b). Thus the total equivalent series impedance referred to the primary is

$$Z_{1\,eq} = R_{1\,eq} + jX_{1\,eq} \qquad (9.56)$$

where

$$R_{1\,eq} = R_1 + R_2' = R_1 + \left(\frac{N_1}{N_2}\right)^2 R_2 \qquad (9.57)$$

and

$$X_{1eq} = X_1 + X_2' = X_1 + \left(\frac{N_1}{N_2}\right)^2 X_2 \qquad (9.58)$$

Fig. 9.16(c) is the complexor diagram corresponding to the equivalent circuit shown at (b). The referred value of load voltage, V_2', is chosen as the reference complexor. The referred value of load current, I_2', is shown lagging behind V_2' by a phase angle ϕ_2. For a given value of V_2', both I_2' and ϕ_2 are determined by the applied load. The voltage drop $I_2'R_{1eq}$ is in phase with I_2', and the voltage drop $I_2'X_{1eq}$ leads I_2' by 90°. When these voltage drops are added to V_2' the input voltage, V_1, is obtained.

I_w is in phase with V_1 ($V_1 I_w$ is approximately equal to the core loss); I_m, the magnetizing current, lags behind V_1 by 90°. The complexor sum of I_w and I_m is I_0. I_0 will flow even when the secondary terminals are open-circuited and is therefore called the *no-load current*. The complexor sum of I_0 and I_2' is the input current, I_1. For the sake of clarity the size of I_0 relative to I_1 is exaggerated in the diagram.

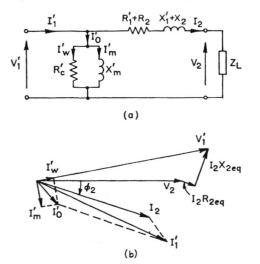

(a)

(b)

Fig. 9.17 TRANSFORMER EQUIVALENT CIRCUIT REFERRED TO SECONDARY

9.12 Equivalent Circuit referred to Secondary

A method similar to that adopted in Section 9.11 to obtain the transformer equivalent circuit referred to the primary may be used to obtain an equivalent circuit referred to the secondary. This is shown in Fig. 9.17(a), and the corresponding complexor diagram is shown at (b).

In Fig. 9.17(a) the primed symbols represent primary quantities referred to the secondary:

$$V_1' = \frac{N_2}{N_1} V_1 \tag{9.59}$$

$$I_1' = \frac{N_1}{N_2} I_1 \tag{9.60}$$

Any primary impedance is referred to the secondary by multiplying its value in the primary by $(N_2/N_1)^2$.

EXAMPLE 9.2 A 4,000/400 V 10 kVA transformer has primary and secondary winding resistances of 13 Ω and 0·15 Ω respectively. The leakage reactance referred to the primary is 45 Ω. The magnetizing impedance referred to the primary is 6 kΩ, and the resistance corresponding to the core loss is 12 kΩ.

Determine the total resistance referred to the primary and the values of all impedances referred to the secondary. Determine the input current (a) when the secondary terminals are open-circuited, and (b) when the secondary load current is 25 A at a power factor of 0·8 lagging.

Turns ratio, $\dfrac{N_1}{N_2} = \dfrac{4,000}{400} = \underline{\underline{10}}$

$R_{1eq} = R_1 + R_2' = R_1 + R_2 \left(\dfrac{N_1}{N_2}\right)^2 = 13 + 0\cdot15 \times 10^2 = \underline{\underline{28\,\Omega}}$

$R_{2eq} = R_{1eq} \left(\dfrac{N_2}{N_1}\right)^2 = 28 \times 0\cdot1^2 = \underline{\underline{0\cdot28\,\Omega}}$

$X_{2eq} = X_{1eq} \left(\dfrac{N_2}{N_1}\right)^2 = 45 \times 0\cdot1^2 = \underline{\underline{0\cdot45\,\Omega}}$

$R_c' = R_c \left(\dfrac{N_2}{N_1}\right)^2 = 12,000 \times 0\cdot1^2 = \underline{\underline{120\,\Omega}}$

$X_m' = X_m \left(\dfrac{N_2}{N_1}\right)^2 = 6,000 \times 0\cdot1^2 = \underline{\underline{60\,\Omega}}$

Core-loss component of current, $I_w = \dfrac{V}{R_c} = \dfrac{4,000}{12,000} = \underline{\underline{0\cdot333\,A}}$

Magnetizing current, $I_m = \dfrac{V}{X_m} = \dfrac{4,000}{6,000} = \underline{\underline{0\cdot667\,A}}$

Input current on no-load, $I_0 = 0\cdot333 - j0\cdot667 = \underline{\underline{0\cdot745\underline{/-63\cdot5°}\,A}}$

When the secondary load current is 25 A at a power factor of 0·8 lagging,

$I_2 = 25\underline{/-36\cdot9°}$

$I_2' = 2\cdot5\underline{/-36\cdot9} = (2\cdot0 - j1\cdot5)\,A$

Input current, $I_1 = I_2' + I_0 = (2\cdot0 - j1\cdot5) + (0\cdot333 - j0\cdot667)$

$\qquad\qquad\qquad = \underline{\underline{3\cdot18\underline{/-43°}\,A}}$

9.13 Regulation

The *voltage regulation* of a transformer is the change in the terminal voltage between no load and full load at a given power factor. This is often expressed as a percentage of the rated voltage or in per-unit form, using rated voltage as base (see Section 9.16).

Consider the equivalent circuit referred to the secondary, Fig. 9.17. On no-load the secondary terminal voltage will be

$$V_2 = V_1' = V_1 \frac{N_2}{N_1}$$

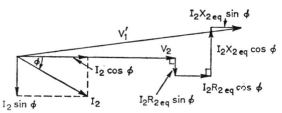

Fig. 9.18   REGULATOR OF A TRANSFORMER WITH LAGGING
POWER FACTOR

When the load is applied the current flowing through the equivalent impedance produces a voltage drop (IZ), and for a fixed value of V_1 this will cause V_2 to change. The equivalent secondary complexor diagram is shown in Fig. 9.18.

The load current I_2 may be represented by its in-phase component $I_2 \cos \phi$, and its quadrature component $I_2 \sin \phi$, and these components each give rise to voltage drops in the total equivalent resistance, R_{2eq}, and the total equivalent reactance, X_{2eq}. These voltage drops are shown in Fig. 9.18, from which it will be seen that, for a lagging power factor,

$$V_1' = \surd\{(V_2 + I_2 R_{2eq} \cos \phi + I_2 X_{2eq} \sin \phi)^2 + (I_2 X_{2eq} \cos \phi - I_2 R_{2eq} \sin \phi)^2\} \quad (9.61)$$

This exact expression for V_1' is rather cumbersome and it is usually sufficient to adopt the following approximate expression for regulations of less than 20 per cent (or 0·2 p.u.):

$$V_1' = V_2 + I_2 R_{2eq} \cos \phi + I_2 X_{2eq} \sin \phi \quad (9.62)$$

The justification for this expression will be clear from Fig. 9.18, where it is apparent that the voltage drops represented by quadrate complexors have little effect on the total magnitude of V_1'.

$$\text{Voltage regulation} = V_1' - V_2 = I_2 R_{2eq} \cos \phi + I_2 X_{2eq} \sin \phi \quad (9.63)$$

If the circuit is referred to the primary, then V_1 will remain constant and V_2' will vary as the load changes. It can be shown, in the same manner as before, that for lagging power factors,

$$V_1 - V_2' \approx I_1(R_1 + R_2') \cos \phi + I_1(X_1 + X_2') \sin \phi \qquad (9.64)$$

The regulation of the transformer is then this voltage multiplied by N_2/N_1.

If the power factor is leading, the regulation is

$$V_1' - V_2 \approx I_2 R_{2\,eq} \cos \phi - I_2 X_{2\,eq} \sin \phi \qquad (9.65)$$

For a given load kVA, the current I_2 will be approximately constant. Hence the regulation may be differentiated with respect to the load phase angle ϕ to determine the way in which the power factor affects the regulation. Thus, from eqn. (9.63),

$$\frac{d}{d\phi}(V_1' - V_2) = -I_2 R_{2\,eq} \sin \phi + I_2 X_{2\,eq} \cos \phi$$

This will be zero when

$$I_2 R_{2\,eq} \sin \phi = I_2 X_{2\,eq} \cos \phi$$

i.e. when

$$\tan \phi = \frac{X_{2\,eq}}{R_{2\,eq}} \qquad \text{or} \qquad \phi = \tan^{-1} \frac{X_{2\,eq}}{R_{2\,eq}} \qquad (9.66)$$

Hence the regulation has a maximum value at that value of load phase angle which is equal to the internal total phase angle of the transformer itself.

Also from eqn. (9.63) the regulation will be zero when

$$I_2 R_{2\,eq} \cos \phi = -I_2 X_{2\,eq} \sin \phi$$

i.e. when

$$\phi = \tan^{-1}\left(-\frac{R_{2\,eq}}{X_{2\,eq}}\right) \qquad (9.67)$$

The minus sign denotes a leading power factor. For leading power factors smaller than this the regulation will be negative, denoting a voltage rise on load.

EXAMPLE 9.3 The primary and secondary windings of a 40 kVA 6,600/250 V single-phase transformer have resistances of $10\,\Omega$ and $0.02\,\Omega$ respectively. The leakage reactance of the transformer referred to the primary is $35\,\Omega$. Calculate (*a*) the primary voltage required to circulate full-load current when the secondary is short-circuited, (*b*) the full-load regulation at (i) unity (ii) 0·8 lagging power factor. Neglect the no-load current.   (*H.N.C.*)

It is obviously convenient to refer impedances to the primary side, as shown in Fig. 9.16. Then, from eqn. (9.52),

$$R_2' = R_2 \left(\frac{N_1}{N_2}\right)^2 = 0.02 \times \left(\frac{6,600}{250}\right)^2 = 13.9\,\Omega$$

since the voltage ratio quoted is equal to the transformation ratio.
The full-load primary current is

$$I_{f1} = \frac{40,000}{6,600} = 6.06\,\text{A}$$

(a) If the secondary is short-circuited, the impedance of the load reflected in the primary is zero, so that the total impedance is simply

$$Z_1 = R_1 + R_2' + jX_{1eq}$$

Hence the primary voltage required to circulate full-load current is

$$V_{sc} = 6.06\sqrt{(23.9^2 + 35^2)} = \underline{256\,\text{V}}$$

(b) In this problem the parallel magnetizing circuit is neglected on full load, to simplify the solution. At unity p.f. and full load, from eqn. (9.64),

$$V_1 - V_2' = (6.06 \times 23.9 \times 1) + 0 = 145\,\text{V}$$

Therefore

$$\text{Actual regulation} = 145 \times \frac{250}{6,600} = 5.5\,\text{V}$$

On no-load $V_2 = 250\,\text{V}$. Therefore

$$\text{Regulation} = \frac{5.5}{250} \times 100 = \underline{2.2 \text{ per cent}}$$

At 0.8 p.f. lagging, from eqn. (9.64),

$$V_1 = V_2' = (6.06 \times (23.9) \times 0.8) + 6.06 \times 35 \times 0.6 = 244\,\text{V}$$

Therefore

$$\text{Actual regulation} = 244 \times \frac{250}{6,600} = 9.25$$

and

$$\text{Percentage regulation} = \frac{9.25}{250} \times 100 = \underline{3.7 \text{ per cent}}$$

9.14 Transformer Losses and Efficiency

Transformer losses may be divided into two main parts, (a) the losses which vary with the load current, and (b) the losses which vary with the core flux. Since the load current is not constant during normal operation the winding I^2R losses will vary. Under normal conditions, however, the core flux will remain approximately constant, so that the losses which vary with the core flux (core losses) will be approximately constant, independent of the load. These losses include stray losses due to e.m.f.s induced by stray fields in adjacent conductors. A further small source of loss is dielectric loss in the insulation, but this will be neglected.

There will be a loss due to the resistance of each winding, the total winding loss being

$$P_c = I_1{}^2 R_1 + I_2{}^2 R_2 = I_2{}^2(R_2 + R_1') \quad \text{watts} \tag{9.68}$$

The core losses are divided into two parts, the hysteresis loss, and the eddy-current loss.

HYSTERESIS LOSS (P_h)

In a specimen of steel subject to an alternating flux, the hysteresis loss per cycle is proportional to the area of the hysteresis loop, and therefore to $B_m{}^n$:

$$\text{Hysteresis loss} = k_h f B_m{}^n \quad \text{watts/metre}^3 \tag{9.69}$$

where B_m is the maximum flux density in teslas, and k_h is a constant. The coefficient n is empirically found to be in the range 1·6 to 2.

EDDY-CURRENT LOSS (P_e)

This is due to the flow of eddy currents in the core. Thin high-resistivity laminations effectively reduce the eddy-current loss to small proportions. Since essentially a transformer action is involved (considering each lamination as a single short-circuited secondary), the induced e.m.f. in the core will be proportional to $f B_m$ (since $E = 4·44 f N \Phi_m$). This causes the flow of an eddy current,

$$I_e = \frac{\text{E.M.F.}}{\text{Impedance of core path}}$$

The impedance of the core path may be assumed constant and independent of frequency for low power frequencies and thin laminations. Thus

$$\text{Eddy current loss} \propto I_e{}^2 \propto f^2 B_m{}^2 = k_e f^2 B_m{}^2 \quad \text{watts/m}^3 \tag{9.70}$$

where k_e is a constant. The eddy current loss is proportional to the square of the frequency. The total core loss, P_i, is

$$P_i = P_h + P_e = k_h f B_m{}^n + k_e f^2 B_m{}^2 \tag{9.71}$$

Provided that B_m and f are constant the core losses should be constant. In order that B_m shall be constant, the magnetizing current and hence the applied voltage must be constant.

EFFICIENCY

The rating of a transformer is an output rating, and hence the efficiency is calculated in terms of the output in kilowatts.

$$\text{Efficiency, } \eta = \frac{\text{Output}}{\text{Input}} = \frac{\text{Output}}{\text{Output} + \text{Losses}} = 1 - \frac{\text{Losses}}{\text{Input}}$$

$$= \frac{V_2 I_2 \cos \phi}{V_2 I_2 \cos \phi + I_2{}^2 R_{2\,eq} + P_i} \qquad (9.72)$$

where $R_{2\,eq} = R_2 + R_1' = R_2 + R_1 \left(\dfrac{N_2}{N_1}\right)^2$, and P_i is the total constant core loss. Therefore

$$\eta = \frac{V_2 \cos \phi}{V_2 \cos \phi + I_2 R_{2\,eq} + \dfrac{P_i}{I_2}}$$

For operation at constant voltage, constant power factor and variable load current, the efficiency will be a maximum when $I_2 R_{2\,eq} + \dfrac{P_i}{I_2}$ is a minimum. This will happen when

$$\frac{d}{dI_2}\left\{I_2 R_{2\,eq} + \frac{P_i}{I_2}\right\} = 0 = R_{2\,eq} - \frac{P_i}{I_2{}^2}$$

ı.e. when

$$P_i = I_2{}^2 R_{2\,eq}$$

Thus the efficiency at any given power factor is a maximum when the load is such that the $I^2 R$ losses are equal to the constant core losses.

EXAMPLE 9.4 The required no-load voltage ratio in a 150kVA single-phase 50 Hz core-type transformer is 5,000/250 V.

(a) Find the number of turns in each winding for a core flux of about 0·06 Wb.
(b) Calculate the efficiency at half-rated kVA and unity power factor.
(c) Determine the efficiency at full load and 0·8 p.f. lagging. The full-load I^2R loss is 1,800 W, and the core loss is 1,500 W.
(d) Find the load kVA for maximum efficiency.

(a) From eqn. (9.45a),

$$E_2 = 4\cdot44 f N_2 \Phi_m$$

Therefore

$$N_2 = \frac{250}{4\cdot44 \times 50 \times 0\cdot06} = 18\cdot8, \text{ or say } \underline{\underline{19 \text{ turns}}}$$

Since only a whole number of turns is possible it is most common, as above, to calculate the number of turns on the low-voltage winding first. Then the number of turns on the high-voltage winding is found to give the correct ratio. Thus

$$N_1 = 19 \times \frac{5,000}{250} = 380 \text{ turns}$$

(*b*) At half-rated kVA, unity p.f. (i.e. at half full-load current).

Winding loss = $(\tfrac{1}{2})^2 \times 1,800 = 450 \text{W} = 0.45 \text{kW}$

Core loss = constant = $P_i = 1,500 \text{W} = 1.5 \text{kW}$

Power output = $\tfrac{1}{2} \times 150 = 75 \text{kW}$

Therefore

$$\text{Efficiency} = 1 - \frac{0.45 + 1.5}{75 + 0.45 + 1.5} = 97.2 \text{ per cent}$$

(*c*) At full-load kVA, 0.8 p.f. lagging.

Power output = $150 \times 0.8 = 120 \text{kW}$

and

$$\text{Efficiency} = 1 - \frac{1.8 + 1.5}{120 + 1.8 + 1.5} = 97.3 \text{ per cent}$$

(*d*) Maximum efficiency.

Let x be the fraction of the full-load kVA at which maximum efficiency occurs. For maximum efficiency the core loss is equal to the I^2R loss, i.e. $P_i = x^2 P_c$, whence

$$x = \sqrt{\frac{P_i}{P_c}} = \sqrt{\frac{1,500}{1,800}} = 0.913$$

Therefore

Load kVA for maximum efficiency = $0.913 \times 150 = 137 \text{kVA}$

9.15 Transformer Tests

OPEN-CIRCUIT TEST

The circuit is shown in Fig. 9.19. The transformer is connected

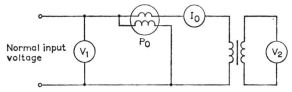

Fig. 9.19   OPEN-CIRCUIT TEST

(in its tank) and the normal rated voltage V_1 is applied. The secondary voltage V_2, the current I_0 and the power P_0 are measured. Then, since the I^2R loss on open-circuit may be neglected (or allowed for, since I_0 is known), the no-load input will be the normal core loss.

The values of R_c and X_m in the parallel exciting circuit can then be calculated. The open-circuit power factor is $\cos\phi_0 = P_0/V_1 I_0$, so that

$$I_w = I_0\cos\phi_0 \quad \text{and} \quad I_m = I_0\sin\phi_0$$

Then

$$R_c = \frac{V}{I_w} \tag{9.73}$$

and

$$X_m = \frac{V}{I_m} \tag{9.74}$$

These are shown on the diagram of Fig. 9.16. The open-circuit ratio V_1/V_2 will be almost equal to the turns ratio N_1/N_2.

SHORT-CIRCUIT TEST (Fig. 9.20)

One winding is short-circuited and the voltage applied to the other is gradually raised from zero until full-load current flows.

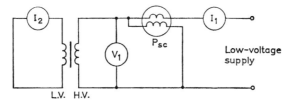

Fig. 9.20 SHORT-CIRCUIT TEST

The readings of the wattmeter (P_{sc}), ammeter (I_1), and voltmeter (V_1) are noted. Since the exciting voltage is small the core flux will be small, and the core losses will be negligible. If the impedance is referred to the primary, then

$$R_{1\,eq} = R_1 + R_2' \quad \text{and} \quad X_{1\,eq} = X_1 + X_2'$$

and these values may be found from the instrument readings.

The short-circuit power factor is

$$\cos\phi_{sc} = \frac{P_{sc}}{V_1 I_1}$$

and

$$Z_{1\,eq} = \frac{V_1}{I_1} = \frac{V_1}{I_1}\cos\phi_{sc} + j\frac{V_1}{I_1}\sin\phi_{\,sc} \tag{9.75}$$

$$= (R_1 + R_2') + j(X_1 + X_2') \tag{9.76}$$

The resistances of the windings can be measured separately by a d.c. test. It is, however, impossible to separate the leakage reactances. .

Note that P_{sc} gives the full-load I^2R loss P_c, so that the efficiency at any load may be calculated from the open- and short-circuit tests.

EXAMPLE 9.5 A 10 kVA 200/400 V 50 Hz single-phase transformer gave the following test results. O.C. test: 200 V, 1·3 A, 120 W, on l.v. side. S.C. test: 22 V, 30 A, 200 W on h.v. side.

(a) Calculate the magnetizing current and the component corresponding to core loss at normal frequency and voltage.

(b) Calculate the magnetizing-branch impedances.

(c) Find the percentage regulation when supplying full load at 0·8 p.f. leading.

(d) Determine the load which gives maximum efficiency, and find the value of this efficiency at unity p.f.

(a) O.C. test

$$\text{Open-circuit power factor} = \frac{120}{200 \times 1\cdot3} = 0\cdot462 = \cos \phi_0$$

Therefore

$$\text{Magnetizing current} = I_0 \sin \phi_0 = 1\cdot3 \times 0\cdot886 = \underline{\underline{1\cdot15\,\text{A}}}$$

and

$$\text{Component of current corresponding to core loss} = I_0 \cos \phi_0$$
$$= 1\cdot3 \times 0\cdot462 = \underline{\underline{0\cdot6\,\text{A}}}$$

(b)

$$R_c = \frac{V_1}{I_0 \cos \phi_0} = \frac{200}{0\cdot6} = \underline{\underline{333\,\Omega}}$$

$$X_m = \frac{V_1}{I_0 \sin \phi_0} = \frac{200}{1\cdot15} = \underline{\underline{174\,\Omega}}$$

(c) Percentage regulation at 0·8 leading p.f.

Total impedance referred to h.v. side, $Z_{2\,eq} = \dfrac{22}{30} = 0\cdot733\,\Omega$

Total resistance referred to h.v. side, $R_{2\,eq} = \dfrac{200}{30^2} = 0\cdot222\,\Omega$

Total reactance referred to h.v. side, $X_{2\,eq} = \sqrt{(0\cdot733^2 - 0\cdot222^2)} = 0\cdot698\,\Omega$

Full load current on h.v. side $= \dfrac{10{,}000}{400} = 25\,\text{A}$

Thus

Regulation at 0·8 leading
$$\approx R_{2\,eq}I_2 \cos \phi - X_{2\,eq}I_2 \sin \phi \qquad \text{(from eqn. (9.65))}$$
$$= (0\cdot222 \times 25 \times 0\cdot8) - (0\cdot698 \times 25 \times 0\cdot6)$$
$$= -6\cdot0\,\text{V} \qquad \text{(voltage rise due to leading power factor)}$$

Therefore

$$\text{Regulation} = -\frac{6 \cdot 0}{400} \times 100 = \underline{\underline{-1 \cdot 5 \text{ per cent}}}$$

(*d*) Maximum efficiency

From o.c. test, full-voltage core loss, $P_i = 120\,\text{W}$

From s.c. test, full-load I^2R loss, $P_c = 200 \times \left(\dfrac{25}{30}\right)^2 = 140\,\text{W}$

Let x be the fraction of full-load kVA at which maximum efficiency occurs. At maximum efficiency the I^2R loss is equal to the core loss, so that

$$x^2 P_c = P_i$$

Therefore

$$x = \sqrt{\frac{P_i}{P_c}} = \sqrt{\frac{120}{140}} = 0 \cdot 925$$

and

kVA for maximum efficiency $= 0 \cdot 925 \times 10 = 9 \cdot 25\,\text{kVA}$

When the load has unity power factor,

Load for maximum efficiency $= \underline{\underline{9 \cdot 25\,\text{kW}}}$

$$\text{Maximum efficiency} = \frac{9{,}250}{9{,}250 + 120 + 120} \times 100 = \underline{\underline{97 \cdot 4 \text{ per cent}}}$$

SEPARATION OF HYSTERESIS AND EDDY-CURRENT LOSSES

From eqn. (9.71) it can be seen that, with sinusoidal flux in the core, the hysteresis loss increases with the frequency, while the eddy-current loss increases as the square of the frequency. This is the

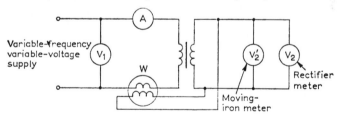

Fig. 9.21 CIRCUIT FOR SEPARATION OF CORE LOSS

basis of one method whereby these losses may be separated (Fig. 9.21). The transformer primary is connected to a variable-frequency and variable-voltage sinusoidal supply. Since it is necessary to ensure a sinusoidal core flux, the secondary is connected to a moving-iron voltmeter (measuring r.m.s. values), with a rectifier voltmeter in parallel (measuring average values × 1·11). If the flux is sinusoidal.

the form factor is 1·11 and the instruments will record the same voltage. If the flux is not sinusoidal, the readings will differ, so that for this test it is essential that both instruments should give the same reading. This reading will be $4·44fN_2\Phi_m$ volts, where N_2 is the number of secondary turns.

The hysteresis and eddy-current losses are separated by their different variations with frequency. The total loss should be measured at various frequencies while the other factors upon which the core losses depend are maintained constant. Thus it is necessary to maintain Φ_m constant as the frequency is varied. Now,

$$\Phi_m \propto \frac{V_1}{f} \quad \text{(from eqn. (9.44}(a)\text{))}$$

so that as the supply frequency is varied the supply voltage must also be varied in such a way that the ratio V_1/f, and hence Φ_m, is maintained constant throughout the test.

Since the voltage coil of the wattmeter is energized from the transformer secondary, the primary I^2R loss is eliminated from the reading. The core loss is obtained by multiplying the wattmeter reading by the turns ratio, and subtracting from this the power loss in the voltmeters and in the voltage coil of the wattmeter.

Hysteresis loss $= Af$ watts, where A is a constant

Eddy current loss $= Cf^2$ watts, where C is a constant

Total core loss $= P_i = Af + Cf^2$

whence

$$\frac{P_i}{f} = A + Cf$$

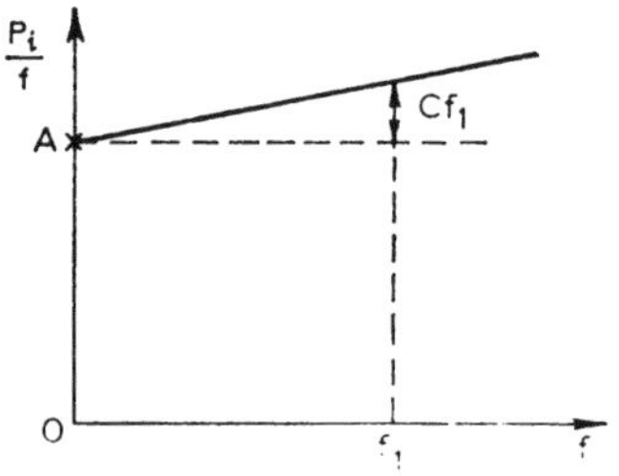

Fig. 9.22   SEPARATION OF CORE LOSSES

If a graph of P_i/f is plotted to a base of frequency, the graph will be a straight line intersecting the P_i/f axis at A and with a slope C (Fig. 9.22). Hence A and C, and the hysteresis and eddy-current losses at a given frequency and peak core flux density may be found.

9.16 The Per-unit System

It is often convenient to carry out calculations relating to transformers and other electrical plant using the *per-unit system*. In this method all the relevant quantities are expressed in per-unit form, i.e. as fractions of chosen base values.

There are two main advantages in using the per-unit system. First, the constants of transformers and other plant fall within narrow limits when expressed in per-unit form based on their rating. Second, in computations involving transformers the need to refer quantities from one side of the transformer to the other is eliminated.

The per-unit value of any quantity may be defined as

$$\text{Quantity in per-unit form} = \frac{\text{Actual quantity}}{\text{Base value of quantity}} \qquad (9.77)$$

For any quantity A,

$$A_{pu} = \frac{A}{A_{base}} \qquad (9.78)$$

The quantity A may be voltage, current, volt-amperes, impedance, admittance or any electrical quantity.

The base values of voltage and current, V_{base}, I_{base}, may be chosen arbitrarily though they will usually be chosen to correspond to *rated voltage* and *rated current*. Once these have been chosen the base values of all other electrical quantities are automatically fixed. For a single-phase system, the power, reactive volt-ampere and volt-ampere bases are

$$P_{base}, \ Q_{base} \quad \text{and} \quad S_{base} = V_{base}I_{base} \qquad (9.79)$$

The resistance, reactance and impedance bases are

$$R_{base}, \ X_{base} \quad \text{and} \quad Z_{base} = \frac{V_{base}}{I_{base}} \qquad (9.80)$$

The conductance, susceptance and admittance bases are

$$G_{base}, \ B_{base} \quad \text{and} \quad Y_{base} = \frac{I_{base}}{V_{base}} \qquad (9.81)$$

In practice it is more usual to choose the voltage base and the volt-ampere base and then to determine the current base from eqn. (9.79).

EXAMPLE 9.6 A 500V 10kVA single-phase generator has an open-circuit voltage of 500V. When the load current is 25A at a certain power factor the terminal voltage falls to 480V. Determine in per-unit form (*a*) the output voltage, (*b*) the output current, (*c*) the output volt-amperes, and (*d*) the voltage regulation.

Choosing rated voltage and rated volt-amperes as bases, and using the subscript B to represent base values,

$$V_B = 500 \text{V} \qquad \text{and} \qquad S_B = 10,000 \text{VA}$$

From eqn. (9.79),

$$I_B = \frac{S_B}{V_B} = \frac{10,000}{500} = 20 \text{A}$$

$$\text{Output voltage p.u.} = \frac{480}{500} = \underline{\underline{0 \cdot 96}}$$

$$\text{Output current p.u.} = \frac{25}{20} = \underline{\underline{1 \cdot 25}}$$

(i.e. an overload since full load is represented by 1 p.u.)

$$\text{Output VA p.u.} = \frac{480 \times 25}{10,000} = 1 \cdot 2$$

or, more directly,

$$\text{Output VA p.u.} = V_{pu} I_{pu} = 0 \cdot 96 \times 1 \cdot 25 = \underline{\underline{1 \cdot 2}}$$

$$\text{Voltage regulation p.u.} = \frac{500 - 480}{500} = \underline{\underline{0 \cdot 04}}$$

or, more directly,

$$\text{Voltage regulation p.u.} = 1 - 0 \cdot 96 = \underline{\underline{0 \cdot 04}}$$

With transformers, provided that rated primary voltage is used as base with primary referred impedances, and rated secondary voltage with secondary referred impedances, the same per-unit values of impedance are obtained. As a result calculations in per-unit form are the same for the primary and the secondary. The actual primary or secondary values are obtained by multiplying the per-unit values by the appropriate base quantities.

EXAMPLE 9.7 A 5 kVA 200/400 V 50 Hz single-phase transformer has an equivalent circuit consisting of shunt admittance $(1 \cdot 5 \times 10^{-3} - j3 \cdot 15 \times 10^{-3})$ S mho and series leakage impedance $(0 \cdot 12 + j0 \cdot 32)\Omega$, both referred to the low-voltage side. Determine the per-unit values of the shunt admittance and the leakage impedance using first the primary referred values and then the secondary referred values. The voltage base is to be the rated value.

Determine also the per-unit value of the core loss when the transformer is excited at rated voltage, the full-load winding loss and the full-load voltage regulation if the load power factor is $0 \cdot 8$ lagging, and express each of these values in actual quantities.

(a) Per-unit values using primary data

Base VA, $S_B = 5,000 \text{VA}$

Base voltage, $V_{B1} = \text{Rated primary voltage} = 200 \text{V}$

$$\text{Base current, } I_{B1} = \frac{S_B}{V_{B1}} = \frac{5,000}{200} = 25 \text{A}$$

Base impedance, $Z_{B1} = \dfrac{V_B}{I_B} = \dfrac{200}{25} = 8\Omega$

Base admittance, $Y_{B1} = \dfrac{I_B}{V_B} = \dfrac{25}{200} = 0.125\text{S}$

Shunt admittance, $Y_{pu} = \dfrac{Y_{1eq}}{Y_B} = \dfrac{1.5 \times 10^{-3} - j3.15 \times 10^{-3}}{0.125}$

$$= (12 \times 10^{-3} - 25.2 \times 10^{-3})\text{p.u.}$$

Leakage impedance, $Z_{pu} = \dfrac{0.12 + j0.32}{8}$

$$= (0.015 + j0.04)\text{p.u.}$$

(b) Actual referred values of equivalent circuit constants

$$Y_{2eq} = Y_{1eq}\left(\frac{200}{400}\right)^2 = \frac{1.5 \times 10^{-3} - j3.15 \times 10^{-3}}{4}$$

$$= (0.375 \times 10^{-3} - j0.787 \times 10^{-3})\,\text{S}$$

$$Z_{2eq} = Z_{1eq}\left(\frac{400}{200}\right)^2 = 4(0.12 + j0.32) = (0.48 + j1.28)\Omega$$

(c) Per-unit values using secondary data

Base VA, $S_B = 5{,}000\,\text{VA}$

Base voltage, $V_{B2} = $ Rated secondary voltage $= 400\,\text{V}$

Base current, $I_{B2} = \dfrac{5{,}000}{400} = 12.5\,\text{A}$

Base impedance, $Z_{B2} = \dfrac{I_{B2}}{V_{B2}} = \dfrac{400}{12.5} = 32\Omega$

Base admittance, $Y_{B2} = \dfrac{I_{B2}}{V_{B2}} = \dfrac{12.5}{400} = 0.03125\text{S}$

Shunt admittance, $Y_{pu} = \dfrac{Y_{2eq}}{Y_{B2}} = \dfrac{0.375 \times 10^{-3} - (j0.787 \times 10^{-3})}{0.03125}$

$$= (12 \times 10^{-3} - j25.2 \times 10^{-3})\text{p.u.}$$

Leakage impedance, $Z_{pu} = \dfrac{Z_{2eq}}{Z_{B2}} = \dfrac{0.48 + j1.28}{32}$

$$= (0.015 + j0.04)\text{p.u.}$$

It will be noted that, where the rated voltage and volt-amperes are used as bases, the per-unit values of the constants of the transformer equivalent circuit are the same whether primary or secondary data are used.

In the transformer equivalent circuit the core loss is represented by power dissipated in G_c the real part of the shunt admittance. Since the transformer is excited at rated voltage, in per-unit values the voltage will be unity.

Core loss p.u., $P_{i\ pu} = V_{pu}^2 G_{pu} = 1^2 \times 12 \times 10^{-3} = 12 \times 10^{-3}\text{p.u.}$

It will be noted that the reference part of the per-unit shunt admittance is numerically equal to the per-unit core loss at normal voltage.

Base power, $P_B = V_{B1}I_{B1} = V_{B2}I_{B2} = 5{,}000$
Actual core loss, $P_i = P_{ipu}P_B = 12 \times 10^{-3} \times 5{,}000 = 60\,\text{W}$

The winding loss may be calculated by determining the power dissipated in the reference part of the leakage impedance. Since the winding loss on full load is required the per unit current will be unity.

Winding loss p.u. $P_{cpu} = I_{pu}{}^2R_{pu} = 1^2 \times 0{\cdot}015 = 0{\cdot}015\,\text{p.u.}$

It will be noted that the reference part of the per-unit leakage impedance is numerically equal to the full-load winding loss.

Actual winding loss, $P_c = P_{cpu}P_B = 0{\cdot}015 \times 5{,}000 = 75\,\text{W}$

The actual voltage regulation is given by eqn. (9.63).

Voltage regulation p.u. $= I_{2pu}R_{pu}\cos\phi + I_{2pu}X_{pu}\sin\phi$

since the full-load voltage regulation is required $I_{2pu} = 1$.

Voltage regulation $= (1 \times 0{\cdot}015 \times 0{\cdot}8) + (0{\cdot}04 \times 0{\cdot}6) = 0{\cdot}036\,\text{p.u.}$
Actual voltage regulation $= 0{\cdot}036V_B = 0{\cdot}036 \times 400 = 14{\cdot}4\,\text{V}$

Since the secondary voltage base is used this gives the actual change in voltage at the secondary terminals.

This last result may be checked by direct substitution into eqn (9.63), which gives

Voltage regulation $= (12{\cdot}5 \times 0{\cdot}48 \times 0{\cdot}8) + (12{\cdot}5 \times 1{\cdot}28 \times 0{\cdot}6) = 14{\cdot}4\,\text{V}$

If calculations relating to two or more transformers, or other plant, of different ratings are to be undertaken, then the per-unit values must all be referred to the same voltage and volt-ampere bases. In such a situation the base values chosen will not be the rated value of volt-amperes of some of the plant involved.

Following eqn. (9.77), the per-unit value of any quantity A to A_{base1} is

$$A_{pu1} = \frac{A}{A_{base1}} \qquad (9.77(a))$$

Similarily the per-unit value of A to a second base value is

$$A_{pu2} = \frac{A}{A_{base2}} \qquad (9.77(b))$$

Combining eqns. (9.77(a)) and (9.77(b)) gives

$$A = A_{pu1}A_{base1} = A_{pu2}A_{base2}$$

i.e.

$$A_{pu2} = A_{pu1} \times \frac{A_{base1}}{A_{base2}} \qquad (9.82)$$

Eqn. (9.82) in combination with eqns. (9.79)–(9.81) may be used to change the base to which any electrical quantity is referred.

For 3-phase plant the rated phase voltage and rated volt-amperes per phase would normally be chosen as base values, and after this has been done eqns. (9.90), (9.91) and (9.92) apply and the problem may be treated as a single-phase problem.

EXAMPLE 9.8 A 60 MVA 3 phase 33/11 kV mesh/star-connected transformer supplies a 10 MVA feeder. The leakage impedance per phase of the transformer is $(0\cdot015 + j0\cdot04)$ p.u. and the impedance per phase of the feeder is $(0\cdot06 + j0\cdot07)$ p.u. The p.u. impedances are based on the nominal ratings per phase of the transformer and feeder respectively. When the load on the distant end of the feeder is 10 MVA at a power factor of 0·8 lagging and the load voltage is 11 kV determine:

    (a) The line current in the feeder.
    (b) The transformer secondary phase current.
    (c) The transformer primary phase current.
    (d) The transformer primary line current.
    (e) The transformer output line voltage.
    (f) The transformer input line voltage.

The per-unit impedances of the feeder and transformer are based on their respective nominal ratings. It will be necessary to express these relative to a common base, and the nominal rating of the transformer is chosen as the common base.

From eqn. (9.82), the per-unit impedance of the feeder on the new base is

$$Z_{pu\,2} = Z_{pu\,1} \frac{Z_{base\,1}}{Z_{base\,2}}$$

$$= Z_{pu\,1} \frac{S_{base\,2}}{S_{base\,1}}$$

since the impedance base is inversely proportional to the volt-ampere base. The feeder impedance referred to this base is

$$Z_F = (0\cdot06 + j0\cdot07)\frac{60}{10} = (0\cdot36 + j0\cdot42)\,\text{p.u.} = 0\cdot551\underline{/49\cdot4^\circ}\,\text{p.u.}$$

The base values for the feeder and the l.v. side of the transformer are

$$S_B = \frac{60 \times 10^6}{3} = 20 \times 10^6\,\text{VA}$$

$$V_B = \frac{11 \times 10^3}{\sqrt{3}} = 6\cdot35 \times 10^3\,\text{V}$$

$$I_B = \frac{20 \times 10^6}{6\cdot35 \times 10^3} = 3\cdot15 \times 10^3\,\text{A}$$

The base values for the h.v. side of the transformer are

$$S_B = \frac{60 \times 10^6}{3} = 20 \times 10^6\,\text{VA}$$

$$V_B = 33 \times 10^3\,\text{V}$$

$$I_B = \frac{20 \times 10^6}{33 \times 10^3} = 0\cdot605 \times 10^3\,\text{A}$$

The actual load on the feeder is $10/3$ MVA per phase at a power factor of 0.8 lagging and a phase voltage of 6.35×10^3 V.

$$\text{Load VA, } S_{pu} = \frac{\text{Actual VA/phase}}{\text{Base VA}} = \frac{10/3}{20} = 0.167 \text{p.u.}$$

$$\text{Load voltage, } V_{pu} = \frac{6.35 \times 10^3}{6.35 \times 10^3} = 1 \text{p.u.}$$

$$\text{Load current, } I_{pu} = \frac{S_{pu}}{V_{pu}} = 0.167 \text{p.u.}$$

(a) Feeder current $= I_{pu} \times I_{base} = 0.167 \times 3.15 \times 10^3 = \underline{\underline{525\,\text{A}}}$

(b) Transformer secondary phase current $= \underline{\underline{525\,\text{A}}}$

(c) Transformer primary phase current $= 0.167 \times 0.605 \times 10^3 = \underline{\underline{101\,\text{A}}}$

(d) Transformer primary line current $= \sqrt{3} \times 101 = \underline{\underline{175\,\text{A}}}$

The transformer secondary voltage is

$$V_2 = V + IZ_F$$

where V = load voltage and I = feeder current.

$$V_{pu2} = 1\underline{/0°} + (0.167\underline{/-36.9°} \times 0.551\underline{/49.4°})$$
$$= (1.09 + j0.0199) = 1.09\underline{/1.04°} \text{p.u.}$$

Transformer secondary phase voltage $= 1.09 \times 6.35 \times 10^3 = 6{,}920\,\text{V}$

(e) Transformer secondary line voltage $= \sqrt{3} \times 6{,}920 = 12{,}000\,\text{V}$

$$\text{Total series impedance, } Z_{puT} = (0.015 + j0.04) + (0.36 + j0.42)$$
$$= 0.375 + j0.46 = 0.594\underline{/50.8°} \text{p.u.}$$

The transformer input line voltage is

$$V_{pu1} = V_{pu} + I_{pu}Z_{puT}$$
$$= 1\underline{/0°} + (0.167\underline{/-36.9°} \times 0.594\underline{/50.8°}$$
$$= 1.0963 + j0.0238 = 1.096\underline{/1.24°} \text{p.u.}$$

(f) Transformer primary line voltage $= 1.096 \times 33 \times 10^3 = \underline{\underline{36{,}200\,\text{V}}}$

9.17 Three-phase Transformers

For 3-phase working it is possible to have either a bank of three single-phase transformers, or a single 3-phase unit (Fig. 9.12). Single-phase construction has the advantage that where single units are concerned only one spare single-phase transformer is needed as a standby, instead of a complete spare 3-phase transformer.

The single 3-phase unit takes up less space and is somewhat cheaper. Technically the difference between the single 3-phase unit and the three single-phase units lies in the fact that there is a direct magnetic coupling between the phases in the first case but not in the second. Star, delta or zigzag windings are possible in

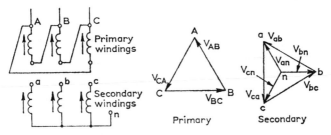

Fig. 9.23 DELTA-STAR 3-PHASE TRANSFORMER

both primary and secondary, giving many possible pairs of connexion. The complexor diagram for any connexion is drawn by observing that the e.m.f.s induced in all windings on the same limb are in phase and in direct ratio to the numbers of turns.

Two cases will be considered in detail, the others being easily followed by similar methods. In Fig. 9.23 the connexions and

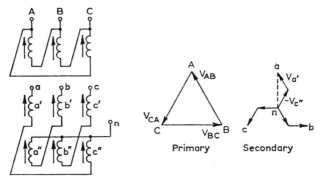

Fig. 9.24 DELTA-ZIGZAG 3-PHASE TRANSFORMER

complexor diagrams for a Δ-Y transformer are shown. The primary complexor diagram is shown on the left, with primary line voltages V_{AB}, V_{BC} and V_{CA}. The secondary is Y-connected, so that the phase voltages V_{an} will be in phase with the primary line voltage V_{AB}, and the ratio V_{an}/V_{AB} will be almost equal to the turns ratio N_2/N_1, etc. The secondary line voltages V_{ab}, etc., will be the complexor differences between successive phase voltages, as shown in the right-hand complexor diagram. In this case the secondary line voltages will lead the primary line voltages by 30°, and will be $\sqrt{3}\,V_p(N_2/N_1)$ in magnitude, where V_p is the primary line voltage.

In Fig. 9.24 are shown the circuit and complexor diagrams of a Δ-zigzag connexion. The secondary is divided into two equal halves on each limb, the top half on one limb being connected in opposition to the lower half on the preceding limb. This connexion is used if the load on the secondary is far out of balance, since each secondary phase is divided between two primary windings. Each secondary phase voltage is thus the difference between the e.m.f.s induced in windings on successive limbs. Thus

$$V_{an} = V_{a'} - V_{c''}$$

where $V_{a'}$ is in phase with V_{AB}, and $V_{c''}$ is in phase with Z_{CA}. The secondary line voltages in this case are in phase with the primary line voltages. If the magnitude of these line voltages is V_l, then

$$V_{a'} = \frac{V_l}{2}\frac{N_2}{N_1}$$

where N_2 is the total number of secondary turns on each limb, and

$$V_{c''} = \frac{V_l}{2}\frac{N_2}{N_1}$$

Hence, from the complexor diagram,

$$V_{an} = \frac{\sqrt{3}}{2}\,V_l\frac{N_2}{N_1}$$

and the secondary line voltages are

$$\frac{3}{2}\,V_l\frac{N_2}{N_1}$$

9.18 Star-star Connexion

It is possible to use a 3-phase transformer with both the primary and the secondary connected in star as follows.

THREE SINGLE-PHASE UNITS (Fig. 9.25)

If the primaries of a bank of three single-phase transformers are star connected to a 3-phase, 4-wire system, then a constant voltage is applied to each primary. The three transformers act independently of one another, the load on each secondary phase being reflected into the corresponding primary phase. This is a perfectly practical connexion, but has the disadvantage that there is no tendency for the primary currents to be balanced when the load on the secondary is unbalanced.

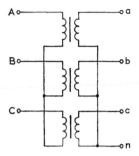

Fig. 9.25 STAR-STAR CONNEXION OF THREE SINGLE-PHASE TRANSFORMERS

If the primaries are connected to a 3-wire system, the primary line voltages are constant, but the primary phase voltages may be unbalanced, since the star-point potential is not fixed. The most extreme case is when one secondary phase only is loaded. The loaded transformer will have a low input impedance while the two unloaded transformers will act as high-impedance chokes. The voltage across the loaded phase will fall to a low value, while the voltages across the other two phases will rise almost to the line voltage value. This connexion is therefore unsuitable if there is any possibility of unbalanced loads.

FIVE-LIMB CORE-TYPE THREE-PHASE TRANSFORMER (Fig. 9.26)

If the primary windings are star connected to a 4-wire system, the primary phase voltages are constant, so that the primary phase e.m.f.s must also be approximately constant. Hence the flux in each core must be approximately constant, and the primary ampere-turns will balance the secondary ampere-turns on each limb. Thus the output will remain approximately constant, and the primary current in any line will be a reflection of the secondary current in the same line.

If there is no primary neutral connexion, the primary line voltages (but not necessarily the primary phase voltages) must remain constant. Hence between line terminals the primary ampere-turns must balance the secondary ampere-turns, but there need not necessarily be ampere-turn balance on each core. Consider, for simplicity, the 1:1 turns-ratio 3-wire star-star transformer which has only one secondary phase loaded, as shown in Fig. 9.26. Let the primary phase currents be I_A, I_B, and I_C, where I_B is the primary current on the loaded limb. By Kirchhoff's first law,

$$I_A + I_B + I_C = 0 \qquad (9.83)$$

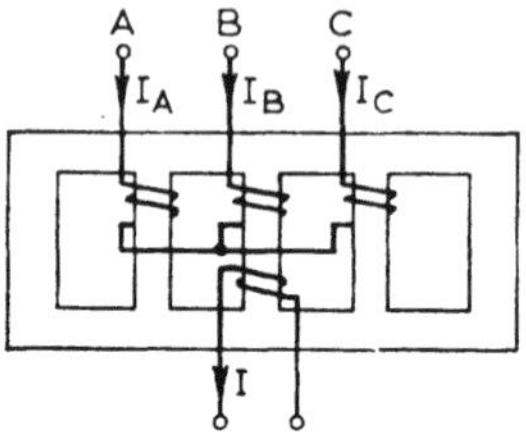

Fig. 9.26   STAR-STAR CONNEXION OF 3-PHASE TRANSFORMER

Suppose that there are N turns on each phase winding; then between terminals A and B:

Primary ampere-turns $= I_A N - I_B N$

$= $ Secondary ampere-turns $= IN$ (9.84)

Also between terminals A and C,

Primary ampere-turns $= I_A N - I_C N$

$= $ Secondary ampere-turns $= 0$

Therefore $I_A = I_C$, and

$$I_B = -(I_A + I_C) = -2I_A \qquad (9.85)$$

Hence, from eqn. (9.84),

$$I_A + 2I_A = I$$

Thus

$I_A = I_C I/3$, and $I_B = -2I/3$.

Examination of the cores will show that on each there are resultant unbalanced ampere-turns of $IN/3$. Further, all these unbalanced ampere-turns are in phase and set up a flux through each of the wound cores in parallel with the path completed through the unwound cores. This flux linking the phase windings will unbalance the phase e.m.f.s with the result that the voltage across the loaded phase tends to fall while the voltages across the two other phases tend to rise. The result is similar to that with three single-phase transformers.

THREE-LIMB CORE-TYPE THREE-PHASE TRANSFORMER

Basically the 3-limb transformer will behave in the same way as the 5-limb transformer. When a 3-limb transformer is supplied from a 3-wire system and feeds an unbalanced load, the phase voltages will remain approximately balanced since there is no low-reluctance path through which the unbalanced ampere-turns on each core can set up a flux. The actual flux path is completed through the air or through the steel tank of enclosed transformers, in which case undesirable heating may occur.

Star-star connexion is not generally satisfactory and should not be used with unbalanced loads, unless an additional delta-connected winding is provided. This *tertiary winding* does not usually feed any load, but if the secondary load is unbalanced, the out-of-balance flux will give rise to a circulating current in the closed tertiary winding, whose ampere-turns will then cancel out the unbalanced ampere-turns due to the load. The phase voltages will then tend to remain balanced.

9.19 Other Three-phase Transformer Connexions

The following brief notes illustrate the applications and limitations of some of the other possible connexions for three-phase transformers.

DELTA-DELTA

This is useful for high-current low-voltage transformers, and can supply large unbalanced loads without disturbing the magnetic equilibrium. There is, however, no available star point.

DELTA-STAR

This arrangement is used to supply large powers at high voltage, and for distribution at low voltages. A large unbalanced load does not

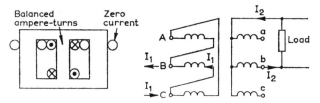

Fig. 9.27   DELTA-STAR CONNEXION WITH UNBALANCED LOAD

disturb the magnitude equilibrium since the primary current will flow in the corresponding winding only, and the primary and secondary ampere-turns will be balanced in each limb (Fig. 9.27).

STAR-DELTA

Used for substation transformers supplied from the grid but not for distribution purposes owing to the absence of a neutral.

DELTA-ZIGZAG

Used for supplying smaller powers with large out-of-balance neutral currents. The zigzag winding establishes magnetic equilibrium.

9.20 Scott Connexion

The Scott connexion is a method of connecting two single-phase transformers to give a 3-phase to 2-phase conversion. The method

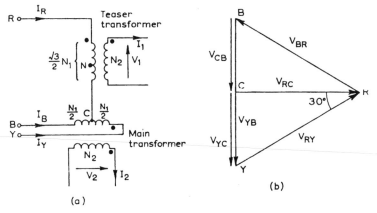

Fig. 9.28   SCOTT CONNEXION

is frequently used to obtain, from a 3-phase supply, a balanced 2-phase supply for a.c. control systems.

Referring to Fig. 9.28(*a*), the centre-tapped primary of the *main transformer* has the line voltage V_{YB} applied to its terminals. The secondary terminal voltage of the main transformer is

$$V_2 = \frac{N_2}{N_1} V_{YB} = \frac{N_2}{N_1} V_l$$

Fig. 9.28(*b*) is the relevant complexor diagram. The line voltages of the 3-phase system, V_{RY}, V_{YB} and V_{BR}, which are balanced, are shown on the complexor diagram as a closed equilateral triangle. The voltages across the two halves of the centre-tapped primary of the main transformer, V_{CB} and V_{YC}, are equal and in phase with

V_{YB}. Evidently V_{RC} leads V_{YB} by 90°. This voltage is applied to the primary of the *teaser transformer*, and so the secondary voltage of the teaser transformer, V_1, will lead the secondary terminal voltage of the main transformer by 90°. However,

$$V_{RC} = V_{RY} \cos 30° = \frac{\sqrt{3}}{2} V_L$$

To make V_1 equal in magnitude to V_2 the primary of the teaser transformer requires to have $\frac{\sqrt{3}}{2} N_1$ turns. Then

$$V_1 = \frac{N_2}{\dfrac{\sqrt{3}}{2} N_1} V_{RC} = \frac{N_2 \sqrt{3}}{\dfrac{\sqrt{3}}{2} N_1} \frac{V_L}{2} = \frac{N_2}{N_1} V_L = V_2$$

The voltages V_1 and V_2 thus constitute what is usually regarded as a balanced 2-phase system comprising two voltages of equal magnitude having a phase difference of 90°.

The primaries of the two transformers may have a 4-wire connexion to the 3-phase supply if a tapping point N is provided on the primary of the teaser transformer such that $V_{RN} = V_l/\sqrt{3}$.

If n is the number of turns in the section RN of the primary winding of the teaser transformer,

$$\frac{n}{\dfrac{\sqrt{3}}{2} N_1} = \frac{V_l/\sqrt{3}}{\dfrac{\sqrt{3}}{2} V_l}$$

Therefore

$$n = \frac{2}{3} \frac{\sqrt{3}}{2} N_1 = 0\cdot577 N_1$$

Number of turns in section NC ⎱ $= 0\cdot866 N_1 - 0\cdot577 N_1$
of teaser transformer primary ⎰ $= 0\cdot299 N_1$

Frequently identical interchangeable transformers are used for the Scott connexion, in which case each transformer has a primary winding of N_1 turns and is provided with tapping points at $0\cdot299 N_1$, $0\cdot5 N_1$ and $0\cdot866 N_1$.

If the 2-phase currents are balanced, the 3-phase currents are also balanced (neglecting magnetizing currents). This may be shown as follows.

Let I_1 be the reference complexor so that

$$I_1 = I\underline{/0°} \quad \text{and} \quad I_2 = I\underline{/-90°} = -jI$$

For m.m.f. balance of the teaser transformer,

$$I_R \frac{\sqrt{3}}{2} N_1 = IN_2 \underline{/0^\circ}$$

so that

$$I_R = \frac{2}{\sqrt{3}} \frac{N_2}{N_1} I \underline{/0^\circ} \tag{9.86}$$

For m.m.f. balance for the main transformer,

$$I_Y \frac{N_1}{2} - I_B \frac{N_1}{2} = -jIN_2$$

or

$$I_Y - I_B = -j \frac{2N_2}{N_1} I \tag{9.87}$$

For 3-wire connexion,

$$I_R + I_Y + I_B = 0 \tag{9.88}$$

Substituting for I_R from eqn. (9.86) in eqn. (9.88),

$$I_B + I_Y = \frac{-2}{\sqrt{3}} \frac{N_2}{N_1} I \tag{9.89}$$

Adding eqns. (9.87) and (9.89),

$$2I_Y = \frac{-2}{\sqrt{3}} \frac{N_2}{N_1} I - j2 \frac{N_2}{N_1} I$$

or

$$I_Y = \frac{2}{\sqrt{3}} \frac{N_2}{N_1} I \underline{/-120^\circ} \tag{9.90}$$

Similarly,

$$I_B = \frac{2}{\sqrt{3}} \frac{N_2}{N_1} I \underline{/+120^\circ} \tag{9.91}$$

That is, the 3-phase currents are balanced if the 2-phase currents are balanced, neglecting the effect of magnetizing current. Since the 3-wire connexion gives balanced currents there will be no neutral current if a neutral wire is connected.

9.21 Transformer Types

POWER TRANSFORMERS

These have a high *utilization factor*, i.e. it is arranged that they run with an almost constant load which is equal to their rating.

The maximum efficiency is designed to be at full load. This means that the full-load winding losses must be equal to the core losses.

DISTRIBUTION TRANSFORMERS

These have an intermittent and variable load which is usually considerably less than the full-load rating. They are therefore designed to have their maximum efficiency at between $\frac{1}{2}$ and $\frac{3}{4}$ of full load.

AUTO-TRANSFORMERS

Consider a single winding AC, on a magnetic core as shown in Fig. 9.29. If this winding is tapped at a point B, and a load is connected between B and C, then a current will flow, under the influence of the e.m.f. E_2, between the two points. The current I_2 will produce an m.m.f. in the core which will be balanced by a current I_1 flowing in the complete winding. This is called the auto-transformer action, and has the advantage that it effects a saving in winding material (copper or aluminium), since the secondary winding is now merged into the primary. The disadvantages of the auto-connexion are:

1. There is a direct connexion between the primary and secondary.
2. Should an open-circuit develop between points B and C, the full mains voltage would be applied to the secondary.
3. The short-circuit current is much larger than for the normal two-winding transformer.

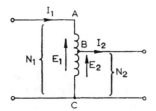

Fig. 9.29 AUTO-TRANSFORMER (STEP DOWN)

It can be seen from Fig. 9.29 that a short-circuited secondary causes part of the primary also to be short-circuited, reducing the effective resistance and reactance.

Applications are—(*a*) Boosting or bucking of a supply voltage by a small amount. (The smaller the difference between the output and input voltages the greater is the saving of winding material.) (*b*) Starting of a.c. machines, where the voltage is raised in two or more steps from a small value to the full supply voltage. (*c*) Continuously variable a.c. supply voltages. (*d*) Production of very high voltages. Auto-transformers are used in the 275 kV and 400 kV grid systems.

The connexion for increasing (boosting) the output voltage by auto-connexion is shown in Fig. 9.30. The operation will be better

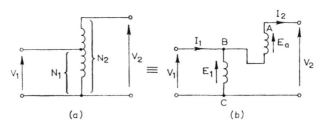

(a)   (b)

Fig. 9.30   AUTO-TRANSFORMER FOR BOOSTING THE OUTPUT VOLTAGE

understood by considering Fig. 9.30(*b*). Neglecting losses, the output voltage E_2 will be the sum of the input voltage E_1 and the e.m.f. E_a induced in the additional winding.

Continuously variable auto-transformers are constructed by arranging for one of the output terminals to be connected to a sliding contact which moves over the whole range of a single-layer winding. The transformer is usually overwound so that voltages in excess of the supply voltage may be obtained.

INSTRUMENT TRANSFORMERS

Current (series) and voltage (shunt) transformers are used for extending the range of a.c. instruments in preference to shunts and series resistors for the following reasons: (i) to eliminate errors due to stray inductance and capacitance in shunts, multipliers and their leads; (ii) the measuring circuit is isolated from the mains by the transformer and may be earthed; (iii) the length of connecting leads from the transformer to the instrument is of lesser importance, and the leads may be of small cross-sectional area; (iv) the instrument ranges may be standardized (usually 1A or 5A for ammeters and 110V for voltmeters); (v) by using a clip-on type of transformer core the current in a heavy-current conductor can be measured without breaking the circuit.

The current transformer has the secondary effectively short-circuited through the low impedance of the ammeter (Fig. 9.31(*a*)).

The voltage across the primary terminals will thus be very small, so that there will only be a very small flux in the core (since $E \propto f\Phi N$). This means that both the magnetizing and the core-loss components of the primary current will be small. Also, the exact value of the secondary load (called the *burden*) will have a negligible effect on the primary current which is to be measured. The current transformation ratio I_p/I_s will not be quite equal to N_s/N_p, and will depend on the ratio of magnetizing current to ammeter current.

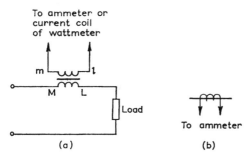

Fig. 9.31 CURRENT TRANSFORMERS

(a) With wound primary (for currents up to about 1,000 A)
(b) With primary consisting of either a single bar on the line itself (for larger currents)

Normally the correction is small. The presence of the magnetizing and loss components of current introduces a phase angle error, due to the fact that the secondary current is not exactly in phase with the primary current. This is of importance only when the transformer is to be used in conjunction with a wattmeter.

The current transformer must never be operated on open-circuit, for two reasons. Firstly, there will be no secondary demagnetizing ampere-turns, and since the primary current is fixed, the core flux will increase enormously. This will cause large eddy-current and hysteresis losses, and the resulting temperature rise may damage the insulation. Even in the absence of evident damage the core may be left with a high value of remanent magnetism which can lead to a large undetected error in subsequent use. Secondly, a very high voltage will be induced in the multi-turn secondary, being dangerous both to life and to the insulation.

Voltage transformers operate with their primaries at the full supply voltage and their secondaries connected to the high impedance of a voltmeter or the voltage coil of a wattmeter (giving a secondary phase angle of almost unity). The secondary current will be very small (of the same order as the magnetizing current) so that transformer may be regarded in the same way as a power transformer on no load. The voltage ratio is effectively the turns ratio, and the phase angle error (due to the fact that the secondary voltage will not be quite in phase with the primary voltage) will generally be negligible.

9.22 Short Transmission Lines in Parallel

Preliminary to the consideration of the operation of two transformers in parallel the simpler case of two transmission lines in parallel will be first considered.

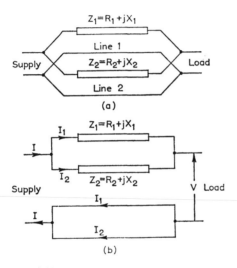

Fig. 9.32 TRANSMISSION LINES IN PARALLEL

Fig. 9.32(a) shows two single-phase short transmission lines connected in parallel. The total series impedance of each pair has been equivalently concentrated in one conductor of each. Fig. 9.32(b) is a simple redraft of the system. It is evident that the voltage drops in the two transmission lines are identical. If the total current I divides between the two lines so that I_1 flows through line 1 (of impedance Z_1) and I_2 flows through line 2 (of impedance Z_2), then

$$I_1 Z_1 = I_2 Z_2 \tag{9.92}$$

Also

$$I = I_1 + I_2 \tag{9.93}$$

Therefore

$$I_1 = \frac{Z_2}{Z_1 + Z_2} I \tag{9.94}$$

Similarly,

$$I_2 = \frac{Z_1}{Z_1 + Z_2} I \tag{9.95}$$

This result will also apply to two balanced 3-phase systems operating in parallel if Z_1 and Z_2 are the equivalent impedances per phase of the 3-phase transmission lines.

In many cases the division of the load between the two lines is required. This may be determined as follows.

Let V be the receiving-end voltage. Multiplying eqns. (9.94) and (9.95) by V for a single-phase system or by $\sqrt{3}\ V$ for a 3-phase system,

$$S_1 = \frac{Z_2}{Z_1 + Z_2} S_T \tag{9.96}$$

and

$$S_2 = \frac{Z_1}{Z_1 + Z_2} S_T \tag{9.97}$$

where the total volt-amperes (S_T) and the volt-amperes delivered by each line $(S_1$ and $S_2)$ are in complexor form, with the system voltage V as reference.

Eqns. (9.96) and (9.97) may be expressed using impedance in per-unit form.

$$S_1 = \frac{Z_{2pu}}{Z_{1pu} + Z_{2pu}} S_T \tag{9.98}$$

$$S_2 = \frac{Z_{1pu}}{Z_{1pu} + Z_{2pu}} S_T \tag{9.99}$$

If the lines have different volt-ampere ratings the p.u. impedances for both lines will have to be based on the volt-ampere rating of one line. If the impedances of the lines are given in per-unit form, each based on the individual volt-ampere rating of the line, the per-unit impedance of one line may be converted to a new base in accordance with eqn. (9.82).

EXAMPLE 9.9 A 3-phase cable A supplies a load of 2,000 kW at 6,600 V and p.f. 0·8 lagging. A second cable B of impedance $(3 + j4·5)\Omega$/phase is connected in parallel with A, and it is found that for the same load as before, A carries 140 A and delivers 1,200 kW at a lagging p.f. What is the impedance of cable A?

$$\text{Total load, } S_T = \frac{2,000}{0\cdot8} \; \underline{/-\cos^{-1} 0\cdot8} = 2,500\underline{/-36\cdot9^\circ} \text{kVA}$$

Power delivered by A when in parallel with B = 1,200 kW
kVA delivered by A when in parallel with B

$$= \frac{\sqrt{3} \times 6,600 \times 140}{1,000} \; \underline{/-\cos^{-1} \frac{1,200}{1,600}} = 1,600\underline{/-41\cdot4^\circ}$$

By eqn. (9.96),

$$\frac{Z_B}{Z_A + Z_B} = \frac{S_A}{S_T} = \frac{1,600\underline{/-41\cdot4^\circ}}{2,500\underline{/-36\cdot9^\circ}} = 0\cdot64\underline{/-4\cdot5^{\,\prime}}$$

and

$$\frac{Z_A + Z_B}{Z_B} = 1\cdot56\underline{/4\cdot5^\circ} = 1\cdot56 + j0\cdot123$$

Therefore

$$\frac{Z_A}{Z_B} = 0\cdot56 + j0\cdot123 = 0\cdot574\underline{/12\cdot4^\circ}$$

and

$$Z_A = Z_B \times 0\cdot574\underline{/12\cdot4^\circ} = 5\cdot41\underline{/56\cdot3^\circ} \times 0\cdot574\underline{/12\cdot4^\circ} = 3\cdot11\underline{/68\cdot7^\circ}\Omega$$

9.23 Single-phase Equal-ratio Transformers in Parallel

The correct method of connecting two single-phase transformers in parallel is shown in Fig. 9.33(a). The wrong method is shown at (b). At (a) it will be seen that round the loop formed by the secondaries, E_1 and E_2 oppose and there will be no circulating current, while at (b) it will be seen that round the loop formed by the two secondaries, E_1 and E_2 are additive, and will give rise to a short-circuit current.

Fig. 9.33(c) shows the two transformer equivalent circuits with the leakage impedances referred to the secondary sides. The two ideal transformers must now have identical secondary e.m.f.s since they have the same turns ratio and have their primaries connected to the same supply. The potentials at A and C and at B and D must then be identical so that these pairs of points may be joined without affecting the circuit. The imaginary joining of these points is shown in Fig. 9.33(d). From this diagram it is clear that two equal-ratio transformers connected in parallel will share the total load in the same way as two short transmission lines; all the previous equations are therefore applicable.

It is noteworthy that the per-unit impedance of a transformer is the same whether the actual impedance is referred to the primary or secondary.

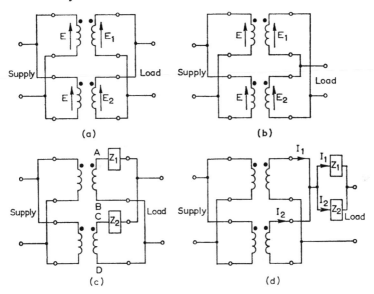

(a)　(b)

(c)　(d)

Fig. 9.33 EQUAL-RATIO TRANSFORMERS IN PARALLEL

From Fig. 9.33(d) it is evident that

$$I_1 Z_1 = I_2 Z_2 \qquad (9.100)$$

and

$$\frac{I_1}{I_2} = \frac{Z_2}{Z_1}$$

The transformer currents (and hence the volt-ampere loads) are in the inverse ratio of the transformer impedances. This is important for parallel operation for it is usually desirable that both transformers be fully loaded simultaneously. If the full-load volt-ampere ratings are S_{fl1} and S_{fl2}, then this condition will be fulfilled if

$$\frac{S_{fl1}}{S_{fl2}} = \frac{Z_2}{Z_1}$$

Thus it is desirable that two transformers for parallel operation should have their rated full-load volt-amperes in the inverse ratio to their impedances. It is also desirable that the two transformers should operate at the same power factor to give the largest resultant volt-ampere rating. To achieve this the two impedances should have the same phase angle.

9.24 Single-phase Transformers with Unequal Ratios in Parallel

If the single-phase transformers connected as in Fig. 9.33(*a*) have unequal ratios, then E_1 and E_2 will be unequal and there will be a circulating current given by

$$I_c = \frac{E_1 - E_2}{Z_1 + Z_2} \qquad (9.101)$$

in the secondary loop.

Since Z_1 and Z_2 will be small, the difference $E_1 - E_2$ must be small or the circulating current will be very large. It is possible

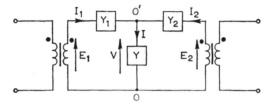

Fig. 9.34   UNEQUAL-RATIO TRANSFORMERS IN PARALLEL

to operate the transformers in parallel satisfactorily only if E_1 and E_2 are very nearly equal. If E_1 and E_2 are the secondary e.m.f.s when the impedances are referred to the secondary, they will be in phase with each other, since they are the e.m.f.s of ideal transformers connected to the same supply.

When the e.m.f.s are unequal they have an important bearing on the load sharing between the two transformers. Fig. 9.34 shows the usual equivalent circuit for the analysis in this case. E_1 and E_2 are in phase; Y_1 and Y_2 are the equivalent secondary admittances and Y is the load admittance. Then by Millman's theorem the voltage $V_{0'0}$ across the load is

$$V_{0'0} = \frac{\Sigma V_{k0} Y_k}{\Sigma Y_k} = \frac{E_1 Y_1 + E_2 Y_2}{Y_1 + Y_2 + Y} = V \qquad (9.102)$$

Hence the load current is

$$I = VY = \frac{E_1 Y_1 Y + E_2 Y_2 Y}{Y_1 + Y_2 + Y} = \frac{E_1 Z_2 + E_2 Z_1}{Z Z_1 + Z_1 Z_2 + Z Z_2} \qquad (9.103)$$

where $Z = 1/Y$, etc.

Also the current through transformer 1 is

$$I_1 = (E_1 - V)Y_1 = \frac{(E_1 - E_2)Y_2Y_1 + E_1YY_1}{Y_1 + Y_2 + Y} \tag{9.104}$$

and that through transformer 2 is

$$I_2 = (E_2 - V)Y_2 = \frac{(E_2 - E_1)Y_2Y_1 + E_2YY_2}{Y_1 + Y_2 + Y} \tag{9.105}$$

9.25 Three-phase Transformers in Parallel

In order that 3-phase transformers may operate in parallel the following conditions must be strictly observed:

(a) The secondaries must have the same phase sequence.

(b) All corresponding secondary line voltages must be in phase.

(c) The secondaries must give the same magnitude of line voltage.

In addition it is desirable that

(d) The impedances of each transformer, referred to its own rating, should be the same, i.e. each transformer should have the same per-unit resistance and per-unit reactance.

If conditions (a), (b) and (c) are not complied with, the secondaries will simply short-circuit one another and no output will be possible. If condition (d) is not complied with the transformers will not share the total load in proportion to their ratings, and one transformer will become overloaded before the total output reaches the sum of the individual ratings. It is difficult to ensure that transformers in parallel have identical per-unit impedances, and this affects the load sharing in the same manner as was indicated for single-phase transformers in parallel.

It is relatively simple to ensure that the phase sequence of all transformer secondaries is the same before connecting them in parallel. In transformers constructed in accordance with B.S. 171 the terminals of the h.v. and l.v. sides are labelled for the conventional positive phase sequence. It is then only necessary to ensure that correspondingly lettered terminals are connected together.

The main difficulty arising from the parallel connexion of 3-phase transformers is to ensure that condition (b) is satisfied. This is because of the phase shift which is possible between primary and secondary line voltages in such transformers.

Three-phase transformers are divided into four groups according to the phase displacement between the primary and secondary line voltages. These groups are

1. No phase displacement
2. 180° phase displacement
3. −30° phase displacement
4. +30° phase displacement

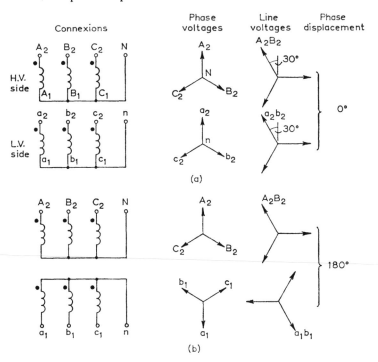

Fig. 9.35   STAR-STAR CONNEXION IN GROUPS 1 AND 2

Only transformers in the same groups may be connected in parallel.

Groups 1 and 2 contain (i) star-star, (ii) delta-delta, and (iii) delta-zigzag combinations.

The connexion and complexor diagrams for a star-star transformer belonging to group 1 are shown in Fig. 9.35(*a*). In this case it is immediately obvious that there is no phase displacement between the primary and secondary phase and line voltages. The essential point to observe is that all windings on the same limb of a transformer must give voltages which are either in phase or in antiphase, according to the relative winding directions. The line voltages are derived from the three phase voltages in the usual manner.

The effect of reversing the connexions to the l.v. winding is shown in Fig. 9.35(*b*). The directions of the phase e.m.f.s in the secondary are reversed, so that there is now 180° phase displacement between the primary and secondary line voltages. This connexion therefore belongs to group 2.

Fig. 9.36 shows diagrams for the delta-delta connexion in group 1, the complexor diagrams indicating that there is no phase shift

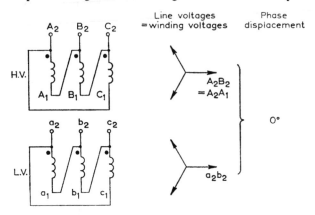

Fig. 9.36 DELTA-DELTA CONNEXION IN GROUP 1

between primary and secondary line voltages. If the connexions to either the primary or the secondary windings are reversed, there will be a 180° phase shift. The delta-zigzag connexion can similarly be shown to belong to either group 1 or group 2.

The connexions in groups 3 and 4, giving −30° and +30° phase displacements respectively, are: (i) delta-star, (ii) star-delta and (iii) star-zigzag. Fig. 9.37(*a*) shows the delta-star connexion which will give −30° phase shift. It should be noted that in this case the line voltage between terminals A_2 and B_2 is actually in anti-phase to the voltage across the h.v. B-winding, so that the voltage induced in the *b*-phase of the l.v. winding will also be in antiphase to the line voltage A_2B_2. If the connexions to the h.v. side are reversed, as shown in Fig. 9.37(*b*), then the phase shift produced will be +30°. Similar connexion and complexor diagrams may be constructed for the star-delta and star-zigzag connexions.

The load-sharing properties of two 3-phase transformers with equal voltage ratios are governed by the same equations (9.98 and 9.99), as single-phase transformers, when the impedances are expressed as per-unit impedances, i.e. irrespective of the methods of connexion used for the transformers.

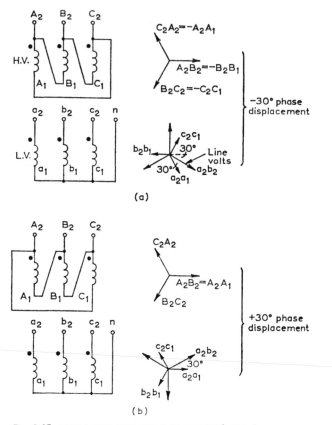

Fig. 9.37   DELTA-STAR CONNEXION IN GROUPS 3 AND 4

9.26 Transformers for High Frequencies

At radio frequencies and above, steel laminations cannot be used for transformers and coils because of the excessive eddy-current and hysteresis losses. One solution to this problem is to form cores of iron dust in an insulating binder. Such dust cores can be used up to radio frequencies, and have an effective relative permeability of about 10.

An alternative solution has become available with the development of homogeneous non-metallic materials called *ferrites*. These have the form $XO.Fe_2O_3$, where X stands for a divalent metallic atom. Ferrites crystallize in a cubic spinel structure, and are characterized by a high initial relative permeability (between 10 and 3,000) and a very high resistivity (typically $10^6 \Omega$-m compared to about $10^{-7} \Omega$-m for iron). Owing to the high resistivity, eddy-current losses are virtually non-existent, so that ferrites can be used up to frequencies in excess of 10^9 Hz. They are not suitable, however, for power-frequency applications owing to their relatively high cost and fairly low saturation flux density (about 0·2 T). Their mechanical properties are similar to those of insulating ceramics—they are hard and brittle and not amenable to mechanical working. It is interesting to note that a naturally occurring ferrite known as lodestone or magnetite was the first material in which magnetic effects were observed.

The magnetic properties of ferrites depend on the metallic atom that occupies the position X in the ferrite formula. In magnetite this happens to be a divalent iron atom, so that magnetite is a double oxide of iron ($FeO . Fe_2O_3$). The manufactured ferrites are generally mixed crystals of two or more single ferrites.

Manganese zinc ferrite ($MnO . Fe_2O_3$, $ZnO . Fe_2O_3$) and nickel zinc ferrite ($NiO . Fe_2O_3$, $ZnO . Fe_2O_3$) have very narrow hysteresis loops and are suitable for high-Q coils, wideband transformers, radio-frequency and pulse transformers and aerial rods. The material is supplied in the form of extrusions or preformed rings. Various grades are available depending on the application and frequency range required.

Magnesium manganese ferrite ($MgO . Fe_2O_3$, $MnO . Fe_2O_3$) exhibits a relatively square hysteresis loop which makes the material suitable for switching and storage applications. A typical B/H characteristic is shown in Fig. 9.38(a).

For use as a storage element, the ferrite is formed in a ring, as shown at (b). With no currents in the windings the magnetic state of the core will be represented by either point X or point Y in (a)—i.e. with the residual flux directed either clockwise or anti-clockwise round the ring. Clockwise flux may arbitrarily designated as the "1" condition and anti-clockwise flux as the "0" condition, so that the core can be used as a storage element for "ones" or "zeros", i.e. as an element in a binary store. Note that no energy is required to maintain the core in either state.

In order to change the state of a core from "1" to "0" it is necessary momentarily to supply a current I in a winding in such a direction as to give rise to demagnetizing ampere-turns. The flux density then changes from X to Z, and falls back to the residual value Y when the current is removed, so changing the core state from "1" to "0". Similarly, if the initial state is "0", then supplying the appropriate magnetizing ampere-turns will cause a change to the "1" state.

In order to determine the state of a core (i.e. to "read" the core) it is also necessary to pass a current I through a winding in a standard direction. Thus, suppose the core state is represented by X at (a). The "read" signal may be chosen in the demagnetizing direction so that when a pulse of "read" current, *I*, passes, the core changes state. An output pulse will then appear in a further winding ("sense" winding) on the core due to the e.m.f. induced by the changing flux.

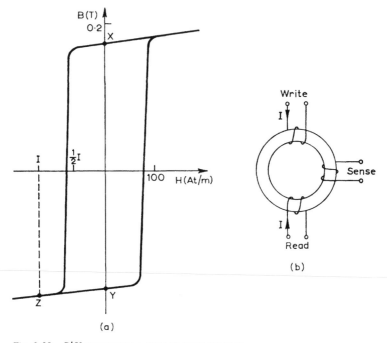

Fig. 9.38 B/H CURVE FOR A SQUARE-LOOP FERRITE

If the state of the core is represented by Y, however, the "read" current pulse causes a change from Y to Z and back to Y. The change in flux will be negligible and only a very small output will appear in the "sense" winding. Hence the size of the "sense" winding output will indicate the state of the core. Note, however, that the "read" signal will itself change the state of the core from "1" to "0", and hence a "rewrite" circuit is required if the "read" signal is not to destroy the "information" stored (i.e. the state of the core).

PROBLEMS

9.1   An air-cored mutual inductor has the following constants:
Resistance of primary winding, $R_1 = 48\,\Omega$
Self-inductance of primary winding, $L_{11} = 80\,\mathrm{mH}$

Resistance of secondary winding, $R_2 = 58\,\Omega$
Self-inductance of secondary winding, $L_{22} = 60\,\text{mH}$
Coupling coefficient, $k = 0.60$

Determine for $\omega = 10{,}000\,\text{rad/s}$ the input impedance to the primary when the secondary terminals are (a) open-circuited, and (b) short-circuited. Find also for $\omega = 10{,}000\,\text{rad/s}$ the input impedance to the coils when they are joined (a) in series aiding, and (b) in series opposing.

Ans. $(48 + j800)\Omega$; $(73.4 + j515)\Omega$; $(106 + j2{,}230)\Omega$; $(106 + j568)\Omega$.

9.2 The primary winding of an air-cored mutual inductor has 1,500 turns and the secondary winding 2,000 turns. When measurements were made at an angular frequency of 5,000 rad/s the following results were obtained:

Input impedance to primary (secondary open-circuited) . $(40 + j325)\Omega$
Input impedance to primary (secondary short-circuited) . $(60 + j165)\Omega$
Input impedance to secondary (primary open-circuited) . $(80 + j650)\Omega$

Determine the self-inductance of each winding, the mutual inductances and the coupling coefficient. Draw the equivalent-T circuit.

Determine also the primary and secondary leakage and magnetizing inductances and draw the transformer equivalent circuit.

What is the input impedance to the primary for a secondary load of $(200 + j200)\Omega$ at an angular frequency $\omega = 5{,}000\,\text{rad/s}$?

Ans. 65 mH; 130 mH; 65 mH; 0.707; 16.2 mH; 43.3 mH; 48.8 mH; 86.7 mH; $(77 + j213)\Omega$

9.3 Deduce an expression for the cross-sectional area of a transformer core in terms of the primary voltage, turns, frequency and flux density. A 50 Hz 3-phase core-type transformer is to be built for a 10,000/400 V ratio connected star-mesh. The cores are to have a square section. Assuming a maximum flux density of 1·1 T and an induced e.m.f. of 10 V per turn, determine the cross-sectional dimensions of the core and the number of turns per phase in each winding.

(*H.N.C.*)

Ans. 20·2 cm × 20·2 cm; 578 turns/phase; 40 turns/phase.

9.4 A 3,200/400 V single-phase transformer has winding resistances and reactances of $3\,\Omega$ and $13\,\Omega$ respectively in the primary and $0.02\,\Omega$ and $0.065\,\Omega$ in the secondary. Express these in terms of (a) primary alone, (b) secondary alone.

Ans. $4.28\,\Omega$, $17.16\,\Omega$; $0.067\,\Omega$, $0.268\,\Omega$.

9.5 Explain with a diagram, how a transformer can be represented by an equivalent circuit. Derive an expression for the equivalent resistance and reactance referred to the primary winding.

A 50 Hz single-phase transformer has a turns ratio of 6. The resistances are $0.9\,\Omega$ and $0.03\,\Omega$ and the reactances $5\,\Omega$ and $0.13\,\Omega$ for high-voltage and low-voltage windings respectively.

Find (a) the voltage to be applied to the high-voltage side to obtain full-load current of 200 A in the low-voltage winding on short-circuit, (b) the power factor on short-circuit. (*H.N.C.*)

Ans. 330 V; 0·2 lagging.

9.6 The primary and secondary windings of a 30 kVA 6,000/230 V transformer have resistances of $10\,\Omega$ and $0.016\,\Omega$ respectively. The reactance of the transformer referred to the primary is $34\,\Omega$. Calculate (a) the primary voltage required to circulate full-load current when the secondary is short-circuited, (b) the percentage voltage regulation of the transformer for a load of 30 kVA having a p.f. of 0·8 lagging.   (*L.U.*)

  Ans. 200 V; 3·1 per cent.

9.7 Calculate (*a*) the full-load efficiency at unity power factor, and (*b*) the secondary terminal voltage when supplying full-load secondary current at power factors (i) 0·8 lagging, (ii) 0·8 leading for the 4 kVA 200/400 V 50 Hz single-phase transformer, of which the following are the test figures:

  Open-circuit with 200 V applied to the primary winding—power 60 W. Short-circuit with 16 V applied to the high-voltage winding—current 8 A, power 40 W.

  Show a complexor diagram in both cases.   (*H.N.C.*)

  Ans. 0·97; 383 V, 406 V.

9.8 A 12 kVA 220/440 V 50 Hz single-phase transformer gave the following test figures:

  No-load: primary data—220 V, 2 A, 165 W.
  Short-circuit: secondary data—12 V, 15 A, 60 W.

  Draw the equivalent circuit, considered from the low-voltage side, and insert appropriate values. Find the secondary terminal voltage on full load at a power factor of 0·8 lagging.   (*H.N.C.*)

  Ans. 422 V.

9.9 The following results were obtained from a 125 kVA 2,000/400 V 50 Hz single-phase transformer:

  No-load test h.v. data—2,000 V, 1 A, 1,000 W.
  Short-circuit tests l.v. data—13 V, 200 A, 750 W.
Calculate:

  (*a*) the magnetizing current and the component corresponding to core loss at normal voltage and frequency;
  (*b*) the efficiency on full load at p.f.s of unity, 0·8 lagging, and 0·8 leading;
  (*c*) the secondary voltage on full load at the above p.f.s.   (*H.N.C.*)
  Ans. 0·866 A; 0·5 A; 0·98; 0·976; 0·976; 394 V; 384 V; 406 V.

9.10 A 5 kVA 200/400 V 50 Hz single-phase transformer gave the following results:

  Open-circuit test: 200 V, 0·7 A, 60 W—low-voltage side.
  Short-circuit test: 22 V, 16 A, 120 W—high-voltage side.

  (*a*) Find the percentage regulation when supplying full load at 0·9 power factor lagging.
  (*b*) Determine the load which gives maximum efficiency and find the value of this efficiency at unity power factor.   (*H.N.C.*)

  Ans. 3·08 per cent, 4·54 kVA, 0·974.

9.11 Enumerate the losses in a transformer and explain how each loss varies with the load when the supply voltage and frequency are constant. Describe how the components of the losses at no load may be determined.

  A transformer having a rated output of 100 kVA has an efficiency of 98 per cent at full-load unity p.f. and maximum efficiency occurs at $\frac{2}{3}$ full load (unity p.f.). Calculate (i) the core losses, and (ii) the maximum efficiency.   (*L.U.*)

  Ans. 0·62 kW; 98·4 per cent.

9.12 Calculate the efficiencies at half-full, full, and $1\frac{1}{4}$ full load of a 100kVA transformer for power factors of (a) unity, (b) 0·8. The winding loss is 1,000 W at full load and the core loss is 1,000 W.

Ans. $\frac{1}{2}$ full, 0·975, 0·969; full, 0·98, 0·975; $1\frac{1}{4}$ full, 0·979, 0·974.

9.13 A 2-phase 240 V supply is to be obtained from a 3-phase 3-wire 440 V supply by means of a pair of Scott-connected single-phase transformers. Determine the turns ratios of the main and teaser transformers.

Find the input current in each of the 3-phase lines (a) when each of the 2-phase currents is 1 A lagging behind the respective phase voltage by 36·9°, and (b) when the secondary phase on the main transformer is open-circuited the other secondary phase being loaded as in (a). Magnetizing current may be neglected.

Ans. 1·83; 1·59; 0·63A; $I_R = 0.63\underline{/0°}\,$A; $I_Y = I_B = 0.315\underline{/180°}\,$A.

9.14 Two transmission lines of impedance $(1 + j2)\Omega$ and $(2 + j2)\Omega$ respectively feed in parallel a load of 7,500kW at 0·8 p.f. lagging.

Determine the power output of each line and its power factor.

Ans. 3,750kW, 0·707 lagging; 3,750kW, 0·894 lagging.

9.15 A 400kVA transformer of 0·01 per unit resistance and 0·05 per unit reactance is connected in parallel with a 200kVA transformer of 0·012 per unit resistance and 0·04 per unit reactance. Find how they share a load of 600kVA at 0·8 p.f. lagging.

Ans. $373\underline{/-39°}$ kVA; $227\underline{/-33.6°}$ kVA.

9.16 Two 3-phase transformers operating in parallel deliver 500 A at a p.f. of 0·8 lagging. The resistances and reactances of the transformers are $R_1 = 0.02\Omega$, $X_1 = 0.2\Omega$; $R_2 = 0.03\Omega$, $X_2 = 0.3\Omega$. Calculate the current delivered by the first transformer and its phase angle with respect to the common terminal voltage.

In this example $R_1/X_1 = R_2/X_2$. Discuss, with reasons, whether or not this is desirable for parallel operation. (*H.N.C.*)

Ans. 300 A, 0·8 lagging.

9.17 A small 3-phase substation receives power from a station some distance away by two feeders which follow different routes. The impedances per phase of the feeders are (a) cable $(3 + j2)\Omega$ and (b) overhead line $(2 + j6)\Omega$. If the power delivered by the line is 4,000kW at 11 kV and p.f. 0·8 lagging, find the total power delivered by the cable and its phase angle. (*H.N.C.*)

Ans. 8,760kW; 1° leading.

9.18 Two transformers A and B are connected in parallel to supply a load having an impedance of $(2 + j1.5)\Omega$. The equivalent impedances referred to the secondary windings are $(0.15 + j0.5)\Omega$ and $(0.1 + j0.6)\Omega$ respectively. The open-circuit e.m.f. of A is 207 V and of B is 205 V. Calculate (i) the voltage at the load, (ii) the power supplied to the load, (iii) the power output of each transformer, and (iv) the kVA input to each transformer.

Ans. (i) $189\underline{/-3.8°}$ V, (ii) 11·5kW, (iii) 6·5kW, 4·95kW, (iv) 8·7kVA, 6·87kVA.

9.19 Explain clearly the essential conditions to be satisfied when two 3-phase transformers are connected in parallel. Give two sets of possible connexions, explaining how these are satisfactory or unsatisfactory.

Two transformers of equal voltage ratios but with the following ratings and impedances,

Transformer A—1,000 kVA, 1 per cent resistance, 5 per cent reactance,
Transformer B—1,500 kVA, 1·5 per cent resistance, 4 per cent reactance,

are connected in parallel to feed a load of 1,000 kW at 0·8 p.f. lagging. Determine the kVA in each transformer and its power factor.   (*H.N.C.*)
(*Note.* Impedances may be expressed in per-unit form by dividing the percentage impedances by 100.)

Ans. A: 448 kVA, 0·73 lagging. B: 804 kVA, 0·834 lagging.

9.20 Two single-phase transformers work in parallel on a load of 750 A at 0·8 p.f. lagging. Determine the secondary voltage and the output and power factor of each transformer.

Test data are:

Open-circuit: 11,000 V/3,300 V for each transformer
Short-circuit with h.v. winding short-circuited:

Transformer A: secondary input 200 V, 400 A, 15 kW
Transformer B: secondary input 100 V, 400 A, 20 kW   (*L.U.*)

Ans. 3,190 V; A: 807 kVA, 0·65 lagging. B: 1,615 kVA, 0·86 lagging.

Chapter 10

GENERAL PRINCIPLES
OF ROTATING MACHINES

Rotating machines vary greatly in size, ranging from a few watts to 600 MW and above—a ratio of power outputs of over 10^7. They also vary greatly in type depending on the number and inter-connexion of their windings and the nature of electrical supply to which they are to be connected. Despite these differences of size and type their general principles of operation are the same, and it is the purpose of this chapter to examine these common principles. Three succeeding chapters give a more detailed treatment of parti-cular types of machine.

10.1 Modes of Operation

There are three distinguishable ways or modes of operation of rotat-ing machines and these are illustrated in the block diagrams of Fig. 10.1. The three modes, motoring, generating and braking, are specified below.

MOTORING MODE

Electrical energy is supplied to the main or *armature winding* of the machine and a mechanical energy output is available at a rotating shaft. This mode of operation is illustrated in Fig. 10.1(*a*), which takes the form of a 2-port representation of a machine, one port being electrical and the other mechanical.

An externally applied voltage v drives a current i through the armature winding against an internally induced e.m.f. e. The process of induction of e.m.f. is discussed in Section 10.4. The winding is thus enabled to absorb electrical energy at the rate ei. At least some of this energy is available for conversion (some may be stored in associated magnetic fields). The armature winding gives rise to an instantaneous torque T_A'* which drives the rotating member of the machine (the *rotor*) at an angular velocity ω_r, and mechanical energy is delivered at the rate of $\omega_r T_A'$. The process of torque production is discussed in Section 10.5. An externally applied load torque

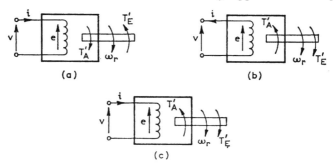

(a)   (b)

(c)

Fig. 10.1   MODES OF OPERATION OF ROTATING MACHINES
(a) Motoring mode
(b) Generating mode
(c) Braking mode

T_E' acting in a direction opposite to that of rotation enables the load to absorb mechanical energy:

$$T_A' - T_E' = J\frac{d\omega_r}{dt} \tag{10.1}$$

J being the moment of inertia of the rotor and its mechanical load.

If the windage and friction torque is included in the external applied load torque, then, following eqn. (10.1), if T_A' and T_E' are equal and opposite, $d\omega_r/dt = 0$ and the machine will rotate at a steady speed.

When steady-state operation prevails, provided a sufficient period of time is considered,

$$(\omega_r T_A')_{mean} = (ei)_{mean} \tag{10.2}$$

Since the armature winding must develop torque and have an e.m.f. induced in it, a magnetic field is required. In very small

---

* To avoid confusion with t for time, instantaneous torque will be represented by T'.

machines this may be provided by permanent magnets, but in most machines it is provided electromagnetically.

Some machines have a separate *field winding* to produce the required magnetic field. For such machines the block diagram of Fig. 10.1(*a*) would require a second electrical port. For the sake of simplicity this has been omitted. The energy fed to the field winding is either dissipated as loss in the field winding or is stored in the associated magnetic field and does not enter into the conversion process.

GENERATING MODE

Mechanical energy is supplied to the shaft of the machine by a prime mover and an electrical energy output is available at the armature-winding terminals. This mode of operation is illustrated in Fig. 10.1(*b*). The shaft of the machine is driven at an angular velocity ω_r in the direction of the applied external instantaneous torque $T_E{}'$ and in opposition to the torque $T_A{}'$ due to the armature winding, enabling the machine to absorb mechanical energy. The armature winding has an e.m.f. *e* induced in it which drives a current through an external load of terminal voltage *v*. Eqns. (10.1) and (10.2) apply equally to generator action.

BRAKING MODE

In this mode of operation the machine has both a mechanical energy input and an electrical energy input. The total energy input is dissipated as loss in the machine. This mode is of limited practical application but occurs sometimes in the operation of induction and other machines.

10.2 Rotating Machine Structures

Rotating electrical machines have two members, a stationary member called the *stator* and a rotating member called the *rotor*. The stator and rotor together constitute the magnetic circuit or core of the machine and both are made of magnetic material so that magnetic flux is obtained for moderate values of m.m.f. The rotor is basically a cylinder and the stator a hollow cylinder. The rotor and stator are separated by a small air-gap as shown in Fig. 10.2. Compared with the rotor diameter the radial air-gap length is small. The stator and rotor magnetic cores are usually, but not invariably,

built up from laminations (typically 0·35 mm thick) in order to reduce
eddy-current loss.

If the rotor is to rotate, a mutual torque has to be sustained between
the rotor and stator. A winding capable of carrying current and of
sustaining torque is required on at least one member and usually,
but not always, on both. One method of arranging windings in a
rotating machine is to place coils in uniformly distributed slots on
both the stator and the rotor. This method is illustrated in Fig.

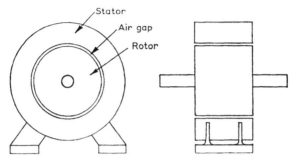

Fig. 10.2   BASIC ARRANGEMENT OF A ROTATING MACHINE

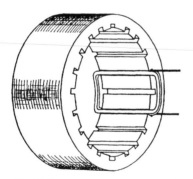

Fig. 10.3   STATOR AND ONE OF ITS COILS

10.3, where, for clarity, only the stator is shown. An arrangement of
this sort is commonly used in induction machines. The distance
between the coil sides is usually about one pole pitch.

To make a complete winding, similar coils are placed in other
pairs of slots and all the coils are then connected together in groups.
The groups of coils may then be connected in series or in parallel,
and in 3-phase machines in star or mesh.

Some windings may be double-layer windings. In such windings
each slot contains two coil sides, one at the top and the other at the

bottom of a slot. Each coil has one coil side at the top of the slot and the other at the bottom.

An alternative arrangement to having uniform slotting on both sides of the air-gap is to have salient poles around which are wound concentrated coils to provide the field winding. The salient poles may be on either the stator or the rotor, and such arrangements are illustrated in Figs. 10.4 and 10.5(*a*).

The salient-pole stator arrangement is commonly used for direct-current machines and occasionally for small sizes of synchronous

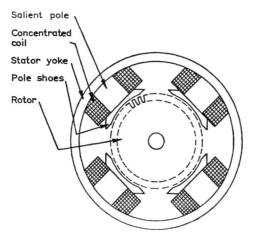

Fig. 10.4 SALIENT-POLE STATOR

machine. As far as d.c. machines are concerned the stator is most often referred to as the field and the rotor as the armature. The main winding in such a machine is on the rotor and is called the armature winding.

The salient pole rotor arrangement is most often used for synchronous machines. In such machines the main winding is on the stator but it is often called the armature winding.

In general, rotating machines can have any even number of poles. The concentrated coil windings surrounding the poles are excited so as to make successive poles of alternate north and south polarity.

The salient-pole rotor structure is unsuitable for large high-speed turbo-alternators used in the supply industry because of the high stress in the rotor due to centrifugal force. In such machines a cylindrical rotor is used as shown in Fig. 10.5(*b*). Uniform slotting occupies two thirds of the rotor surface, the remaining third being unslotted. Such rotors are usually solid steel forgings.

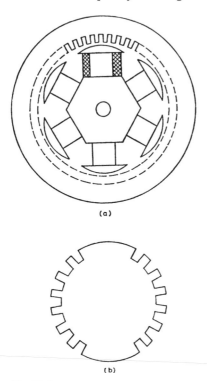

(a)

(b)

Fig. 10.5   ROTORS
(a) Salient pole   (b) Cylindrical

10.3 Self- and Mutual Inductance of Stator and Rotor Windings

The simplest rotating machine structure is a 2-pole machine with a uniform air-gap as shown in Fig. 10.2 which does not exhibit "saliency" (i.e. does not have salient poles) on either side of the air-gap. In this and all succeeding sections of this chapter only 2-pole machines will be considered.

Fig. 10.6(a) shows such a machine. The stator winding axis is chosen to correspond with a horizontal angular reference axis called the *direct axis* (*d*-axis) at which $\theta = 0$.

A convention for positive current in a coil must be established and this is done in the following way. Consider a winding such as the stator winding of Fig. 10.6(a) whose axis corresponds with the *d*-axis. Positive current is taken to produce an m.m.f. acting in the positive direction of the *d*-axis. Thus the stator winding in Fig. 10.6(a) is excited by positive current. The angular position, θ, of a

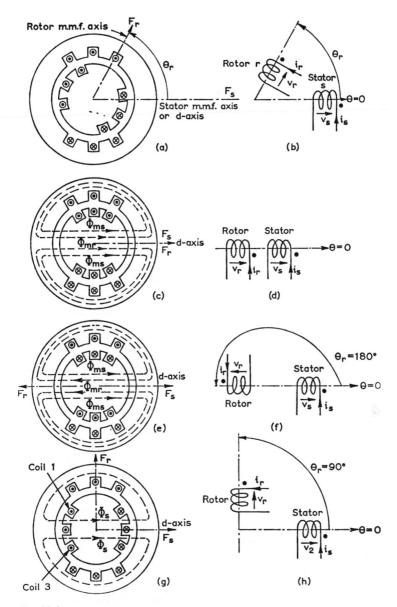

Fig. 10.6 MUTUAL COUPLING OF STATOR AND ROTOR COILS

winding whose axis does not correspond with the d-axis is the magnitude of the clockwise angle through which the coil must be rotated so that its m.m.f. acts along the positive direction of the d-axis. Fig. 10.6(a) shows a rotor winding at an angle θ_r to the d-axis. If the current in this coil were reversed, represented in the diagram by interchange of the dots and crosses of the rotor winding, the winding position would be taken as $180° + \theta_r$.

Fig. 10.6(b) is a circuit representation of the configuration shown at (a) which uses the dot notation. The dot notation for coils capable of rotation can be expressed as "currents entering the dotted end of a winding give rise to an m.m.f. which acts towards the dotted coil end".

Evidently the mutual inductance between the stator and rotor windings is a positive maximum in the configuration of Fig. 10.6(c), where $\theta_r = 0$, and a negative maximum for that of Fig. 10.6(e), where $\theta_r = 180°$. Further, in the configuration of Fig. 10.6(g), where the winding axes are at right angles, the mutual inductance between the windings is zero. If the rotor is considered to have diametral coils (i.e. coil sides in diametrically opposite slots) then the current of coil 1 links the stator flux in the opposite direction to that of the current in coil 3 so that the net current-flux linkage is zero.

The mutual coupling between the stator and rotor coils depends on the angular separation of their m.m.f. axes θ_r. When an inductance is a function of θ in this way it will be denoted by the symbol $\mathscr{L}$. Where an inductance is not a function of θ it is written L.

The mutual coupling between the stator and rotor windings is evidently a cosine-like or even function of the form

$$\mathscr{L}_{sr} = L_{sr}(\cos \theta_r + k_3 \cos 3\theta_r + k_5 \cos 5\theta_r \ . \ . \ .) \tag{10.3}$$

If all terms except the fundamental are ignored,

$$\mathscr{L}_{sr} = L_{sr} \cos \theta_r \tag{10.4}$$

Since the air-gap is uniform, the permeance of the stator and rotor magnetic circuits is unaffected by rotor position and the stator and rotor winding self-inductances are constants.

If the stator has salient poles, the mutual inductance between the stator and rotor windings is still given by eqn. (10.3) though the space-harmonic coefficients, k_3, k_5, k_7, etc. will be different. Since the space harmonics are ignored, the mutual inductance is given by eqn. (10.4). The self-inductance of the stator winding will be a constant independent of θ_r, but the self-inductance of the rotor coil will depend on the rotor position (see Fig. 10.7). When the rotor m.m.f. axis is lined up with the d-axis, its self-inductance will be a maximum, L_{dd}, say, but when it is lined up with an axis at right angles to the d-axis, the *quadrature axis* (q-axis), the self-inductance will have fallen to a minimum value L_{qq}, say, because of the much lower permeance of the magnetic circuit centred on this axis.

After the rotor has turned through 180° the rotor m.m.f. axis again corresponds with the d-axis so that its self-inductance again

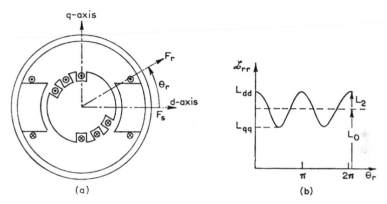

Fig. 10.7 VARIATION OF SELF-INDUCTANCE DUE TO SALIENCY

has the maximum value L_{dd}. Fig. 10.7(*b*) shows $\mathscr{L}_{rr}$ to a base of θ_r, from which approximately

$$\mathscr{L}_{rr} = L_0 + L_2 \cos 2\theta_r \tag{10.5}$$

or, from Fig. 10.7(*b*),

$$\mathscr{L}_{rr} = \tfrac{1}{2}(L_{dd} + L_{qq}) + \tfrac{1}{2}(L_{dd} - L_{qq}) \cos 2\theta_r \tag{10.6}$$

When the saliency occurs on the rotor, on the other hand, the rotor self-inductance becomes constant and the stator self-inductance changes. In such a case the d- and q-axis are deemed to rotate, the d-axis coinciding with the salient-pole axis and the q-axis being at right angles to the d-axis. Eqn. (10.6) then gives the variation of stator-winding self-inductance without modification as

$$\mathscr{L}_{ss} = \tfrac{1}{2}(L_{dd} + L_{qq}) + \tfrac{1}{2}(L_{dd} - L_{qq}) \cos 2\theta_r \tag{10.7}$$

10.4 General Expression for Induced E.M.F.

Consider a stator winding s, and a rotor winding r that rotates at a steady angular velocity $\omega_r = d\theta_r/dt$ with respect to the stator winding, as shown in Fig. 10.8. Let the windings carry instantaneous currents i_s, i_r which, in general, will be functions of time. The mutual inductance of the windings is a function of θ_r as also will be one of the self-inductances if saliency exists on either side of the air-gap. Since the position of the rotor winding θ_r is a function of time,

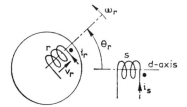

Fig. 10.8 INDUCED E.M.F. AND TORQUE IN A ROTATING MACHINE

such inductances are implicitly functions of time. The e.m.f. induced in the rotor winding is

$$e_r = \frac{d}{dt}(\mathcal{L}_{rr}i_r + \mathcal{L}_{sr}i_s)$$

Differentiating each term as a product,

$$e_r = \frac{\partial \mathcal{L}_{rr}}{\partial \theta_r}\frac{d\theta_r}{dt}i_r + \mathcal{L}_{rr}\frac{di_r}{dt} + \frac{\partial \mathcal{L}_{sr}}{\partial \theta_r}\frac{d\theta_r}{dt}i_s + \mathcal{L}_{sr}\frac{di_s}{dt}$$

$$e_r = \mathcal{L}_{rr}{}'\omega_r i_r + \mathcal{L}_{sr}{}'\omega_r i_s + \mathcal{L}_{rr}\frac{di_r}{dt} + \mathcal{L}_{sr}\frac{di_s}{dt} \qquad (10.8)$$

$$\underbrace{\qquad\qquad\qquad\qquad}_{\text{Rotational voltages}}\quad\underbrace{\qquad\qquad\qquad\qquad}_{\text{Transformer voltages}}$$

where $\mathcal{L}_{rr}{}' = \partial\mathcal{L}_{rr}/\partial\theta$ etc. and $\omega_r = d\theta_r/dt$.

It will be seen that the expression for e_r contains terms of two distinct types: (*a*) voltages proportional to the rotor angular velocity and called *rotational voltages*, and (*b*) voltages proportional to the rate of change of the winding currents. These latter are often called *transformer voltages*.

Including the voltage drop in the rotor winding resistance the voltage applied to that winding is

$$v_r = r_r i_r + \mathcal{L}_{rr}{}'\omega_r i_r + \mathcal{L}_{sr}{}'\omega_r i_s + \mathcal{L}_{rr}\frac{di_r}{dt} + \mathcal{L}_{sr}\frac{di_s}{dt} \qquad (10.9)$$

Similarly, the voltage applied to the stator winding is

$$v_s = r_s i_s + \mathcal{L}_{ss}{}'\omega_r i_s + \mathcal{P}_{sr}{}'\omega_r i_r + \mathcal{L}_{ss}\frac{di_s}{dt} + \mathcal{L}_{sr}\frac{di_r}{dt} \qquad (10.10)$$

In general, for n coupled windings the voltage applied to the jth winding is

$$v_j = r_j i_j + \sum_{k=1}^{k=n} \frac{d}{dt}(\mathscr{L}_{jk} i_k) \tag{10.11}$$

When there are additional stator or rotor windings the expression for a winding voltage may usually be inferred by extension of eqn. (10.9) or eqn. (10.10).

10.5 General Expression for Torque

Consider again a machine consisting of a stator winding s and a rotor winding r rotating at a steady angular velocity ω_r, as shown in Fig. 10.8. The total instantaneous power fed into the machine is

$$p_e = v_r i_r + v_s i_s$$

Substituting for v_s and v_r in terms of eqns. (10.9) and (10.10),

$$p_e = r_r i_r^2 + \mathscr{L}_{rr}{}' \omega_r i_r^2 + \mathscr{L}_{sr}{}' \omega_r i_r i_s + \mathscr{L}_{rr} i_r \frac{di_r}{dt} + \mathscr{L}_{sr} i_r \frac{di_s}{dt}$$

$$+ r_s i_s^2 + \mathscr{L}_{ss}{}' \omega_r i_s^2 + \mathscr{L}_{sr}{}' \omega_r i_r i_s + \mathscr{L}_{ss} i_s \frac{di_s}{dt} + \mathscr{L}_{sr} i_s \frac{di_r}{dt} \tag{10.12}$$

In this equation the terms $r_r i_r^2$ and $r_s i_s^2$ represent power loss in the winding resistances.

The energy stored in the magnetic fields associated with the two coils is

$$w_f = \tfrac{1}{2}\mathscr{L}_{rr} i_r^2 + \tfrac{1}{2}\mathscr{L}_{ss} i_s^2 + \mathscr{L}_{sr} i_s i_r$$

The rate at which energy is stored in the magnetic field is

$$\frac{dw_f}{dt} = \tfrac{1}{2} i_r^2 \mathscr{L}_{rr}{}' \omega_r + \mathscr{L}_{rr} i_r \frac{di_r}{dt} + \tfrac{1}{2} i_s^2 \mathscr{L}_{ss}{}' \omega_r + \mathscr{L}_{ss} i_s \frac{di_s}{dt}$$

$$+ i_s i_r \mathscr{L}_{sr}{}' \omega_r + \mathscr{L}_{sr} i_s \frac{di_r}{dt} + \mathscr{L}_{sr} i_r \frac{di_s}{dt} \tag{10.13}$$

where again $\mathscr{L}_{rr}{}' = \partial \mathscr{L}_{rr}/\partial \theta_r$, etc., and $\omega_r = d\theta_r/dt$.

There is an instantaneous mechanical power output corresponding to that portion of the instantaneous electrical power input which is neither dissipated in the winding resistances nor used to store energy in the magnetic field. If the instantaneous torque on the rotor is T', then

$$p_m = \omega_r T' = p_e - r_r i_r^2 - r_s i_s^2 - \frac{dw_f}{dt}$$

$$= \tfrac{1}{2}\mathscr{L}_{rr}{}' \omega_r i_r^2 + \mathscr{L}_{sr}{}' \omega_r i_r i_s + \tfrac{1}{2}\mathscr{L}_{ss}{}' \omega_r i_s^2$$

or

$$T' = \tfrac{1}{2}\mathscr{L}_{rr}{}' i_r^2 + \mathscr{L}_{sr}{}' i_r i_s + \tfrac{1}{2}\mathscr{L}_{ss}{}' i_s^2 \tag{10.14}$$

When there are additional stator or rotor windings the total torque acting on the rotor or stator may be inferred by extension of eqn. (10.14).

EXAMPLE 10.1 A torque motor has a uniform air-gap. The stator and rotor each carry windings and the axis of the rotor coil may rotate relative to that of the stator coil. The mutual inductance between the coils is such that

$$\mathscr{L}_{sr} = L_{sr} \cos \theta_r$$

(a) Show that, when the axes of the coils are lined up on the d-axis and each coil carries conventionally positive current, the coils are in a position of stable equilibrium.

(b) If with the coils so aligned the current in either coil is reversed show that the position is one of unstable equilibrium.

(c) In such an arrangement the rotor and stator coils are in series, the rotor coil axis is at $\theta_r = 135°$ and the maximum mutual inductance is 1 H. Calculate the coil currents if the mutual torque on the rotor is to be 100 N-m in the $-\theta$ direction and the coils are excited with direct current.

The instantaneous torque is given by eqn. (10.14) as

$$T' = \tfrac{1}{2}\mathscr{L}_{rr}{}' i_r^2 + \mathscr{L}_{sr}{}' i_r i_s + \tfrac{1}{2}\mathscr{L}_{ss}{}' i_s^2 \tag{10.14}$$

Since the air-gap is uniform the rotor and stator winding self-inductances are constants and their angular rates of change are zero, i.e. $\mathscr{L}_{rr}{}' = 0$ and $\mathscr{L}_{ss}{}' = 0$.

$$\mathscr{L}_{sr}{}' = \frac{d}{d\theta_r}(\mathscr{L}_{sr}) = \frac{d}{d\theta_r}(L_{sr} \cos \theta_r) = -L_{sr} \sin \theta_r$$

Substituting these conditions into the expression for instantaneous torque,

$$T' = -L_{sr} i_r i_s \sin \theta_r \tag{10.15}$$

With the rotor coil axis aligned with the stator coil axis $\theta_r = 0$ and the torque is zero. If the rotor coil is given a small deflection $+\delta\theta_r$ from this position, with positive currents flowing in both windings the torque takes on a negative value and acts in the $-\theta_r$ direction to restore the rotor to its initial position. Similarly, if the rotor is given a small deflection $-\delta\theta_r$ the torque takes on a positive value and acts in the $+\theta_r$ direction to restore the rotor to its initial position.

If one of the coil currents is reversed, however, the opposite result occurs and a small deflection in either direction leads to a torque acting so as to increase the deflection. If the rotor is free to move in this case it will take up an equilibrium position at $\theta_r = 180°$, where the torque is again zero.

It will be noted that in both cases the tendency is for the coils to align themselves in the positive maximum mutual inductance configuration.

From eqn. (10.15) for $i_s = i_r = I$, the steady torque when the axis of the rotor winding is at any angle θ_r is

$$T = -L_{sr}I^2 \sin \theta_r$$

As the torque on the rotor is to act in the $-\theta$ direction, $T = -100$ N-m. This gives

$$I = \sqrt{\frac{100}{1 \times \sin 135°}} = \underline{\underline{11{\cdot}9\,\text{A}}}$$

EXAMPLE 10.2 An electrodynamic wattmeter has a fixed current coil and a rotatable voltage coil. The magnetic circuit of the device does not exhibit saliency. The following are details of a particular wattmeter:

Full-scale deflection	. 110°
Control-spring constant . . .	. 10^{-7} N-m/deg
Maximum current-coil current (r.m.s.) .	. 10 A
Maximum voltage-coil voltage (r.m.s.) .	. 60 V
Voltage-coil resistance	. 600 Ω

The mutual inductance between the coils varies cosinusoidally with the angle of separation of the coil axes. The zero on the instrument corresponds to a voltage-coil position of $\theta_r = 145°$.

(a) Determine the direct current flowing in the current coil when a direct voltage of 60 V is applied to the voltage coil and the angular deflection is 100° from the instrument scale zero.

(b) For a sinusoidally varying current-coil current of 6 A (r.m.s.) and a voltage-coil voltage of 60 V (r.m.s.) of the same frequency as the current, determine the phase angle by which the current lags the voltage when the voltage-coil deflection is 60° from the instrument scale zero.

The reactance of the voltage coil is negligible compared with its resistance. A diagram of the arrangement is given in Fig. 10.9.

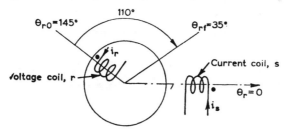

Fig. 10.9

Since there is no saliency and $\mathscr{L}_{sr} = L_{sr} \cos \theta_r$, the expression for the instantaneous torque is the same as that found in Example 10.2, namely

$$T' = -L_{sr}i_r i_s \sin \theta_r \qquad (10.15)$$

If the conditions for full-scale deflection are substituted the value of L_{sr}, the maximum possible mutual inductance between the coils, is found. Since the deflection is in the $-\theta_r$ direction the torque is negative.

$$T = -110 \times 10^{-7}\,\text{N-m}, \; i_r = \frac{60}{600} = 0\cdot1\,\text{A}, \; i_s = 10\,\text{A}, \theta_{rf} = 145° - 110° = 35°$$

whence

$$L_{sr} = \frac{110 \times 10^{-7}}{0\cdot1 \times 10 \times \sin 35°} = 192 \times 10^{-7}\,\text{H}$$

(a) When the angular deflection is 100° the position of the voltage coil is $\theta_r = 145 - 100 = 45°$. This gives, in eqn. (10.15),

$$-100 \times 10^{-7} = -192 \times 10^{-7} \times 0\cdot1 \times I_s \sin 45°$$

so that

$$I_s = \frac{100}{192 \times 0\cdot1 \times 0\cdot707} = \underline{\underline{7\cdot37\,\text{A}}}$$

(b) If the voltage-coil current is taken as reference, the current-coil current is

$$i_s = I_{sm} \cos(\omega t - \phi)$$

The instantaneous torque is, from eqn. (10.15),

$$T' = -L_{sr}I_{rm} \cos \omega t I_{sm} \cos(\omega t - \phi) \sin \theta_r$$
$$= -\tfrac{1}{2}L_{sr}I_{rm}I_{sm} \sin \theta_r[\cos \phi + \cos(2\omega t - \phi)]$$

This expression shows that the instantaneous torque consists of two components: (a) a steady component, and (b) an alternating component which oscillates at twice the frequency of the currents in the two coils. The inertia of the rotating system will prevent its responding to the alternating component. The average torque is therefore

$$T = -\tfrac{1}{2}L_{sr}I_{rm}I_{sm} \sin \theta_r \cos\phi$$

whence

$$\cos \phi = -\frac{2T}{L_{sr}I_{rm}I_{sm} \sin \theta_r}$$

When the angular deflection is 60° the position of the voltage coil is $\theta_r = 145 - 60 = 85°$:

$$\cos \phi = -\frac{-2 \times 60 \times 10^{-7}}{192 \times 10^{-7} \times \dfrac{\sqrt{2 \times 60}}{600} \times \sqrt{2} \times 6 \times 0\cdot995} = 0\cdot525$$

so that

$$\phi = \underline{\underline{58\cdot3°}}$$

10.6 The Alignment Principle

Example 10.1 has shown that the torque acting on the rotor of a simple rotating machine structure consisting of a stator and a rotor coil is such as to tend to align the coils in their maximum positive mutual inductance position. The mutual torque on the system is then zero. If a continuously rotating machine is to be produced, therefore, some method must be found of maintaining a constant angular displacement of the axes of the rotor and stator winding m.m.f.s under steady conditions despite the rotation of the rotor and its winding. Several different methods exist for achieving this; the particular method chosen determines the type of machine. The rest of the chapter is devoted to considering how this constant angular displacement of the axes of the winding m.m.f.s is brought about in some common types of machine.

10.7 The Commutator

In some types of machine the stator winding is excited with direct current and so the axis of the stator m.m.f., F_s, is fixed. If a constant angle is to be maintained between the axes of the stator and rotor winding m.m.f.s the rotor winding m.m.f. must be stationary despite

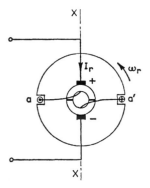

Fig. 10.10   A TWO-SEGMENT COMMUTATOR

the rotation of that winding. This may be achieved by exciting the rotor winding with direct current supplied through a commutator.

Fig. 10.10 shows a simple 2-segment commutator which consists essentially of a hollow cylinder of copper split in half, each being insulated from the other and from the shaft. One end of a rotor coil is joined to each commutator segment. Two brushes, fixed in space, make alternate contact with each segment of the commutator as it rotates. Although the current in the coil reverses twice in each revolution, it will be seen that whichever of the coil sides, a or a', lies to the left of the brush axis XX will carry current in the direction indicated by ⊙, whereas whichever coil side lies to the right of XX carries current in the direction indicated by ⊕.

A rotor winding consisting of many coils wound into uniformly distributed slots may also be supplied through a commutator. Each of the two ends of each coil is connected to two different commutator segments. The rotor coils are connected in series, the ends of successive coils being joined at the commutator as shown in Fig. 10.11(*a*). Such windings are double-layer windings.

Fig. 10.11(*b*) is a conventional representation of such a winding where the commutator is not shown and the brushes are thought

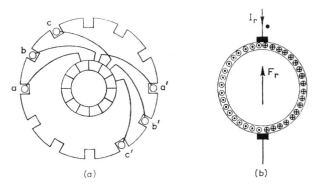

(a)   (b)

Fig. 10.11   A MULTI-SEGMENT COMMUTATOR

of as bearing directly on the conductors. All the conductors to the left of the brush axis carry current in the direction $\odot$ and all those to its right carry current in the direction $\oplus$. It will be seen, therefore, that supplying the rotor winding through a commutator has the effect of fixing a certain current pattern in space despite rotation of the winding. As a result the axis of the rotor winding m.m.f., F_r, is fixed in space and coincides with the brush axis. The positive brush at which the current enters the winding corresponds to the dotted end of the winding.

10.8 Separately Excited D.C. Machine

The d.c. machine has almost invariably a salient pole structure on the stator and a non-salient pole rotor. The stator has a concentrated coil winding; the rotor winding is distributed in slots. Fig. 10.4 shows the structure commonly adopted for the d.c. machine. The stator winding is excited with direct current, and the rotor winding is supplied with direct current through a commutator, thus maintaining a constant angular displacement between the axes of the stator and rotor winding m.m.f.s as is required for torque maintenance.

Since the rotor is not salient pole the stator winding self-inductance does not vary with the angular position of the rotor as explained in Section 10.3, i.e.

$$\mathscr{L}_{ss} = L_{ss} \tag{10.16}$$

so that

$$\mathscr{L}_{ss}' = \frac{\partial \mathscr{L}_{ss}}{\partial \theta} = 0 \tag{10.17}$$

The mutual inductance between the stator and rotor windings is

$$\mathscr{L}_{sr} = L_{sr} \cos \theta_r \tag{10.4}$$

so that

$$\mathscr{L}_{sr}' = -L_{sr} \sin \theta_r \tag{10.18}$$

Since there is saliency on the stator the rotor self-inductance varies with the angular position of the rotor and is given by eqn. (10.6) as

$$\mathscr{L}_{rr} = \tfrac{1}{2}(L_{dd} + L_{qq}) + \tfrac{1}{2}(L_{dd} - L_{qq}) \cos 2\theta_r \tag{10.6}$$

so that

$$\mathscr{L}_{rr}' = -(L_{dd} - L_{qq}) \sin 2\theta_r \tag{10.19}$$

Eqn. (10.9) gives the voltage to the rotor winding as

$$v_r = r_r i_r + \mathscr{L}_{rr}' \omega_r i_r + \mathscr{L}_{sr}' \omega_r i_s + \mathscr{L}_{rr} \frac{di_r}{dt} + \mathscr{L}_{sr} \frac{di_s}{dt} \tag{10.9}$$

Consider a separately excited d.c. machine with steady, direct voltages, V_s and V_r, applied to the stator and rotor windings. Let the steady, direct currents in the stator and rotor windings be I_s and I_r. The time rates of change of these steady currents are zero so that the voltage applied to the rotor winding is

$$V_r = r_r I_r + \mathscr{L}_{rr}' \omega_r I_r + \mathscr{L}_{sr}' \omega_r I_s \tag{10.20}$$

or

$$V_r = r_r I_r - (L_{dd} - L_{qq}) \sin 2\theta_r \omega_r I_r - L_{sr} \sin \theta_r \omega_r I_s \tag{10.21}$$

In this equation the terms involving ω_r are rotational voltages, and the larger these terms are, for a given angular velocity and for given winding currents, the more effective the machine will be as an energy convertor. For the commutation of the rotor winding current to take place without sparking between brushes and commutator, the brush axis must be approximately at right angles to the stator winding m.m.f. This condition is represented by substituting the value $\theta_r = -\pi/2$ in eqn. (10.21) and has the effect of making the

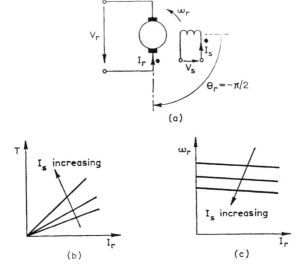

(a)

(b)

(c)

Fig. 10.12   SEPARATELY EXCITED D.C. MOTOR

rotational voltage involving the term $(L_{dd} - L_{qq})$ zero but the rotational voltage involving the the term L_{sr} a maximum. Substituting $\theta_r = -\pi/2$ in eqn. (10.21) has the advantage of removing from the equations minus signs which could be a source of confusion. Assigning the value $\theta_r = -\pi/2$ means that the dotted end of the rotor winding (i.e. the positive commutator brush) is placed at

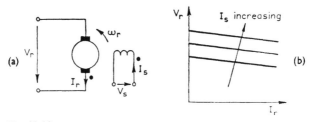

Fig. 10.13   SEPARATELY EXCITED D.C. GENERATOR

$\theta_r = -\pi/2$, as shown in Figs. 10.12, 10.13 and 10.14. Carrying out this substitution gives

$$V_r = r_r I_r + L_{sr} \omega_r I_s \qquad (10.22)$$

The machine configuration is shown in Fig. 10.12(*a*).

The voltage $L_{sr}\omega_r I_s$ is a steady, direct rotational voltage due to the rotation of the rotor winding in a magnetic field set up by the stator winding. The fact that it is a direct voltage is due to the effect of the commutator.

When attention is directed to the voltage applied to the stator winding it will be realized that no rotational voltage will appear in the winding. This is so because, in spite of rotor rotation, the rotor winding m.m.f. axis is fixed in space by the action of the commutator, and so the rotor flux does not change its linkage with the stator winding even when the rotor rotates. The voltage applied to the stator winding is therefore given by eqn. (10.10):

$$V_s = r_s I_s \tag{10.23}$$

The instantaneous torque developed is given by eqn. (10.14) as

$$T' = \tfrac{1}{2}\mathscr{L}_{rr}' i_r{}^2 + \mathscr{L}_{sr}' i_r i_s + \tfrac{1}{2}\mathscr{L}_{ss}' i_s{}^2 \tag{10.14}$$

As previously noted,

$$\mathscr{L}_{ss}' = 0 \tag{10.17}$$

$$\mathscr{L}_{sr}' = -L_{sr} \sin \theta_r \tag{10.18}$$

$$\mathscr{L}_{rr}' = -(L_{dd} - L_{qq}) \sin 2\theta_r \tag{10.19}$$

The positive rotor brush or terminal is located at $\theta_r = -\pi/2$, and the rotor and stator windings carry steady currents I_r and I_s. The steady torque T is then given by substitution in eqn. (10.14) as

$$\begin{aligned}
T &= -\tfrac{1}{2}(L_{dd} - L_{qq}) \sin (-\pi) I_r{}^2 - L_{sr} \sin (-\pi/2) I_r I_s \\
&= L_{sr} I_r I_s \tag{10.24}
\end{aligned}$$

The torque/rotor-winding-current characteristic is shown in Fig. 10.12(*b*).

The result shown in eqn. (10.24) may be confirmed by multiplying eqn. (10.22) by I_r, which gives

$$V_r I_r = r_r I_r{}^2 + L_{sr}\omega_r I_r I_s$$

The term $V_r I_r$ represents the input power to the rotor winding, and $L_{sr}\omega_r I_r I_s$ the portion which is available for conversion to mechanical power. Therefore

$$\omega_r T = L_{sr}\omega_r I_r I_s$$

or the torque on the rotor is

$$T = L_{sr} I_r I_s \tag{10.24}$$

From eqn. (10.22),

$$\omega_r = \frac{V_r - r_r I_r}{L_{sr} I_s} \tag{10.25}$$

Eqns. (10.22), (10.23) and (10.24) have been set up for conventionally positive current entering the windings corresponding to electrical power input and therefore motoring mode operation. Eqn. (10.24) shows the torque developed as positive, i.e. acting in the $+\theta_r$ direction. For motor operation it is to be expected that rotation will take place in the same direction as that in which the torque acts, and this is confirmed by the positive sign of ω_r given by eqn. (10.25).

For a constant applied rotor-winding voltage and a constant stator current, eqn. (10.25) shows that the speed of the separately excited d.c. machine operating in the motoring mode will remain almost constant as the rotor winding current varies with load, since the internal voltage drop $r_r I_r$ will be small compared with V_r in any efficient machine. The speed/rotor-winding-current characteristic is shown in Fig. 10.12(c).

Eqns. (10.22), (10.23) and (10.24) apply equally to generator action. In this I_r will be taken to emerge from the dotted end of the rotor winding and will be negative. As a result the torque due to the rotor winding will be negative and will act in the $-\theta_r$ direction. Therefore the rotor must be assumed to be driven in the $+\theta_r$ direction by the prime mover, to be consistent with this assumed current direction. Changing the sign of I_r in eqn. (10.22),

$$V_r = -r_r I_r + L_{sr} \omega_r I_s \tag{10.26}$$

The V_r/I_r characteristic of the separately excited d.c. generator is shown in Fig. 10.13(b).

EXAMPLE 10.3   A separately excited d.c. machine is rotated at 500 rev/min by a prime mover. When the field (stator winding) current is 1 A the armature (rotor winding) generated voltage is 125 V with the armature open-circuited. The armature resistance is $0.1\,\Omega$ and the field resistance is $250\,\Omega$. Determine:

    (a) The rotational voltage coefficient, $L_{sr}\omega_r$.
    (b) The maximum mutual inductance between the stator and rotor windings.
    (c) The armature terminal voltage if the machine acts as a generator delivering a current of 200 A at a speed of 1,000 rev/min and the field current is 2 A.
    (d) The input current and speed if the machine acts as a motor and develops a gross torque of 1,000 N-m. The armature and field windings are each excited from a 500 V supply.

Neglect iron loss and the effect of magnetic saturation.

Using the assumptions previously made the steady-state operating equations are

$$V_r = r_r I_r + L_{sr}\omega_r I_s \tag{10.22}$$

$$V_s = r_s I_s \tag{10.23}$$

$$T = L_{sr} I_r I_s \tag{10.24}$$

(*a*) Adhering to the previous sign conventions and considering the armature winding open-circuit test, $V_r = 125\,V$, $I_r = 0$, $I_s = 1\,A$. Substituting in eqn. (10.22),

$$125 = (0 \cdot 1 \times 0) + (L_{sr}\omega_{r1} \times 1)$$

Therefore

$$L_{sr}\omega_{r1} = \underline{\underline{125\,V/A}}$$

(*b*) $\quad L_{sr} = \dfrac{125}{\omega_{r1}} = \dfrac{125}{2\pi 500/60} = \underline{\underline{2 \cdot 39\,H}}$

(*c*) When the speed of the machine is doubled, this will double the value of the voltage coefficient. Substituting the given data for generator action in (10.22),

$$V_r = \{0 \cdot 1 \times (-200)\} + \left(125 \times \dfrac{1{,}000}{500} \times 2\right) = \underline{\underline{480\,V}}$$

Note that $I_r = -200\,A$ for generator action.

(*d*) From eqn. (10.25),

$$I_s = \dfrac{V_s}{r_s} = \dfrac{500}{250} = \underline{\underline{2\,A}}$$

In eqn. (10.24),

$$1{,}000 = 2 \cdot 39 \times I_r \times 2 \qquad \text{so that} \qquad I_r = \dfrac{1{,}000}{2 \cdot 39 \times 2} = \underline{\underline{209\,A}}$$

If ω_{r2} is the new angular velocity, then from (10.22) the new voltage coefficient, $L_{sr}\omega_{r2}$, is

$$L_{sr}\omega_{r2} = \dfrac{V_r - r_r I_r}{I_s} = \dfrac{500 - (0 \cdot 1 \times 209)}{2} = \underline{\underline{240\,V/A}}$$

The new speed is

$$n_2 = n_1 \left(\dfrac{L_{sr}\omega_{r2}}{L_{sr}\omega_{r1}}\right) = 500 \times \dfrac{240}{125} = \underline{\underline{960\,rev/min}}$$

10.9 Shunt and Series D.C. Machines

The stator winding of a d.c. machine is usually excited from the same supply as the rotor winding. The stator winding may be connected in parallel with the rotor winding across the supply to form a d.c. *shunt machine* or in series with the rotor winding to form a d.c. *series machine*.

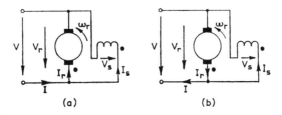

Fig. 10.14   D.C. SHUNT MACHINE

Fig. 10.14(a) shows the connexion diagram for a d.c. shunt machine operating in the motoring mode, and Fig. 10.14(b) shows it operating in the generating mode. The operating equations for the shunt machine may be obtained from those of the separately excited machine. In eqns. (10.22) and (10.23), putting $V_r = V$ and $V_s = V$ gives

$$V = r_r I_r + L_{rs} \omega_r I_r \tag{10.27}$$

$$V = r_s I_s \tag{10.28}$$

The torque equation remains unchanged as

$$T = L_{sr} I_r I_s \tag{10.24}$$

Referring to Fig. 10.14(a) for motor-mode operation,

$$I = I_r + I_s \tag{10.29}$$

The equations for generating action are obtained by putting $I_r = -I_r$ in eqns. (10.27) and (10.24). In addition, for generator action,

$$I_r = I + I_s \tag{10.30}$$

The characteristics of the shunt machine are similar to those of the separately excited machine shown in Figs. 10.12 and 10.13. The establishment of a stable output voltage for shunt generator operation requires some saturation of the magnetic circuit.

Fig. 10.15(a) shows the connexion diagram for a d.c. series machine operating in the motoring mode. The operating equations for the series machine may also be obtained from those of the separately excited machine. Substituting $I_r = I$ and $I_s = I$ in eqns. (10.22), (10.23) and (10.24) gives

$$V_r = r_r I + L_{sr} \omega_r I \tag{10.31}$$

$$V_s = r_s I \tag{10.32}$$

$$T = L_{sr} I^2 \tag{10.33}$$

The torque/current characteristic of the d.c. series motor is shown in Fig. 10.15(b). From Fig. 10.15(a),

$$V = V_r + V_s = r_r I + L_{sr}\omega_r I + r_s I$$

$$= (r_r + r_s)I + L_{sr}\omega_r I \qquad (10.34)$$

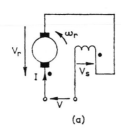

(a)

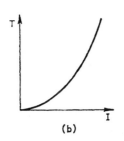

(b)

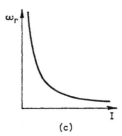

(c)

Fig. 10.15   D.C. SERIES MOTOR

From this equation,

$$\omega_r = \frac{V - (r_r + r_s)I}{L_{sr}I} \qquad (10.35)$$

Since $(r_r + r_s)I$ is very much smaller than V, the speed of the d.c. series motor is approximately inversely proportional to the input current. Therefore, on light loads dangerously high speeds could be reached. In practical applications of the motor, protective devices are used to guard against this contingency. The speed/current characteristic is shown in Fig. 10.15(c).

The output voltage of a d.c. series generator is approximately proportional to the output current. The establishment of this output voltage also is dependent upon there being some saturation of the magnetic circuit.

10.10 Universal Motor

The universal motor is a series connected motor suitable for operation on either a.c. or d.c. supplies.

As previously, the inductance coefficients are

$$\mathscr{L}_{ss} = L_{ss} \tag{10.16}$$

$$\mathscr{L}_{sr} = L_{sr} \cos \theta_r \tag{10.4}$$

$$\mathscr{L}_{rr} = \tfrac{1}{2}(L_{dd} + L_{qq}) + \tfrac{1}{2}(L_{dd} - L_{qq}) \cos 2\theta_r \tag{10.6}$$

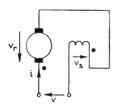

Fig. 10.16   A.C. SERIES MOTOR

To consider a.c. operation of the motor using the series connexion shown in Fig. 10.16, let the supply current be

$$i = i_r = i_s = I_m \cos \omega t$$

Following eqn. (10.9), the instantaneous rotor-winding voltage is

$$v_r = r_r i + \mathscr{L}_{rr}{}' \omega_r i + \mathscr{L}_{sr}{}' \omega_r i + \mathscr{L}_{rr} \frac{di}{dt} + \mathscr{L}_{sr} \frac{di}{dt} \tag{10.9}$$

The dotted end of the rotor winding is at $\theta_r = -\pi/2$; when substituted in the above equations this gives $\mathscr{L}_{rr}{}' = 0$, $\mathscr{L}_{sr}{}' = L_{sr}$, $\mathscr{L}_{rr} = \tfrac{1}{2}(L_{dd} + L_{qq}) = L_{rr}$, say, and $\mathscr{L}_{sr} = 0$. Eqn. (10.9) then becomes

$$v_r = r_r i + L_{sr}\omega_r i + L_{rr} \frac{di}{dt} \tag{10.36}$$

This equation may be written in complexor form. Let V_r be the complexor corresponding to v_r and I the complexor corresponding to i. Then

$$V_r = r_r I + L_{sr}\omega_r I + j\omega L_{rr} I \tag{10.37}$$

Due to the action of the commutator in fixing the axis of the rotor-winding m.m.f., no rotational voltages appear in the stator winding whether operation is from a d.c. or an a.c. supply. The instantaneous stator-winding voltage is therefore, from eqn. (10.10),

$$v_s = r_s i + \mathcal{L}_{ss} \frac{di}{dt} + \mathcal{L}_{sr} \frac{di}{dt} \qquad (10.38)$$

$\mathcal{L}_{ss} = L_{ss}$, and with the rotor winding at $\theta_r = -\pi/2$, $\mathcal{L}_{sr} = 0$. This gives

$$v^s = r_s i + L_{ss} \frac{di}{dt} \qquad (10.39)$$

Eqn. (10.38) written in complexor form gives

$$V_s = r_s I + j\omega L_{ss} I \qquad (10.40)$$

If V is the complexor representing the supply voltage, then

$$V = V_r + V_s$$

i.e.

$$V = L_{sr}\omega_r I + [(r_r + r_s) + j\omega(L_{rr} + L_{ss})]I \qquad (10.41)$$

The instantaneous torque developed is, from eqn. (10.14),

$$T' = \tfrac{1}{2}\mathcal{L}_{rr}' i_r{}^2 + \mathcal{L}_{sr}' i_r i_s + \tfrac{1}{2}\mathcal{L}_{ss}' i_s{}^2 \qquad (10.14)$$

Substituting the conditions previously found ($\mathcal{L}_{rr}' = 0$, $\mathcal{L}_{sr}' = L_{sr}$, $\mathcal{L}_{ss}' = 0$, $i_r = i_s = i$) in eqn. (10.14) gives

$$T' = L_{sr} i^2 = L_{sr} I_m{}^2 \cos^2 \omega t = L_{sr} \frac{I_m{}^2}{2} (\cos 2\omega t + 1) \qquad (10.42)$$

This equation therefore shows that for a.c. operation the torque developed by the machine consists of two components, a steady torque and one that pulsates at twice the supply frequency. The average torque is

$$T = L_{sr} I^2 \qquad (10.43)$$

where I is the r.m.s. value of i. The torque/current characteristic is therefore the same as for d.c. operation.

Universal motors usually have a compensating winding on the stator with its m.m.f. axis coinciding with the rotor brush axis. The compensating winding is connected in series opposition with the rotor winding and serves to reduce the voltage drop in the internal reactance as well as assisting commutation. It has been neglected in the above analysis. A laminated stator construction is essential for a.c. operation.

EXAMPLE 10.4   A 0·1 kW series motor has the following constants:

Armature resistance . . . . $r_r = 12\,\Omega$
Series field resistance . . . . . . $r_s = 36\,\Omega$
Effective armature inductance . . . . . $L_{rr} = 0\cdot3\,\mathrm{H}^*$
Series field inductance . . . . . . $L_{ss} = 0\cdot34\,\mathrm{H}$
Maximum mutual inductance between rotor and stator windings $L_{sr} = 0\cdot71\,\mathrm{H}$

Determine the input current and speed when the load torque applied to the motor is 0·18 N-m (*a*) when connected to a 200 V d.c. supply, and (*b*) when connected to a 200 V 50 Hz a.c. supply.
Neglect windage and friction and all core losses.

(*a*) Considering first d.c. operation, the applied voltage is

$$V = (r_r + r_s)I + L_{sr}\omega_r I \tag{10.34}$$

and the torque developed is

$$T = L_{sr}I^2 \tag{10.33}$$

From eqn. (10.33),

$$I = \sqrt{\frac{T}{L_{sr}}} = \sqrt{\frac{0\cdot18}{0\cdot71}} = \underline{\underline{0\cdot504\,\mathrm{A}}}$$

From eqn. (10.34),

$$\omega_r = \frac{V - (r_r + r_s)I}{L_{sr}I} = \frac{200 - (48 \times 0\cdot504)}{0\cdot71 \times 0\cdot504} = 492\,\mathrm{rad/s}$$

$$n_r = \frac{\omega_r}{2\pi} \times 60 = \frac{492}{2\pi} \times 60 = \underline{\underline{4{,}700\,\mathrm{rev/min}}}$$

(*b*) Considering a.c. operation, the r.m.s. current is, from eqn. (10.43),

$$I = \sqrt{\frac{T}{L_{sr}}} = \sqrt{\frac{0\cdot18}{0\cdot71}} = \underline{\underline{0\cdot504\,\mathrm{A}}}$$

The applied voltage is

$$V = L_{sr}\omega_r I + [(r_r + r_s) + j\omega(L_{rr} + L_{ss})]I$$

Taking *I* as the reference complexor,

$$200\underline{/\theta} = 0\cdot71 \times 0\cdot504\omega_r\underline{/0^\circ} + [48 + j2\pi \times 50(0\cdot3 + 0\cdot34)]0\cdot504\underline{/0^\circ}$$

$$200\cos\theta + j200\sin\theta = 0\cdot358\omega_r + 24\cdot2 + j101$$

Equating quadrate parts in this equation,

$$200\sin\theta = 101$$

whence $\sin\theta = 0\cdot505$, $\cos\theta = 0\cdot864$.

Equating reference parts,

$$200\cos\theta = 0\cdot358\omega_r + 24$$

$$\omega_r = \frac{200 \times 0\cdot864 - 24\cdot2}{0\cdot358} = 415\,\mathrm{rad/s}$$

$$n_r = \frac{415}{2\pi} \times 60 = \underline{\underline{3{,}970\,\mathrm{rev/min}}}$$

---

* The actual armature inductance $\approx L^2_{sr}/L_{ss} \approx 1\cdot48\,\mathrm{H}$. The effective value is reduced to 0·3 H due to the effect of a compensating winding connected in series opposition with the armature.

10.11 Rotating Field due to a Three-phase Winding

Fig. 10.17 shows a stator winding with three diametral coils aa', bb' and cc', each having N_s turns. The dots and crosses indicate the direction of conventionally positive current in each coil as explained

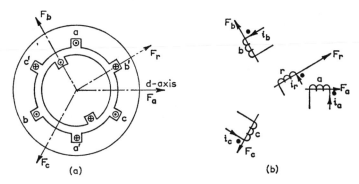

Fig. 10.17   M.M.F. DUE TO A 3-PHASE WINDING

in Section 10.3. The axes of the coil m.m.f.s are therefore mutually displaced by $2\pi/3$ radians, as shown in Fig. 10.17.

Suppose the three coils are supplied with balanced 3-phase currents, i_a, i_b and i_c, such that

$$i_a = I_{sm} \cos \omega t = \frac{I_{sm}}{2} (e^{j\omega t} + e^{-j\omega t}) \tag{10.44}$$

$$i_b = I_{sm} \cos (\omega t - 2\pi/3) = \frac{I_{sm}}{2} (e^{j(\omega t - 2\pi/3)} + e^{-j(\omega t - 2\pi/3)}) \tag{10.45}$$

$$i_c = I_{sm} \cos (\omega t + 2\pi/3) = \frac{I_{sm}}{2} (e^{j(\omega t + 2\pi/3)} + e^{-j(\omega t + 2\pi/3)}) \tag{10.46}$$

The m.m.f. of coil a is directed in the reference direction when i_a is positive. The instantaneous value of this m.m.f. is therefore

$$F_a' = \frac{I_{sm}N_s}{2} (e^{j\omega t} + e^{-j\omega t})e^{j0} \tag{10.47}*$$

This expression has been multiplied by $e^{j0} (= 1)$ to indicate that it acts in the space reference direction.

---

* To avoid confusion with f for frequency, instantaneous m.m.f. will be represented by F'.

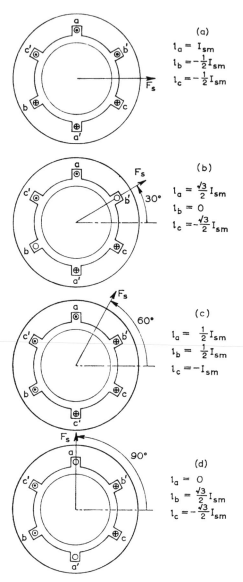

(a)

$1_a = I_{sm}$

$1_b = -\frac{1}{2}I_{sm}$

$1_c = -\frac{1}{2}I_{sm}$

(b)

$1_a = \frac{\sqrt{3}}{2}I_{sm}$

$1_b = 0$

$1_c = -\frac{\sqrt{3}}{2}I_{sm}$

(c)

$1_a = \frac{1}{2}I_{sm}$

$1_b = \frac{1}{2}I_{sm}$

$1_c = -I_{sm}$

(d)

$1_a = 0$

$1_b = \frac{\sqrt{3}}{2}I_{sm}$

$1_c = -\frac{\sqrt{3}}{2}I_{sm}$

Fig. 10.18 M.M.F. DUE TO A 3-PHASE WINDING AT DIFFERENT INSTANTS

The m.m.f. of coil b is directed along an axis $+2\pi/3$ radians from the reference direction when i_b is positive. The instantaneous value of this m.m.f. is therefore

$$F_b' = \frac{I_{sm}N_s}{2}\left(e^{j(\omega t - 2\pi/3)} + e^{-j(\omega t - 2\pi/3)}\right)e^{j2\pi/3}$$

$$= \frac{I_{sm}N_s}{2}\left(e^{j\omega t} + e^{-j(\omega t - 4\pi/3)}\right) \tag{10.48}$$

Similarly the m.m.f. due to coil c at any instant is

$$F_c' = \frac{I_{sm}N_s}{2}\left(e^{j(\omega t + 2\pi/3)} + e^{-j(\omega t + 2\pi/3)}\right)e^{-j2\pi/3}$$

$$= \frac{I_{sm}N_s}{2}\left(e^{j\omega t} + e^{-j(\omega t + 4\pi/3)}\right) \tag{10.49}$$

The resultant stator m.m.f. due to all three coils is

$$F_s' = F_a' + F_b' + F_c'$$

$$= \frac{I_{sm}N_s}{2}\left[e^{j\omega t} + e^{-j\omega t} + e^{j\omega t} + e^{-j(\omega t - 4\pi/3)} + e^{j\omega t}\right.$$
$$\left. + e^{-j(\omega t + 4\pi/3)}\right]$$

Since $e^{-j\omega t} + e^{-j(\omega t - 4\pi/3)} + e^{-j(\omega t + 4\pi/3)} = 0$,

$$F_s' = \tfrac{3}{2} I_{sm}N_s e^{j\omega t} \tag{10.50}$$

This equation shows that, when three coils are so positioned that their m.m.f. axes are mutually displaced by $2\pi/3$ radians and are then supplied with balanced 3-phase currents, an m.m.f. of constant magnitude results and the m.m.f. axis rotates at an angular velocity of ω radians per second.

For the coil configuration and phase sequence chosen the direction of rotation is in the $+\theta$ direction. It will be found that, if the phase sequence is reversed, the direction of rotation of the resultant m.m.f. axis is also reversed.

Fig. 10.18 shows the m.m.f. due to a 3-phase winding supplied with balanced 3-phase currents for a number of different instants. At (a) the current in phase a is positive maximum value and the currents in the two other phases are half the negative maximum value. The negative currents are indicated by showing the current in the cross direction in coil sides b and c, and in the dot direction in coil sides b' and c'. F_s is shown acting along the stator m.m.f. axis.

Figs. 10.18(b), (c) and (d) show successive instants in the 3-phase cycle corresponding to 30° rotations of the complexor diagram.

It will be seen that the axis of the stator m.m.f. is also displaced by successive steps of 30° in the $+\theta$ direction, so that F_s completes one revolution in each cycle and thus must rotate with an angular velocity of ω radians per second. This is in agreement with eqn. (10.50).

Eqn. (10.50) also shows that the m.m.f. due to a 3-phase winding when excited by balanced 3-phase currents could be represented as the m.m.f. of a single winding of N_s turns and excited with a direct current of value $\frac{3}{2}I_{sm}$, where the winding is considered to rotate at an angular velocity ω and N_s represents the number of turns of each stator phase.

In Chapter 11 the resultant m.m.f. due to 3-phase distributed windings is considered, and the effect of space harmonics is discussed. These are ignored in the present treatment.

10.12 Three-phase Synchronous Machine

In the previous section it has been shown that, when three coils have their m.m.f. axes mutually displaced by $2\pi/3$ and are then supplied with balanced 3-phase currents, an m.m.f. of constant magnitude results, the m.m.f. axis rotating at ω radians per second. If a constant angular displacement is to be maintained between the resultant stator and rotor m.m.f.s, as is required for the continuous production of torque, the rotor m.m.f. must also rotate at ω in the same direction as the stator m.m.f.

This rotation of the rotor m.m.f. may be brought about in a number of different ways. In the synchronous machine the rotor winding is excited with direct current supplied through slip rings. The axis of the rotor m.m.f. then rotates at the same speed as the rotor itself, so that the condition for continuous torque production is that the rotor should rotate at ω in the same direction as the resultant stator m.m.f. axis.

The rotors of synchronous machines are often of the salient-pole type shown in Fig. 10.5(a), but for simplicity only the non-salient pole type of rotor as shown in Fig. 10.5(b) will be considered. Instead of the stator phase windings consisting of the single coils considered in Section 10.11, the phase windings consist of several coils distributed in slots and occupying the whole stator periphery as shown in Fig. 10.19. The effect of this distribution of the winding is to introduce constants called *distribution factors* into equations relating to the operation of the machine. These constants are ignored here; in most practical cases they have numerical values close to unity.

The coupling between the d.c. excited rotor winding r (see Fig. 10.17) and the stator reference phase a is a cosine-like or even function which, ignoring space harmonics and taking θ_r as the instantaneous angle of the axis of the rotor winding with respect to phase *a*, is

$$\mathcal{L}_{ar} = L_{sr} \cos \theta_r \qquad (10.51)$$

Since it is assumed that there are no salient poles, all the self-inductances are constant and their angular rates of change are zero.

As explained in the previous section, the 3-phase stator winding may be considered to be replaced by a representative stator windings of N_s turns, excited by a direct current of $\frac{3}{2} I_{sm}$ and rotating at an angular velocity of ω radians per second. The axes of both the stator

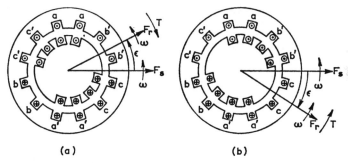

(a)   (b)

Fig. 10.19 SYNCHRONOUS MACHINE
(a) Generating   (b) Motoring

and rotor windings therefore rotate at ω. If the angular displacement between these axes is ϵ as shown in Fig. 10.19, then, from eqn. (10.51),

$$\mathcal{L}_{sr} = L_{sr} \cos \epsilon \qquad (10.52)$$

When ϵ is positive the rotor m.m.f. axis is displaced anticlockwise from the stator m.m.f. axis. The angular rate of change of this mutual inductance is

$$\mathcal{L}_{sr}' = -L_{sr} \sin \epsilon \qquad (10.53)$$

The instantaneous torque on the rotor is given by eqn. (10.14) as

$$T' = \tfrac{1}{2}\mathcal{L}_{rr}'i_r^2 + \mathcal{L}_{sr}'i_r i_s + \tfrac{1}{2}\mathcal{L}_{ss}'i_s^2 \qquad (10.14)$$

The rotor winding current i_r is I_r, a steady, direct current, and the representative stator winding carries a current of $\frac{3}{2}I_{sm}$, where I_{sm} is the maximum current per phase in the actual 3-phase winding. Substituting for these currents and for the angular rates of change of inductance in eqn. (10.14),

$$T' = -\tfrac{3}{2}L_{sr} I_r I_{sm} \sin \epsilon \qquad (10.54)$$

For a steady angular displacement ϵ_s between the axes of the stator and rotor m.m.f.s, the mean torque on the rotor is

$$T = -\tfrac{3}{2} L_{sr}\, I_r I_{sm} \sin \epsilon_s \tag{10.55}$$

It is to be noted that under steady conditions the 3-phase machine, unlike the single-phase machine, does not produce an oscillating component of torque.

At starting, as a motor, the rotor angular velocity ω_r is zero, so that the displacement between the rotor and stator m.m.f. axes is $\epsilon = \omega t$, and the instantaneous torque on the rotor given by eqn. (10.54) is

$$T' = -\tfrac{3}{2} L_{sr} I_r I_{sm} \sin \omega t \tag{10.56}$$

The mean value of the torque given by this equation is zero, and since the inertia of the rotating system is too large to allow the rotor to respond to a torque which oscillates at mains frequency, the synchronous motor is not self-starting.

The currents in the actual 3-phase stator winding may be taken to be the same as those of Section 10.11:

$$i_a = I_{sm} \cos \omega t \tag{10.44}$$

$$i_b = I_{sm} \cos (\omega t - 2\pi/3) \tag{10.45}$$

$$i_c = I_{sm} \cos (\omega t + 2\pi/3) \tag{10.46}$$

For the stator phase currents so chosen, the stator m.m.f. axis at $t = 0$ is along the positive direction of the d-axis, and therefore ϵ_s is the angle of separation of the rotor and stator m.m.f. axes as shown in Fig. 10.19. When ϵ_s is positive the rotor m.m.f. axis is displaced anticlockwise from the stator m.m.f. axis, and as shown by eqn. (10.55), the torque on the rotor acts in the $-\theta$ direction, i.e. in the direction opposite to rotation. Under such circumstances the machine acts as a generator, the rotor being driven against the direction of the torque developed on it by a prime mover. When ϵ_s is negative the torque acts in the $+\theta$ direction, i.e. in the same direction as rotation, and the machine acts as a motor.

When the machine is unloaded $\epsilon_s = 0$, corresponding to the alignment of the rotor and stator m.m.f. axes. As load is imposed the value of ϵ_s increases, the rotor and stator m.m.f. axes are displaced and the appropriate torque is developed.

The operation of the synchronous machine is illustrated in Figs. 10.19(a) and (b). In both diagrams the stator current distribution is drawn for the instant in the 3-phase cycle when $i_a = I_{sm}$ and $i_b = i_c = -\tfrac{1}{2} I_{sm}$, so that the axis of the stator m.m.f. is along the positive direction of the *d*-axis. Fig. 10.19(*a*) illustrates generator action and Fig. 10.19(*b*) motor action.

If the stator phases *b* and *c* are assumed to be open-circuited, then the voltage applied to stator phase *a* may be obtained by adapting the subscripts of eqn. (10.10) as

$$v_a = ri_a + \mathscr{L}_{ar}{}'\omega_r i_r + \mathscr{L}_{aa}\frac{di_a}{dt} + \mathscr{L}_{ar}\frac{di_r}{dt} \tag{10.57}$$

$$\mathscr{L}_{ar} = L_{sr}\cos\theta_r \tag{10.51}$$

$$\mathscr{L}_{ar}{}' = -L_{sr}\sin\theta_r = -L_{sr}\sin(\omega_r t + \epsilon) \tag{10.58}$$

where ϵ is the position of the rotor winding axis at $t = 0$. Since the rotor winding is excited with direct current, $i_r = I_r$ and $di_r/dt = 0$. The current in phase *a* is, from eqn. (10.44), $i_a = I_{sm}\cos\omega t$. Substituting in eqn. (10.57),

$$v_a = rI_{sm}\cos\omega t - \omega_r L_{sr}I_r\sin(\omega_r t + \epsilon) + L_{aa}\frac{d}{dt}(I_{sm}\cos\omega t)$$

$$= rI_{sm}\cos\omega t + \omega_r L_{sr}I_r\cos(\omega_r t + \epsilon + \pi/2)$$

$$+ L_{aa}\frac{d}{dt}(I_{sm}\cos\omega t) \tag{10.59}$$

Under normal operating conditions all three phases carry current, and under balanced conditions this has the effect of increasing the effective inductance per phase by approximately 50 per cent because of the mutual inductance between phases. If the effective inductance per phase is L_{ss}, eqn. (10.59) becomes

$$v_a = rI_{sm}\cos\omega t + \omega_r L_{sr}I_r\cos(\omega_r t + \epsilon + \pi/2)$$

$$+ L_{ss}\frac{d}{dt}(I_{sm}\cos\omega t) \tag{10.60}$$

In complexor form eqn. (10.60) becomes

$$V_s e^{j\phi} = \frac{\omega_r L_{sr}I_r}{\sqrt{2}}e^{j(\epsilon + \pi/2)} + (r + j\omega L_{ss})\frac{I_{sm}}{\sqrt{2}}e^{j0} \tag{10.61}$$

or

$$V_s = E_s + Z_s I \tag{10.62}$$

EXAMPLE 10.5 A 2-pole 1,000 V 50 Hz synchronous machine has a 3-phase star-connected stator winding each phase of which has an effective inductance of 0·01 H and negligible resistance. The maximum mutual inductance between the rotor winding and a stator phase is 0·4 H.

(*a*) Determine the developed torque, the stator phase current, the rotor winding current, the angle between the stator and rotor m.m.f. axes and the induced rotational voltage per phase when the machine acts as a motor with an output power of 224 kW. The input power factor is unity and the stator line voltage is 1,000 V. Neglect all losses.

(b) Determine the load current and output power when the machine acts as a generator if the rotor current is 12 A, the output power factor is 0·8 lagging and the stator line terminal voltage is 1,000 V. Find also the angle between the rotor and stator m.m.f. axes and the phase angle between the stator phase terminal voltage and the stator phase induced rotational voltage.

Neglect the effects of distribution of the windings and of magnetic saturation.

(a) Rotor angular velocity, $\omega_r = 2\pi f = 2\pi \times 50 = 314 \, \text{rad/s}$.

Developed torque, $T = \dfrac{P}{\omega_r} = \dfrac{224{,}000}{314} = 712 \, \text{N-m}$

Neglecting all losses,

$\sqrt{3} V_L I_L \cos\phi = \text{Power output}$

For the star connexion,

$$I_p = I_L = \frac{224{,}000}{\sqrt{3} \times 1{,}000 \times 1} = 129 \, \text{A}$$

From eqn. (10.61) the impedance per phase is

$Z_s = r + j\omega L_{ss} = 0 + (j314 \times 0{\cdot}01) = j3{\cdot}14\,\Omega$

From eqn. (10.62),

$V = E + ZI$

Taking the stator current as the reference complexor, and remembering that the input power factor is unity,

$$\frac{1{,}000}{\sqrt{3}} \underline{/0°} = E\underline{/\epsilon + 90°} + (j3{\cdot}14 \times 129)\underline{/0°}$$

Therefore

$E\underline{/\epsilon + 90°} = 577 - j405 = 706\underline{/-35°}$

Thus the magnitude of the induced rotational voltage per phase is 706 V

The angle between the stator and rotor m.m.f. axes is

$\epsilon = -35 - 90 = -125°$

From eqn. (10.55) the mean torque is

$$T = -\tfrac{3}{2} L_{sr} I_r I_{sm} \sin\epsilon \tag{10.55}$$

Thus the rotor current is

$$I_r = -\frac{712}{\tfrac{3}{2} \times 0{\cdot}4 \times \sqrt{2} \times 129 \sin(-125°)} = 7{\cdot}95 \, \text{A}$$

As a check, from eqn. (10.73),

$$E = \frac{L_{sr}\omega_r I_r}{\sqrt{2}} = \frac{0{\cdot}4 \times 314 \times 7{\cdot}95}{\sqrt{2}} = 706 \, \text{V}$$

(b) From eqn. (10.62),

$$I = \frac{V - E}{Z}$$

Taking the stator current as the reference complexor, then for an output power factor of 0·8 lagging, the phase voltage will *lead* the current by $\cos^{-1} 0·8 = 36·9°$
Therefore

$$I\underline{/0°} = \dfrac{\dfrac{577\underline{/+36·9°} + 0·4 \times 314 \times 12\underline{/\epsilon + 90°}}{\sqrt{2}}}{j3·14}$$

and

$$I = 184\underline{/-53·1°} - 339\underline{/\epsilon}$$
$$= 110 - j147 - (339 \cos \epsilon + j339 \sin \epsilon)$$

The quadrate part of the complex expression for the current is zero; hence

$$-147 - 339 \sin \epsilon = 0$$
$$\sin \epsilon = -\frac{147}{339} = -0·433 \quad \epsilon = -25·7° \quad \text{and} \quad \cos \epsilon = 0·901$$

Thus

$$I = 110 - (339 \times 0·901) = -196\,\text{A}$$

and

$$I = 196\underline{/180°}\,\text{A}$$

The negative value of current corresponds to generating action and the reversal of current with respect to the terminal voltage. The axis of the stator m.m.f. at $t = 0$ is therefore at $180°$ whereas that of the rotor m.m.f. is at $\epsilon = -25·7°$. The axis of the rotor m.m.f. is thus displaced from that of the stator m.m.f. by $180 - 25·7 = 154·3°$ in the $+\theta$ direction as is to be expected for generator action.

$$\text{Output power} = \frac{3 \times 577 \times 196 \times 0·8}{1,000} = 271\,\text{kW}$$

The phase angle between the induced rotational voltage per phase and the terminal voltage is

$$90 - 25·7 - 36·9 = 27·4°$$

10.13 Three-phase Induction Machine

The 3-phase induction machine has a uniformly slotted stator and rotor. The stator has a 3-phase winding like that of the synchronous machine. Unlike that of the synchronous machine the rotor winding is not excited with direct current and may be supposed to consist of short-circuited coils. The effects of distribution are again neglected, and the three stator phase windings and three rotor phase windings are each treated as if they were single, concentrated coils. The arrangement is shown in Fig. 10.20(a).

The stator winding is excited with balanced 3-phase currents, which, as explained in Section 10.11, set up an m.m.f. of constant magnitude the axis of which rotates at ω radians per second in the $+\theta$ direction for the 2-pole configuration shown in Fig. 10.20(a) and for the stator phase currents given by

$$i_a = I_{sm} \cos \omega t \tag{10.44}$$

$$i_b = I_{sm} \cos (\omega t - 2\pi/3) \tag{10.45}$$

$$i_c = I_{sm} \cos (\omega t + 2\pi/3) \tag{10.46}$$

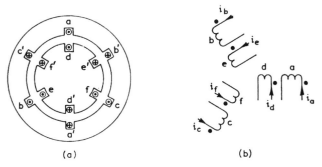

(a)   (b)

Fig. 10.20 MUTUAL INDUCTANCES OF THE 3-PHASE INDUCTION MACHINE

If the machine is to produce torque there must be a rotor winding current. Since the resultant stator m.m.f. is constant in magnitude and rotates at constant speed, a rotor winding current will only be obtained if the angular velocity of the rotor, ω_r, differs from that of the stator m.m.f., ω, because only in this way can a rotational voltage be obtained in the rotor winding.

The magnitude and frequency of the induced currents in the rotor windings are clearly proportional to $\omega - \omega_r$, the rotor angular velocity relative to that of the axis of the stator m.m.f. It is usual to analyse the action of induction machines in terms of the *per-unit slip*, s, which is defined as

$$s = \frac{\omega - \omega_r}{\omega} \tag{10.63}$$

The voltage applied to any winding is

$$v_j = r_j i_j + \sum_{k=1}^{k=n} \frac{d}{dt} (\mathscr{L}_{jk} i_k) \tag{10.11}$$

Each of the rotor coils is short-circuited so that the above equation may be used to obtain expressions for the rotor currents. For the configuration of Fig. 10.20, however, there are six self-inductances and nine mutual inductances. The resulting algebraic work though not difficult is extremely tedious. More sophisticated methods of analysis exist which greatly simplify the algebraic manipulation but are outside the scope of the present chapter.

Without carrying out the actual algebraic work, however, it is not difficult to anticipate the form of the expression for the current in any rotor phase.

Rotor current/phase $\propto$ Induced rotor e.m.f. per phase

Rotor e.m.f. per phase $\propto$ Relative angular velocity of stator m.m.f., $s\omega$

$\propto$ Maximum mutual inductance between a rotor and stator phase winding, L_{sr}

$\propto$ Maximum stator current per phase, I_{sm}.

The angular frequency of the rotor current per phase is also directly proportional to the relative angular velocity of the stator m.m.f., $s\omega$.

The current in phase d of the rotor winding is thus of the form

$$i_d = \frac{ks\omega L_{sr}I_{sm}}{Z_r} \cos(s\omega t - \phi_r + \alpha) \tag{10.64}$$

where k and α are constants, and Z_r is the rotor impedance per phase:

$$Z_r = r_r + js\omega L_{rr} \tag{10.65}$$

$$\phi_r = \tan^{-1} \frac{s\omega L_{rr}}{r_r} \tag{10.66}$$

L_{rr} is the total effective rotor inductance per phase. This includes a contribution due to mutual coupling with the two other rotor phases.

Analysis shows that $k = \frac{3}{2}$ and that if rotor phase d is at $\theta_r = 0$ at $t = 0$, then $\alpha = -\pi/2$. Therefore

$$i_d = \frac{\frac{3}{2}s\omega L_{sr}I_{sm}}{Z_r} \cos(s\omega t - \phi_r - \pi/2) \tag{10.67}$$

The currents in the two other rotor phases are such as to form, with that in rotor phase d, a balanced 3-phase system of currents. Therefore, as shown by eqn. (10.50), the 3-phase rotor winding will give rise to an m.m.f. of constant magnitude rotating at an angular velocity $s\omega$ in the $+\theta$ direction. Following eqn. (10.50), the rotor m.m.f. is

$$F_r = \tfrac{3}{2} I_{rm} N_r e^{j(s\omega t - \phi_r - \pi/2)} \qquad (10.68)$$

where I_{rm} is the maximum rotor current per phase and N_r is the number of turns per rotor phase:

$$I_{rm} = \frac{\tfrac{3}{2} s\omega L_{sr} I_{sm}}{Z_r} \qquad (10.69)$$

Since the rotor winding itself is rotating at $\omega_r = (1 - s)\omega$, the axis of the rotor m.m.f. has an angular velocity in space given by

Absolute angular velocity of rotor m.m.f.
$$= (1 - s)\omega + s\omega = \omega$$

i.e. the angular velocity of the rotor m.m.f. is constant and independent of rotor speed. The axes of the stator and rotor m.m.f.s both rotate at the same angular velocity with an angular displacement $-\phi - \pi/2$ (obtained by comparing eqns. (10.50) and (10.68)). The machine will therefore produce a steady torque.

Just as the m.m.f. due to a 3-phase winding when excited by balanced 3-phase currents can be represented as the m.m.f. of a single winding of N_s turns excited by a direct current of $3I_{sm}$, where the winding is considered to rotate at ω radians per second, so also may the m.m.f. due to a 3-phase rotor winding when excited by balanced 3-phase currents be represented as the m.m.f. of a winding of N_r turns carrying a direct current $\tfrac{3}{2} I_{rm}$ and rotating at an angular velocity ω. The mutual inductance between a stator phase a and a rotor phase r is

$$\mathscr{L}_{ar} = L_{sr} \cos \theta_r$$

The mutual inductance between the two equivalent windings of N_s and N_r turns carrying direct currents $\tfrac{3}{2} I_{sm}$ and $\tfrac{3}{2} I_{rm}$ respectively is

$$\mathscr{L}_{sr} = L_{sr} \cos \epsilon \qquad \text{whence} \qquad \mathscr{L}_{sr}{}' = -L_{sr} \sin \epsilon$$

The angular displacement of the axes of the stator and rotor m.m.f.s is

$$\epsilon = -\phi_r - \pi/2$$

The instantaneous torque developed on the rotor may be obtained from an application of eqn. (10.14):

$$
\begin{aligned}
T' &= \tfrac{1}{2}\mathscr{L}_{rr}{}' i_r^2 + \mathscr{L}_{sr}{}' i_r i_s + \tfrac{1}{2}\mathscr{L}_{ss}{}' i_s^2 \\
&= -L_{sr} \sin\left(-\phi_r - \pi/2\right)\tfrac{3}{2} I_{rm} \tfrac{3}{2} I_{sm} \qquad (10.14)
\end{aligned}
$$

Substituting for I_{rm} from eqn. (10.69),

$$T' = -\left(\frac{3}{2}\right)^3 \frac{s\omega L_{sr}^2 I_{sm}^2}{Z_r} \sin\left(-\phi_r - \pi/2\right)$$

Since none of these terms varies with time the mean torque is

$$T = \left(\frac{3}{2}\right)^3 \frac{s\omega L_{sr}^2 I_{sm}^2 \cos \phi_r}{Z_r} \tag{10.70}$$

The per-unit slip, s, is positive when $\omega > \omega_r$, and when this condition obtains the torque developed on the rotor is positive and acts in the $+\theta$ direction, the direction of assumed rotor rotation. The machine therefore acts in the motoring mode for positive values of s. If the rotor is coupled to a prime mover and driven so that $\omega_r > \omega$ the slip and torque become negative and the machine acts in the generating mode.

Eqn. (10.70) shows that, like the 3-phase synchronous motor, the 3-phase induction motor produces a non-oscillatory torque. Unlike the 3-phase synchronous motor, however, the 3-phase induction motor is self-starting. At starting $\omega_r = 0$, and from eqn. (10.63),

$$s = \frac{\omega - 0}{\omega} = 1$$

Eqn. (10.70) shows that a torque will be developed on the rotor for this value of s.

PROBLEMS

10.1 An electrodynamic ammeter consists of a fixed coil and a moving coil connected in series. The self-inductance of the fixed coil is $400\,\mu$H and that of the moving coil $200\,\mu$H. The mutual inductance between the coils is

$$\mathscr{L}_{sr} = 100 \times 10^{-6} \cos \theta_r \quad \text{henry}$$

where θ_r is the position of the axis of the moving coil relative to that of the fixed one. The zero on the instrument scale corresponds to a position of the axis of the moving coil $\theta_r = 145°$. Full-scale deflection is $110°$ from the scale zero. The control constant is $5 \cdot 22 \times 10^{-5}$ N-m per degree of deflection. Determine the direct current required for full-scale deflection.

If an alternating current of 5A r.m.s. and of frequency 50 Hz passes through the ammeter coils, what is the voltage drop across the instrument terminals? The resistance of the windings may be neglected.

Ans. 10A; $0 \cdot 91$V. (The angular deflection of the moving coil is approximately $48°$ from the instrument zero.)

10.2 A rotating relay consists of a stator coil of self-inductance $2 \cdot 0$H and a rotor coil of self-inductance $1 \cdot 0$H. The axes of the rotor and stator coils are displaced by $30°$, and in this configuration the mutual inductance of the coils is $1 \cdot 0$H. Neither stator nor rotor has salient poles. A current $i_s = 14 \cdot 14 \cos 314t$

amperes is passed through the stator coil. The rotor coil is short-circuited. Draw a diagram to show the relative directions of the stator and rotor coil currents and calculate the r.m.s. value of the rotor current. Determine also the r.m.s. value of voltage applied to the stator coil. Describe the direction in which the torque acts. The resistance of each winding can be neglected.

Ans. 10 A; 3,140 V. The torque acts so as to tend to align the coils.

10.3 A 2-pole d.c. machine has a field (stator) winding resistance of 200 Ω and an armature (rotor) resistance of 0·1 Ω. When operating as a generator the output voltage is 240 V when the armature winding current is 100 A, the field winding current 2 A, and the speed 500 rev/min.

Determine the armature current and speed when the machine is connected as a shunt motor to a 400 V d.c. supply and the total load torque imposed on the motor is 1,000 N-m. Assume that the machine is linear.

Ans. 209 A; 758 rev/min.

10.4 A 93 W 2-pole series motor has the following constants:

Armature resistance	$r_r = 12 \Omega$
Series field resistance	$r_s = 36 \Omega$
Effective armature inductance	$L_{rr} = 0\cdot3\,\text{H}$
Series field inductance	$L_{ss} = 0\cdot34\,\text{H}$
Maximum mutual inductance between rotor and stator windings	$L_{sr} = 0\cdot71\,\text{H}$

Determine the speed, output power, input current and power factor when the motor is connected to a 200 V 50 Hz supply and the load torque applied is 0·25 N-m. Neglect windage and friction and iron losses.

Ans. 3,000 rev/min; 78·5 W; 0·594 A; 0·802 lagging.

10.5 A 100 V 3-phase 9 kVA 2-pole 50 Hz star-connected alternator has a total effective self-inductance per phase of 1·5 mH. The maximum mutual inductance between the rotor (field winding) and a stator phase winding is 90 mH. Determine the field current required to give an open-circuit line voltage of 100 V when the machine is driven at 3,000 rev/min. Find also, for this speed and field current, the terminal line voltage when the synchronous generator delivers rated full load current at (*a*) a power factor of 0·8 lagging; (*b*) unity power factor; (*c*) a power factor of 0·8 leading. The armature resistance per phase is negligible.

Ans. 2·89 A; 68·6 V; 89·1 V; 123 V.

10.6 The synchronous machine of Problem 10·5 is run as a motor connected to 3-phase 100 V 50 Hz busbars. Determine the field current required if the input power factor is to be unity when the gross torque imposed is 20 N-m. Neglect all losses.

Ans. 3·01 A.

Chapter 11

THREE-PHASE WINDINGS AND FIELDS

In an a.c. machine the armature (or main) winding may be either on the stator (i.e. the stationary part of the machine) or on the rotor, the same form of winding being used in each case. The simplest form of 3-phase winding has concentrated coils each spanning one pole pitch, and with the starts of each spaced 120° (electrical) apart on the stator or rotor. These coils may be connected in star or delta as required.

In most machines the coils are not concentrated but are distributed in slots over the surface of the stator or rotor, and it is this type of winding which will now be considered. The same type of winding is common to both synchronous and asynchronous (induction) machines.

11.1 Flux Density Distributions

In all a.c. machines an attempt is made to secure a sinusoidal flux density distribution in the air-gap. This may be achieved approximately by the distribution of the winding in slots round the air-gap or by using salient poles with shaped pole shoes.

In Fig. 11.1(a) a section of a multipolar machine is shown. If the flux density in the air-gap is to be sinusoidally distributed, the flux density must be zero on the inter-polar axes such as OA, OC and OE, and maximum on the polar axes OB and OD. Since successive poles are of alternate north and south polarities, the maximum flux densities along OB and OD are oppositely directed. Thus a complete cycle of variation of the flux density takes place in a

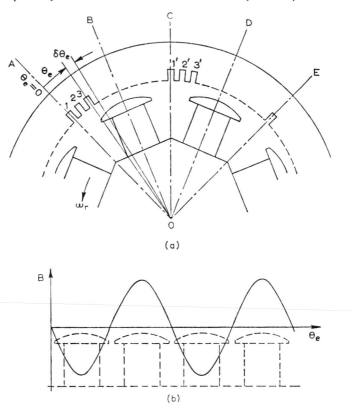

Fig. 11.1   SINUSOIDAL FLUX DENSITY DISTRIBUTION

double pole pitch from the axis OA to the axis OE. This is shown in Fig. 11.1(*b*).

Taking axis OA as the datum for angular measurements, the flux density at any point in the air-gap is

$$B = B_m \sin \theta_e \tag{11.1}$$

where θ_e is the angle from the origin measured in electrical radians or electrical degrees. Since one cycle of variation of the flux density occurs in a double pole pitch,

1 double pole pitch $\equiv 2\pi$ electrical radians or 360 electrical degrees

If the machine has $2p$ poles or p double pole pitches,

$$\theta_e = p\theta_m \qquad\qquad (11.2)$$

where θ_m is the angular measure in mechanical radians or degrees.

11.2 Three-phase Single-layer Concentric Windings

The two sides of an armature coil must be placed in slots which are approximately a pole pitch (180 electrical degrees) apart so that the e.m.f.s in the coil sides are cumulative. In addition, in 3-phase machines the starts of each phase winding must be 120 electrical degrees apart.

In single-layer windings one coil side occupies the whole of a slot. As a result, difficulty is experienced in arranging the end connectors, or overhangs. In concentric and split-concentric windings differently shaped coils having different spans are necessary. To preserve e.m.f. balance in each of the phases, each phase must contain the same number of each shape of coil.

Fig. 11.2(a) represents a developed stator with 24 stator slots, and it is desired to place a 4-pole 3-phase concentric winding in them:

$$\text{Number of slots per pole} = \frac{24}{4} = 6$$

$$\text{Number of slots per pole and phase} = \frac{24}{4 \times 3} = 2$$

Fig. 11.2(a) shows the coil arrangement for the red phase as a thin full line. The start and finish (marked S and F respectively) of the phase winding are brought out, all the coils in the one phase being connected in series. For a phase sequence RYB, the yellow phase (shown dotted) must start 120 electrical degrees after the red phase. One pole pitch contains six slots and is equivalent to 180 electrical degrees. Hence a slot pitch is equivalent, in this case, to 30 electrical degrees.

The red phase starts in slot 1 and therefore the yellow phase must start in slot 5. In the same way the blue phase is 240 electrical degrees out of space phase with the red phase. The blue phase must therefore start in slot 9.

In Fig. 11.2 the finishes of the three phases have been commoned, making a star-connected winding. It would have been equally correct to common the three starts. The winding might also have been mesh-connected, in which case the finish of the red phase would have been connected to the start of the yellow phase, the finish of the yellow to the start of the blue, the finish of the blue to the start of the red, three connectors to the three junctions being brought out to terminals.

It will be observed that each phase has coils of each of the four different sizes used, thus maintaining balance between the phases.

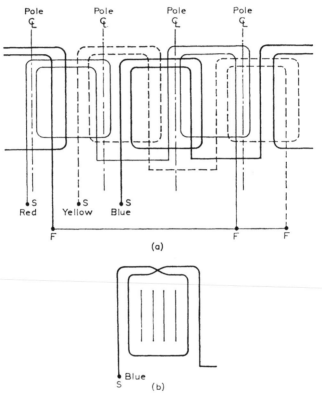

Fig. 11.2   FOUR-POLE 3-PHASE SINGLE-LAYER CONCENTRIC WINDING

It will also be seen that a coil group of any one phase consists of two coils per double pole pitch, one coil being greater than a pole pitch by one slot pitch and the other being less than a pole pitch by the same amount. If the end connexions of these two coils were crossed over as shown in Fig. 11.2(*b*) two full-pitch coils (i.e. having a span of exactly one-pole pitch) would be formed. Therefore each such coil group is the equivalent, electrically, of two full-pitch coils joined in series. All single-layer windings are effectively composed of full-pitch coils.

11.3 Three-phase Single-layer Mush Winding

Fig. 11.3 shows a 4-pole 3-phase single-layer mush winding. The distinctive feature of the mush winding is the utilization of constant-span coils. The overhangs are arranged in a similar manner to those of a conventional double-layer winding.

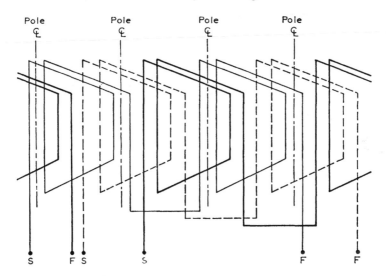

Fig. 11.3 FOUR-POLE 3-PHASE SINGLE-LAYER MUSH WINDING

11.4 Three-phase Double-layer Windings

The double-layer windings used in 3-phase machines are essentially similar to those used in d.c. machines except that no connexions to a commutator are required.

Since each phase must be balanced, all must contain equal numbers of coils and the starts of each phase must be displaced by 120 electrical degrees. If a number of groups of coils are to be connected in parallel, then similar parts in the winding at equal potentials must be available, a condition obtainable only in machines having a number of poles divisible by three when a wave winding is used.

On the other hand, tooth ripple, which arises where there are an integral number of slots per pole, resulting in the same relative positions of equivalent slots under each pole, may be avoided in double-layer windings by the use of winding pitches different from the pole pitch, thus giving a fractional number of slots per pole. A further advantage of the double-layer winding is the possibility of using constant-span coils. Only single-layer windings are considered in the rest of this chapter.

11.5 E.M.F. Induced in a Full-pitch Coil

Consider a full-pitch coil C with coil sides lying in slots 3 and 3' as shown in Fig. 11.1. Let the coil side in slot 3 lie at θ_e so that the coil side in slot 3' lies at $\theta_e + \pi$ electrical radians, where θ_e is measured from the interpolar axis OA. Let the stator diameter be D and the effective stator length L. Assume that the flux density distribution is sinusoidal, i.e. that

$$B = B_m \sin \theta_e \tag{11.1}$$

The flux in the stator segment between θ_e and $\theta_e + \delta\theta_e$ is

$$\delta\phi = BL \frac{D}{2} \delta\theta_m = B_m \sin \theta_e \, L \frac{D}{2} \frac{\delta\theta_e}{p}$$

The total flux linked with coil C is

$$\phi = \frac{B_m L D}{2p} \int_{\theta_e}^{\theta_e + \pi} \sin \theta_e \, d\theta_e$$

$$= + \frac{B_m L D}{2p} 2 \cos \theta_e \tag{11.3}$$

If a coil lies with its sides on the interpolar axes, as, for example, the coil lying in slots 1 and 1' of Fig. 11.1, then the coil links the total flux per pole, Φ:

$$\Phi = \frac{B_m L D}{2p} \int_0^\pi \sin \theta_e \, d\theta_e$$

$$= + \frac{B_m L D}{2p} 2 \tag{11.4}$$

The flux linked with coil C is therefore, by substitution in eqn. (11.3),

$$\phi = \Phi \cos \theta_e \tag{11.5}$$

Suppose the pole system rotates in the direction shown at a uniform angular velocity

$$\omega_r = 2\pi n_0 \quad \text{radians/second} \tag{11.6}$$

where n_0 is the rotor speed in revolutions per second. The position of any coil such as C at any instant, in electrical radians, is

$$\theta_e = \omega t + \theta_0$$

where θ_0 is the position of the coil at $t = 0$, and

$$\omega = p\omega_r = 2\pi n_0 p \quad \text{electrical radians/second} \qquad (11.7)$$

Substituting for θ_e in eqn. (11.5), the flux linking any coil such as C at any time t is

$$\phi = \Phi \cos(\omega t + \theta_0) \qquad (11.8)$$

The e.m.f. induced in any coil of N_c turns is

$$e = N_c \frac{d\phi}{dt}$$

$$= N_c \frac{d}{dt}\{\Phi \cos(\omega t + \theta_0)\}$$

$$= -\omega \Phi N_c \sin(\omega t + \theta_0)$$

The r.m.s. coil e.m.f. is therefore

$$E_c = \frac{\omega \Phi N_c}{\sqrt{2}} \qquad (11.9)$$

Thus a sinusoidal flux density distribution in space may give rise to an e.m.f. induced in a coil which varies sinusoidally with time. This is achieved by giving the coil and the flux density distribution a constant relative angular velocity.

The frequency of the induced e.m.f. is

$$f = \frac{\omega}{2\pi} = \frac{2\pi n_0 p}{2\pi} = n_0 p \qquad (11.10)$$

n_0 is called the *synchronous speed*. In this equation it is measured in revolutions per second.

11.6 Distribution (or Breadth) Factor and E.M.F. Equation

Suppose that under each pole pair each phase of the winding has g coils connected in series, each coil side being in a separate slot. The e.m.f. per phase and pole pair is the complexor sum of the coil voltages. These will not be in time phase with one another since successive coils are displaced round the armature, and hence will not be linked by the same value of flux at the same instant. E_1, E_2, $E_3 \ldots E_g$ (as shown in Fig. 11.4(a)) represent the r.m.s. values of the e.m.f.s in successive coils. The phase displacement between successive e.m.f.s is ψ, which depends on the electrical angular displacement between successive slots on the armature.

Suppose the machine has a total of S slots and $2p$ poles. Then

$$\text{Number of slots per pole} = \frac{S}{2p}$$

The slot pitch (electrical angle between slot centre lines) is

$$\psi = \frac{180°_e}{S/2p} \quad \text{(since 1 pole pitch} = 180°_e) \tag{11.11}$$

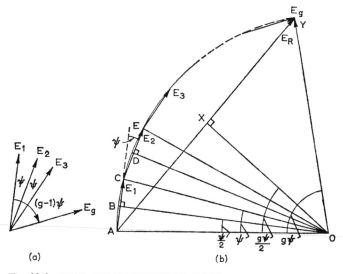

Fig. 11.4   DERIVATION OF DISTRIBUTION FACTOR

(a) Complexor diagram of slot e.m.f.s
(b) Resultant of slot e.m.f.s

The e.m.f. complexors E_1, E_2, E_3 . . . E_g are placed end to end in order in Fig. 11.4(b). The resultant complexor E_R, represents the complexor sum of the e.m.f.s of the g coils connected in series.

Since the complexors E_1, E_2, E_3 . . . E_g are all of the same length and are displaced from one another by the same angle, they must be successive chords of the circle whose centre is O in Fig. 11.4(b). The complexor sum AY may be found as follows.

Join OA, OC, OE, etc., draw the perpendicular bisectors of each chord (i.e. OB, OD, etc.) and also the perpendicular bisector OX of the chord AY.

In the triangle AOX,

$$AX = AO \sin AOX = AO \sin g\frac{\psi}{2}$$

Therefore

$$AY = 2AO \sin g \frac{\psi}{2}$$

In the triangle AOB,

$$AB = AO \sin AOB = AO \sin \frac{\psi}{2}$$

$$AC = 2AB = 2AO \sin \frac{\psi}{2}$$

Therefore

$$\frac{AY}{AC} = \frac{E_R}{E_1} = \frac{\sin g \frac{\psi}{2}}{\sin \frac{\psi}{2}}$$

Thus the *distribution factor* is

$$K_d = \frac{\text{Complexor sum of coil e.m.f.s}}{\text{Arithmetic sum of coil e.m.f.s}}$$

$$= \frac{E_R}{gE_1} = \frac{\sin g \frac{\psi}{2}}{g \sin \frac{\psi}{2}} \tag{11.12}$$

The product $g\psi$ represents the electrical angle over which the conductors of one phase are spread under any one pole and is referred to as the *phase spread*. In a 3-phase single-layer winding each phase has two phase spreads under each pole pair. Therefore, for a single-layer 3-phase winding,

$$g\psi = \frac{360}{2 \times 3} = 60°_e \quad \text{or} \quad \pi/3 \text{ electrical radians}$$

Clearly the highest value which the distribution factor K_d can have is unity, corresponding to a situation where there is one coil per pole pair and phase. A lower limit for the value of K_d also exists. Thus, if the number of separate slots g in the phase spread $g\psi$ is considered to increase without limit, then

$$\psi \to 0 \quad \text{and} \quad \sin \frac{\psi}{2} \to \frac{\psi}{2}$$

A 3-phase winding with a phase spread of $60°_e$ is said to be *narrow spread*.

For a narrow-spread 3-phase winding ($g\psi = \pi/3$),

$$\lim_{\psi \to 0} K_d = \frac{\sin \dfrac{g\psi}{2}}{g \dfrac{\psi}{2}} = \frac{\sin \pi/6}{\pi/6} = \frac{3}{\pi} \qquad (11.13)$$

A winding having this limiting condition is called a *uniform winding*, and in such winding the phase spreads may be thought of as current sheets with the effect of the slotting eliminated.

The lower limit of K_d for a 3-phase narrow-spread winding ($3/\pi = 0\cdot955$), corresponding to a very large number of slots per pole and phase, shows that the distribution of the winding will have little effect on the magnitude of the fundamental e.m.f. per phase.

Ideally the flux density distribution linking the winding should be sinusoidal. In practice this ideal is not usually achieved; the air-gap flux density distribution is then of the form

$$B = B_{m1} \sin \theta_e + B_{m3} \sin (3\theta_e + \epsilon_3)$$
$$+ \ldots B_{mn} \sin (n\theta_e + \epsilon_n) \quad (11.14)$$

In this expression the first term on the right-hand side is called the *fundamental space distribution*. The other terms are referred to as *space harmonics*. The nth space harmonic goes through n cycles of variation for one cycle of variation of the fundamental. Only odd space harmonics are present since the flux density distribution repeats itself under each pole and is therefore symmetrical.

Just as the fundamental flux density gives rise to a fundamental e.m.f. induced in a coil, so the nth space harmonic in the flux density distribution will give rise to an nth time harmonic in the coil e.m.f. The distribution factor for the nth harmonic is

$$K_{dn} = \frac{\sin \dfrac{gn\psi}{2}}{g \sin \dfrac{n\psi}{2}} \qquad (11.15)$$

Although the distribution of the winding has little effect on the magnitude of the fundamental, it may cause considerable reduction in the magnitude of harmonic e.m.f.s compared with those occurring in a winding for which $g = 1$, i.e. one coil per pole pair and phase.

11.7 Coil-span Factor

The e.m.f. equation of Section 11.5 has been deduced on the assumption of full-pitch coils, i.e. coils whose sides are separated by one pole pitch. As has been pointed out, the coils in double-layer wind-ings are often made either slightly more or slightly less than a pole pitch. Fig. 11.5 illustrates coils with various pitches.

If the coil has a pitch of exactly one pole pitch, it will at some instant link the entire flux of a rotor pole. If the coil pitch is less than one pole pitch, it will never link the entire flux of a rotor pole and the maximum coil e.m.f. will be reduced. If the coil pitch is greater than one pole pitch, the coil must always be linking flux

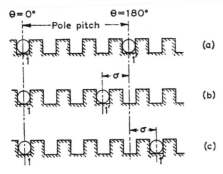

Fig. 11.5 COIL SPANS

(a) Full pitch
(b) Short pitch
(c) Over-full pitch

from at least two adjacent rotor poles so that the net flux linked will be less than the flux of one pole and the maximum coil e.m.f. will again be reduced.

The factor by which the e.m.f. per coil is reduced is called the *coil span factor*, K_s:

$$K_s = \frac{\text{E.M.F. in the short or long coil}}{\text{E.M.F. in a full-pitched coil}} \qquad (11.16)$$

The magnitude of the coil span factor may most readily be obtained by considering the e.m.f. induced in each coil side, namely

$$e = Blv \text{ volts}$$

where B = air-gap flux density, l = active conductor length and v = conductor velocity at right angles to the direction of B.

This e.m.f. will have the same waveform as the flux density in the air-gap, since l and v are constant, and hence if the flux density is sinusoidally distributed the e.m.f. in each conductor will be sinu-soidal so that the resultant coil e.m.f. will also be sinusoidal. If the pitch is short or long by an electrical angle σ, then, assuming a sinusoidal flux density distribution, the e.m.f.s in each side of the coil will differ in phase by σ but will have the same r.m.s. value. The resultant coil e.m.f. will be the complexor sum of the e.m.f.s in each coil side, as shown in Fig. 11.6.

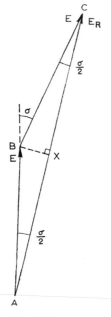

Fig. 11.6   DERIVATION OF COIL SPAN FACTOR

Resultant e.m.f. $= AC = 2AB \cos \dfrac{\sigma}{2}$

E.M.F. for a full-pitch coil $= 2AB$

Therefore

$$K_s = \frac{2AB \cos \dfrac{\sigma}{2}}{2AB} = \cos \frac{\sigma}{2} \tag{11.17}$$

If the flux density distribution contains space harmonics, the coil span factor for the nth harmonic e.m.f. is

$$K_{sn} = \cos \frac{n\sigma}{2} \tag{11.18}$$

All single-layer windings are effectively made up of full-pitch coils, but double-layer windings usually have short-pitched or short-chorded coils. The nth harmonic coil e.m.f. is reduced to zero if the *chording angle*, σ, is such that

$$\cos \frac{n\sigma}{2} = 0$$

or

$$\frac{n\sigma}{2} = 90°_e \tag{11.19}$$

This enables windings to be designed which will not permit specified harmonics to be generated (e.g. if $\sigma = 60°_e$ there can be no third-harmonic generation).

11.8 E.M.F. Induced per Phase of a Three-phase Winding

Following eqn. (11.9) the r.m.s. e.m.f. induced in a full-pitch coil of N_c turns due to its angular velocity relative to the pole system is

$$E_c = \frac{\omega \Phi N_c}{\sqrt{2}} \tag{11.9}$$

For a coil-span factor, K_s, due to chording,

$$E_c = K_s \frac{\omega \Phi N_c}{\sqrt{2}}$$

Further, if there are g coils in a phase group under a pole pair the resultant complexor sum is

$$E_g = K_d g E_c = K_d K_s g \frac{\omega \Phi N_c}{\sqrt{2}}$$

Assuming that the e.m.f.s of coil groups of the same phase under successive pole pairs are in phase and connected in series, the e.m.f. per phase is

$$E_p = pEg = pK_d K_s g \frac{\omega \Phi N_c}{\sqrt{2}}$$

or

$$L_p = K_d K_s \frac{\omega \Phi N_p}{\sqrt{2}} \tag{11.20}$$

where the number of turns per phase, N_p, is pgN_c.

This equation is sometimes written in the form

$$E_p = 4.44 \, K_d K_s f \Phi N_p \tag{11.21}$$

since $\omega = 2\pi f$ and $2\pi/\sqrt{2} = 4.44$.

Sometimes the conductors per phase rather than the turns per phase are specified, in which case eqn. (11.21) becomes

$$E_p = 2 \cdot 22 \, K_s K_d f \Phi Z_p \qquad (11.22)$$

since $N_p = Z_p/2$.

The line voltage will depend on whether the winding is star or delta connected.

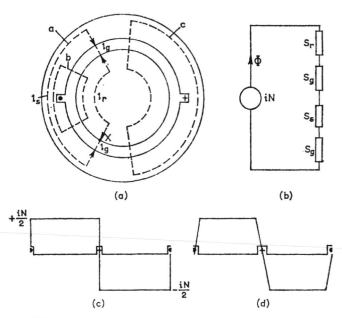

Fig. 11.7 M.M.F. DUE TO A FULL-PITCH COIL

11.9 M.M.F. due to a Full-pitch Coil

Fig. 11.7(a) shows a stator and rotor separated by a uniform air-gap, i.e. one whose radial length l_g is constant. The stator has two diametrically opposite slots in which one stator coil of N turns carries a current i. The slot opening is assumed to be very small compared with the internal circumference of the stator.

Consider the closed path a of Fig. 11.7(a) and let distances be measured from point X on this path. Now,

$$\text{M.M.F.} = iN = \oint H \, dl$$

Assuming that the reluctance of the rotor and stator core paths is zero, and that the magnetic field strength in the gap, H_g, is constant along the radial length l_g, then

$$iN = 2H_g l_g$$

or

$$H_g l_g = \frac{iN}{2} \tag{11.23}$$

Fig. 11.7(b) shows the equivalent magnetic circuit with path reluctances S_r (rotor), S_s (stator), S_g (air-gap). It will readily be confirmed that, if $S_r = 0$ and $S_s = 0$, the magnetic potential difference across each of the two equal reluctances S_g is $\frac{1}{2}iN$.

It is clear that, adhering to the assumptions of zero reluctance in the rotor and stator core and constant field strength in the air gap, the same result as that of eqn. (11.23) is obtained for other paths of integration such as b or c in Fig. 11.7(a). Indeed the magnetic potential drop across the air gap is $\frac{1}{2}iN$ at all points. The magnetic potential drop across the air-gap is, for the direction of coil current chosen, directed from rotor to stator for the upper half of the stator, and from stator to rotor for the lower half of the stator in this case.

Fig. 11.7(c) shows a graph of air-gap magnetic potential difference plotted to a base of the developed stator surface. The magnetic potential difference has been arbitrarily assumed positive when it is directed from rotor to stator and shown above the datum line. It is therefore taken to be negative when directed from stator to rotor and shown below the datum line. The magnetic potential difference is shown as changing abruptly from $+\frac{1}{2}iN$ to $-\frac{1}{2}iN$ opposite the slot opening. This corresponds to the situation where the slot is extremely thin.

Although Fig. 11.7(c) is properly described as showing the variation of air-gap magnetic potential difference to a base of the developed stator surface, such a diagram is often called an m.m.f. wave diagram, and the quantity $\frac{1}{2}iN$ is often called the m.m.f. per pole.

Where the width of the slot opening is not negligible the m.m.f. wave for a coil may be taken to be trapezoidal as shown in Fig. 11.7(d).

11.10 M.M.F. due to One Phase of a Three-phase Winding

Fig. 11.8(a) shows the coil for a double pole pitch of one phase of a 3-phase concentric winding of the type illustrated in Fig. 11.2. The position of the stator slots and coils is indicated on a developed diagram of the stator slotting. The start of the red phase winding is shown with a current emerging from the start end of the winding. This is taken as a conventionally positive current for generator

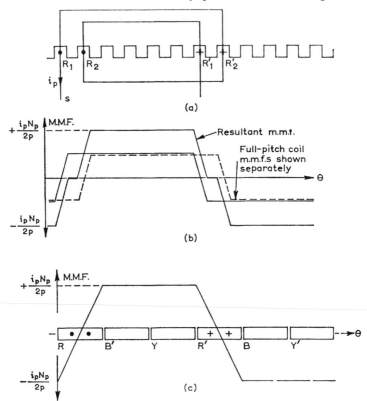

Fig. 11.8   M.M.F. DUE TO ONE PHASE OF A 3-PHASE WINDING

action. Conventionally positive current for motor action enters the start end of the winding.

As has been explained previously, the concentric coil arrangement shown is equivalent to an arrangement of two full-pitch coils where the coil sides in slots R_1 and R_1' and the coil sides in slots R_2 and R_2' are joined. Fig. 11.8(*b*) shows the m.m.f. for each of such coils separately and also their resultant obtained by adding together the two separate m.m.f. waves. The convention regarding positive m.m.f. explained in the previous section has been adhered to.

The resultant m.m.f. shown in Fig. 11.8(*b*) is stepped, owing to the effect of the discrete coils. Fig. 11.8(*c*) shows the m.m.f. per phase when the effect of discrete coils is ignored. The rectangular blocks represent the phase spreads, and these are considered to extend over both the regions previously occupied by slots and by teeth. The phase spread containing the start of the phase winding is identified by the unprimed letter R. The other phase spread of the same phase is marked R'. The phase current is considered to be uniformly distributed in the block representing the phase spread. Such a winding is a uniformly distributed winding as described in Section 11.6, and the m.m.f. per phase for such a winding is of the trapezoidal shape shown.

The maximum value of the m.m.f. wave at any instant is the m.m.f. per pole for the phase considered. For N_p total turns per phase the m.m.f. per phase and pole is $i_p N_p / 2p$.

If a sinusoidal alternating current $i_p = I_{pm} \sin \omega t$ flows in the phase winding, the maximum value of the m.m.f. wave will vary sinusoidally:

$$\frac{i_p N_p}{2p} = \frac{I_{pm} N_p \sin \omega t}{2p}$$

In subsequent work the m.m.f. due to uniform windings only will be considered.

11.11 M.M.F. due to a Three-phase Winding (graphical treatment)

Fig. 11.9 shows the m.m.f.s for each phase of a 3-phase winding carrying balanced 3-phase currents for two different instants in the current cycle. The resultant m.m.f., due to the combined action of the separate phases, is also shown in each diagram.

Fig. 11.9(a) is drawn for the instant when the instantaneous currents in the three phases are

$$i_r = I_{pm}$$
$$i_y = -\tfrac{1}{2} I_{pm}$$
$$i_b = -\tfrac{1}{2} I_{pm}$$

The current in the red phase is positive, so according to the convention for positive current explained in Section 11.10, phase spread R has the current direction indicated by a dot and phase spread R' has the current direction indicated by a cross. The red phase m.m.f., F_r, therefore has the trapezoidal distribution shown having a maximum value of

$$\frac{i_p N_p}{2p} = \frac{I_{pm} N_p}{2p} = F_{pm}$$

where F_{pm} is the maximum m.m.f. per phase and pole.

The currents in the yellow and blue phases are both negative so that the Y and B phase spreads have crosses, and the phase spreads Y' and B' have dots, to show the current direction. The m.m.f.

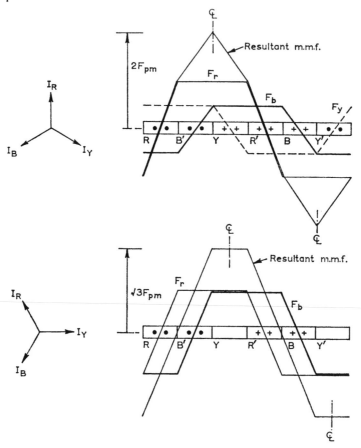

Fig. 11.9 M.M.F. DUE TO A 3-PHASE WINDING
(GRAPHICAL TREATMENT)

waves for these phases, F_y and F_b, are also trapezoidal and at th instant shown in the diagram have a maximum value of

$$\frac{i_p N_p}{2p} = \frac{1}{2}\frac{I_{pm} N_p}{2p} = \tfrac{1}{2}F_{pm}$$

The resultant stator m.m.f., F_A, is obtained by finding the sur of the separate phase m.m.f.s, F_r, F_y and F_b. This resultant m.m. has a maximum value of $2F_{pm}$.

Fig. 11.9(*b*) has been drawn for an instant $\frac{1}{12}$th of a cycle later than Fig. 11.9(*a*). The instantaneous phase currents are

$$i_r = \frac{\sqrt{3}}{2} I_{pm}$$

$$i_y = 0$$

$$i_b = -\frac{\sqrt{3}}{2} I_{pm}$$

The red and blue phase m.m.f.s occupy the same positions as in Fig. 11.9(*a*), but the maximum value of the red phase m.m.f., F_r, has fallen to $(\sqrt{3}/2)F_{pm}$, whereas the maximum value of the blue phase m.m.f., F_b, has risen to this value. Since the yellow phase current is zero, the yellow phase m.m.f., F_y, is zero.

The resultant stator m.m.f. is the sum of F_r and F_b, and has a maximum value of $\sqrt{3}F_{pm}$.

Comparing the resultant m.m.f.s of Figs. 11.9(*a*) and (*b*) it will be seen that the centre-line of the resultant m.m.f. has moved $30°_e$ in the $+\theta$ direction, and that the shape of the distribution has become trapezoidal. The maximum value of the resultant m.m.f. has fallen slightly from $2F_{pm}$ to $\sqrt{3}_{pm}$.

After the next $\frac{1}{12}$th cycle the waveshape will be found to be the same as in Fig. 11.9(*a*), but displaced a further $30°_e$ round the armature. Hence the following points may be noted.

1. The m.m.f. wave is continually changing shape between the limits of the peaked wave of Fig. 11.9(*a*) and the flat-topped wave of Fig. 11.9(*b*).
2. The wave may be approximated to by a sinusoidal wave of constant maximum value. It is shown in Section 11.12 that this value is $\dfrac{18}{\pi^2} F_{pm}$.
3. The m.m.f. wave moves past the coils as the alternating currents vary throughout their cycle.
4. The m.m.f. wave moves by $\frac{1}{12}$th of one pole pair in $\frac{1}{12}$th cycle, i.e. the m.m.f. wave moves through one pole pair in one cycle.

If the frequency of the 3-phase currents is f and the speed of rotation of the field is n revolutions per second,

$$\text{Time to move through 1 pole pair} = \frac{1}{f} = \frac{1}{np}$$

Therefore

$$n = \frac{f}{p} = n_0$$

i.e. the field rotates at synchronous speed as defined by eqn. (11.10). Summarizing these points it may be said that a 3-phase current in a 3-phase winding produces a rotating magnetic field in the air-gap of the machine, the speed of rotation being the synchronous speed for the frequency of the currents and the number of pole pairs in the machine.

The production of the rotating field is the significant difference between a 3-phase and a single-phase machine. Due to its rotating field a 3-phase machine gives a constant, non-pulsating torque in a direction independent of any subsidiary gear or auxiliary windings.

EXAMPLE 11.1   Compare the e.m.f.s at 50 Hz of the following 20-pole alternator windings wound in identical stators having 180 slots:

(a) a single-phase winding with 5 adjacents slots per pole wound, the remaining slots being unwound.
(b) a single-phase winding with all slots wound,
(c) a 3-phase star-connected winding with all slots wound.

All the coils in each phase are connected in series, and each slot accommodates 6 conductors. The total flux per pole is 0·025 Wb.

Assuming a single-layer winding with full-pitch coils there will be 6 turns per coil and the coil span factor will be unity.

There are 9 slots per pole, and thus the slot pitch, ψ, is given by

$$\psi = \frac{180}{9} = 20°_e$$

(a) Number of coils per pole pair and phase, $g = 5$

$$\text{Distribution factor} = \frac{\sin 5 \times \dfrac{20°}{2}}{5 \sin \dfrac{20°}{2}} = \frac{0·766}{0·868} = 0·883$$

E.M.F. per phase $= 4·44 K_d K_s f \Phi N_p$ (11.21)

$$= 4·44 \times 0·883 \times 1 \times 50 \times 0·025 \times 5 \times 6 \times 10$$

$$= 1,470\,\text{V}$$

(b) Number of coils per pole pair and phase, $g = 9$

$$\text{Distribution factor} = \frac{\sin 9 \times \dfrac{20'}{2}}{9 \sin \dfrac{20°}{2}} = \frac{1}{1·563} = 0·64$$

E.M.F. per phase $= 4{\cdot}44K_dK_sf\Phi N_p$

$\qquad = 4{\cdot}44 \times 0{\cdot}64 \times 1 \times 50 \times 0{\cdot}025 \times 9 \times 6 \times 10$

$\qquad = 1{,}920\,\mathrm{V}$

(c) Number of coils per pole pair and phase, $g = 3$

$$\text{Distribution factor} = \frac{\sin 3 \times \dfrac{20°}{2}}{3 \sin \dfrac{20°}{2}} = \frac{0{\cdot}5}{0{\cdot}521} = 0{\cdot}96$$

E.M.F. per phase $= 4{\cdot}44K_dK_sf\Phi N_p$

$\qquad = 4{\cdot}44 \times 0{\cdot}96 \times 1 \times 50 \times 0{\cdot}025 \times 3 \times 6 \times 10$

$\qquad = 960\,\mathrm{V}$

(Line voltage for star connexion $= 1{,}600\,\mathrm{V}$)

Comparing (a) and (b) it will be seen that the e.m.f. in case (b) is only 30 per cent greater than that in case (a), while the amount of winding material is 80 per cent greater.

The winding losses for the same current would also be 80 per cent greater for case (b). Thus it is common practice to omit some coils in each pole pair in a single-phase winding.

Supposing that with the above e.m.f.s there is a current of I amperes in the coils.

In case (a), armature power $= 1{,}470I$ watts

In case (b), armature power $= 1{,}920I$ watts

In case (c), armature power $= \sqrt{3} \times 1{,}660I = 2{,}800I$ watts

Comparing (b) and (c) above it will be realized that for the same frame size with the same winding and core losses the output from a 3-phase machine is about 1·5 times greater than that from a single-phase machine.

11.12 M.M.F. due to a Three-phase winding (analytical treatment)

In Fig. 11.10 the m.m.f. due to one phase acting separately is shown as a trapezoidal wave. This trapezoidal wave can be represented by using an appropriate Fourier series consisting of a fundamental and a series of space harmonics. In the analysis below all space harmonics are neglected, and the m.m.f. due to each phase acting separately is assumed to be of sinusoidal form having a maximum value equal to the maximum value of the fundamental in the Fourier series.

In Fig. 11.10 the axis $\theta = 0$ is the centre-line of the positive half-wave of the m.m.f., F'_r, due to the red phase only when the red phase carries conventionally positive current (i.e. emerging from the start end of the winding). The Fourier series of a trapezoidal wave having this origin is

$$F(\theta) = \frac{8A}{\pi(\pi - 2\beta)} \sum_n \frac{1}{n^2} \cos n\beta \cos n\theta \quad (n \text{ is odd}) \tag{11.24}$$

where A and β are as indicated on Fig. 11.10.

A is the maximum value of the trapezoid, and since the red phase is excited by alternating current, this maximum value varies sinusoidally with time so that

$$A = F_{pm} \cos \omega t \quad \left(F_{pm} = \frac{I_{pm} N_p}{2p} \right)$$

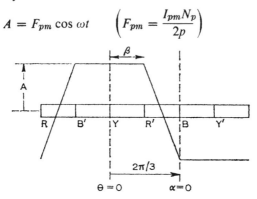

Fig. 11.10   M.M.F. DUE TO A 3-PHASE WINDING
(ANALYTICAL TREATMENT)

The angle β corresponds to a phase spread, so that $\beta = \pi/3$. Substituting in eqn. (11.24), the red phase m.m.f. at any time t and at any position θ is

$$F'_r = \frac{8F_{pm} \cos \omega t}{\pi(\pi - 2\pi/3)} \cos \frac{\pi}{3} \cos \theta$$

$$= \frac{12F_{pm}}{\pi^2} \cos \omega t \cos \theta \tag{11.25}$$

In Fig. 11.10 the axis $\alpha = 0$ is the centre-line of the positive half-wave of the m.m.f. due to the yellow phase. Since the yellow phase current lags in time behind the red phase current by $2\pi/3$ radians, the variation of the maximum value of the yellow phase m.m.f. will lag behind that of the red phase m.m.f. by the same amount.

The yellow phase m.m.f. is

$$F'_y = \frac{8F_{pm} \cos (\omega t - 2\pi/3)}{\pi(\pi - 2\pi/3)} \cos \frac{\pi}{3} \cos \alpha$$

Evidently

$$\theta = \alpha + 2\pi/3 \quad \text{so that} \quad \alpha = \theta - 2\pi/3$$

and

$$F'_y = \frac{12F_{pm}}{\pi^2} \cos \left(\omega t - \frac{2\pi}{3} \right) \cos \left(\theta - \frac{2\pi}{3} \right) \tag{11.26}$$

Similarly the blue phase m.m.f. is

$$F_b' = \frac{12F_{pm}}{\pi^2} \cos\left(\omega t + \frac{2\pi}{3}\right) \cos\left(\theta + \frac{2\pi}{3}\right) \quad (11.27)$$

The resultant m.m.f. of the 3-phase winding is

$$F_A' = F_r' + F_y' + F_b'$$

$$= \frac{12F_{pm}}{\pi^2} \left\{ \begin{matrix} \cos \omega t \cos \theta + \cos(\omega t - 2\pi/3)\cos(\theta - 2\pi/3) \\ + \cos(\omega t + 2\pi/3)\cos(\theta + 2\pi/3) \end{matrix} \right\}$$

On simplifying this gives

$$F_A' = \frac{18}{\pi^2} F_{pm} \cos(\theta - \omega t) \quad (11.28)$$

This is the fundamental in the space distribution of m.m.f. It has a maximum value at $\theta - \omega t = 0$, i.e. when $\theta = \omega t$. That is, the position of the maximum value travels in the $+\theta$ direction with an angular velocity ω radians per second.

Thus an m.m.f. of the form $F = F_m \cos(\theta - \omega t)$ represents an m.m.f. wave cosinusoidally distributed in space and travelling in the $+\theta$ direction at ω radians per second. The above equation is therefore the equation of a *travelling wave* and is referred to as a *retarded function*.

11.13 Three-phase Rotating Field Torques

Consider a rotating field, derived from the rotor, say, linking a 3-phase winding in which a 3-phase current is flowing; i.e. the rotating field is due to the m.m.f. of rotating poles, not to the 3-phase current. Suppose the speed of the rotating field is such that the frequency of the e.m.f. induced in the 3-phase winding is the same as that of the currents in the 3-phase winding. Unless this is the case there will be no mean torque since the direction of the torque will be alternating.

Let E_{ph} = R.M.S. value of e.m.f. induced in each phase of 3-phase winding

ϕ = Phase angle between induced e.m.f. and winding current

I_{ph} = R.M.S. phase current

The machine may be acting as either a generator or a motor:

Mean phase power = $E_{ph}I_{ph} \cos\phi$ watts

This, by the law of conservation of energy and neglecting losses, must be the mechanical power required to drive the rotor if the machine is acting as a generator, or the mechanical power developed if the machine is acting as a motor.

Total mechanical power developed $= 3E_{ph}I_{ph}\cos\phi = 2\pi n_0 T$

where T is the total torque developed (newton-metres) and n_0 is the speed of the rotor (f/p revolutions per second). Therefore

$$T = \frac{3E_{ph}I_{ph}\cos\phi}{2\pi n_0} \tag{11.29}$$

$$= \frac{3}{2\pi\dfrac{f}{p}} \frac{2\pi f N_{ph}\Phi_m K_d K_s}{\sqrt{2}} I_{ph}\cos\phi$$

$$= \frac{3p}{\sqrt{2}} N_{ph}\Phi_m K_d K_s I_{ph}\cos\phi \quad \text{newton-metres} \tag{11.30}$$

When the machine is motoring, the torque will act on the rotor in the direction of rotation and react on the stator in the opposite direction. These directions will interchange when the machine is generating.

11.14 Non-pulsating Nature of the Torque in a Three-phase Machine

It has been shown that the 3-phase currents in the stator of a 3-phase machine produce a magnetic field of effectively constant amplitude rotating round the air-gap at synchronous speed. The torque developed is due to the magnetic forces between the rotor poles and the rotating field, so that so long as the rotor poles move at synchronous speed there will be a constant magnetic force between stator and rotor. Hence the 3-phase machine will develop a constant torque which does not pulsate in magnitude. (Note that this differs from the case of the single-phase machine.)

The above conclusion may also be derived by considering that the total power delivered to a balanced 3-phase load is non-pulsating, so that, if the load is a machine which is running at a constant speed, the torque developed must also be non-pulsating. On this basis the single-phase machine has a pulsating torque since the power supplied pulsates at twice the supply frequency.

PROBLEMS

11.1 Derive an expression for the e.m.f. induced in a full-pitched coil in an alternator winding, assuming a sinusoidal distribution of flux in the air-gap.

Show how the voltage of a group of such coils, connected in series, may be found.

Calculate the speed and open-circuit line and phase voltages of a 4-pole 3-phase 50 Hz star-connected alternator with 36 slots and 30 conductors per slot. The flux per pole is 0·0496 Wb, sinusoidally distributed.

Ans. 1,500 rev/min, 3,300 V, 1,910 V.

11.2 Derive the expression for the voltage in a group of *m* full-pitch coils each having an electrical displacement of *ψ*.

An 8-pole 3-phase star-connected alternator has 9 slots per pole and 12 conductors per slot. Calculate the necessary flux per pole to generate 1,500 V at 50 Hz on open-circuit. The coil span is one pole pitch.

With the same flux per pole and speed, what would be the e.m.f. when the armature is wound as a single-phase alternator using two-thirds of the slots?

(*H.N.C.*)

Ans. 0·0283 Wb; 1,480 V.

11.3 A 6-pole machine has an armature of 90 slots and 8 conductors per slot and revolves at 1,000 rev/min, the flux per pole being 5×10^{-2} Wb. Calculate the e.m.f. generated (*a*) as a d.c. machine if the winding is lap-connected; (*b*) as a 3-phase star-connected machine if the winding factor is 0·96 and all the conductors in each phase are in series. Deduce the expression used in each case.

(*L.U.*)

Ans. 600 V, 2,200 V.

11.4 Derive an expression for the e.m.f. induced in each phase of a single-layer distributed polyphase winding assuming the flux density distribution to be sinusoidal.

A 4-pole 3-phase 50 Hz star-connected alternator has a single-layer armature winding in 36 slots with 30 conductor per slot. The flux per pole is 0·05 Wb. Determine the speed of rotation. Draw, to scale, the complexor diagram of the phase e.m.f.s,

1. when the phase windings are symmetrically star-connected,
2. when the phase windings are asymmetrical star-connected, the yellow phase winding being reversed with respect to the red and blue phase windings.

Give the numerical values of all the line voltages in each case. The phase sequence for the symmetrical star connection is RYB.

Ans. 1,500 rev/min; $V_{RY} = V_{YB} = B_{BR} = 3,320$ V; $V_{RY} = 1,920$ V; $V_{YB} = 1,920$ V; $V_{BR} = 3,320$ V.

11.5 An 8-pole rotor, excited to give a steady flux per pole of 0·01 Wb, is rotated at 1,200 rev/min in a stator containing 72 slots. Two 100-turn coils A and B are accommodated in the stator slotting as follows:

Coil A. Coil sides lie in slots 1 and 11,
Coil B. Coil sides lie in slots 2 and 10.

Calculate the resultant e.m.f. of the two coils when they are joined (*a*) in series aiding and (*b*) in series opposing. Assume the flux density distribution to be sinusoidal.

Ans. 700 V; 0 V.

11.6 A rotor having a d.c. excited field winding is rotated at *n* revolutions per second in a stator having uniformly distributed slots. If the air-gap flux density is sinusoidally distributed, derive an expression for the e.m.f. induced in a coil of N turns the sides of which lie in any two slots.

In such an arrangement the flux per pole is 0·01 Wb, the rotor speed is 1,800 rev/min, the stator has 36 slots and the rotor has 4 poles. Calculate the frequency and r.m.s. value of the induced e.m.f. in the coil if the coil sides lie in slot 1 and slot 9 and the coil has 100 turns.

An exactly similar coil is now placed with its coil sides in slots adjacent to the first coil. Determine the resultant e.m.f.s when the coils are connected in series.

Ans 60 Hz; 263 V; 518 V or 91·3 V.

Chapter 12

THE THREE-PHASE
SYNCHRONOUS MACHINE

A synchronous machine is an a.c. machine in which the rotor moves at a speed which bears a constant relationship to the frequency of the current in the armature winding. As a motor, the shaft speed must remain constant irrespective of the load, provided that the supply frequency remains constant. As a generator, the speed must remain constant if the frequency of the output is not to vary. The field of a synchronous machine is a steady one. In very small machines this field may be produced by permanent magnets, but in most cases the field is excited by a direct current obtained from an auxiliary generator which is mechanically coupled to the shaft of the main machine.

12.1 Types of Synchronous Machine

The armature or main winding of a synchronous machine may be on either the stator or the rotor. The difficulties of passing relatively large currents at high voltages across moving contacts have made the stator-wound armature the common choice for large machines. When the armature winding is on the rotor, the stator carries a salient-pole field winding excited by direct current and very similar to the stator of a d.c. machine (Fig. 12.1(*a*)). Stator-wound armature machines fall into two classes: (*a*) salient-pole rotor machines, and (*b*) non-salient-pole, or cylindrical-rotor, machines

(Fig. 12.1(*b*) and (*c*)). The salient-pole machine has concentrated field windings and generally is cheaper than the cylindrical-rotor machine when the speed is low (less than 1,500 rev/min). Salient-pole alternators are generally used when the prime mover is a water

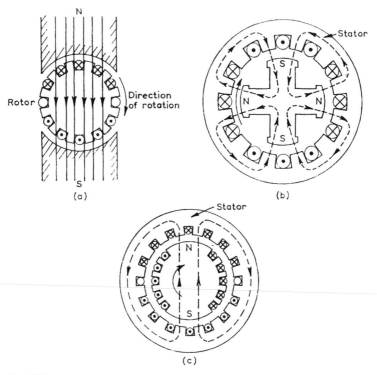

Fig. 12.1   SYNCHRONOUS MACHINES

    (a) Armature on rotor
    (b) Salient-pole rotor (4 poles)
    (c) Cylindrical rotor (2 poles)

turbine or a reciprocating engine. The cylindrical rotor has a distributed winding in the rotor slots and is most suitable for high speeds: steam-turbine-driven alternators are generally high-speed machines (3,000 rev/min) and have the cylindrical-rotor construction.

For an alternator to generate a 50 Hz voltage the speed n_0 will be given by eqn. (11.10) as

$$n_0 = \frac{60 \times 50}{p} = \frac{3,000}{p} \quad \text{rev/min}$$

where p is the number of pole-pairs on the machine. n_0 would also be the speed of a synchronous motor with p pole-pairs operating from a 50 Hz supply.

12.2 M.M.F. Wave Diagrams of the Synchronous Generator

The operation of a synchronous machine may be understood by a consideration of its m.m.f. waves. There are three m.m.f. waves to be considered: that due to the field winding, F_F, which is separately excited with direct current; that due to the 3-phase armature winding, F_A; and their resultant, F_R.

In the first stages of the explanation the following assumptions will be made.

1. Magnetic saturation is absent so that the machine is linear.
2. The field and armature m.m.f.s are sinusoidally distributed.
3. The air-gap is uniform, i.e. the machine does not exhibit saliency on either side of the air-gap.
4. The reluctance of the magnetic paths in the stator and rotor is negligible.
5. The armature-winding leakage inductance and resistance are negligible.

The last assumption will be removed at a convenient stage in the development. The machine considered will be a cylindrical-rotor machine with the 3-phase winding on the stator and the d.c.-excited field winding on the rotor. The generating mode of action will first be considered.

The method of drawing the m.m.f. waves will be that used in Chapter 11, which should be read in conjunction with this chapter. The same conventions for positive m.m.f. and positive current are adopted. Positive current is assumed to emerge from the start ends of the phase windings, so that the unprimed phase spreads R,Y,B are dotted when the current is positive, and the primed phase spreads R′Y′B′ are crossed when the current is positive. Positive e.m.f. may now be defined in the same way. Positive m.m.f. is assumed to be directed from rotor to stator and is shown above the θ axis in the m.m.f. wave diagrams. These are shown superimposed on a representation of a double pole-pitch of the stator winding and of the rotor.

GENERATING-MODE OPERATION ON OPEN-CIRCUIT

Fig. 12.2(*a*) shows the relevant m.m.f. wave for the synchronous generator on open-circuit. The instant chosen in the e.m.f. cycle, indicated by the complexor diagram, is such that

$$E_R = E_{pm} \qquad E_Y = -\tfrac{1}{2}E_{pm} \qquad E_B = -\tfrac{1}{2}E_{pm}$$

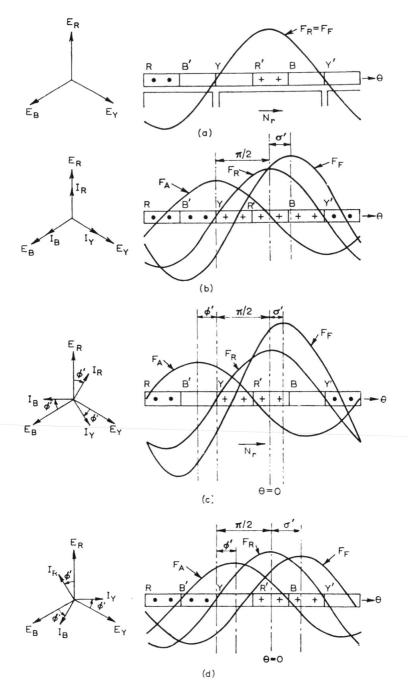

Fig. 12.2 M.M.F. WAVE DIAGRAMS FOR SYNCHRONOUS GENERATOR

The phase spreads of the red phase, R and R', have been dotted and crossed in accordance with the convention for positive e.m.f. The rotor rotates in the direction shown; the direction of the stator conductors relative to the rotor is in the opposite direction, and this latter direction must be used in applying the right-hand rule to find the direction of the field flux and m.m.f. This is as shown in Fig. 12.2(a), the maximum rotor m.m.f.s occurring opposite the centres of the red phase spreads, since this phase has maximum e.m.f. induced in it. The field m.m.f., F_F, shown in Fig. 12.2(a) is also the resultant m.m.f., F_R, since on open-circuit there is no armature current and consequently F_A is zero at all times and at all points in the air-gap.

The field m.m.f. is stationary with respect to the rotor winding, which is excited with direct current but moves, with the rotor, at synchronous speed past the stator winding.

GENERATING-MODE OPERATION AT UNITY POWER FACTOR

Fig. 12.2(b) represents the m.m.f. waves when the armature winding is supplying current at unity power factor.

To obtain comparability between Figs. 12.2(a) and (b), both diagrams have been drawn for the same instant in the e.m.f. cycle, and the magnitudes of the e.m.f.s are the same in each case as indicated by the e.m.f. complexor diagram. Since the e.m.f. is caused by the resultant m.m.f., F_R, this will have the same magnitude and position in Fig. 12.2(b) as it has in Fig. 12.2(a).

However, since in this case armature current flows, there will be an armature m.m.f., F_A. The resultant m.m.f., F_R, is the sum of F_A and F_F, so in this case F_F and F_R are different.

The instant in the current cycle is such that

$$i_R = I_{pm} \qquad i_Y = -\tfrac{1}{2}I_{pm} \qquad i_B = -\tfrac{1}{2}I_{pm}$$

This is the instant in the 3-phase cycle for which Fig. 11.9(a) was drawn. The armature m.m.f. in Fig. 12.2(b), therefore, is in the same position as the armature m.m.f. in Fig. 11.9(a) and lags behind the resultant m.m.f. wave by $\pi/2$ radians, but the space harmonics which give the armature m.m.f. wave its distinctive peaked shape are ignored, and this m.m.f. is shown as a sine distributed wave. The armature m.m.f. moves at synchronous speed, so that the m.m.f.s F_A and F_F and their resultant F_R all move at the same speed and in the same direction under steady conditions.

At any time and at any point in the air-gap,

$$F_R = F_A + F_F \tag{12.1}$$

Therefore $F_A = F_R - F_F$.

The field m.m.f., F_F, in Fig. 12.2(*b*) is therefore obtained by point-by-point subtraction of the resultant and armature m.m.f. waves. Comparing the field m.m.f. at Fig. 12.2(*b*) with that for Fig. 12.2(*a*) two changes may be observed:

1. To maintain the e.m.f. constant, the separate excitation has had to be increased in value as shown by the higher maximum value of F_F. The effect of armature m.m.f. is therefore the same as that of an internal voltage drop.
2. The axis of the field m.m.f. has been displaced by an angle σ' in the direction of rotation, and as a result a torque is exerted on the rotor in the direction opposite to that of rotation. The rotor must be driven by a prime mover against this torque, so that machine absorbs mechanical energy and is therefore able to deliver electrical energy.

GENERATING-MODE OPERATION AT POWER FACTORS OTHER THAN UNITY

Fig. 12.2(*c*) shows the m.m.f. waves for the same instant in the e.m.f. cycle but with the phase currents lagging behind their respective e.m.f.s by a phase angle ϕ'. As compared with its position in Fig. 12.2(*b*), therefore, the armature m.m.f. wave is displaced by an angle ϕ' in the direction opposite to that of rotation as compared with its unity-power-factor position. This change in the relative position of the armature m.m.f. wave brings it more into opposition with the field m.m.f., so that the latter must be further increased to maintain the resultant m.m.f. and e.m.f. constant.

Fig. 12.2(*d*) shows the m.m.f. waves for the same instant in the e.m.f. cycle but with the phase currents leading their respective e.m.f.s by a phase angle ϕ'. As compared with its position for unity power factor, the armature m.m.f. wave is displaced in the direction of rotation by the angle ϕ'. In this case the change in the relative position of the armature m.m.f. wave gives it a component which aids the field m.m.f., which must then be reduced for a constant resultant m.m.f. and e.m.f.

If the power factor is zero lagging, the armature and field m.m.f. waves are in direct opposition, whereas if the power factor is zero leading, the armature and field m.m.f. waves are directly aiding.

MOTORING-MODE OPERATION

Positive current in a motor conventionally enters the positive terminal and circulates in the windings in the direction opposite to that in which the e.m.f. acts. In Fig. 12.3 the phase of the current with

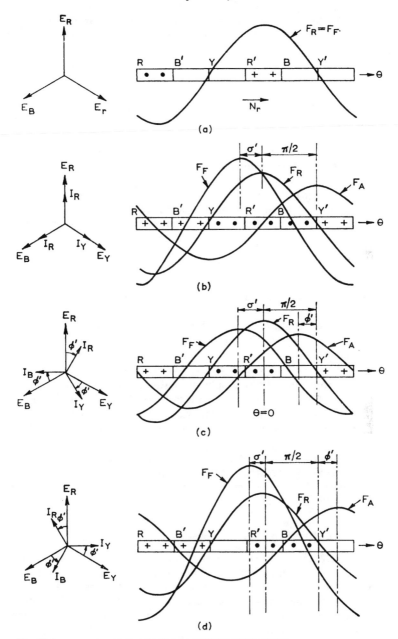

Fig. 12.3 M.M.F. WAVE DIAGRAMS FOR SYNCHRONOUS MOTOR

respect to the induced e.m.f. is kept the same as for generator action, but the opposite convention for positive current is used in drawing the m.m.f. wave; i.e. when the phase current is positive it is assumed to enter the start ends of the phase windings, so that the unprimed phase spreads are crossed when the current is positive, and the primed phase spreads are dotted when the current is positive. The convention for e.m.f. is the same as that used for generator action in the construction of Fig. 12.2.

Fig. 12.3(*a*) shows the m.m.f. waves for the synchronous motor on no-load. It has been assumed that the no-load current is negligible and the armature m.m.f. zero. The field and resultant m.m.f.s are then identical as shown, and Fig. 12.3(*a*) corresponds exactly to Fig. 12.2(*a*).

Fig. 12.3(*b*) shows the m.m.f. wave diagram corresponding to motor operation at unity power factor. The armature m.m.f. is reversed at all points compared with the corresponding armature m.m.f. wave in Fig. 12.2(*b*). The field m.m.f. is found by carrying out the subtraction $F_R - F_A$ point by point round the air-gap. It will be noted that, compared with the no-load condition, the field m.m.f. is displaced by an angle σ' in the direction opposite to rotation, thus giving rise to a torque on the rotor acting in the direction of rotation. The machine thus delivers mechanical power, having absorbed electrical power from the supply.

Figs. 12.3(*c*) and (*d*) show the m.m.f. for operation at lagging and leading power factors respectively. Synchronous motors are operated from a constant-voltage supply, so that variation in the field excitation cannot affect the machine e.m.f. It follows that the power factor of the armature current must alter with variation of field excitation (this effect also occurs in generators connected in large constant-voltage systems).

From Fig. 12.3(*c*) it will be seen that the input power factor becomes leading when the excitation is increased above the unity p.f. condition. Similarly when the excitation is reduced the p.f. becomes lagging (Fig. 12.3(*d*)).

12.3 M.M.F. Travelling-wave Equations

Figs. 12.2 and 12.3 are the m.m.f. wave diagrams of a synchronous machine for both the generating and motoring modes of operation at various power factors. Each of the waves travels in the $+\theta$ direction at synchronous speed. Therefore, each of these m.m.f.s may be represented by a travelling-wave equation, or retarded function, as explained in Section 11.12.

For example, the equation of the cosinusoidally distributed m.m.f.

shown in Fig. 12.4, which is travelling in the $+\theta$ direction at ω electrical radians per second is

$$F' = F_m \cos(\omega t - \theta) \tag{12.2}*$$

Considering generator action at a lagging power factor, the resultant m.m.f. (Fig. 12.2(c)) is

$$F_R' = F_{Rm} \cos(\omega t - \theta) \tag{12.3}$$

The field m.m.f. is

$$F_F' = F_{Fm} \cos(\omega t - \theta + \sigma') \tag{12.4}$$

and the armature m.m.f. is

$$F_A' = F_{Am} \cos(\omega t - \theta - \pi/2 - \phi') \tag{12.5}$$

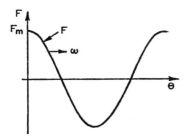

Fig. 12.4 TRAVELLING-WAVE M.M.F.

When the generator works at a leading power factor, eqns. (12.3) and (12.4) still apply for the resultant and separate field m.m.f.s. It will be seen, by examination of Fig. 12.2(d), that the equation for the armature m.m.f. becomes

$$F_A' = F_{Am} \cos(\omega t - \theta - \pi/2 + \phi') \tag{12.6}$$

Considering now the motoring mode at a lagging power factor, to which Fig. 12.3(c) refers, the resultant m.m.f. is the same as for generator action, i.e.

$$F_R' = F_{Rm} \cos(\omega t - \theta) \tag{12.7}$$

The separate field m.m.f. is

$$F_F' = F_{Fm} \cos(\omega t - \theta - \sigma') \tag{12.8}$$

and the armature m.m.f. is

$$F_A' = F_{Am} \cos(\omega t - \theta + \pi/2 - \phi') \tag{12.9}$$

* A prime (') is used to indicate instantaneous values of m.m.f.

When the power factor is leading, the armature m.m.f. becomes

$$F_A' = F_{Am} \cos(\omega t - \theta + \pi/2 + \phi') \tag{12.10}$$

These results may be summarized as follows: The resultant m.m.f. for any mode of operation is

$$F_R' = F_{Rm} \cos(\omega t - \theta) \tag{12.3}$$

The separate field m.m.f. is

$$F_F' = F_{Fm} \cos(\omega t - \theta \pm \sigma') \tag{12.8}$$

where $+\sigma'$ is used for the generating mode and $-\sigma'$ for the motoring mode.

The armature m.m.f. is

$$F_A' = F_{Am} \cos(\omega t - \theta - \pi/2 \pm \phi') \tag{12.11}$$

for the generating mode and

$$F_A' = F_{Am} \cos(\omega t - \theta + \pi/2 \pm \phi') \tag{12.12}$$

for the motoring mode. For both cases $+\phi'$ refers to operation at a leading power factor and $-\phi'$ to operation at a lagging power factor.

12.4 M.M.F. Complexor Diagrams

The sum (or difference) of two sinusoidally space-distributed m.m.f.s may be found using the same methods as are used for time-varying sinusoidal quantities. Although the sinusoidally space-distributed m.m.f.s of the synchronous machine are all travelling at synchronous speed (in a stator-wound machine), they may still be dealt with by means of complexor diagrams, since, under steady-state conditions, the relative positions of the waves do not alter.

The m.m.f. complexor diagrams may be deduced either directly from the m.m.f. wave diagrams of Figs. 12.2 and 12.3 or from the travelling-wave equations derived in Section 12.3.

Adopting the latter method for generator mode operation at a lagging power the travelling-wave equations are

$$F_R' = F_{Rm} \cos(\omega t - \theta) \tag{12.2}$$

$$F_F' = F_{Fm} \cos(\omega t - \theta + \sigma') \tag{12.4}$$

$$F_A' = F_{Am} \cos(\omega t - \theta - \pi/2 - \phi') \tag{12.5}$$

At any particular point in the air-gap denoted by $\theta = \theta_0$ the m.m.f.s are

$$F_R' = F_{Rm} \cos(\omega t - \theta_0) \tag{12.13}$$

$$F_F' = F_{Fm} \cos(\omega t - \theta_0 + \sigma') \tag{12.14}$$

$$F_A' = F_{Am} \cos(\omega t - \theta_0 - \pi/2 - \phi') \tag{12.15}$$

Since θ_0 is a particular value of θ and therefore not a variable, the above equations represent, not travelling waves, but quantities varying sinusoidally with time.

The corresponding complexor diagram is shown in Fig. 12.5(a), where the m.m.fs. F_R', F_F' and F_A' are represented by the complexors

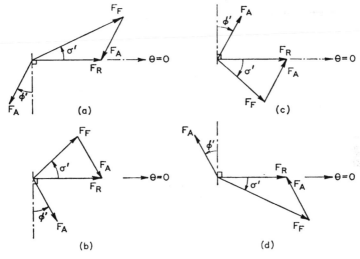

Fig. 12.5 M.M.F. COMPLEXOR DIAGRAMS FOR THE SYNCHRONOUS MACHINE

(a) Generator operation at a lagging power factor
(b) Generator operation at a leading power factor
(c) Motor operation at a lagging power factor
(d) Motor operation at a leading power factor

F_R, F_F and F_A. For simplicity the diagram has been drawn for the air-gap position $\theta_0 = 0$. The chain-dotted line indicates the unity-power-factor position of the m.m.f. F_A. Figs. 12.5(b), (c) and (d) are similar diagrams for different power factors and different operating modes.

12.5 E.M.F. Complexor Diagram

Assuming that the reluctance of the magnetic paths in the stator and rotor is negligible and that the air-gap is uniform, the sinusoidally distributed travelling-wave m.m.f.s F_F and F_A and their resultant F_R may be assumed to give rise to separate sinusoidally distributed flux densities. Each of these flux density distributions will travel at synchronous speed, its maximum value occurring at the same place as that of the corresponding m.m.f. and travelling with it at synchronous speed. The relative motion between these flux density distributions and the phase windings will induce e.m.f.s in the windings.

Since only the air-gap reluctance is taken into account, the magnetic circuit is linear and the principle of superposition may be applied. For a particular phase winding let

E_F = E.M.F. due to field m.m.f., F_F
E_A = E.M.F. due to armature m.m.f., F_A
E_R = Resultant e.m.f. due to the resultant m.m.f., F_R

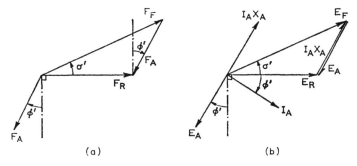

Fig. 12.6 M.M.F. AND E.M.F. COMPLEXOR DIAGRAMS FOR THE SYNCHRONOUS GENERATOR

The relative phase angles of E_F, E_A and E_R will be the same as those of F_F, F_A and F_R.

The m.m.f. complexor diagram, as explained in Section 12.4, is drawn for a particular position round the air-gap $\theta_0 = 0$.

The m.m.f. and e.m.f. complexor diagrams are shown in Figs. 12.6(a) and (b) respectively for the case of a generator working at a lagging power factor. The resultant e.m.f. E_R is shown as the complexor sum of E_F and E_A:

$$E_R = E_F + E_A \qquad (12.16)$$

For a fixed value of separate excitation, E_F will be constant. An examination of Fig. 12.6(b) reveals that any change of either the armature current or the load power factor would alter the resultant e.m.f. E_R. It is more convenient to work with a constant-voltage source in an equivalent circuit, so that E_F is customarily regarded as the e.m.f. since it does not alter with load and is also the terminal voltage on open-circuit (when $E_A = 0$). The effect of the armature m.m.f. is treated, not as a contribution to the available e.m.f. but as an internal voltage drop, and the phase opposite of E_A is subtracted from E_F.

Examination of Fig. 12.6(b) shows that E_A will always lag I_A by 90° irrespective of the power-factor angle. The phase opposite of E_A, namely $-E_A$, leads I_A by 90° for all conditions.

The peak value of the e.m.f. due to the armature m.m.f. is E_{Am}:

$$E_{Am} \propto F_{Am} \propto I_{Am}$$

Therefore

$$\frac{E_{Am}}{I_{Am}} = \frac{E_A}{I_A} = \text{constant}$$

Since the quotient of E_A and I_A is a constant, and since the phase opposite of E_A leads I_A by 90°, this voltage may be represented as an inductive voltage drop, and the quotient as an inductive reactance, i.e.

$$-E_A = I_A X_A$$

where

$$X_A = \frac{E_A}{I_A} \tag{12.17}$$

Substituting for E_A in eqń. (12.16) and rearranging,

$$E_F = E_R + I_A X_A \tag{12.18}$$

This complexor summation is shown in Fig. 12.6(b).

In Section 12.7 an expression for X_A is found in terms of the physical dimensions of the machine.

12.6 Equivalent Circuit of the Synchronous Machine

The preceding section has shown that the equivalent circuit of a synchronous machine must contain a voltage source E_F which is constant for a constant excitation current I_F and a series-connected reactance X_A. In addition, an actual machine winding will have resistance R and (in the same way as a transformer) leakage reactance X_L.

Fig. 12.7(a) shows the full equivalent circuit of the synchronous machine in which the current flows in the conventionally positive direction for generator-mode operation (a source), i.e. emerging from the positive terminal. Applying Kirchhoff's law to this circuit,

$$E_F = V + IR + jIX_L + jIX_A \tag{12.19}$$

Fig. 12.7(*b*) is the corresponding complexor diagram. The resultant e.m.f. E_R is shown for the sake of completeness but will be omitted in subsequent diagrams.

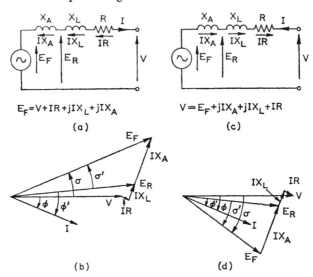

$$E_F = V + IR + jIX_L + jIX_A$$

(*a*)

$$V = E_F + jIX_A + jIX_L + IR$$

(*c*)

(*b*)

(*d*)

Fig. 12.7   EQUIVALENT CIRCUITS AND FULL COMPLEXOR DIAGRAMS
FOR THE SYNCHRONOUS MACHINE
(*a*), (*b*) Generator   (*c*), (*d*) Motor

Eqn. (12.19) may be rewritten as

$$E_F = V + IZ_s \tag{12.20}$$

where Z_s is the *synchronous impedance*.

$$Z_s = R + j(X_L + X_A) \tag{12.21}$$

or

$$Z_s = R + jX_s \tag{12.22}$$

where X_s is the *synchronous reactance*:

$$X_s = X_L + X_A \tag{12.23}$$

In polar form the synchronous impedance is

$$Z_s = Z_s\underline{/\psi} \tag{12.24}$$

where

$$\psi = \tan^{-1} \frac{X_s}{R} \tag{12.25}$$

and

$$Z_s = \sqrt{(R^2 + X_s^2)} \tag{12.26}$$

Frequently in synchronous machines $X_s \gg R$, in which case eqn. (12.26) becomes

$$Z_s = X_s \underline{/90°} = jX_s \tag{12.27}$$

Fig. 12.7(c) shows the full equivalent circuit of the synchronous motor, in which the current flows in the conventionally positive direction for motor-mode operation (a load), i.e. entering the position terminal. Applying Kirchhoff's law to this circuit gives

$$V = E_F + jIX_A + jIX_s + IR \tag{12.28}$$

Fig. 12.7(d) is the corresponding complexor diagram.

EXAMPLE 12.1 A 3-phase 11·8kV 75MVA, 50Hz 2-pole star-connected synchronous generator requires a separate field m.m.f. having a maximum value of 3.0×10^4 At/pole to give normal rated voltage on open-circuit. The flux per pole on open-circuit is approximately 5·3Wb.

Determine (a) the maximum armature m.m.f. per pole corresponding to rated full-load current, and (b) the synchronous reactance if the leakage reactance of the armature winding is 0.18Ω. Find also the p.u. value of the synchronous reactance.

Neglect the effect of space harmonics in the field and armature m.m.f.s. Assume the flux per pole to be proportional to the m.m.f. and the armature winding to be uniform and narrow spread.

The e.m.f. per phase, from eqn. (11.20), is

$$E_p = \frac{K_d K_s \omega \Phi N_p}{\sqrt{2}}$$

and the distribution factor for a uniform narrow-spread winding is

$$K_d = \frac{3}{\pi} = 0.955 \tag{11.13}$$

$$E_p = \frac{11.8 \times 10^3}{\sqrt{3}} = 6,800 \, \text{V}$$

Taking $K_s = 1$, the number of turns per phase is

$$N_p = \frac{\sqrt{2}E_p}{K_s K_d \omega \Phi} = \frac{2 \times 6,800}{0.955 \times 2\pi 50 \times 5.3} = 6.05$$

Since the number of turns per phase must be an integer, take $N_p = 6$. The maximum armature m.m.f., from eqn. (11.28), is

$$F_{Am} = \frac{18 F_{pm}}{\pi^2} = \frac{18}{\pi^2} \frac{\sqrt{2} I_p N_p}{2}$$

$$\text{Rated current per phase} = \frac{75 \times 10^6}{3 \times 6,800} = 3,680 \, \text{A}$$

so that

$$F_{Am} = \frac{18}{\pi^2} \times \sqrt{2} \times \frac{3,680 \times 6}{2} = 2.85 \times 10^4 \, \text{At/pole}$$

Since the flux per pole is proportional to the m.m.f.,

$$\frac{E_A}{E_F} = \frac{F_{Am}}{F_{Fm}}$$

$$E_A = 6{,}800 \times \frac{2 \cdot 85 \times 10^4}{3 \cdot 0 \times 10^4} = 6{,}450\,\text{V}$$

$$X_A = \frac{E_A}{I_A} = \frac{6{,}450}{3{,}680} = 1 \cdot 75\,\Omega$$

$$X_s = X_A + X_L = 1 \cdot 75 + 0 \cdot 18 = \underline{\underline{1 \cdot 93\,\Omega}}$$

Taking rated phase voltage and current as bases,

$$X_{spu} = \frac{3{,}680 \times 1 \cdot 93}{6{,}800} = \underline{\underline{1 \cdot 04\,\text{p.u.}}}$$

12.7 Synchronous Reactance in Terms of Main Dimensions

It is assumed that (a) magnetic saturation is absent; (b) the armature m.m.f. is sinusoidally distributed; (c) the air-gap is uniform; and (d) the reluctance of the magnetic paths in the stator and rotor is negligible.

Let D = Internal stator diameter
L = effective stator (or core) length
l_g = radial gap length

The synchronous reactance X_s is

$$X_s = X_L + X_A \tag{12.23}$$

The reactance X_A is

$$X_A = \frac{E_A}{I_A} \tag{12.17}$$

where E_A is the armature e.m.f. per phase due to the armature m.m.f. F_A, and I_A is the armature current per phase.

The value of E_A may be found by using eqn. (11.20), which gives the e.m.f. per phase of a polyphase winding:

$$E_A = K_d K_s \frac{\omega \Phi_A N_p}{\sqrt{2}} \tag{12.30}$$

where the distribution factor for a narrow-spread uniform winding is, from eqn. (11.13), $K_d = 3/\pi$; the coil span factor, $k_s = 1$; and Φ_A is the flux per pole due to the armature m.m.f. F_A.

From eqn. (11.28), and assuming that the armature m.m.f. is sinusoidally distributed,

$$\left.\begin{array}{l}\text{Maximum armature} \\ \text{m.m.f. per pole}\end{array}\right\} \quad F_{Am} = \frac{18 F_{pm}}{\pi^2} = \frac{18}{\pi^2} \frac{\sqrt{2} I_A N_p}{2p}$$

$$\left.\begin{array}{l}\text{Maximum air-gap} \\ \text{field strength}\end{array}\right\} \quad H_{gm} = \frac{F_{Am}}{l_g} = \frac{1}{l_g} \frac{18}{\pi^2} \frac{\sqrt{2} I_A N_p}{2p}$$

$$\left.\begin{array}{l}\text{Maximum air-gap} \\ \text{flux density}\end{array}\right\} \quad B_{gm} = \mu_0 H_{gm} = \frac{\mu_0}{l_g} \frac{18}{\pi^2} \frac{\sqrt{2} I_A N_p}{2p}$$

Since the air-gap flux density is sinusoidally distributed,

$$\left.\begin{array}{l}\text{Average air-gap} \\ \text{flux density}\end{array}\right\} \quad B_{av} = \frac{2}{\pi} B_{gm}$$

$$\left.\begin{array}{l}\text{Flux per pole due} \\ \text{to armature m.m.f.}\end{array}\right\} \quad \Phi_A = B_{av} \times \text{Pole area}$$

$$= \frac{2}{\pi} \frac{\mu_0}{l_g} \frac{18}{\pi^2} \frac{\sqrt{2} I_A N_p}{2p} \frac{\pi DL}{2p}$$

Substituting for K_d, K_s and Φ_A in eqn. (12.30), and then substituting the resulting expression for E_A in eqn. (12.17),

$$X_A = \frac{E_A}{I_A} = \frac{\dfrac{3}{\pi} \omega \dfrac{2}{\pi} \dfrac{\mu_0}{l_g} \dfrac{18}{\pi^2} \dfrac{\pi DL}{2p} \dfrac{\sqrt{2} I_A N_p}{2p} N_p}{\sqrt{2} I_A}$$

$$= \omega \left(\frac{18}{\pi^2}\right)^2 \frac{\mu_0}{3 l_g} \pi DL \left(\frac{N_p}{2p}\right)^2 \qquad (12.31)$$

To obtain the synchronous reactance, an allowance for the leakage reactance X_L must be added to X_A. The leakage reactance is mainly due to (a) slot leakage flux, which links individual slots and is not, therefore, part of the main flux, and (b) end-turn leakage flux, which links the end turns of the stator winding following mainly air paths. The evaluation of these leakage fluxes, particularly the latter, presents some difficulties and is beyond the scope of the present volume.

The e.m.f. per phase, E_A, is due to the armature current itself and is therefore an e.m.f. of self-induction. The reactance X_A is therefore a *magnetizing reactance*.

Eqn. (12.31) shows that the value of this reactance may be reduced by increasing the gap length l_g.

Looked at in another way, for a given machine rating the rated armature current is fixed as is also, as a consequence, the maximum armature m.m.f. per pole. This fixed value of armature m.m.f. has a progressively smaller effect as the air-gap is lengthened.

EXAMPLE 12.2 A 3-phase 13·8 kV 100 MVA 50 Hz 2-pole star-connected cylindrical-rotor synchronous generator has an internal stator diameter of 1·08 m and an effective core length of 4·6 m. The machine has a synchronous reactance of 2 p.u. and a leakage reactance of 0·16 p.u. The average flux density over the pole area is approximately 0·6 Wb/m². Estimate the gap length.

Assume that the radial air-gap is constant and the armature winding uniform. Neglect the reluctance of the iron core and the space harmonics in the armature m.m.f.

With the above assumptions the reactance X_A is

$$X_A = \omega \left(\frac{18}{\pi^2}\right)^2 \frac{\mu_0}{3 l_g} \pi DL \left(\frac{N_p}{2p}\right)^2 \tag{12.31}$$

Base voltage, $V_B = V_p = \dfrac{13·8 \times 10^3}{\sqrt{3}} = 7,960 \,\text{V}$

Base current, $I_B = \dfrac{\text{VA/phase}}{V_B} = \dfrac{100 \times 10^6}{3 \times 7,960} = 4,180 \,\text{A}$

Base impedance, $Z_B = \dfrac{V_B}{I_B} = \dfrac{7,960}{4,180} = 1·91 \,\Omega$

$X_{A\,pu} = X_{s\,pu} - X_{L\,pu} = 2·00 - 0·16 = 1·84 \,\text{p.u.}$

$X_A = X_{A\,pu} Z_B = 1·84 \times 1·91 = 3·52 \,\Omega$

Flux per pole, $B_{av} \times \text{Pole area} = B_{av} \dfrac{\pi DL}{2} = \dfrac{0·6 \times \pi \times 1·08 \times 4·6}{2}$

$$= 4·68 \,\text{Wb}$$

$$E_p = K_d K_s \frac{\omega \Phi N_p}{\sqrt{2}} \tag{11.20}$$

For a uniform winding, $K_d = 3/\pi$ and $K_s = 1$, so that

$$N_p = \frac{\sqrt{2} E_p}{K_d K_s \omega \Phi} = \frac{\sqrt{2} \times 7,960}{3/\pi \times 2\pi \times 50 \times 4·68} = 8·02$$

The number of turns per phase must be an integer, say 8. This will require a slightly higher flux per pole and average value of flux density. From eqn. (12.31),

$$l_g = 2\pi \times 50 \times \left(\frac{18}{\pi^2}\right)^2 \times \frac{4\pi \times 10^{-7}}{3 \times 3·52} \times \pi \times 1·08 \times 4·6 \times \left(\frac{8}{2}\right)^2$$

$$= 3·10 \times 10^{-2} \,\text{m}$$

12.8 Determination of Synchronous Impedance

The ohmic value of the synchronous impedance, at a given value of excitation may be determined by open-circuit and short-circuit tests (Fig. 12.8).

On open-circuit the terminal voltage depends on the field excitation and the magnetic characteristics of the machine. Fig. 12.9 includes a

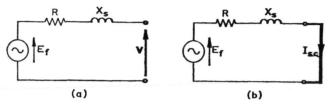

Fig. 12.8 DETERMINATION OF SYCHRONOUS IMPEDANCE
(a) Open-circuit test   (b) Short-circuit test

typical open-circuit characteristic showing the usual initial linear portion and subsequent saturation portion of a magnetization curve.
  On short-circuit the current in an alternator winding will normally lag behind the induced voltage by approximately 90° since the leakage

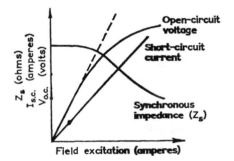

Fig. 12.9 VARIATION OF SYNCHRONOUS IMPEDANCE WITH EXCITATION

reactance of the winding is normally much greater than the winding resistance. The complexor diagram for short-circuit conditions is shown in Fig. 12.10. It is found that the armature and field m.m.f.s

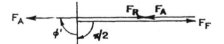

Fig. 12.10 COMPLEXOR DIAGRAM FOR SHORT-CIRCUIT
CONDITIONS

are directly in opposition, so that a surprisingly large excitation is required to give full-load short-circuit current in the windings. The resultant m.m.f. and flux are small since the induced voltage is only required to overcome resistance and leakage reactance voltage drops in the windings. Since the flux is small, saturation effects will be negligible and the short-circuit characteristic is almost straight.

The synchronous impedance Z_s may be found by dividing the open-circuit voltage by the short-circuit current at any particular value of field excitation (Fig. 12.9). Over the range of values where the open-circuit characteristic is linear the synchronous impedance is constant, but when the open-circuit characteristic departs from linearity the value of the synchronous impedance falls. It is often difficult to estimate the most appropriate value of Z_s to use for a particular calculation.

12.9 Voltage Regulation

The voltage regulation of an alternator is normally defined as the rise in terminal voltage when a given load is thrown off. Thus, if E_F is the induced voltage on open-circuit and V is the terminal voltage at a given load, the voltage regulation is given by

$$\text{Per-unit regulation} = \frac{E_F - V}{V} \tag{12.32}$$

There are a number of methods of predicting the voltage regulation of an alternator. None are completely accurate. Only the synchronous impedance method is considered here.

12.10 Synchronous Impedance Method

Using a suitable value for Z_s,

$$E_F = V + IZ_s \tag{12.20}$$

EXAMPLE 12.3 A 3-phase star-connected alternator has a resistance of $0.5\,\Omega$ and a synchronous reactance of $5\,\Omega$ per phase. It is excited to give $6,600\,\text{V}$ (line) on open circuit. Determine the terminal voltage and per-unit voltage regulation on full-load current of $130\,\text{A}$ when the load power factor is (*a*) 0.8 lagging, (*b*) 0.6 leading.

It is best to take the phase terminal voltage V as the reference complexor since the phase angle of the current is referred to this voltage. (The magnitude of V is, however, not known): i.e.

Phase terminal voltage, $V = V\underline{/0^\circ}$

The magnitude of the e.m.f E_F is known but not its phase with respect to V; i.e.

$$E_F = E_F\underline{/\sigma^\circ} = \frac{6,600}{\sqrt{3}}\,\underline{/\sigma^\circ} = 3,810\underline{/\sigma^\circ}$$

where σ° is the phase of E_F with respect to V as reference.

(*a*) The phase current *I* lags behind *V* by a phase angle corresponding to a power factor of 0·8 lagging, i.e.

$$I = 130/\!-\!\cos^{-1} 0.8 = 130/\!-\!36.9°\,\text{A}$$

The synchronous impedance per phase is

$$Z_s = (0.5 + j5)\Omega = 5.02/84.3°\,\Omega$$

In eqn. (12.20),

$$3{,}810/\sigma° = V/0° + (130/\!-\!36.9° × 5.02/84.3°)$$
$$= V/0° + 653/47.4°$$

Expressing all the terms in rectangular form,

$$3{,}810 \cos \sigma + j\, 3{,}810 \sin \sigma = V + j0 + 442 + j482$$

Equating quadrate parts,

$$3{,}810 \sin \sigma = 482$$

whence $\sin \sigma = 0.127$ and $\cos \sigma = 0.992$
Equating reference parts,

$$3{,}810 \cos \sigma = V + 442$$
$$V = (3{,}810 × 0.992) - 442 = \underline{\underline{3{,}340\,\text{V}}}$$

and

$$\text{Per-unit regulation} = \frac{3{,}810 - 3{,}340}{3{,}340} = \underline{\underline{0.141}}$$

(*b*)   Phase current = 130 A at 0·6 **leading** with respect to *V*
$$= 130/\!+\!53.1°$$

Following the same procedure as in part (*a*) it will be found that there is an on-load phase terminal voltage of $\underline{\underline{4{,}260\,\text{V}}}$. Hence the per-unit regulation, since

there is a voltage rise, is given by

$$\frac{3810 - 4260}{4{,}260} = \underline{\underline{-0.106\,\text{p.u.}}}$$

12.11 Synchronous Machines connected to Large Supply Systems

In Britain, electrical energy is supplied to consumers from approximately 100 generating stations. These stations vary considerably in size, the declared gross capability of the largest exceeding 3,000MW. A small number of stations have a declared gross capability of 100 MW or less.

The generating stations do not operate as isolated units but are interconnected by the national *grid*, which consists of almost 15,000 km of main transmission line, for the most part overhead lines operating at 132, 275 and 400 kV. The total generating capacity interconnected by the grid system is over 60,000 MW. The output of any single machine is therefore small compared with the total interconnected capacity. The biggest single generator has a rating of 660 MW. For this reason the performance of a single machine is unlikely to affect appreciably the voltage and frequency of the whole system. A machine connected to such a system, where the capacity of any one machine is small compared with the total interconnected capacity, is often said to be connected to *infinite busbars*. The outstanding electrical characteristics of such busbars are that they are constant-voltage constant-frequency busbars.

When the machine is connected to the infinite busbars the terminal voltage and frequency becomes fixed at the values maintained by the rest of the system. Unless the machine is grossly overloaded or under-excited, no change in the mechanical power supply, load or excitation will alter the terminal voltage or frequency. If the machine is acting as a generator and the mechanical driving power is increased the power output from the machine to the busbars must increase, assuming that the efficiency does not greatly change. In the same way, a decrease of mechanical driving power or the application of a mechanical load (motoring) will produce a decrease in output power or the absorption of power from the busbars.

12.12 Synchronizing

The method of connecting an incoming alternator to the live busbars will now be considered. This is called *synchronizing*.

A stationary alternator must not be connected to live busbars, or, since the induced e.m.f. is zero at standstill, a short-circuit will result. The alternator induced e.m.f. will prevent dangerously high switching currents only if the following conditions are almost exactly complied with:

1. The frequency of the induced voltages in the incoming machine must equal the frequency of the voltages of the live busbars.
2. The induced voltages in the incoming machine must equal the live busbar voltages in magnitude and phase.
3. The phase sequence of the busbar voltages and the incoming-machine voltages must be the same.

In modern power stations alternators are synchronized automatically. The principles may be illustrated by the three-lamp method, which, along with a voltmeter, may be used for synchronizing low-power machines.

Fig. 12.11(*a*) is the connexion diagram from which it will be noted that one lamp is connected between corresponding phases while the two others are cross-connected between the other two phases. In the complexor diagram at (*b*) the machine induced voltages are represented by E_R, E_Y and E_B, while the live supply voltages are represented by V_R, V_Y, and V_B. The lamp symbols have been added to the complexor diagram to indicate the instantaneous lamp voltages. It will be realized that the speed of rotation of the complexors will correspond to the frequencies of the supply and the machine—if these are the same then the lamp brilliancies will be constant. The speed of the machine should be adjusted until the machine frequency is nearly that of the supply, but exact equality is inconvenient for there would then be, in all probability, a permanent phase difference between corresponding voltages. The machine excitation should now be varied until the two sets of voltages

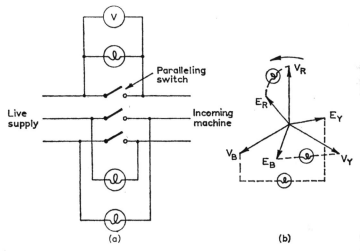

Fig. 12.11 SYNCHRONIZING BY CROSS-CONNECTED LAMP METHOD

are equal in magnitude. The correct conditions will be obtained at an instant when the straight-connected lamp is dark and the cross-connected lamps are equally bright. If the phase sequence is incorrect no such instant will occur as the cross-connected lamps will, in effect, be straight-connected and all the lamps will be dark simultaneously. In this event the direction of rotation of the incoming machine should be reversed or two lines of the machine should be interchanged. Since the dark range of a lamp extends over a considerable voltage range it is advisable to connect a voltmeter across the straight-connected lamp and to close the paralleling switch when the voltmeter reading is zero. It should be noted that the lamps and the voltmeter must be able to withstand twice the normal phase voltage.

12.13 Effect of Variation of Excitation of Synchronous Machine connected to Infinite Busbars

Consider an alternator connected to infinite busbars. Fig. 12.12(a) is the complexor diagram of such a machine when operating at unity power factor. The voltage drop I_aR is in phase with I_a and the voltage drop I_aX_s leads I_a by $90°$. The complexor sum of I_aR and I_aX_s is I_aZ_s, and the e.m.f. (E_F) of the machine is obtained by adding I_aZ_s to V, the constant busbar voltage. R, X_s and Z_s refer to the winding resistance, reactance and impedance respectively. I_aZ_s makes an angle $\psi = \tan^{-1}(X_s/R)$ with V.

Suppose that the excitation of the alternator is reduced while its power input is not altered. The power output will thus remain unchanged. As a result the active component of current, and the voltage drop I_aZ_s due to this current, will remain unchanged. I_aZ_s is shown separately in Fig. 12.12(b).

However, when the excitation of the alternator is reduced the e.m.f. (E_F) of the machine must fall, so there must be a difference between this new, lower value of E_F and the complexor sum of V and I_aZ_s, both of which remain unchanged. This difference is made up by a leading reactive current (which contributes nothing to the power output of the alternator) which sets up the voltage drop I_rZ_s which leads I_aZ_s by $90°$. The complexor diagram illustrating this condition is shown in Fig. 12.12(b). If it is said that the alternator is normally excited when it is working at unity power factor, then when the alternator is under-excited it will work at a leading power factor.

In a similar way, if the excitation of the alternator is increased from the normally excited unity-power-factor condition, the e.m.f. (E_F) of the machine will increase, so that there must be a difference between this new, higher value of E_F and the complexor sum of V and I_aZ_s, both of which remain unchanged. This difference is made up by a lagging reactive current (which contributes nothing to the power output of the alternator) which sets up the voltage drop I_rZ_s lagging behind I_aZ_s by $90°$. The complexor diagram illustrating this condition is shown in Fig. 12.12(c). Thus when the alternator is over-excited it will work with a lagging power factor.

Fig. 12.12(d), (e) and (f), give the corresponding diagrams for the synchronous motor connected to infinite busbars. These are essentially similar to those of the alternator. It will be noted that when the motor is under-excited it works with a lagging power factor, whereas the alternator under similar conditions of excitation works with a leading power factor; and that when the synchronous motor is over-excited it works with a leading power factor, whereas the alternator works with a lagging power factor.

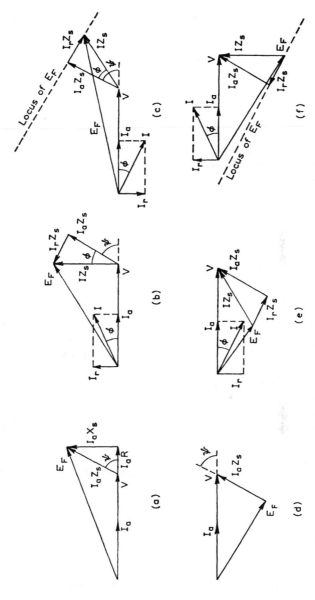

Fig. 12.12 SYNCHRONOUS GENERATOR DELIVERING CONSTANT POWER TO INFINITE BUSBARS

417

For a constant power output as a generator the voltage drop I_aZ_s will be constant. If the excitation is now varied, I_a remains unchanged but I_r, and hence the I_rZ_s drop, will change in size. This voltage drop is always directed at right angles to the I_aZ_s voltage drop, however, so that the locus of E_F as the excitation varies at constant power is the straight line perpendicular to I_aZ_s, as shown in Fig. 12.12(c).

In the same way the locus of E_F for constant load in a motor is the straight line perpendicular to I_aZ_s as shown in Fig. 12.12(f).

In each case illustrated in Fig. 12.12, the line joining the end point of the complexor V to the end point of the complexor E_F represents the overall internal voltage drop in the machine, i.e. IZ_s. It should be noted that the angle between complexors I_aZ_s and IZ_s is in every case the phase angle, ϕ, of the resultant load current.

12.14 General Load Diagram

Fig. 12.13 shows the general load diagram of a synchronous machine connected to infinite busbars. In the diagram OV, represents the constant busbar voltage; VC, displaced from OV by the angle $\psi = \tan^{-1}(X_s/R)$, the phase angle of the synchronous impedance of the machine, represents a voltage drop, I_aZ_s, caused by the full (100 per cent) load active component of current when the machine acts as a generator. If the machine is working at unity power factor, the e.m.f. E_F, is represented by OC. If the machine works at other than unity power factor, the voltage drop caused by the reactive component of current (I_rZ_s) must be at right angles to VC, since I_r is at right angles to I_a. Hence the locus of E_F for constant full (100 per cent) load is AD, drawn at right angles to VC, passing through C.

The line ZZ is VC produced in both directions.

At 50 per cent load the active component of current will be half its value at 100 per cent load. The I_aZ_s drop at 50 per cent load is therefore half the value corresponding to 100 per cent load. Thus the locus of E_F for constant 50 per cent load is the straight line drawn parallel to AD and passing through the mid-point of VC. A series of constant-power lines for operation both as a generator and a motor may be drawn using the same principle and are shown in Fig. 12.13. The no-load line is the one which passes through the extremity of vector OV as shown.

If the machine is working at unity power factor there is no reactive current and hence no I_rZ_s voltage drop. The line ZZ therefore represents the locus of E_F for unity power factor at any load. Z_1Z_1 is a similar locus for power factor 0·866 leading, and Z_2Z_2 for power factor 0·866 lagging.

The locus of E_F for constant excitation is evidently a semicircle with centre O. Loci for 50 per cent, 100 per cent, 150 per cent and 200 per cent normal excitation are shown.

Referring to Fig. 12.13, suppose the machine has 100 per cent excitation and is on no-load; E_F is then coincident with V. Suppose the mechanical power input to the machine is increased; it must now act as a generator delivering electrical power to the busbars.

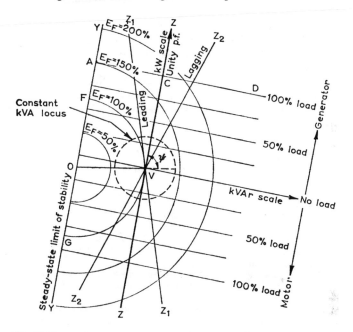

Fig. 12.13 GENERAL LOAD DIAGRAM FOR A SYNCHRONOUS
MACHINE CONNECTED TO INFINITE BUSBARS

Physically the rotor is displaced slightly forward in the direction of rotation with respect to the instantaneous pole-centres of the stator field. In other words, E_F advances in phase on V along the circular locus marked "100 per cent excitation". As the mechanical power input is increased, there is an increasing phase displacement between E_F and V until a stage is reached where E_F has the position along the line YY, which is parallel to ZZ through O. If the mechanical power input were further increased E_F would tend to swing beyond YY. However, the extremity of E_F would then approach the 50 per cent load line instead of moving further away from it and the electrical power fed to the busbars would tend to decrease. The excess mechanical energy fed into the machine must then be absorbed by the machine increasing its speed and breaking from synchronism with the constant-frequency system. YY therefore represents the steady-state limit of stability of the machine. It gives the maximum load which the machine may deliver for a given excitation when the load is very gradually applied. For suddenly applied loads the limit is somewhat lower (see Section 15.6).

When the machine acts as a motor, the rotor poles are displaced backwards against the direction of rotation and E_F lags behind V when the load is increased. Thus the steady-state limit of stability is reached at OG along YY when the excitation is 100 per cent.

The general load diagram is based on the assumption that the synchronous impedance Z_s is a constant. Since the value of Z_s is affected by saturation, numerical results obtained from the diagram are only approximate.

The broken circle (Fig. 12.13) whose centre is the end point of V and whose radius is the internal voltage drop (IZ_s to scale) will represent the locus of E_F for a constant volt-ampere value.

EXAMPLE 12.4   An 11 kV 3-phase star-connected turbo-alternator delivers 200A at unity power factor when connected to constant-voltage constant-frequency busbars.

   (a) Determine the armature current and power factor at which the machine works when the mechanical input to the machine is increased by 100 per cent, the excitation remaining unchanged.

   (b) Determine the armature current and power factor at which the machine works when the excitation is raised by 20 per cent, the power input remaining doubled.

   (c) Determine the maximum power output and corresponding armature current and power factor at this new value of excitation, i.e. as in (b).

   The armature resistance is 0·4Ω/phase and the synchronous reactance is 8Ω/phase.

   Assume that the efficiency remains constant.

   The problem is best solved graphically; an analytical solution is tedious in this case where the armature resistance is not neglected. The graphical solution is shown in Fig. 12.14.

   Using phase values,

$$V_{ph} = \frac{11,000}{\sqrt{3}} = 6,350\,\text{V}$$

OV is drawn as reference complexor 6,350V in length to a suitable scale.

$$Z_s = 0·4 + j8 = 8·00\underline{/87·1°}\,\Omega$$

V*b* is drawn making an angle of 87·1° with OV as shown. This is the unity-power-factor line.

$$I_a Z_s = 200 \times 8 = 1,600\,\text{V}$$

V*a* is cut off along the unity-power-factor line 1,600 V in length to scale. O*a* represents the e.m.f. under the initial conditions stated.

$$E_{F1} = Oa = 6,600 \text{ V to scale}$$

The line at right angles to V*a* passing through *a* is the constant-power line corresponding to an active component of 200 A.

(*a*) When the mechanical power input is doubled the power output must be doubled and therefore the active component of current, $I_a{}'$, is doubled.

$$I_a{}'Z_s = 400 \times 8 = 3,200 \text{ V}$$

V*b* is cut off along the unity-power-factor line 3,200 V in length to scale.

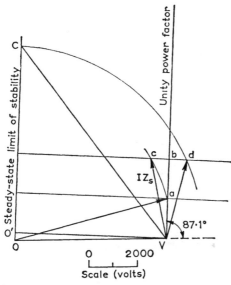

Fig. 12.14

A line at right angles to V*b* passing through *b* is drawn. This is the constant-power line. With centre O and radius O*a* an arc is drawn to intersect this constant-power line in *c*. Let the resultant current be *I*. Then

$$IZ_s = Vc = 3,300 \text{ V to scale}$$

Thus

$$I = \frac{3,300}{8} = 412 \text{ A}$$

Power factor, $\cos \phi = \dfrac{I_a{}'Z_s}{IZ_s} = \dfrac{3,200}{3,300} = 0.97 \text{ leading}$

(*b*) The new value of e.m.f. is $1.2 \times 6,600 = 7,920 \text{ V}$.

With centre O and radius $E_{F2} = 7,920 \text{ V}$ to scale an arc is struck to cut the constant-power line in *d*, and OC, the steady-state limit of stability line, in C.

$$Vd = 3,300 \text{ V} \quad \text{(by measurement)}$$

New current, $I_3 = \dfrac{3,300}{8} = 412\text{A}$

Power factor, $\cos \phi = \dfrac{I_4'Z_s}{I_3Z_s} = \dfrac{3,200}{3,300} = 0\text{-}97 \text{ lagging}$

(c) At the steady-state limit of stability for this excitation (point C), the current will be I_4, where $I_4Z_s = \text{CV} = 9,900\text{V}$. Thus

$$I_4 = \dfrac{9,900}{8} = 1,235\text{A}$$

$I_{a4}Z_s = \text{CO}' = 7,600\text{V}$

Power factor, $\cos \phi = \dfrac{I_{a4}Z_s}{I_4Z_s} = \dfrac{7,600}{9,900} = 0\text{-}768 \text{ leading}$

Maximum power output $= \sqrt{3}VI \cos \phi$

$$= \dfrac{\sqrt{3} \times 11,000 \times 1,235 \times 0\text{-}768}{1,000} = 18,100\text{kW}$$

12.15 Power/Angle Characteristic of a Synchronous Machine

Fig. 12.15(a) is part of the general load diagram for a synchronous machine and shows the complexor diagram corresponding to generation into infinite busbars at a lagging power factor. Fig. 12.15(b) is

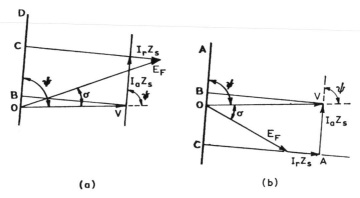

(a)   (b)

Fig. 12.15   POWER TRANSFER FOR A SYNCHRONOUS MACHINE
(a) Generator   (b) Motor

the corresponding complexor diagram for motor operation also at a lagging power factor. The power transfer is

$$P = 3VI \cos \phi = 3VI_a \qquad\qquad (12.33)$$

where V is the phase voltage and I is the phase current.

The projection of the complexors of Fig. 12.15(a) on the steady-state limit of stability line OD gives

$$I_a Z_s = E_F \cos (\psi - \sigma) - V \cos \psi \tag{12.34}$$

Substituting the expression for I_a obtained from eqn. (12.34) in eqn. (12.33) gives

$$P = \frac{3V}{Z_s} \{E_F \cos (\psi - \sigma) - V \cos \psi\} \tag{12.35}$$

Following the same procedure for motor action and using Fig. 12.15 (b) the power transfer is found to be

$$P = \frac{3V}{Zs} \{V \cos \psi - E_F \cos (\psi + \sigma)\} \tag{12.36}$$

Evidently eqn. (12.36) will cover both generator action and motor action if the power transfer P and the load angle σ are taken, conventionally, to be positive for generator action and negative for motor action.

Since, for steady-state operation, the speed of a synchronous machine is constant, the torque developed is

$$T = \frac{P}{2\pi n_0} = \frac{3}{2\pi n_0} \frac{V}{Z_s} \{E_F \cos (\psi - \sigma) - V \cos \psi\} \tag{12.37}$$

In many synchronous machines $X_s \gg R$, in which case $Z_s/\psi \approx X_s/90°$. When this approximation is permissible eqn. (12.35) becomes

$$P = \frac{3V}{Z_s} \{E_F \cos (90° - \sigma) - V \cos 90°\}$$

$$= \frac{3VE_F}{X_s} \sin \sigma \tag{12.38}$$

Similarly eqn. (12.37) becomes

$$T = \frac{3}{2\pi n_0} \frac{VE_F}{X_s} \sin \sigma \tag{12.39}$$

The power/load-angle (or torque/load-angle) characteristic is shown in Fig. 12.16. The dotted parts of this characteristic refer to operation beyond the steady-state limit of stability. Usually stable operation cannot be obtained beyond this limit, so that if the load angle exceeds ±90° the operation is dynamic with the machine either

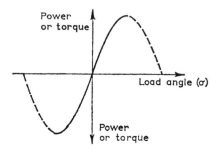

Fig. 12.16   POWER/LOAD-ANGLE AND TORQUE/LOAD-ANGLE
CHARACTERISTICS OF A SYNCHRONOUS MACHINE
CONNECTED TO INFINITE BUSBARS

accelerating or decelerating. In this case eqns. (12.38) and (12.39) are only approximately true.

12.16 Synchronizing Power and Synchronizing Torque Coefficients

A synchronous machine, whether a generator or a motor, when synchronized to infinite busbars has an inherent tendency to remain synchronized.

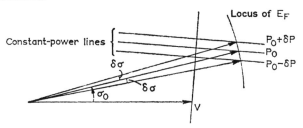

Fig. 12.17   DETERMINATION OF SYNCHRONIZING POWER
COEFFICIENT

In Fig. 12.17, which applies to generator operation at a lagging power factor, the complexor diagram is part of the general load diagram. At a steady load angle σ_0 the steady power transfer is P_0.

Suppose that, due to a transient disturbance, the rotor of the machine accelerates, so that the load angle increases by $\delta\sigma$. This alters the operating point of the machine to a new constant-power line and the load on the machine increases to $P_0 + \delta P$. Since the steady power input remains unchanged, this additional load retards the machine and brings it back to synchronism.

Similarly, if owing to a transient disturbance, the rotor decelerates so that the load angle decreases, the load on the machine is thereby reduced to $P_0 - \delta P$. This reduction in load causes the rotor to accelerate and the machine is again brought back to synchronism.

Clearly the effectiveness of this inherent correcting action depends on the extent of the change in power transfer for a given change in load angle. A measure of this effectiveness is given by the *synchronizing power coefficient*, which is defined as

$$P_s = \frac{dP}{d\sigma} \tag{12.40}$$

From eqn. (12.35),

$$P = \frac{3V}{Z_s} \{E_F \cos{(\psi - \sigma)} - V \cos{\psi}\} \tag{12.35}$$

so that

$$P_s = \frac{dP}{d\sigma} = \frac{3VE_F}{Z_s} \sin{(\psi - \sigma)} \tag{12.41}$$

Similarly the synchronizing torque coefficient is defined as

$$T_s = \frac{dT}{d\sigma} = \frac{1}{2\pi n_0} \frac{dP}{d\sigma} \tag{12.42}$$

From eqn. (12.42), therefore,

$$T_s = \frac{3}{2\pi n_0} \frac{VE_F}{Z_s} \sin{(\psi - \sigma)} \tag{12.43}$$

In many synchronous machines $X_s \gg R$, in which case eqns. (12.42) and (12.43) become

$$P_s = \frac{3VE_F}{X_s} \cos{\sigma} \tag{12.44}$$

$$T_s = \frac{3}{2\pi n_0} \frac{VE_F}{X_s} \cos{\sigma} \tag{12.45}$$

Eqns. (12.44) and (12.45) show that the restoring action is greatest when $\sigma = 0$, i.e. on no-load. The restoring action is zero when $\sigma = \pm 90°$. At these values of load angle the machine would be at the steady-state limit of stability and in a condition of unstable equilibrium. It is impossible, therefore, to run a machine at the steady-state limit of stability since its ability to resist small changes is zero unless the machine is provided with a special fast-acting excitation system.

EXAMPLE 12.5 A 2 MVA 3-phase 8-pole alternator is connected to 6,000 V 50 Hz busbars and has a synchronous reactance of 4Ω per phase. Calculate the synchronizing power and synchronizing torque per mechanical degree of rotor displacement at no-load. (Assume normal excitation.)

The synchronizing power coefficient is

$$P_s = \frac{3VE_F}{X_s} \cos \sigma \tag{12.44}$$

On no-load the load angle $\sigma = 0$.

Since there are 4 pole-pairs, 1 mechanical degree of displacement is equivalent to 4 electrical degrees; therefore

$$P_s = 3 \times \frac{6,000}{\sqrt{3}} \times \frac{6,000}{\sqrt{3} \times 4} \times \frac{4}{1,000} \times \frac{\pi}{180} = \underline{\underline{627\,kW/mech.\ deg}}$$

Synchronous speed of alternator, $n_0 = \dfrac{f}{p} = 12\cdot5$ rev/s

Thus

$$2\pi n_0 T_s = 627 \times 10^3$$

and

Synchronizing torque, $T_s = \underline{\underline{8,000\,N\text{-}m/mech.\ deg}}$

12.17 Oscillation of Synchronous Machines

In the previous sections, transient accelerations or decelerations of an alternator rotor were assumed in order to investigate the synchronizing power and synchronizing torque. Such transients may be caused by irregularities in the driving torque of the prime mover or, in the case of a motor, by irregularities in the load torque, or by irregularities in other machines connected in parallel, or by sudden changes in load.

Normally the inherent stability of alternators when running in parallel quickly restores the steady-state condition, but if the effect is sufficiently marked, the machine may break from synchronism. Moreover, if the disturbance is cyclic in effect, recurring at regular intervals, it will produce forced oscillations in the machine rotor. If the frequency of this cyclic disturbance approaches the value of the natural frequency of the rotor, when connected to the busbar system, the rotor may be subject to continuous oscillation and may eventually break from synchronism. This continuous oscillation of the rotor (periods of acceleration and deceleration) is sometimes known as *phase swinging* or *hunting*.

Fig. 12.18 shows the torque/load-angle characteristic of a synchronous generator. The steady input torque is T_0, corresponding to a steady-state load angle σ_0. Suppose a transient disturbance occurs such as to make the rotor depart from the steady state by σ'. Let σ' be sufficiently small to assume that the synchronizing

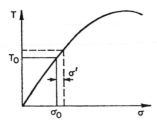

Fig. 12.18   OSCILLATION OF A SYNCHRONOUS MACHINE CONNECTED
TO INFINITE BUSBARS

torque is constant; i.e. the torque/load-angle characteristic is assumed to be linear over the range of σ' considered.

Let T_s = Synchronizing torque coefficient (N-m/mech. rad)
σ' = Load angle deviation from steady-state position (mech. rad)
J = Moment of inertia of rotating system (kg-m^2)

Assuming that there is no damping,

$$J\frac{d^2\sigma'}{dt^2} = -T_s\sigma' \tag{12.46}$$

The solution of this differential equation is

$$\sigma' = \sigma_m' \sin\left(\sqrt{\frac{T_s}{J}}\, t + \psi\right) \tag{12.47}$$

From eqn. (12.47), the frequency of undamped oscillation is

$$f = \frac{1}{2\pi}\sqrt{\frac{T_s}{J}} \tag{12.48}$$

Synchronous machines intended for operation on infinite busbars are provided with damping windings in order to prevent the sustained oscillations predicted by eqn. (12.48).

In salient-pole machines the damping winding takes the form of a short-circuited cage consisting of copper bars of relatively large cross-section embedded in the rotor pole-face. In cylindrical-rotor machines the solid rotor provides considerable damping, but a cage winding may also be provided. This consists of copper fingers inserted in the rotor slots below the slot wedges and joined together at each end of the rotor. The currents induced in the damping bars give a damping torque which prevents continuous oscillation of the rotor.

EXAMPLE 12.6   A 3-phase 3·3 kV 2-pole 3,000 rev/min 934 kW synchronous motor has an efficiency of 0·95 p.u. and delivers full-load torque with its excitation adjusted so that the input power factor is unity. The moment of inertia of the motor and its load is 30 kg-m², and its synchronous impedance is $(0 + j11·1)\Omega$. Determine the period of undamped oscillation on full-load for small changes in load angle.

$$\text{Input current, } I = \frac{934 \times 10^3}{\sqrt{3} \times 3·3 \times 10^3 \times 0·95} = 172\,\text{A}$$

Taking the phase voltage as reference,

$$E_F = V - IX_s$$
$$= \frac{3·3 \times 10^3}{\sqrt{3}}\,\underline{/0°} - (172\underline{/0°} \times 11·1\underline{/90°}) = 2,700\underline{/-45°}\,\text{V}$$

The synchronizing torque coefficient is

$$T_s = \frac{3}{2\pi n_0}\frac{VE_F}{X_s}\cos\sigma \tag{12.45}$$
$$= \frac{3}{2\pi 50} \times \frac{3·3 \times 10^3}{\sqrt{3}} \times \frac{2,700}{11·1} \times 0·707 = 3·14 \times 10^3\,\text{N-m/rad}$$

The undamped frequency of oscillation is

$$f = \frac{1}{2\pi}\sqrt{\frac{T_s}{J}}$$

The period of oscillation is

$$T = \frac{1}{f} = 2\pi\sqrt{\frac{30}{3·14 \times 10^3}} = \underline{\underline{0·612\,\text{s}}}$$

12.18   Synchronous Motors

A synchronous motor will not develop a driving torque unless it is running at synchronous speed, since at any other speed the field poles will alternately be acting on the effective N and S poles of the rotating field and only a pulsating torque will be produced. For starting either (*a*) the induction motor principle or (*b*) a separate starting motor must be used. If the latter method is used the machine must be run up to synchronous speed and synchronized as an alternator. To obviate this trouble, synchronous motors are usually started as induction motors, and have a squirrel-cage winding embedded in the rotor pole faces to give the required action. When the machine has run up to almost synchronous speed the d.c. excitation is switched on to the rotor, and it then pulls into synchronism. The induction motor action then ceases (see Chapter 13).

The starting difficulties of a synchronous motor severely limit its usefulness—it may only be used where the load may be reduced for starting and where starting is infrequent. Once started, the motor has the advantage of running at a constant speed with any desired power factor. Typical applications of synchronous motors are the driving of ventilation or pumping machinery where the machines run almost continuously. Synchronous motors are often run with no load to utilize their leading power factor characteristic for power factor correction or voltage control. In these applications the machine is called a synchronous phase modifier.

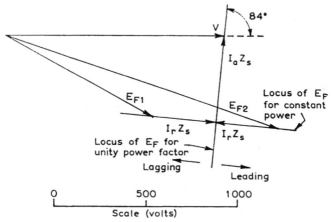

Fig. 12.19

EXAMPLE 12.7 A 2,000 V 3-phase 4-pole star-connected synchronous machine has resistance and synchronous reactance per phase of $0.2\,\Omega$ and $1.9\,\Omega$ respectively.

Calculate the e.m.f. and the rotor displacement when the machine acts as a motor with an input of 800 kW at power factors of 0.8 lagging and leading.

If a field current of 40 A is required to produce an e.m.f. per phase equal to rated phase voltage, determine also the field current for each condition.

Synchronous impedance, $Z_s = 0.2 + j1.9 = 1.91\underline{/84°}\,\Omega/\text{phase}$

Constant phase terminal voltage, $V = \dfrac{2,000}{\sqrt{3}} = 1,150\,\text{V}$

Total phase current in both cases $= \dfrac{800 \times 10^3}{\sqrt{3} \times 2,000 \times 0.8} = 288\,\text{A}$

Active component of current in both cases, $I_a = 288 \times 0.8 = 230\,\text{A}$

Reactive component of current in both cases, $I_r = 288 \times 0.6 = 173\,\text{A}$

$I_a Z_s = 230 \times 1.91 = 440\,\text{V}$

$I_r Z_s = 173 \times 1.91 = 330\,\text{V}$

Fig. 12.19 is now drawn to scale for the motoring condition.

At the lagging power factor the excitation voltage is measured from the complexor diagram as $E_{F1} = 880 \text{V/phase.}$

Field current required, $I_{F1} = 40 \times \dfrac{880}{1,150} = 30.5 \text{A}$

The rotor displacement is the phase angle between E_{F1} and V with the rotor lagging for motor action as previously described. Therefore at the lagging p.f.

Rotor angle $= 27°_e = 13.5°_m$ for a 4-pole machine

At the leading p.f. the excitation voltage, $E_{F2} = 1,520 \text{V/phase}$

Field current required, $I_{F2} = 52.9 \text{A}$

Rotor angle $= 17°_e = 8.5°_m$

EXAMPLE 12.8 A 2,000 V 3-phase 4-pole star-connected synchronous motor runs at 1,500 rev/min. The excitation is constant and gives an e.m.f. per phase of 1,150 V. The resistance is negligible compared with the synchronous reactance of 3Ω per phase.

Determine the power input, power factor and torque developed for an armature current of 200 A.

Synchronous impedance $= j3 = 3\underline{/90°}\,\Omega/\text{phase}$

Phase voltage, $V = \dfrac{2,000}{\sqrt{3}} = 1,150 \text{V}$

E.M.F./phase, $E_F = 1,150 \text{V}$

$IZ_s = 200 \times 3 = 600 \text{V}$

In Fig. 12.20 V represents the phase voltage taken as reference complexor.

A circular arc whose radius represents the open-circuit voltage of 1,150 V is the locus of E_F for constant excitation.

AB is the locus of E_F for unity power factor operation; in this case AB is perpendicular to V since the phase angle of Z is 90°.

A circle whose radius represents 600 V is the locus of E_F for constant kVA operation. For the actual operating conditions E_F must lie at the intersection of the two circles.

From the diagram,

$I_aZ_s = 580 \text{V}$

Active component of current, $I_a = 193 \text{A}$

Therefore

Total power input $= \dfrac{3VI_a}{1,000} = \dfrac{3 \times 1,150 \times 193}{1,000} = 666 \text{kW}$

Operating power factor $= \dfrac{I_a}{I} = \dfrac{193}{200} = 0.96 \text{ lagging}$

Torque developed, $T = \dfrac{3VI_a}{2\pi n_0}$

$$= \dfrac{3 \times 1150 \times 193}{2\pi \times 1{,}500} \times 60$$

$$= 4{,}250 \text{ N-m}$$

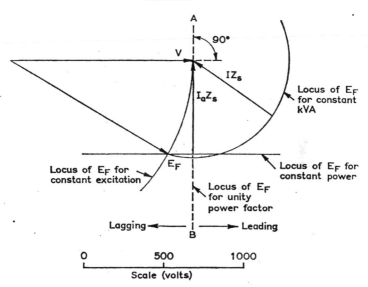

Fig. 12.20

PROBLEMS

12.1 A 3-phase 11 kV star-connected alternator has an effective armature resistance of 1 Ω and a synchronous reactance of 20 Ω per phase. Calculate the percentage regulation for a load of 1,500 kW at p.f.s of (*a*) 0·8 lagging, (*b*) unity, (*c*) 0·8 leading.

Ans. 22 per cent, 4·25 per cent, −13·4 per cent.

12.2 Describe the tests carried out in order that the synchronous impedance of an alternator can be obtained. By means of diagrams show how the synchronous impedance can be used to determine the regulation of an alternator at a particular load and power factor.

A 6,600 V 3-phase star-connected alternator has a synchronous impedance of $(0·4 + j6)$ Ω/phase. Determine the percentage regulation of the machine when supplying a load of 1,000 kW at normal voltage and p.f. (i) 0·866 lagging, (ii) unity, (iii) 0·866 leading, giving complexor diagrams in each case.   (*H.N.C.*)

Ans. 9·7 per cent, 1·84 per cent, −6·03 per cent.

12.3 Explain, with the aid of complexor diagrams, the effect of varying the excitation of a synchronous motor driving a constant-torque load.

A 3-phase 4-pole 50 Hz 2,200 V 1,870 kW star-connected synchronous motor has a synchronous impedance of $(0.06 + j0.6\,\Omega)$/phase. The motor is to be run in parallel with an inductive load of 1,000 kVA having a power factor of 0.707 lagging, and is to be so excited that the power factor of the combined loads is 0.9 lagging. If the motor output is 1,870 kW and its efficiency is 0.9 p.u., determine (*a*) the kVAr, kW and kVA input to the motor, (*b*) the input current and power factor to the motor, (*c*) the load angle of the motor in mechanical degrees, and (*d*) the field current of the motor.

Ans. 64.8 kVAr; 2,080 kW; 2,170 kVA; 570 A; 0.955 lagging; 7.78 mechanical degrees.

12.4 A 2,200 V 3-phase star-connected synchronous motor has a resistance of 0.6 Ω/phase and a synchronous reactance of 6 Ω/phase. Find graphically or otherwise the generated e.m.f. and the angular retardation of the rotor when the input is 200 kW at (*a*) a power factor of unity, (*b*) a power factor of 0.8 leading.
(*C. & G.*)

Ans. 2,200 V, 15°$_e$; 2,640 V, 13.5°$_e$.

12.5 A 400 V 3-phase 50 Hz star-connected synchronous motor has a synchronous impedance per phase of $(1 + j5)\,\Omega$. It takes a line current of 10 A at unity power factor when operating with a certain field current. If the load torque is increased until the line current is 40 A, the field current remaining unchanged, find the new power factor and the gross output power.
(*H.N.C.*)

Ans. 0.957 lagging, 25 kW.

12.6 A 150 kW 3-phase induction motor has a full-load efficiency and power factor of 0.91 and 0.89 respectively. A 3-phase star-connected synchronous motor, connected to the same mains, is to be over-excited in order to improve the resultant power factor to unity. The synchronous motor also drives a constant load, its power input being 100 kW. The line voltage is 415 V and the synchronous reactance per phase of the synchronous motor is 0.5 Ω, the resistance being negligible. Determine the induced e.m.f. per phase of the synchronous motor.
(*H.N.C.*)

Ans. 306 V.

12.7 A synchronous generator operates on constant-voltage constant-frequency busbars. Explain the effect of variation of (*a*) excitation and (*b*) steam supply on power output, power factor, armature current and load angle of the machine.
An 11 kV 3-phase star-connected synchronous generator delivers 4,000 kVA at unity power factor when running on constant-voltage constant-frequency busbars. If the excitation is raised by 20 per cent determine the kVA and power factor at which the machine now works. The steam supply is constant and the synchronous reactance is 30 Ω/phase. Neglect power losses and assume the magnetic circuit to be unsaturated.
(*L.U.*)

Ans. 4,280 kVA; 0.935 lagging.

12.8 Describe briefly the procedure for synchronizing and connecting a 3-phase alternator to constant-voltage constant-frequency busbars. How is the output of the machine adjusted?
A single-phase alternator operates on 10 kV 50 Hz busbars. The winding resistance is 1 Ω and the synchronous impedance 10 Ω. If the excitation is adjusted to give an open-circuit e.m.f. of 12 kV, what is the maximum power

output of the machine? Find the armature current and power factor for this condition. (*H.N.C.*)

Ans. 10·9 MW; 1,480 A; 0·737 leading.

12.9 Show that the maximum power that a synchronous generator can supply when connected to constant-voltage constant-frequency busbars increases with the excitation.

An 11 kV 3-phase star-connected turbo-alternator delivers 240 A at unity power factor when running on constant voltage and frequency busbars. If the excitation is increased so that the delivered current rises to 300 A, find the power factor at which the machine now works and the percentage increase in the induced e.m.f. assuming a constant steam supply and unchanged efficiency. The armature resistance is 0·5 Ω per phase and the synchronous reactance 10 Ω per phase. (*H.N.C.*)

Ans. 0·802 lagging; 24 per cent.

12.10 An 11 kV 300 MVA 3-phase alternator has a steady short-circuit current equal to half its rated value. Determine graphically or otherwise the maximum load the machine can deliver when connected to 11 kV constant-voltage constant-frequency busbars with its field excited to give an open-circuit voltage of 12·7 kV/phase. Find also the armature current and power factor corresponding to this load. Ignore armature resistance. (*H.N.C.*)

Ans. 300 MW; 17·4 kA; 0·895 leading.

12.11 An alternator having a synchronous impedance of $R + jX$ ohms/phase is supplying constant voltage and frequency busbars. Describe, with the aid of complexor diagrams, the changes in current and power factor when the excitation is varied over a wide range, the steam supply remaining unchanged. The complexor diagrams should show the locus of the induced e.m.f.

A star-connected alternator supplies 300 A at unity power factor to 6,600 V constant voltage and frequency busbars. If the induced e.m.f. is now reduced by 20 per cent, the steam supply remaining unchanged, determine the new values of the current and power factor. Assume the synchronous reactance is 5 Ω/phase, the resistance is negligible and the efficiency constant. (*H.N.C.*)

Ans. 350 A; 0·85 leading.

12.12 Deduce an expression for the synchronizing power of an alternator.

Calculate the synchronizing power in kilowatts per degree of mechanical displacement at full load for a 1,000 kVA 6,600 V 0·8 power-factor 50 Hz 8-pole star-connected alternator having a negligible resistance and a synchronous reactance of 60 per cent. (*L.U.*)

Ans. 158 kW per mechanical degree.

12.13 A 40 MVA 50 Hz 3,000 rev/min turbine-driven alternator has an equivalent moment of inertia of 1,310 kg-m², and the machine has a steady short-circuit current of four times its normal full-load current.

Deducing any formula used, estimate the frequency at which hunting may take place when the alternator is connected to an "infinite" grid system.

(*H.N.C.*)

Ans. 3·14 Hz.

12.14 An 11 kV 3-phase star-connected turbo-alternator is connected to constant-voltage constant-frequency busbars. The armature resistance is negligible and the synchronous reactance is 10 Ω. The alternator is excited to deliver 300 A at unity power factor. Determine the armature current and power factor when the excitation is increased by 25 per cent, the load on the machine being unchanged.

(*H.N.C.*)

Ans. 356A; 0·843 lagging.

12.15 Derive an expression for the synchronous reactance per phase of a 3-phase cylindrical-rotor synchronous machine assuming the armature m.m.f. is sinusoidally distributed, the armature winding uniform and the air-gap uniform. Neglect armature leakage flux and all reluctance except that of the air-gap.

Using these assumptions, calculate the approximate synchronous reactance per phase of a 3-phase 200 MVA 11 kV 2-pole 50 Hz star-connected synchronous generator having a stator internal diameter of 1 m, a core length of 4·6 m and an air-gap length of 4·7 cm. A flux per pole of 5 Wb is required to give an e.m.f. equal to rated voltage.

Ans. 1·21 Ω.

12.16 If a synchronous generator operating on infinite busbars has a fast-acting field excitation system, it is possible to operate the machine beyond the steady-state limit of stability at load angles in excess of 90 electrical degrees.

A 3-phase 75 MVA 11 kV star-connected synchronous generator has such an excitation system. Determine the maximum leading MVAr the machine can supply to the infinite busbars to which the machine is synchronized without exceeding its rating. The load on the machine is 30 MW. What is the load angle for this condition?

The synchronous reactance is 3 Ω/phase and the resistance is negligible.

Ans. 68·5 MVAr leading; 128°.

12.17 Explain why a synchronous machine synchronized to infinite busbars tends to remain synchronized.

A 3-phase 75 MVA 11·8 kV 50 Hz 2-pole synchronous generator has a period of oscillation of 1·3 s when synchronized to infinite busbars, the excitation and steam supply to the prime mover being so adjusted that there is no current transfer under steady conditions.

Neglecting the effect of damping, calculate the moment of inertia of the rotating system. The synchronous reactance per phase may be taken as 3 Ω and the resistance per phase to be negligible.

Ans. 5×10^3 kg-m².

Chapter 13

INDUCTION MACHINES

It was shown in Chapter 11 that, when a polyphase stator winding is excited from a balanced polyphase supply, a stator m.m.f. distribution is set up and travels at synchronous speed given by eqn. (11.10) as

$$n_0 = \frac{f}{p} \tag{13.1}$$

Associated with the stator m.m.f. distribution is a flux density distribution which also travels at synchronous speed and is often referred to as a "rotating field".

The stator field induces voltages in the rotor phase windings so that a rotor m.m.f. distribution and an associated flux density distribution are set up. The rotor distributions travel at the same speed as the stator distribution. The axes of the stator and rotor distributions have an angular displacement, and as a result a torque acts on the rotor and causes it to accelerate in the same direction as the stator field.

The steady-state rotor speed is normally slightly less than synchronous so that the motor runs with a *per-unit slip*, s, defined as

$$s = \frac{n_0 - n_r}{n_0} \tag{13.2}$$

where n_r is the rotor speed.

At standstill, $n_r = 0$ and $s = 1$. For the rotor to reach synchronous speed ($n_r = n_0$ and $s = 0$), an external drive is necessary,

435

since for this condition there is no rotor e.m.f. and hence no rotor current or torque. If the rotor is driven so that $n_r > n_0$, the slip becomes negative, the rotor torque opposes the external driving torque and the machine acts as an induction generator.

In all cases the slip speed is

$$n_s = n_0 - n_r \tag{13.3}$$

From eqn. (13.2),

$$n_s = sn_0 \tag{13.4}$$

and

$$n_r = (1 - s)n_0 \tag{13.5}$$

The frequency of the rotor e.m.f.s and currents is proportional to the difference in speed between the rotating field and the rotor, so that

$$f_r = (n_0 - n_r)p = sn_0 p = sf$$

where p is the number of pole pairs. Hence

$$\frac{f_r}{f} = s \tag{13.6}$$

Similarly for the angular frequencies corresponding to f_r and f:

$$\frac{\omega_r}{\omega_0} = \frac{2\pi f_r}{2\pi f} = s \tag{13.7}$$

The 3-phase induction motor has a torque characteristic similar to that of the d.c. shunt motor, is robust, and is low in initial cost. Other forms of asynchronous machine are the a.c. commutator motor, which gives a wide range of speed control, and various types of single-phase motor, which are employed for fractional-horsepower drives, in individual units, and in traction.

13.1 Construction

The induction machine consists essentially of a stator, which carries a 3-phase winding, and a rotor. The stator winding is a 3-phase winding of one of the types described in Chapter 11, often being a narrow-spread mesh-connected closed winding. The winding is laid in open or half-closed slots in a laminated silicon-steel core.

The rotor winding is placed in half-closed or closed slots, the air-gap between stator and rotor being reduced to a minimum. There are two main types of rotor, the *wound rotor* and the *squirrel-cage rotor*. In the squirrel-cage rotor, solid conducting rods are inserted into closed slots, and at each end the rods are connected to a heavy short-circuiting ring. This forms a permanently short-circuited winding which is practically indestructible. In some smaller machines the conductors, end rings and fan are cast in one piece in aluminium. The cage rotor is cheap and robust, but suffers from the disadvantage of a low starting torque.

The wound rotor has a 3-phase winding with the same number of poles as the stator; the ends of the rotor winding may be brought out to three slip rings. The advantage of the wound-rotor machine is that an external starting resistance can be connected to the slip rings to give a large starting torque. This resistance is reduced to zero as the machine runs up to speed.

13.2 Equivalent Circuit of Induction Machine at Any Slip

The approximate equivalent circuit per phase of a polyphase induction machine at standstill ($s = 1$) is shown in Fig. 13.1. The equivalent circuit takes the same form as that adopted for the power transformer, since at standstill the induction machine consists of two polyphase windings linked by a common flux.

Unlike that of a power transformer, the magnetic circuit of the induction machine has an air-gap, and this makes the per-unit

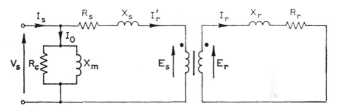

Fig. 13.1 EQUIVALENT CIRCUIT OF THE POLYPHASE INDUCTION
MACHINE AT STANDSTILL

value of magnetizing current much higher than that of the power transformer. As a result the approximation of showing the shunt magnetizing branch of the equivalent circuit at the input terminals is less close than for the power transformer. The approximation is nevertheless acceptable for large machines, but not for small machines. To keep the magnetizing current as small as possible, the air-gap length of induction machines is made as short as is consistent with mechanical considerations.

A further difference between the polyphase induction machine and the power transformer is that in the former the windings are distributed, and this affects the effective turns ratio.

In this and subsequent sections it is assumed that the rotor has a 3-phase winding. A cage rotor is, in effect, a rotor with a large number of short-circuited phases. Such an arrangement may be represented by an equivalent 3-phase winding; I_r is not then the current in an actual rotor phase, but the stator current I_s is preserved as the true stator current.

The induced stator e.m.f. per phase when connected to a supply of frequency f hertz is, from eqn (11.20),

$$E_s = K_{ds}K_{ss}\frac{\omega_0\Phi N_s}{\sqrt{2}} \tag{13.8}$$

where $\omega_0 = 2\pi f = 2\pi n_0 p$.

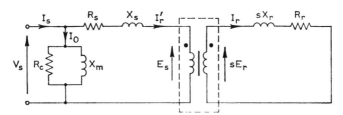

Fig. 13.2   EQUIVALENT CIRCUIT OF THE POLYPHASE INDUCTION MACHINE AT ANY SLIP, s

At standstill the frequency of the rotor e.m.f. per phase is the supply frequency and the rotor e.m.f. per phase is, from eqn (11.20),

$$E_r = K_{dr}K_{sr}\frac{\omega_0\Phi N_r}{\sqrt{2}} \tag{13.9}$$

When the rotor rotates the rotor e.m.f. per phase is altered both in size and frequency.

$$\left.\begin{array}{c}\text{Rotor e.m.f. per phase} \\ \text{at any slip } s\end{array}\right\} = K_{dr}K_{sr}\frac{\omega_r\Phi N_r}{\sqrt{2}}$$

$$= K_{dr}K_{sr}\frac{s\omega_0\Phi N_r}{\sqrt{2}}$$

$$= sE_r$$

The rotor reactance per phase at standstill is X_r. At any slip s, therefore, the rotor reactance per phase will be sX_r, since reactance is proportional to frequency. The equivalent circuit per phase of a polyphase inductor motor at any slip s is shown in Fig. 13.2.

13.3 Slip Ratios

The element enclosed by the dotted box in Fig. 13.2 represents an ideal or lossless induction machine. This differs from the ideal transformer considered in Chapter 9 in that (a) the current and voltage transformation ratios differ, and (b) the frequencies of the voltages and currents at the input and output terminal pairs of the ideal element also differ.

From eqns. (13.8) and (13.9) the effective turns ratio, k_t, at standstill ($s = 1$) is

$$k_t = \frac{E_s}{E_r} = \frac{K_{ds}K_{ss}N_s}{K_{dr}K_{sr}N_r} \tag{13.10a}$$

At any slip s the voltage ratio is

$$\frac{E_s}{sE_r} = \frac{k_t}{s} \tag{13.10b}$$

At any slip s, m.m.f., balance must exist between the stator and rotor phase windings so that

$$I_r' \, K_{ds}K_{ss}N_s = I_r K_{dr}K_{sr} \, N_r$$

or

$$\frac{I_r'}{I_r} = \frac{K_{dr}K_{sr}N_r}{K_{ds}K_{ss}N_s} = \frac{1}{k_t} \tag{13.11}$$

Assuming there are three phases on both the stator and rotor,

$$\left.\begin{array}{l}\text{Power absorbed by ideal}\\ \text{stator winding}\end{array}\right\} P_0 = 3E_s I_r' \cos \phi_r \tag{13.12}$$

This power is obtained from the supply when the machine acts as a motor, and from the prime mover driving the rotor when it acts as a generator.

$$\left.\begin{array}{l}\text{Power dissipated in}\\ \text{the rotor circuit}\end{array}\right\} P_r = 3sE_r I_r \cos \phi_r$$

$$= 3s \frac{E_s}{k_t} k_t I_r' \cos \phi_r$$

$$= 3sE_s I_r' \cos \phi_r \tag{13.13}$$

Dividing eqn. (13.12) by eqn. (13.13),

$$\frac{P_0}{P_r} = \frac{1}{s} \tag{13.14}$$

The power dissipated in the rotor circuit consists of winding loss in the rotor circuit and core loss in the rotor magnetic circuit. Since the core loss varies with frequency this implies that the equivalent circuit-element, R_r, is frequency dependent. Under normal running conditions, for plain induction motors, however, the rotor frequency and rotor core loss are low and the latter may usually be neglected. The power dissipated in the rotor is obtained from the ideal stator winding when the machine acts as a motor and from the prime mover when the machine acts as a generator.

When the machine acts as a motor the power absorbed by the ideal stator windings is greater than that dissipated in the rotor circuit except when the rotor is stationary (standstill) when they are equal. The difference in these two powers appears as gross mechanical power output:

$$\text{Mechanical power, } P_m = P_0 - P_r = P_0 - sP_0$$

or

$$P_m = P_0(1 - s) \tag{13.15}$$

Combining eqns. (13.14) and (13.15),

$$P_0:P_r:P_m = 1:s:(1 - s) \tag{13.16}$$

When the machine acts as a generator the net mechanical power input is the sum of the stator and rotor powers:

$$\text{Mechanical power, } P_m = P_0 + P_r$$

or

$$P_m = P_0(1 + s) \tag{13.17}$$

For generator action, therefore,

$$P_0:P_r:P_m = 1:s:(1 + s) \tag{13.18}$$

Eqn. (13.16) will serve for both motor and generator action if the slip s, the power absorbed by the ideal stator winding P_0 and the mechanical power P_m are taken to be negative for generator action, and it is remembered that P_m is the gross mechanical power output for the motoring mode and the net power input for the generating mode. Figs. 13.3(a) and (b) are block diagrams representing the power transfer in a plain induction machine for motor and generator action.

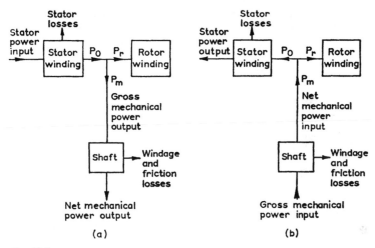

Fig. 13.3 **POWER TRANSFER IN A PLAIN INDUCTION MACHINE**

(a) Motoring mode (b) Generating mode

EXAMPLE 13.1 A 37·3 kW 4-pole 50 Hz induction machine has a friction and windage torque of 22 N-m. The stator losses equal the rotor circuit loss. Calculate:

(a) The input power to the stator when delivering full-load output at a speed of 1,440 rev/min.

(b) The gross input torque and stator output power when running at a speed of 1,560 rev/min. The stator losses are as in (a) and the windage and friction torque is unchanged.

(a) Synchronous speed $= \dfrac{f}{p} \times 60 = \dfrac{50 \times 60}{2} = 1{,}500 \, \text{rev/min}$

Per-unit slip, $s = \dfrac{n_0 - n_r}{n_r} = \dfrac{1{,}500 - 1{,}440}{1{,}500} = 0{\cdot}04$

Windage and friction loss $= 2\pi n_r T = \dfrac{2\pi \times 1{,}440 \times 22}{60} = 3{,}320 \, \text{W}$

Gross mechanical power output, $P_m = 37{,}300 + 3{,}320 = 40{,}620 \, \text{W}$

Power absorbed by ideal stator winding, $P_0 = \dfrac{P_m}{1-s} = \dfrac{40{,}620}{0{\cdot}96} = 42{,}300 \, \text{W}$

Stator losses $=$ Rotor loss $= sP_0 = 0{\cdot}04 \times 42{,}300 = 1{,}690 \, \text{W}$

Stator input power $= 42{,}300 + 1{,}690 = \underline{\underline{44{,}000 \, \text{W}}}$

(b) Per-unit slip $= \dfrac{1{,}500 - 1{,}560}{1{,}500} = -0{\cdot}04$

That is, the machine is now operating as an induction generator. Since the rotor circuit loss and the stator losses are equal and the latter are unchanged,

Rotor circuit loss, $P_r = 1{,}600\,\mathrm{W}$

Net mechanical power input, $P_m = \dfrac{1+s}{s}\,P_r = \dfrac{1{\cdot}04}{0{\cdot}04} \times 1{,}690 = 44{,}000\mathrm{W}$

Torque corresponding to net mechanical power input $= \dfrac{44{,}000 \times 60}{2\pi \times 1{,}560}$

$$= 269\,\mathrm{N\text{-}m}$$

Gross input torque $= 269 + 22 = \underline{\underline{291\,\mathrm{N\text{-}m}}}$

Power absorbed by ideal stator winding, $P_0 = \dfrac{P_m}{1+s} = \dfrac{44{,}000}{1{\cdot}04} = 42{,}300\,\mathrm{W}$

Stator output power $= 42{,}300 - 1{,}690 = \underline{\underline{40{,}300\,\mathrm{W}}}$

13.4 Transformer Equivalent Circuit of the Induction Machine

Referring to the equivalent circuit of Fig. 13.2, the rotor current per equivalent phase is

$$I_r = \frac{sE_r}{R_r + jsX_r}$$

If the numerator and denominator are divided by s this gives

$$I_r = \frac{E_r}{\dfrac{R_r}{s} + jX_r} \tag{13.19}$$

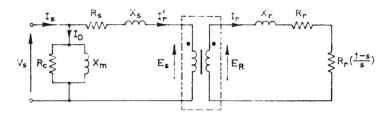

Fig. 13.4 TRANSFORMER EQUIVALENT CIRCUIT OF THE INDUCTION MACHINE

This latter expression for the rotor current per equivalent phase is consistent with the rotor equivalent circuit shown in Fig. 13.4. Although the value of I_r is unchanged this equivalent circuit is significantly different from that of Fig. 13.2 in that both the induced voltage and the reactance per equivalent rotor phase have their standstill ($s = 1$) values. Nor do the voltage and current ratios now differ as they did in the equivalent circuit of Fig. 13.1. Both are now equal to the virtual turns ratio, k_t. The element enclosed in the dotted box of Fig. 13.4 represents an ideal transformer. Therefore the power absorbed by the ideal stator winding must be equalled by the power delivered by the ideal rotor winding to R_r/s. Thus the power dissipated in the rotor equivalent circuit of Fig. 13.4 must be both the rotor loss and the gross mechanical power output. This may be shown to be so as follows.

$$\left.\begin{array}{l}\text{Added resistance per equivalent}\\ \text{rotor phase}\end{array}\right\} = \frac{R_r}{s} - R_r = R_r\left(\frac{1-s}{s}\right)$$

$$\left.\begin{array}{l}\text{Power dissipated in}\\ \text{added rotor resistance}\end{array}\right\} = 3I_r{}^2 R_r\left(\frac{1-s}{s}\right) = P_r\left(\frac{1-s}{s}\right) = P_m$$

That is, the equivalent circuit has additional resistance of $R_r\left(\dfrac{1-s}{s}\right)$ in each phase and the power dissipated in these additional resistances is equal to the gross mechanical power output. Fig. 13.4 shows the equivalent circuit of an induction machine which is also the equivalent circuit of a transformer the power dissipated in whose secondary load is equal to the gross mechanical power output of the induction machine when operating in the motoring mode.

Since the equivalent circuit of Fig. 13.4 is that of a transformer, the secondary values may be referred to the primary by multiplying them by $k_t{}^2$, the square of the virtual turns ratio. The turns ratio is defined in eqn. (13.10a). Fig 13.5 shows the referred equivalent circuit in which

$$\frac{R_r{}'}{s} = k_t{}^2\frac{R_r}{s} \quad \text{and} \quad X_r{}' = k_t{}^2 X_r$$

It should be noted that the equivalent circuits are valid only if the variations in speed or slip are relatively slow. Usually the moment of inertia of the rotor is sufficiently large for this condition to be realized.

13.5 Torque developed by an Induction Machine

An expression for the torque developed by an induction machine may be obtained by reference to the equivalent circuit of Fig. 13.5. Assuming that the stator winding has three phases,

Rotor circuit loss, $P_r = 3(I_r{}')^2 R_r{}'$ (13.20)

$$\left.\begin{array}{l}\text{Power absorbed by ideal}\\ \text{stator winding}\end{array}\right\} P_0 = \frac{P_r}{s} = 3(I_r{}')^2\frac{R_r{}'}{s} \qquad (13.21)$$

Gross mechanical power output $\left.\right\}$ $P_m = 3(I_r')^2 R_r' \dfrac{1-s}{s}$ (13.22)

Gross torque developed, $T = \dfrac{3}{2\pi n_r} (I_r')^2 R_r' \dfrac{1-s}{s}$ (13.23)

where $n_r = (1-s)n_0$ (13.5)

Substituting for n_r in eqn. (13.23),

$$T = \frac{3}{2\pi n_0} (I_r')^2 \frac{R_r'}{s}$$ (13.24)

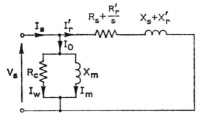

Fig. 13.5 REFERRED TRANSFORMER EQUIVALENT CIRCUIT OF : ··. INDUCTION MACHINE

Comparing eqn. (13.23) with eqn. (13.21) it will be seen that

$$T = \frac{P_0}{2\pi n_0}$$ (13.25)

Thus the torque developed is proportional to the power, P_0, absorbed by the ideal stator winding . The quantity P_0 is sometimes referred to as the torque measured in "synchronous watts", which presumably implies that, if this power is divided by the synchronous angular velocity, the torque is obtained.

Referring to the equivalent circuit of Fig. 13.5, evidently

$$I_r' = \frac{V_s}{\sqrt{\left\{\left(R_s + \dfrac{R_r'}{s}\right)^2 + (X_s + X_r')^2\right\}}}$$

Substituting for $(I_r')^2$ in eqn. (13.24),

$$T = \frac{3}{2\pi n_0} \frac{V_s^2}{\left(R_s + \dfrac{R_r'}{s}\right)^2 + (X_s + X_r')^2} \frac{R_r'}{s}$$ (13.26)

Multiplying numerator and denominator by s gives

$$T = \frac{3V_s^2}{2\pi n_0} \frac{sR_r'}{(sR_s + R_r')^2 + s^2(X_s + X_r')^2} \tag{13.27}$$

If the stator impedance is neglected, this equation reduces to

$$T = \frac{3V_s^2}{2\pi n_0} \frac{sR_r'}{(R_r')^2 + s^2(X_r')^2} \tag{13.28}$$

Eqns. (13.27) and (13.28) have been obtained by considering motor action, but they apply equally to generator action. If the torque is taken to be positive when the machine acts as a motor it will be negative for generator action since the slip then becomes negative. For motor action eqn. (13.28) gives the gross torque developed, but for generator action it gives the net input torque, as will be evident from a consideration of Figs. 13.2(a) and (b), the torque in each case being $P_m/2\pi n_r$.

13.6 Slip/Torque Characteristics of the Induction Machine

Referring to the expression for torque given by eqn. (13.27), since the slip s is positive in both the numerator and the denominator, the

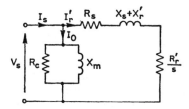

Fig. 13.6   PERTAINING TO MAXIMUM DEVELOPED TORQUE

torque will be zero both at $s = 0$ and $s = \infty$. Therefore, the torque will be a maximum at some intermediate value of slip.

From eqn. (13.24),

$$T = \frac{3}{2\pi n_0} (I_r')^2 \frac{R_r'}{s} \tag{13.24}$$

Therefore the torque will be a maximum when there is maximum power transfer into the load R_r'/s. As shown in Fig. 13.6, the source impedance is $R_s + j(X_s + X_r')$ assuming the supply itself to have zero internal impedance. According to Section 2.4 the maximum power is transferred to R_r'/s when

$$\frac{R_r'}{s} = \pm\sqrt{\{R_s^2 + (X_s + X_r')^2\}}$$

or

$$s = \pm \frac{R_r'}{\sqrt{\{R_s^2 + (X_s + X_r')^2\}}} \tag{13.29}$$

The plus sign refers to motor action, the minus sign to generator action.

An expression for maximum torque may be obtained by substituting in eqn. (13.27) the value of s given in eqn. (13.29). The algebra is a little simplified if in the first instance a substitution for $s^2(X_s + X_r')^2$ is made. From eqn. (13.29),

$$s^2(X_s + X_r')^2 = (R_r')^2 - s^2 R_s^2$$

In eqn. (13.27),

$$T_{max} = \frac{3V_s^2}{2\pi n_0} \frac{sR_r'}{(sR_s + R_r')^2 + (R_r')^2 - s^2 R_s^2}$$

$$= \frac{3V_s^2}{2\pi n_0} \frac{1}{2} \frac{s}{(sR_s + R_r')}$$

Substituting for s and simplifying further,

$$T_{max} = \pm \frac{3V_s^2}{2\pi n_0} \frac{1}{2}\left[\frac{1}{\sqrt{\{R_s^2 + (X_s + X_r')^2\}} \pm R_s}\right] \tag{13.30}$$

Here again the plus signs refer to motor action, the minus signs to generator action. It is to be noted that the maximum net input torque for generator action is greater than the maximum gross output torque for motor action. This inequality would disappear if the stator resistance were negligible. Although, as eqn. (13.29) shows, the slip at which the maximum torque occurs is proportional to the referred value of rotor resistance per stator phase, R_r', the actual value of the maximum torque in independent of R_r'. Therefore variation of R_r' changes the slip at which maximum torque occurs without affecting its value. The maximum torque is sometimes called the *pull-out torque*.

If stator impedance is neglected (i.e. $R_s = 0$, $X_s = 0$) eqn. (13.29) becomes

$$s = \pm \frac{R_r'}{X_r'} \tag{13.31}$$

and eqn. (13.30) becomes

$$T_{max} = \pm \frac{3V_s^2}{2\pi n_0} \frac{1}{2X_r'}$$ (13.32)

It will be appreciated that, to obtain a high starting torque and a high maximum torque, the combined rotor and stator leakage reactance must be small. The shorter the air-gap is made the more the leakage flux is reduced. This is an additional reason for minimizing the air-gap.

Fig. 13.7 is a typical slip/torque characteristic of an induction machine. The hatched areas in the region of $s = 0$ show the normal operating range of the machine for motor and generator action. Referring to motor action, AB represents the torque at standstill or starting torque. Provided the load torque is less than this the motor will accelerate until the developed motor torque and load torque come into equality at a speed close to but less than synchronous speed. The machine operates stably as a motor over the range indicated. If the machine is operating in this region and a load torque in excess of the maximum motoring torque, CD, is imposed on the machine, it will decelerate to standstill or stall. The range DB represents unstable motor action.

Fig. 13.7 shows positive values of slip greater than unity. To achieve such values the rotor must be coupled to a prime mover and driven in the opposite direction to that of the stator rotating field, the stator still being connected to the 3-phase supply. In such conditions of operation the machine acts as neither a motor nor a generator as it receives both electrical and mechanical input power, all the power input being dissipated as loss. This mode of operation is referred to as *brake action*.

For the machine to operate as an induction generator a prime mover must drive the rotor in the same direction as the stator rotating field but at a higher speed with the stator connected to a pre-existing supply. In Fig. 13.7 OF represents the range of stable generator action. If the input torque to the generator exceeds the maximum generating torque, EF, the machine accelerates and passes into the region of unstable generator action, FG. Unless the input torque is removed, the speed may rise dangerously.

At some negative value of slip the machine will pass from unstable generator action to brake action. This will occur when the stator I^2R loss exceeds the power absorbed by the ideal stator winding, since the machine must then draw power from the electrical supply to meet completely the stator loss as well as having mechanical power input.

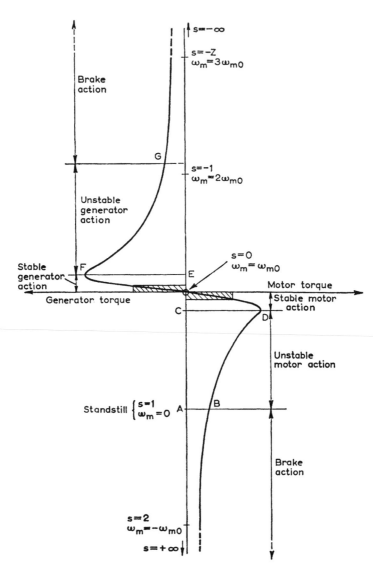

Fig. 13.7 SLIP/TORQUE CHARACTERISTIC OF THE INDUCTION MACHINE

Stator I^2R loss $= 3(I_r')^2 R_s$ (13.33)

Power absorbed by ideal stator, $P_0 = -3(I_r')^2\dfrac{R_r'}{s}$ (13.21)

P_0 is taken to be negative when the input to the ideal stator winding is derived from a mechanical source. Equating eqns. (13.33) and (13.21),

$$s = -\frac{R_r'}{R_s} \tag{13.34}$$

At this value of slip the machine changes from unstable generator to brake action. Since R_r' and R_s will be of the same order this value of slip will be approximately -1.

The slip/torque characteristic shown in Fig. 13.7 has its maximum value at a relatively small value of slip. This is typical of induction machines and is desirable, since when the machine operates as a motor the speed regulation with load is small, giving the machine a speed/torque characteristic over its operating region similar to that of the d.c. shunt motor. This matter is discussed further in Section 13.7. Further, since from eqn. (13.14) the slip is equal to $s = P_r/P_0$, an induction machine operating with a large value of slip would have a large rotor loss and consequently a low efficiency. A disadvantage of having the maximum torque occur at a low value of slip is that, as Fig. 13.7 shows, this arrangement makes the starting torque low.

EXAMPLE 13.2 A 440 V 4-pole 3-phase 50 Hz slip-ring induction motor has its stator winding mesh connected and its rotor winding star connected. The standstill voltage measured between slip rings with the rotor open-circuited is 218 V. The stator resistance per phase is $0\cdot6\,\Omega$ and the stator reactance per phase is $3\,\Omega$. The rotor resistance per phase is $0\cdot05\,\Omega$ and the rotor reactance per phase is $0\cdot25\,\Omega$. Calculate the maximum torque and the slip at which it occurs. If the ratio of full-load to maximum torque is $1:2\cdot5$ find the full-load slip and the power output.

All values must be referred to either the stator or the rotor. It is usual to refer to the stator. Since the rotor is star connected:

Induced standstill rotor voltage, $E_r = \dfrac{218}{\sqrt3} = 126\,V$

From eqn. (13.10a) the standstill turns ratio is

$k_t = \dfrac{E_s}{E_r} = \dfrac{440}{126} = 3\cdot49$

Rotor resistance/phase referred to stator, $R_r' = 0\cdot05 \times 3\cdot49^2 = 0\cdot61\,\Omega$
Rotor reactance/phase referred to stator, $X_r' = 0\cdot25 \times 3\cdot49^2 = 3\cdot05\,\Omega$

From eqn. (13.29) the slip for maximum torque is

$$s = \frac{R_r'}{\sqrt{\{(R_s')^2 + (X_s \times X_r')^2\}}} = \frac{0\cdot61}{\sqrt{(0\cdot6^2 + 6\cdot05^2)}} = \underline{\underline{0\cdot1}}$$

Synchronous speed, $n_0 = \dfrac{f}{p} = \dfrac{50}{2} = 25\,\text{rev/s}$

From eqn. (13.30) the maximum torque is

$$T_{max} = \frac{3V_s^2}{2\pi n_0}\frac{1}{2}\left[\frac{1}{\sqrt{\{R_s^2 + (X_s + X_r')^2\}} + R_s}\right]$$

$$= \frac{3 \times 440^2}{2\pi 25}\frac{1}{2}\frac{1}{6\cdot06 + 0\cdot6} = 278\,\text{N-m}$$

Full-load torque $= \dfrac{278}{2\cdot5} = 111\,\text{N-m}$

The slip for full-load torque may be obtained from eqn. (13.27):

$$T = \frac{3V_s^2}{2\pi n_0}\frac{sR_r'}{(sR_s + R_r')^2 + s^2(X_s + X_r')^2} \qquad (13.27)$$

i.e.

$$111 = \frac{3 \times 440^2}{2\pi \times 25}\frac{0\cdot61s}{(0\cdot6s + 0\cdot61)^2 + 6\cdot05^2s^2}$$

This gives

$$s^2 - 0\cdot53s + 0\cdot01 = 0$$

so that

$$s = \frac{0\cdot53 \pm \sqrt{(0\cdot53^2 - 4 \times 0\cdot01)}}{2} = 0\cdot02 \text{ or } 0\cdot51$$

Since the slip for maximum torque is $0\cdot1$, $s = 0\cdot02$ is on the stable part of the slip/torque characteristic and $s = 0\cdot51$ is on the unstable part. Selecting the value of s giving stable operation,

Power output, $P_m = 2\pi n_r T = 2\pi \times 25(1 - 0\cdot02) \times 111 = 17\cdot1\,\text{kW}$

13.7 Starting

SLIP-RING MACHINES

To obtain a satisfactory operating characteristic giving a reasonable efficiency and a small speed regulation with load, the slip for the maximum torque developed by an induction motor must have a value in the range from $0\cdot1$ to $0\cdot2$, as explained at the end of Section 13.6. From eqn. (13.29) the slip for maximum torque is

$$s = \frac{R_r'}{\sqrt{\{R_s^2 + (X_s + X_r')^2\}}} = \text{from } 0\cdot1 \text{ to } 0\cdot2$$

At starting $s = 1$, and to obtain maximum torque on starting,

$$\frac{R_r'}{\sqrt{\{R_s^2 + (X_s + X_r')^2\}}} = 1$$

In the plain-cage-rotor induction motor these conflicting requirements cannot well be met, though specially shaped rotor slots or a
double-cage rotor may make this possible. In slip-ring machines,

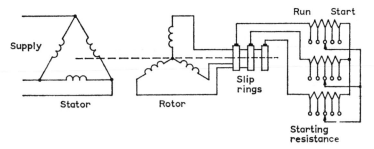

Fig. 13.8 STARTING OF WOUND-ROTOR MACHINES

however, the rotor resistance per phase is such as to give a satisfactory operating characteristic.

Slip-ring machines are invariably started by means of external
resistances connected through the slip rings to the rotor circuit
(Fig. 13.8). The machine is started with all the resistances in,
giving a high starting torque. As the machine runs up to speed the

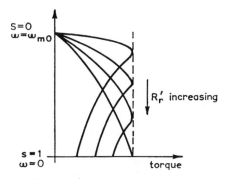

Fig. 13.9 SLIP/TORQUE CURVES FOR ROTOR-RESISTANCE STARTING

external resistance is reduced until the machine attains full speed
with no external resistance.

Fig. 13.9 shows the slip/torque curves of a slip-ring induction
motor corresponding to various positions of the starting resistance.

NON-SLIP-RING MACHINES

Stator starting must be used for cage-rotor machines, since no
connexion can be made to the rotor, and direct switching of large

machines would cause huge starting currents, which must be avoided. The cage-rotor machine suffers from the disadvantage that the starting torque is low if the resistance is low, while the efficiency is reduced if the rotor resistance is high. Various methods of starting will be examined.

Direct-on-line starting. In small workshops, direct-on-line starting may be restricted to motors of 2 kW or less, but in large industrial premises the tendency is to "direct switch" whenever possible, direct-on starting of motors of 40 kW at medium voltages being common.

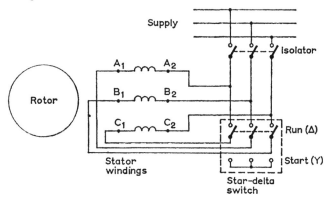

Fig. 13.10 STAR-DELTA STARTING

Reduced-voltage starters, described below, may be used to limit the initial starting torque and thus to reduce the shock to the driven machine.

Star-delta starting. In this type of starter the stator is star-connected for running up to speed, and is then delta connected. The applied voltage per phase in star is only $1/\sqrt{3}$ of the value which would be applied if the windings were connected in delta; hence, from eqn. (13.27), the starting torque is reduced to 1/3. The phase current in star is $1/\sqrt{3}$ of its value in delta, so that the line current for star connexion is 1/3 of the value for delta. Fig. 13.10 shows a connexion diagram for a star-delta starter.

Auto-transformer starting. In auto-transformer starting the transformer has at least three tappings giving open-circuit voltages of not less than 40, 60 and 75 per cent of line voltage for starting, and the stator is switched directly to the mains when the motor has run up to speed (Fig. 13.11). If the fractional tapping is x, then the applied voltage per phase on starting is xV_1 (where V_1 is the mains voltage), and the starting torque is reduced by x^2. The starting current from the mains will also be reduced by approximately x^2.

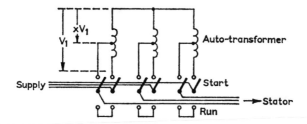

Fig. 13.11 AUTO-TRANSFORMER STARTING

EXAMPLE 13.3   A 3-phase squirrel-cage induction motor has a stator resistance per phase of $0.5\,\Omega$ and a rotor resistance per phase referred to the stator of $0.5\,\Omega$. The total standstill reactance per phase referred to the stator is $4.92\,\Omega$. If the ratio of maximum torque to full-load torque is $2:1$, find the ratio of actual starting to full-load torque for (*a*) direct starting, (*b*) star-delta starting and (*c*) auto-transformer starting with a tapping of 75 per cent.

The maximum torque is given by eqn. (13.30):

$$T_{max} = \pm \frac{3V_s^2}{2\pi n_0} \frac{1}{2} \left[\frac{1}{\sqrt{\{R_s^2 + (X_s + X_r')^2\}} \pm R_s} \right]$$

For motor action

$$T_{max} = k \frac{V_s^2}{2} \frac{1}{\sqrt{(0.5^2 + 4.92^2)} + 0.5} = \frac{kV_s^2}{10}$$

Full-load torque, $T_{FL} = \tfrac{1}{2}T_{max} = \dfrac{kV_s^2}{20}$

(*a*) *Direct-on-line starting.* The starting torque is obtained by substituting $s = 1$ in eqn. (13.27), which is

$$T = \frac{3V_s^2}{2\pi n_0} \frac{sR_r'}{(sR_s + R_r')^2 + s^2(X_s + X_r')^2} \qquad (13.27)$$

The starting torque is

$$T_0 = kV_s^2 \frac{0.5}{(0.5 + 0.5)^2 + 4.92^2} = \frac{kV_s^2}{2 \times 25.2}$$

Therefore

$$\frac{\text{Starting torque}}{\text{Full-load torque}} = \frac{T_0}{T_{FL}} = \frac{kV_s^2}{2 \times 25.2} \times \frac{20}{kV_s^2} = \underline{\underline{0.397}}$$

(*b*) *Star-delta starting.* The effective phase voltage is reduced to $1/\sqrt{3}$ of its original value. Therefore

$$T_0 = \left(\frac{1}{\sqrt{3}} \right)^2 \text{ of } T_0 \text{ for direct starting, and}$$

$$\frac{\text{Starting torque}}{\text{Full-load torque}} = \frac{0.397}{3} = \underline{\underline{0.132}}$$

(c) *Auto-transformer starting.* The effective phase voltage is reduced to 0·75 of its original value. Thus

$$T_0 = 0·75^2 \text{ of } T_0 \text{ for direct starting}$$

and

$$\frac{\text{Starting torque}}{\text{Full-load torque}} = 0·397 \times 0·75^2 = \underline{\underline{0·223}}$$

Where a high starting torque is required from a squirrel-cage motor, it may be achieved by a double-cage arrangement of the rotor conductors, as shown in Fig. 13.12(a). The equivalent electrical

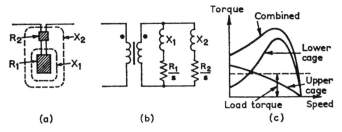

Fig. 13.12   IMPROVEMENT IN STARTING TORQUE OF CAGE ROTORS
(a) Double-cage rotor
(b) Equivalent circuit
(c) Combined torque/speed characteristics

rotor circuit is shown at (b), where X_1 and X_2 are leakage reactances. This equivalent circuit neglects mutual inductance between the cages. For the upper cage the resistance is made intentionally high, giving a high starting torque, while for the lower cage the resistance is low, and the leakage reactance is high, giving a low starting torque but high efficiency on load. The resultant characteristic will be roughly the sum of these two as shown in Fig. 13.12(c).

If a 3-phase induction motor starts in the wrong direction, this can be remedied by interchanging any two of the three supply leads to the stator.

13.8 Stability and Crawling

Curve (a) in Fig. 13.13 is the torque/speed curve for a typical induction motor. Consider that this motor is required to drive a constant-torque load having the torque/speed characteristic illustrated by curve (b). T_0, the starting torque with direct-on switching, is greater than the load torque T_b; thus there will be an excess starting torque (T_0-T_b) which will accelerate the motor and the load. The acceleration at any speed will be proportional to the torque difference $(T-T_b)$

so that the acceleration will be a maximum when the driving torque, T, is a maximum; and the acceleration will be zero, i.e. a steady speed will be obtained, when the speed corresponds to the operating point A. This is a stable operating point, since if the speed rose by a small amount Δn_r from its value at A, the load torque would exceed the driving torque and there would be a deceleration back to the

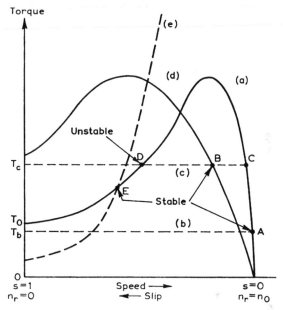

Fig. 13.13 PERTAINING TO STABILTY

     - - - Load characteristics
     —— Motor characteristics

speed at A, and vice versa for a decrease in speed. The conclusions from this argument are:

(a) The operating point must be at the intersection of the two torque/speed characteristics.

(b) The slope of the load torque/speed curve must be greater than that of the driving-motor torque/speed curve for the operating point to be stable; i.e.

$$\frac{dT}{dn_r} \text{ for the load} > \frac{dT}{dn_r} \text{ for the drive}$$

Curve (c) in Fig. 13.13 represents a second load. In this case the load torque T_c is greater than the starting torque T_0 of the motor.

and with direct-on switching the motor would fail to start. The motor could be started by the use of additional rotor resistance sufficient to give the motor the characteristic of curve (*d*). The motor would drive the load at the speed corresponding to point C when the additional rotor resistance is short-circuited. Operation at point C may be unsatisfactory, since C is relatively near the maximum torque point and fluctuations in the load might too easily stall the motor. If the fluctuation does not cause the speed to fall

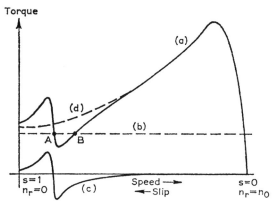

Fig. 13.14   TORQUE/SPEED CHARACTERISTICS TO ILLUSTRATE
CRAWLING

    (*a*) Resultant of (*c*) and (*d*)
    (*b*) Constant-torque characteristic
    (*c*) Characteristic due to 7th harmonic flux density
    (*d*) Characteristic due to fundamental flux density

below that at D the motor should accelerate the load back to its speed at C when the load torque returns to normal. Since dT/dn_r for the load is not greater than dT/dn_r for the drive at D, this is an unstable operating point; i.e. if some random cause makes the speed fall slightly the load torque will exceed the driving torque and cause a further reduction in speed, or vice versa. The portion of the normal characteristic curve (*a*) which lies to the left of the maximum-torque point is called the *nominally unstable* portion. Though it is not normally possible for a motor to operate at a point on the nominally unstable portion of its characteristic, this may be arranged if a load, such as a fan, with a steep rising characteristic is chosen. A particular case is represented by curve (*e*); the motor would drive this load at a speed corresponding to the point E.

An induction motor may sometimes run in a stable manner at a low speed on a constant-torque load. This can be the result of a kink in the normal torque/speed characteristic. In Fig. 13.14

± ∞, $R_r'/s = 0$, so that any point P_s on the impedance

, $R_r'/s = R_r' \equiv P_\infty P_1$, and P_s is at P_1.
sitive slip s,

P_s

$$P_\infty P_s - P_\infty P_1 = \frac{R_r'}{s} - R_r' = R_r'\left(\frac{1-s}{s}\right)$$

$$P_1:P_1P_s = R_s:R_r':R_r'\left(\frac{1-s}{s}\right) \tag{13.36}$$

X_r' is the minimum value of the impedance $Z + R_r'/s$
corresponding maximum value of current $V_s/(X_s + X_r')$.
OB as diameter is the locus of $V_s/(Z + R_r'/s)$ with
when V_s is taken as reference complexor. Points Q_∞,
Q_s' on this locus corresponding to points P_∞, P_1, P_s
ectively. The subscripts refer to the values of slip to
ints correspond.

= +∞, a condition not practically attainable since it
the rotor is driven at infinite speed, any point P_s on the
ocus is at P_∞, and any point Q_s on the admittance locus
dingly at Q_∞. As the value of s decreases, P_s moves to
P_1, corresponding to $s = 1$ (standstill), and Q_s moves
he range between P_∞ and P_1, the mode of operation is
n as explained in Section 13.6. As the slip decreases
P_s moves to the right of P_1 and correspondingly Q_s
d the circular locus towards O. If the slip were to become
uld be an infinite distance to the right of P_1 and Q_s would
For the range of slip from +1 to almost zero the mode of
s motor action.

= −∞, a condition also not practically realizable, P_s
d Q_s at Q_∞. As s takes up smaller and smaller negative
moves to the left of P_∞ and Q_s moves anticlockwise
circular locus towards O, which it would reach when
nfinite distance to the left of P_∞.

L POWER LINE

.16 Q_sT is drawn parallel to the reference axis cutting
, OQ_∞ in S and meeting OB in T. Q_1V is also drawn
the reference axis, cutting OQ_∞ in U and meeting OB in
tly,

$$UQ_1 = AP_\infty:P_\infty P_1 = R_s:R_r' \tag{13.37}$$

curve (a) shows a torque/speed curve with such a kink, and curve
(b) represents a constant-torque load. The intersection A of curves
(a) and (b) represents a stable operating point, so that the machine
would not run up to full speed but merely drive the load at the speed
corresponding to point A. This is termed *crawling*. To make the
motor run up to full speed the load would have to be reduced to a
value less than that of the minimum occurring between A and B. The
kinks are due to irregularities (such as teeth) in the machine which
accentuate the effect of space harmonics in the flux density distribution. Curve (c) in Fig. 13.14 shows the slip/torque characteristic due to the 7th space harmonic, which rotates at a speed of
$n_0/7$ in the same direction as the fundamental. When the rotor speed
n_r is less than $n_0/7$ this space harmonic produces a motoring torque,
but when n_r is greater than $n_0/7$ it produces a generating torque.
When this torque/slip characteristic is added to the torque/slip
characteristic due to the fundamental (curve (d)) a kink in the
resultant torque/slip characteristic occurs.

13.9 Stator Current Locus of the Induction Machine

Fig. 13.15(a) shows the approximate referred equivalent circuit of
the induction machine. The impedance Z_s is

$$Z_s = R_s + \frac{R_r'}{s} + j(X_s + X_r') \tag{13.35}$$

As the slip s varies the locus of Z_s is that of an impedance of fixed
reactance $X_s + X_r'$ and of variable resistance. In Fig. 13.15(b)
W'AW represents the locus of Z_s. For any positive slip s, OP_s
represents the impedance Z_s in modulus and phase, OA represents
the fixed reactance $X_s + X_r'$, and AP_s represents the particular
resistance $R_s + R_r'/s$. For negative values of slip points such
as P_s' fall to the left of A and represent induction generator
action.

As shown in Section 1.7, if the complexor representing the current
I_r' in the impedance Z_s, is drawn from the origin O and V_s is taken as
reference complexor, the extremity of I_r' must lie on a circle of diameter $V_s/(X_s + X_r')$. This locus is shown in Fig. 13.15(b). OB is
the current $V_s/(X_s + X_r')$ when the slip takes up a value such as to
make $R_s + R_r'/s = 0$.

For any value of Z_s (such as OP_s making an angle θ with the first
reference axis), the corresponding value of I_r' is found by drawing
OQ_s making an angle $-\theta$ with the first reference axis, as shown in

Fig. 13.15(*b*). OQ_s' is the current complexor corresponding to the impedance OP_s' drawn for a negative slip.

To obtain the total stator current I_s the fixed current I_0 must be added to I_r', whose value varies as s varies. This may be done most

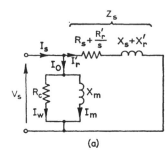

(a)

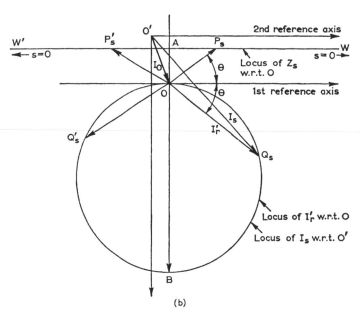

(b)

Fig. 13.15   STATOR CURRENT LOCUS OF THE INDUCTION MACHINE

(*a*) Equivalent circuit at any slip, s
(*b*) Impedance and current loci

conveniently by shifting the origin of measurement for the current by an amount equal to I_0 (to scale) from O to O' as shown in Fig. 13.15(*b*). The locus of I_s is then the same circle as the locus of I_r', but the origin of measurement for I_s is O', while that for I_r' is O.

13.10 Torque and Mecha
Circle Diagram

Fig. 13.16 again shows th

$$Z_s = R_s + \frac{R_r'}{s} + j($$

and of the current, I_r', fl

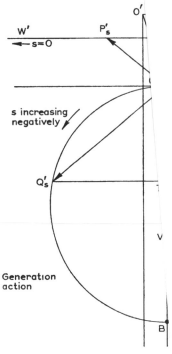

Fig. 13.16   TORQUE AND MECHANICAL
INDUCTION-MACHINE CIRC

sake of clarity the size of th
no-load current I_0 has been exa

OP_∞ represents the fixed im
that $AP_\infty = R_s$ and $OA = X_s$ +
$Z + R_r'/s$ as the slip s varies fr
W'P_∞W parallel to the reference
lies to the right of P_∞ for posi
and to the left of P_∞ for negativ

When $s = $
locus is at P_∞
When $s = $
For any po

$$\frac{R_r'}{s} = P$$

$$P_1P_s = $$

$$AP_\infty : P_\infty$$

$OA = X_s + $
and OB is the
The circle o
respect to O
Q_1, Q_s and
and P_s' resp
which the p

When $s = $
implies that
impedance l
is correspon
the right to
to Q_1. In t
brake actio
from unity,
moves roun
zero, P_s wo
move to O.
operation i

When s
is at P_∞ an
values, P_s
round the
P_s was an

MECHANICA

In Fig. 13
OQ_1 in R
parallel to
V. Evide

VU:

Also

$$\cot \theta = \frac{VQ_1}{OV} = \frac{AP_1}{OA} = \frac{R_s + R_r'}{X_s + X_r'} \qquad (13.38)$$

$$I_r' = OQ_s = \frac{V_s}{R_s + \dfrac{R_r'}{s} + j(X_s + X_r')}$$

V_s is the reference complexor, i.e. $V_s = V\underline{/0°}$.

$$OQ_s = \frac{V_s}{Z_s^2}(R_s + R_r'/s) - j\frac{V_s}{Z_s^2}(X_s + X_r')$$

or

$$OQ_s = TQ_s - jOT$$

$$TR = OT\frac{VQ_1}{OV} = \frac{V_s}{Z_s^2}(X_s + X_r')\frac{R_s + R_r'}{X_s + X_r'} = \frac{V_s}{Z_s^2}(R_s + R_r')$$

$$RQ_s = TQ_s - TR = \frac{V_s}{Z_s^2}(R_s + R_r'/s) - \frac{V_s}{Z_s^2}(R_s + R_r')$$

$$= \frac{V_s}{Z_s^2} R_r'\left(\frac{1-s}{s}\right)$$

The mechanical power is

$$P_m = 3(I_2')^2 R_r'\left(\frac{1-s}{s}\right) \qquad (13.22)$$

$$= 3\frac{V_s^2}{Z_s^2} R_r'\left(\frac{1-s}{s}\right)$$

$$= 3V_s . RQ_s \qquad (13.39)$$

Thus for any point Q_s on the current locus the distance RQ_s parallel to the reference axis is proportional to the mechanical power, so that OQ_1 is called the *mechanical power line*. For generator action $R'Q_s'$ represents the mechanical power input.

TORQUE LINE

$$SR = RQ_s \times \frac{P_\infty P_1}{P_1 P_s} = \frac{V_s}{Z_s^2} R_r'\left(\frac{1-s}{s}\right) \times \frac{R_r'}{R_r'\left(\frac{1-s}{s}\right)} = \frac{V_s}{Z_s^2} R_r'$$

Rotor loss, $P_r = 3(I_r')^2 R_r' \qquad (13.20)$

$$= 3\frac{V_s^2}{Z_s^2} R_r'$$

ı.e.

$$P_r = 3V_s . SR \tag{13.40}$$

Ideal stator power transfer, $P_0 = 3(I_r')^2 \dfrac{R_r'}{s}$ (13.21)

$$= 3 \frac{V_s^2}{Z_s^2} \left\{ R_r' + R_r' \left(\frac{1-s}{s} \right) \right\} = 3V_s(SR + RQ_s)$$

i.e.

$$P_0 = 3V_s . SQ_s \tag{13.41}$$

Torque developed, $T = \dfrac{3}{2\pi n_0} V_s . SQ_s$ (13.42)

Thus for any point Q_s on the current locus the distance SQ_s parallel to the reference axis is proportional to the developed torque, so that OQ_∞ is called the *torque line*. For generator action the torque is $S'Q_s'$.

13.11 Determination of Equivalent Circuit and Locus Diagram

The current locus may be directly determined by test, and this is usually achieved by means of a no-load test and a short-circuit test.

NO-LOAD TEST

In Fig. 13.16, $O'O$ represents the current I_0, which may be determined by operating the machine in such a way that $I_r' = 0$. Referring to Fig. 13.15(a) it will be seen that this condition could be achieved exactly by coupling the machine to a prime mover and rotating the rotor at exactly synchronous speed so that the slip was zero. In these circumstances R_r'/s is infinite and I_r' is zero. In practice it is usually sufficiently accurate to take I_0 as the input current with the machine running unloaded, when the slip will be very small and I_r' negligible compared with I_0. The test is usually performed with full voltage applied to the stator.

Let $V_0 = $ input voltage per phase on no-load
    $I_0 = $ Input current per phase on no-load
    $P_0 = $ Input power per phase on no-load

In the approximate conditions of the test it should be noted that the power P_0 is made up of stator core loss, stator I^2R loss, rotor loss, and windage and friction loss. The stator I^2R loss and the rotor loss are usually negligible, but the windage and friction loss may well be comparable to the stator core loss.

$$\cos \phi_0 = \frac{P_0}{V_0 I_0} \tag{13.43}$$

Hence the complexor I_0 of Fig. 13.17 can be drawn.

The quantities in the shunt arm of the equivalent circuit of Fig. 13.15(a) are

$$I_w = I_0 \cos \phi_0 \qquad I_m = I_0 \sin \phi_0$$

and

$$R_c = \frac{V_0}{I_w} = \frac{V_0}{I_0 \cos \phi_0} \tag{13.44}$$

$$X_m = \frac{V_0}{I_m} = \frac{V_0}{I_0 \sin \phi_0} \tag{13.45}$$

It should be noted that the power dissipated in R_c, namely $I_w^2 R_c$, represents stator core loss together with windage and friction loss for this method of testing.

LOCKED-ROTOR TEST

In Fig. 13.16, $O'Q_1$ represents the total input current I_s at $s = 1$, and this can be measured with the rotor at standstill. Since the machine produces a torque at standstill it is necessary to lock the rotor to prevent its accelerating. The test is usually performed with reduced voltage.

Let V_{sc} = Input voltage per phase at standstill
$\quad\ I_{sc}$ = Input current per phase at standstill
$\quad\ P_{sc}$ = Input power per phase at standstill

The power input P_{sc} includes stator I^2R loss, stator core loss and rotor loss. The stator core loss will be negligible, however, since V_{sc} will be a small fraction of the rated value, whereas I_{sc} will be of the same order as the rated current.

$$\cos \phi_{sc} = \frac{P_{sc}}{V_{sc} I_{sc}} \tag{13.46}$$

The standstill stator current if full voltage were applied would be

$$I_s = I_{sc} \frac{V_s}{V_{sc}} \tag{13.47}$$

Hence the complexor I_s $(O'Q_1)$ of Fig. 13.17 may be drawn.

The input current I_s is $I_0 + I_r'$. Since I_0 is normally negligible compared with I_r' at standstill, the input impedance at standstill may be taken as

$$Z_1 = R_s + R_r' + j(X_s + X_r')$$

$$= \frac{V_{sc}}{I_{sc}} \cos \phi_{sc} + j \frac{V_{sc}}{I_{sc}} \sin \phi_{sc}$$

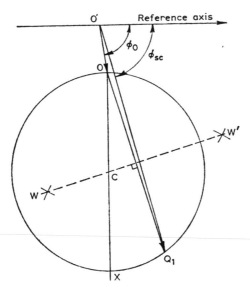

Fig. 13.17   CONSTRUCTION OF THE CURRENT LOCUS DIAGRAM
FROM TEST RESULTS

Therefore

$$R_s + R_r' = \frac{V_{sc}}{I_{sc}} \cos \phi_{sc} \tag{13.48}$$

$$X_s + X_r' = \frac{V_{sc}}{I_{sc}} \sin \phi_{sc} \tag{13.49}$$

R_s may be separated from R_r' by performing a d.c. resistance test on the stator winding.

In Fig. 13.17 $O'O$ represents I_0 and $O'Q_1$ represents the value of I_s at standstill with normal voltage applied. Both O and Q_1 lie on the circular locus, so OQ_1 is joined and represents a chord of the circle. WW' is the perpendicular bisector of the chord OQ_1. The centre, C, of the circular locus lies along WW′ where it intersects OX, the line drawn through O perpendicular to the reference axis.

EXAMPLE 13.4   A 2·25 kW 440 V 4-pole 50 Hz 3-phase mesh-connected induction motor gave the following test results at the stator terminals.

(i) Standstill test (rotor locked): 142 V, 5 A, 205 W.
(ii) No-load test: 440 V, 2·21 A, 122 W.

The ratio of stator: rotor I^2R loss at standstill is 1·2 : 1.
From these data construct the circle diagram for the machine, and hence determine

(a) Full-load stator current, power factor, slip, speed and efficiency.
(b) Pull-out torque, and the slip at which it occurs for both motor and generator action.
(c) The maximum mechanical power for both motor and generator action.
(d) The starting torque for direct-on-line starting and star/delta starting.

From the standstill test the standstill current is

$$I_s = I_{sc} \times \frac{V_s}{V_{sc}} = 5 \times \frac{440}{142} = 15\cdot5 \,\text{A}$$

and the power factor is

$$\cos\phi_{sc} = \frac{P_{sc}}{\sqrt{3}V_{sc}I_{sc}} = \frac{205}{\sqrt{3}\times142\times5} = 0\cdot167 \text{ lagging}$$

whence

$$\phi_{sc} = 80°\ 30' \text{ lagging}$$
$$I_s = 15\cdot5\underline{/-80°\ 30'}\,\text{A}$$

From the no-load test,

$$\cos\phi_0 = \frac{122}{\sqrt{3}\times440\times2\cdot21} = 0\cdot0725 \text{ lagging}$$

whence

$$\phi_0 = 85°\ 50' \text{ lagging}$$
Open-circuit line current, $I_0 = 2\cdot21\underline{/-85°\ 50'}\,\text{A}$

The circle diagram of the machine is shown in Fig. 13.18. The diagram is drawn for line values of current, and the circular locus is obtained by the method explained in Section 13.11.
I_s and I_0 are drawn as the complexors O′O and O′Q₁. The perpendicular bisector of the chord OQ₁ is drawn to cut OX in C, the centre of the circular locus. OX is the line through O perpendicular to the reference axis. A point U is chosen in VQ₁ according to eqn. (13.37) such that

$$\frac{VU}{UQ_1} = \frac{\text{Stator winding loss}}{\text{Rotor winding loss}} = \frac{R_s}{R_r'} = 1\cdot2$$

OQ₁ is the mechanical power line, and OU is the torque line. OZ is an axis drawn through O parallel to the reference axis.

(a) Full-load power output = 2·25 kW
Active component of current corresponding to full load output

$$= \frac{2,250}{\sqrt{3} \times 440} = 2·93\,\text{A}$$

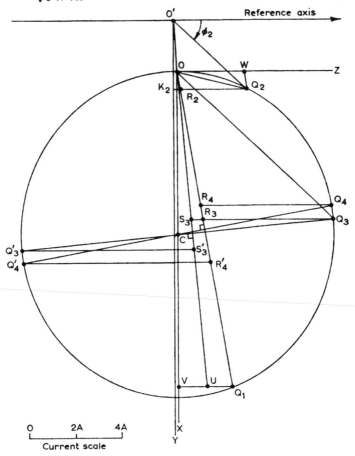

Fig. 13.18

In Fig. 13.18, OW is drawn equal to 2·93A to scale. WQ₂ is drawn parallel to the mechanical power line to cut the circle in Q₂. OQ₂ is joined Q₂K₂ is drawn parallel to the reference axis to meet O′Y in K₂ and to cut the mechanical power line in R₂.

Full-load stator current = $O'Q_2$ = 4·30 A

Input power factor = $\cos \phi_2 = \cos 41·5 = 0·748$ lagging

From eqn. (13.14) the slip is

$$s = \frac{P_r}{P_0}$$

Fig. 13.19 shows an enlargement of the area at the origin of Fig. 13.18. In that diagram

$$s = \frac{S_2R_2}{S_2Q_2} = \frac{0.12}{5.94} = 0.02\,\text{p.u.}$$

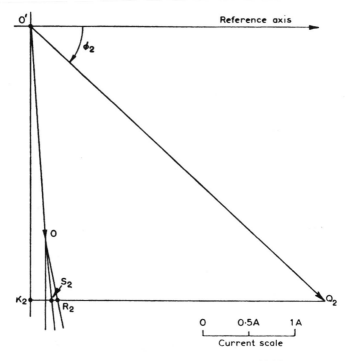

O′

Reference axis

ϕ_2

O

S_2

K_2

R_2

Q_2

O   0·5A   1 A

Current scale

Fig. 13.19   ENLARGEMENT OF AREA AT ORIGIN OF FIG. 13.18

The speed is

$$n_r = (1 - s)n_0 = (1 - 0.02) \times \tfrac{50}{2} = 24.5\,\text{rev/s or } 1{,}470\,\text{rev/min}$$

It should be realized that Q_2K_2 is the reference or active component of the input current $O'Q_2$. Of the current Q_2K_2, the component R_2Q_2 is equal to OW, which corresponds to the output. Hence the component R_2K_2 corresponds to losses, and Q_2K_2 corresponds to the electrical power input. Therefore

$$\text{Efficiency} = \frac{Q_2R_2}{Q_2K_2} = \frac{1.47}{1.60} = 0.92\,\text{p.u.}$$

(b) The pull-out torque is the maximum torque. Hence two points, Q_3 for motor action and Q_3' for generator action, must be found on the circular locus which make the horizontal distance from the locus to the torque line a maximum. This is most easily accomplished by drawing a line from the centre of the circle, perpendicular to the torque line, to meet the circle at Q_3 and Q_3'. $Q_3 S_3$ and $Q_3' S_3'$ are then drawn in the reference direction to meet the torque line in S_3 and S_3'.

$$\text{Pull-out torque, motoring} = \frac{3 V_s \cdot S_3 Q_3}{2 \pi n_0} \tag{13.42}$$

or, when line values of voltage and current are used,

$$\text{Pull-out torque, motoring} = \frac{\sqrt{3} \times 440 \times 3 \cdot 09 \times 2}{2 \pi \times 50/2} = \underline{\underline{30 \, \text{N-m}}}$$

$$\text{Slip for pull-out torque} = \frac{S_3 R_3}{S_3 Q_3} = \frac{0 \cdot 24}{3 \cdot 09} = \underline{\underline{0 \cdot 078}}$$

Since $S_3 Q_3$ is now known to represent a torque of 30 N-m, all other torques may be found by simple proportion.

$$\text{Pull-out torque, generating} = 30 \times \frac{S_3' Q_3'}{S_3 Q_3} = 30 \times \frac{3 \cdot 73}{3 \cdot 09} = \underline{\underline{36 \cdot 2 \, \text{N-m}}}$$

The slip for maximum torque, generating, will be numerically the same as that for maximum torque, motoring, but will be negative.

(c) To find the maximum mechanical power, the points Q_4 and Q_4' must be found on the circular locus which make the horizontal distance from the locus to the mechanical power line a maximum. A similar method is used as was used to obtain the maximum torque. A line is drawn from the centre of the circle at right angles to the mechanical power line to meet the circular locus in Q_4 and Q_4'. $Q_4 R_4$ and $Q_4' R_4'$ are then drawn in the reference direction to meet the mechanical power line in R_4 and R_4'.

$$\text{Maximum mechanical power, motoring} = 3 V_s \cdot R_4 Q_4 \tag{13.39}$$

or, when line values of voltages and current are used,

$$\text{Maximum mechanical power motoring} = \sqrt{3} \times 440 \times 2 \cdot 84 \times 2$$
$$= \underline{\underline{4{,}320 \, \text{W}}}$$

Otherwise, since it is already known that $Q_2 R_2$ represents 2,240 W (full load), all other mechanical powers may be found by ratio.

$$\text{Maximum mechanical power, motoring} = 2{,}240 \times \frac{Q_4 R_4}{Q_2 R_2}$$

$$= 2{,}240 \times \frac{2 \cdot 84}{1 \cdot 47} = \underline{\underline{4{,}320 \, \text{W}}}$$

$$\text{Maximum mechanical power generating} = 2{,}240 \times \frac{Q_4' R_4'}{Q_2 R_2}$$

$$= 2{,}240 \times \frac{4 \cdot 03}{1 \cdot 47} = \underline{\underline{6{,}140 \, \text{W}}}$$

(d) The starting torque is the torque when $s = 1$ and is therefore represented by $U Q_1$ on the diagram.

Starting torque for direct-on starting $= 30 \times \dfrac{UQ_1}{S_3Q_3}$

$$= 30 \times \frac{0 \cdot 55}{3 \cdot 09} = \underline{\underline{5 \cdot 34 \, \text{N-m}}}$$

When a star-delta starter is used, the voltage applied to each stator phase is $1/\sqrt{3}$ of the normal running value. Eqn. (13.27) shows that the torque is proportional to the square of the voltage applied to each stator phase. Therefore,

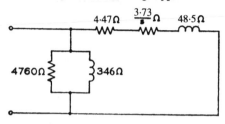

Fig. 13.20 (EXAMPLE 13.4)

when a star-delta starter is used the starting torque will be $(1/\sqrt{3})^2$ or $1/3$ of that for direct-on-line starting.

Starting torque for star-delta starting $= \dfrac{5 \cdot 34}{3} = \underline{\underline{1 \cdot 78 \, \text{N-m}}}$

The equivalent circuit calculated from the test results using the method given in Section 13.11 is shown in Fig. 13.20.

EXAMPLE 13.5 In the approximate equivalent circuit of one phase of a 3-phase mesh-connected induction motor shown in Fig. 13.21, $V_s = 415\,\text{V}$, $R_c = 200\,\Omega$,

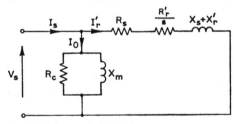

Fig. 13.21 (EXAMPLE 13.5)

$R_s = 0 \cdot 1\,\Omega$, $R_r' = 0 \cdot 2\,\Omega$, $X_m = 25\,\Omega$, and $X_s + X_r' = 1 \cdot 0\,\Omega$. Determine the input current, power factor, output power and efficiency if the full-load slip is $0 \cdot 03$ when the machine is connected to a 3-phase 415 V 50 Hz supply.

This problem could be solved by finding the input current on no load (when $s \approx 0$) and on short-circuit (when $s = 1$) and then constructing the circle diagram as in Example 13.4. An alternative solution by calculation is given below.

$$Z_s = R_s + \frac{R_r'}{s} + j(X_s + X_r') = 0 \cdot 1 + \frac{0 \cdot 2}{+0 \cdot 03} + j1 \cdot 0 = 6 \cdot 85 \underline{/8 \cdot 4^\circ}$$

$$Y_s = \frac{1}{Z_s} = 146 \times 10^{-3} / \underline{-8\cdot4^\circ} = 144 \times 10^{-3} - j21\cdot3 \times 10^{-3}$$

$$Y_0 = \frac{1}{R_c} - j\frac{1}{X_m} = \frac{1}{200} - j\frac{1}{25} = 5 \times 10^{-3} - j40 \times 10^{-3}$$

Total input admittance, $Y_T = Y_0 + Y_s = 149 \times 10^{-3} - j61\cdot3 \times 10^{-3}$
$$= 161 \times 10^{-3} / \underline{-22\cdot4^\circ}$$

Input current per phase $= VY_T = 415 \times 161 \times 10^{-3} = \underline{\underline{66\cdot8\,A}}$

Input power factor $= \cos 22\cdot4^\circ = \underline{0\cdot925\ \text{lagging}}$

Input power $= 3 \times 415 \times 66\cdot8 \times 0\cdot925 = 77{,}100\,W$
$I_r' = VY_s = 415 \times 146 \times 10^{-3} = 60\cdot5\,A$
Rotor loss, $P_r = 3(I_r')^2 R_r'$

Mechanical power output, $P_m = P_r \left(\dfrac{1 - s}{s} \right)$

$$= 3 \times 60\cdot5^2 \times 0\cdot2 \times \frac{1 - 0\cdot03}{0\cdot03}$$

$$= \underline{\underline{71{,}100\,W}}$$

Efficiency $= \dfrac{71{,}100}{77{,}100} = \underline{\underline{0\cdot923\,\text{p.u.}}}$

13.12 Speed Control

In the majority of applications the speed of a driving motor is required to be almost constant, and hence the plain induction motor is very suitable. Sometimes, however, it is desirable to be able to control the speed of the motor, and this may be achieved in three main ways. From eqn. 13.5,

$$n_r = (1 - s)n_0 = (1 - s)\frac{f}{p} \tag{13.50}$$

It follows that the speed n_0, may be varied by varying the number of stator poles or varying the supply frequency. It is rarely that variation of the supply frequency is used.

In wound-rotor machines the speed for a given load torque may be varied by varying the rotor resistance. This (as shown in Fig. 13.22) gives a range of speeds near full speed, four speeds being possible for the same load torque, with a four-position resistance. The disadvantages of this method are the heat lost in the regulating resistor, and the dependence of the speed variation on the load torque.

If two or three different operating speeds are required near the synchronous speeds of 3,000, 1,500, 1,000, 750, 500, etc., rev/min. these may be achieved by having two or more stator windings each having a different number of pole-pairs. The required speed is obtained by switching on the appropriate winding. This is only possible with cage-rotor machines, since wound-rotor machines must have a fixed number of rotor poles. To avoid the added cost of separate windings with different numbers of pole-pairs, pole-changing windings were introduced. In these, by a series or parallel grouping of the coils (achieved by switching), the number

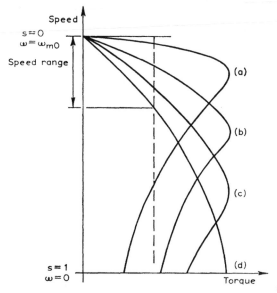

Fig. 13.22 SPEED CONTROL BY VARYING ROTOR RESISTANCE

(a) $R_r' = 0.15\sqrt{[R_s^2 + (X_s + X_r')^2]}$
(b) $R_r' = 0.4\sqrt{[R_s^2 + (X_s + X_r')^2]}$
(c) $R_r' = 0.7\sqrt{[R_s^2 + (X_s + X_r')^2]}$
(d) $R_r' = \sqrt{[R_s^2 + (X_s + X_r')^2]}$

of stator poles may be altered. Speed ratios of 2:1 and 1.5:1 may be achieved in this manner.

EXAMPLE 13.6 An 8-pole 50 Hz 3-phase slip-ring induction motor has a total leakage impedance of $(2.0 + j5.4)\,\Omega$ per phase referred to the stator. The stator resistance per phase is $1.1\,\Omega$. When 415 V is applied to the mesh-connected stator winding the voltage between any pair of open-circuited slip rings to which the star-connected rotor winding is connected is 239 V. The motor develops full-load torque with a slip of 0.04 with the slip rings short-circuited. Calculate the approximate speed of the machine if a 3-phase non-inductive resistor of $0.5\,\Omega$ per phase is connected in series with the slip rings when full-load torque is applied.

Since the rotor winding is star connected,

Induced rotor voltage/phase on open-circuit, $E_r = \dfrac{239}{\sqrt{3}}$ V

Since the stator winding is mesh connected,

Induced stator voltage/phase on open-circuit, $E = 415\,\text{V}$

Voltage drop due to any magnetizing current is neglected.

Effective phase turns ratio $= \dfrac{415}{239/\sqrt{3}} = 3$

Rotor resistance/phase referred to stator, $R_{r1}' = 2\cdot0 - 1\cdot1 = 0\cdot9\,\Omega$

Added rotor resistance/phase referred to stator $= 0\cdot5 \times 3^2 = 4\cdot5\,\Omega$

Total rotor resistance/phase referred to stator, $R_{r2}' = 0\cdot9 + 4\cdot5 = 5\cdot4\,\Omega$

$$\text{Torque developed, } T = \frac{3V_s^2}{2\pi n_0}\frac{s_1 R_{r1}'}{(s_1 R_s + R_{r1}')^2 + s_1^2(X_s + X_r')^2} \tag{13.27}$$

$$= \frac{3V_s^2}{2\pi n_0}\frac{s_2 R_{r2}'}{(s_2 R_s + R_{r2}')^2 + s_2^2(X_s + X_r')^2}$$

where $s_1 =$ slip for full-load torque with slip rings short-circuited, and $s_2 =$ slip for full-load torque with added rotor resistance.

$$\frac{0\cdot04 \times 0\cdot9}{(0\cdot04 \times 1\cdot1 + 0\cdot9)^2 + 0\cdot04^2 \times 5\cdot4^2} = \frac{5\cdot4 s_2}{(1\cdot1 s_2 + 5\cdot4)^2 + 5\cdot4^2 s_2^2}$$

Therefore

$$s_2^2 - 4\cdot22 s_2 + 0\cdot96 = 0$$

whence

$$s_2 = 0\cdot235 \quad \text{or} \quad 3\cdot92$$

The second result refers to brake action and may be neglected here.

Speed $= (1 - s) = (1 - 0\cdot235) \times \tfrac{50}{4} \times 60 = 573\,\text{rev/min}$

13.13 Single-phase Induction Machines

The construction of the single-phase induction motor is similar to that of the 3-phase type: a single-phase winding replaces the 3-phase winding. For the same output, the size of the single-phase machine is about 1·5 times that of the corresponding 3-phase machine.

A single-phase current in a single-phase winding produces a pulsating, not a rotating, magnetic field. However, the theory of single-phase motors may be placed on the same basis as that of 3-phase motors by representing a single-phase pulsating m.m.f. by two fields of constant amplitude rotating in opposite directions. This representation is valid so long as the original pulsating field has a sinusoidal distribution round the armature and varies sinusoidally with time.

If the current in the single-phase stator winding is

$$i_s = I_{sm}\sin\omega t$$

then the fundamental stator m.m.f. at any position θ is

$$F'_s = \frac{I_{sm}N_s}{2p} \sin \omega t \sin \theta$$

$$= \frac{I_{sm}N_s}{4p} \{\cos (\omega t - \theta) - \cos (\omega t + \theta)\} \qquad (13.51)$$

Thus the pulsating stator m.m.f. may be represented by two oppositely rotating cosinusoidally distributed m.m.f.s each travelling at synchronous speed. Each of these component stator m.m.f. distributions gives rise to corresponding rotor m.m.f. distributions. The resulting instantaneous torque developed has four clearly distinguishable components, which are:

1. A torque due to the interaction of the forward-travelling stator and rotor m.m.f. distributions.
2. A torque due to the interaction of the backward-travelling stator and rotor m.m.f. distributions.
3. A torque due to the forward-travelling stator m.m.f. distribution and the backward-travelling rotor m.m.f. distribution.
4. A torque due to the backward-travelling stator m.m.f. distribution and the forward-travelling rotor m.m.f. distribution.

The first component gives a steady non-pulsating torque acting on the rotor in the forward direction and gives rise to a component slip/torque characteristic of the form obtained from a polyphase induction machine as shown in Fig. 13.23. The second component gives rise to a similar slip/torque characteristic, the torque acting in the opposite, backward direction.

The third and fourth components give rise to torques which pulsate at twice supply frequency and do not contribute to the mean torque. Thus, unlike the polyphase induction machine, the single-phase induction machine has a pulsating component of developed torque. Examination of the component slip/torque curves of Fig. 13.23 reveals that at starting the forward- and backward-directed mean torques are equal, so that the single-phase machine is not self-starting. If the machine is started in either direction it will run up to speed and run stably in that direction.

Following eqn. (13.2), the slip when the rotor runs in the forward direction is

$$s_f = \frac{n_0 - n_r}{n_0} = 1 - \frac{n_r}{n_0} \qquad (13.52)$$

The corresponding value of slip for the rotor measured with reference to the backward-travelling m.m.f. is

$$s_b = \frac{-n_0 - n_r}{-n_0} = 1 + \frac{n_r}{n_0} \qquad (13.53)$$

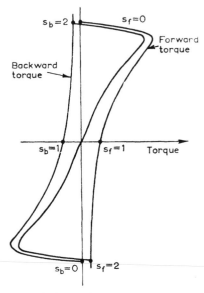

Fig. 13.23 SLIP/TORQUE CHARACTERISTIC OF A SINGLE-PHASE INDUCTION MOTOR

Adding eqns. (13.52) and (13.53),

$$s_f + s_b = 2$$

or

$$s_b = 2 - s_f \qquad (13.54)$$

13.14 Equivalent Circuit of Single-phase Induction Machine

An equivalent circuit for the single-phase induction machine may be obtained by considering it as a 3-phase machine with one supply line disconnected as shown in Fig. 13.24. Evidently, from Fig. 13.24,

$$I_R = I \qquad I_Y = -I \qquad I_B = 0$$
$$V_{RS} = \tfrac{1}{2}V \qquad V_{YS} = -\tfrac{1}{2}V$$

Also, $V_{BS} = 0$ since the voltages induced in the blue phase due to its couplings with the red and yellow phases cancel out.

Considering first the symmetrical components of the voltage applied, the zero-sequence voltage is

$$V_{R0} = \tfrac{1}{3}(V_{RS} + V_{YS} + V_{BS}) = 0$$

so that

$$V_{R0} = V_{Y0} = V_{B0} = 0$$

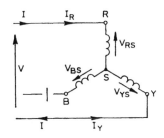

Fig. 13.24 SINGLE-PHASE OPERATION OF A 3-PHASE INDUCTION MACHINE

The applied voltage is

$$V = V_{RS} - V_{YS}$$
$$= (V_{R+} + V_{R-} + V_{R0}) - (V_{Y+} + V_{Y-} + V_{Y0})$$
$$= (1 - a^2)V_{R+} + (1 - a)V_{R-} \tag{13.55}$$

Considering now the symmetrical components of the current, the zero-sequence current is

$$I_{R0} = \tfrac{1}{3}(I_R + I_Y + I_B) = 0$$

The positive phase-sequence current is

$$I_{R+} = \tfrac{1}{3}(I_R + aI_Y + a^2 I_B)$$
$$= \tfrac{1}{3}I(1 - a) \tag{13.56}$$
$$I_{R-} = \tfrac{1}{3}(I_R + a^2 I_Y + aI_B)$$
$$= \tfrac{1}{3}I(1 - a^2) \tag{13.57}$$

Thus from eqns. (13.56) and (13.57),

$$I = \frac{3I_{R+}}{1 - a} = \frac{3I_{R-}}{1 - a^2} \tag{13.58}$$

The total input impedance is

$$Z = \frac{V}{I} = \frac{(1 - a^2)V_{R+}}{3I_{R+}} + \frac{(1 - a)V_{R-}}{3I_{R-}}$$
$$\frac{1 - a}{\phantom{3I_{R+}}} \qquad \frac{1 - a^2}{\phantom{3I_{R-}}}$$

$$= \frac{V_{R+}}{I_{R+}}\frac{(1 - a^2)(1 - a)}{3} + \frac{V_{R-}}{I_{R-}}\frac{(1 - a)(1 - a^2)}{3}$$

$$= \frac{V_{R+}}{I_{R+}} + \frac{V_{R-}}{I_{R-}}$$

i.e.

$$Z = Z_+ + Z_- \qquad\qquad (13.59)$$

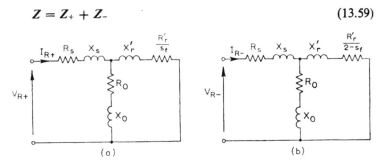

Fig. 13.25   PHASE-SEQUENCE NETWORKS OF THE 3-PHASE INDUCTION MACHINE

(a) Positive   (b) Negative

where Z_+ and Z_- are the positive phase-sequence and negative phase-sequence impedance operators of the 3-phase induction machine respectively.

The positive phase-sequence equivalent circuit of the induction machine, as shown in Fig. 13.25(a), is evidently of the same form as the normal equivalent circuit per phase, except that the parallel magnetizing branch has been replaced by its equivalent series circuit. No simplification is obtained by the approximation of showing the magnetizing branch at the input terminals. It should be noted that the parameters in the magnetizing branch will have different values for single-phase excitation than for 3-phase excitation owing to different mutual effects.

The negative phase-sequence equivalent circuit, as shown in Fig. 13.25(b), is the equivalent circuit for the machine when a negative phase-sequence 3-phase supply is applied to the stator and the rotor is driven in the opposite direction to the stator field. For any given rotor speed, if s_f is the slip for the positive phase-sequence network, the corresponding slip for the negative phase-sequence network is, according to eqn. (13.54),

$$s_b = 2 - s_f \qquad (13.54)$$

Evidently the positive phase-sequence currents give rise to the forward-travelling component wave and the negative phase-sequence currents to the backward-travelling wave, so that the positive phase-sequence network may be used to calculate the forward torque,

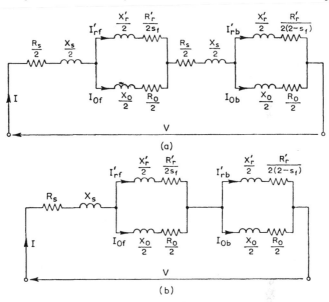

Fig. 13.26 EQUIVALENT CIRCUITS OF THE SINGLE-PHASE INDUCTION MACHINE

and the negative phase-sequence network to predict the backward torque.

Eqn. (13.59) shows that, to obtain the equivalent circuit of the single-phase induction machine, the positive and negative phase-sequence networks must be joined in series. Since one phase of the 3-phase machine corresponds to half of the single-phase winding, the equivalent circuits of the single-phase machine are as shown in Figs. 13.26(*a*) and (*b*).

Following eqn. (13.24) for the 3-phase machine, the forward torque component is

$$T_f = \frac{(I_{rf}')^2 R_r'/2s_f}{2\pi n_0} \qquad (13.60)$$

and the backward torque component is

$$T_b = \frac{(I_{rb}')^2 R_r'/2(2 - s_f)}{2\pi n_0} \qquad (13.61)$$

EXAMPLE 13.7 A single-phase 230 V 4-pole 50 Hz 0·5 kW induction motor gave the following test results:

Locked rotor test	60 V	1·5 A	Power factor, 0·6 lagging
No load test	230 V	0·535 A	Power factor, 0·174 lagging

Determine the approximate equivalent circuit for the machine. Find also the torque developed, the power output, the input current and power factor when the machine runs with a fractional slip of 0·05.

Assume that the stator and rotor $I^2 R$ losses at standstill are equal, and that the rotor leakage reactance referred to the stator and the stator leakage reactance are equal.

For the locked-rotor test, $s_f = 1$. The referred value of rotor impedance, $R_r' + jX_r'$, will be much smaller than the magnetizing impedance, $R_0 + jX_0$; therefore, to a good approximation the equivalent circuit for $s = 1$ is as shown in Fig. 13.27(a).

Input impedance with rotor locked, $Z_{sc} = R_s + R_r' + j(X_s + X_r')$

$$= \frac{60/0°}{1·5/-\cos^{-1} 0·6} = (24 + j32)\Omega$$

$R_s + R_r' = 24\Omega$ so that $R_s = R_r' = 12\Omega$

$X_s + X_r' = 32\Omega$ so that $X_s = X_r' = 16\Omega$

For no-load conditions $s_f \to 0$, in which case the magnetizing impedance, $\frac{R_0}{2} + j\frac{X_0}{2}$, will be much smaller than $\frac{R_r'}{2s_f} + j\frac{X_r'}{2}$ which tends to infinity when $s_f \to 0$. On the other hand, $\frac{R_r'}{2(2 - s_f)} + j\frac{X_r'}{2}$, which tends to $\frac{R_r'}{4} + \frac{X_r'}{2}$ when $s_f \to 0$, will be much smaller than $\frac{R_0}{2} + j\frac{X_0}{2}$. A good approximation for the equivalent circuit of the single-phase machine on no-load is therefore as shown in Fig. 13.27(b).

Input impedance on no-load, $Z_{nl} = R_s + \frac{R_0}{2} + \frac{R_r'}{4} + j\left(X_s + \frac{X_0}{2} + \frac{X_r'}{2}\right)$

$$= \frac{230/0°}{0·535/-\cos^{-1} 0·174} = (75 + j424)\Omega$$

Therefore

$$\frac{R_0}{2} = 75 - R_s - \frac{R_r'}{4} = 75 - 12 - 3 = 60\Omega$$

$$\frac{X_0}{2} = 424 - X_s - \frac{X_r'}{2} = 424 - 16 - 8 = 400\Omega$$

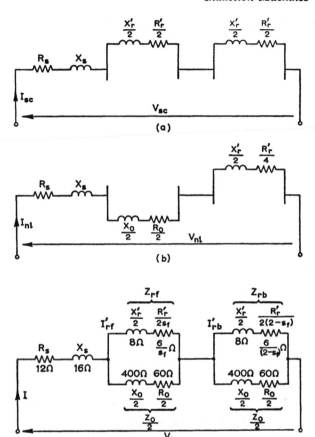

Fig. 13.27

The full equivalent circuit is shown in Fig. 13.27(c). When $s_f = 0.05$,

$$Z_{rf} = \frac{6}{0.05} + j8 = 120 + j8 = 120\underline{/3.8°}\,\Omega$$

$$Z_{rb} = \frac{6}{(2 - 0.05)} + j8 = 3.08 + j8 = 8.55\underline{/69°}\,\Omega$$

$$\frac{Z_0}{2} = 60 + j400 = 404\underline{/81.5}\,\Omega$$

$$Z_{rf} + \frac{Z_0}{2} = 180 + j408 = 446\underline{/66.2°}\,\Omega$$

$$Z_{rb} + \frac{Z_0}{2} = 63.1 + j408 = 413\underline{/81.3°}\,\Omega$$

The total impedance is

$$Z_T = 12 + j16 + \frac{120\underline{/3\cdot8^\circ} \times 404\underline{/81\cdot5^\circ}}{446\underline{/66\cdot2^\circ}} + \frac{8\cdot55\underline{/69^\circ} \times 404\underline{/81\cdot5^\circ}}{413\underline{/81\cdot3^\circ}}$$

$$= (118 + j59\cdot6)\Omega = 132\underline{/26\cdot9}\,\Omega$$

Input current, $I = \dfrac{230\underline{/0^\circ}}{132\underline{/26\cdot9^\circ}} = 1\cdot74\underline{/-26\cdot9^\circ}\,\text{A}$

Input power factor $= \cos 26\cdot9^\circ = 0\cdot892$ lagging

Synchronous speed, $n_0 = \dfrac{f}{p} = \dfrac{50}{2} = 25\,\text{rev/s}$

$I_{rf}' = I\dfrac{Z_0/2}{Z_{rf} + Z_0/2} = 1\cdot74 \times \dfrac{404}{446} = 1\cdot57\,\text{A}$

$I_{rb}' = I\dfrac{Z_0/2}{Z_{rb} + Z_0/2} = 1\cdot74 \times \dfrac{404}{413} = 1\cdot7\,\text{A}$

From eqns. (13.60) and (13.61),

$$T_f = \frac{(I_{rf}')^2 R_r'/2s_f}{2\pi n_0} = \frac{1\cdot7^2 \times 120}{2\pi 25} = 1\cdot88\,\text{N-m}$$

$$T_b = \frac{(I_{rb}')^2 R_r'/2(2 - s_f)}{2\pi n_0} = \frac{1\cdot7^2 \times 3\cdot08}{2\pi 25} = 0\cdot0566\,\text{N-m}$$

Net forward torque $= 1\cdot88 - 0\cdot0566 = 1\cdot82\,\text{N-m}$

Power output $= 2\pi n_r T = 157(1 - 0\cdot05) \times 1\cdot82 = 0\cdot27\,\text{kW}$

13.15 Starting

One of the chief disadvantages of the single-phase induction motor is the fact that special arrangements must be made for starting, so increasing the cost and limiting the motor size which is practical. A second starting winding of short time rating is used. This winding is connected to the single-phase supply through a capacitor or an inductor, producing a phase shift which causes the machine to start as a 2-phase induction motor. A centrifugal switch disconnects the starting winding from the supply when the machine runs up to speed. Fig. 13.28 shows connexion diagrams and a torque/speed curve for this type of motor. It will be observed that the starting direction is reversed with inductance starting.

13.16 Synchronous Induction Motor

A disadvantage of the induction motor, as has been seen, is the fact that it operates at a lagging power factor. It is, however, self-starting. On the other hand, the synchronous motor, which can be

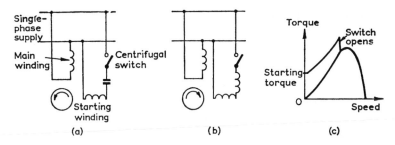

Fig. 13.28   STARTING OF SINGLE-PHASE INDUCTION MOTORS

(*a*) Capacitor start
(*b*) Inductor start
(*c*) Torque/speed characteristics

operated with a leading power factor, has no starting torque. The synchronous induction motor combines the high starting torque of the induction motor with the leading power factor of the synchronous motor. The machine consists essentially of a wound-rotor induction motor, which has a longer air-gap than the normal induction motor to reduce the effect of armature m.m.f. in the machine when it is running as a synchronous motor. It is started by

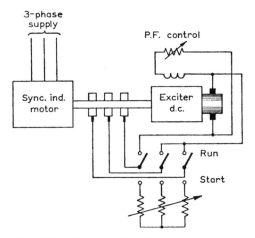

Fig. 13.29   SYNCHRONOUS INDUCTION MOTOR

resistance starting as an induction motor, and when it has run up to speed the starting resistance is disconnected, and direct current from a small exciter on the same shaft as the motor is fed into the rotor (Fig. 13.29). The machine then runs as a synchronous motor, the power factor being varied by controlling the direct current in the rotor.

The most common method of connecting the rotor for direct current is shown in Fig. 13.30, from which it will be seen that one phase carries the total current I_d, while the two other phases carry half each in the opposite direction. Since the rotor runs at synchronous speed, the rotor currents will move with the rotating field, and will produce a synchronously rotating field. Thus the rotor current may be represented on the stator complexor diagram. For the connexion shown the direct currents correspond to the instant when a 3-phase system has positive maximum current in one phase, and half

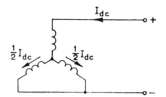

Fig. 13.30   CONNEXION OF A 3-PHASE WINDING FOR DIRECT CURRENT

the negative maximum current in the two other phases. The equivalent r.m.s. rotor current will thus be

$$I_{ac} = \frac{1}{\sqrt{2}} I_{dc} \tag{13.62}$$

This current must be multiplied by the rotor/stator turns ratio, to obtain the equivalent stator current.

13.17 Circle Diagram for a Synchronous Induction Motor

For starting and running up to speed the current locus will be the same as the circle diagram for the induction motor. When the direct current is switched into the rotor, there will be an additional stator current (at supply frequency) due to transformer action. The resultant stator current will be the complexor sum of the no-load current and this additional current, so that, for a constant direct current, the locus of the stator current I_1 (Fig. 13.31), will be a circle, centre D, and of a radius representing the rotor current; i.e.

$$I_1 = \frac{I_d}{\sqrt{2}} \times \frac{\text{Rotor turns}}{\text{Stator turns}}$$

The operating point for a given power output may be found (approximately) by the following procedure.

1. Calculate the active component of the total current which will correspond to the required output power. Mark this component from D parallel to the voltage complexor:

$$DQ' = \frac{\text{Power output (watts)}}{3V_{ph}}$$

(The entire diagram should preferably be drawn for phase values of voltage and current.)

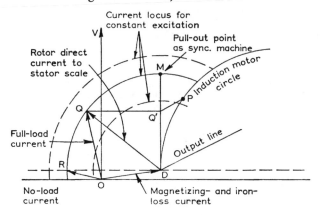

Fig. 13.31 CIRCLE DIAGRAM FOR SYNCHRONOUS INDUCTION MOTOR

2. Draw a line at right angles to DQ' from Q' to cut the circle representing the rotor current in use at the point Q. Q is then the operating point and OQ represents the total stator phase current.

It will be seen that for this load and rotor excitation current the machine has a leading power factor. DM represents the maximum possible load which the motor could supply with this particular d.c. excitation without falling out of step as a synchronous machine.

PROBLEMS

13.1 A 4-pole induction motor runs from a 50 Hz supply at 1,450 rev/min. Determine the frequency of the rotor current and the percentage slip.

Ans. 1·67, 3·3 per cent.

13.2 The rotor of a 3-phase 440 V 50 Hz 4-pole induction motor has a resistance of 0·3 Ω/phase and an inductance of 0·008 H/phase. The ratio of stator to rotor turns is 2:1. Determine the voltage between the rotor slip-rings on open-circuit, if the stator is delta connected and the rotor is star-connected. Also calculate the standstill rotor current and the rotor current when running with 3 per cent slip.

Ans. 381 V, 87 A; 21·4 A.

13.3   A 6-pole 3-phase 440 V 50 Hz induction motor has an output of 14·9 kW the rotor e.m.f. making 90 cycles per minute. Calculate the motor input power and efficiency if the rotational losses absorb a torque of 16·4 N-m, and the stator losses are 800 W.   (*H.N.C.*)

Ans. 17·9 kW, 84 per cent.

13.4   The power input to a 3-phase induction motor is 40 kW. The stator losses total 1 kW, and the mechanical losses are 1·6 kW. Determine the brake output power, the rotor winding loss per phase, and the efficiency if the motor runs with a slip of 3·5 per cent.   (*H.N.C.*)

Ans. 36 kW; 445 W; 0·9.

13.5   A 3-phase wound-rotor induction motor has six poles, and operates on a 50 Hz supply. The rotor resistance is 0·1 Ω/phase and the standstill reactance 0·5 Ω/phase. Draw on graph paper the torque/slip curve of the motor. Calculate:

   1. The speed for maximum torque.

   2. The torque, expressed as a percentage of maximum torque, when the slip is 4 per cent.

   3. The value of the external rotor starter resistance to obtain maximum torque on starting.   (*H.N.C.*)

Ans. 800 rev/min; 38·5 per cent; 0·4 Ω/phase.

13.6   A 200 V 50 Hz 3-phase induction motor has a 4-pole star-connected stator winding. The rotor resistance and standstill reactance per phase are 0·1 Ω and 0·9 Ω respectively. The ratio of rotor to stator turns is 2:3. Calculate the total torque developed when the slip is 4 per cent. Neglect stator resistance and leakage reactance.   (*H.N.C.*)

Ans. 39·8 N-m.

13.7   A 3-phase 50 Hz induction motor with its rotor star connected gives 500 V (r.m.s.) at standstill between slip rings on open circuit. Calculate the current and power factor in each phase of the rotor winding at standstill when joined to a star-connected circuit each limb of which has a resistance of 10 Ω and an inductance of 0·6 H. The resistance per phase of the rotor is 0·2 Ω and its inductance per phase is 0·03 H. Calculate also the current and power factor in each rotor phase when the slip rings are short-circuited and the motor is running with a slip of 4 per cent. Neglect the impedance of the stator.   (*L.U.*)

Ans. 1·46 A; 0·0515 lagging; 27·1 A; 0·4 lagging.

13.8   Explain the advantage gained by using a slip-ring rotor instead of a cage rotor for a 3-phase induction motor.

   A 3-phase 4-pole induction motor works at 200 V, 50 Hz on full load of 7·5 kW; its speed is 1,440 rev/min. (Frictional losses total 373 W.) Determine approximately (*a*) its speed at 200 V and half load, and (*b*) its speed with an output of 7·5 kW at 190 V, 50 Hz.   (*L.U.*)

Ans. 1,470 rev/min; 1,433 rev/min.

13.9   A 3-phase 4-pole induction motor connected to a 50 Hz supply has a constant flux per pole of 0·05 Wb.

The 3-phase star-connected rotor winding has 48 slots and 6 conductors per slot, all the conductors of each phase being connected in series. The resistance of each phase is $0.2\,\Omega$ and the inductance $0.006\,H$.

Assuming the flux to remain constant in magnitude, determine the rotor current and the torque in newton-metres (*a*) when the rotor is stationary and short-circuited, and (*b*) when the rotor is running at 5 per cent slip. Assume that the distribution factor for the rotor winding is 0.96.

(*L.U.*)

Ans. 270 A, 279 N-m; 116 A, 1,030 N-m.

13.10   Explain the action of a 3-phase induction motor (i) at starting, (ii) when running. Deduce expressions for the rotor current, torque and slip when running, assuming the resistance of the rotor circuit to be constant. Sketch a typical slip/torque characteristic from standstill to synchronous speed.

A 6-pole 3-phase induction motor runs at a speed of 960 rev/min when the shaft torque is 136 N-m and the frequency 50 Hz. Calculate the rotor I^2R loss if the friction and windage losses are 150 W. (*L.U.*)

Ans. 574 W.

13.11   Explain the action of a 3-phase induction motor when running loaded. Derive the expression for the torque as a function of the slip assuming the impedance of the stator winding to be negligible.

If the star-connected rotor winding of such a motor has a resistance per phase of $0.1\,\Omega$ and a leakage reactance per phase at standstill of $0.9\,\Omega$, calculate: (*a*) the slip at which maximum torque occurs, and (*b*) the additional resistance required to obtain maximum torque at starting. (*L.U.*)

Ans. 0.111; 0.8 Ω.

13.12   A 33.5 kW 4-pole 440 V 3-phase 50 Hz mesh-connected induction motor gave the following results on test:

(*a*) No load—440 V, 22 A, 1,500 W
(*b*) Standstill—220 V, 140 A, 12,000 W

Draw the circle diagram and determine the power factor, load current, and efficiency on full load.

If the rotor winding loss is 45 per cent of the total winding loss, find the slip, and the ratio of starting to full load torque. (*H.N.C.*)

Ans. 0.85, 56 A, 95 per cent; 0.022, 0.69.

13.13   Tests on a 440 V 50 Hz 3-phase 4-pole induction motor showed that on no-load the motor took 22 A at a power factor of 0.2, while at standstill the current was 135 A at a power factor of 0.4 with the applied voltage reduced to 200 V. Determine (*a*) the maximum power output, (*b*) the maximum torque, (*c*) the starting torque on normal voltage. Assume a star-connected stator of resistance $0.164\,\Omega$/phase.

Ans. 75 kW, 585 N-m, 280 N-m.

13.14   A 6-pole 50 Hz 3-phase induction motor develops a useful full-load torque of 163 N-m when the rotor e.m.f. makes 120 complete cycles per minute. Calculate the brake power. If the mechanical torque lost in friction is 17.6 N-m and the stator loss is 750 W, find the winding loss in the rotor circuit, the input to the motor, and its efficiency. (*E.E.P.*)

Ans. 16.4 kW, 770 W, 20 kW, 82 per cent.

13.15 A 75 kW 440 V 3-phase 50 Hz 2-pole synchronous induction motor takes a no-load current of 30 A at a power factor of 0·2 lagging when operating as an induction motor. If the stator-to-rotor turns ratio is 1 : 2, determine the pull-out torque for a direct current of 99 A. What is the power factor on full load if the direct current is maintained constant?   (*H.N.C.*)

Ans. 340 N-m, 0·84 leading.

13.16 Use the test data of Problem 13.12 to determine the parameters of the approximate equivalent circuit of the induction motor.

Ans. $R_c = 388 \, \Omega; \quad X_m = 35 \cdot 2 \, \Omega; \quad X_s + X_r' = 2 \cdot 65 \, \Omega;$
$$R_s = 0 \cdot 34 \, \Omega \, R_r' = 0 \cdot 278 \, \Omega.$$

13.17 Tests on a 3-phase 4-pole 50 Hz mesh-connected induction motor established the following values for the parameters of the approximate equivalent circuit (per phase) as shown in Fig. 13.21: $V = 415$ V, $R_c = 1{,}600 \Omega$, $X_s + X_r' = 5 \cdot 4 \Omega$, $X_m = 120 \Omega$, $R_s = 0 \cdot 48 \Omega$, $R_r' = 0 \cdot 41 \Omega$. Determine the input current and power factor and the gross torque and mechanical power developed when the slip is 0·05 p.u. The stator is mesh connected.

Ans. 74 A; 0·812 lagging; 260 N-m; 38·8 kW.

13.18 For the 3-phase induction machine whose equivalent circuit parameters are given in Problem 13.17, (*a*) draw the impedance locus, and (*b*) draw the current locus. (Shift the origin of the current locus by I_0 as explained in Section 13.9.)

(*c*) Use the locus diagram to find the input current and power factor at a slip of 0·05. Compare the values obtained with those found in Problem 13.17. (*Hint.* Find the point on the impedance locus corresponding to $R_s + R_r'/0 \cdot 05$.)

13.19 A single-phase 230 V 4-pole 50 Hz 0·4 kW induction motor gave the following test results:

Locked-rotor test	60 V	1 A	36 W
No-load test	230 V	0·35 A	16 W

Determine the approximate equivalent circuit for the machine. Find also the torque developed, the power output, the input current and power factor when the machine runs with a fractional slip of 0·04. Estimate the value of slip when full-load torque is applied. Assume the stator and rotor winding losses at standstill equal and the rotor leakage reactance referred to the stator and the stator leakage reactance equal.

Ans. 1·03 N-m; 155 W; 0·99 A; 0·89 lagging.

13.20 A 3-phase 440 V 4-pole 50 Hz mesh-connected double-cage induction motor has a stator leakage impedance of $(1 + j5) \Omega$/phase. The impedance of the inner cage referred to the stator is $(1 + j10) \Omega$/phase at standstill, while that of the outer cage is $(5 + j0) \Omega$/phase at standstill. Determine (*a*) the starting torque for full stator voltage applied, and (*b*) the gross torque developed when the fractional slip is 0·04.

Ans. 105 N-m; 123 N-m.

13.21 Tests on a 3-phase 440 V 33·5 kW synchronous induction motor gave the following results:

No-load	440 V	50 A	3,400 W
Locked-rotor	220 V	152 A	13,000 W

When running synchronously at full load the excitation is adjusted to give unity power factor.

If $\quad \dfrac{\text{Stator turns per phase}}{\text{Rotor turns per phase}} = \dfrac{1}{2}$

Calculate the d.c. field current required.

For synchronous operation the rotor is connected with one phase joined in series with two in parallel. Draw the locus of stator current for both synchronous and induction running. From the diagram determine the maximum power output and the corresponding input currents and power factors for (a) synchronous running at the above excitation, and (b) induction running.

Ans. 47·6 A; 51·5 kW, 87·2 A, 0·83 lagging; 77·6 kW, 199 A, 0·67 lagging.

Chapter 14

THE D.C. CROSS-FIELD MACHINE

The d.c. cross-field machine is a two-axis machine consisting in effect of two machines connected in cascade. The armature winding plays a double role, acting as a field winding on one axis and as an output winding on the second axis. The machine as a whole operates as a two-stage d.c. amplifier suitable for use as a relatively fast-acting exciter to control large machines. The general arrangement of d.c. machines is described in Chapter 10 and is not repeated here. To reduce the effects of eddy currents in slowing response times, the entire magnetic circuit of a cross-field machine is often laminated.

14.1 E.M.F. and Torque Equations of a Single-axis D.C. Machine

Let E_A = Average e.m.f. induced in any one parallel path in the armature winding

Φ = Flux per pole

Z = Number of armature conductors

$2p$ = Number of poles

$2a$ = Number of parallel paths in the armature winding

n_r = Speed, rev/s

N = Number of turns in series in any one of the parallel paths in the armature winding

I_A = Armature winding current

T_A = Torque developed by armature

$$E_A = N\frac{\Delta\Phi}{\Delta t} \tag{14.1}$$

where $\Delta\Phi$ is the change in the flux linked with a coil in time Δt.

A coil whose coil sides are at A and A′ in Fig. 14.1 links the flux per pole, Φ. When the coil moves through a pole-pitch from this position it links the flux per pole in the reverse direction so that the

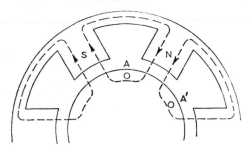

Fig. 14.1 INDUCTION OF E.M.F. IN A SINGLE-AXIS D.C. MACHINE

change in flux linked with coil is 2Φ, i.e.

$$\Delta\Phi = 2\Phi$$

This change takes place in the time required for the coil to move through one pole-pitch, i.e.

$$\Delta t = \frac{1}{2pn_r}$$

Since it requires two conductors to make a turn,

$$N = \frac{1}{2a}\frac{Z}{2}$$

Substituting in eqn. (14.1),

$$E_A = \frac{1}{2a}\frac{Z}{2}\frac{2\Phi}{\dfrac{1}{2pn_r}} = \Phi Z n_r \frac{p}{a} \tag{14.2}$$

This is the average e.m.f. induced in any one of the parallel paths in the armature winding, and is the voltage measured between the positive and negative brushes when they are positioned to give maximum output voltage and the machine acts as a generator on open-circuit.

Since energy is conserved,

$$2\pi n_r T_A = E_A I_A$$

Therefore the torque developed by the armature is

$$T_A = \frac{1}{2\pi n_r} \, \Phi Z n_r \frac{p}{a} \, I_A$$

$$= \frac{\Phi Z I_A}{2\pi} \frac{p}{a} \tag{14.3}$$

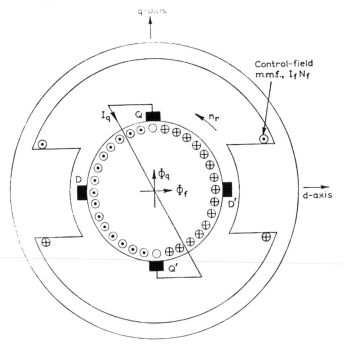

Fig. 14.2   TWO-POLE SINGLE-AXIS D.C. MACHINE

14.2 Performance Equations of a Generalized D.C. Cross-field Generator

Consider a 2-pole single-axis d.c. machine driven by a prime mover at constant speed as shown in Fig. 14.2. It is assumed that the magnetic circuit of the machine is unsaturated and that the flux per pole is proportional to the m.m.f. causing it.

A main field or control winding m.m.f., $I_f N_f$, gives rise to a direct-axis flux

$$\Phi_f = \Lambda_d I_f N_f \tag{14.4}$$

where Λ_d is the permeance of the direct-axis magnetic circuit. The control winding is sometimes called the *variator winding*.

When the field is excited, voltages are induced in the armature conductors in the direction indicated by the dots and crosses in Fig. 14.2 for anticlockwise rotation. A voltage will be obtained at a pair of diametral brushes, and this brush voltage has its greatest value when the brushes are in the position shown in Fig. 14.2. Brushes in the position DD', on the other hand, have zero generated voltage and would be absent in a conventional single-axis machine.

If the quadrature axis brushes QQ' are joined together, an armature current I_q will flow which will be large relative to the control-field current, I_f. The distribution of the armature current I_q corresponds to the dots and crosses of Fig. 14.2. This armature current gives rise to an armature flux Φ_q in the direction shown in Fig. 14.2, where

$$\Phi_q = \Lambda_q I_q N_A' \tag{14.5}$$

and Λ_q is the permeance of the quadrature-axis magnetic circuit.

The effective number of turns in the armature winding, N_A', is the number of turns which when multiplied by the total armature flux, Φ_q, gives the total flux linkage. The effective number of turns is less than the actual number, since not all the armature turns link all the armature flux.

The armature flux, Φ_q, resulting from a given armature current, I_q, may be increased by increasing the value of Λ_q. This may be done by providing poles on the quadrature axis as shown in Fig. 14.3.

In a conventional d.c. machine an interpole winding would be provided. This would be excited so as to reduce Φ_q to zero or even to produce a commutating flux in the opposite direction. In a cross-field machine, where a relatively large flux is required, commutation may be critical. To improve commutation the pole may be bifurcated as shown in Fig. 14.4, leaving only a small flux in the commutating zone. High-resistance brushes are employed, and the current to be commutated is reduced by providing a stator quadrature winding g excited so as to aid Φ_q. Thus for a given quadrature axis flux the current I_q required is reduced by increasing the effective number of turns. Thus

$$\Phi_q + \Phi_g = \Lambda_q I_q (N_A' + N_g) \tag{14.6}$$

The stator quadrature winding is sometimes called the *ampliator winding*.

The total quadrature axis flux, $\Phi_q + \Phi_g$, will give rise to induced voltages in the armature conductors. The directions of these e.m.f.s are indicated by the dots and crosses placed beside the armature conductors in Fig. 14.3; the dots and crosses within the armature conductors indicate the induced voltages due to the control flux, Φ_f.

Due to the existence of the quadrature-axis flux, a voltage will occur at the direct-axis brushes DD' (but the quadrature-axis flux does not, of course, give an e.m.f. at the brushes QQ').

When the direct-axis brushes are joined to a load resistance, R_L, a direct-axis armature current, I_d, flows. The distribution of this

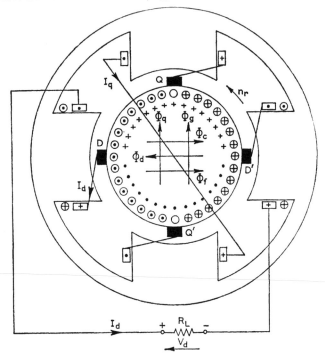

Fig. 14.3   TWO-POLE TWO-AXIS D.C. CROSS-FIELD MACHINE

current in the armature winding is also given by the dots and crosses placed beside the armature conductors. It will be noted that the m.m.f., $I_d N_A'$, due to this load current flowing through the armature winding is in opposition to the input m.m.f., $I_f N_f$, due to the control field. This means that the m.m.f. constitutes negative feedback that is proportional to the output current. In this mode of operation the machine will act as a constant-current generator. The component flux due to the direct-axis armature current is

$$\Phi_d = \Lambda_d I_d N_A' \tag{14.7}$$

The negative feedback due to the armature m.m.f. may be reduced or entirely eliminated by leading the output current I_d through a compensating winding that provides an m.m.f. I_dN_c acting in the opposite direction to that due to the armature. The component flux due to the compensating winding is

$$\Phi_c = \Lambda_d I_c N_c \tag{14.8}$$

The machine is exactly compensated when

$$I_d N_A' = I_d N_c \tag{14.9}$$

The compensating winding is more effective if it is distributed in a similar manner to the armature winding. In some cross-field machines

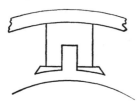

Fig. 14.4 BIFURCATED POLE STRUCTURE

the stator field windings are distributed in slots that are fewer and of a larger size than those on the armature.

The e.m.f. induced in the armature winding due to the component fluxes on the direct axis is

$$E_q = (\Phi_f - \Phi_d + \Phi_c)Zn_r\frac{p}{a} \tag{14.10}$$

The e.m.f. induced in the armature winding due to the component fluxes on the quadrature axis is

$$E_d = (\Phi_q + \Phi_g)Zn_r\frac{p}{a} \tag{14.11}$$

Substituting for the component fluxes in terms of the winding m.m.f.s in eqns. (14.10) and (14.11),

$$E_q = \Lambda_d\{I_fN_f - I_d(N_A' - N_c)\}Zn_r\frac{p}{a} \tag{14.12}$$

$$E_d = \Lambda_q I_q(N_A' + N_g)Zn_r\frac{p}{a} \tag{14.13}$$

Cross-field machines are commonly designed with the permeances of their direct and quadrature axes equal. Assuming this to be so, let

$$\lambda = \Lambda_d Zn_r\frac{p}{a} = \Lambda_q Zn_r\frac{p}{a}$$

Eqns. (14.12) and (14.13) then become

$$E_q = \lambda I_f N_f - \lambda(N_A{}' - N_c)I_d \tag{14.14}$$

$$E_d = \lambda(N_A{}' + N_g)I_q \tag{14.15}$$

These e.m.f.s are equal to the sums of the voltage drops in the quadrature-axis and direct-axis circuits respectively. Therefore, for generator action,

$$\lambda I_f N_f - \lambda(N_A{}' - N_c)I_d = I_q(R_A + R_g) \tag{14.16}$$

$$\lambda I_q(N_A{}' + N_g) = V_d + I_d(R_A + R_c) \tag{14.17}$$

where R_A = Armature winding resistance
$\quad\quad\ R_g$ = Stator quadrature-winding resistance
$\quad\quad\ R_c$ = Compensating-winding resistance

If the directions of the armature currents, I_d and I_q, are considered with respect to the total component fluxes on each of the axes in Fig. 14.3, it will be seen that:

1. The direct-axis armature current I_d produces no resultant torque with the direct-axis component fluxes.
2. The quadrature-axis armature current I_q produces no resultant torque with the quadrature-axis component fluxes.
3. The component torques due to the interaction of I_d and quadrature-axis component fluxes, and due to the interaction of I_q and the direct-axis component fluxes, are additive.

The torque due to the quadrature-axis armature current and the direct-axis flux is

$$T_q = \frac{\Phi_f - \Phi_d + \Phi_c}{2\pi}\, I_q Z \frac{p}{a} \tag{14.18}$$

and the torque due to the direct-axis armature current and the quadrature-axis flux is

$$T_d = \frac{\Phi_q + \Phi_g}{2\pi}\, I_d Z \frac{p}{a} \tag{14.19}$$

Substituting for the component fluxes in terms of the winding m.m.f.s,

$$T_q = \frac{\Lambda_d Z \frac{p}{a} I_q}{2\pi}\, (I_f N_f - I_d N_A{}' + I_d N_c)$$

$$T_d = \frac{\Lambda_q Z \frac{p}{a} I_d}{2\pi}\, (I_q N_A{}' + I_q N_g)$$

The total torque is

$$T = T_q + T_d$$

$$= \frac{\lambda}{2\pi n_r} \{I_q I_f N_f - I_q I_d (N_A' - N_c) + I_q I_d N_A' + I_q I_d N_g\}$$

$$= \frac{\lambda}{2\pi n_r} \{I_q I_f N_f + I_q I_d (N_c + N_g)\} \qquad (14.20)$$

14.3 Fully Compensated Cross-field Generator

The cross-field machine is said to be fully compensated when the m.m.f. of the compensating winding is equal and opposite to the armature winding m.m.f. due to the direct-axis armature current, I_d. To ensure exact compensation, the compensating winding is usually overwound (i.e. provided with more turns than necessary), and fine adjustment is then made by connecting an adjustable diverter resistance in parallel with the compensating winding. When the machine is fully compensated, $I_d N_A' = I_d N_c$.

Substituting this condition in eqn. (14.16),

$$\lambda I_f N_f = I_q (R_A + R_g) \qquad (14.21)$$

As previously,

$$\lambda I_q (N_A' + N_g) = V_d + I_d (R_A + R_c) \qquad (14.17)$$

If the machine is driven at constant speed as a generator and is excited with a constant control-field current, the quadrature-axis armature current, from eqn. (14.21), is

$$I_q = \frac{\lambda I_f N_f}{R_A + R_g} = \text{constant}$$

From eqn. (14.17), therefore,

$$V_d + I_d (R_A + R_c) = \text{constant}$$

In this mode of operation the machine will act as an approximately constant-voltage source, provided that the internal voltage drop, $I_d(R_A + R_c)$, is small compared with the output voltage V_d, a condition which usually obtains since internal losses must be small for the machine to have reasonable efficiency. *Amplidyne generators* are fully compensated cross-field machines.

14.4 Uncompensated Cross-field Generator

When the cross-field machine has no compensating winding, $N_c = 0$ and $R_c = 0$, and eqns. (14.16), (14.17) and (14.20) become

$$\lambda I_f N_f - \lambda I_d N_A' = I_q(R_A + R_g) \tag{14.22}$$

$$\lambda I_q(N_A' + N_g) = V_d + I_d R_A \tag{14.23}$$

$$T = \frac{\lambda}{2\pi n_r} (I_q I_f N_f + I_q I_d N_g) \tag{14.24}$$

If the machine is driven at constant speed as a generator and is excited with a constant control-field current, then

$$\lambda I_f N_f = \text{constant}$$

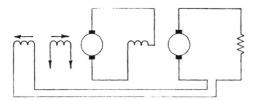

Fig. 14.5   TWO-MACHINE EQUIVALENT OF AN UNCOMPENSATED
CROSS-FIELD GENERATOR

Provided that the internal voltage drop, $I_q(R_A + R_g)$, is small compared with either of the generated voltages, then

$$\lambda I_f N_f - \lambda I_d N_A' \to 0$$

or

$$\lambda I_d N_A' \to \lambda I_f N_f$$

i.e.

$$I_d \to \frac{\lambda I_f N_f}{N_A'} \to \text{constant}$$

In this mode of operation, therefore, the machine acts as an approximately constant-current source, a result already anticipated becau~~ of the existence of the negative-feedback term proportional to output current. *Metadyne generators* are uncompensated or undercompensated cross-field generators.

Fig. 14.5 shows the two-machine equivalent of the uncompensated cross-field generator. An external feedback loop is shown providing an m.m.f. proportional to output current at the input in opposition to the separately excited control-field m.m f

14.5 Determination of Winding Transfer Constants by Test

In order to predict the operation of a cross-field machine usi
eqns. (14.16) and (14.17) it is necessary to determine the values of t
winding resistances R_A, R_g and R_c, and the values of the ter
λN_f, $\lambda N_A'$, λN_c and λN_g. These latter quantities may be called t
winding transfer constants. They can be determined by performing
a series of open-circuit tests on the machine. All the winding interconnexions are removed, and with the machine running at rated
speed one winding only is excited with a known current and the
appropriate open-circuit brush voltage is measured.

For example, with the control-field winding only excited, the
quadrature-axis brush voltage is, from eqn. (14.14),

$$E_q = \lambda I_f N_f \qquad \text{so that} \qquad \lambda N_f = \frac{E_q}{I_f}$$

The other winding transfer constants are found in a similar
manner. To determine $\lambda N_A'$, a known current must be fed to one set
of brushes and the open-circuit voltage measured at the other set.
If the permeances of the direct and quadrature axes are equal, the
same result will be obtained whether the current is fed in at the
direct or quadrature-axis brushes.

EXAMPLE 14.1 A fully compensated d.c. cross-field generator has the following winding resistances:

Armature winding	$R_A = 0{\cdot}05\,\Omega$
Compensating winding	$R_c = 0{\cdot}05\,\Omega$
Quadrature winding	$R_g = 0{\cdot}15\,\Omega$

Separate excitation open-circuit tests gave the following constants at rated
speed:

O.C. quadrature-axis brush voltage	750 V/control-winding ampere
O.C. direct-axis brush voltage	1·5 V/quadrature-winding ampere

The number of stator quadrature-axis turns is equal to the effective number
of armature turns. The permeances of the direct and quadrature axes are equal.
Magnetic saturation is negligible.
Determine for steady-state operation:

(i) The output voltage on open-circuit and for a load current of 50 A when
the machine runs as a fully compensated generator at rated speed with a
control current of 3·43 mA.

(ii) The output current on short-circuit and for an output voltage of 50 V
when the machine runs without the compensating winding connected at
rated speed and with a control current of 73·3 mA.

Sketch the output-voltage/output-current characteristics in each case.

Winding transfer constants are $\lambda N_f = 750\,\text{V/A}$ and $\lambda N_A' = 1\cdot5\,\text{V/A}$. Since the stator quadrature-winding turns are equal to the effective armature turns,

$$\lambda N_g = \lambda N_A' = 1\cdot5\,\text{V/A}$$

For the fully compensated machine,

$$\lambda N_c = \lambda N_A' = 1\cdot5\,\text{V/A}$$

For the uncompensated machine,

$$\lambda N_c = 0$$

(i) Fully compensated machine
Substituting the known constants in eqns. (14.16) and (14.17),

$$750 I_f = 0\cdot2 I_q \tag{i}$$

$$3 I_q = V_d + 0\cdot1 I_d \tag{ii}$$

Substituting the value $I_f = 3\cdot43 \times 10^{-3}$ in eqn. (i),

$$I_q = \frac{750 \times 3\cdot43 \times 10^{-3}}{0\cdot2} = 12\cdot9\,\text{A}$$

When the machine is open-circuited, $I_d = 0$ and from eqn. (ii),

$$V_d = 3 I_q = 3 \times 12\cdot9 = \underline{\underline{38\cdot6\,\text{V}}}$$

When the load current is 50 A, from eqn. (ii),

$$V_d = 3 I_q - 0\cdot1 I_d = (3 \times 12\cdot9) - (0\cdot1 \times 50) = \underline{\underline{33\cdot6\,\text{V}}}$$

(ii) Uncompensated machine
Substituting the known constants in eqns. (14.16) and (14.17),

$$750 I_f - 1\cdot5 I_d = 0\cdot2 I_q \tag{iii}$$

$$3 I_q = V_d + 0\cdot05 I_d \tag{iv}$$

Substituting the value $I_f = 73\cdot3 \times 10^{-3}$ in eqn. (iii),

$$55 - 1\cdot5 I_d = 0\cdot2 I_q \tag{v}$$

When the machine is short-circuited, $V_d = 0$. From eqn. (iv),

$$I_q = \frac{0\cdot05 I_d}{3}$$

Substituting for I_q in eqn. (v),

$$55 - 1\cdot5 I_d = \frac{0\cdot2 \times 0\cdot05 I_d}{3}$$

so that

$$I_d = \frac{55}{1\cdot503} = \underline{\underline{36\cdot5\,\text{A}}}$$

When the load voltage $V_d = 50\,\text{V}$, then from eqn. (iv),

$$I_q = \frac{50 - 0.05I_d}{3}$$

Substituting for I_q in eqn. (v),

$$55 - 1.5I_d = 0.2\left(\frac{50 - 0.05I_d}{3}\right)$$

This gives

$$I_d = \frac{55 - 3.33}{1.503} = \underline{\underline{34.3\,\text{A}}}$$

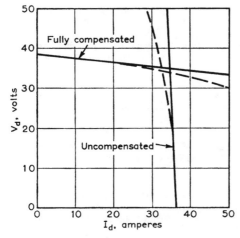

Fig. 14.6 V_d/I_d CHARACTERISTICS OF FULLY COMPENSATED AND
UNCOMPENSATED CROSS-FIELD GENERATORS

It will be noted that I_q is constant in the operation of the fully compensated machine, and is approximately proportional to the output voltage in the uncompensated machine. The V_d/I_d characteristic for each mode of operation is shown in Fig. 14.6.

The V_d/I_d characteristics shown in Fig. 14.6 are straight lines since the operating equations were derived assuming that the machine was linear and that, in particular, magnetic saturation was absent.

In practical machines the magnetic circuits are subjected to magnetic saturation, and this causes the characteristics to be non-linear. As magnetic saturation grows the values of the direct-axis and quadrature-axis permeances, Λ_d and Λ_q, will fall, and this causes the V_d/I_d characteristics of the fully compensated and uncompensated machines to be modified in the way indicated by the dotted lines in Fig. 14.6.

PROBLEMS

14.1 A 220 V 0·8 kW fully compensated cross-field generator has the following winding resistances:

Armature winding	$R_A = 0\cdot5\,\Omega$
Compensating winding	$R_c = 0\cdot5\,\Omega$
Quadrature winding	$R_g = 1\cdot5\,\Omega$
Control field winding	$R_f = 340\,\Omega$

Separate excitation open-circuit tests at rated speed gave the following constants:

O.C. quadrature-axis brush voltage	1,000 V/control-winding ampere
O.C. direct-axis brush voltage	250 V/quadrature-winding ampere

The stator quadrature-axis turns are equal to the effective armature turns. The permeances of the direct and quadrature axes are equal. Magnetic saturation is negligible.

Determine for steady-state operation (*a*) the control-winding current required to give rated voltage output on open-circuit, and (*b*) the terminal voltage at rated full-load current and with a control-field current as calculated in (*a*).

Ans. 0·88 mA; 216 V.

14.2 The cross-field generator of Problem 14.1 is operated with only half the compensating winding connected ($R_c = 0\cdot25\,\Omega$ and $N_c = 0\cdot5N_A'$). Determine, for steady-state operation,

    (*a*) The control-winding current required to give rated output current on short-circuit.

    (*b*) The output current at rated full-load voltage and with a control-field current as calculated in (*a*).

Ans. 0·455 A; 3·64 A.

14.3 A cross-field machine has an armature resistance of R_A when measured either at the direct-axis or the quadrature-axis brushes. The open-circuit quadrature-axis brush voltage is $\lambda N_A'$ per ampere of the direct-axis brush current. The machine has no stator windings. The permeances of the direct- and quadrature-axis magnetic circuits are equal.

Show that, if a constant direct voltage, V_q, is applied to the quadrature axis, and the direct- and quadrature-axis brush currents are I_d and I_q, the operating equations of the machine are

$$\lambda N_A'I_d = V_q - I_qR_A$$
$$\lambda N_A'I_q = V_d + I_dR_A$$

Assume that magnetic saturation is negligible and that the speed is constant.

Find the value of I_d when $R_A = 0\cdot50\,\Omega$, $V_q = 500$ V and $\lambda N_A' = 50$ V/A (*a*) if $R_L = 0$, and (*b*) if $R_L = 60\,\Omega$.

The subject of this question is a simple example of a machine called a *metadyne transformer*.

Ans. 10 A; 9·88 A.

14.4 Starting from eqns. (14.16) and (14.17), show, by eliminating I_q between the equations, that the control-field current of a cross-field generator is related to the output voltage and current by the equation

$$I_f = \frac{R_A + R_g}{\lambda^2 N_f(N_A' + N_g)} V_d + \left\{ \frac{N_A'(1 - k)}{N_f} + \frac{(R_A + R_c)(R_A + R_g)}{\lambda^2 N_f(N_A' + N_g)} \right\} I_d$$

where the compensation ratio, $k = N_c/N_A'$.

14.5 Using the equation derived in Problem 14.4, show that the control-field current of a cross-field generator is related to the output voltage and current by the equation

$$I_f = c_o V_d + \{c_s(1 - k) + c_o(R_A + R_c)\} I_d$$

where

$$c_o = \frac{\text{Control-field current}}{\text{Output voltage on open-circuit}} = \frac{\lambda^2 N_f(N_A' + N_g)}{R_A + R_g}$$

and

$$c_s = \frac{\text{Control-field current}}{\text{Output current on short-circuit with no compensation}}$$

$$= \frac{I_f}{I_d} = \frac{N_A'}{N_F} \quad \text{when } k = 0$$

For the machine to be efficient $N_A'/N_f \gg c_o(R_A + R_c)$, and this should be assumed.

14.6 Open- and short-circuit tests on a cross-field generator running at rated speed gave the following results:

Open-circuit tests: voltage gain, $V_d/V_f = 25$
Short-circuit test (no compensation): current gain $I_d/I_f = 0.95$

The armature winding resistance is $5\,\Omega$, the compensating winding resistance is $5\,\Omega$ and the control-field winding resistance is $100\,\Omega$.
Use the equation found in Problem 14.5 to determine the output voltage and current of the generator when it is driven at constant speed with a control-field current of 4 mA and a load resistance of $100\,\Omega$, when the machine has a compensation ratio of 0.99.

Ans. 73.5 V; 0.735 A.

Chapter 15

INTERCONNECTED SYSTEMS

Fig. 15.1 is a line diagram of two power stations A and B joined by an interconnector, the interconnector being connected to the busbars

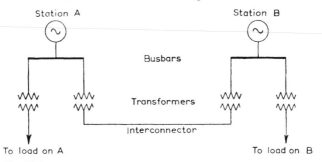

Fig. 15.1 INTERCONNEXION OF POWER STATIONS

of each station through transformers. Each station also has a feeder load connected through a transformer to its busbars.

The power sent across the interconnector will depend, ultimately, on the steam supply to the turbines of each station. For example, if the feeder loads on the busbars of A and B are each 50 MW and the output of the generators on A's busbars is 30 MW, the output of the generators on B's busbars must then be 70 MW, and 20 MW must be transmitted across the interconnector from B to A. As

502

was shown in Chapter 12, the output of the generators depends only on the power supply to their prime movers. Thus the power transmitted from point to point in an interconnected network depends ultimately on the steam supplies to the prime movers. Where more than one path is available between interconnected points, the proportion of power transmitted by each path may be controlled, but the total remains dependent on the load conditions.

The control of the power transmitted over the National Grid in this country is centralized in the control rooms of the Generating Divisions and in the National Control Room. These maintain communication with the generating stations coming under their control and issue instructions to station engineers to increase or reduce station loadings. The control room engineers thus control the frequency and the loading of transmission links in the network.

Apart from the question of the control of the power flow there is the question of voltage regulation. When power is transmitted across the interconnector there will be a voltage drop in the interconnector, the magnitude of which will depend on the impedance of the interconnector and on the power factor at which the power is transmitted. This voltage drop may be accommodated in a number of ways. Assuming that power is being transmitted from B to A (Fig. 15.1) these are as follows.

1. The busbar voltage at B or at A may be so adjusted that the difference in the busbar voltages is equal to the voltage drop in the interconnector and associated transformers. The disadvantage of this method is that it affects the voltages at which the loads connected to the station busbars are supplied.

2. The interconnector transformers may be equipped with on-load tap-changing gear. The voltage drop in the interconnector may then be supplied by adjusting the secondary e.m.f.s of the interconnector transformers, and the busbar voltages may be maintained constant. This method is commonly used where main transformers are, in any case, necessary.

3. A voltage boost in the appropriate direction may be injected into the interconnector either by an induction regulator or by a series boosting transformer. The latter is now of less importance due to the modern practice of incorporating on-load tap-changing gear in main transformers which, in effect, performs the same function as the series boosting transformer.

4. The secondary terminal voltages of the interconnector transformers may be held constant and the voltage drop in the interconnector may be accommodated by adjusting the relative phase of the voltages at the sending and receiving ends of the interconnector by means of a synchronous phase modifier. Synchronous phase

modifiers are used only on transmission links some hundreds of miles in length.

A further use of voltage regulating equipment is to control the division of power between two or more feeders or transmission lines operating in parallel. In the absence of voltage regulating equipment the division of the load between two lines is determined by their respective impedances. This division of the load may be modified by the introduction of a voltage boost in one line.

The control of the power division between lines in parallel by voltage boosting has the important advantage that both lines may be utilized to maximum capacity. It was shown in Chapter 9 that, when lines are operated in parallel, one may become fully loaded before the other has taken up its full load because of disproportionate impedances. A voltage boost of the appropriate magnitude and direction in such under-loaded lines may allow them to take up their full load.

15.1 Tap-changing Transformers

Fig. 15.2(*a*) shows a transformer having variable tappings in the secondary winding. As the position of the tap is varied, the effective

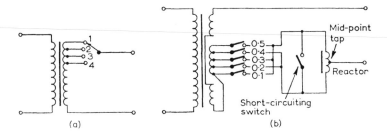

(a)   (b)

Fig. 15.2 TAP-CHANGING TRANSFORMER

number of secondary turns is varied, and hence the e.m.f. and output voltage of the secondary can be altered.

In supply networks, however, tap-changing has normally to be performed on load (that is, without causing an interruption to supply). The arrangement shown in Fig. 15.2(*a*) is unsuitable for this purpose. Suppose that the tapping is to be altered from position 1 to position 2. If contact with position 1 is broken before contact with position 2 is made, an open-circuit results. If, on the other hand contact with position 2 is made before contact with position 1 is broken, the coils connected between these two tapping points are short-circuited, and will carry damagingly heavy currents. Moreover, in both cases, switching would be accompanied by excessive arcing.

Fig. 15.2(*b*) shows diagrammatically one type of on-load tap-changing transformer. With switch 5 closed, all the secondary turns are in circuit. If the reactor short-circuiting switch is also closed, half the total current flows through each half of the reactor—since the currents in each half of the reactor are in opposition, no resultant flux is set up in the reactor and there is no inductive voltage-drop across it.

Suppose now it is desired to alter the tapping point to position 4. The reactor short-circuiting switch is opened. The load current now flows through one-half of the reactor coil only so that there is a voltage drop across the reactor. Switch 4 is now closed, so that the coils between tapping points 4 and 5 are now connected through the whole reactor winding. A circulating current will flow through this local circuit, but its value will be limited by the reactor. Switch 5 is now opened and the reactor short-circuiting switch is closed, thus completing the operation.

The tapping coils are placed physically in the centre of the transformer limb to avoid unbalanced axial forces acting on the coils, as would arise if they were placed at either end of the limb. Electrically, the tapped coils are at one end of the winding, the practice being to connect them at the earth-potential end.

15.2 Three-phase Induction Regulator

In construction, the 3-phase induction regulator resembles a 3-phase induction motor with a wound rotor. In the induction regulator, the rotor is locked, usually by means of a worm gear, to prevent its revolving under the action of the electromagnetic force operating on it. The position of the rotor winding relative to the stator winding is varied by means of the worm gear.

If the stator winding is connected to a constant-voltage constant-frequency supply, a rotating magnetic field is set up and will induce an e.m.f. in each phase of the rotor winding. The magnitude of the induced rotor e.m.f. per phase is independent of the rotor position, since the e.m.f. depends only on the speed of the rotating field and the strength of the flux, neither of which varies with rotor position. However, variation of the position of the rotor will affect the phase of the induced rotor e.m.f. with respect to that of the applied stator voltage.

Fig. 15.3(*a*) shows the star-connected stator winding of a 3-phase induction regulator with each of the rotor phase windings in series with one line of an interconnector. In Fig. 15.3(*b*) O*a*, O*b*, O*c* represent the input values of the line-to-neutral voltages of the interconnector. The circles drawn at the extremities *a*, *b*, *c* of these complexors represent the loci of the rotor phase e.m.f.s as the rotor position is varied with respect to the stator. The complexors *aa'*, *bb'* and *cc'* represent the voltage boosts introduced by the induction

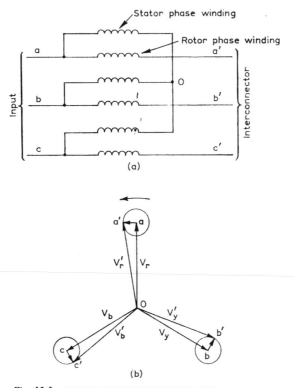

Fig. 15.3 POLYPHASE INDUCTION REGULATOR

regulator when the rotor position is such as to cause these voltage boosts to lead on their respective line-to-neutral voltages by 90°. O*a'*, O*b'* and O*c'* represent the resultant voltages V_r', V_y' and V_b'.

It will at once be seen that the induction regulator has altered the phase of the voltages as well as introducing a voltage boost.

To eliminate this phase displacement, a double polyphase induction regulator is employed, in which two rotors are assembled on a common shaft. The connexion diagram is shown in Fig. 15.4(*a*). The rotor windings of each regulator are connected in series with the interconnector. The stator windings are star-connected, but the phase sequence of one regulator stator is reversed with respect to the other. This reversal has the effect of eliminating any phase displacement in the resultant voltage boost in the interconnector. Thus, when the shaft of the double regulator is displaced, both rotors move by the same angular amount, but if the e.m.f. induced in one leads its former value, then the e.m.f. induced in the other lags by the same amount since the rotating fields in the regulators rotate in

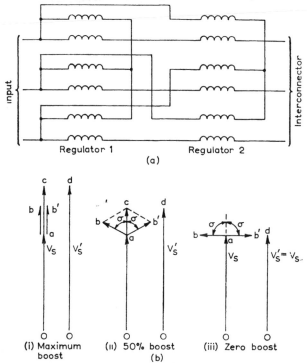

Fig. 15.4 DOUBLE POLYPHASE INDUCTION REGULATOR

opposite directions. Fig. 15.4(*b*) is a complexor diagram for various rotor positions. One phase only is shown for clarity. O*a* represents the unboosted input-end voltage, V_s, *ab* and *ab'* represent the voltage boosts supplied by each rotor, *ac* represents the resultant voltage boost, and O*d* represents the resultant voltage V_s'.

It is often convenient to reverse the functions of the stator and the rotor windings in induction regulators used for boosting. The rotor then carries the primary winding. This has the advantage of requiring only three connexions to the rotor instead of six, and the interconnector current flows in the stator instead of the rotor.

15.3 Synchronous Phase Modifier

In Chapter 12, it was shown that variation in the excitation of a synchronous motor alters the power factor at which the machine works. As the excitation of the machine is increased, the power factor passes from a lagging, through unity, to a leading power factor.

Use is made of this characteristic of the synchronous motor to correct the power factors of loads taking a lagging current. When so used the motor always acts with a leading power factor and is often

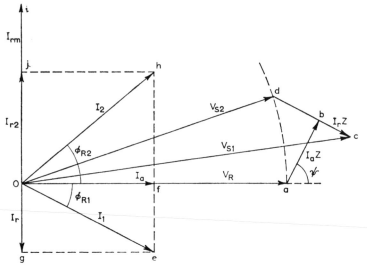

Fig. 15.5   VOLTAGE REGULATION BY SYNCHRONOUS PHASE
MODIFIER

called a *synchronous capacitor*. When the synchronous motor is used as a means of controlling the voltage of a transmission line the term *synchronous phase modifier*, or *synchronous compensator* is usually preferred, since, in this application, the machine may be adjusted to take either a leading or a lagging current. The machine is connected in parallel with the load at the receiving end of the line.

The action of the synchronous phase modifier in controlling the voltage of a transmission line is best understood by reference to the complexor diagram shown in Fig. 15.5. For simplicity the diagram is that of a short line where the effects of capacitance are neglected, but it should be understood that this method of control is mostly applied to long lines where, with other methods of control, the voltage drop along the line would be excessive.

In Fig. 15.5, Oa represents the receiving-end voltage V_R, and Oe represents the receiving-end current, I_1, lagging behind the receiving-end voltage by a phase angle ϕ_{R1}. Of and Og represent the active and reactive components (I_a and I_r) of current, respectively, ab represents the voltage drop I_aZ caused by the active component of current, which leads V_R by the phase angle of the line impedance, ψ ($\tan^{-1} X_L/R$), Z being the line impedance. bc represents the voltage drop I_rZ caused by the reactive component of current. bc lags I_aZ by 90°, since I_r lags behind I_a by 90° when the load power factor is lagging. In an unregulated line the sending-end voltage, V_{s1}, is the complexor sum of V_R, I_aZ and I_rZ.

Suppose now that the sending-end and receiving-end voltages are to be held constant at the same value; then the extremity of Od representing the new value of the sending-end voltage V_{s2} must be at some point along the arc ad, whose centre is O and radius is OD. Moreover, if the same power is to be sent along the line as previously, the I_aZ drop will remain the same since the active component of current must remain the same. However, if the excitation of a synchronous phase modifier connected to the receiving end is adjusted so that it takes a leading current—the current will lead by almost 90° since the modifier works on no-load—then as this leading current is increased the lagging reactive current drawn along the line will be reduced and the voltage drop I_rZ will be reduced. The extremity of the complexor representing the sending-end voltage will move along the line cb towards b. When the leading reactive current taken by the modifier is equal to the lagging reactive current of the load, there will be no reactive current drawn along the line and no I_rZ drop, and hence the extremity of the complexor representing the sending-end voltage will be at b. If the leading current taken by the modifier is further increased, the overall power factor of the load and the modifier together becomes leading and the extremity of the complexor representing the sending-end voltage lies along the line bd between b and d. Thus if the leading current taken by the modifier is made sufficiently great the sending-end voltage complexor takes up the position Od.

The synchronous phase modifier may therefore be used to control the voltage drop of a transmission line. If the sending-end voltage is maintained constant, then on full-load at a lagging power factor (the usual condition) the modifier will be over-excited to take a leading current. The receiving-end voltage will thus increase compared with its value had the line been unregulated. On no-load, on the other hand, the modifier would be under-excited and would take a lagging current in order to offset the voltage rise which occurs at the receiving end of a long unregulated line when the load is removed.

The power which may be sent along a transmission line is limited by either the power loss in the line reaching its permissible maximum value or by the voltage drop along the line reaching the maximum value which can be conveniently dealt with. On long transmission lines it is the voltage drop which limits the power which can be sent. Thus, if synchronous phase modifiers are used to regulate the voltage, more power can be dealt with by the line. Moreover, since voltage drop in the line and associated plant is not the first consideration when synchronous phase modifiers are used to control the voltage, current-limiting reactors may be incorporated in the system to reduce the maximum short-circuit current should a fault occur.

The principal disadvantage, apart from cost, of using synchronous phase modifiers is the possibility of their breaking from synchronism and causing an interruption to the supply.

15.4 Sending-end Voltage

In constant-voltage transmission systems using synchronous phase modifiers, the sending-end and receiving-end voltages are held constant, but they do not necessarily have to be equal. There is, indeed, an advantage in having the sending-end voltage higher than the receiving-end voltage, particularly with short lines, since under such conditions a smaller synchronous phase modifier capacity will satisfactorily regulate the voltage. For example, referring to Fig. 15.5, if the sending-end voltage had been greater than the receiving-end voltage, the reactive voltage drop *cd*, due to the reactive current of the synchronous phase modifier, would have been smaller and a synchronous phase modifier of smaller capacity would have been sufficient.

On longer lines, the capacitive effect tends to cause a voltage rise on light loads and no load, and the synchronous phase modifier has to work with a lagging power factor in order to hold the voltage constant. Thus the longer the line the less is the advantage of having the sending-end voltage higher than the receiving-end voltage.

EXAMPLE 15.1 A 3-phase transmission line has a resistance of $8 \cdot 75\,\Omega$ and an inductive reactance of $15\,\Omega$ per phase. The line supplies a load of $10\,\text{MW}$ at $0 \cdot 8$ power factor lagging and $33\,\text{kV}$. Determine the kVA rating of a synchronous phase modifier operating at zero power factor such that the sending-end voltage may be maintained at $33\,\text{kV}$. The effect of the capacitance of the line may be neglected. Determine, also, the maximum power which may be transmitted by the line at these voltages.

The problem is most easily tackled graphically.

The line current drawn by the load is

$$I = \frac{10^7}{\sqrt{3} \times 33 \times 10^3 \times 0\cdot8} \, \underline{/-\cos^{-1} 0\cdot8}$$

$$= 219\underline{/-36\cdot9°} = (175 - j132)\,\text{A}$$

Phase impedance $= 8\cdot75 + j15 = 17\cdot4\underline{/59\cdot7}\,\Omega$

The voltage drop caused by the active component of the load current is

$$I_a Z = 175 \times 17\cdot4 = 3\cdot04 \times 10^3\,\text{V}$$

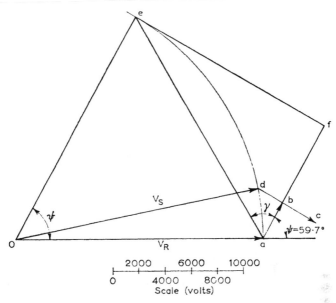

Fig. 15.6

The voltage drop caused by the reactive component of the load current is

$$I_r Z = 132 \times 17\cdot4 = 2\cdot3 \times 10^3\,\text{V}$$

Receiving-end voltage (phase value), $V_R = 19\cdot1 \times 10^3\,\text{V}$

Phase angle of line impedance, $\psi = \tan^{-1}\dfrac{X_L}{R} = \tan^{-1}\dfrac{15}{8\cdot75} = 59\cdot7°$

The graphical construction is shown in Fig. 15.6.

Oa is drawn to represent V_R to a suitable scale.

ab, equivalent in length to $3\cdot04 \times 10^3\,\text{V}$ to scale, is drawn making an angle ψ ($=59\cdot7$) with Oa to represent $I_a Z$.

bc, equivalent in length to $2\cdot3 \times 10^3\,\text{V}$ to scale, is drawn lagging behind ab by 90° to represent $I_r Z$.

The line joining O and c would then represent the necessary sending-end voltage in an unregulated line. With centre O and radius Oa (equivalent in length to 19·1 × 10³V) an arc *ae* is drawn. The extremity of the complexor representing the sending-end voltage must lie on this arc. *cb* is now produced until it cuts arc *ae* in *d*. *Od* represents the sending-end voltage.

cd represents the voltage drop due to the reactive current of the modifier alone.

It will be observed that the resultant power factor is leading and the resultant voltage drop, *bd*, due to the reactive current leads the voltage drop, I_aZ, due to the active component of current by 90°.

If I_r' is the reactive current of the synchronous phase modifier,

$$I_r'Z = dc = 4·46 \times 10^3 \text{ V to scale}$$

Therefore

$$I_r' = \frac{4,460}{17·4} = 257 \text{ A}$$

$$\text{Reactive MVA drawn by modifier} = 3 \times \frac{V_R I_r'}{10^6}$$

$$= \frac{3 \times 19·1 \times 10^3 \times 257}{10^6}$$

$$= 14·6 \text{ MVA}$$

To find the maximum power which can be sent along the line it is necessary to determine the maximum value which the active component of current, I_a, may have. This may be done from the diagram by determining the maximum value of the voltage drop I_aZ.

With centre O and radius O*a* (equal in magnitude to the sending-end voltage V_S), an arc, *ae*, is described. O*e* is drawn parallel to *ab* to cut this arc in *e*. From *e* a perpendicular to *ab* is drawn meeting *ab* in *f*. *af* is the maximum value which voltage drop I_aZ may have; *fe* is the corresponding voltage drop caused by the reactive component of current.

The maximum power is transmitted when the phase angle between the receiving-end and sending-end voltages is equal to that of the line impedance. If the sending-end voltage V_S were to lead the receiving-end voltage V_R by more than ψ the power sent would decrease, since the projection of the sending-end voltage on *ab* would decrease. From Fig. 15.6,

$$I_aZ = af = 9·6 \text{ kV} \quad \text{so that} \quad I_a = \frac{9·6 \times 10^3}{17·4} = 552 \text{ A}$$

$$\text{Power sent} = \frac{\sqrt{3}VI_a}{10^6} = \frac{\sqrt{3} \times 33,000 \times 552}{10^6} = 31·5 \text{ MW}$$

EXAMPLE 15.2 A 3-phase 50 Hz 132 kV transmission line 100 km long has the following constants per km: resistance 0·2 Ω, inductance 2 mH, capacitance 0·015 μF.

The sending-end voltage is 132 kV and the receiving-end voltage is held constant at 132 kV by means of a synchronous phase modifier. Determine the reactive kVA of the synchronous phase modifier when the load at the receiving end is 50 MW at a power factor of 0·8 lagging.

One method of taking the effect of line capacitance into account is to use a nominal-π equivalent circuit. The advantage of doing this is that the graphical method developed for the short line is still applicable if, instead of the receiving-end current, I_R, the current in the mid-section of the π-circuit, I', is used.

Fig. 15.7(*a*) shows the nominal-π equivalent circuit, and Fig. 15.7(*b*) shows the complexor diagram of this circuit. The voltage drop in the line impedance is shown as the sum of $I_a'Z$ (the voltage drop caused by the active component of I' and $I_r'Z$ (the voltage drop caused by the reactive component of I'). The voltage diagram showing the relationship between V_R and V_S is now similar to that for the short line.

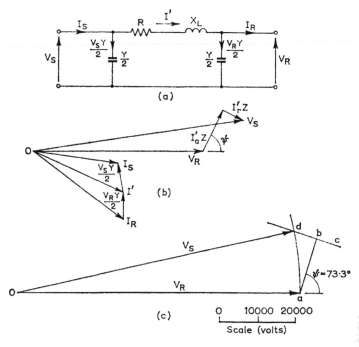

Fig. 15.7

Taking the receiving-end voltage as reference complexor and working with phase values,

$$V_R = \frac{132 \times 10^3}{\sqrt 3} \underline{/0^\circ} = 76 \cdot 2 \times 10^3 \underline{/0^\circ} \, \text{V}$$

$$I_R = \frac{50 \times 10^6}{\sqrt 3 \times 132 \times 10^3 \times 0 \cdot 8} \underline{/-\cos^{-1} 0 \cdot 8} = 273 \underline{/-36 \cdot 9^\circ} \, \text{A}$$

$$Z = 0 \cdot 2 \times 100 + j2\pi \times 50 \times 2 \times 10^{-3} \times 100$$

$$= 20 + j62 \cdot 8 = 66 \cdot 2 \underline{/72 \cdot 4^\circ} \, \Omega$$

$$\frac{Y}{2} = j \frac{2\pi \times 50 \times 0 \cdot 015 \times 10^{-6} \times 100}{2}$$

$$= j0 \cdot 236 \times 10^{-3} = 0 \cdot 236 \times 10^{-3} \underline{/90^\circ} \, \text{S}$$

Before proceeding to the normal graphical construction, I' should first be calculated with its active and reactive components.

$$I' = I_R + \frac{V_R Y}{2}$$

$$= (273\underline{/-36\cdot9^\circ}) + (76\cdot2 \times 10^3 \times 0\cdot236 \times 10^{-3}\underline{/90^\circ})$$

$$= 219 - j146\,\text{A}$$

$$I_a'Z = 219 \times 66\cdot2 = 14\cdot5 \times 10^3\,\text{V}$$

$$I_r'Z = 146 \times 66\cdot2 = 9\cdot65 \times 10^3\,\text{V}$$

The graphical construction shown in Fig. 15.7(c) is the same as that described in Example 15.1.

If I_r'' is the reactive current of the modifier, then

$$I_r''Z = cd = 15\cdot7 \times 10^3\,\text{V} \qquad \text{so that} \qquad I_r'' = \frac{15\cdot7 \times 10^3}{66\cdot2} = 236\,\text{A}$$

$$\text{Modifier MVAr} = \frac{\sqrt{3}VI_r''}{10^6} = \frac{\sqrt{3} \times 132 \times 10^3 \times 236}{10^6}$$

$$= 54\,\text{MVAr}$$

15.5 Power/Angle Diagram for a Short Line

In a constant-voltage system of transmission where the sending-end and receiving-end voltages, V_S and V_R, are maintained constant,

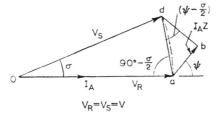

$$V_R = V_S = V$$

Fig. 15.8 COMPLEXOR DIAGRAM FOR A SHORT LINE

the power transmitted depends on the phase displacement between these voltages. It is possible to derive an expression for the power sent as a function of the phase displacement between V_S and V_R. To simplify the derivation it will be assumed (i) that the line is a short one, and (ii) that $V_S = V_R = V$ (phase values).

Fig. 15.8 shows the complexor diagram in which the receiving-end voltage V_R is taken as the reference. By geometry,

$$\angle adb = (\psi - \sigma/2) \qquad \text{so that} \qquad ab = ad \sin(\psi - \sigma/2)$$

But $ad = 2V \sin \sigma/2$; therefore

$$ab = 2V \sin (\sigma/2) \sin (\psi - \sigma/2) = V\{\cos (\sigma - \psi) - \cos \psi\}$$

Therefore

$$I_A = \frac{V}{Z} \{\cos (\sigma - \psi) - \cos \psi\}$$

and

$$\text{Power transmitted per phase} = VI_A = \frac{V^2}{Z} \{\cos (\sigma - \psi) - \cos \psi\}$$

When $\sigma = 0$ the power transmitted is zero. When $\sigma = \psi$ the power transmitted is a maximum (compare with Example 15.1). The power/angle diagram is shown in Fig. 15.9.

15.6 Stability of Operation of Synchronous Systems

Experience in operating transmission lines with synchronous machinery at both ends has shown that there are definite limits

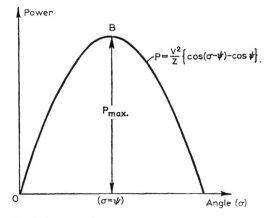

Fig. 15.9 POWER/ANGLE DIAGRAM FOR A SHORT LINE WITH EQUAL SENDING- AND RECEIVING-END VOLTAGES

beyond which operation becomes unstable, resulting in a loss of synchronism between the sending and receiving ends. It is possible to distinguish two limits of stability, a static limit and a dynamic or transient limit.

For given constant values of sending-end and receiving-end voltages, the load on a transmission line can be gradually increased until a condition is reached corresponding to B in Fig. 15.9. At this point the power transmitted is a maximum and corresponds to an angle of phase difference between the sending-end and receiving-end voltages of ψ ($\tan^{-1} X_L/R$). This point represents the static limit of stability (i.e. for a gradually applied load), and any attempt to impose further load on the line results in loss of synchronism between the ends of the line.

Short lines, where constant-voltage transmission using synchronous phase modifiers is not used, cannot be operated near the limit of stability, since the voltage drop along the line would become excessive. On long lines, however, where synchronous phase modifiers are used to control the voltage this limit may be approached and becomes of practical importance. Because of the high capital cost of transmission lines, the more power which can be transmitted over a given line the more economical becomes the operation.

The limit of stability of transmission lines is analogous to the limit of stability of a synchronous motor, where, as load is imposed on the machine, the rotor shifts backwards relative to the rotating field of the stator, and the angle of phase difference between the applied voltage and the e.m.f. increases. The limit of stability is reached when the e.m.f. lags behind the applied voltage by an angle equal to $\tan^{-1} X_s/R$.

This analogy may be made use of in examining the transient stability of a transmission line by considering a line loaded by a synchronous motor. Fig. 15.10 shows the power/angle diagram of a transmission line with the power axis scaled in units of the torque output of the synchronous motor at the receiving end of the line. This may be done conveniently since the synchronous motor is a constant-speed machine and consequently the torque output is proportional to the power transmitted.

When a load is suddenly applied to a synchronous motor, the inertia of the rotor prevents it from immediately falling back by the appropriate electrical angle. Suppose that, in Fig. 15.10, a motor is operating stably at a point A, where the torque developed, T_A, is equal to the load torque. If the load torque is suddenly increased to T_B, the angular displacement of the rotor remains momentarily at σ_A so that the electrically developed torque remains at T_A. The difference torque ($T_B - T_A$) must therefore slow down the rotor causing its angular displacement to increase. The rotor, in slowing down, loses an amount of kinetic energy proportional to the area of the triangle DBA (energy = torque × angular displacement), this energy being transferred to the load.

At B the rotor is running below synchronous speed, so that the angular displacement will continue to increase. However, for angular displacements greater than σ_B the generated torque will exceed the load torque and the rotor will be accelerated; it will regain synchronous speed at an angular displacement σ_C. During this time the load torque remains constant at T_B, so that the torque represented by the difference between the curve BC and the line BE must be that which accelerates the rotor. The area BCE is then proportional to the energy stored in the rotor due to its acceleration. The areas ADB and BCE must be equal if the kinetic energy taken from the rotor during deceleration is all to be returned during the accelerating

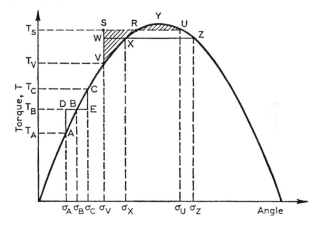

Fig. 15.10   TRANSIENT STABILITY OF A TRANSMISSION LINE

period. This is necessary for the operation to be again stable. This is called the *equal-area criterion for stability*.

At C the electrically developed torque, T_C, exceeds the load torque, so that the machine will continue to accelerate and its speed will rise above synchronous speed. The angular displacement will therefore be reduced, and the electrically developed torque will fall. Beyond B the rotor will experience a synchronizing torque tending to decelerate it. It will fall in speed until it again runs at synchronous speed at A. The whole cycle of events will then be repeated, giving rise to phase swinging or hunting between A and C.

In actual fact the damping which takes place means that the swing of the rotor will becomes less and less, and the machine will eventually run stably at B. The frequency of the oscillation may be determined from eqn. (12.48).

Consider now the motor operating at V with a load torque T_V. If the load torque suddenly increases to T_S, then the shaded area VSR represents the energy which is taken from the rotor as it slows down to accommodate the increased torque. The energy returned to the rotor while it accelerates under the influence of the synchronizing torque is the shaded area RUY. This is less than area VSR, so that the rotor cannot have as much energy returned to it as has been taken from it. It must therefore fall out of step (i.e. lose synchronism).

The maximum load which can be suddenly applied at V is represented by VW, where area VWX is just equal to area XYZ. Note that in this case it is possible for the rotor to swing through an angle which is greater than the maximum angular displacement for static stability. It is, of course, not possible for the operating point to lie beyond the peak point Y on the power/angle diagram.

PROBLEMS

15.1 Explain, with the aid of a complexor diagram, how the power factor of the load influences the voltage drop in a transmission line.

A 3-phase load of 10 MW at 0·8 p.f. lagging is supplied at 33 kV by an overhead line, each conductor of which has a resistance of 2·9 Ω and an inductive reactance of 6·5 Ω. A 3-phase bank of capacitors is connected to the load end of the line so that the voltage at the sending end is equal to that at the load. Calculate the MVA rating of the capacitors. (*L.U.*)

Ans. 12·2 MVA.

15.2 A 3-phase transmission line 25 km in length supplies a load of 10 MW at 0·8 p.f. lagging at a voltage of 33 kV. The resistance and reactance per km per conductor are 0·35 Ω and 0·6 Ω respectively. Neglecting the capacitance of the line, determine the rating of a synchronous capacitor, operating at zero power factor, connected at the load end of the line such that the sending-end voltage may be 33 kV. (*L.U.*)

Ans. 14·7 MVA.

15.3 A 3-phase transmission line is automatically regulated to zero voltage regulation by means of a synchronous phase modifier at the load end. If the full-load output is 50 MW at 0·8 p.f. lagging delivered at 200 kV and the line-to-neutral impedance is $(20 + j60)\,\Omega$, find the input to the synchronous set under these conditions. Deduce any formula employed. (*H.N.C.*)

Ans. 60·3 MVAr.

15.4 A 3-phase overhead line has resistance and reactance of 12 Ω and 40 Ω respectively per phase. The supply voltage is 132 kV and the load-end voltage is maintained constant at 132 kV for all loads by an automatically controlled synchronous phase modifier. Determine the kVAr of the modifier when the load at the receiving end is 120 MW at power factor 0·8 lagging. (*H.N.C.*)

Ans. 145 MVAr.

15.5 A 3-phase transmission line has a resistance of 6 Ω/phase and a reactance of 20 Ω/phase. The sending-end voltage is 66 kV and the voltage at the receiving end is maintained constant by a synchronous phase modifier.

Determine the MVAr of the synchronous phase modifier when the load at the receiving end is 75 MW at 0·8 p.f. lagging, and also the maximum load which can be transmitted over the line, the voltage being 66 kV at both ends.   (*L.U.*)

Ans. 96·8 MVAr; 148 MW.

15.6 A 3-phase transmission line has a resistance per phase of 5 Ω and an inductive reactance per phase of 12 Ω, and the line voltage at the receiving end is 33 kV.

(*a*) Determine the voltage at the sending end when the load at the receiving end is 20 MVA at 0·8 p.f. lagging.

(*b*) The voltage at the sending end is maintained constant at 36 kV by means of a synchronous phase modifier at the receiving end, which has the same rating at zero load at the receiving end as for the full load of 16 MW. Determine the power factor of the full-load output and the rating of the synchronous phase modifier.   (*L.U.*)

Ans. 40·1 kV; 0·90 lagging; 7·92 MVAr.

15.7 The "constants" per kilometre per conductor of a 150 km 3-phase line are as follows:

Resistance, 0·25 Ω; inductance, 2 × 10⁻³ H; capacitance to neutral, 0·015 μF.

A balanced 3-phase load of 40 MVA at 0·8 p.f. lagging is connected to the receiving end, and a synchronous capacitor operating at zero power factor leading, is connected to the mid-point of the line. The frequency is 50 Hz.

If the voltage at the load is 120 kV, determine the MVA rating of the synchronous capacitor in order that the voltage at the sending end may be equal in magnitude to that at the mid-point. The nominal-T circuit is to be used for the calculations.   (*L.U.*)

Ans. 31·9 MVA.

15.8 A 3-phase 50 Hz transmission line has the following values per phase per km: *R* = 0·25 Ω;   *L* = 2·0 mH;   *C* = 0·014 μF. The line is 50 km long, the voltage at the receiving end is 132 kV and the power delivered is 80 MVA at 0·8 power factor lagging.

If the voltage at the sending end is maintained at 140 kV by a synchronous phase modifier at the receiving end, determine the kVAr of this machine (i) with no load, (ii) with full load at the receiving end.   (*L.U.*)

Ans. 30·4 MVAr lagging; 42·6 MVAr leading.

15.9 A 3-phase transmission line has a resistance of 10 Ω per phase and a reactance of 30 Ω per phase.

Determine the maximum power which could be delivered if 132 kV were maintained at each end.

Derive a curve showing the relationship between the power delivered and the angle between the voltage at the sending and receiving ends, and explain how this curve could be used to determine the maximum additional load which could be suddenly switched on without loss of stability if the line were already carrying, say, 50 MW.   (*L.U.*)

Ans. 380 MW.

15.10 Develop an expression connecting the power received with the angle between the sending-end and receiving-end voltages of a regulated transmission line in terms of these voltages and the line constants, ignoring capacitance between lines.

Sketch the curve which this equation represents and use it to describe briefly what is meant by (*a*) static, and (*b*) transient stability.

Find the maximum power which can be transmitted over the following line:

Impedance per conductor: $(24 + j45)\,\Omega$
Receiving-end voltage: 110 kV
Regulation: Zero   (*H.N.C.*)

Ans. 126 MW.

15.11 Describe the necessary conditions under which power can be transmitted between two interconnected power stations.

Two power stations are linked by a 3-phase interconnector the impedance per line of which (including transformers and reactors) is $(10 + j40)\,\Omega$. The busbar voltage of each station is 66 kV. Calculate the angular displacement between the two station voltages in order to transmit 8 MW from one station to the other.

(*H.N.C.*)

Ans. 4·8°.

Chapter 16

TRAVELLING VOLTAGE SURGES

A voltage surge in a transmission system consists of a sudden voltage rise at some point in the system, and the transmission of this voltage to other parts of the system, at a velocity which depends on the medium in which the voltage wave is travelling.

The initiation of voltage surges on overhead transmission lines is frequently caused by lightning discharges. The voltage may be induced in the line due to a lightning discharge in the vicinity without the discharge occurring directly to the line. A most severe voltage rise may be caused where the lightning discharge is direct to a line conductor; such direct strokes, however, are not common.

Voltage surges may also be initiated by switching. Such transient disturbances are due to the rapid redistribution of the energy associated with electric and magnetic fields. For example when the current in an inductive circuit is interrupted the energy stored in the magnetic field must be rapidly transferred to the associated electric field and will give rise to a sudden increase in voltage.

Fig. 16.1 shows the waveform of a typical surge voltage. This is, in effect, a graph of the build-up of voltage at a particular point to a base of time. The steepness of the wavefront is of great importance, since the steeper the wavefront the more rapid is the build-up of voltage at any point on the network, and the properties of insulators depend on the rate of rise of voltage. In most cases the build-up is comparatively rapid, being of the order of 1–5 μs.

Surge voltages are usually specified in terms of the rise time and the time to decay to half maximum value. For example, a $1/50\ \mu s$ wave is one which reaches its maximum value in 1 μs and decays to

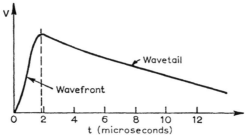

Fig. 16.1   WAVEFORM OF A VOLTAGE SURGE

half its maximum value in 50 μs. Impulse voltage tests are usually carried out with a wave of this shape.

16.1   Velocity of Propagation of a Surge

In the circuit of Fig. 16.2 the high-voltage source is assumed to give a constant high voltage E, which is applied to one end of the

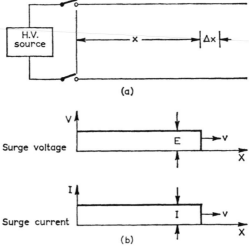

Fig. 16.2   CONSTANT VOLTAGE SURGE

line when the switch is closed. This will simulate a lightning stroke with a rapid rise of voltage and a long tail. It will also represent 50 Hz phenomena on short lines for reasons which will be discussed later. The lines will be assumed loss-free, i.e. no conductor resistance or leakage conductance. This greatly simplifies the theory and errs on the right side, since the effect of losses is to reduce the size of a surge and the rate of rise of voltage. Real surges then will not be quite so severe as those calculated by the present theory.

Let C = Line capacitance per unit length (F/m)
L = Line inductance per unit length (H/m)

When the switch is closed the whole line will not immediately become charged to a voltage E, for raising the voltage of a length of line entails charging the capacitance of that length and this entails a current flow from the source. The current cannot immediately flow due to the inductance of the line and therefore the line voltage cannot immediately rise to the voltage E, though in time it will do so.

Suppose that at an instant t after the closing of the switch a length x of the line has become charged to voltage E, and that those parts of the line beyond a further distance Δx are not yet charged at all. The intermediate region of length Δx will be called the "disturbance" since it is only in this region that the voltage is changing—in front of the disturbance the voltage is zero; behind it the voltage is constant at E. The disturbance will move along the line from the switch with a uniform velocity since the line is uniform.

Let v be the velocity of the disturbance; i.e. the *surge velocity*, or *speed of propagation*.

Since the velocity is uniform equal lengths (v units) of the line will be charged up each second as the disturbance moves along the line. Therefore

Charge required per second = E × capacitance of length v

$$= ECv$$

i.e.

Charging current flowing along line, $I = ECv$ (16.1)

I is called the *surge current*.

As the surge proceeds along the line a new length v will carry the current I each second, i.e. a length which has an inductance Lv has the current in it changed from zero to I in 1 sec.

Potential required to increase current

= Inductance × Rate of change of current

$$= LvI \text{ volts}$$

The potential applied to increase the current is E since in front of

the disturbance the potential is zero and behind it the potential
is E. Therefore

$$E = Lvl \tag{16.2}$$

From eqns. (16.1) and (16.2),

$$\text{Surge velocity, } v = \frac{1}{\sqrt{(LC)}} \tag{16.3}$$

Consider, for example, the velocity of propagation in a concentric
cable.

$$\text{Inductance/unit length} = \frac{\mu}{2\pi} \log_e \frac{b}{a} \text{ henrys/metre} \tag{7.29}$$

$$\text{Capacitance/unit length} = 2\pi \frac{\epsilon}{\log_e \dfrac{b}{a}} \text{ farads/metre} \tag{7.7}$$

where b and a are the sheath and core radii respectively, and μ and ϵ
are the permeability and permittivity of the dielectric material
respectively. The internal sheath linkages are not considered since
a surge is a high-speed effect equivalent to a high-frequency effect
and skin effect will reduce the internal linkages. If the resistance is
zero, the depth of penetration is also necessarily zero.

$\mu = \mu_0$ if the dielectric is non-magnetic
$\epsilon = \epsilon_r \epsilon_0$ where ϵ_r is the relative permittivity

Therefore

$$v = \frac{1}{\sqrt{(LC)}} = \frac{1}{\sqrt{(\mu_0 \epsilon_0 \epsilon_r)}} \tag{16.4}$$

i.e. the velocity is independent of the size and spacing of the conductors. This applies to all configurations of lossless conductors,
e.g. a core that is not concentric, or a twin-line system. Substituting
numerical values for μ_0 and ϵ_0,

$$v = \frac{1}{\sqrt{\left(4\pi \times 10^{-7} \times \dfrac{\epsilon_r}{36\pi \times 10^9}\right)}} = \frac{3 \times 10^8}{\sqrt{\epsilon_r}} \text{ metres/second} \tag{16.5}$$

This is the velocity of electromagnetic waves in the medium concerned.

The surge velocity is seen to be extremely fast (3×10^8 m/s in air).
Even in a cable with a relative permittivity of 9 (fairly high) the
velocity is 10^8 m/s. If, say, $5\,\mu$s is the time of rise of the surge voltage
(length of the wavefront), then the length of the disturbance, is

$5 \times 10^{-6} \times 10^8 = 500$ m, which is short for power transmission-line distances (not, of course, for high-frequency transmission-line distances). For power transmission lines, surges are often represented by a "vertical front block" as in Fig. 16.2(*b*) since the length of the disturbance is small.

In $\frac{1}{50}$th second a surge will travel $\frac{1}{50} \times 3 \times 10^8 = 6 \times 10^6$ m (3,730 miles). Thus in $\frac{1}{10}$th of a cycle, for a 50 Hz system, a surge will travel the whole length of a 600 km line (about 400 miles). In this period the voltage of the 50 Hz source will not have greatly changed, so that if a 50 Hz source is suddenly connected to a line the surge may be examined by the present theory taking the surge voltage as the instantaneous voltage when the switch is closed. Naturally the peak voltage should be considered, since this will be at the most dangerous instant at which the switch might be closed.

16.2 Surge Impedance

Dividing and simplifying eqns. (16.1) and (16.2),

$$\frac{E}{I} = \sqrt{\frac{L}{C}} \tag{16.6}$$

$\sqrt{(L/C)}$ is called the *surge impedance*, or *characteristic impedance*, Z_0, of a line. For a lossless line the surge impedance is evidently a pure resistance.

It should be noted that the surge current I, consequent on a voltage surge E, is related to the voltage surge by eqn. (16.6), i.e. by the properties of the line in which the surge travels.

The magnitude of the surge impedance for a particular conductor configuration may be determined from

$$Z_0 = \sqrt{\frac{L}{C}} \text{ ohms} \tag{16.7}$$

For an overhead power line the surge impedance is usually about $300\,\Omega$; for a power cable it usually is about $50\,\Omega$. The inductance per unit length of a line increases with the spacing of the conductors while the capacitance per unit length decreases as the spacing is increased. Thus, by eqn. (16.7), the surge impedance will increase with the spacing.

16.3 Power Input and Energy Storage

When the switch of Fig. 16.2(*a*) is closed, the high-voltage source maintains a potential E volts across the input to the line and supplies a current I amperes:

Power input to line $= EI$ watts

This energy input must become the energy stored in the line, for until the surge reaches the termination, there can be no output of energy from the line.

If v is the surge velocity, then in one second a length v stores electrostatic energy $\frac{1}{2}CvE^2$ and electromagnetic energy $\frac{1}{2}LvI^2$.

Energy input per second = Energy stored per second

i.e.

$$EI = \frac{1}{2}CvE^2 + \frac{1}{2}LvI^2 \tag{16.8}$$

Now,

$$\frac{1}{2}CvE^2 = \frac{1}{2}E\frac{C}{\sqrt{(LC)}}E = \frac{1}{2}E\frac{E}{\sqrt{\dfrac{L}{C}}} = \frac{1}{2}EI$$

and

$$\frac{1}{2}LvI^2 = \frac{1}{2}I\frac{L}{\sqrt{(LC)}}I = \frac{1}{2}I\sqrt{\frac{L}{C}}I = \frac{1}{2}EI$$

Thus the electrostatic and electromagnetic stored energies are equal.

16.4 Terminations

Consider a surge, of voltage E_i, impinging on the termination of a transmission line; if the characteristic or surge impedance of the line is Z_0, then the surge current I_i in the transmission line is given by

$$I_i = \frac{E_i}{Z_0} \tag{16.6a}$$

This surge impinging on the termination is called the *incident surge*; E_i is the *incident surge voltage* and I_i is the *incident surge current*.

Power conveyed to termination with incident surge = E_iI_i

Consider the particular case of a line terminated in a pure resistor equal to the surge impedance of the line. When the surge arrives the current in the resistor is E_i/Z_0.

Power absorbed by resistor = $E_i \times \dfrac{E_i}{Z_0} = E_iI_i$

= Power transmitted by the surge

In this case the surge power is exactly absorbed by the terminating impedance and there will be no further changes, i.e. the line will continue to be charged to potential E_i, and will continue to carry a current I_i.

If the terminating resistor has a resistance R, either higher or lower than the surge impedance Z_0, then changes in both voltage and current will occur. For instance if $R > Z_0$, then, on the arrival of the surge at the termination, the current through the terminating resistor will be E_i/R, which will be less than E_i/Z_0. The incident current on the line is in excess of the current which can be absorbed by the terminating resistor at the surge voltage. Since the current cannot instantaneously decrease due to the inductance of the line, the excess current will increase the charge on the capacitance at the end of the line. This increases the voltage at the termination to a value higher than the incident voltage E_i.

Let the voltage at the termination rise to E_T. Then

$$\text{Current through terminating resistor} = \frac{E_T}{R} = I_T$$

$$\text{Excess voltage appearing at termination} = E_T - E_i$$

$$\text{Excess current at termination} = I_T - I_i$$

The whole line is now charged to a potential E_i, but there has been a further rise of potential to E_T at the end of the line, i.e. an excess potential suddenly appears at the termination of the line. This is a similar condition to the initial closure of the switch which produced the incident surge, and so, in a similar manner, a surge will now travel from the termination back along the line. This is called the *reflected surge*.

$$\text{Reflected surge voltage, } E_r = E_T - E_i$$

Therefore

$$E_i + E_r = E_T \tag{16.9}$$

$$\text{Reflected surge current, } I_r = I_T - I_i$$

Thus

$$I_i + I_r = I_T \tag{16.10}$$

where $I_i = E_i/Z_0$, $I_r = -E_r/Z_0$, $I_T = E_T/R$.

The positive current direction is the direction of the current in the incident surge. Since the reflected current has the opposite direction the minus sign is necessary.

Substituting in eqn. (16.10),

$$\frac{E_i}{Z_0} - \frac{E_r}{Z_0} = \frac{E_T}{R}$$

Substituting for E_T from eqn. (16.9),

$$\frac{E_i}{Z_0} - \frac{E_r}{Z_0} = \frac{E_i + E_r}{R}$$

Therefore

$$RE_i - RE_r = Z_0 E_i + Z_0 E_r$$

and

$$E_r = \frac{R - Z_0}{R + Z_0} E_i \tag{16.11}$$

Substituting for E_r in eqn. (16.9),

$$E_i + \frac{R - Z_0}{R + Z_0} E_i = E_T$$

Thus

$$E_T = \frac{2R}{R + Z_0} E_i \tag{16.12}$$

Also

$$I_r = -\frac{E_r}{Z_0} = -\frac{1}{Z_0} \frac{R - Z_0}{R + Z_0} E_i \tag{16.13}$$

$$I_T = \frac{E_T}{R} = \frac{2}{R + Z_0} E_i \tag{16.14}$$

Graphs of the voltage and current distributions along the line are shown in Fig. 16.3 for instants before and after the incident surge reaches the termination. Since $R > Z_0$, the current at the termination is reduced and the reflected current surge is negative.

If the terminating resistance R is less than Z_0, then on the arrival of the incident surge the current, E_i/R, in the terminating resistor will be greater than the surge current $I_i = E_i/Z_0$. Due to the line inductance the line current cannot suddenly increase. The capacitance at the end of the line is then discharged, so reducing the voltage at the termination. The reduction of voltage at the termination is equivalent to the sudden application of a negative voltage surge at the termination which travels back along the line. This is the *reflected surge*.

Since the reflected voltage is negative the reflected current will be a negative current in the negative direction, i.e. equivalent to a positive current.

All the previous equations apply without change. The voltage and current distributions are shown in Fig. 16.4.

For an open-circuited line, $R \to \infty$, and hence the terminating current tends to zero. Therefore, by eqn. (16.11),

Reflected voltage, $E_r = \dfrac{\infty - Z_0}{\infty + Z_0} E_i = E_i$

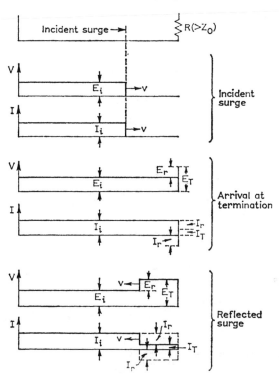

Fig. 16.3   REFLECTION AT A TERMINATION WHERE $R > Z_0$

and

Termination voltage, $E_T = E_i + E_i = 2E_i$

Reflected current, $I_r = -\dfrac{E_r}{Z_0} = -\dfrac{E_i}{Z_0} = -I_i$

Thus

Termination current, $I_T = I_r + I_i = 0$

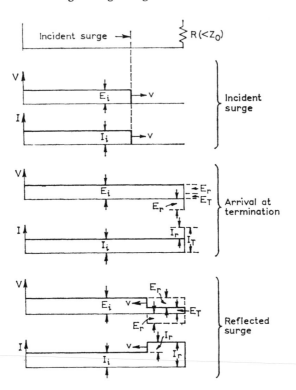

Fig. 16.4   REFLECTION AT A TERMINATION WHERE $R < Z_0$

For a short-circuited line, $R = 0$; therefore by eqn. (16.11),

Reflected voltage, $E_r = \dfrac{0 - Z_0}{0 + Z_0} E_t = -E_t$

and

Termination voltage, $E_T = E_t - E_t = 0$

Reflected current, $I_r = -\dfrac{E_r}{Z_0} = \dfrac{E_t}{Z_0} = I_t$

Thus

Termination current, $I_T = I_t + I_t = 2I_t$

Obviously there can be no voltage at the short-circuit. It should be noted that the terminating voltage and current are often called the *transmitted voltage* and *current*.

To summarize, if $R > Z_0$ the voltage surge is reflected unchanged in sign, while the current surge is reflected with the opposite sign. If $R < Z_0$ the sign of the reflected voltage surge is reversed, while that of the current surge is unchanged.

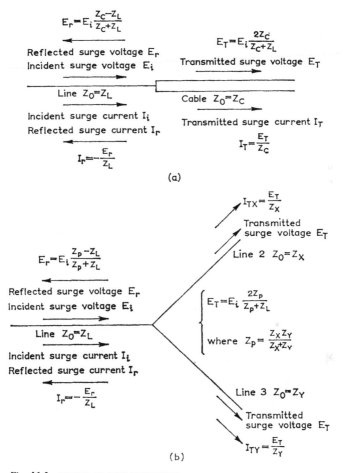

$$E_r = E_i \frac{Z_C - Z_L}{Z_C + Z_L}$$

Reflected surge voltage E_r
Incident surge voltage E_i

$$E_T = E_i \frac{2Z_C}{Z_C + Z_L}$$

Transmitted surge voltage E_T

Line $Z_0 = Z_L$

Cable $Z_0 = Z_C$

Incident surge current I_i
Reflected surge current I_r

Transmitted surge current I_T

$$I_T = \frac{E_T}{Z_C}$$

$$I_r = -\frac{E_r}{Z_L}$$

(a)

$$I_{TX} = \frac{E_T}{Z_X}$$

Transmitted surge voltage E_T

Line 2 $Z_0 = Z_X$

$$E_r = E_i \frac{Z_p - Z_L}{Z_p + Z_L}$$

Reflected surge voltage E_r
Incident surge voltage E_i

$$E_T = E_i \frac{2Z_p}{Z_p + Z_L}$$

where $Z_p = \frac{Z_X Z_Y}{Z_X + Z_Y}$

Line $Z_0 = Z_L$

Incident surge current I_i
Reflected surge current I_r

$$I_r = -\frac{E_r}{Z_L}$$

Line 3 $Z_0 = Z_Y$

Transmitted surge voltage E_T

$$I_{TY} = \frac{E_T}{Z_Y}$$

(b)

Fig. 16.5 SURGES AT LINE JUNCTIONS

16.5 Junctions of Lines having Different Characteristic Impedances

If a voltage surge travels along a line towards a junction where the characteristic impedance of the line changes, reflexion will take place at the junction and the surge transmitted into the section beyond the junction will be modified in value (Fig. 16.5(a)).

Consider the case of a junction between an overhead line and a cable. Let the characteristic impedance of the line be Z_L and that of the cable be Z_C. Suppose a surge voltage E_i is initiated in the line and travels down the line to the junction between the line and the cable. The cable presents a terminal impedance of Z_C to the line, since the cable voltage will be Z_C times the cable current under surge conditions.

Initial surge voltage in line $= E_i$

Initial surge current in line $= I_i = \dfrac{E_i}{Z_L}$

The final voltage in the line after reflexion from the junction is the voltage surge transmitted into the cable:

$$E_T = \frac{2Z_C}{Z_C + Z_L}\, E_i \tag{16.12}$$

The final current in the line after reflexion from the junction is the current surge transmitted into the cable:

$$I_T = \frac{2E_i}{Z_C + Z_L} \tag{16.14}$$

Reflected voltage in line, $E_r = \dfrac{Z_C - Z_L}{Z_C + Z_L}\, E_i \tag{16.11}$

Reflected current in line, $I_r = -\dfrac{1}{Z_0}\dfrac{Z_C - Z_L}{Z_C + Z_L}\, E_i \tag{16.13}$

It should be noted that, since the characteristic impedance of the cable Z_C is likely to be much less than the characteristic impedance of the line Z_L, the magnitude of the voltage surge in the cable will be much less than that in the line.

If there is a junction between three lines or a "tee" junction on one line (Fig. 16.5(b)), the reflected and transmitted surges may be calculated by the above equations with Z_p replacing Z_C, where

$$Z_p = \frac{Z_X Z_Y}{Z_X + Z_Y}$$

EXAMPLE 16.1 An underground cable having an inductance of 0·3 mH/km and a capacitance of 0·4 μF/km is connected in series with an overhead line having an inductance of 2·0 mH/km and a capacitance of 0·014 μF/km.

Calculate the values of the transmitted and reflected waves of voltage and current at the junction, due to a voltage surge of 100 kV travelling to the junction (a) along the cable, and (b) along the line.   (*L.U.*)

Characteristic impedance of cable, $Z_C = \sqrt{\dfrac{L}{C}}$

$$= \sqrt{\frac{0.3 \times 10^{-3}}{0.4 \times 10^{-6}}} = 27.4\,\Omega$$

Characteristic impedance of line, $Z_L = \sqrt{\dfrac{L}{C}}$

$$= \sqrt{\frac{2.0 \times 10^{-3}}{0.014 \times 10^{-6}}} = 378\,\Omega$$

(a) 100 kV surge initiated in cable

Initial value of surge voltage, $E_i = 100\,\text{kV}$

Initial value of surge current, $I_i = \dfrac{E_i}{Z_C} = \dfrac{100}{27.4} = 3.65\,\text{kA}$

Surge voltage transmitted into line, $E_T = \dfrac{2Z_L}{Z_L + Z_C}\,E_i$

$$= \frac{2 \times 378}{378 + 27.4} \times 100 = \underline{\underline{186\,\text{kV}}}$$

Surge current transmitted into line, $I_T = \dfrac{E_T}{Z_T} = \dfrac{E_T}{Z_L}$

$$= \frac{186}{378} = \underline{\underline{0.492\,\text{kA}}}$$

Reflected surge voltage in cable, $E_r = E_i \dfrac{Z_L - Z_C}{Z_L + Z_C}$

$$= 100 \times \frac{378 - 27.4}{378 + 27.4} = \underline{\underline{86\,\text{kV}}}$$

Reflected surge current in cable, $I_r = -\dfrac{E_r}{Z_C} = -\dfrac{86}{27.4} = \underline{\underline{-3.16\,\text{kA}}}$

(b) 100 kV surge initiated in line

Initial value of surge voltage, $E_i = 100\,\text{kV}$

Initial value of surge current, $I_i = \dfrac{E_i}{Z_L} = \dfrac{100}{378} = 0.264\,\text{kA}$

Surge voltage transmitted into cable, $E_T = \dfrac{2Z_C}{Z_C + Z_L}\,E_i$

$$= \frac{2 \times 27.4}{27.4 + 378} = 100 = \underline{\underline{13.5\,\text{kV}}}$$

Surge current transmitted into cable, $I_T = \dfrac{E_T}{Z_C} = \dfrac{13.5}{27.4} = \underline{\underline{0.494\,\text{kA}}}$

Reflected surge voltage in line, $E_r = E_T - E_i = \underline{\underline{-86.5\,\text{kV}}}$

Reflected surge current in line $= -\dfrac{E_r}{Z_L} = +\dfrac{86.5}{378} = \underline{\underline{0.23\,\text{kA}}}$

534 Travelling Voltage Surges

EXAMPLE 16.2 A single-phase overhead line is 50 km long and has a surge impedance of 300 Ω. The line has a circuit-breaker at the input end. A dead

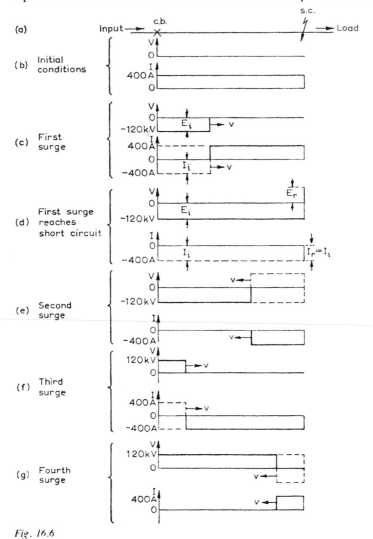

Fig. 16.6

short-circuit occurs at the terminating end and the circuit-breaker suddenly interrupts the short-circuit current when it has an instantaneous value of 400 A. Describe the surge phenomena which will occur in the line.

The line is shown in Fig. 16.6(a). Under the initial short-circuit conditions the voltage along the line may be taken as zero and the current as uniform at 400 A.

When the circuit-breaker opens, the current of 400 A cannot immediately cease due to the line inductance; thus the line capacitance at the circuit-breaker end must become negatively charged by the instantaneous continuation of the 400 A current. A negative voltage then arises at the circuit-breaker end while the rest of the line is uncharged; hence a negative voltage surge travels from the circuit-breaker end. The current associated with this surge will be −400 A, since there can be no resultant current at the open-circuited circuit-breaker end. Therefore

$$\text{Voltage of surge} = -400 \times Z_0 = -400 \times 300 \, \text{V} = -120 \text{kV}$$

i.e. a surge of −120 kV and −400 A travels down the line to the short-circuit termination.

At the short-circuit the termination voltage is necessarily zero; thus a reflected surge arises with a voltage of +120 kV and a current of −400 A. (At a short-circuit the surge voltage is reflected with change of sign and the surge current is reflected without change of sign, see Section 16.4.) When the reflected surge reaches the open-circuited circuit-breaker end it will in turn be reflected back down the line. The third surge voltage will be +120 kV and the third surge current will be +400 A. (At an open-circuit the surge voltage is reflected without change of sign and the surge current is reflected with change of sign; see Section 16.4.) These reflexions obey the rules: (i) the resultant voltage at a short-circuit must be zero, and (ii) the resultant current at an open-circuit must be zero.

The fourth surge stage is shown in Fig. 16.6(*g*).

It will be seen that after the fourth stage the resultant voltage and current are the same as the initial voltage and current. The fifth surge would then be the same as the first surge. In all real lines the losses will continually be reducing the magnitude of the surges so that the later surges are much smaller than the first ones.

Since this line is an overhead line, the surge velocity will approach 3×10^8 m/s. Thus the time required for a surge to travel the length of the line will be $50,000/(3 \times 10^8)$ s, i.e. 0·167 ms.

16.6 Surges of Short Duration

The previous theory has been developed on the assumption of a sudden rise of voltage followed by the steady application of the same voltage. In practice the sudden rise of voltage is usually followed by a slow fall of voltage back to normal. If the surge front reaches the far end of the line before the input voltage falls appreciably, then the previous theory gives a good representation of the surge conditions on the line. If the sudden rise of voltage is followed by a rapid fall so that the voltage surge becomes a voltage pulse, the previous theory does not give a clear representation of the conditions on the line. The theory is, however, still applicable and a further simple assumption gives a good representation of the actual conditions.

Fig. 16.7(*a*) shows a pulse or short-duration-surge source connected to a line. The output voltage of the source is shown in Fig. 16.7(*b*).

This output voltage may be considered as (i) a positive voltage surge of infinite duration followed at a discrete interval by (ii) a negative voltage surge of infinite duration and the same magnitude. The negative surge will then cancel the "tail" of the positive surge. Since both the positive and negative long-duration surges will obey the previous equations, the pulse voltage will also obey them.

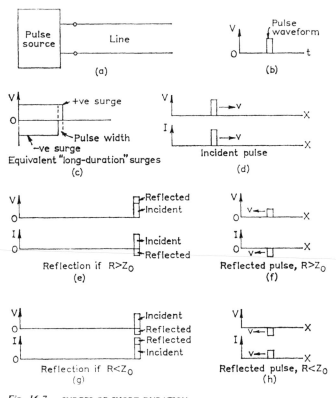

Fig. 16.7.   SURGES OF SHORT DURATION

Incident and reflected pulses are shown in Figs. 16.7(*d*)–(*h*). Losses along the line will tend to decrease the pulse magnitude and to broaden the pulse front and pulse tail.

EXAMPLE 16.3   A short-duration pulse of magnitude 10 kV travels along a very long line of characteristic impedance 300 Ω. The line is joined to a similar very long line by a cable of characteristic impedance 30 Ω and with a relative permittivity of 4. If the cable is 1·5 km long, calculate the magnitude of the first and second pulses entering the second line. What is the time interval between the two pulses?

$$\text{Magnitude of pulse transmitted into cable} = 10 \times \frac{2 \times 30}{30 + 300} \qquad (16.12)$$

$$= 1 \cdot 82\,\text{kV}$$

Part of the input pulse will be reflected back along the first very long line and thus conveyed away from the cable. Assuming no losses in the cable, the pulse of $1 \cdot 82\,\text{kV}$ will be incident on the junction of the cable with the second length of line.

Magnitude of pulse transmitted into second line from cable

$$= 1 \cdot 82 \times \frac{2 \times 300}{300 + 30} = 3 \cdot 31\,\text{kV}$$

This is the first pulse to travel along the second length of line.

Magnitude of pulse reflected back into cable from junction with second length of line

$$= 1 \cdot 82 \times \frac{300 - 30}{300 + 30} = 1 \cdot 49\,\text{kV} \qquad (16.11)$$

This reflected pulse will travel back along the cable to the junction with the first length of line. Part of the reflected pulse will be transmitted into the first length of line and part of it will be reflected and will again pass down the cable toward the second length of line.

Magnitude of pulse reflected back into cable from junction with first length of line

$$= 1 \cdot 49 \times \frac{300 - 30}{300 + 30} = 1 \cdot 22\,\text{kV} \qquad (16.11)$$

This pulse will form a second pulse incident on the junction of the cable and the second length of line. Therefore

Magnitude of second pulse transmitted into second line

$$= 1 \cdot 22 \times \frac{2 \times 300}{300 + 30} = 2 \cdot 22\,\text{kV}$$

The time interval between the two pulses will be the time required for a pulse to travel first back and then forward along the length of cable, i.e. a total distance of 3 km in the cable.

$$\text{Velocity of pulse in cable} = \frac{3 \times 10^8}{\sqrt{4}} = 1 \cdot 5 \times 10^8\,\text{m/s}$$

Therefore

$$\text{Time interval} = \frac{3 \times 10^3}{1 \cdot 5 \times 10^8} = 2 \times 10^{-5}\,\text{s}$$

Note. The surge pulses in the second line were much smaller than those in the first. Equipment in stations at the ends of overhead lines is sometimes protected from overvoltage surges by bringing the overhead lines through a short length of cable before reaching the station.

16.7 Mitigation of High-voltage Surges

High-voltage surges are mainly due to either (*a*) lightning discharges, or (*b*) switching. The voltages set up by lightning discharges are reduced by stringing one or two earth wires above the main conductors. The voltages set up by switching are reduced by using *resistance switching*.

A direct lightning stroke to a line causes enormous voltages and there is little possibility of preventing these. Fortunately direct lightning strokes are rare. ʼIt is more common for high-voltage surges to be caused by the charge induced on the conductors of an overhead line when a charged cloud passes over or near to the line. The charge induced on the conductors will have the opposite polarity to that in the cloud. If the cloud passes slowly away, the charges induced on the conductors will gradually flow to earth and no disturbance will be caused. If, however, the cloud is suddenly discharged by a lightning stroke to earth or another cloud, then the induced charge on the line will be suddenly released and a surge voltage will travel along the line in either direction.

The charge induced on an overhead transmission line system by a charged cloud is mainly concentrated in the uppermost conductor, since the other conductors are to some extent electrostatically shielded by the uppermost one. If the uppermost conductor is made an earth wire and not one of the system conductors, then the charges induced on the system conductors are considerably reduced. This reduces the surges in the system conductors. It also affords some mitigation of the effects of a direct lightning stroke.

Where an earth wire is present the resistance of the tower footings must be kept low or *back flashover* from the earth wire may occur. For example, in the extreme case of the resistance to earth of the tower footing being infinite, any surge voltage wavefront reaching the tower base is doubled and reflected back to the earth wire. Eventual build-up of earth-wire potential to a value well above that of the system conductors may result in a discharge from the earth wire to the system conductors.

Transient currents and voltages naturally occur with most switching operations. Generally switching-in and disconnecting can be performed without dangerous disturbances arising. The interruption of a high short-circuit current by an efficient circuit-breaker does, however, tend to give high-voltage surges. These can be mitigated by arranging that the circuit-breaker will be opened in stages. During the stages one or more resistance sections carry the current which is being interrupted and part, at least, of the energy stored in the line inductance is dissipated in the switch resistors.

16.8 Protection of Insulation

The insulators which support an overhead line and the insulation of cables, switches or transformers will, under some surge conditions, have voltages impressed on them which are greater than the breakdown strength of the insulators or insulation. To prevent the breakdown of these costly units and to prevent the interruption of the supply which would result from their breakdown, the insulation is usually protected by an air-gap so arranged that the surge voltage will produce breakdown of the air-gap rather than of the insulation. A string of insulators for an overhead line, or the bushing of a transformer, has frequently a rod gap across it (Fig. 16.8), so that a

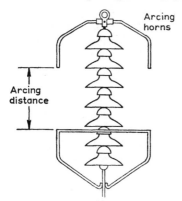

Fig. 16.8   ROD GAP PROTECTING AN INSULATOR STRING

spark or an arc will jump across the rod gap rather than down the insulator or the bushing.

Alternatively a metal ring concentric with the insulator string and about level with the third insulator shed may be used as the lower electrode in place of a rod electrode.

When setting the rod gap two factors must be taken into account: (*a*) impulse ratio and (*b*) time factor.

Impulse ratio

$$= \frac{\text{Breakdown voltage under surge conditions}}{\text{Breakdown voltage under low-frequency conditions}}$$

It is found that the breakdown voltage under surge, i.e. rapidly changing or high-frequency conditions, is often higher than the breakdown voltage under steady or low-frequency conditions. The *impulse ratio* is a measure of this difference. Supposing the breakdown voltage of a string of insulators is, say, 300 kV at 50 Hz and that the string is protected by a rod gap with a breakdown voltage of, say, 200 kV at 50 Hz. If the impulse ratio for the insulators is, say, 1·3, then the surge breakdown voltage for the insulators will be 390 kV; and if the impulse ratio for the rod gap is, say, 2·1, then the surge breakdown voltage for the rod gap will be 420 kV. The rod gap does not then protect the insulators under surge conditions. Either the impulse ratio for the rod gap must be improved or the 50 Hz setting for the rod gap must be reduced. The impulse ratio is found to depend on the geometry of the air-gap. A sphere gap, with relatively close spacing, has an impulse ratio of unity—a needle gap may have an impulse ratio of between 2 and 3.

Time Factor. The breakdown of insulation or an air-gap does not occur instantaneously on the application of the excess voltage. The time for the complete breakdown to develop depends (i) on the magnitude of the excess voltage, (ii) on the material in the breakdown path, and (iii) on the shape and spacing of the electrodes. Naturally the greater the excess voltage the shorter is the time required for breakdown to develop—a typical characteristic is shown in Fig. 16.9(*a*).

Fig. 16.9(*b*) compares the characteristics of a rod gap and an insulator which are used in conjunction with one another. If the voltage across them were slowly increased, the rod gap would correctly break down first, i.e. at the lower voltage. If a voltage greater than V_c were suddenly applied, the insulator would break down first and thus, under a steep-wavefront surge condition, the rod gap does not protect the insulator. The correct relative characteristics for a rod gap to protect the insulation for all surge voltages is shown in Fig. 16.9(*c*).

The time delay is relatively short for sphere gaps and relatively long for needle gaps.

16.9 Surge Diverters

Rather than permit a surge to impinge on the terminal apparatus it is advantageous to eliminate the surge if possible. The elimination may be carried out in two ways, either (*a*) the surge may be diverted to earth, i.e. short-circuited, or (*b*) the surge energy may be absorbed. The latter method is not now used.

Modern surge diverters consist essentially of elements having a non-linear volt/ampere characteristic, and made of a ceramic material consisting of silicon carbide bonded with clay.

To protect plant successfully against high-voltage travelling waves the surge diverter must operate, as far as possible, simultaneously with the incidence of the surge. Since the surge is travelling at

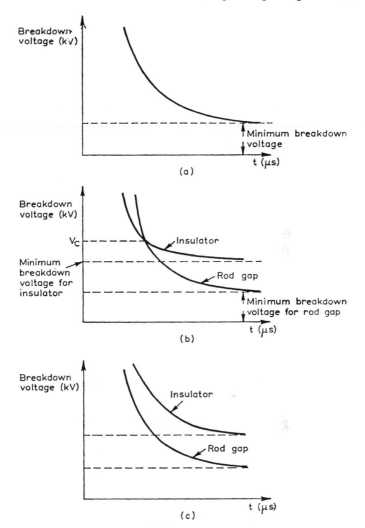

Fig. 16.9 BREAKDOWN-VOLTAGE/TIME CHARACTERISTICS

(*a*) Typical rod gap characteristic
(*b*) Unsuitable combination of insulator and rod gap
(*c*) Correct relative characteristics for insulator and rod gap protection

approximately 3×10^8 m/s a short delay will permit the surge to pass the diverter and be transmitted into the plant which the diverter is intended to protect. Moreover, the surge diverter should be placed as close as possible to the plant to be protected to obviate the risk of a surge being initiated in the line between the horn gap and the plant. The line between the surge diverter and the plant should be protected by an earth wire or wires.

Fig. 16.10 shows approximately a typical characteristic of a Metrosil disc suitable for incorporation in a surge diverter designed

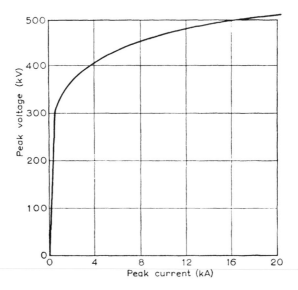

Fig. 16.10   TYPICAL METROSIL VOLT/AMPERE CHARACTERISTIC

to operate on a high-voltage transmission system to protect transformers and other plant from high-voltage surges.

The law connecting the applied voltage and the current is of the form

$$V = kI^\beta$$

where k is a constant depending on the geometrical form and β is a constant depending on the composition and treatment of the substance. Ideally β should be zero, so that whatever the value of the surge current the voltage would be constant. In practice values for β of the order of 0·2 are achieved.

The principle of operation is that a stack of Metrosil discs is connected between line and earth close to the transformer (or other plant) to be protected. Because of the nature of the volt/ampere characteristic, at normal voltage the diverter passes only a very small current to earth, but when a high over-voltage occurs the resistance

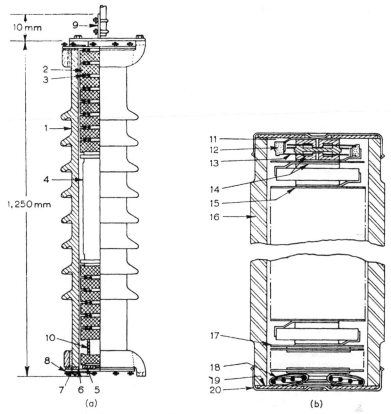

Fig. 16.11 METROSIL DIVERTER AND SPARK-GAP ASSEMBLY

(AEI Ltd.)

(a) Surge diverter

1. Glazed porcelain housing	6. Inner sealing gasket
2. Metrosil disc	7. Outer sealing gasket
3. Metallic spacers	8. Sealing plate
4. Spark-gap assembly	9. Terminal assembly
5. Compression spring	10. Spacing tube

(b) Section through spark-gap assembly

11. Contact clips	16. Porcelain housing
12. Metrosil grading ring	17. Metal spacers
13. Mica disc	18. Compression spring
14. Electrode	19. Contact plate
15. Locating disc	20. Sealing cap

of the diverter falls and the diverter passes a high current, diverting the surge energy to earth.

In practice, in a diverter suitable for operation on a 132kV system, a stack of Metrosil discs 6in. in diameter is assembled in a glazed porcelain housing which is provided on its exterior with rain sheds, which may be of a special shape for operation in dirt-laden atmospheres. Spark gaps are incorporated to prevent current flow to earth under normal-voltage conditions. The air in the interior of the porcelain housing containing the spark-gap assembly is evacuated and then the housing is filled with nitrogen. Fig. 16.11 shows the details of the arrangement of a 33 kV Metrosil diverter and spark-gap assembly.

The diverters are mounted vertically, mechanical support being also provided at the top of the assembly for larger ratings. External stress rings are provided for ratings above 110 kV.

The diverters are normally set to operate on twice normal voltage, it being undesirable for them to operate on small over-voltages. Operation is extremely rapid, taking less than a microsecond. The impulse ratio is practically unity.

PROBLEMS

16.1 An overhead line of surge impedance 500 Ω terminates in a transformer of surge impedance 3,500 Ω. Find the amplitudes of the current and voltage surge transmitted to the transformer due to an incident voltage of 30 kV.   (*H.N.C.*)

Ans. 52·5 kV; 0·015 kA.

16.2 Derive an expression for the surge impedance of a transmission line.

A transmission line has a capacitance of 0·012 μF per km and an inductance of 1·8 mH per km. This overhead line is continued by an underground cable with a capacitance of 0·45 μF per km and an inductance of 0·3 mH per km. Calculate the maximum voltage occurring at the junction of line and cable when a 20 kV surge travels along the cable towards the overhead line.   (*H.N.C.*)

Ans. 37·5 kV.

16.3 Obtain an expression for the surge impedance of a transmission line and for the velocity of propagation of electric waves in terms of the line inductance and capacitance.

A cable having an inductance 0·3 mH per km and a capacitance of 0·4 μF per km is connected in series with a transmission line having an inductance of 1·5 mH per km and a capacitance of 0·012 μF per km. A surge of peak value 50 kV originates in the line and progresses towards the cable. Find the voltage transmitted into the cable. Use the result to explain the practice sometimes adopted of terminating a line by a short length of cable before connecting to reactive apparatus.   (*H.N.C.*)

Ans. 7·2 kV.

16.4 An overhead transmission line 300 km long, having a surge impedance of 500 Ω is short-circuited at one end and a steady voltage of 3 kV is suddenly applied at the other end.

Neglecting the resistance of the line explain, with the aid of diagrams, how the current and voltage change at different parts of the line, and calculate the current at the end of the line 0·0015 s after the voltage is applied.

Ans. 0.

16.5 Two stations are connected together by an underground cable having a capacitance of 0·15 μF/km and an inductance of 0·35 mH/km joined to an overhead line having a capacitance of 0·01 μF/km and an inductance of 2·0 mH/km.

If a surge having a steady value of 100 kV travels along the cable towards the junction with an overhead line, determine the values of the reflected and transmitted waves of voltage and current at the junction.

State briefly how the transmitted waves would be modified along the overhead line if the line were of considerable length.   *(L.U.)*

Ans. 81 kV; 181 kV; 1·57 kA; 0·404 kA.

16.6 Derive an expression for the velocity with which a disturbance will be transmitted along a transmission line.

A disturbance, due to lightning, travels along an overhead line of characteristic impedance 200 Ω. After travelling 30 km along the line the disturbance reaches the end of the line where it is joined to a cable of surge impedance 50 Ω and dielectric constant [relative permittivity] 6. Calculate the relative magnitude of the energy of the disturbance in the cable and the time taken between initiation and arrival at a point 15 km along the cable from the junction.   *(H.N.C.)*

Ans. 0·64; 225 μs.

16.7 An overhead transmission line having a surge impedance of 500 Ω is connected at one end to two underground cables, one having a surge impedance of 40 Ω and the other one of 60 Ω. A rectangular wave having a value of 100 kV travels along the overhead line to the junction.

Deduce expressions for, and determine the magnitude of, the voltage and current waves reflected from and transmitted beyond the junction.

If the rectangular wave originated at a long distance from the junction, state how and why it would be modified in its passage along the line.   *(L.U.)*

Ans. 90·8 kV; 0·182 kA; 9·16 kV; 0·229 kA; 0·153 kA.

16.8 Two single transmission lines A and B with earth return are connected in series and at the junction a resistance of 2,000 Ω is connected between the lines and earth. The surge impedance of line A is 400 Ω and of B 600 Ω. A rectangular wave having an amplitude of 100 kV travels along line A to the junction.

Develop expressions for and determine the magnitude of the voltage and current waves reflected from and transmitted beyond the junction. What value of resistance at the junction would make the magnitude of the transmitted wave 100 kV?   *(L.U.)*

Ans. 7 kV; 0·018 kA; 107 kV; 0·178 kA; 1,200 Ω.

Chapter 17

SHORT-CIRCUIT PROTECTION

The possibility of a short-circuit to earth or between the phases of any transmission system due to mechanical and/or electrical break-down cannot be ignored. Hence it is essential that the maximum fault currents which could exist in such a system be calculated and that adequate measures be taken to reduce to a minimum the effects which these currents would have. It is also necessary to ensure that the circuit-breakers which are installed in the system will be capable of interrupting these fault currents.

Modern switchgear ratings are approximately:

35,000 MVA for the 400 kV system
15,000 MVA for the 275 kV system
7,000 MVA for the 132 kV system
2,500 MVA for the 66 kV system

Usually the internal leakage reactances of the interconnected generators and transformers are sufficient to limit the prospective fault MVA within these ratings, but in exceptional cases separate reactors may be used to limit fault levels. In older power stations the use of external reactors was standard practice. The methods of reactor connection are discussed below.

17.1 Reactor Control of Short-circuit Currents

The circuit-breakers connected in a transmission network must

546

be capable of dealing with the maximum possible short-circuit current that can occur at their points of connexion. If no steps are taken to limit the value of the possible short-circuit currents, not only will the duty required of circuit-breakers be excessively heavy, but damage to the lines, cables and interconnected plant will almost certainly occur.

Reactors may be connected in power station circuits to limit the maximum possible short-circuit current occurring at any point to a value which will not cause damage to plant for the short time during which it flows, and to relieve the duty required of circuit-breakers. The use of reactors may also increase the chance of continuity of supply by making it possible for healthy sections of the power station busbars to continue in operation.

There are three types of reactor in general use.

IRON-CORED MAGNETICALLY-SHIELDED REACTOR (Fig. 17.1)

The turns of the reactor are surrounded by a number of iron cores providing a complete magnetic path. The cross-sectional area of the cores is made as generous as possible to limit the change in reactance due to magnetic saturation of the cores. The disposition of the cores gives almost complete magnetic shielding. The whole unit is reinforced mechanically and immersed in oil to enable it to withstand high voltages, to permit of good cooling, and to make outdoor operation possible.

AIR-CORED RING-SHIELDED REACTOR (Fig. 17.2)

A reactance coil having an air core is surrounded by short-circuited shielding rings mounted sufficiently far from the coil not to reduce the inductance unduly. The shielding achieved is sufficiently good to permit the assembly to be inserted in an oil-filled tank. The chief advantage of this type is that saturation difficulties are avoided, but, compared with the iron-cored type, the L/R ratios obtainable are smaller.

AIR-CORED NON-SHIELDED REACTOR (Fig. 17.3)

In this type of reactor, the air-cored coil is supported by cast concrete pillars. The reactor is mounted on porcelain pedestal-type insulators. The construction is cheap and robust, but since no shielding is provided the reactor is not suitable for immersion in an oil tank. The reactor is not suitable for outdoor operation and may not be placed near metal objects.

17.2 Location of Current-limiting Reactors

Current-limiting reactors may be connected in series with each generator, in series with each feeder, or between busbar sections.

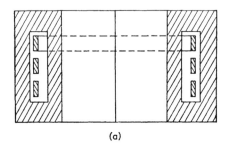

(a)

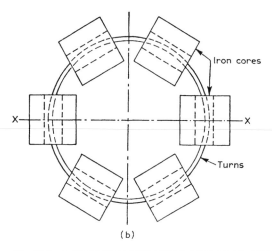

(b)

Fig. 17.1   IRON-CORED MAGNETICALLY-SHIELDED REACTOR
(a) Elevation on XX
(b) Plan

The connexion of reactors in series with each generator is not common, since modern power station generators have sufficient leakage reactance to enable them to withstand a symmetrical short-circuit across their terminals. The disadvantages of connecting reactors in series with the generators are (i) that there is a relatively large power loss and voltage drop in the reactor, (ii) that a busbar or feeder fault close to the busbar will reduce the busbar voltage to a low value thereby causing the generators to fall out of step, and (iii) that a fault on one feeder is likely to affect the continuity of supply to others.

When the reactors are connected in series with each feeder, a feeder fault will not seriously affect the busbar voltage so that

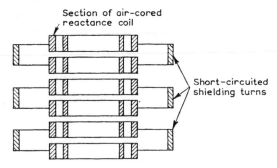

Fig. 17.2 AIR-CORED RING-SHIELDED REACTOR

there is little tendency for the generators to lose synchronism, other feeders will be little affected and the effects of the fault will be localized. The disadvantages of this method of connexion are (i) that

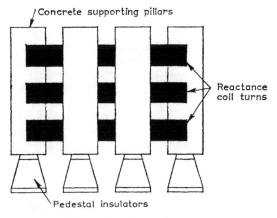

Fig. 17.3 AIR-CORED NON-SHIELDED REACTOR

there is a relatively large power loss and voltage drop in each reactor, since the reactors are in series with the feeder currents, (ii) if the number of generators is increased so the size of the feeder reactors will also have to be increased to keep the short-circuit current within the rating of the feeder circuit-breakers, and (iii) no protection is given against busbar faults.

The most common system of connecting reactors is between busbar sections. There are two methods, the ring system and the tie-bar system, and these are shown in Fig. 17.4. Under normal operating conditions, each generator will supply its own section of the load and no current will flow through the reactors, thus reducing power loss and voltage drop in the reactors. Even when power has to be transferred from one busbar section to another, a large voltage difference between sections is not necessary, since the voltage drop

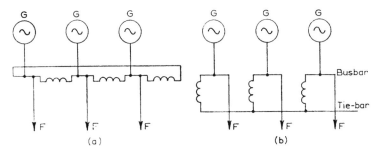

Fig. 17.4   BUSBAR REACTORS
(*a*) Ring reactors
(*b*) Tie-bar reactors

in the reactor will be almost in quadrature with the busbar voltage. The transfer of reactive current between sections is, however, difficult and would necessitate large voltage differences between sections, so that each generator must supply the reactive current for its own load.

The presence of the reactors between busbar sections has the effect of tending to localize any fault which occurs. Thus a feeder fault is likely to affect only the busbar section to which it is connected, the other sections being able to continue in normal operation.

Comparing the ring system with the tie-bar system, it will be seen that in the tie-bar system there are effectively two reactors in series between sections so that the reactors must have approximately half the reactance of those used in a comparable ring system. The tie-bar system has the disadvantage of requiring an additional busbar, but has the advantage that additional generators may be added to the system without necessitating changes in the existing reactors. This is because there is a limiting value of short-circuit current which can be fed into a busbar fault when the tie-bar system is used (see Section 17.5).

17.3 Per-unit Reactance and Resistance*

The per-unit value of a quantity A was defined in Chapter 9 as

$$A_{pu} = \frac{A}{A_{base}} \tag{9.78}$$

The per-unit reactance is therefore

$$X_{pu} = \frac{X}{X_{base}} \tag{17.1}$$

From eqn. (9.80) the base reactance is

$$X_{base} = \frac{V_{base}}{I_{base}} \tag{9.80}$$

The rated full-load values are usually chosen as bases, so that

$$X_{base} = \frac{V_{fl}}{I_{fl}} \tag{17.2}$$

Substituting for X_{base} in eqn. (17.1),

$$X_{pu} = X \frac{I_{fl}}{V_{fl}} \tag{17.3}$$

Consider a supply system of reactance X ohms and negligible resistance supplying a load; if the load is suddenly short-circuited, the current I_{sc} which will flow is

$$I_{sc} = \frac{V_{fl}}{X} = \frac{V_{fl}}{X_{pu}\dfrac{V_{fl}}{I_{fl}}} = \frac{I_{fl}}{X_{pu}} \tag{17.4}$$

If the rated full-load volt-amperes is S_{fl}, then multiplying eqn. (17.4) by V_{fl} gives

$$V_{fl}I_{sc} = \frac{V_{fl}I_{fl}}{X_{pu}} = \frac{S_{fl}}{X_{pu}}$$

i.e.

$$\text{Short-circuit volt-amperes} = \frac{S_{fl}}{X_{pu}} \tag{17.5}$$

* Reactances and resistances are often quoted in percentage values. The corresponding per-unit values are obtained by dividing the percentage values by 100.

Per-unit resistance is dealt with in an analogous manner; thus

$$R_{pu} = R \frac{I_{fl}}{V_{fl}} \qquad (17.6)$$

and

$$\text{Short-circuit volt-amperes} = \frac{S_{fl}}{R_{pu}} \qquad (17.7)$$

In a supply system containing both series resistance and reactance,

$$\text{Short-circuit volt-amperes} = \frac{S_{fl}}{R_{pu} + jX_{pu}} = \frac{S_{fl}}{Z_{pu}} \qquad (17.8)$$

17.4 Reference of Per-unit Impedance to a Common Base

Where a system contains units of different volt-ampere ratings the per-unit impedance of each unit may be based on one of the several ratings of the units involved. Before any calculations can be undertaken the per-unit impedances must all be referred to the same base volt-amperes.

Suppose the following details refer to two units A and B.

	kVA	Per-unit impedance	Full-load current
A	S_A	Z_{Apu}	$I_A = \dfrac{S_A}{V_{fl}}$
B	S_B	Z_{Bpu}	$I_B = \dfrac{S_B}{V_{fl}}$

Following eqn. (9.77) the per-unit value of any quantity A to $A_{base\,1}$ is

$$A_{pu\,1} = \frac{A}{A_{base1}} \qquad (9.77a)$$

Similarly the per-unit value of A to a second base value is

$$A_{pu\,2} = \frac{A}{A_{base2}} \qquad (9.77b)$$

Combining the above equations,

$$A_{pu\,2} = A_{pu\,1} \frac{A_{base1}}{A_{base2}} \qquad (9.82)$$

Let Z_{Apu}' be the per-unit value of Z_A referred to base C; i.e.

$$Z_{Apu}' = Z_{Apu} \frac{Z_{baseA}}{Z_{baseC}}$$

$$= Z_{Apu} \frac{V_{fl}}{I_A} \frac{I_C}{V_{fl}}$$

or

$$Z_{Apu}' = Z_{Apu} \frac{S_C}{S_A} \qquad (17.9)$$

Similarly the per-unit impedance of B referred to base C is

$$Z_{Bpu}' = Z_{Bpu} \frac{S_C}{S_A} \qquad (17.10)$$

EXAMPLE 17.1 A generating station is laid out as shown in Fig. 17.5(*a*). The ratings and per-unit reactances of the different elements are as indicated. Calculate the volt-amperes and the current fed into the following symmetrical 3-phase short circuits: (*a*) at a busbar section, e.g. at D; (*b*) at the distant end of a feeder, e.g. at I.

Refer to a base of 5 MVA.

Reference value of generator per-unit reactance $= 0.30 \times \dfrac{5}{10} = 0.15$ p.u.

Reference value of reactor per-unit reactance $= 0.10 \times \dfrac{5}{10} = 0.05$ p.u.

(*a*) The equivalent reactance diagram for a fault on a busbar section is shown in Fig. 17.5(*b*).

Equivalent per-unit reactance of parallel branch $= \dfrac{0.2 \times 0.2}{0.2 + 0.2} = 0.1$ p.u.

Equivalent per-unit reactance of branch $1 = 0.1 + 0.05 = 0.15$ p.u.

Total equivalent per-unit reactance $= \dfrac{0.15 \times 0.15}{0.15 + 0.15} = 0.075$ p.u.

Short-circuit MVA $= \dfrac{5}{0.075} = 66.7$ MVA

Short-circuit current $= \dfrac{66.7 \times 10^6}{\sqrt{3} \times 6.6 \times 10^3} = 5.83$ kA

(*b*) The equivalent impedance diagram for a fault at the distant end of a feeder is shown in Fig. 17.5(*c*).

Total equivalent per-unit impedance $= (0.06 + j0.205)$ p.u.

Magnitude of per-unit impedance $= 0.214$ p.u.

Short-circuit MVA $= \dfrac{5}{0.214} = 23.4$ MVA

Short-circuit current $= \dfrac{23.4 \times 10^6}{\sqrt{3} \times 11 \times 10^3} = 1.24$ kA

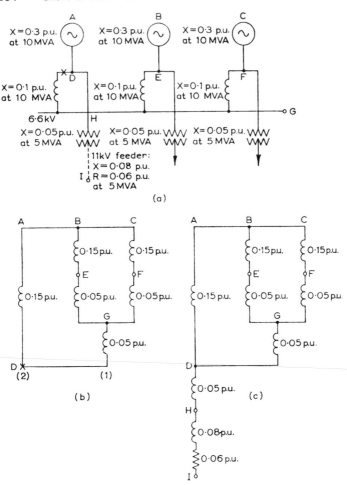

Fig. 17.5

17.5 Limiting Value of Short-circuit VA using Tie-bar Reactors

Consider a generating station having N busbar sections connected on the tie-bar system. Let each section have a generating capacity of S volt-amperes with internal per-unit reactances $X_{g\,pu}$ on a basis of S volt-amperes. Let each tie-bar reactor have a per-unit reactance of $X_{r\,pu}$ also on a basis of S volt-amperes. Assume a symmetrical 3-phase short-circuit occurs in a feeder connected to the busbars of section 1, as indicated by X in Fig. 17.6.

Per-unit reactance to fault from 1st generator $= X_{gpu}$

Short-circuit VA fed into fault by 1st generator $= \dfrac{S}{X_{gpu}}$

Per-unit reactance to fault from remaining $(N - 1)$ generators

$$= X_{rpu} + \frac{X_{rpu} + X_{gpu}}{N - 1} = \frac{X_{gpu} + NX_{rpu}}{N - 1}$$

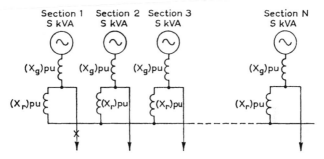

Fig. 17.6   CALCULATION OF SHORT-CIRCUIT VOLT-AMPERES ON
TIE-BAR SYSTEM

Short-circuit VA fed into fault by $(N - 1)$ generators

$$= \frac{S(N - 1)}{X_{gpu} + NX_{rpu}}$$

Total short-circuit VA $= S\left(\dfrac{1}{X_{gpu}} + \dfrac{N - 1}{X_{gpu} + NX_{rpu}}\right)$

Therefore

$$\underset{N \to \infty}{\mathrm{Lt}}\,(\text{s.c. VA}) = S\left(\frac{1}{X_{gpu}} + \frac{1}{X_{rpu}}\right) \tag{17.11}$$

This is the short-circuit volt-amperes at the fault if N is very large. A smaller number of generators would reduce the short-circuit volt-amperes. If the circuit-breaker on the faulty feeder has the rating derived from eqn. (17.11), then no matter how many extra generators and reactors are added, the circuit-breaker rating will remain adequate. This is a useful property of the tie-bar arrangement. With the ring arrangement extra generators may not be added without changing the existing circuit-breakers or increasing the existing reactance.

EXAMPLE 17.2   The busbars of a generating station are to be divided into three sections by the use of three reactors. A 60 MVA generator having 0·15 p.u.

leakage reactance is to be connected to each busbar section. Determine the minimum value of reactor reactance, in ohms, if the maximum MVA fed into a symmetrical 3-phase short-circuit at a section is to be 500, (a) if the three reactors are connected to a common tie-bar, and (b) if the three reactors are ring connected. The busbar voltage is 22 kV.

(a) Tie-bar Reactors (Fig. 17.7(a)).

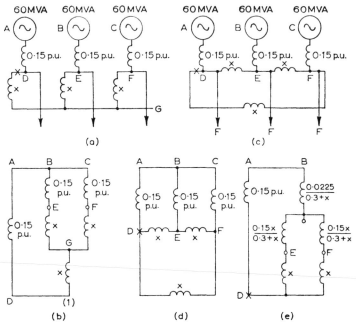

Fig. 17.7

The equivalent reactance diagram is shown in Fig. 17.7(b).
Let x be the per-unit reactance of each reactor on a 60 MVA base.

$$\text{Equivalent per-unit reactance of parallel branch} = \frac{0.15 + x}{2} \text{ p.u.}$$

$$\text{Equivalent per-unit reactance of branch 1} = x + \frac{0.15 + x}{2}$$

$$= \frac{0.15 + 3x}{2} \text{p.u.}$$

$$\text{Total equivalent per-unit reactance} = \frac{0.15 \times \dfrac{0.15 + 3x}{2}}{0.15 + \dfrac{0.15 + 3x}{2}} \text{ p.u.}$$

$$= \frac{0.15(0.05 + x)}{0.15 + x} \text{p.u.}$$

$$\text{Short-circuit MVA} = \frac{0 \cdot 15 + x}{0 \cdot 15(0 \cdot 05 + x)} = 500$$

since the short-circuit MVA is not to exceed 500 MVA. Solving gives

Per-unit reactance of each reactor, $x = 0 \cdot 35$ p.u.

$$\text{Full-load current, } I_{fl} = \frac{60 \times 10^6}{\sqrt{3} \times 22 \times 10^3} = 1{,}580 \text{ A}$$

$$\text{Reactance of each reactor, } X = \frac{V_{ph}}{I_{fl}} = \frac{22 \times 10^3}{\sqrt{3} \times 1{,}580} \times 0 \cdot 35 = \underline{\underline{2 \cdot 8 \Omega}}$$

(b) Ring Reactors (Fig. 17.7(c)).
The equivalent reactance diagram is shown in Fig. 17.7(d).
Let x be the equivalent per-unit reactance of each reactor on a 60 MVA base. To obtain an expression for the total equivalent per-unit reactance, the mesh BCFE may be replaced by the equivalent star.
Let the star-point of the equivalent star be O.

$$X_{BOpu} = \frac{0 \cdot 15 \times 0 \cdot 15}{0 \cdot 15 + 0 \cdot 15 + x} = \frac{0 \cdot 0225}{0 \cdot 30 + x} \text{ p.u.}$$

$$X_{FOpu} = \frac{0 \cdot 15x}{0 \cdot 15 + 0 \cdot 15 + x} = \frac{0 \cdot 15x}{0 \cdot 30 + x} \text{ p.u.}$$

$$X_{EOpu} = \frac{0 \cdot 15x}{0 \cdot 15 + 0 \cdot 15 + x} = \frac{0 \cdot 15x}{0 \cdot 30 + x} \text{ p.u.}$$

The equivalent reactance diagram is redrawn in Fig. 17.7(e).

$$X_{BDpu} = \frac{1}{2} \left(\frac{0 \cdot 15x}{0 \cdot 30 + x} + x \right) + \frac{0 \cdot 0225}{0 \cdot 30 + x} = \frac{x + 0 \cdot 15}{2}$$

$$\text{Total equivalent per-unit reactance} = \frac{0 \cdot 15 \times \dfrac{x + 0 \cdot 15}{2}}{0 \cdot 15 + \dfrac{x + 0 \cdot 15}{2}}$$

$$= \frac{0 \cdot 15(x + 0 \cdot 15)}{x + 0 \cdot 45} \text{ p.u.}$$

$$\text{Short-circuit MVA} = \frac{60(x + 0 \cdot 45)}{15(x + 0 \cdot 15)} = 500 \text{ MVA}$$

since the short-circuit MVA is not to exceed 500 MVA. Hence

Per-unit reactance of each reactor, $x = 1 \cdot 05$ p.u.

As before, the full-load current is 1,580 A; hence

$$\text{Reactance of each reactor} = \frac{22 \times 10^3}{\sqrt{3} \times 1{,}580} \times \frac{105}{100} = \underline{\underline{8 \cdot 4 \Omega}}$$

Note. In this particular problem, it is not absolutely necessary to use the delta-star transformation, since the symmetry of paths BED and CFD in Fig. 17.14(d) will mean that there is no current in the reactor EF.

17.6 Principles of Arc-extinction in Circuit-breakers

The calculation of the currents fed into symmetrical 3-phase short-circuits made in the previous sections, while giving an indication of the duty to which a particular circuit-breaker may be subjected, ignores the fact that there may be considerable asymmetry in the short-circuit current due to the presence of a d.c. component. It was seen in Section 6.4 that, when a short-circuit occurs in a circuit whose resistance is negligible compared with the inductive reactance, which is usually the case in transmission networks, the resulting short-circuit current has a d.c. component except when the short-circuit occurs at the instant at which the circuit voltage is a maximum. This d.c. component has a maximum value when the short-circuit occurs at the instant at which the circuit voltage is zero. Since in a 3-phase system there are six voltage zeros per cycle, it is certain that there will be considerable asymmetry in the current flowing in at least one of the phases. In considering the operation of circuit-breakers it is therefore necessary to take account of this asymmetry.

Fig. 17.8 indicates the process of arc extinction in a circuit-breaker. Fig. 17.8(*a*) shows the alternator line-to-neutral voltage. This voltage is shown as having constant amplitude, but it may be subjected to some decrement due to alternator armature reaction until the arc is extinguished, when the alternator voltage will slowly recover. Whether or not this effect is noticeable depends on whether the short-circuit endures for a sufficient length of time and on the distance of the fault from the alternator. If this distance is great (say at the distant end of a feeder), the alternator voltage decrement will not be marked.

In Fig. 17.8 it is assumed that the short-circuit occurs when the alternator phase voltage is passing through zero. As has been seen, this gives rise to maximum asymmetry when the circuit reactance is much greater than the resistance. The maximum peak current is about 1·8 times the peak symmetrical current, i.e. about 1·8 × √2, or 2·55 times the r.m.s. value of the symmetrical short-circuit current. The d.c. component rapidly dies away, a typical value of decrement factor being 0·8 per half-cycle. Thus, under this decrement factor, the d.c. component will have fallen to about 30 per cent of its initial value after 2½ cycles.

Fig. 17.8(*c*) shows the voltage across the contacts of the circuit-breaker. This is zero until the instant of separation. The voltage across the contacts after separation is the arc drop. The arc extinguishes each half-cycle and the principle of arc extinction in an a.c. circuit-breaker is to permit the arc to interrupt itself at a current zero. Whether or not the arc will restrike after a current zero depends on whether the insulation strength of the medium between the contacts (usually the medium is air or oil) builds up more rapidly or less rapidly than the voltage across the contacts. The build-up of insulation strength depends largely on the speed and thoroughness

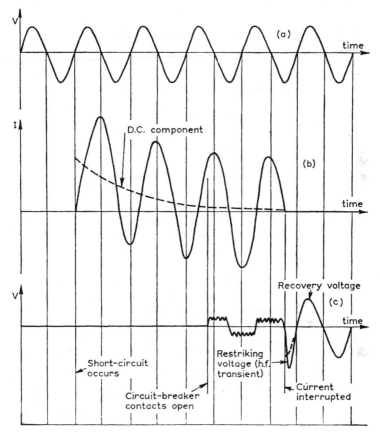

Fig. 17.8   INTERRUPTION OF AN ASYMMETRICAL SHORT-CIRCUIT
CURRENT

(a) Alternator line-to-neutral voltage
(b) Short-circuit current
(c) Voltage across circuit-breaker contacts

with which the ionized gas (caused by the heat and electronic bombardment in the arc) is removed and replaced by un-ionized air or oil.

When the arc is finally extinguished, there is an extremely rapid rise of voltage across the circuit-breaker contacts. This voltage, which is called the *restriking voltage*, is a high-frequency transient voltage and is caused by the rapid redistribution of energy between the magnetic and electric fields associated with the plant and transmission lines of the system.

The *recovery voltage* is the 50 Hz voltage that appears across the circuit-breaker contacts when final extinction takes place and is approximately equal to the alternator phase voltage. Since the power factor of the fault circuit is low, the instantaneous recovery voltage will be almost equal to the maximum value of the alternator phase voltage.

If the alternator voltage has been subjected to decrement due to armature reaction, then after final extinction of the arc has taken place there will be a relatively slow growth in the value of recovery voltage. This is not shown in Fig. 17.8(*c*).

17.7   Rating of Circuit-breakers

It was seen in Section 17.6 that, when a symmetrical 3-phase short-circuit occurs in a transmission system, there will be a considerable asymmetry in the short-circuit current of at least one phase. The more rapidly a circuit-breaker operates after the occurrence of a short-circuit, therefore, the more onerous may be its duty. If the operation of the circuit-breaker is delayed for a few cycles, the asymmetry will be considerably reduced. On the other hand, the longer the short-circuit persists, the greater is the chance of synchronous plant losing synchronism and causing a serious interruption to the supply. Therefore circuit-breaker action is made as rapid as possible even though this may tend to make their duty more onerous.*

It is normal practice to specify the *rupturing capacity* of circuit breakers in kilovolt-amperes or megavolt-amperes. This practice is well established but may be criticized as not being logical, since the breaking capacity in megavolt-amperes is obtained from the product of short-circuit current and recovery voltage. While the short-circuit current is flowing, however, there is only a small voltage across the circuit-breaker contacts (the recovery voltage does not appear across them until final extinction has taken place). Thus the MVA rating is the product of two quantities which do not exist simultaneously in the circuit-breaker. It would appear more logical to have a current rather than an MVA rating for circuit-breakers. The agreed international standard method of specifying circuit-breaker rupturing capacity is defined as a rated symmetrical current at a rated voltage.

---

* Asymmetry in the short-circuit current may actually relieve circuit-breaker duty since current zeros occur closer to voltage zeros so that the value of the fundamental component of recovery voltage is reduced.

The symmetrical breaking current of a circuit-breaker is the current which the circuit-breaker will interrupt at a power factor of 0·15 for ratings up to 500 MVA and a power factor of 0·3 for ratings of 750 MVA or upwards with a recovery voltage of 95 per cent normal voltage.

The asymmetrical breaking current is the current the circuit-breaker will interrupt when there is asymmetry in one of the phases. It is assumed, for purposes of proving the capacity of a circuit-breaker, that the asymmetrical d.c. component is 50 per cent of the maximum value of the a.c. component. Normal British practice is to make the a.c. component equal to the rated symmetrical current. With a decrement factor of 0·8 per half-cycle, the maximum asymmetrical d.c. component, which is initially 80 per cent of the maximum value of the a.c. component, will have fallen to 50 per cent of this value in a little over one cycle.

The r.m.s. value of an asymmetrical current having a 50 per cent d.c. component is approximately 1·25 times that of the a.c. component. If the a.c. component (I_{ac}) is equal to the rated symmetrical component, then the asymmetrical breaking current (I) will be equal to 1·25 times the rated symmetrical breaking current. This may be shown from eqn. (5.19), since

$$I = \sqrt{\left(I_{dc}{}^2 + \frac{I_m{}^2}{2}\right)}$$

where I_m is the peak value of the a.c. component. Therefore

$$I = \sqrt{(0{\cdot}25I_m{}^2 + 0{\cdot}5I_m{}^2)} = 0{\cdot}866I_m \approx 1{\cdot}25I_{ac}$$

17.8 Types of A.C. Circuit-breaker

It is impossible here to give more than an outline of the many types of a.c. circuit-breaker in use. Circuit-breakers may be divided into (i) those which do not incorporate some form of arc control, and (ii) those which do. The latter class may be subdivided into (ii*a*) self-blast arc-control circuit-breakers, and (ii*b*) circuit-breakers in which the arc control is provided by mechanical means external to the circuit-breaker. Fig. 17.9 shows a classification of circuit-breakers in the form of a diagram.

Circuit-breakers which do not have a form of arc control may be either plain air-break or plain oil-break. In these types the contacts separate either in air or in oil. In the plain oil-break circuit-breakers some assistance to arc extinction is afforded by the gas bubble generated around the arc. The gas bubble, by setting up turbulence in the oil, helps to eliminate ionized arc-products from the arc path. Long and inconsistent arcing times are obtained with these types of circuit-breaker and they are only suitable for low-current low-voltage operation.

Self-blast circuit-breakers are oil circuit-breakers in which the pressure of the gas bubble set up in the oil by the arc is utilized to force fresh, un-ionized oil into the arc path, thus materially increasing the rate of rise of insulation resistance in the circuit-breaker.

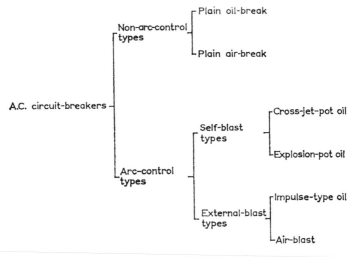

Fig. 17.9   TYPES OF CIRCUIT-BREAKER

Fig. 17.10 shows the explosion pot of a cross-jet oil circuit-breaker, which incorporates this principle of arc control. In this type of circuit-breaker, as the moving contact is withdrawn, the gas generated by the arc exerts pressure on the oil in the back passage, and as a result, when the moving contact uncovers the arc-splitting jets, fresh oil is forced across the arc path.

Circuit-breakers of this type are made with rupturing capacities of up to 2,500 MVA at 66 kV. For higher voltages and capacities multiple-break units have been developed. In these, two or more sets of cascaded contacts open simultaneously. The main difficulty associated with multiple-break units is to ensure uniform voltage distribution over the breaks. On open-circuit the voltage distribution is mainly controlled by the self-capacitance of the breaks, the insulation resistance being extremely high. Fig. 17.11 shows the opening sequence for a multi-break circuit-breaker with four cross-jet pots per phase. Four Metrosil resistors are used to control the voltages across the breaks and to eliminate over-voltages.

A difficulty experienced with self-blast oil circuit-breakers is that

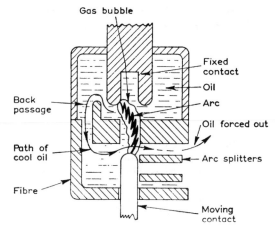

Fig. 17.10   EXPLOSION POT OF A CROSS-JET OIL CIRCUIT-BREAKER

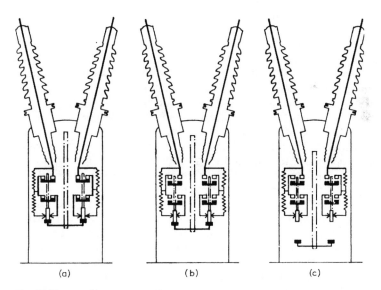

(a)   (b)   (c)

Fig. 17.11   OPENING SEQUENCE FOR A MULTI-BREAK CIRCUIT-BREAKER

(AEI Ltd.)

(a) Switch closed
(b) Resistors in circuit
(c) Switch completely open

arcing times tend to be long and inconsistent when operating against current considerably less than rated current—say about 30 per cent of rated current. This is so because the gas pressure generated is much reduced compared with that generated at rated current. Indeed the most onerous duty of the self-blast circuit-breaker is when it is called upon to break fault currents of the order of 30 per cent of its rated current.

This particular difficulty is overcome in circuit-breakers which utilize a form of arc control in which the blast is provided by external means and thus is independent of the value of fault current to be broken. Circuit-breakers using this form of arc control may be either impulse-oil circuit-breakers or air-blast circuit-breakers.

In impulse oil circuit-breakers, oil is forced across the arc-path, the necessary pressure, being produced by external mechanical means, does not in any way depend on the strength of current to be broken. Such circuit-breakers are usually multi-break and have capacitance or resistance shunts to control the voltage across the cascaded breaks. They are suitable for very-high-voltage systems of 200 kV and over.

A disadvantage attaching to all oil circuit-breakers is the risk of fire due to the inflammability of the oil. For this reason, and others, the air-blast circuit-breaker has been developed.

Air-blast circuit-breakers are similar to impulse oil circuit-breakers in that arc control, which takes the form of an air blast which may be across the arc (cross-blast), along the arc (axial blast), or radial to the arc (radial-blast), is provided by an external air compressor and is independent of the current to be interrupted.

Fig. 17.12 indicates one type of air-blast circuit-breaker. The type shown has four series breaks shunted by non-linear resistors.

Arc control is used basically to remove ionized gas, which acts as a conductor rather than an insulator, from the arc path. A recent development is to surround the circuit-breaker contacts with either sulphur hexafluoride (SF_6) or vacuum which makes it possible to dispense with external arc control.

In an SF_6 circuit-breaker the contacts are surrounded by SF_6 gas at a pressure of 45 lb/in.^{2} when the breaker is quiescent. Movement of the contacts is synchronized with the opening of a valve which allows additional gas to flow into the interrupter from a receiver containing gas at about 250 lb/in.^{2} The increased pressure assists in arc quenching.

The large SF_6 molecules rapidly absorb free electrons produced in the arc path to form heavy negative ions of low mobility which are ineffective as charge carriers. Rapid restrike-free arc interruption is thus achieved.

A typical SF$_6$ circuit-breaker consists of interrupter units each capable of dealing with currents of up to 60 kA and voltages of 50 to 80 kV. A number of units is connected in series according to the system voltage.

SF$_6$ circuit-breakers for lower voltages sometimes operate in the same manner as oil impulse breakers where the movement of the contacts operates a piston which produces a gas flow across the arc.

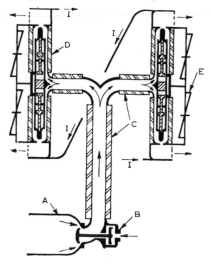

Fig. 17.12 AIR-BLAST CIRCUIT-BREAKER

(*AEI Ltd.*)

A. Air reservoir
B. Blast valve at earth potential
C. Porcelain blast tubes
D. Porcelain interrupting chamber
E. Non-linear resistor
I. Current paths

The advantageous features of SF$_6$ circuit-breakers are that SF$_6$ is odourless, non-toxic, inert and non-inflammable, and should any decomposition take place, the products (fluorine powders) possess high electric strength and are not a maintenance problem. The circuit-breakers are extremely quiet in operation, and maintenance costs are low.

Vacuum circuit-breakers are based upon the employment of a number of vacuum interrupter units in series. Each unit consists of a pair of separable contacts in a sealed, evacuated envelope of borosilicate glass. The moving contact is operated by flexible metal bellows, and it is essential that all occluded gases be removed from all components within the envelope during manufacture to maintain

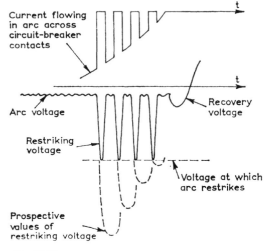

Fig. 17.13   PREMATURE EXTINCTION OF ARC-CURRENT CHOPPING
(SIMPLIFIED ACTION)

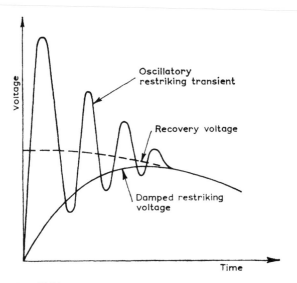

Fig. 17.14   CRITICAL DAMPING OF RESTRIKING VOLTAGE

vacuum. The high electric strength of the vacuum allows a short contact separation (0·25 in.) and rapid restrike-free interruption of the arc is achieved. A tubular shield around the contacts condenses any metallic vapours expelled from the contacts during operation, and effective interruption depends on the thoroughness and speed with which these arc products are condensed. The interrupter units are sealed into an outer envelope containing SF_6 gas. A typical 132 kV 3,500 MVA vacuum circuit-breaker uses eight interrupter units in series in each phase, each shunted by a capacitor to secure uniform voltage distribution across the units.

The advantageous features of vacuum circuit-breakers are that vacuum is non-inflammable and non-toxic, that the circuit-breaker has a light, simple, short-stroke mechanism giving high-speed performance. The relatively rapid rate of recovery of electric strength in vacuum circuit-breakers make them ideal for the interruption of large leading-power-factor currents encountered in underground cable systems. These circuit-breakers require little maintenance and are extremely quiet in operation.

17.9 Current Chopping

Current chopping is the name given to a phenomenon associated with circuit-breakers when they extinguish the arc before the natural current zero. This occurs when the circuit-breaker is breaking against a fault current considerably below its rated current. When the current is chopped in this manner the energy associated with the inductance of the circuit ($\frac{1}{2}Li^2$, where i is the instantaneous value of the current chopped) is rapidly transferred to the circuit capacitance and gives rise to a high voltage of high frequency, the frequency being determined by the inductance and capacitance of the circuit.

Current chopping occurs in all types of circuit-breaker but is probably more marked in those in which the arc control is not proportional to the fault current, i.e. in impulse oil circuit-breakers and air circuit-breakers. In air circuit-breakers, where the electric strength is usually lower than in oil circuit-breakers, the rise in voltage across the contacts may cause the arc to restrike, thus relieving the electric stress on the system insulation and preventing the transmission of a high-voltage surge. Indeed the arc may restrike several times as indicated in Fig. 17.13. In oil circuit-breakers, where electric strengths are normally higher than in air circuit-breakers, restriking may not take place and other methods of protection against the effects of current chopping such as resistance switching are necessary. Such methods may also be applied to air circuit-breakers.

17.10 Resistance Switching

The value which the restriking voltage reaches after arc extinction may be limited by the use of a shunt resistance connected across the main contacts of the circuit-breakers. The effect of the resistance is to prevent the oscillatory growth of voltage and to cause the voltage to grow exponentially up to the recovery value. Fig. 17.14 shows oscillatory growth and exponential growth when the circuit is critically damped. The dotted line represents the generated e.m.f., which is approximately at maximum value when the current zero occurs. The value of resistance required to obtain critical damping is $\frac{1}{2}\sqrt{(L/C)}$. When this technique is used some form of auxiliary contact is required to break the current through the resistor.

PROBLEMS

17.1 Explain what is meant by the percentage leakage reactance of an alternator

Two 18 MVA alternators each with a 25 per cent leakage reactance, run in parallel on a section busbar A which is connected through a 36 MVA reactor having 30 per cent reactance to a busbar B having two alternators similar to those on A. A 9 MVA feeder having 10 per cent reactance is connected to busbar A. Estimate the initial fault MVA if a short-circuit occurs between the three conductors at the far end of the feeder.   (*L.U.*)

Ans. 62·8 MVA.

17.2 The busbars of a station are in two sections, A and B, separated by a reactor. Connected to section A are two 15 MVA generators of 12 per cent reactance each, and to B one 8 MVA generator of 10 per cent reactance. The reactor is rated at 10 MVA and 15 per cent reactance. Feeders are connected to the busbar A through transformers, each rated at 5 MVA and 4 per cent reactance. Determine the maximum short-circuit kVA with which oil-switches on the outgoing side of the transformer have to deal.   (*H.N.C.*)

Ans. 87·1 MVA.

17.3 The 3-phase busbars of a station are divided into two sections, A and B, joined by a reactor having a reactance of 10 per cent at 20 MVA.

A 60 MVA generator with 12 per cent reactance is connected to section A and a 50 MVA generator with 8 per cent reactance is joined to section B. Each section supplies a transmission line through a 50 MVA transformer with 6 per cent reactance which steps up the voltage to 66kV.

If a 3-phase short-circuit occurs on the high-voltage side of the transformer connected to section A, calculate the maximum initial fault current which can flow into the fault.

Explain how you would estimate the current to be interrupted by a circuit-breaker opening after 0·3 s, and show why this value would differ from the maximum value calculated.   (*L.U.*)

Ans. 3,210 A.

17.4 Explain the object of (i) sectionalizing the busbars of a large generating station, and (ii) connecting reactance coils between the sections. Sketch alternative arrangements of connecting the sections and reactance coils, and discuss their relative advantages and disadvantages.

A station has three busbar sections A, B and C, which are interconnected by reactance coils (one coil between A and B, and one coil between B and C), each coil being rated at 14 per cent reactance on a basis of 60 MVA. Generators are connected as follows: one 50 MVA of 18 per cent reactance to A; one 60 MVA of 20 per cent reactance to B; one 75 MVA of 20 per cent reactance to C. Calculate the MVA which would be fed into a short-circuit on section B when all the generators are running. (*L.U.*)

Ans. 668 MVA.

17.5 A generating station has three section busbars with the following plant:

Busbar section	Plant	Percentage reactance
1	10 MVA generator	10
2	7 MVA generator	5
3	8 MVA grid transformer	8

Sections 1 and 2 are connected through a 6 MVA, 5 per cent busbar reactor, and sections 2 and 3 by a 6 MVA, 6 per cent reactor. Calculate the MVA fed into a short-circuit at the distant end of one of the outgoing feeders on section 2. The reactance of the feeder may be taken as 10 per cent on a basis of 5 MVA. Treat the grid networks as having infinite capacity. (*H.N.C.*)

Ans. 40·2 MVA.

17.6 Write a short account of the use of reactors in busbar layouts.

Generators aggregating 10 MVA and 20 per cent reactance are connected to each section busbar of a system consisting of three sections which later are connected in ring formation by a 4 MVA, 8 per cent reactor between each section. A feeder having a rating (including transformer) of 8 MVA and 10 per cent reactance is connected to one section. If a fault occurs at the distant end of the feeder, find the percentage of normal voltage at each busbar section. (*H.N.C.*)

Ans. 45 per cent; 90·1 per cent; 90·1 per cent.

17.7 Describe the "star" or "tie-bar" method of interconnecting the busbar sections in a generating station and compare it with other busbar arrangements.

A station contains 4 busbar sections to each of which is connected a generating unit of 30 MVA having 12 per cent leakage reactance, the busbar ·reactors having a reactance of 10 per cent.

Calculate the maximum MVA fed into a fault on any busbar section and also the maximum MVA if the number of similar busbar sections were increased to infinity. Deduce any formula used. (*L.U.*)

Ans. 422 MVA; 550 MVA.

17.8 What are the objects of sectionalizing the busbars in a large power station?

The busbars of a generating station are divided into three sections to each of which is connected a 16 MVA generator having 30 per cent reactance. The busbars are connected to a common tie-bar through 12 MVA reactors having 15 per cent reactance. To each busbar is connected a 10 MVA transformer having 10 per cent reactance and a 20 km 10 MVA feeder having a reactance of 0·7 Ω/km and negligible resistance is connected to the 33 kV secondary of one transformer.

Find the current fed into a symmetrical 3-phase short-circuit at the distant end
of the feeder.   (*H.N.C.*)

Ans. 513 A.

17.9 In a generating station there are four busbar sections with a 60 MVA 33 kV
3-phase generator having 15 per cent leakage reactance connected to each section.
The sections are connected through a 10 per cent reactor to a common tie-bar.
A 1 MVA 3-phase feeder joined to one of the busbar sections has a resistance
of 60 Ω/phase and a reactance of 70 Ω/phase.

If a symmetrical 3-phase short-circuit occurs at the receiving end of the feeder,
determine the MVA and also the voltage on the four busbar sections.

(*L.U.*)

Ans. 11·6 MVA, 3·83 MVA; 32·3 kV, 32·7 kV.

17.10 Define the terms "symmetrical breaking current" and "asymmetrical
breaking current" as applied to oil circuit-breakers. Show how these quantities
are determined from oscillograms of short-circuit tests on a circuit-breaker.
Explain why, on symmetrical 3-phase short-circuit tests, some initial asymmetry
in the current always occurs.

A 50 MVA generator of 18 per cent leakage reactance and a 60 MVA generator
of 20 per cent leakage reactance are connected to separate busbars which are
interconnected by a 50 MVA reactor. Calculate the percentage reactance this
reactor must possess in order that switches rated at 500 MVA may be employed
on feeders connected to each of the busbars.   (*L.U.*)

Ans. 7 per cent, or 0·07 p.u.

17.11 Enumerate the positions in which current-limiting reactors may be
connected. What advantage does the tie-bar system have over the ring system?

A station busbar has three sections A, B and C, to each of which is connected
a 20 MVA generator of reactance 8 per cent. Two similar reactors are to be
connected, one between the busbar sections A and B and one between the busbar
sections B and C. Calculate the percentage reactance of these reactors if the
MVA fed into a symmetrical short-circuit on section A busbar is not to exceed
400. The reactors are to be rated at 10 MVA.   (*H.N.C.*)

Ans. 4 per cent, or 0·04 p.u.

17.12 Three 11 kV 40 MVA alternators are connected to three sets of 33 kV
busbars A, B and C by means of three 11/33 kV 40 MVA transformers. The busbars are joined by two similar reactor sets, one set being connected between A
and B and the other between B and C. The reactance of each alternator is 20 per
cent and that of each transformer is 6 per cent at 40 MVA.

Determine the percentage reactance of each set of reactors at 10 MVA in
order that the symmetrical short-circuit on busbars A should be limited to
350 MVA.   (*L.U.*)

Ans. 1·5 per cent, or 0·015 p.u.

17.13 The 33 kV busbars of a generating station are divided into three sections
A, B and C, which are connected to a common tie-bar by a reactance *X* ohms.
To section A is connected a 60 MVA generator having a leakage reactance of
15 per cent, to B a 40 MVA generator of leakage reactance 12 per cent, and to C
a 30 MVA generator of leakage reactance 10 per cent.

If the breaking capacity of the circuit-breaker connected to section A is not to
exceed 500 MVA, determine the minimum value of the reactance *X*.

Ans. 6·12 Ω.

Chapter 18

CLOSED-LOOP CONTROL SYSTEMS

In almost every sphere of human endeavour there is a need to exercise control of physical quantities. Manual control, involving a human operator, suffers from several disadvantages among which may be numbered fatigue, slow reaction time (some 0·3 s), lack of exact reproducibility, limited power, tendency to step-by-step action and variations between one operator and another. The demand for precision control of physical quantities has led to the development of *automatic control systems* or *servo systems*. It is the purpose of this chapter to examine some simple servo systems, as an introduction to a subject which is of ever-growing importance.

All precision control involves the feedback of information about the controlled quantity, in such a way that if the controlled quantity differs from the desired value an error is observed, and the control system operates to reduce this error. This type of control is called *closed-loop control* and can be either manual or automatic. In simple regulating systems there is no feedback of information, and precise control cannot be achieved. This is called *open-loop control*, since there is no feedback loop.

18.1 Open-loop Control

The operation of an open-loop regulating system may be understood by considering one or two illustrations of such systems. For example, the flow of water in a pipe may be controlled by a valve. The opening of the valve can be measured on a scale, but for any one setting the actual flow of water will depend on the available head at the inlet to the valve, and on the loading at the outlet, as well as on the valve setting. The accuracy of the setting is thus dependent on external disturbances.

Again, consider the speed control of a d.c. shunt motor by field resistance. Increasing the field resistance increases the motor speed, but at any setting of the field rheostat the actual speed will depend on the supply voltage and on the load on the machine. The rheostat cannot be calibrated accurately in terms of speed, and the system is an open-loop control. If now we connect a tachometer to the shaft, and mark on its scale the desired speed, then a human operator may adjust the field rheostat as required to keep the actual speed as near as he can to the desired speed. The operator then acts as the feedback loop, and the system has become a closed-loop system, where the control action depends on the observed error between actual and desired speed. An increased accuracy of control is thus achieved. An automatic control is achieved by replacing the operator by an error-measuring device and output controller.

18.2 Basic Closed-loop Control

The basic elements of a simple closed-loop control system are illustrated in Fig. 18.1. In this case the output of some industrial

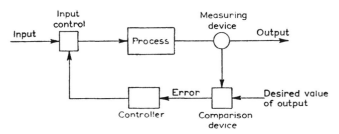

Fig. 18.1 SIMPLE CLOSED-LOOP PROCESS CONTROL

process is being controlled by a control element. The actual output from the process is measured and compared with a desired value in the comparison device. The magnitude and sense of any difference between the desired and actual values of the output is fed as an error signal to the controller which in turn actuates the correcting device in such a way as to tend to reduce the error. Note that correction

takes place no matter how the error arises, e.g. by external disturbances or changes in input conditions. The control gear forms a
closed loop with the process.

The essential elements of the automatic control system are thus
(*a*) a measuring device, which can often be combined with (*b*) a
comparison device to produce an error signal, (*c*) a controller,
which normally incorporates power amplification, and (*d*) a correction
device.

One very important type of control system is that in which the
angular position of a shaft has to be controlled from some remote
position with great accuracy. Such a system is called a *remote
position control* (or r.p.c.) servo, and has applications including the
automatic control of gun positions, servo-assisted steering of vehicles
and ships, positioning of control rods in nuclear reactors, and automatic control of machine tools. In the following sections an electrical
r.p.c. servo will be considered in more detail. In such a system, the
shaft position is measured electrically, an electrical error signal is
generated, amplified, and used to control an electric positioning
motor.

18.3 The Summing Junction

In electrical servos it is often required to apply the sum of or difference between two or more signals to an amplifier. This can conveniently be done by means of a summing junction, as illustrated

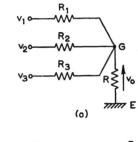

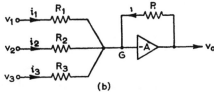

Fig. 18.2 THE SUMMING JUNCTION

in Fig. 18.2(a). Thus, if the free ends of the input resistors R_1, R_2, R_3 have voltages v_1, v_2, v_3 to earth, then by Millman's theorem,

$$v_0 = v_{GE} = \frac{\Sigma v_{ne} Y_n}{\Sigma Y_n} = \frac{\dfrac{v_1}{R_1} + \dfrac{v_2}{R_2} + \dfrac{v_3}{R_3}}{\dfrac{1}{R_1} + \dfrac{1}{R_2} + \dfrac{1}{R_3} + \dfrac{1}{R}} \tag{18.1}$$

The output voltage is thus dependent on the sum of the input voltages taken in proportions which depend on the ratios of the resistors. Note that, if (as is not uncommon) $R_1 = R_2 = R_3$ and $R \gg R_1$, then

$$v_0 = \tfrac{1}{3}(v_1 + v_2 + v_3) \tag{18.2}$$

For two inputs (i.e. if R_3 were disconnected) this would reduce to

$$v_0 = \tfrac{1}{2}(v_1 + v_2)$$

A more sophisticated version of the summing junction is obtained by using a high-gain d.c. amplifier as shown in Fig. 18.2(b). If the gain of the amplifier is $-A$ and its input impedance is very high, then

$$i = -(i_1 + i_2 + i_3)$$

and the potential of point G is $-v_0/A$ which will be very small if A is large (typically A can be of the order of 10^7). Thus G can be considered to be almost at earth potential, and is called a *virtual earth*, so that

$$i_1 = \frac{v_1}{R_1} \qquad i_2 = \frac{v_2}{R_2} \qquad i_3 = \frac{v_3}{R_3} \qquad i = \frac{v_0}{R}$$

Hence

$$\frac{v_0}{R} = -\left(\frac{v_1}{R_1} + \frac{v_2}{R_2} + \frac{v_3}{R_3}\right)$$

or

$$v_0 = -\left\{v_1 \frac{R}{R_1} + v_2 \frac{R}{R_2} + v_3 \frac{R}{R_3}\right\} \tag{18.3}$$

In the case where $R_1 = R_2 = R_3 = R$,

$$v_0 = -(v_1 + v_2 + v_3) \tag{18.4}$$

This arrangement is called an *operational amplifier*, and finds an application in analogue computers as well as in servo systems.

18.4 Measurement of Shaft Position Error by Voltage Dividers

Two methods of obtaining an electrical signal which will give a measure of the size and sense of the difference between the actual angular position of a shaft and the desired angular position will be considered. The first of these methods involves voltage dividers whose sliders are fixed to a reference and an actual output shaft respectively.

Consider the two linearly wound voltage dividers shown in Fig. 18.3, connected to a summing junction, G, though resistors which are of such high values that they give negligible loading on the voltage dividers. The slider of the reference voltage divider is set at the desired angular shaft position θ_i, so that, assuming the divider to be wound over 300°,

$$v_1 = \frac{\theta_i}{300} \times (-V)$$

The slider of the output voltage divider is connected to the output shaft, and for an output shaft angular position of θ_o,

$$v_2 = \frac{\theta_o}{300} \times (V)$$

It follows from eqn. (18.4) that

$$v_e = -\left(\frac{-\theta_i V}{300} + \frac{\theta_o V}{300}\right) = \frac{V}{300}(\theta_i - \theta_o) \tag{18.5}$$

The voltage v_e is thus proportional to the shaft error (i.e. the difference between the desired and actual shaft positions $(\theta_i - \theta_o)$ and is called the *error voltage*.

18.5 Synchros as Shaft Position Error Detectors

The synchro gives an a.c. error voltage whose amplitude is proportional to the shaft error and whose phase depends on the sense of the error. The principle of operation can be seen from Fig. 18.4. The stator houses three windings whose centre-lines are 120° apart. The rotors may be of the salient-pole type or of the wound-rotor type.

Two units are used. The transmitter has its rotor supplied from an a.c. source, the rotor shaft being set to the desired angular position. The rotor current sets up an air-gap flux which links the three stator windings and induces e.m.f.s in them according to the relative position of the rotor. The stator windings of the transmitter are linked to those of the receiver as shown in Fig. 18.4,

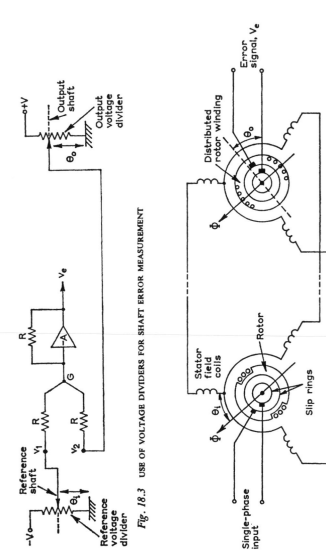

Fig. 18.3 USE OF VOLTAGE DIVIDERS FOR SHAFT ERROR MEASUREMENT

Fig. 18.4 USE OF SYNCHROS FOR SHAFT ERROR DETECTION

so that the induced e.m.f.s will set up circulating currents, which in turn cause a flux in the air-gap of the receiver. This receiver flux will have the same direction relative to the receiver stator as the transmitter air-gap flux has relative to the transmitter stator.

When the centre-line of the receiver rotor is at 90° to this flux, the e.m.f. induced in the rotor winding will be zero. For any deviation θ from the 90° position, the alternating e.m.f. induced in the receiver rotor will be

$$V_e = V \sin \theta = V \sin (\theta_i - \theta_o) \qquad (18.6)$$

where V is the r.m.s. voltage induced in the receiver rotor when it links all the stator flux. For small deviations from exact quadrature between receiver flux and rotor centre-line, $\sin (\theta_i - \theta_o) \approx \theta_i - \theta_o$ and

$$V_e \approx V(\theta_i - \theta_o) \qquad (18.7)$$

i.e. the error voltage is linearly related to the difference between input and output shaft positions (after allowing for the initial 90° displacement). The phase of this error voltage will be 180° different for the case when $\theta_o < \theta_i$ than it is when $\theta_o > \theta_i$, thus giving a measure of the sense of the error.

Shaft error detection is only one of several applications of these devices. As error detectors they have the advantage over voltage dividers of negligible wear, greater accuracy, and error detection over a full 360° rotation.

18.6 Small Servo Motors and Motor Drives

Electrical servos may be (*a*) entirely d.c. operated, using voltage dividers, d.c. amplifiers, and d.c. driving motors, (*b*) entirely a.c. operated using synchros, a.c. amplifiers and 2-phase a.c. driving motors, or (*c*) a.c./d.c. operated using a.c. error detection, phase-sensitive rectification and d.c. driving motors.

SPLIT-FIELD MOTOR

Small d.c. servo motors are generally of the split-field type illustrated in Fig. 18.5. Neglecting saturation and assuming a constant armature current, the output torque will be proportional to the net field current, and will reverse when this net field current reverses. The field may be fed from a push-pull amplifier stage. If the armature is not fed from a constant-current source, the build-up of armature e.m.f. with speed causes a falling torque/speed characteristic, which is equivalent to viscous-friction damping in the servo system. Approximately constant armature current can be achieved by feeding the armature through a high resistance from a high-voltage d.c. supply.

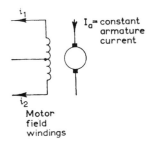

Fig. 18.5   THE D.C. SERVO MOTOR

PHASE-SENSITIVE RECTIFIER

An a.c. error signal may be used with a *phase-sensitive rectifier* (p.s.r.) to produce a d.c. error voltage. The basic operation of a p.s.r. is shown in Fig. 18.6. When there is no error voltage, each diode conducts during positive half-cycles of the reference voltage and there is no net voltage between A and B. The peak error signal,

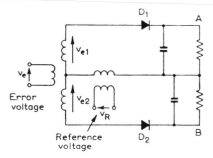

Fig. 18.6   PHASE-SENSITIVE RECTIFICATION

v_e, is arranged to be smaller than the reference voltage, v_R. When the error signal is in phase with the reference voltage, the voltage applied to D_1 during the positive half-cycles of v_R, i.e. $(v_R + v_{e1})$, is greater than that applied to D_2 $(v_R - v_{e2})$, and hence A is positive with respect to B. During the negative half-cycles of v_R both diodes remain non-conducting (since $v_R > v_e$). If the error voltage is now changed in phase by 180°, D_2 has a larger voltage applied to it during the positive half-cycles than D_1 and B is positive with respect to A. Hence the voltage between A and B gives the magnitude and sense of the error, and may be applied direct to the bases of a long-tailed pair (Section 22.11). The capacitors provide smoothing of the output signal.

TWO-PHASE SERVO MOTOR

In an a.c. servo the error signal from a synchro is fed to an a.c. amplifier, whose output feeds one phase of 2-phase motor. The phase of the voltage applied to this winding is arranged to be in quadrature with that applied from a constant reference source to the second winding of the motor (the reference winding), as shown in Fig. 18.7(*a*).

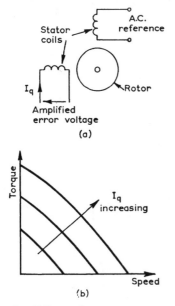

Fig. 18.7 THE 2-PHASE SERVO MOTOR

When there is no error voltage, only the reference winding is energized and the rotor is locked in position. The motor is designed to give maximum torque at standstill. The size of the output torque will depend on the magnitude of the error signal, and the direction of the torque will depend upon whether the error signal lags or leads the reference voltage by 90°. Typical characteristics are shown in Fig. 18.7(*b*).

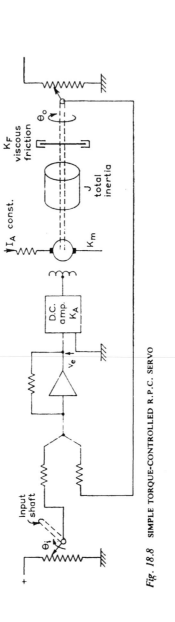

Fig. 18.8 SIMPLE TORQUE-CONTROLLED R.P.C. SERVO

The a.c. servo is not used where large power is required, because the electronic amplifier has a limited power output. In such cases rotating d.c. amplifiers (amplidyne or metadyne type) or magnetic amplifiers are used with d.c. driving motors. High-power servos will not be dealt with in this text.

Note that in both d.c. and a.c. servo motors the armatures are usually long and of small diameter in order to give a high torque/inertia ratio.

18.7 Simple Torque-controlled R.P.C. Servo

The main components of an electrical r.p.c. servo system having been briefly discussed, a simple control system can now be considered. Such a closed-loop system is shown schematically in Fig. 18.8, where the servo motor drives a load shaft (the inertia of the load and the motor being J kg-m^2), and where there is viscous-friction damping (i.e. a frictional force proportional to the angular velocity of the output shaft). The driving torque produced by the motor is directly proportional to the error voltage, v_e, and is given by

$$T_D = K_A K_m v_e \quad \text{newton-metres}$$

where K_A is the amplifier *transconductance* in amperes per volt, and K_m is the motor torque constant in newton-metres per ampere of field current.

The error voltage is related to the shaft error by the equation

$$v_e = K_S(\theta_i - \theta_o) \tag{18.12}$$

where K_S is the voltage divider and summing junction constant in volts per radian of error.

Neglecting any load torque, the motor driving torque must be sufficient to overcome the inertia torque, $J(d^2\theta_o/dt^2)$, and the viscous friction torque, $K_F(d\theta_o/dt)$ (where K_F is the friction torque per unit of angular velocity), so that

$$J\frac{d^2\theta_o}{dt^2} + K_F\frac{d\theta_o}{dt} = K_A K_m v_e = K_A K_m K_S(\theta_i - \theta_o) = K(\theta_i - \theta_o)$$

$$\tag{18.13}$$

where $K = K_A K_m K_S$ newton-metres per radian. Rearranging,

$$\frac{d^2\theta_o}{dt^2} + \frac{K_F}{J}\frac{d\theta_o}{dt} + \frac{K}{J}\theta_o = \frac{K}{J}\theta_i \tag{18.14}$$

Obviously, at standstill, when $d^2\theta_o/dt^2 = d\theta_o/dt = 0$, then $\theta_o = \theta_i$ and there is no error.

The response of the system (which is known as a *second-order system*) to a step change in input, θ_i, may be computed in exactly the same way as was used for the double-energy transient in Chapter 7. Rewriting eqn. (18.14) in standard form,

$$\frac{d^2\theta_o}{dt^2} + 2\zeta\omega_n \frac{d\theta_o}{dt} + \omega_n{}^2\theta_o = \omega_n{}^2\theta_i \qquad (18.14a)$$

where

$$\omega_n = \sqrt{(K/J)} = \text{undamped natural angular frequency} \qquad (18.15)$$

and ζ (zeta) is the *damping ratio given by*

$$\zeta = \frac{\text{Actual damping constant}}{\text{Damping constant for critical damping}} = \frac{K_F}{2\sqrt{(JK)}} \qquad (18.16)$$

As shown in Fig. 18.9, there are four solutions to eqn. (18.14a).

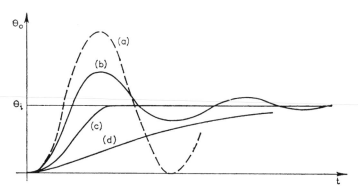

Fig. 18.9   RESPONSE OF A SECOND-ORDER SERVO TO A STEP CHANGE OF INPUT

(a) *Zero damping* ($\zeta = 0$; i.e. $K_F = 0$)
The response to a step input change is given by

$$\theta_o = \theta_i(1 - \cos \omega_n t) \qquad (18.17)$$

The output oscillates continuously, the amplitude of the oscillations corresponding to the input step change. This condition should obviously be avoided.
(b) *Underdamping* ($\zeta < 1$; i.e. $K_F < 2\sqrt{(JK)}$)
The output oscillates but finally settles down to a steady value of $\theta_o = \theta_i$. The response is characterized by an overshoot (or several

overshoots) but gives a fast rise to around the value of θ_i. The actual equation of the response is

$$\theta_o = \theta_i \left\{ 1 - e^{-\zeta\omega_n t}(\cos \omega t + \frac{\zeta}{\sqrt{(1 - \zeta^2)}} \sin \omega t) \right\} \qquad (18.18)$$

where

$$\omega = \omega_n \sqrt{(1 - \zeta^2)} \qquad (18.19)$$

The time constant of the decaying oscillation is $\tau = 1/\zeta\omega_n$. The slightly underdamped response is the type usually employed for fast-acting servos, damping ratios of the order of 0·6 being common.

(c) *Critical damping* ($\zeta = 1$; i.e. $K_F = 2\sqrt{(JK)}$)
This condition marks the transition between the oscillatory and the overdamped solution, and the response to a step input change is

$$\theta_o = \theta_i(1 - \zeta\omega_n t\, e^{-\zeta\omega_n t} - e^{-\zeta\omega_n t}) \qquad (18.20)$$

(d) *Overdamping* ($\zeta > 1$; i.e. $K_F > 2\sqrt{(JK)}$)
This represents a condition of slow response and is normally avoided in practice. The mathematical expression for the response is

$$\theta_o = \theta_i \left\{ 1 - e^{-\zeta\omega_n t} \left(\cosh \beta t + \frac{\zeta}{\sqrt{(1 - \zeta^2)}} \sinh \beta t \right) \right\} \qquad (18.21)$$

where

$$\beta = \omega_n \sqrt{(\zeta^2 - 1)} \qquad (18.22)$$

It should be noted that exactly the same results are obtained with an a.c. servo, or a d.c. servo with synchro error detection and a phase-sensitive rectifier.

In all servos there is a lower limit to the size of error signal which will just cause a correcting action to take place. If the system is made too sensitive, spurious operation may result from random noise inputs to the amplifier. The *dead zone* of an r.p.c. servo is the range of shaft errors over which no correcting action will take place, since the motor torque will not be sufficient to overcome stiction (static friction). This dead zone will give the limit of accuracy of the system.

18.8 Gearing

It is usually economic to design servo motors which run at a much higher speed than that required for the output shaft. A reduction gear is then used to connect the motor to the load shaft (Fig. 18.10).

If the gear ratio is $n:1$, the shaft output angular rotation, velocity and acceleration will each be $1/n$ of the corresponding input quantities. Also, assuming that there is no power loss in the gearing,

Input power = Output power   i.e. $T_1\omega_1 = T_2\omega_2$

or

$$T_2 = nT_1 \tag{18.23}$$

where T represents torque, ω is the angular velocity, and the subscripts 1 and 2 refer to the input and output sides of the gearbox.

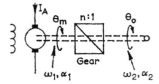

Fig. 18.10   SERVO MOTOR WITH GEAR TRAIN

The torque at the motor shaft required to overcome the motor inertia (J_m) is $J_m\alpha_1$ where α_1 is the angular acceleration at the motor. This torque at the motor shaft gives a torque of $nJ_m\alpha_1$ at the output shaft.

Hence

Output shaft torque $= nJ_m\alpha_1 = n^2J_m\alpha_2$

where α_2 is the output shaft acceleration; i.e. the motor inertia is equivalent to an inertia *at the output shaft* of

$$J_m' = n^2J_m \tag{18.24}$$

If the load inertia is J_L, the load acceleration, α_2, due to a motor torque T_m will be

$$\alpha_2 = \frac{nT_m}{J_L + n^2J_m}$$

where the total inertia referred to the load shaft is

$$J = J_L + n^2J_m \tag{18.25}$$

α_2 is a maximum as n varies when $n = \sqrt{(J_L/J_m)}$, and this gear ratio is said to match the inertias.

18.9 Velocity-feedback Damping

Normal viscous friction of the mechanical system is usually insufficient to provide enough damping for the satisfactory operation

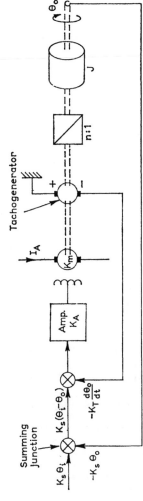

Fig. 18.11 D.C. SERVO WITH VELOCITY-FEEDBACK DAMPING

of an r.p.c. servo, and any increase in the mechanical viscous friction would involve additional power loss. The damping term in eqn. (18.14) can be readily increased, however, by *velocity feedback*, in which a second feedback loop is provided to give a negative signal at the amplifier input proportional to the shaft velocity. This signal can be obtained from a d.c. tachogenerator connected to the motor shaft, usually before the gearing, as shown in Fig. 18.11. In a.c. servos an a.c. tachogenerator can be used in the same way.

SYNCHRO ERROR DETECTION

The dynamic equation of the system is obtained as before by equating motor driving torque to the opposing torques. Neglecting mechanical viscous friction and load torque, and assuming a total inertia at the output shaft of J,

$$nK_AK_m \left\{ K_S(\theta_i - \theta_o) - K_Tn \frac{d\theta_o}{dt} \right\} = J \frac{d^2\theta_o}{dt^2} \tag{18.26}$$

where K_T is the tachogenerator constant in volts per rad/s *at the motor shaft*, K_S is the error constant in volt/rad error, and it is assumed that the summing junction adds the input voltages (algebraically) directly. Rearranging,

$$\frac{d^2\theta_o}{dt^2} + \frac{n^2K_AK_mK_T}{J} \frac{d\theta_o}{dt} + \frac{nK_AK_mK_S}{J} \theta_o = \frac{nK_AK_mK_S}{J} \theta_i \tag{18.27}$$

This equation is of exactly the same form as eqn. (18.14) and will thus yield similar results. The damping can be readily varied by including a voltage divider in the tachogenerator feedback path.

EXAMPLE 18.1 An r.p.c. servo uses voltage dividers with a 300° travel and a total voltage of 30 V across them for error detection. Damping is provided by a d.c. tachogenerator, whose output is added to the shaft error voltage in an operational summing amplifier. The amplifier has a transconductance of 250 mA/V, and the motor has a torque constant of 4×10^{-4} N-m/mA and an inertia of 50×10^{-6} kg-m². The motor is coupled to the load, whose moment of inertia is 40×10^{-2} kg-m², through a 100:1 reduction gear. Calculate (a) the undamped natural frequency of the system, and (b) the tachogenerator constant in volts per 1,000 rev/min to give a damping ratio of 0·8. Neglect viscous friction.

From eqn. (18.25),

Total inertia referred to load shaft $= (40 \times 10^{-2}) + (10^4 \times 50 \times 10^{-6})$
$$= 90 \times 10^{-2} \text{kg-m}^2$$

Voltage divider constant $= \dfrac{30}{300} \times \dfrac{360}{2\pi} = 5\cdot73 \text{ V/rad}$

Hence eqn. (18.27) can be written

$$\frac{d^2\theta_0}{dt^2} + 1{,}110\,K_T\,\frac{d\theta_0}{dt} + 63{\cdot}7\theta_0 = 63{\cdot}7\theta_i$$

Thus from eqn. (18.15), $\omega_n = \sqrt{63{\cdot}7} = 7{\cdot}98$, so that

$$f_n = \frac{\omega_n}{2\pi} = 1{\cdot}27\mathrm{Hz}$$

Comparison with eqn. (18.14a) yields

$$2\zeta\omega_n = 1{,}110\,K_T$$

Hence

$$K_T = \frac{2 \times 0{\cdot}8 \times 7{\cdot}98}{1{,}110} = 0{\cdot}0115 \text{ V per rad/s}$$

$$= 1{\cdot}20\mathrm{V} \text{ per } 1{,}000\,\mathrm{rev/min}$$

18.10 Velocity Lag

In a second-order r.p.c. servo with viscous friction damping, suppose that the input shaft is rotated at a constant angular velocity $\omega_i = d\theta_i/dt$. The output shaft will continue to accelerate until it is rotating at the same angular velocity as the input. Then, since $d^2\theta_0/dt^2$ will be zero under these conditions, the steady-state equation of motion will be, from eqn. (18.13),

$$K_F\,\frac{d\theta_0}{dt} = K_A K_m K_S(\theta_i - \theta_o) = K_A K_m K_S \varepsilon$$

where ε is the angular difference between input and output shafts. Hence

$$\varepsilon = \frac{K_F \omega_i}{K_A K_m K_S} \tag{18.28}$$

This constant error is required to give the motor torque needed to drive the output shaft against the viscous friction loading, and is called the *velocity lag*.

For an r.p.c. servo which is stabilized by negative velocity feedback in addition to viscous friction damping, the dynamic equation is

$$J\,\frac{d^2\theta_0}{dt^2} + K_F\,\frac{d\theta_0}{dt} = nK_A K_m \left\{ K_S(\theta_i - \theta_o) - K_T n\,\frac{d\theta_0}{dt} \right\}$$

(from eqn. (18.26)), where J is the total inertia referred to the output shaft. Under conditions of steady velocity input ($d^2\theta_0/dt^2 = 0$, $d\theta_0/dt = \omega_i$), the velocity lag is

$$\varepsilon = \theta_i - \theta_o = \frac{(K_F + n^2 K_T K_A K_m)\omega_i}{nK_A K_m K_S} \approx \frac{nK_T\omega_i}{K_S} \tag{18.29}$$

if (as is usually the case) $K_F \ll n^2 K_T \cdot K_A K_m$.

It is possible to eliminate the velocity lag in an r.p.c. servo with negative velocity feedback by arranging that the velocity feedback is removed when steady-state conditions are achieved. This is done by connecting a large capacitor in series in the velocity feedback loop. Essentially the capacitor passes any changing voltage conditions, but acts as a d.c. block to steady voltages. Thus the velocity feedback is effective only when the velocity of the output shaft is changing. This is termed *transient velocity feedback*.

EXAMPLE 18.2. An r.p.c. servo with velocity-feedback damping uses synchros as error detectors. The output of the phase-sensitive rectifier is 1.5 V/deg error, and is fed to the summing junction of an operational amplifier through a $1\,\text{M}\Omega$ resistor, the feedback resistor being also $1\,\text{M}\Omega$. The amplifier transductance is $400\,\text{mA/V}$, and the motor torque at standstill is $5 \times 10^{-2}\,\text{N-m}$ when the full field current of $80\,\text{mA}$ flows in one half of the split field and zero in the other half. The tachogenerator output is 0.3 V per rev/s and is fed through a $2\,\text{M}\Omega$ resistor to the summing amplifier, the tacho generator being on the motor shaft. The output shaft is coupled to the motor through an $80:1$ reduction gear, the total inertia at this shaft being $50 \times 10^{-2}\,\text{kg-m}^2$. Determine the system damping ratio, the magnitude of the first overshoot when the input shaft is given a sudden displacement of $10°$, and the velocity lag when the input shaft is rotated at $5\,\text{rev/min}$.

The motor torque constant is $(5 \times 10^{-2})/80 = 6.25 \times 10^{-4}\,\text{N-m/mA}$; K_T is $0.3/2\pi$ V per rad/s; and K_S is $1.5 \times 360/2\pi$ V/rad.

The dynamic equation of the system is

$$J\frac{d^2\theta_0}{dt^2} = nK_AK_m \left\{ K_S(\theta_i - \theta_0) - \tfrac{1}{2}nK_T \frac{d\theta_0}{dt} \right\}$$

The factor of one-half is present since the tachogenerator output is fed to the summing junction through a $2\,\text{M}\Omega$ resistor. This gives

$$\frac{d^2\theta_0}{dt^2} + 76.2\frac{d\theta_0}{dt} + 3,440\theta_0 = 3,440\theta_i$$

Comparing with eqn. (18.14a),

$$\omega_n = \sqrt{3,440} = 58.5 \quad \text{and} \quad 2\zeta\omega_n = 76.2$$

Hence the damping ratio, ζ, is $\underline{0.65}$.

The system is thus underdamped and the response to a step input of $10°$ is given by eqn. (18.18) as

$$\theta_0 = 10\{1 - e^{-38.1t}(\cos \omega t + 0.86 \sin \omega t)\}$$

$$= 10\left\{1 - \frac{e^{-38.1t}}{\sqrt{(1 + 0.86^2)}} \cos(\omega t - \tan^{-1} 0.86)\right\}$$

where $\omega = \omega_n\sqrt{(1 - \zeta^2)} = 58.5 \times 0.76 = 44.5$.

Thus θ_0 has its first maximum when $\cos(\omega t - 40.5°) = -1$, i.e. when

$$\omega t - \frac{40.5 \times 2\pi}{360} = \pi \quad \text{or} \quad t = \frac{1.23\pi}{44.5} = \underline{\underline{0.087\,\text{s}}}$$

Hence

$$\theta_{0_{max}} = 10\{1 + e^{-3\cdot3}1\cdot32\} = \underline{\underline{10\cdot5°}}$$

so that the magnitude of the first overshoot is $\underline{\underline{0\cdot5°}}$.

The velocity lag is given by eqn. (18.29) as

$$\varepsilon = \frac{nK_T\omega_i}{K_S} \text{ rad } = \frac{80 \times 0\cdot3 \times 5}{1\cdot5 \times 60} = \underline{\underline{1\cdot3°}}$$

18.11 Effect of Load Torque on a Simple R.P.C. Servo

So far, position control systems where the load torque is negligible have been considered. If there is a constant load torque, the motor must supply this even at standstill, and must therefore have an input. This in turn presumes that there is an error between input and output shafts. The dynamic equation for the system of Fig. 18.11 will be

$$J\frac{d^2\theta_o}{dt^2} + T_L = nK_AK_m\left\{K_S(\theta_i - \theta_o) - nK_T\frac{d\theta_o}{dt}\right\} \qquad (18.30)$$

where T_L is the constant load torque.

Under steady-state conditions when the input is set at some fixed value (θ_i = costant), $d^2\theta_o/dt^2 = d\theta_o/dt = 0$, and the expression for the error is

$$\varepsilon = \theta_i - \theta_o = \frac{T_L}{nK_AK_mK_S} \text{ radians} \qquad (18.31)$$

This error is often called the *offset*. Offset may be eliminated by altering the input signal to the servo amplifier so that it corresponds, not only to the error, but also to the error plus the time integral of the error.

18.12 Simple Speed Control—the Velodyne

The speed of the output shaft of a small servo motor can be controlled by a closed loop system similar to that used for position control. Such a system is shown in Fig. 18.12 and is commonly known as a

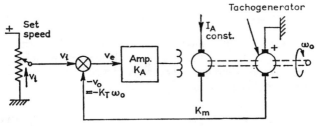

Fig. 18.12 VELODYNE SPEED CONTROL

velodyne. The output shaft drives a d.c. tachogenerator, and the voltage which it produces is compared with a set value derived from a reference voltage-divider, which can be calibrated in terms of output speed. With the polarities shown, the error voltage v_e is

$$v_e = v_t - K_T\omega_0 \qquad (18.32)$$

where K_T is the tachometer constant in volts per rad/s and ω_0 ($= d\theta_0/dt$) is the angular velocity of the output shaft.

The motor torque (assuming a constant armature current, an amplifier transconductance K_A, and a motor torque constant K_m) is $K_A K_m v_e$, so that, neglecting friction and loading, the dynamic equation for the system is

$$J\frac{d\omega_0}{dt} = K_A K_m v_e = K_A K_m(v_t - K_T\omega_0) \qquad (18.33)$$

where J is the total inertia at the motor shaft.

Under conditions of steady output speed, $d\omega_0/dt = 0$, and this condition is fulfilled only when $v_t = K_T\omega_0$, so that the output speed is

$$\omega_0 = \frac{v_t}{K_T} \qquad (18.34)$$

If there is a constant load torque, T_L, coupled through an $n:1$ reduction gear, with the tachogenerator direct on the motor shaft, then the tachogenerator output will be $nK_T\omega_0$ (where ω_0 is the output velocity). If the total inertia referred to the output shaft is J', the dynamic equation becomes

$$J'\frac{d\omega_0}{dt} + T_L = K_A K_m(v_t - nK_T\omega_0) \qquad (18.35)$$

In this case the condition for steady output-shaft angular velocity (i.e. when $d\omega_0/dt = 0$) is

$$T_L = K_A K_m(v_t - nK_T\omega_0) = K_A K_m(nK_T\omega_t - nK_T\omega_0) \qquad (18.36)$$

since the reference input voltage, v_t, can be calibrated in terms of the desired speed, ω_t, by eqn. (18.34), taking the gear ratio into account in this case. The difference between the desired speed, ω_t, and the actual speed, ω_0, is given by

$$\varepsilon = \omega_t - \omega_0 = \frac{T_L}{nK_T K_A K_m} \qquad (18.37)$$

This error is called the *droop*, and it should be noted that it is independent of the output speed. Thus at high shaft speeds it will represent a smaller percentage error than at low shaft speeds. It may be eliminated by the use of *integral-of-error compensation* as in the r.p.c. servo.

EXAMPLE 18.3 In the velodyne speed control shown in Fig. 18.12 the constants are as follows: amplifier transconductance, 200 mA/V; motor torque constant, 5×10^{-3} N-m/mA; tachogenerator constant 10 V per 1,000 rev/min. Determine the input voltage to give a speed of 2,000 rev/min. If the input setting is at half this value, find the droop when a load torque of 6×10^{-2} N-m is applied.

The tachogenerator constant in volts per rad/s is

$$K_T = \frac{10 \times 60}{1,000 \times 2\pi}$$

Hence, since there is no gearing, eqn. (18.34) gives

$$v_i = \omega_0 K_T = \underline{\underline{20\,\text{V}}}$$

When the input is set at 10 V, the no-load shaft speed will be 1,000 rev/min. If the load torque is now applied there must be an amplifier input voltage given by

$$v_e = \frac{T_L}{K_A K_m} = \frac{6 \times 10^{-2}}{200 \times 5 \times 10^{-3}} = 6 \times 10^{-2}\,\text{V}$$

Hence the tachogenerator output is $10 - (6 \times 10^{-2})$ V, and the actual shaft speed is $(10 - 6 \times 10^{-2} \times 1,000/10)$ rev/min, i.e. the droop is

$$6 \times 10^{-2} \times 100 = \underline{\underline{6\,\text{rev/min}}}$$

Note that this result can also be obtained by applying eqn. (18.37). The speed regulation is $6/1,000 \times 100 = \underline{\underline{0.6\,\text{per cent.}}}$

18.13 Some Limitations of the Simple Theory

In the simple theory developed in the preceding sections no account has been taken of any non-linearities in the system, such as the saturation of the amplifier, backlash in gearing, and stiction. Nor have the effects of actual servo-motor characteristics on the system damping, the mechanical limit of system acceleration (to keep acceleration stresses within reason), or the effect on dynamic response of the introduction of integral compensation been considered. Only velocity-feedback stabilization has been dealt with, and other forms of stabilization have been omitted. Such subjects, together with the important questions of system stability and harmonic response are the concern of full courses on servo-mechanism, as are considerations of large power systems using magnetic or rotating amplifiers.

PROBLEMS

18.1 A summing junction has three inputs A, B and C connected through resistors of $1\,\text{M}\Omega$, $2\,\text{M}\Omega$ and $250\,\text{k}\Omega$ to a star point, G. A resistor of $1\,\text{k}\Omega$ is connected between the star point and earth. The voltages of A, B and C with respect to earth are $20\,\text{V}$, $-19\,\text{V}$ and $-2\,\text{V}$ respectively. Find the voltage between G and earth.

Ans. $2\cdot49\,\text{mV}$.

18.2 An operational amplifier has a feedback resistor of $1\,\text{M}\Omega$ and three inputs A, B and C with input resistors of $1\,\text{M}\Omega$, $1\,\text{M}\Omega$ and $2\,\text{M}\Omega$ respectively. If the input voltages on A, B and C are $3\cdot5\,\text{V}$, $-3\cdot7\,\text{V}$ and $-1\cdot2\,\text{V}$ what is the output voltage?

Ans. $-0\cdot8\,\text{V}$.

18.3 A shaft-error-measuring system employs voltage dividers with 330° of travel, having $20\,\text{V}$ d.c across them. What is the voltage per radian error?

Ans. $3\cdot48$.

18.4 The transient action of an r.p.c. servo can be described by the differential equation

$$J\frac{d^2\theta_0}{dt^2} + F\frac{d\theta_0}{dt} + K\theta_0 = 0$$

Define the terms used, and use the equation to find expressions for (a) the undamped natural frequency, and (b) the critical viscous friction coefficient, F_{crit}. Verify that these expressions are dimensionally correct. (It may be assumed that the solution to the differential equation is of the form $\theta_0 = Ae^{\alpha_1 t} + Be^{\alpha_2 t}$).

An r.p.c. servo has a moment of inertia of $0\cdot024\,\text{kg-m}^2$ referred to the output shaft. The error-measuring element gives a signal of $1\,\text{V}$ per degree of misalignment. The motor torque constant referred to the output shaft is $10^{-2}\,\text{N-m/mA}$, and the amplifier transconductance is $50\,\text{mA/V}$. Determine: (a) the undamped natural frequency of the system, and (b) the viscous friction coefficient in newton-metre-seconds [i.e. N-m per rad/s] as measured at the output shaft, which will give a damping ratio of $0\cdot6$.   (*H.N.C.*)

Ans. $5\cdot5\,\text{Hz}$; $0\cdot99\,\text{N-m-s}$.

18.5 An r.p.c. servo uses synchros for error detection, giving an output of $1\,\text{V}$ per degree error. The motor is coupled through a $100:1$ reduction gear to the load. Given that amplifier transductance $= 400\,\text{mA/V}$; motor torque constant $= 5 \times 10^{-4}\,\text{N-m/mA}$; motor inertia $= 30 \times 10^{-6}\,\text{kg-m}^2$; load inertia $= 0\cdot5\,\text{kg-m}^2$; calculate (a) the viscous friction coefficient in N-m-s at the output shaft to give a damping ratio of $0\cdot6$, (b) the undamped natural frequency, (c) the steady-state error in degrees when the input shaft is rotated at $12\,\text{rev/min}$.   (*H.N.C.*)

Ans. $36\cdot3\,\text{N-m-sec}$; $6\,\text{Hz}$; $2\cdot3°$.

18.6 What do you understand by the term "dead zone" as applied to a servo-mechanism?

A servo system has a motor with inertia $10^{-6}\,\text{kg-m}^2$ matched through a gearbox to a load of inertia $0\cdot01\,\text{kg-m}^2$. The viscous friction coefficient measured at the load is $0\cdot64\,\text{N-m-s}$. The motor is excited from an amplifier and produces a torque of $10^{-4}\,\text{N-m/mA}$ of amplifier output. The error-measuring element gives $1\,\text{V}$ per degree of misalignment of input and output shafts.

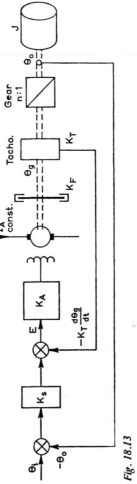

Fig. 18.13

If the amplifier transconductance is 56 mA/V calculate (a) the undamped natural frequency of the system, (b) the damping factor, and (c) the steady-state error in degrees for an input speed of 10 rev/min.   (*H.N.C.*)

Ans. 6·4 Hz; 0·4; 1·2°.

18.7 The following data relates to an r.p.c. servo with the output connected direct to the motor shaft:

Error transducer constant, 57·4 V/rad
Motor torque constant, 10^{-2} N-m/mA of amplifier output
Moment of inertia at motor shaft, 0·143 kg-m²
Viscous friction coefficient at motor shaft, 0·13 N-m-s
Damping ratio, 1·03

Determine (a) the amplifier transductance, and (b) the undamped natural frequency of oscillation.   (*H.N.C.*)

Ans. 60·7 mA/V; 2·49 Hz.

18.8 A servo motor of moment of inertia J_M drives a load of moment of inertia J_L through a reduction gear-box of ratio $n:1$. Deduce an expression for acceleration in terms of these quantities and the motor torque. Hence, determine the gear ratio for maximum acceleration of the load.

A motor is required to drive a load whose moment of inertia is 5 kg-m² through a gear-box of ratio 10:1. Two motors, A and B having moments of inertia 5×10^{-2} kg-m² and 3×10^{-2} kg-m² respectively are available. If both motors have the same torque-to-inertia ratio of 0·06 N-m/kg-m², which will give the maximum angular acceleration of the load, and what is the value of this acceleration?   (*H.N.C.*)

Ans. A; 0·3 rad/s².

18.9 Fig. 18.13 is a block diagram of a simple remote-position servo-mechanism which drives a load through a gear-box of ratio $n:1$. The moment of inertia of the moving parts is J; K_F is the viscous friction coefficient; $K_A K_m$ is the motor torque per volt; and K_S is the error constant in volts per radian of misalignment. Establish the differential equation of the system when it is stabilized by velocity feedback, the tachogenerator having a constant of K_T volts per rad/s.

In this control system the total moment if inertia referred to the output shaft is 10 kg-m², the motor torque is 2 N-m/V and the error constant is 25 V/rad. If the gear-box ratio is 20 : 1 and the tachogenerator constant is 0·03 V per rad/s referred to the output shaft, calculate the natural angular frequency. Find the coefficient of viscous damping if the system is critically damped by combined viscous friction and velocity feedback and the steady-state velocity error if the input velocity is 0·2 rad/s.

Ans. 10 rad/s; 176 N-m per rad/s; 0·04 rad

18.10 A velodyne speed control uses an amplifier with a transconductance of 300 mA/V, a motor with a torque constant of 2×10^{-4} N-m/mA, and a tachogenerator with a constant of 2 V per 1,000 rev/min. The reference input and the tachogenerator feedback voltage are fed into the amplifier through equal resistors. Find (a) the reference input voltage for a shaft speed of 2,500 rev/min on no load, and (b) the droop when a constant load torque of 8×10^{-3} N-m is applied.

Ans. 5 V; 70 rev/min.

Chapter **19**

ELECTRON DYNAMICS

In the following chapters it is intended to cover some of the basic concepts of the subject of electronics but it is assumed that the student is already familiar with the more elementary ideas. For this reason only a brief summary of the characteristics of electronic devices will be given.

In considering the dynamics of electron motion in a vacuum the electron will be visualized as a particle which has a negative electric charge, $-e$ coulomb. The electron is associated with a mass m, which increases as the velocity of the electron increases towards the velocity of light, but may be assumed constant at $9 \cdot 1 \times 10^{-31}$ kg up to one-third of the velocity of light. Since the negative charge is $-1 \cdot 6 \times 10^{-19}$ C, the ratio of charge to mass for a single electron is $-1 \cdot 76 \times 10^{11}$ C/kg. This huge ratio accounts for the fact that when an electron is placed in an electrostatic field the gravitational forces on it may be neglected compared with the electrostatic forces.

19.1 Acceleration of an Electron

Newton's second law of motion gives the force F required to produce an acceleration of a metres per second per second on a mass of m kilogrammes as

$$F = ma \quad \text{newtons} \tag{19.1}$$

If an electron is placed in an electrostatic field of strength E volts per metre (or newtons per coulomb), it will be acted upon by a force given by

$$F = -eE \quad \text{newtons} \tag{19.2}$$

595

the minus sign indicating that the force acts in the opposite direction to that of the field. The electron will, if free to move, be accelerated towards the positive electrode (Fig. 19.1). Combining eqns. (19.1)

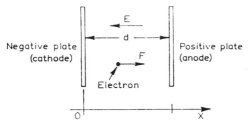

Fig. 19.1   FORCE ON AN ELECTRON BETWEEN TWO PARALLEL PLATES
E = Electric field strength = $-Fe$ = $-Vd$ F = Force on electron

and (19.2) gives the acceleration produced as

$$a = -\frac{eE}{m} \quad \text{metres/second}^2 \tag{19.3}$$

19.2 Motion of an Electron in a Plane Electrostatic Field

In considering the motion of an electron between the large plane electrodes shown in Fig. 19.1 it is assumed (i) that the space between the plates is completely evacuated, (ii) that the electrodes have a large area compared with their distance (d metres) apart so that the field strength at all points between the plates is V/d volts per metre, and (iii) that an electron leaves the negative plate (cathode) with zero initial velocity. If the positive X-direction is taken as shown from the cathode to the anode, then the following relations exist.

$$\text{Electric field strength} = -\frac{V}{d} \quad \text{volts/metre}$$

(negative since the cathode is more negative than the anode).

$$\text{Force on electron} = (-e) \times \left(-\frac{V}{d}\right) = e\frac{V}{d} \quad \text{newtons}$$

$$\text{Electron acceleration} = \frac{e}{m}\frac{V}{d} \quad \text{metres/second}^2 \tag{19.4}$$

Considering that the electron travels from the cathode with this uniform acceleration for a time t seconds, then

$$\text{Final velocity attained} = \frac{e}{m}\frac{V}{d}t \quad \text{metres/second} \tag{19.5}$$

$$\text{Average velocity} = \frac{1}{2}\frac{e}{m}\frac{V}{d}t \quad \text{metres/second} \tag{19.6}$$

and

$$\text{Distance moved} = \frac{1}{2}\frac{e}{m}\frac{V}{d}t^2 \quad \text{metres} \tag{19.7}$$

When the electron reaches the positive plate it will have moved a distance d metres. Let t_1 be the time taken to traverse the whole distance (transit time). Then

$$\frac{1}{2}\frac{e}{m}\frac{V}{d}t_1{}^2 = d \tag{from 19.7}$$

Therefore

$$t_1 = \sqrt{\frac{d^2}{\frac{1}{2}\frac{e}{m}V}}$$

and

$$\text{Final velocity} = \frac{e}{m}\frac{V}{d}t_1 = \frac{e}{m}\frac{V}{d}\sqrt{\frac{d^2}{\frac{1}{2}\frac{e}{m}V}} = \sqrt{2\frac{e}{m}V} \tag{19.8}$$

This is seen to depend only on the potential difference V, through which the electron has moved, and on the ratio e/m.

19.3 Energy Relationships for an Electron in an Electrostatic Field

At a point in an electrostatic field an electron will possess (*a*) kinetic energy by virtue of its motion (i.e. $\frac{1}{2}mv^2$ joules), and (*b*) potential energy by virtue of its position in the field and the electrostatic forces acting upon it. When the electron is at the negative plate, or cathode, its potential energy is a maximum. The potential energy declines as the electron moves toward the positive plate, or anode.

If the electrostatic potential at a point is V volts, then the work done in bringing a positive charge of 1 coulomb to this point from a point of zero potential is V joules. The work done in bringing an electron (of charge $-e$ coulomb) up to the point will be $-eV$ joules. This is the energy stored by virtue of the electron's position.

$$\text{Potential energy of electron} = -eV \quad \text{joules} \tag{19.9}$$

If the electron moves under the action of the electrostatic forces only, it will accelerate towards the anode, losing potential energy and gaining kinetic energy but having a constant total energy, provided there are no collisions with stray gas particles.

If at one point an electron has velocity v_1 metres per second, and the point potential is V_1 volts, then

Total electron energy $= \frac{1}{2}mv_1^2 - eV_1$ joules

If, at a second point, the electron velocity is v_2, and the potential is V_2, then the total energy at this point is $\frac{1}{2}mv_2^2 - eV_2$. But since the total energy must remain constant,

$$\frac{1}{2}mv_1^2 - eV_1 = \frac{1}{2}mv_2^2 - eV_2$$

Therefore

$$v_1^2 - v_2^2 = \frac{2e}{m}(V_1 - V_2) \tag{19.10}$$

Often an electron starts with effectively zero velocity from a cathode with effectively zero potential, i.e. $v_1 = 0$, $V_1 = 0$. Then

$$v_2 = \sqrt{\frac{2e}{m}V_2} \quad \text{metres/second} \tag{19.8a}$$

This gives the same result as was derived in the previous section.

These particular conditions are so often met with that it is important to remember the above equation and also to realize that, in the "zero-initial-velocity earthed-cathode" case, the magnitude of the electron velocity at any point depends on the potential of that point only, e and m being assumed constant.

19.4   The Electron-volt

Compared with the energy usually associated with a single electron, the joule is inconveniently large. The more common unit in electronics is the *electron-volt* (eV).

One electron-volt is defined as the kinetic energy gained or lost by an electron when it passes, without collision, through a p.d. of 1 V.

Now, the work done when a charge of 1 C moves through a p.d. of 1 V is 1 J. Thus

Work done when 1 electron moves through a p.d. of 1 V
$$= eV \text{ joules} = 1{\cdot}6 \times 10^{-19} \times 1 \text{ joule}$$

Therefore
$$1 \text{ eV} = 1{\cdot}6 \times 10^{-19} \text{ J}$$

If an electron "falls" through a p.d. of one million volts the work done is 1 million electron-volts (1 MeV).

19.5   Electronic Conduction through a Vacuum

The dynamics of electron motion which have been considered so far has assumed that the movement of the electrons depends only on the electrode potentials and on the properties of individual electrons. This is true only so long as the number of electrons which are involved is small (e.g. the beam current of a cathode-ray tube). For most valves the simple theory is inadequate as it takes no account of the forces which the electrons in the valve exert on each other.

Consider the simple diode circuit shown in Fig. 19.2(*a*). With the

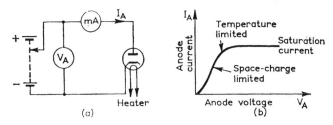

Fig. 19.2   CONDUCTION IN A THERMIONIC DIODE

heater current at a constant value the cathode will attain a constant temperature and electrons will be emitted from it at a constant rate. If the anode voltage is gradually increased, the anode current does not immediately increase to the saturation value corresponding to the cathode temperature, but rises to this value in the manner indicated in Fig. 19.2(*b*). Once saturation has been reached any alteration in the cathode temperature will alter the valve current. The operation is said to be *temperature limited*.

The more important region of the characteristic is the initial part where the anode current increases with the anode voltage. The explanation for this lies in the effect on each other of the electrons in the space between the cathode and the anode. These electrons are said to form a *space charge*, a sample space charge being shown in Fig. 19.3(*a*). For an electron on the cathode side of the space charge the electrostatic force will be reduced since the repulsion of the negative space charge counters the attraction of the positive anode. On the other hand, the force (and therefore the field strength) towards the anode will be increased for the same reason. This gives rise to the voltage distribution between cathode and anode as shown in Fig. 19.3(*b*). Unless the field strength at the cathode is reduced to zero by the space charge, all the emitted electrons must be attracted to the anode. Hence the initial portion of the characteristic is due to the field strength at the cathode being reduced to zero, and the operation in this region is called *space-charge-limited* operation. The action in the space-charge-limited region may be viewed as follows. When the cathode is heated, electrons are emitted and form a "cloud" or negative space charge round the cathode so that any further emitted electrons are repelled and fall back into the cathode. Some electrons may, however, be emitted with velocities

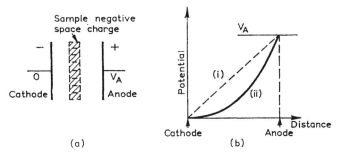

(a)

Fig. 19.3   EFFECT OF SPACE CHARGE ON VOLTAGE DISTRIBUTION
IN A DIODE

(a) Voltage distribution without space charge
(b) Actual voltage distribution

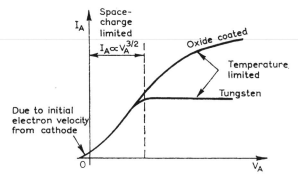

Fig. 19.4   DIODE CHARACTERISTICS

sufficient to overcome the repulsion of the space charge, and these will reach the anode, giving the small anode current for zero anode voltage shown in Fig. 19.4. When a positive potential is applied to the anode some of the electrons from the space charge are attracted to the anode, so that temporarily the space charge is reduced and further electrons will leave the cathode to join the space charge. It can be shown that in this region the anode current is proportional to the three-halves power of the anode voltage, i.e.

$$I_A \propto V_A^{3/2} \tag{19.11}$$

If the anode voltage is sufficient to draw the electrons from the space charge at the maximum rate at which they are emitted from the cathode, then the space charge disappears and the current reaches its temperature limited saturation value.

The space charge may be assumed to collect very close to the cathode.

19.6 Cathode-ray Tubes

In the cathode-ray tube a pencil-like beam of electrons is directed at a fluorescent screen, where it produces a visible spot. The beam, and hence the spot, may be directed at any part of the screen by a beam deflecting system. The construction of a typical tube with electrostatic deflexion and focusing is shown in Fig. 19.5. It con-

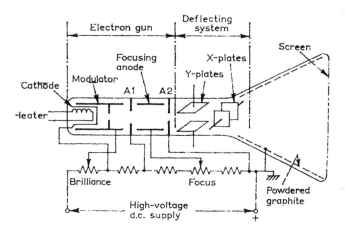

Fig. 19.5 CONSTRUCTION OF A CATHODE-RAY TUBE WITH ELECTROSTATIC FOCUSING AND DEFLEXION

sists of (*a*) an *electron gun* to produce the beam of electrons, (*b*) the *deflecting system*, and (*c*) the *fluorescent screen*. The whole is mounted in an evacuated glass envelope.

The source of electrons for the electron gun is an indirectly heated cathode in the form of a nickel cylinder, the end cap of which is oxide coated. Surrounding and extending beyond the cathode is a second cylinder (*modulator*) which is kept at a negative potential with respect to the cathode. A constriction at the end of this cylinder serves to concentrate the electrons into a rough beam before they pass through the first or accelerating anode (A1). If the modulator is sufficiently negative the beam will be cut off, so that by varying the modulator potential a brilliancy control is achieved.

The anodes A1 and A2 are in the form of discs with small central holes through which the electrons pass. Between them is a third anode in the form of a cylinder, the whole arrangement serving to focus the electrons into a narrow beam and being termed an *electron lens*.

It is usual to have the final anode A2 at earth potential so that the deflecting plates and the screen may also be at earth potential. This in turn means that the cathode is considerably negative with respect to earth, and explains why the filament supply must be well insulated from earth.

After leaving the electron gun the beam passes between two pairs of parallel deflecting plates which are mutually perpendicular. If no potential is applied between the plates, then the beam strikes the centre of the fluorescent screen. A potential applied between the *X*-plates produces a horizontal deflexion of the spot, and a potential between the *Y*-plates produces a vertical deflexion. Both deflexions may take place at the same time.

The funnel-shaped part of the tube leading to the screen is usually coated with graphite and earthed to form a shield. The screen itself is coated with a fluorescent powder such as zinc sulphide (blue glow) or zinc orthoscilicate (blue-green glow). Special powders have been developed to give black-and-white and coloured pictures for television.

For television tubes, magnetic focusing and deflexion have been widely used; electrostatic focusing and magnetic deflexion are now common; and electrostatic focusing and deflexion are coming into favour. For oscilloscopes, electrostatic focusing and deflexion are universal (apart from large-screen display osilloscopes).

19.7 Electrostatic Deflexion of an Electron Beam

Consider a beam from an electron gun passing between two charged parallel plates, *d* metres apart and impinging on a screen which is *D* metres from the centre of the plates, as shown in Fig. 19.6. If the potential of the final anode in the electron gun is V_A volts relative to the cathode, then the velocity of the electrons emerging from the gun will be, by eqn. (19.8),

$$v_z = \sqrt{\frac{2e}{m} V_A} \quad \text{metres/second}$$

This velocity will be called the *z*-velocity.

If the p.d. between the parallel plates is V volts, then the field strength between them is V/d volts/metre, directed at right angles to the original direction of the electron beam. The field is assumed to be uniform between the plates; hence while the electrons are travelling between the plates they are acted on by a force of eV/d newtons (from Section 19.2), which causes an acceleration in the y-direction of $\dfrac{e}{m}\dfrac{V}{d}$ metres per second per second (eqn. (19.9)). The velocity in the z-direction is unaffected, so that the transit time (i.e. the time an electron remains between the plates) will be

$$t_1 = \frac{l}{v_z} \quad \text{seconds}$$

where $l =$ length of plates in metres.

Hence the final velocity in the y-direction will be

$$v_y = (\text{acceleration in } y\text{-direction}) \times (\text{transit time})$$

$$= \frac{e}{m}\frac{V}{d}\frac{l}{v_z} \tag{19.12}$$

The resultant velocity, v, after deflexion will be obtained by the vector addition of v_y and v_z:

$$v = \sqrt{(v_y{}^2 + v_z{}^2)} \tag{19.13}$$

This velocity will make an angle $\theta = \tan^{-1}(v_y/v_z)$ with v_z. The path of the beam between the plates will be parabolic (shown dotted in Fig. 19.6), but there is no loss of accuracy in assuming that the

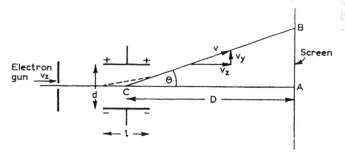

Fig. 19.6 ELECTROSTATIC DEFLEXION OF AN ELECTRON BEAM

beam suddenly deflects by an angle θ at the centre of the plates.

Assuming that the screen is flat, the resultant deflexion of the spot is AB, where

$$\text{AB} = D \tan \theta = D\frac{v_y}{v_z} = D\frac{e}{m}\frac{d}{V}\frac{l}{v_z{}^2} = \tfrac{1}{2}D\frac{l}{d}\frac{V}{V_A} \tag{19.14}$$

EXAMPLE 19.1 In a given cathode-ray tube the anode voltage is 2kV. The *y*-deflecting plates are 3cm long and 0·5cm apart, the distance from the centre of the plates to the screen being 20cm. Determine the deflexion sensitivity in volts per centimetre.

From eqn. (19.14) the deflexion δ for a voltage V between the plates is

$$\delta = \tfrac{1}{2} \times 20 \times \frac{3}{0\cdot 5} \times \frac{V}{2{,}000} = \frac{V}{33\cdot 3}\text{cm}$$

Thus

$$\text{Sensitivity} = \frac{V}{\delta} = 33\cdot 3\,\text{V/cm}$$

Note that in this example the transit time for an electron between the *y*-plates is

$$t_1 = \frac{l}{v_t} = \frac{0\cdot 03}{\sqrt{2\dfrac{e}{m}V_A}} = 1\cdot 13 \times 10^{-9}\,\text{s}$$

Thus the electron passes between the deflecting plates in an exceedingly short time. If an alternating voltage with a frequency of less than, say 1MHz is applied to the plates, then each electron will pass through the plates in such a small fraction of a cycle that during the passage of the electron the deflecting voltage may be regarded as constant at its instantaneous value.

19.8 Magnetic Deflexion of an Electron

When an electron moves across a magnetic field it is found that the path of the electron is affected by the field. The deflexion produced depends on the velocity of the electron and may be calculated by assuming that the moving electron is equivalent to a current-carrying element. Consider an electron of charge $-e$ coulombs moving with velocity v along an element of length Δl:

$$\text{Time to traverse element} = \Delta l/v \quad \text{seconds}$$

Therefore

$$\text{Rate of passage of charge} = -\frac{e}{\Delta l/v} = -\frac{ev}{\Delta l} \quad \text{coulombs/second}$$

Hence

$$\text{Equivalent current, } i = -\frac{ev}{\Delta l} \quad \text{amperes}$$

Therefore

$$i\Delta l = -ev$$

The force acting on the electron is the same as the force on the current element $i\Delta l$. If the electron is moving at right angles to the field, which has a flux density of B teslas, then

$$\text{Force on electron} = Bi\Delta l = -Bev \quad \text{newtons} \tag{19.15}$$

The minus sign is included since the direction of positive current is opposite to the direction of electron motion. The force will act in a direction which is mutually perpendicular to the electron path and to the magnetic field. The relative directions are illustrated in Fig. 19.7.

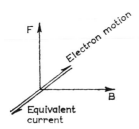

Fig. 19.7   FORCE ON A MOVING ELECTRON IN A MAGNETIC FIELD
F. Force   *B.* Magnetic flux density

19.9   Motion of an Electron in a Uniform Magnetic Field

Two cases will be considered.

(*a*) *Electron passing completely through the field.* Consider an electron entering a magnetic field with a linear velocity v metres per second. Let the path of the electron be directly across the magnetic field. From eqn. (19.15) the force on the electron is Bev newtons. Since this force acts perpendicular to the direction of motion of the electron, the linear velocity of the electron will be unaffected, but the electron will describe a circular path as shown in Fig. 19.8. If the radius of the path is R metres, then since the inward magnetic force is equal to the outward centrifugal force,

$$Bev = \frac{mv^2}{R}$$

Thus

$$R = \frac{mv}{Be} \quad \text{metres} \tag{19.16}$$

By geometry the angle θ through which the electrons are deflected is the angle subtended at the centre of curvature. Hence, from Fig. 19.8,

$$\text{Deflexion of beam, } \theta = \sin^{-1} \frac{l}{R} \tag{19.17}$$

where l is the width of the magnetic field.

(*b*) *Electron moving in a large field*. Since an electron projected into a magnetic field moves with constant linear velocity along a circular path, it follows that if the extent of the field is large enough the electron will describe a complete circle. This is the basis of operation of some types of particle accelerator such as the cyclotron

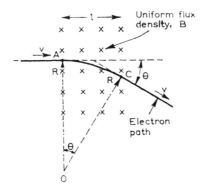

Fig. 19.8   PATH OF AN ELECTRON IN A MAGNETIC FIELD

B. Uniform magnetic field   O. Centre of curvature

and the betatron, and also of some types of ultra-high-frequency valves, particularly the magnetron. It should be noted that, if the electron enters the magnetic field at an angle other than 90°, its component of velocity in the direction of the field remains unaltered, and it will describe a spiral path.

PROBLEMS

19.1 If an electron is accelerated from rest through a potential difference of 4 kV, determine the final energy of the electron and the time required for the electron to pass along an evacuated tube 10 in. long after it has attained its final velocity.

Ans. 6.4×10^{-16} J; 6.73×10^{-9} s.

19.2 Two large plane parallel electrodes are placed 1 cm apart in a vacuum and a p.d. of 100 V is applied between them. If an electron is emitted from the negative plate with zero initial velocity, calculate the time required for this electron to reach the positive plate.

Ans. 3.37×10^{-9} s.

19.3 Define the electron-volt.

400 V electrons are introduced at A (Fig. 19.9) into a uniform electric field of intensity 150 V/cm. If the electrons emerge at B 5.0×10^{-9} s later, determine (*a*) the distance AB; (*b*) the angle θ.   (*H.N.C.*)

Ans. 4.95 cm; 33.8°.

19.4 Find the path of an electron moving perpendicular to a uniform magnetic field.

A 15 cm cathode-ray tube is to be magnetically deflected. The final anode voltage is 2 kV, the deflecting coil is 2½ cm long and the distance from its centre to the screen is 15 in. Find the flux density required to deflect the spot to the edge of the screen. (*H.N.C.*)

Ans. 1·06 × 10⁻³T.

19.5 A cathode-ray tube has an accelerating potential of 2 kV. It is fitted with two pairs of deflecting plates, the pairs being mutually perpendicular. Each plate is 1·5 cm long, the spacing between the plates being 1 cm and the distance between

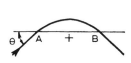

Fig. 19.9

centres of pairs being 2·5 cm. Neglecting fringing effects, calculate the length of the line produced on a fluorescent screen 30 cm distant from the centre of the nearer pair of plates when (*a*) a 50 V (r.m.s.) voltage is applied between the further pair of plates, (*b*) a 50 V (r.m.s.) voltage is applied between the nearer pair of plates, (*c*) both voltages are applied simultaneously.

Ans. 1·72 cm; 1·59 cm; 2·34 cm.

19.6 Electrons are projected with a velocity *v* centimetres per second into a magnetic field of strength *B* teslas and are observed to travel in a circular path of 10 cm radius. An electrostatic field, existing between a pair of plates 3 cm apart is now superimposed on the magnetic field at right angles to both the magnetic field and to the initial direction of motion of the electrons. If the voltage between the plates is adjusted until the electron beam passes through the combined fields without deviation, and is found to be 22·5 V, calculate the strength of the magnetic field and the velocity of the electrons at their time of entry into the field. (*H.N.C.*)

Ans. 2·06 × 10⁻⁴T; 3·64 × 10⁶ m/s.

Chapter 20

ELECTRONIC DEVICE CHARACTERISTICS

In order to understand electronic circuits it is necessary to know something about the characteristics of the devices which are employed in these circuits. In this chapter the operating characteristics of valves and transistors will be described. The physical basis of operation of these devices will be discussed very briefly. There will be no attempt to develop the theory of physical electronics in great detail.

Transistors and valves are called active devices, because their operation can be described in terms of equivalent constant-voltage or constant-current sources whose output is controlled by input voltages or currents. The essential basic feature of these active devices is that the controlled current or voltage can be much larger than the input current or voltage. Thus the output current or voltage variation can be an amplified version of the input signal variation. The output power is derived from an external supply and not from the controlling input signal.

20.1 The Triode

In the vacuum *triode* a heated *cathode* is surrounded by a metal *anode*, A, and a fine wire spiral, G, called the *grid* is placed between the cathode, K, and the anode, to act as the control element. The assembly is housed in an evacuated envelope. The graphical symbol for a triode is shown in Fig. 20.1(*a*). For a given positive anode voltage, the more negative the grid with respect to the cathode, the smaller will be the anode current.

When the grid–cathode voltage V_{GK} is zero, the static anode characteristic (Fig. 20.1(*b*)) is similar to that of a space-charge-limited diode. The greater the negative value of V_{GK} (i.e. the larger the negative grid bias) the larger is the anode–cathode voltage

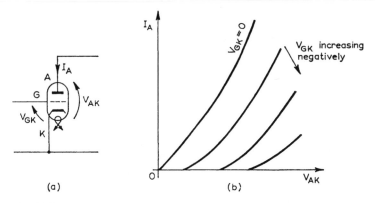

(a)   (b)

Fig. 20.1   ANODE CHARACTERISTICS OF A TRIODE

V_{AK} required to give the same anode current. Hence the static curves for various values of negative grid–cathode voltage are displaced to the right as shown in the diagram.

So long as the grid is maintained at a negative potential with respect to the cathode, relatively few electrons will hit it—i.e. the grid current will be negligibly small. Most of the electrons moving towards the anode will pass through the spaces in the grid wire helix.

20.2 The Tetrode—Secondary Emission

The vacuum *tetrode* was originally developed to reduce the capacitive coupling which exists between the grid and anode of a triode. To do this a second grid called the *screen grid* is introduced between the *control grid* and the anode. This grid also consists of a wire helix, through which most of the electrons comprising the cathode current can pass. It is maintained at a fixed potential and so "screens" the electrostatic field at the anode from that of the rest of the valve.

The screen potential must be positive with respect to the cathode, since otherwise there would be no electric field to draw electrons from the space charge, and hence no anode current. The graphical symbol for a tetrode is shown in Fig. 20.2(*a*).

The static anode characteristics are shown in Fig. 20.2(*b*). Because of the constant screen voltage and the screening effect on the anode, the cathode current, I_K, will be nearly constant for any negative value of grid–cathode voltage, V_{GK}. (I_K is shown for $V_{GK} = 0$ at (*b*).) This current divides between screen and anode in a manner

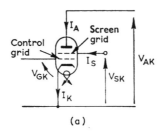

(a)

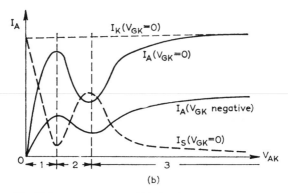

(b)

Fig. 20.2   TETRODE CHARACTERISTICS

which depends on their relative potentials. As the anode–cathode voltage is increased, three distinct stages of anode current, I_A, can be distinguished.

Stage 1. When $V_{AK} = 0$, nearly all of the cathode current passes to the screen grid so that I_A is almost zero. Any electrons passing through the screen helix are attracted back to it by the positive screen potential. As V_{AK} is increased, however, more electrons which pass through the screen helix will reach the anode, so that I_A increases and the screen current, I_S, falls by a corresponding amount.

Stage 2. When V_{AK} exceeds about 20V (the exact voltage depends on the anode material), the electrons which reach the anode bombard it with sufficient energy to cause electrons to be emitted from the anode surface. These *secondary electrons* are relatively slow moving, and will tend to be attracted to whichever electrode is more positive. So long as the screen is more positive than the anode the secondary electrons will tend to be attracted to the screen. Since the bombarding electrons can each cause more than one electron to be emitted from the anode, the anode current will fall and the screen current will correspondingly rise. As V_{AK} rises the secondary emission increases and the anode current falls further.

Stage 3. When V_{AK} exceeds V_{SK} the secondary electrons emitted from the anode will generally return to the anode, so that the anode current rises towards the value of the almost constant cathode current, and the screen current falls to a relatively low value (most of the electron stream will pass through the screen and will reach the anode). The anode current is then almost independent of the anode–cathode voltage.

Note that during stage 2 the anode exhibits a negative resistance characteristic—*an increase* in voltage causes a *decrease* in current. This is an unstable operating region. In practice, simple tetrodes were never popular owing to the negative-slope-resistance "kink" in the characteristics, but it should be noted that the curve in stage 3 gives the possibility of high amplification, since anode current in this region depends on negative grid voltage only and is almost independent of anode voltage (unlike the triode). This is discussed further in Section 20.4.

20.3 The Pentode

The *pentode* gives the advantages of the tetrode (high amplification and low electrostatic coupling between control grid and anode circuits) without the disadvantage of the negative-resistance kink in the anode characteristics. A third open-wire helical grid (the *suppressor grid*) is inserted between the screen grid and the anode. This suppressor grid is connected either internally or externally to the cathode. Hence any secondary electrons from the anode find themselves in an electric field which forces them back to the anode, so that they have no net effect on either the anode or the screen currents. The cathode current is determined almost entirely by the constant screen potential, V_{SK}, and the control grid voltage, V_{GK}. Hence the anode current is largely independent of the anode voltage above some critical value which can lie between 20 V and 100 V. The graphical symbol for a pentode and typical anode characteristics are shown in Fig. 20.3.

Similar "kink-free" characteristics are obtained in the *beam tetrode*, by aligning the control and screen grid helices and using the electron beam itself as a suppressor of secondary emission effects. The alignment of the grids also has the effect of reducing the screen-grid current below the value obtained in the equivalent pentode, which can be very advantageous. Beam tetrodes are often used in

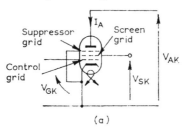

(a)

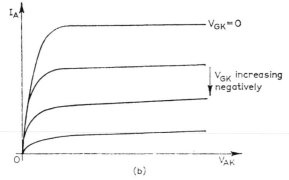

(b)

Fig. 20.3   ANODE CHARACTERISTICS OF A PENTODE

the audio-frequency power-output stages of amplifiers and in radio-frequency power amplifiers. Note that the screen–cathode voltage must be kept constant in order to obtain normal pentode or tetrode characteristics.

20.4 Simple Valve Voltage Amplifier—Load Line

The circuit of a simple amplifier using a pentode is shown in Fig. 20.4(a). A steady negative voltage (the grid-bias voltage V_{GB}) is applied to the grid through the grid-leak resistor, R_g. The grid carries no current, and hence if the input voltage, V_i, is zero there will be no voltage drop across R_g and $V_{gk} = -V_{GB}$. Note that this form of negative grid-bias supply from a separate source is not now used (see Section 20.5). The anode is supplied from a constant-voltage supply, V_B, through the anode load resistor R_L.

When a signal, V_i, is applied across R_g the grid voltage will vary, causing a variation, in the anode current I_A. The voltage drop across R_L will also vary, and hence the output voltage V_o will vary, since

$$V_o = V_{ak} = V_B - I_a R_L \qquad (20.1)$$

If the change in V_{ak} is greater than the voltage, V_i, which caused it, the circuit is a *voltage amplifier*. Note that when V_i becomes more

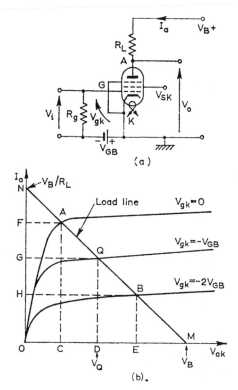

Fig. 20.4   SIMPLE VOLTAGE AMPLIFIER

positive, I_a will increase, and from eqn. (20.1), V_o will decrease, i.e. the output voltage variation is inverted with respect to the input voltage variation.

The relationship between V_{ak} and I_a must satisfy two conditions simultaneously, namely those imposed by the anode characteristics of the valve and those expressed by eqn. (20.1). If $(V_B - I_a R_L)$ is plotted on the anode characteristics, as shown in Fig. 20.4(b), it gives the straight line MN running from M (when $I_a = 0$, and $V_{ak} = V_B$) to N (when $I_a = V_B/R_L$, and $V_{ak} = 0$). This is called the *load line;* its slope is $-1/R_L$ amperes per volt (siemens).

For any value of grid voltage, the operating current and anode–cathode voltage must be represented by a point on that particular grid-voltage curve and also by a point on the load line, i.e. the intersection of these two curves represents the *operating point.*

Thus for $V_{gk} = 0$ the intersection is at point A, and the anode current is given by OF. The voltage across the valve is OC and the voltage across R_L is CM. With no input signal, the grid voltage is $-V_{GB}$ and so the operating point is Q, which gives the quiescent anode current (OG) and anode voltage (OD).

Suppose that an alternating input voltage of peak value $V_{i\,max}$ is applied to the grid (this maximum value is so chosen that the grid will never be positive to the cathode). The operating point moves up and down the load line between points A and B. Hence the output voltage varies between OC and OE and the voltage gain A_v is given by

$$A_v = \frac{\text{Change in output voltage}}{\text{Change in input voltage}} = -\frac{\text{CE}}{2V_{i\,max}} \qquad (20.2)$$

The minus sign indicates the inversion which takes place between input and output voltage variations.

If the static characteristics are equally spaced for equal changes in V_{gk}, the output-voltage waveform will be an exact copy of the alternating input-voltage waveform. If the static characteristics are not equispaced, the output voltage will be a distorted version of the input. The grid-bias voltage is chosen so that a minimum of distortion is produced.

EXAMPLE 20.1   The anode characteristics of a pentode (Mullard EF37A) are given by the curves of Fig. 20.5(a) for a screen grid voltage of 100 V.

If the supply voltage is 400 V plot the load lines for anode resistances of (a) 50 kΩ and (b) 200 kΩ. (i) Choose a suitable grid bias voltage for operation with each of the above resistances. (ii) Estimate the power dissipation at the anode of the valve for each of the grid-bias/anode-resistance conbinations when there is no input grid voltage. (iii) Estimate the voltage amplification obtained with each resistance.

The load lines are drawn on the characteristic curves of Fig. 20.5, as shown. Suitable grid-bias voltages would be, say, -2 V for the 50 kΩ line (Q_1), and -3.3 V for the 200 kΩ line (Q_2).

With these grid bias values and no alternating grid voltage:
(a) For 50 kΩ,

> Steady anode current = 3 mA
> Steady anode voltage = 245 V

Therefore

Power dissipation at anode $= 0.003 \times 245 = 0.735\,\mathrm{W}$

(*b*) For $200\,\mathrm{k}\Omega$,

Steady anode current $= 1.2\,\mathrm{mA}$
Steady anode voltage $= 160\,\mathrm{V}$

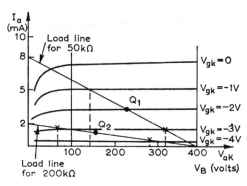

Fig. 20.5

Therefore

Power dissipation at anode $= 0.192\,\mathrm{W}$

To determine the voltage amplification for $50\,\mathrm{k}\Omega$ suppose the grid voltage varies by $\pm 1\,\mathrm{V}$ about the $-2\,\mathrm{V}$ bias.

$$\text{Voltage amplification} = -\frac{325 - 150}{2} = -87.5$$

To determine the voltage amplification for $200\,\mathrm{k}\Omega$ suppose the grid voltage varies by $\pm 0.5\,\mathrm{V}$ about the $-3.3\,\mathrm{V}$ bias.

$$\text{Voltage amplification} = -\frac{275 - 70}{1} = -205$$

Similar conditions apply for a triode amplifier. The construction of the load line is the same as for a pentode. In this case, however, the anode current is not independent of the anode voltage, but rises and falls as the anode voltage rises and falls. Hence, if the grid voltage falls so that the anode current decreases, the anode voltage will rise, and this will tend to counteract the decrease in anode current. The result is a smaller change in anode voltage compared with a pentode, where the anode voltage hardly affects the anode current, i.e. the voltage amplification is less in the triode circuit.

20.5 Cathode Bias

The grid-cathode bias voltage (i.e. the steady or quiescent value of V_{gk}) is commonly obtained by connecting a resistor in the cathode lead as shown in Fig. 20.6.

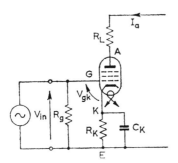

Fig. 20.6   AUTOMATIC CATHODE BIAS

Then, if the quiescent cathode current is I_Q, a direct voltage given by

$$V_{KE} = I_Q R_K \tag{20.3}$$

will be developed across R_K, and hence the grid will be more negative than the cathode by this voltage (G is connected to E through the grid-leak resistor R_g). The capacitor C_K connected across R_K serves to short-circuit the cathode to earth for alternating components of current, and so smooths out voltage variations across R_K due to any alternating component of cathode current. The calculation of the value of C_K is deferred until the next chapter, but as a rule of thumb the reactance $1/\omega C_K$ at the lowest signal frequency is normally less than $0 \cdot 1/g_m$, where g_m is a constant of the valve (see Section 20.6). Note that, if R_K is returned to the earth line, then R_K should be approximately equal to $1/g_m$.

20.6 Linear Valve Parameters

When the a.c. signals which are being dealt with by an amplifier are small, it is usually possible to operate the valve over a range of current and voltage where the static characteristics are linear and equally spaced. In this case the operation of the valve may be represented by constants (or parameters) which are assumed to remain unchanged in value over the whole operating range. This will not be true if large signals are dealt with (see Chapter 24).

Consider a section of the static anode characteristics of a valve as shown in Fig. 20.7. From these the following expressions are obtained.

1. The *anode slope resistance*, r_a, of the valve is defined as the change in anode voltage per unit change in anode current when the grid voltage is held constant. Hence from Fig. 20.7,

$$r_a = \left(\frac{\delta V_{ak}}{\delta I_a}\right)_{V_{gk} \text{ const.}} = \frac{AB(\text{volts})}{BC(\text{amperes})} \tag{20.4}$$

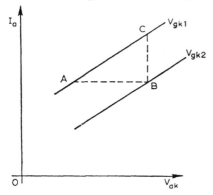

Fig. 20.7 CALCULATION OF LINEAR VALVE PARAMETERS

2. The *mutual conductance*, g_m, is defined as the change in anode current per unit change in control grid voltage when the anode voltage is held constant; i.e.

$$g_m = \left(\frac{\delta I_a}{\delta V_{gk}}\right)_{V_{ak} \text{ const.}} = \frac{BC(\text{amperes})}{V_{gk1} - V_{gk2}} \tag{20.5}$$

Since the anode current is normally measured in milliamperes, g_m is usually expressed in milliamperes per volt.

3. The *amplification factor*, μ, is defined as the change in anode voltage per unit change in grid voltage with the anode current held constant, i.e.

$$\mu = \left(\frac{\delta V_{ak}}{\delta V_{gk}}\right)_{I_a \text{ const.}} = \frac{AB}{V_{gk1} - V_{gk2}} \tag{20.6}$$

From eqns. (20.4)–(20.6), the following important relation can be deduced:

$$\mu = \left(\frac{\delta V_{ak}}{\delta V_{gk}}\right)_{I_a \text{ const.}} = \left(\frac{\delta V_{ak}}{\delta I_a}\right)_{V_{gk} \text{ const.}} \times \left(\frac{\delta I_a}{\delta V_{gk}}\right)_{V_{ak} \text{ const.}}$$

i.e.

$$\mu = r_a g_m \tag{20.7}$$

The mutual conductance depends mainly on the spacing and construction of the control grid and cathode. It has thus approximately the same value for triodes and pentodes, and is usually within the range 1–10 mA/V.

The anode slope resistance is the reciprocal of the slope of the static anode characteristics, and will therefore be much larger for pentodes than for triodes. Typical values range from 1 kΩ–50 kΩ for triodes and 100 kΩ–2 MΩ for pentodes.

From this and by considering eqn. (20.7) it follows that the amplification factor of a triode is much lower than that of a pentode, typical values being 20 for triodes and 1,000 for pentodes.

20.7   Anode Resistors for Voltage Amplifiers

Fig. 20.8(*a*) shows the anode characteristics of a triode with two load lines drawn for the same supply voltage V_B. Consideration of the diagram will show that the load line for the high resistance, R_H, gives a larger voltage amplification than the load line for the lower resistance, R_L. Though the amplification is not markedly different it appears that the larger the load resistance the better. However,

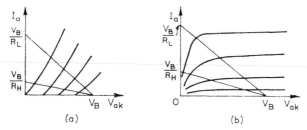

Fig. 20.8   CHOICE OF ANODE RESISTOR FROM VALVE CHARACTERISTICS
(*a*) Triode   (*b*) Pentode

the load line for R_H runs through the curved and less regular parts of the characteristics so that the output voltage variations will be a less faithful reproduction of the input voltage variations than if the load line for R_L were traversed. An intermediate value, usually two or three times r_a, is chosen.

The same arguments for and against a high load resistance apply with even more force for a pentode. Load lines for both high and low resistance in this case are shown in Fig. 20.8(*b*). Since r_a for a pentode valve is very high, a load resistance of the order of r_a would give marked distortion of the output. A suitable value is very often one for which the load line runs through the knee of the $V_{gk} = 0$ characteristic. This is usually about two or three times the value which would be used with a roughly equivalent triode.

The increased amplification obtainable with a high anode resistance is often, in any case, illusory, since this gives a large internal resistance to the amplifier and hence the output voltage falls considerably when the next stage is connected to the amplifier terminals. Also, with a high anode resistance, stray capacitances have a relatively greater shunting effect, giving reduced amplification, particularly at high frequencies.

EXAMPLE 20.2 For the triode, whose anode characteristics are given in Fig. 20.9(a), determine the valve parameters in the linear region. Design an

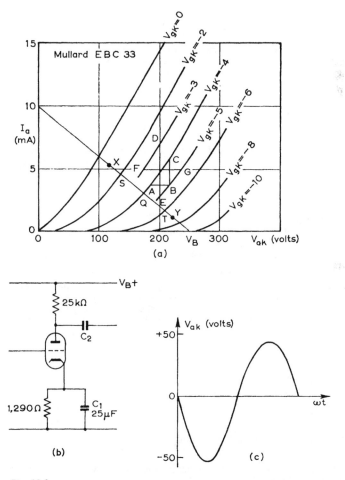

Fig. 20.9

amplifier using this valve and a supply of 250 V to give uniform amplification to audio frequencies above 50 Hz. Include a cathode bias circuit. Estimate the voltage amplification obtainable and draw the output voltage waveform corresponding to a sinusoidal input of 2 V (r.m.s.).

The parameters are taken for the region around the point $I_a = 5$ mA, $V_{ak} = 200$ V in Fig. 20.9(*a*).

$$r_a = \frac{AB}{CB} = \frac{33}{2\cdot3} \times 1,000 = \underline{14\cdot4\,k\Omega}$$

$$g_m = \frac{DE}{5-3} = \frac{4}{2} = \underline{2\,mA/V}$$

$$\mu = \frac{FG}{5-3} = \frac{58}{2} = \underline{29}$$

The -3 V and -5 V characteristics were interpolated.

A suitable anode resistance might be 25 kΩ for a 250 V supply, i.e. approximately $2r_a$ (up to about 50 kΩ may be used without severe distortion).

The load line for 25 kΩ at 250 V is shown, i.e. the line from

$$(V_{ak} = 250\text{V}, \ I_a = 0) \quad \text{to} \quad \left(V_{ak} = 0, I_a = \frac{250}{25,000} = 10\,\text{mA} \right)$$

For an input of 2 V r.m.s., i.e. 2·83 V peak, the grid-bias voltage should be at least -3 V. Suppose -4 V is chosen for convenience; the static operating point is then Q in Fig. 20.9. Thus

Mean anode current $= 3\cdot1$ mA

And by eqn. (20.3),

Cathode bias resistance, $R_1 = \frac{4}{3\cdot1} \times 1,000 = 1,290\,\Omega$

When the internal impedance of the valve is taken into account it can be shown that the effective cathode-to-earth resistance is 1,290 Ω in parallel with $1/g_m$. The capacitor C_1 is chosen so that its reactance at the lowest frequency to be amplified is very much less than the effective cathode-to-earth resistance; $C_1 = 25\mu$F (say). The circuit is shown in Fig. 20.9(*b*).

To estimate the voltage amplification assume that the input voltage is 2 V peak so that the grid voltage varies from -2 V to -6 V, i.e. a variation of 4 V from S to T in Fig. 20.9(*a*).

At S, anode voltage $\doteqdot$ 135 V

At T, anode voltage $=$ 207 V

Therefore

Voltage amplification $= -\dfrac{207 - 135}{4} = \underline{-18}$

It will be appreciated that the voltage amplification varies somewhat with the amplitude of the input voltage.

For a sinusoidal input voltage of 2 V r.m.s., the anode voltage wave may be derived from the load line in Fig. 20.9(a).

ωt	0	45°	90°	135°	180°	225°	270°	315°	360°
Input voltage	0	2	2·83	2	0	−2	−2·83	−2	0
Operating point	Q	S	X	S	Q	T	Y	T	Q
Anode voltage	173	135	120	135	173	207	218	207	173
Variation of V_{ak}	0	−38	−53	−38	0	34	45	34	0

The anode voltage variation, i.e. the output waveform, is shown in Fig. 20.9(c). The waveform will, of course, be nearer to a true sinusoid if the amplitude of the input voltage is reduced.

20.8 Estimating Voltage Gain from Valve Parameters

For a valve amplifier the anode current, I_a, is some function of the grid–cathode and anode–cathode voltages, i.e.

$$I_a = f(V_{gk}, V_{ak})$$

Hence any change in anode current, δI_a, due to changes δV_{gk} and δV_{ak} in the grid–cathode and anode–cathode voltages can be expressed as

$$\delta I_a = \delta V_{gk} \left(\frac{\delta I_a}{\delta V_{gk}}\right)_{V_{ak}\, \text{const.}} + \delta V_{ak} \left(\frac{\delta I_a}{\delta V_{ak}}\right)_{V_{gk}\, \text{const.}}$$

or

$$\delta I_a = g_m \delta V_{gk} + \frac{1}{r_a}\delta V_{ak} \qquad (20.8)$$

Also

$$\delta V_{ak} = -R_L \delta I_a \qquad (20.9)$$

where R_L is the anode circuit resistance, and the minus sign is used to indicate that if I_a increases V_{ak} decreases.

From eqns. (20.8) and (20.9),

$$-\frac{\delta V_{ak}}{R_L} = g_m \delta V_{gk} + \frac{1}{r_a}\delta V_{ak}$$

or

$$\delta V_{ak} = \frac{-g_m r_a R_L \delta V_{gk}}{r_a + R_L} = \frac{-\mu R_L \delta V_{gk}}{r_a + R_L} \qquad (20.10)$$

Let V_{gk} represent the r.m.s. alternating grid–cathode voltage, and V_{ak} the r.m.s. alternating anode–cathode voltage; then, in terms of complexors, eqn. (20.10) can be rewritten as

$$V_{ak} = \frac{-\mu R_L V_{gk}}{r_a + R_L} \tag{20.11}$$

$$= \frac{\mu R_L V_{gk}\underline{/180°}}{r_a + R_L} \quad \text{(since } -1 \equiv 180°\text{)} \tag{20.11a}$$

The complex stage gain, A_v, is thus

$$A_v = \frac{V_{ak}}{V_{gk}} = \frac{-\mu R_L}{r_a + R_L} = \frac{\mu R_L\underline{/180°}}{r_a + R_L} \tag{20.12}$$

Eqns. (20.11) and (20.12) thus represent the operation of the amplifier under small-signal linear conditions. They refer only to the alternating components of voltage.

20.9 Coupling of Multistage Valve Amplifiers

In order to increase the overall gain of an amplifier several stages may be connected in cascade. A two-stage amplifier is shown in Fig. 20.10, in which resistance-capacitance coupling is used to isolate

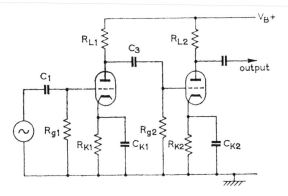

Fig. 20.10   CR-COUPLED TWO-STAGE AMPLIFIER

each stage from the d.c. levels of the previous stage. Thus capacitor C_1 acts as a d.c. block for the input signal source, and C_3 prevents the direct anode voltage of valve V_1 appearing at the grid of valve V_2. The values of C_1 and C_2 must be chosen so that at the lowest frequency of the input signal they are virtually short-circuits to alternating currents. Usually it is sufficient to make the reactance of the capacitors less than one-tenth of the resistance of the following grid resistors, at the lowest frequency that is to be amplified; i.e.

$$\frac{1}{\omega C_1} < 0 \cdot 1 R_{g1} \quad \text{and} \quad \frac{1}{\omega C_3} < 0 \cdot 1 R_{g2}$$

Coupling circuits will be discussed in greater detail in the next chapter.

Note that in a two-stage amplifier the effective resistance, R_P, in the anode circuit at signal frequencies is the parallel combination of the anode resistor, R_{L1}, and the following grid resistor, R_{g2}. In estimating the gain from the characteristics an a.c. load line must be used. This has a slope of $-1/R_P$ and passes through the d.c. operating point, Q.

20.10 The Semiconductor Junction Diode

It is assumed that the reader is aware that in an *intrinsic* semiconducting material (such as pure monocrystalline germanium, silicon, gallium arsenide, indium antimonide, etc.) conduction takes place by the movement of two types of *charge carriers, electrons* and *holes*. These are produced in equal numbers in the intrinsic material by thermal generation or by radiation. The number of *electron-hole pairs* increases rapidly as the temperature rises. Holes may be regarded as positive charge carriers, while electrons are, of course, negative charge carriers. In an *extrinsic* or *doped* semiconductor, impurities are added to the intrinsic material to give a predominance of either electrons (in *n*-type material) or holes (in *p*-type material) as charge carriers. The conductivity of a doped semiconductor is greater than and is less temperature sensitive than that of the intrinsic material up to fairly high temperatures. The charge carriers which predominate in a doped semiconductor are called the *majority carriers*.

Elemental semiconductors (e.g. silicon and germanium, in group IV of the periodic table of elements) have four electrons in the valence band, and form crystals of the diamond lattice structure where each atom shares its valence electrons with four adjacent atoms. This is called covalent bonding. If impurities in group V of the periodic table (e.g. phosphorus or arsenic) are added to the intrinsic semiconductor, these impurity atoms fit into the crystal lattice quite happily, and share four of their five valence electrons with four adjacent elemental semiconductor atoms, the fifth electron moving into the conduction band to contribute to the conductivity of the material. The material is then said to be *n*-type. Impurities which produce this effect are called *donor* impurities.

Intrinsic semiconductors are also formed when elements of group III (e.g. gallium) and group V (e.g. arsenic) form compounds (e.g. gallium arsenide) which have an average of four valence electrons per atom. Similarly some group II–group VI compounds form semiconductor crystals (e.g. cadmium sulphide), and silicon carbide is an example of two elements in Group IV forming a semiconductor crystal. Doping to give an *n*-type material can be achieved in a similar manner to that for elemental semiconductors. Thus if a group VI impurity (e.g. tellurium or sulphur) is added to a III–V compound, the impurity atom can replace one of the group V elements, its extra valence electron then being free to contribute to conduction. The same effect can be achieved by adding an excess of the group V element to the compound. Compound semiconductors are more difficult to purify and dope than elemental semiconductors.

In all *n*-type semiconductors, each added impurity atom gives one extra conduction electron. This increased number of conduction electrons results in a reduction in the mean density of thermally generated holes. This follows since if there are more conduction electrons there is a greater statistical probability of their recombining with holes, so that the mean lifetime of the holes falls. Note that since the donor atom loses one of its valence electrons it becomes a positive ion. These ions do not contribute to conduction since they are fixed in the crystal lattice. The net result of adding an *n*-type impurity is to increase the number of mobile electrons and to reduce the number of holes. Conduction in *n*-type semiconductors is thus mainly by electrons.

Elemental semiconductors become extrinsic *p*-type material when *acceptor* impurities (e.g. indium or boron in group III of the periodic table) are added to the intrinsic material. The impurity atoms fit into the crystal lattice, but since they have only three valence electrons instead of the four of the semiconductor atom that they replace, they each contribute a mobile hole (or space in the valence band that can be filled by an electron from an adjacent atom). Compound semiconductors can be similarly doped (e.g. zinc, in group II is often used as an acceptor to dope Group III–V compounds).

By analogy with the conditions in *n*-type semiconductors the increase in the number of holes in a *p*-type material due to the acceptor impurities gives rise to a reduction in the mean density of thermally generated electrons, and results in a material in which conduction is mainly by the movement of holes. These act as positive charge carriers.

In general, doped semiconductors are of higher conductivity than intrinsic semiconductors, the conductivity being a function of the number of impurity atoms added. The conductivity is also less temperature dependent than that of intrinsic material.

The conductivity of a material can be expressed in terms of the number of mobile charge carriers per unit volume, the size of the charge, and a constant, μ, called the *mobility* of the charge carriers. The mobility simply expresses the mean velocity attained by the charge carriers per unit of applied electric field. Mathematically the conductivity, σ, of a semiconductor may be written as

$$\sigma = en\mu_n + ep\mu_p$$

where e = electronic charge (1.6×10^{-19} C); n = electron density, i.e. the number of mobile electrons per cubic metre; μ_n = electron mobility (m/s per V/m); p = hole density; μ_p = hole mobility.

In intrinsic material $n = p = n_i$ (the intrinsic carrier density), and it may be shown that in doped semiconductors the product np is a constant, equal to $n_i{}^2$.

Generally hole mobility is smaller than electron mobility. Table 20.1 gives some idea of mobilities in different materials at room temperature (mobility varies with temperature—at around room temperature it falls as the temperature rises).

TABLE 20.1

Material	Electron mobility, μ_n	Hole mobility, μ_p	Intrinsic conductivity
	m/s per V/m	m/s per V/m	S/m
Germanium	0·38	0·18	2
Silicon	0·19	0·05	2×10^{-3}
Gallium arsenide	0·5	0·03	
Indium antimonide	10·0	0·3	
Silicon carbide	<0·01	<0·01	

Consider a junction formed between *p*-type and *n*-type materials as shown in Fig. 20.11(*a*). The alloy junction shown can be achieved by fusing an indium pellet into a small *n*-type wafer. This alloying process produces a very heavily doped *p*-type region to the left of the junction as the indium fuses into the *n*-type wafer. The *n*-type region may be lightly doped. Three conditions will be discussed.

1. *No external bias and open-circuit connexion.* At the junction shown schematically at (*b*), holes from the *p*-type material diffuse across into the *n*-type region, where there is an excess of conduction

electrons. The holes recombine with electrons in the *n*-type region so that a positive charge of captured (bound) holes is built up near the junction. Similarly, electrons diffusing into the *p*-type material from the *n*-type combine with the majority carriers (holes) there, to form a bound negative charge just on the *p*-side of the junction. The bound charges build up until the repulsion between them is just sufficient to prevent any further net diffusion of holes from *p* to *n* or electrons from *n* to *p*.

However, the bound charges attract across the junction the *minority carriers* (electrons in *p* and holes in *n*) which are generated thermally

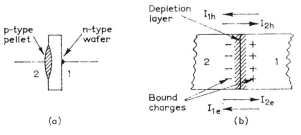

Fig. 20.11   ALLOY JUNCTION DIODE
(*a*) Construction of diode   (*b*) The *p–n* junction

on the opposite side of the junction; i.e. minority holes in *n* are attracted across to the bound negative charge in *p*, and minority electrons in *p* are attracted to the bound positive charge in *n*. Let these currents be

I_{1h} = Minority hole current from *n* to *p*
I_{2e} = Minority electron current from *p* to *n* (see Fig. 20.11(*b*))

Since the *p*-region is assumed to be much more heavily doped than the *n*-region, there will be a *lower concentration* of thermally generated electrons in *p* than of holes in *n*; i.e.

$I_{1h} \gg I_{2e}$ (a ratio of 100:1 can easily be achieved)

A junction of this nature is referred to as a $p^+–n$ junction, to indicate that the *p*-region is more heavily doped than the *n*-region.

In equilibrium these minority-carrier currents must be exactly balanced by the majority-carrier currents crossing the opposite way to maintain the bound-charge distribution; there is no net junction current.

In the region in which bound charges exist there are virtually no free carriers, and this is called the *depletion layer*. Mobile carriers are moved out of this region because of the electrostatic field due to the bound charges. Since the diffusing carriers must penetrate further into a lightly doped region before finding an opposite carrier with which to recombine, the depletion layer extends further into the lightly doped side of the junction than into the heavily doped side. Indeed if one side is much more heavily doped than the other, the depletion layer exists almost entirely on the lightly doped side of the junction.

If the junction is short-circuited by an external connexion, an alternative path for the return of charge carriers exists and there

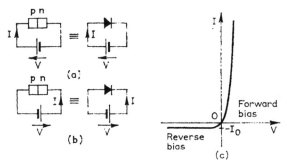

Fig. 20.12   EFFECT OF BIAS ON A *p–n* JUNCTION

(a) Forward bias
(b) Reverse bias
(c) Forward and reverse *I/V* characteristics

can be no net build-up of charge and hence the current will normally be negligible.

It is worth noting that the heavier the doping of the two regions the greater is the charge barrier which is produced to maintain equilibrium.

2. *Forward bias.* If an external d.c. supply is connected across a *p–n* junction as shown in Fig. 20.12(*a*), electrons in the *p*-type material will drift away from the junction towards the positive connexion to the supply, while holes in the *n*-type region will drift towards the negative connexion (where they recombine with electrons from the external circuit). This drift reduces the bound charge density at the junction, and allows majority carriers to flood both ways across the junction, giving rise to a large current. This is the foward bias connexion.

Since the *p*-region is assumed to be more heavily doped than the *n*-region, the junction current will consist mainly of holes crossing from *p* to *n*, i.e. majority carriers from the heavily doped region. The current in the connecting wires is, of course, an electron current. Above about 0·2 to 0·3 V forward bias for germanium and about

0·6 to 0·7 V for silicon, the current is limited only by the bulk resistance of the semiconductor material and the external circuit resistance.

3. *Reverse bias.* If the polarity of the supply is reversed (to give the reverse bias connexion of Fig. 20.12 (*b*)),the potential barrier at the junction is increased so that the majority carrier current falls, and eventually (above about 0·1 V reverse bias) becomes negligibly small. However, the minority carriers continue to be attracted strongly across the junction. The number of these minority carriers is limited by the thermal generation rate of electron-hole pairs on each side of the junction to the equilibrium values in the unbiased junction i.e. to I_{1h} for holes crossing from *n* to *p* and I_{2e} for electrons crossing from *p* to *n*. Hence the reverse bias current becomes constant at a value

$$I_0 = I_{1h} + I_{2e}$$

It has been seen that $I_{1h} \gg I_{2e}$ for a p^+–n junction so that the reverse bias current depends mainly on the flow of minority carriers from the lightly doped region.

The complete characteristic is shown in Fig. 20.12(*c*). The reverse-bias saturation current is typically of the order of a few microamperes for germanium diodes, and at least two orders of magnitude less for similar silicon diodes at room temperature, depending upon fabrication techniques.

Since the reverse bias current depends on the minority carrier concentration in the lightly doped region it increases rapidly with temperature, approximately doubling for every 10°C rise in temperature in germanium and 6°C rise in silicon. The equation for junction current can be shown to be

$$I = I_0(e^{eV/kT} - 1) \tag{20.13}$$

where *e* is the magnitude of the charge of an electron or a hole ($1·6 \times 10^{-19}$ C), *V* is the bias applied to the depletion layer, *k* is Boltzmann's constant ($1·38 \times 10^{-23}$ J/K), and *T* is the absolute temperature of the junction in kelvins.

Similar considerations apply to a junction in which the *n*-region is more heavily doped than the *p*-region (a p–n^- junction). In this case the forward current across the junction is mainly electrons from the *n*-region (majority carriers from the heavily doped region), while the reverse bias current is mainly electrons from the *p*-region (minority carriers from the lightly doped region).

20.11 The Junction Transistor

There are two main types of transistor—*junction* also known as *bipolar* because carriers of both polarities, electrons and holes, take part in the action, and *field-effect,* also known as *unipolar* because carriers of only one polarity are involved. Field-effect transistors are considered in Chapter 26.

The construction of a *p–n–p* germanium alloy junction transistor is shown in Fig. 20.13 together with its graphical symbol. A small indium pellet is fused into one side of a thin, lightly doped *n*-type wafer of germanium (called the *base*) to give a heavily doped *p*-region, called the *emitter.* A *p+–n* junction is thus formed. On the other side of the base wafer a slightly larger indium pellet is fused, to give a second *p–n* junction. This second *p*-region is called the *collector.*

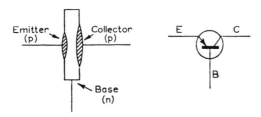

Fig. 20.13 ALLOY JUNCTION *p–n–p* TRANSISTOR

The *p–n–p* alloy junction germanium transistor was one of the earliest types to be produced. It is still widely used for low-frequency applications. Owing to difficulties with the materials used, *n–p–n* transistors (in which the base layer is of *p*-type material and the collector and emitter regions are of *n*-type) are not made by the alloy process; nor are satisfactory devices made of silicon manufactured in this way.

Since silicon has considerable advantages electrically compared to germanium when used to make transistors, considerable effort has been put into the technology of silicon device manufacture, despite the fact that silicon is a more difficult material to handle than germanium. It has been found that *p*- and *n*-regions can be introduced into semiconductor crystals by a diffusion process, and that both silicon and germanium devices can be made using this process. With diffusion methods it is relatively easy to obtain silicon *n–p–n* transistors with higher gain, lower leakage current and much better high-frequency response than is possible with germanium *p–n–p* alloy junction types. In addition it is possible to use diffusion in order to produce complete circuits on a single silicon crystal chip. Such integrated circuits are becoming widely used in present-day technology.

The construction of a planar *n–p–n* transistor is shown in Fig. 20.14(*a*), together with its graphical symbol. A lightly doped wafer of *n*-type silicon is used as the collector (the light doping enables an increased collector voltage to be used compared to the alloy junction type). A thin *p*-type layer is formed by exposing an etched area of the surface to hot gases containing acceptor impurities. The *p*-type impurity diffuses into the *n*-type wafer to form the base of the transistor. A small *n*-type layer is then diffused into a further etched area in the centre of the base layer to give the emitter; this

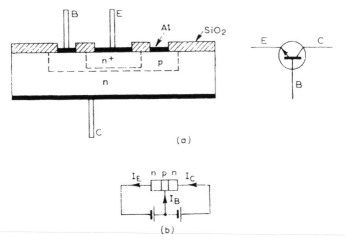

Fig. 20.14   *n–p–n* SILICON PLANAR TRANSISTOR

(*a*) Construction and graphical symbol
(*b*) Method of biasing

is a heavily doped region. Aluminium electrodes are evaporated on, and the terminations of the junctions are protected by a layer of silicon dioxide, which prevents the ingress of moisture and reduces surface leakage current.

Since the diffusion process can be accurately controlled (as distinct from the alloying process), very thin base widths (about 1μm) are possible, and this increases the gain and improves the high-frequency response of the device.

In operation the emitter–base junction is forward biased, and the collector–base junction is reverse biased as shown for an *n–p–n* transistor in Fig. 20.14(*b*). Owing to the forward bias on the emitter-base junction, a large current, I_E, flows. Since the emitter is heavily doped, I_E consists mainly of electrons injected into the base from the emitter. Transistor action depends on these injected majority carriers, and this is why it is essential to have a heavily doped emitter and a lightly doped base (the holes injected from the base into the emitter play no part in transistor action). A ratio of electron current to hole current of more than 100:1 can readily be achieved. The ratio of electron current to hole current is called the *emitter injection ratio*.

The injected electrons diffuse across the base towards the reverse-biased collector–base junction. As soon as they reach the collector-junction depletion layer the electrons are pulled into the collector (or collected) so that the collector current increases by almost the same amount as the injected electron current from the emitter. (Remember that the current across a reverse-biased junction is mainly formed by minority carriers from the lightly doped side, and electrons are the minority carriers in the very lightly doped p-type base). Since the base is very thin and the collector area large, almost all the injected electrons are collected. However, some recombination of injected electrons with the majority holes in the base does take place. To maintain charge neutrality a current of electrons must flow out of the base. This current, I_B, will normally be only a very small fraction of the collector current I_C. Obviously,

$$I_E = I_C + I_B \qquad (20.14)$$

Note that the transistor (unlike a valve which requires a heated filament in order to produce an electron stream) does not require a heate⁻ supply. This is one of its main advantages over valves.

The p–n–p transistor operates in a similar manner, but the injection is of holes at the emitter junction, and it is the hole current which is collected by the collector. The polarities for forward and reverse bias are reversed.

20.12 Static Characteristics—Common Base Connexion

The connexion of the n–p–n transistor illustrated in Fig. 20.15(a) in which the input is applied between emitter and base and the output is taken between collector and base is called *common base* (C.B.) connexion. The input characteristic relating emitter current and base–emitter voltage shown at (b) exhibits the curvature of a forward-biased junction.

The static collector, or output, characteristics are shown at (c). In the first quadrant they have been drawn as horizontal lines, but they should really have positive slopes of less than $1 \mu A/V$, and should therefore have counterparts in a family of current-transfer curves, each for a particular value of V_{CB}. However, corresponding to the horizontal lines is the single current-transfer characteristic at (d). When the extremely small slope of the output characteristics is taken into account, the equation for the collector bias current at a particular quiescent point is

$$I_C = I_{CBO} + \bar{\alpha}_B I_E \qquad (V_{CB} \text{ constant}) \qquad (20.15)$$

I_{CBO} is the collector leakage current, i.e. the current through the reverse-biased collector–base junction when $I_E = 0$, and $\bar{\alpha}_B$ is the mean static current amplification factor, or the inherent static

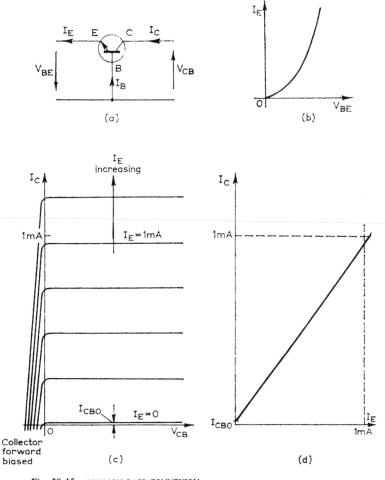

Fig. 20.15   COMMON BASE CONNEXION

(a) Circuit
(b) Input characteristic
(c) Output characteristics

forward current-transfer ratio, for the C.B. connexion. I_{CBO} is small, especially with silicon, and $\bar{\alpha}_B$ may lie between 0·95 and almost unity, so that I_C is very nearly equal to I_E. The ratio I_C/I_E is given the symbol $-h_{FB}$, which, with silicon, means much the same as $\bar{\alpha}_B$.

EXAMPLE 20.3 A silicon transistor has a collector leakage current, I_{CBO}, of 0·02μA at 300K. The leakage current doubles for every 6K rise in temperature. Calculate the base current at (a) 300K, (b) 330K when the emitter current is 1 mA, given that $\bar{\alpha}_B = 0·99$.

(a) At 300K, eqn. (20.15) gives

$$I_C = 0·00002 + 0·99 = 0·99\,\text{mA}$$

Hence, since $I_E = I_C + I_B$,

$$I_B = 1·0 - 0·99 = 0·01\,\text{mA} = \underline{10\mu\text{A}}\ \text{directed towards the base}$$

(b) At 330K, I_{CBO} will be 0·64. Hence

$$I_C = 0·00064 + 0·99 \approx 0·9906\,\text{mA} \qquad \text{so that} \qquad I_B = \underline{9·4\mu\text{A}}$$

If the temperature were to increase sufficiently the base current would reverse in direction.

At any given temperature, germanium transistors have a much larger leakage current than silicon transistors. Hence they are much more affected by temperature variations, despite the fact that the per-unit change in I_{CBO} with temperature is smaller in germanium than in silicon.

Note that when $V_{CB} = 0$ the collector still collects almost all the injected emitter current, since its barrier charge pulls the minority carriers across. This is particularly true of transistors with thin bases. It is only when the collector junction is forward biased that the collector current becomes zero, due to the forward collector-base current cancelling the reverse current. The larger the emitter current the greater must be the collector forward current and hence the greater must be the forward bias voltage, in order to give $I_C = 0$. This explains why the knee of the characteristics is usually in the forward-biased collector-voltage region.

20.13 Static Characteristics—Common Emitter Connexion

The common emitter (C.E.) connexion of a transistor is illustrated in Fig. 20.16(a). This corresponds to a normal common-cathode valve configuration. The input is applied between base and emitter, and the output is taken between collector and emitter. The static collector characteristics give the relation between collector current, I_C, and collector–emitter voltage, V_{CE}, for various values of base current, I_B, as shown at (b).

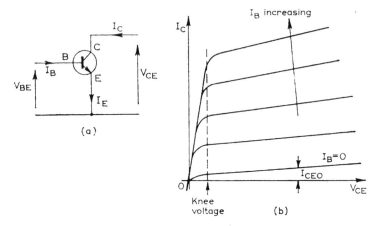

Fig. 20.16   COMMON EMITTER CONNEXION

(a) Circuit   (b) Output characteristics

The collector current, I_C, is obtained from eqns. (20.14) and (20.15):

$$I_E = I_C + I_B \quad \text{and} \quad I_C = I_{CBO} + \bar{\alpha}_B I_E$$

whence

$$I_C = I_{CBO} + \bar{\alpha}_B(I_C + I_B)$$

or

$$I_C = \frac{I_{CBO}}{1 - \bar{\alpha}_B} + \frac{\bar{\alpha}_B I_B}{1 - \bar{\alpha}_B} \tag{20.16}$$

This equation is of the same form as eqn. (20.15), so that, for the C.E. connexion we may write

$$I_C = I_{CEO} + \bar{\alpha}_E I_B \quad (V_{CE} \text{ constant}) \tag{20.17}$$

$I_{CEO} (= I_{CBO}/(1 - \bar{\alpha}_B))$ is the collector leakage current, i.e. the current through the collector–base junction when $I_B = 0$, and $\bar{\alpha}_E (= \bar{\alpha}_B/(1 - \bar{\alpha}_B))$ is the static current amplification factor, for the C.E. connexion.

Since $\bar{\alpha}_B$ is nearly unity $(1 - \bar{\alpha}_B)$ is small, so that I_{CEO} is very much larger than I_{CBO}. Thus any temperature effect which increases I_{CBO} will cause a much greater increase in I_{CEO}, so that if germanium transistors are used it is particularly important to stabilize against temperature changes. Again because $(1 - \bar{\alpha}_B)$ is small, $\bar{\alpha}_E$ can have values ranging from about 20 to 100 or more. The ratio I_C/I_B is given the symbol h_{FE}, which, with silicon, means much the same as $\bar{\alpha}_E$.

The C.E. static characteristics shown at (b) have a greater slope than those for the C.B. connexion, mainly because $\bar{\alpha}_B$ increases slightly with increased collector–base voltage and the small increase in I_{CBO} with collector voltage is magnified $1/(1 - \bar{\alpha}_B)$ times—indeed the slope in the C.E. connexion will be greater by about this factor than that in the C.B. connexion. The knee voltage of a C.E. characteristic occurs at a voltage which is much lower than that of a pentode and in small power transistors is less than a volt. The knee occurs at a positive value of collector–emitter voltage (for *n–p–n* transistors) since the voltage drop of the forward biased emitter junction must be added to V_{CB} in order to obtain the corresponding value of V_{CE}.

In the C.E. connexion the input current/voltage relation is a curve which resembles that relating emitter current to base–emitter voltage in the C.B. connexion. Eqn. (20.16) shows that the change in collector current is a linear function of base current, and it therefore follows that the change in collector current is *not* a linear function of base–emitter voltage. This is why transistors are normally operated with a current drive, as distinct from valves, which have a voltage drive. It should be noted that in fact $\bar{\alpha}_B$ and $\bar{\alpha}_E$ both vary slightly with collector current so that the linear relations expressed by eqns. (20.16) and (20.15) are themselves an approximation.

20.14 The C.E. Amplifier

In order to utilize the transistor as an amplifier in the C.E. connexion, the base–emitter junction is forward biased, and a load resistor is connected in the collector lead. The signal source is connected in series with the base connexion, and the resultant variation of base current causes a change in collector current. This in turn gives an alternating component of voltage across the load resistor. A simple circuit is shown in Fig. 20.17(*a*), using an *n–p–n* transistor.

A constant-current d.c. bias source is used to give a suitable quiescent collector current. When the constant-current a.c. source is connected in parallel with the bias source, the base current will vary about its quiescent value to give a corresponding amplified change in collector current.

The operation of the circuit may be described graphically by drawing a load line on the static collector characteristics in a manner similar to that for the valve amplifier. Thus the operating point of the transistor must satisfy two requirements, namely

1. The collector current must be related to the base current by the static characteristics.
2. The collector current must satisfy the equation

$$V_{CE} = V_o = V_{CC} - I_C R_L$$

This is the equation of the load line shown in Fig. 20.17(b), which runs from ($I_C = V_{CC}/R_L$, $V_{CE} = 0$) to ($I_C = 0$, $V_{CE} = V_{CC}$). Then if the bias is adjusted to give a base current of I_{BO}, the quiescent operating point will be Q. If now the signal causes the base current

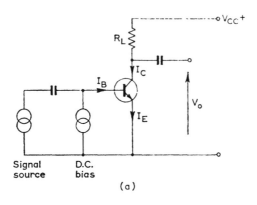

(a)

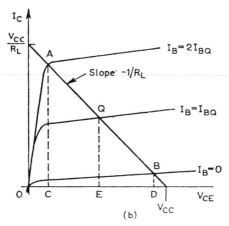

(b)

Fig. 20.17   SIMPLE C.E. AMPLIFIER

(a) Circuit
(b) Output characteristics with load line

to vary sinusoidally with a peak value of I_{BQ}, the operating point will move up and down the load line between A and B, giving an output peak-to-peak swing of CD volts. This simple circuit gives no compensation for temperature variations in I_{CEO}.

Note particularly that when I_B increases, I_C increases, so that the voltage drop across R_L increases and the output voltage (between collector and emitter) *decreases*, i.e. there is inherent phase reversal in the amplifier just as there is in the equivalent valve amplifier.

The steady bias current is chosen so that the quiescent operating point lies about the middle of the load line where the characteristics are most linear. Distortion of the signal is thus minimized.

20.15 Automatic Bias and Stability

The base-bias circuits employed in C.E. amplifiers should not only provide a suitable quiescent base current, but should also provide against changes in temperature, transistor characteristics, and supply voltage from affecting the quiescent operating point. It has been seen that, in an uncompensated transistor any increase in leakage current, ΔI_{CBO}, causes an increase in I_{CEO} of $\Delta I_{CBO}/(1 - \bar{\alpha}_B)$. In germanium transistors a rise or fall in temperature may change the quiescent collector current so much that the transistor is forced to operate over a non-linear range of its characteristics, and distortion will result. The *stability factor*, S, may be defined by the relation

$$S = \frac{\text{Change in } I_C}{\text{Given change in } I_{CBO}} = \frac{\delta I_C}{\delta I_{CBO}} \tag{20.18}$$

For the unstabilized circuit $S = 1/(1 - \bar{\alpha}_B) = \bar{\alpha}_E + 1$ (since $\bar{\alpha}_E = \bar{\alpha}_B/(1 - \bar{\alpha}_B)$).

Two other stability factors may also be defined:

$$M = \frac{\delta I_C}{\delta \bar{\alpha}_B} \tag{20.19}$$

which relates the change in I_C to changes in the C.B. current amplification factor $\bar{\alpha}_B$, and

$$N = \frac{\delta I_C}{\delta V_{CC}} \tag{20.20}$$

which relates the response of I_C to changes in the supply voltage, V_{CC}. In the ideal case S, M and N are all zero. Generally, if N turns out to be unacceptably high, the answer is to provide a supply that is better stabilized, so that $\delta V_{CC} \rightarrow 0$.

20.16 Base Resistor Bias

Perhaps the simplest way to achieve the required base-bias current is that shown in Fig. 20.18. A high-value resistor, R_1, is connected

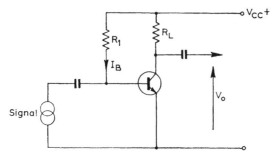

Fig. 20.18   SIMPLE BASE BIAS

from the supply direct to the base. Then if the base–emitter voltage drop is V_{BE}, it follows that the quiescent base current is given by

$$I_B = \frac{V_{CC} - V_{BE}}{R_1} \tag{20.21}$$

The quiescent collector current is

$$I_C = I_{CBO}/(1 - \bar{\alpha}_B) + \bar{\alpha}_B I_B/(1 - \bar{\alpha}_B) = (1 + \bar{\alpha}_E)I_{CBO} + \alpha_E I_B$$
$$= (1 + \bar{\alpha}_E)I_{CBO} + \bar{\alpha}_E(V_{CC} - V_{BE})/R_1 \tag{20.22}$$

so that the stability factor is

$$S = \frac{\delta I_{CQ}}{\delta I_{CBO}} = 1 + \bar{\alpha}_E \tag{20.23}$$

The circuit is not stabilized at all against changes in I_{CBO} or $\bar{\alpha}_E$ caused by changes in temperature or tolerances in transistor parameters, since S has the same value as in the uncompensated circuit.
Also from eqn. (20.22),

$$N = \frac{\delta I_C}{\delta V_{CC}} = \frac{\bar{\alpha}_E}{R_1} \tag{20.24}$$

In a typical case with $\bar{\alpha}_E = 100$ and $R_1 = 100\,\mathrm{k\Omega}$, S would be 101 and N would be $1\,\mathrm{mA/V}$.

20.17   Collector–Base Resistance Bias

A measure of stability can be achieved by connecting the base-bias resistor (R_1 in Fig. 20.18) to the collector instead of direct to the supply rail. This is shown in Fig. 20.19(*a*).

Stabilization against changes in I_{CBO} is achieved since the quiescent base current now depends on the collector potential. If I_{CBO} increases, I_C will also increase; the collector potential will therefore decrease and the base current will decrease, tending to reduce the collector current to its original value.

The emitter resistor R_3 shown at (*b*) further increases stability, since any increase in quiescent collector current, I_C, will now reduce the base–emitter voltage and again cause a reduction in I_B counter-

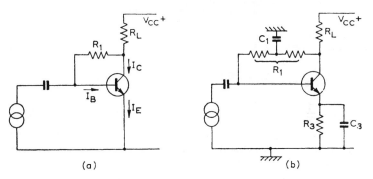

(a)   (b)

Fig. 20.19   COLLECTOR–BASE BIAS

acting the increase in I_C. To prevent this effect from also reducing signal-current changes, R_3 is short-circuited for alternating currents by the capacitor C_3. As far as d.c. conditions are concerned, inserting R_3 is equivalent to increasing the collector load resistance, R_L.

The stability factor, S, is obtained from the following relations.

$$I_E = I_C + I_B \quad \text{(hence the current in } R_L \text{ is } I_E)$$
$$I_C = (1 + \bar{\alpha}_E)I_{CBO} + \bar{\alpha}_E I_B$$

and

$$V_{CC} = I_E R_L + I_B R_1 + V_{BE} + I_E R_3$$

Hence, after some manipulation,

$$I_C = \frac{R_1 + R_L + R_3}{\dfrac{R_1}{(1 + \bar{\alpha}_E)} + R_L + R_3} I_{CBO} + \frac{\bar{\alpha}_E(V_{CC} - V_{BE})}{R_1 + (R_L + R_3)(1 + \bar{\alpha}_E)}$$

so that

$$S = \frac{\delta I_C}{\delta I_{CBO}} = \frac{R_1 + R_L + R_3}{R_1/(1 + \bar{\alpha}_E) + R_L + R_3} \tag{20.25}$$

Since I_B must always be positive in this circuit, the minimum collector current is equal to I_{CEO}, and this occurs when $I_B = 0$.

Note that there will be signal feedback through R_1 which will reduce the overall gain unless R_1 is divided into two sections, the junction being connected through a capacitor to earth as shown at (*b*).

It follows from the equation for I_C that

$$N = \frac{\delta I_C}{\delta V_{CC}} = \frac{\bar{\alpha}_E}{R_1 + (R_L + R_3)(1 + \bar{\alpha}_E)} \qquad (20.26)$$

This factor in turn is obviously very dependent on any changes in $\bar{\alpha}_E$ (and hence in $\bar{\alpha}_B$).

20.18 Base Voltage-divider Bias

The circuit shown in Fig. 20.20 is one of the most commonly used in C.E. amplifiers for base current bias and operating point stabilization. Essentially the resistors R_1 and R_2 fix the base potential. The capacitor C_3 short-circuits the emitter for a.c. signals to prevent reduction of signal gain.

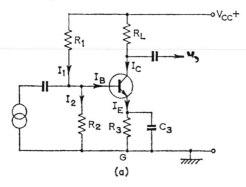

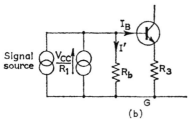

Fig. 20.20   VOLTAGE-DIVIDER BASE BIAS

(*a*) Complete circuit   (*b*) A.C. circuit

The stabilizing action is achieved because the base potential is kept nearly constant, so that any increase in I_{CBO} (and hence increase in I_C and I_E) causes the emitter potential to rise, and hence the base-emitter voltage to fall. The base current is therefore reduced and the collector current falls towards its original value.

Expressions for the stability factors can be obtained by using Norton's theorem to replace the voltage divider R_1, R_2 by the constant-current source V_{CC}/R_1 in parallel with the resistance $R_b = R_1R_2/(R_1 + R_2)$ as shown in Fig. 20.20(b). Then, for the d.c. conditions (zero signal),

$$I_C = (1 + \bar{\alpha}_E)I_{CBO} + \bar{\alpha}_E I_B \tag{20.27}$$

$$V_{BG} = V_{BE} + (I_B + I_C)R_3 = I'R_b \tag{20.28}$$

and

$$I' + I_B = \frac{V_{CC}}{R_1} = \frac{V_{BE}}{R_b} + (I_B + I_C)\frac{R_3}{R_b} + I_B$$

or

$$I_B = \frac{\dfrac{V_{CC}}{R_1} - \dfrac{V_{BE}}{R_b} - I_C\dfrac{R_3}{R_b}}{\dfrac{R_3}{R_b} + 1} \tag{20.29}$$

Substituting in eqn. (20.27) from eqn. (20.29),

$$I_C = (1 + \bar{\alpha}_E)I_{CBO} + \frac{\bar{\alpha}_E R_b}{R_3 + R_b}\left(\frac{V_{CC}}{R_1} - \frac{V_{BE}}{R_b} - I_C\frac{R_3}{R_b}\right)$$

so that

$$I_C = \frac{(1 + \bar{\alpha}_E)I_{CBO}}{1 + \dfrac{\bar{\alpha}_E R_3}{R_3 + R_b}} + \frac{\bar{\alpha}_E R_b}{(R_3 + R_b)\left(1 + \dfrac{\bar{\alpha}_E R_3}{R_3 + R_b}\right)}\left(\frac{V_{CC}}{R_1} - \frac{V_{BE}}{R_b}\right)$$

It follows that

$$S = \frac{\delta I_C}{\delta I_{CBO}} = \frac{1 + \bar{\alpha}_E}{1 + \dfrac{\bar{\alpha}_E R_3}{R_3 + R_b}}$$

$$= \frac{R_3 + R_b}{\dfrac{R_b}{(1 + \bar{\alpha}_E)} + R_3} = \frac{R_3 + R_b}{R_b(1 - \bar{\alpha}_B) + R_3} \tag{20.30}$$

Also

$$N = \frac{\delta I_C}{\delta V_{CC}} = \frac{R_b}{R_1} \frac{\bar{\alpha}_E}{R_b + R_3(1 + \bar{\alpha}_E)} \qquad (20.31)$$

With this circuit the above equations indicate that it is possible to bias so that the quiescent collector current $I_C < I_{CEO}$, the base current then being in the reverse direction. Indeed if $R_b \rightarrow 0$ and $R_3 \rightarrow \infty$ the minimum collector current will tend to I_{CBO}. The quiescent base current I_B will then be negative. Note that the improvement in stability factor S compared with the unstabilized circuit is given by the factor $1/(1 + \bar{\alpha}_E R_3/(R_3 + R_b))$.

From eqn. (20.26) it will be seen that the stabilization is best when $R_b \ll R_3$. In practical circuits this cannot be achieved since if R_b is small a large part of the signal input current will be bypassed by it (so that current gain is lost), while if R_3 is large, a large value of supply voltage is required.

The stability factor S can be used in designing circuits for germanium transistors. Thus if the largest permissible change in I_C is known, and the greatest variation in I_{CBO} due to expected temperature changes is assumed, then S and hence the ratio R_3/R_b is determined.

EXAMPLE 20·4 For the circuit of Fig. 20.20 (a) $R_1 = 33$ kΩ, $R_2 = 5$ kΩ, $R_3 = 2\cdot2$ kΩ and $\bar{\alpha}_B = 0\cdot99$. Determine the stability factors S and N.

$$R_b = \frac{R_1 R_2}{R_1 + R_2} = 4\cdot5 \text{ k}\Omega$$

Hence from eqn. (20.30),

$$S = \frac{\delta I_C}{\delta I_{CBO}} = \frac{(4\cdot5 + 2\cdot2)10^3}{(4\cdot5 \times 0\cdot01 + 2\cdot2)10^3} = \underline{3}$$

From eqn. (20.31),

$$N = \frac{\delta I_C}{\delta V_{CC}} = \frac{4\cdot5 \times 10^3}{33 \times 10^3} \times \frac{99}{(4\cdot5 + 2\cdot2 \times 100)10^3}$$
$$= 6 \times 10^{-5} \text{ A/V} = \underline{60 \ \mu\text{A/V}}$$

Although the foregoing analysis applies equally to germanium and silicon transistors, the value of I_{CBO} in silicon is so small that stabilization against any change of temperature on this account is virtually unnecessary unless the circuit is operating at unusually low values of collector current or at very high temperatures.

Another factor which affects the quiescent operating point in transistor circuits is the variation in $\bar{\alpha}_E$ from one transistor to another. This spread in values is generally greater than the production spread of valve parameters. It could give rise to large variations in the gain of transistor amplifiers.

Moreover, $\bar{\alpha}_E$, like I_{CBO}, has a positive temperature coefficient, and V_{BE} has a negative temperature coefficient. Generally, however, if S is small, the circuit will also be stabilized against changes in $\bar{\alpha}_E$ and V_{BE}.

Stabilization circuits are applications of the negative feedback principles discussed in Chapter 22.

20.19 Reverse Breakdown—the Zener Diode

The forward and reverse bias characteristics of a *p–n* junction were described in Section 20.10, where it was seen that when reverse bias is applied the current is almost constant at the very small reverse-bias

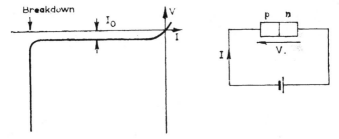

Fig. 20.21 REVERSE CHARACTERISTIC OF A *p–n* JUNCTION

saturation value. If the reverse bias voltage is increased, a point will be reached at which a large reverse current flows, and *breakdown* of the junction is said to have occurred.

The voltage across the junction then remains almost constant as shown in Fig. 20.21. Breakdown is due to two causes: *Zener effect* and *avalanche effect*, both of which arise in the depletion layer.

Avalanche Effect. The reverse bias voltage creates an electric field across the depletion layer. Carriers in the depletion layer are accelerated by the field, and if the electric stress is large enough, they may gain sufficient kinetic energy to produce electron-hole pairs by collision with some of the atoms in the crystal lattice. The electrons released by this ionization may then themselves be accelerated sufficiently to cause further electron-hole generating collisions, so that the process becomes cumulative. This is the avalanche effect. The holes produced by collision are less mobile than the electrons and are unlikely to add to the cumulative process. Avalanche effect gives rise to a very large increase in current, with little change in the reverse bias voltage. Hot spots may form at any imperfection at the junction, causing destructive breakdown.

Zener Effect. This effect occurs when the electric field in the depletion layer is large enough to produce electron-hole pairs by pulling electrons forcibly away from their parent atoms, i.e. by breaking the covalent bonds. This high field effect causes an increase in free charge carriers and hence a large increase in current for small increases in voltage, as in the avalanche effect.

Both effects depend on a high electric field, i.e. a large voltage gradient across the depletion layer. This in turn depends on (*a*) the applied voltage, and (*b*) the width of the depletion layer. Since the depletion layer is wider in a lightly doped material it follows that the *breakdown voltage* depends on the degree of doping of the lightly doped side of the junction, being higher for a more lightly doped material.

In *Zener diodes* the junction is specially made to be as uniform and free from imperfections as possible, so that breakdown will occur uniformly and will not be destructive. The voltage at which breakdown occurs can be chosen by a suitable choice of doping, and this gives a wide range of diodes suitable as voltage stabilizers. With a high level of doping the depletion layer will be narrow so that a small voltage will establish a high enough field in the depletion layer to cause breakdown. In a lightly doped junction, the depletion layer will be wide and the applied voltage will require to be large in order to give the electric field necessary for breakdown.

For breakdown voltages below about 6 V the main effect is Zener breakdown. Above this voltage the main effect is avalanche breakdown. The effects may be distinguished since the temperature coefficient of avalanche breakdown is positive because a higher temperature reduces the mean free path of the electrons and so requires a higher voltage for breakdown; whereas the temperature coefficient of Zener breakdown is negative, because increased thermal vibrations help the electric field to break the covalent bonds in the semiconductor. Diodes which break down at about 6 V show almost zero temperature coefficient—both avalanche and Zener effects are present, and their temperature coefficients almost cancel, being slightly negative for low currents and increasing through zero to become slightly positive at high currents due to the increased avalanche effect. Another distinguishing feature is that, since avalanching is a multiplicative process, the onset of breakdown is much more sudden than in the case of Zener breakdown.

20.20. Depletion Layer Capacitance

In Section 20.10 it was seen that when a *p–n* junction is formed in a semiconductor, holes from *p* diffuse across to *n* while electrons from *n* diffuse across to *p*. These diffusing minority carriers combine with the majority carriers on the other side of the junction to set up a bound charge distribution. The result of this recombination is that a layer exists on both sides of the junction which is depleted of mobile carriers. On the *p*-side of the metallurgical junction this *depletion layer* will have a net negative charge. On the *n*-side it will have an equal positive charge. Hence opposite charges exist across the junction separated by a depletion layer with few free carriers. This constitutes a capacitance called the *depletion layer capacitance*.

The extent of the junction depletion layers is dependent on the doping levels on either side. Thus for a *p⁺n* junction holes must diffuse further into the lightly doped *n*-type region before they

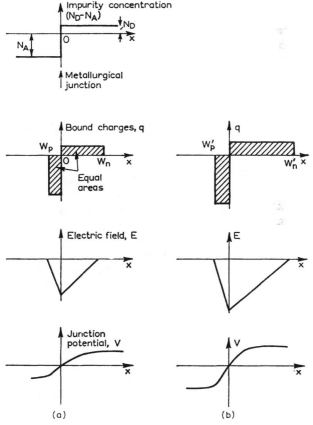

Fig. 20.22 JUNCTION CONDITIONS AT A *p–n* BOUNDARY
(a) Zero external bias.   (b) Applied reverse bias

recombine to form bound charges than do the electrons that diffuse across to the p^+-side. In other words, the depletion layer is wider on the more lightly doped side of a p-n junction. Indeed if the difference in doping levels is large enough it can usually be assumed that the depletion layer exists entirely on the lightly doped side of the junction.

Idealized junction relations are shown in Fig. 20.22(a) for zero bias conditions. The top diagram shows the impurity doping density at an abrupt junction, (N_D donors per cubic metre in n, and N_A acceptors per cubic metre in p). For a p^+n junction $N_A \gg N_D$. The next diagram shows the bound charge density. The relative

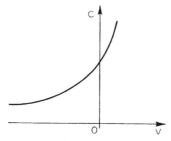

Fig. 20.33   VARIATION OF DEPLETION-LAYER CAPACITANCE WITH VOLTAGE

lengths of the depletion layers, W_p and W_n, may be estimated from the fact that the bound charges must be equal, i.e. $W_p N_A = W_n N_D$. and the shaded areas are thus equal. The charge distribution sets up an electric field across the junction, as shown in the third diagram. The electric field E, has a maximum value at the metallurgical boundary (it can be shown that E is proportional to $\int$ (bound charge density) dx). The junction potential, V, is found (since $V = -\int E \, dx$) by taking the negative of the integral of the curve of E, as shown in the fourth diagram. The total depletion layer width is $W_p + W_n$, which in a heavily doped p^+n junction becomes W_n. If a reverse bias is applied externally the corresponding conditions are shown at (b). The top diagram here shows the bound charge distribution. The reverse bias increases the junction potential, and this in turn increases the electric field and also the total bound charge. Since the height of the bound charge distribution is dependent on the impurity concentration, this will be constant. Hence the width of the depletion layer will increase and the depletion layer capacitance will decrease (the usual capacitor effect—if the dielectric is wider the capacitance is less).

It may be shown that the depletion layer width varies with the square root of the reverse bias voltage, and that the associated capacitance varies inversely as $\sqrt{V}$. Typical curves of capacitance variation are shown in Fig 20.23. This indicates the possibility of a voltage-sensitive capacitive element which finds applications in frequency-modulation circuits and parametric amplifiers.

PROBLEMS

20.1 A region of the anode characteristics of the Mullard EC52 triode is given in the following table. Estimate the values of r_a, g_m and μ for the given region.

$$V_{GK} = -2\text{V} \begin{cases} I_A \text{ (mA)} & 5 & 10 & 15 \\ V_A \text{ (V)} & 150 & 205 & 245 \end{cases}$$

$$V_{GK} = -3\text{V} \begin{cases} I_A \text{ (mA)} & 4 & 10 & 15 \\ V_A \text{ (V)} & 200 & 260 & 305 \end{cases}$$

$$V_{GK} = -4\text{V} \begin{cases} I_A \text{ (mA)} & 4 & 9 & 14 \\ V_A \text{ (V)} & 260 & 310 & 360 \end{cases}$$

Ans. 9,500Ω, 6·0mA/V, 57·5.

20.2 One stage of a low-frequency amplifier employs a tetrode having static characteristics as given in the accompanying table. The anode load of the valve is a resistance of 17 kΩ. Determine the maximum amplification of the stage.

V_A volts	50	100	150	200
V_{GK} volts	I_A milliamperes			
−2·25	11·6	12·4	13·0	
−4·5		7·0	7·5	7·75
−6·75		2·9	3·2	3·4

Ans. 33·5.

20.3 The accompanying table gives anode-current/anode-voltage characteristics for a beam tetrode for two different values of screen voltage. Determine:

    (a) the constants of the equivalent generator for the valve operated normally as a tetrode, and calculate the voltage amplification obtained with a 5,000Ω pure-resistance anode load; and

    (b) the corresponding data for the valve operated as a triode with anode and screen directly connected. The current drawn by the screen under tetrode operation may be neglected.

Assume the static operating point to be at $V_A = 200\,\text{V}$ and $V_{GK} = -5\,\text{V}$.

Screen Voltage 250 V

$V_{GK} = 0\,\text{V}$		$V_{GK} = -5\,\text{V}$		$V_{GK} = -10\,\text{V}$	
V_A	I_A	V_A	I_A	V_A	I_A
100	104 mA	100	76 mA	100	55 mA
200	110	200	82	200	59
300	112	300	85	300	61

Screen Voltage 300 V

$V_{GK} = 0\,\text{V}$		$V_{GK} = -5\,\text{V}$		$V_{GK} = -10\,\text{V}$	
V_A	I_A	V_A	I_A	V_A	I_A
100	125 mA	100	100 mA	100	76 mA
200	135	200	108	200	84
300	141	300	114	300	88

(*L.U. part question*)

Ans. (*a*) 5·1 mA/V, 33 kΩ, 22; (*b*) 5·1 mA/V, 1,600 Ω, 6·2.

20.4 Explain the action of the three-electrode valve when used as an amplifier. In a particular case, with a load resistance of 8,000 Ω the voltage amplification was 5·5 and with 12,000 Ω it was 6·5. What amplification at a frequency of 800 Hz would be expected, using a choke coil of 10 H inductance?   (*L.U.*)
Ans. 10.

20.5 A single-stage triode voltage amplifier has a gain of 20. The gain is reduced to 15 by halving the load resistance. Find the amplification factor of the triode. If the operating voltages are 300 V and $-3\,\text{V}$, calculate the change in static current caused by doubling the bias assuming that the 3/2 power law is adhered to.   (*H.N.C.*)
Ans. 30; 56·7 per cent reduction.

20.6 The static collector characteristics of an *n–p–n* transistor are linear over the range indicated.

$$I_B = 0 \quad \begin{cases} V_{CE} \text{ (volts)} & 1 & 10 \\ I_C \text{ (mA)} & 0{\cdot}04 & 0{\cdot}08 \end{cases}$$

$$I_B = 20\,\mu\text{A} \quad \begin{cases} V_{CE} \text{ (volts)} & 1 & 10 \\ I_C \text{ (mA)} & 0{\cdot}93 & 0{\cdot}98 \end{cases}$$

$$I_B = 40\,\mu\text{A} \quad \begin{cases} V_{CE} \text{ (volts)} & 1 & 10 \\ I_C \text{ (mA)} & 1{\cdot}9 & 1{\cdot}96 \end{cases}$$

The transistor is used as a common-emitter amplifier from a 10 V d.c. supply. The load in the collector circuit is 5 kΩ, and the quiescent base current is 20 μA. Determine the current gain (ratio of change in I_C to change in I_B) when the signal gives a base current variation of ±20 μA. What are the maximum and minimum values of collector voltage?
Ans. 46; 9·6 V; 0·5 V.

20.7 In the CE amplifier circuit shown in Fig. 20.20(*a*) a *p–n–p* germanium transistor is used, the supply voltage is $-6\,\text{V}$, and $R_1 = 25\,\text{k}\Omega$, $R_2 = 5\,\text{k}\Omega$, $R_3 = 0{\cdot}8\,\text{k}\Omega$, $R_L = 2{\cdot}5\,\text{k}\Omega$, and $I_{CBO} = 1\,\mu\text{A}$. Determine the quiescent collector–earth voltage if $\bar{\alpha}_B = 0{\cdot}98$ and the base–emitter voltage is 0·3 V. What is the value of the stability factor S?
Ans. $-3{\cdot}75\,\text{V}$; 5·7.

20.8 A silicon *n–p–n* transistor is used in the circuit of Fig. 20.19.
If $\bar{\alpha}_B = 0.99$, $I_{CBO} = 0.1\,\mu A$, $R_L = 3\,k\Omega$, $R_1 = 330\,k\Omega$ and $R_3 = 0.5\,k\Omega$, determine the quiescent collector voltage if the supply voltage is 10 V and $V_{BE} = 0.7\,V$. Evaluate the stability factor S.

Ans. 6 V; 49.

20.9 In the circuit of Fig. 20.24 show that

$$\frac{\delta I_C}{I_{CBO}} = \frac{R_1 + R_2}{R_1(1 - \bar{\alpha}_B) + R_2} \quad \text{and} \quad \frac{\delta I_C}{\delta V_{CC}} = \frac{\bar{\alpha}_E}{R_1 + (1 + \bar{\alpha}_E)R_2}$$

Neglect the base–emitter voltage drop. Would you expect this circuit to be more stable than that of Fig. 20.18? Explain your answer.

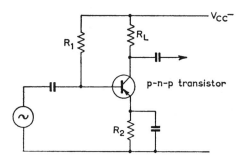

Fig. 20.24

20.10 An *n–p–n* silicon transistor at room temperature has its emitter disconnected. A voltage of 5 V is applied between collector and base, with the collector positive. A current of 0.2 A flows. When the base is disconnected and the same voltage is applied between collector and emitter the current is found to be 20 μA. Explain this effect and calculate the d.c. short-circuit current gain of the transistor, and the base and emitter currents when the collector current is 1 mA.

Ans. 0.99; 10μA; 1,010 μA.

20.11 In the transistor of Problem 20.10 the leakage current may be assumed to double for every 8°C rise in temperature above ambient. Determine the base and emitter currents when the collector current is 1 mA at a temperature 40°C above ambient. Assume that the d.c. short-circuit current gain remains constant.

Ans. 3.6 μA; 1,004 μA.

20.12 Find the values of R_1 and R_L for the simple base-resistor bias circuit of Fig. 20.18 assuming that a *p–n–p* germanium transistor is used and that $V_{CC} = -15$ V, $V_{BE} = -0.3$ V, $I_{CBO} = 5\,\mu A$ and $\bar{\alpha}_E = 50$. The quiescent values of I_C and V_{CE} are to be 2.5 mA and -5 V respectively. Determine the values of the stability factors S and N.

Ans. 330 kΩ; 4kΩ; 51; 0.15 mA/V.

20.13 In the circuit of Fig. 20.18 an additional resistor R_E is connected between the emitter and earth. Determine the values of R_1, R_E and R_L assuming that the transistor used is the same as in Problem 20.12 and that the quiescent values of I_C, V_{CE} and V_E (the voltage across R_E) are 2·5 mA, −3 V and −3 V respectively. Also find the values of the stability factors S and N.

Ans. 260 kΩ; 1·2 kΩ; 3·6 kΩ; 41·5; 0·155 mA/V.

20.14 A silicon *n–p–n* transistor is used with the collector–base resistance bias circuit of Fig. 20.19. Design the circuit (i.e. find suitable values of R_1, R_L and R_3) for the following conditions: $V_{CC} = 20$ V; $\bar{\alpha}_E = 100$; $I_{CBO} = 0·2$ μA; $V_{BE} = 0·7$ V; the quiescent values of I_C, V_{CE} and the voltage across R_3 are 1·5 mA, 8 V and 2 V respectively. Find the values of the stability factors S and N.

Ans. 480 kΩ; 6·66 kΩ; 1·33 kΩ; 38; 78 μA/V.

Chapter 21

SMALL-SIGNAL AMPLIFIERS

Small-signal amplifiers are assumed to operate over the linear range of the active device (valve or transistor) which they employ. They may be either voltage or current amplifiers, and will normally feed a further amplifier stage. The d.c. bias circuits are designed so that the active device operates over the linear parts of its characteristics. The quiescent voltages are found by considering the maximum peak-to-peak currents or voltages required for the signal (the quiescent current may have to be appreciably larger than the peak current swing to ensure linear operation). Under linear operating conditions the a.c. operation of the amplifier may be deduced by representing the active device by an equivalent circuit consisting of linear generators and circuit elements. In this chapter bipolar-transistor small-signal circuits will be considered. Field-effect transistor amplifiers will be dealt with in Chapter 26.

21.1 Transistor Equivalent Circuits

Unlike valves, which are voltage-controlled devices with almost infinite input impedance (the grid normally takes no current), bipolar junction transistors are generally current-controlled devices. The input impedance is not normally very high, and must be taken into account when circuit calculations are made.

Several transistor equivalent circuits can be drawn, and manufacturers quote many different parameters. Only two of these will be considered, namely (*a*) the equivalent-T circuit, and (*b*) the *h*-parameter equivalent circuit. It is assumed that the quiescent point has been chosen in the middle of the linear portion of the transistor characteristics, and that suitable d.c. bias circuits have been selected.

In a.c. (signal) equivalent circuits only those external components that are effective under a.c. conditions are shown. Thus the d.c. power supply is simply a short-circuit to alternating currents. Similarly, at mid-frequencies, decoupling and coupling capacitors will act as short-circuits. The effect of d.c. bias circuits on the signal is usually small so that they too can often be neglected in a.c. equivalent circuits.

EQUIVALENT-T CIRCUIT

The graphical symbol for a *p–n–p* transistor is shown in Fig. 21.1, together with the equivalent-T circuit, which applies to either type of transistor. The parameters of this circuit can be related to the

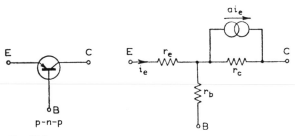

Fig. 21.1   TRANSISTOR EQUIVALENT-T CIRCUIT

physical operation of the transistor. Thus the resistance r_e represents the relatively low resistance of the forward-biased emitter–base junction, r_b represents approximately the resistance of the base layer (which will be appreciable since the base is very lightly doped, thin, and fed from its edge), and r_c represents the very high resistance of the reverse-biased base–collector junction. Physically, the emitter current is injected into the base and is transported across the base-collector junction by diffusion or drift, so that a proportion, α, actually crosses to the collector. This is represented by the constant-current generator, ai_e, across r_c, where $a \approx \alpha$ except at high frequencies.

The small-signal operation of a transistor amplifier stage may be determined by replacing the transistor by its equivalent circuit and solving the resulting circuit by conventional means. It will be noted that in common-base connexion, the resistance r_b is common to both input and output circuit loops. This means that the output load will affect input conditions—i.e. there is internal feedback in the transistor—and this causes considerable complications.

In practice the equivalent-T parameters are now seldom quoted on manufacturers' data sheets since it is rather difficult to measure the parameters r_e, r_b and r_c and the approximation made lead to inaccuracies if the common-base T-circuit is converted for common-emitter connexion. Typical values at a quiescent collector current of 1mA for a small germanium audio-frequency transistor are $r_e = 20\Omega$, $r_b = 700\Omega$, $r_c = 1\text{M}\Omega$ and $a = 0.98$; and for a small silicon transistor, $r_e = 10\Omega$, $r_b = 1,000\Omega$, $r_e = 0.5\text{M}\Omega$ and $a = 0.99$.

THE h-PARAMETER EQUIVALENT CIRCUIT

In the small-signal range the transistor can be regarded as an active two-port "black box", i.e. it has two input terminals and two output terminals (Fig. 21.2(a)). There are four external variables, the

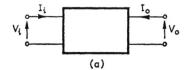

(a)

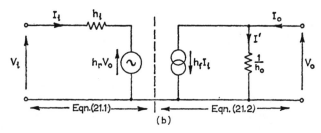

(b)

*Fig. 21.2 h-*PARAMETER EQUIVALENT CIRCUIT

input voltage and current (V_i and I_i) and the output voltage and current (V_o and I_o). Since operation takes place over the linear range of transistor characteristics, V_i, I_i, V_o and I_o can be related to each other by constant parameters. Thus V_i and I_o can be related to I_i and V_o by four constants, called the *h-parameters*, giving the following complexor equations:

$$V_i = h_i I_i + h_r V_o \qquad (21.1)$$

$$I_o = h_f I_i + h_o V_o \qquad (21.2)$$

The directions of the currents and the polarities of the voltage are conventionally chosen as shown in Fig. 21.2(*a*). There are several alternative ways of relating the input and output voltages and currents but these will not be discussed.

Eqns. (21.1) and (21.2) give rise to the equivalent circuit of Fig. 21.2(*b*). Thus the input voltage to the black box is made up of the voltage drop $h_i I_i$ across a resistor and the e.m.f. $h_r V_o$ of a constant-voltage generator (the left-hand half of Fig. 21.2(*b*)). It follows that h_i has the dimensions of a resistance and that h_r is a numerical constant. Also I_o is made up of the current $h_f I_i$ in a constant-current source and the current $V_o h_o$ through an admittance, h_o, which is connected across the output terminals (the right-hand half of Fig. 21.2(*b*)). The parameter h_f is a numerical constant.

The subscripts used for the *h*-parameters indicate their function in the equivalent circuit—h_i is the *input impedance* with short-circuited output (i.e. when $V_o = 0$), h_r is the *reverse voltage transfer constant*, h_o is the *output admittance* when the input is open-circuited (i.e. when $I_i = 0$), and h_f is the *forward short-circuit current-transfer constant* (i.e. $h_f = I_o/I_i$ when $V_o = 0$).

Since the transistor is a three-terminal device, it is possible to choose any one of its terminals as the one which is common to input and output circuits. This gives rise to the three practical connexions, common base, common emitter and common collector. The *h*-parameter equivalent circuit has the same form for each connexion, but the values of the parameters are different for all three. It is usual to indicate the connexion to which the parameters refer by a second subscript, *b* for common base, *e* for common emitter and *c* for common collector. Thus h_{fe} is the forward short-circuit current gain in the common-emitter connexion. Typical values of the small-signal *h*-parameters are indicated in Table 21.1, for a quiescent collector current of 1 mA.

The small-signal *h*-parameters can be evaluated by simple circuit tests. Suppose the output terminals are short-circuited ($V_o = 0$) and a voltage V_i is applied to the input. Then if a small change δV_i in V_i causes a small change δI_i in the input current I_i,

$$h_i = \left. \frac{\delta V_i}{\delta I_i} \right|_{V_o = 0} \tag{21.3}$$

If the corresponding change in the output current is δI_o, then

$$h_f = \left. \frac{\delta I_o}{\delta I_i} \right|_{V_o = 0} \tag{21.4}$$

h_f may be negative or positive, according to the connexion used (see Table 21.1).

TABLE 21.1

Parameter	OC71 Ge $p–n–p$	BCY30 Si $n–p–n$	2N930 Si $n–p–n$
h_{ib} (Ω)	35	30	20
h_{rb}	7×10^{-4}	$1{\cdot}5 \times 10^{-4}$	6×10^{-4}
h_{fb}	$-0{\cdot}98$	$-0{\cdot}98$	$-0{\cdot}995$
h_{ob} (μS)	$1{\cdot}0$	$0{\cdot}5$	$0{\cdot}5$
h_{ie} (kΩ)	$1{\cdot}5$	$1{\cdot}4$	$4{\cdot}0$
h_{re}	8×10^{-4}	6×10^{-4}	4×10^{-4}
h_{fe}	49	49	199
h_{oe} (μS)	50	25	100
h_{ic} (kΩ)	$1{\cdot}5$	$1{\cdot}4$	$4{\cdot}0$
h_{rc}	1	1	1
h_{fc}	-50	-50	-200
h_{oc} (μS)	50	25	100

In order to determine h_r and h_o the input is open-circuited ($I_t = 0$) and a voltage V_o is applied at the output terminals, giving rise to a current I_o and a voltage across the input terminals of V_t. For a small change δV_o in V_o, let the corresponding changes in I_o and V_t be δI_o and δV_t. Then

$$h_r = \left.\frac{\delta V_t}{\delta V_o}\right|_{I_t=0} \tag{21.5}$$

and

$$h_o = \left.\frac{\delta V_o}{\delta I_o}\right|_{I_t=0} \tag{21.6}$$

21.2 Common-emitter Amplifier

In the common-emitter connexion, the input to the transistor is between base and emitter and the output is taken between collector and emitter. The graphical symbols for a $p–n–p$ and an $n–p–n$ transistor, and the common-emitter h-parameter equivalent circuit, are shown in Figs. 21.3(*a*) and (*b*). It is convenient in the analysis to assume that all signal currents are directed towards the transistor as shown, whether the transistor is $n–p–n$ or $p–n–p$. Thus the following equation is always true

$$I_e + I_b + I_c = 0 \tag{21.7}$$

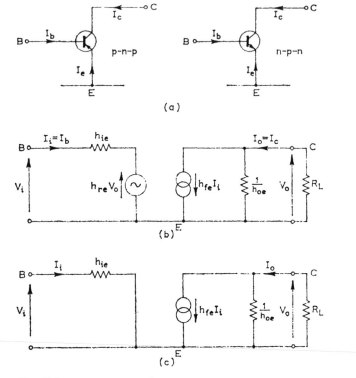

Fig. 21.3   COMMON-EMITTER h-PARAMETER EQUIVALENT CIRCUIT

Comparing Figs. 21.3(a) and (b) it will be seen that

$$I_b = I_i \qquad I_c = I_o \qquad V_{be} = V_i \qquad V_{ce} = V_o$$

Under normal operating conditions, and considering alternating currents and voltages only,* the value of h_{re} is small, so that there is small error in neglecting the generator $h_{re}V_o$ in the input circuit. This assumption considerably simplifies the equivalent circuit calculations.

From the simplified small-signal equivalent circuit of Fig. 21.3(c), it follows that

$$V_i = h_{ie}I_i \tag{21.8}$$

and

$$I_o = h_{fe}I_i + h_{oe}V_o \tag{21.9}$$

* Subscripts in small letters are used with the symbols for signal quantities, and subscripts in capital letters, with the symbols for quiescent d.c. quantities. Thus I_e = emitter signal current and I_E = emitter bias current.

Hence

$$\text{Input resistance, } R_i = \frac{V_i}{I_i} = h_{ie} \qquad (21.10)$$

and

$$\text{Output resistance, } R_o = \frac{1}{h_{oe}} \qquad (21.11)$$

The output impedance of an amplifier is the internal impedance as seen by any load placed upon it and may be defined as the ratio of output voltage to output current when an a.c. source is applied to the output terminals and the input is represented by the internal impedance of any input source.

$$\text{Current gain, } A_i = \frac{V_o/R_L}{I_i} = \frac{-I_o}{I_i}$$

From eqn. (21.9),

$$I_o = h_{fe}I_i + h_{oe}V_o$$
$$= h_{fe}I_i - h_{oe}R_LI_o \quad (\text{since } V_o = -I_oR_L)$$

so that

$$I_o = \frac{h_{fe}I_i}{1 + h_{oe}R_L}$$

and

$$A_i = \frac{-h_{fe}}{1 + h_{oe}R_L} = \frac{h_{fe}\underline{/180°}}{1 + h_{oe}R_L} \qquad (21.12)$$

Since h_{fe} is positive, the minus sign indicates the 180° phase reversal between output and input.

$$\text{Voltage gain, } A_v = \frac{V_o}{V_i} = \frac{-I_oR_L}{I_iR_i}$$

$$= A_i\frac{R_L}{R_i} = \frac{-h_{fe}R_L}{R_i(1 + h_{oe}R_L)} \qquad (21.13)$$

Again the minus sign indicates the 180° phase reversal between output and input voltage.

$$\text{Power gain, } G = \frac{\text{Output power}}{\text{Input power}} = \frac{I_o^2R_L}{I_i^2R_i}$$

i.e.

$$G = A_i^2\frac{R_L}{R_i} = A_iA_v \qquad (21.14)$$

Typical figures for the C.E. circuit are as follows:

	Germanium	Silicon
Input resistance	$1.5\,k\Omega$	$2.5\,k\Omega$
Output resistance	$25\,k\Omega$	$70\,k\Omega$
Current gain	-45	-80
Voltage gain	-80	-120

The actual values depend on the source and load resistances, as can be seen by carrying out an exact analysis of the circuit.

It is left as an exercise for the reader to deduce from the complete h-parameter equivalent circuit (i.e. including the feedback generator) that

$$\text{Input impedance} = h_{ie} - \frac{h_{re}h_{fe}R_L}{1 + h_{oe}R_L} \tag{21.15}$$

$$\text{Output admittance} = h_{oe} - \frac{h_{re}h_{fe}}{h_{ie} + R_g} \tag{21.16}$$

where R_g = resistance of signal source, and

$$\text{Current gain, } A_i = \frac{-I_o}{I_i} = \frac{-h_{fe}}{1 + h_{oe}R_L} \tag{21.17}$$

From these expressions it is seen that the input resistance is h_{ie} if R_L is small (this is the value of input resistance which the simplified circuit gives), and falls towards $h_{ie} - h_{re}h_{fe}/h_{oe}$ as $R_L \to \infty$ (an unusual operating condition). Also the output admittance rises towards h_{oe} as R_g rises.

EXAMPLE 21.1   For the common-emitter amplifier stage shown in Fig. 21.4(a), determine (a) the approximate d.c. conditions, (b) the small-signal mid-frequency current and power gains. The transistor is a BCY30; $R_1 = 60\,k\Omega$; $R_2 = 10\,k\Omega$; $R_3 = 1\,k\Omega$; $R_4 = 5\,k\Omega$; $R_L = 1.4\,k\Omega$; $C_1 = C_2 = 100\,\mu F$; $C_3 = 250\,\mu F$. Neglect the leakage current.

The coupling capacitors, C_1 and C_2, are required to isolate the source and the load from the direct voltages on the transistor; and the emitter bypass capacitor, C_3, effectively connects the emitter to earth for the a.c. signal.

It is assumed that the reactances of the capacitors are negligibly small, and that the d.c. supply is also of negligible impedance. Hence the a.c. equivalent circuit is as shown in Fig. 21.4(b). The parallel combination of R_1 and R_2 shunts some of the input current, I_{in}, and so reduces the gain. Also the transistor output current, I_o, is divided between the load R_L and the collector circuit resistance R_4. On small a.c. signals the input impedance of the stage, R_{in}, is the parallel combination of R_1, R_2 and h_{ie}, i.e.

$$R_{in} = R_1\|R_2\|h_{ie} = 10^3/(1/60 + 1/10 + 1/1.4) = 1{,}200\,\Omega$$

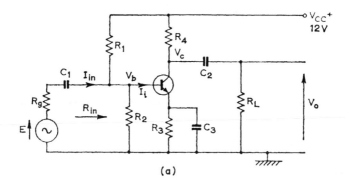

(a)

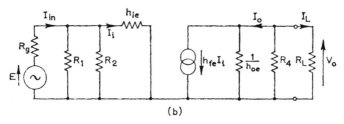

(b)

Fig. 21.4

(a) The d.c. operating conditions may be obtained by assuming that the base bias current, I_B, is very much smaller than the direct current through R_2, so that the direct voltage on the base, V_B, is given approximately

$$V_B = \frac{V_{CC}R_2}{R_1 + R_2} = \frac{12 \times 10,000}{70,000} = 1.7\,\text{V}$$

Since a silicon transistor is used it may be assumed that the base–emitter voltage is $0.7\,\text{V}$, so that the emitter bias voltage is

$$V_E = V_B - 0.7 = 1\,\text{V}$$

The emitter quiescent current is therefore

$$I_E = \frac{V_E}{R_3} = 1\,\text{mA}$$

This is also approximately the collector quiescent current, I_C, so that the quiescent voltage on the collector is

$$V_C = V_{CC} - I_C R_4 \approx 12 - 5 \times 10^{-3} \times 1000 = 7\,\text{V}$$

Note that the quiescent base current is $I_B \approx I_C/h_{FE}$, where h_{FE} is normally approximately the same as h_{fe} (50 in this case). Hence $I_B \approx 1/50 = 0.02\,\text{mA}$. The quiescent current through R_2 is $I_{R2} = V_B/R_2 = 1.7/10^4 = 0.17\,\text{mA}$. This demonstrates that the quiescent base current is indeed small compared to the current through R_2, as was originally assumed.

(*b*) From eqn. (21.12) the mid-frequency current gain of the transistor is given by

$$\frac{-I_0}{I_i} = \frac{h_{fe}\underline{/180^c}}{1 + h_{oe}R'}$$

where R' is the resistance of R_4 and R_L in parallel (i.e. $1 \cdot 1\,k\Omega$).

The actual stage current gain is $A_i = I_L/I_{in}$ and is less than the transistor gain since (i) the input current is partially shunted by R_1 and R_2, and (ii) the output current is divided between R_4 and R_L. Thus

$$I_i = I_{in}\frac{R_p}{R_p + h_{ie}}. \quad \text{where } R_p = \frac{R_1R_2}{R_1 + R_2} = 8\cdot6\,k\Omega$$

and

$$I_L = \frac{-I_0R_4}{R_4 + R_L}$$

Hence the current gain of the stage is

$$A_i = \frac{I_L}{I_{in}} = \frac{-I_0R_4}{(R_4 + R_L)}\frac{R_p}{I_i(R_p + h_{ie})}$$

$$= \frac{-h_{fe}R_4R_p}{(1 + h_{oe}R')(R_4 + R_L)(R_p + h_{ie})}$$

Inserting numerical values,

$$A_i = \frac{-50 \times 5 \times 10^3 \times 8\cdot6 \times 10^3}{\{1 + (25 \times 10^{-6} \times 1\cdot1 \times 10^3)\}(5 + 1\cdot4) \times 10^3(8\cdot6 + 1\cdot4) \times 10^3}$$

$$= -33 = 33\underline{/180^c}$$

The mid-frequency power gain is

$$G = A_i^2\frac{R_L}{R_{in}} = 33^2 \times \frac{1\cdot4}{1\cdot2} = 1,250$$

21.3 Common-base Amplifier

The common-base amplifier was frequently used in the early days of transistors owing to its good high-frequency response. It has a very low input impedance and high output impedance combined with a current gain of less than unity, so that matching transformers are required for cascaded stages. The fractional current gain is the main disadvantage of this connexion. The modern improvement in the high-frequency performance of transistors has meant that the common-base circuit has been largely replaced by the common-emitter circuit, which can be cascaded simply. However, the common-base circuit can be used as a low-to-high impedance-matching stage. This has basically the same buffer action for a current amplifier as the emitter follower (see Chapter 22) has for a voltage amplifier. A typical stage is shown in Fig. 21.5(*a*).

The a.c. small-signal equivalent circuit is shown in Fig. 21.5(*b*), where again the feedback generator has been omitted. The expressions for input and output resistance, current and voltage gain have

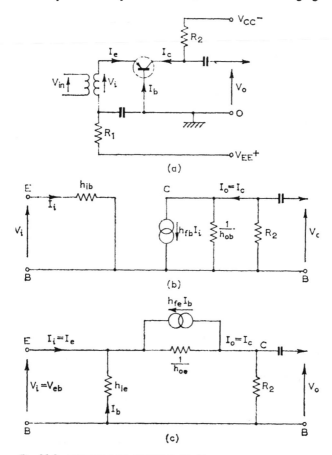

Fig. 21.5 COMMON-BASE AMPLIFIER STAGE

therefore the same form as those derived for the common-emitter stage, but the common base (C.B.) *h*-parameters replace the common-emitter (C.E.) *h*-parameters. Thus

$$V_i = V_{eb} = h_{ib}I_e \tag{21.18}$$

$$I_o = I_c = h_{fb}I_e + h_{ob}V_{cb} \tag{21.19}$$

and

Input resistance $= h_{ib}$ $\qquad$ (21.20)

Output resistance $= 1/h_{ob}$ $\qquad$ (21.21)

Internal current gain $= \dfrac{-I_o}{I_i} = \dfrac{-h_{fb}}{1 + h_{ob}R_L}$ $\qquad$ (21.22)

Voltage gain $= \dfrac{-h_{fb}}{1 + h_{ob}R_L} \dfrac{R_L}{R_{in}}$ $\qquad$ (21.23)

The fact that the current gain is positive is accounted for by describing h_{fb} as a negative constant. There is, of course, no phase reversal through the amplifier—as is apparent since the input current (the emitter current) flows across to the collector to become the output current. Expressions similar to eqns. (21.15)–(21.17) are obtained if the full h-parameter circuit is used. As with the common-emitter circuit, and because of the production spreads of transistor parameters and the small value of h_{rb} it is normally sufficiently accurate to use the simplified circuit.

Frequently only common-emitter parameters are quoted by manufacturers, so that it is often more convenient simply to rearrange the common-emitter equivalent circuit for common-base operation as shown in Fig. 21.5(c). Then, from eqn. (21.9),

$$I_c = h_{fe}I_b + h_{oe}V_{ce}$$
$$= -h_{fe}(I_c + I_e) + h_{oe}V_{ce}$$

since $I_b + I_c + I_e = 0$. Hence

$$I_c = \frac{-h_{fe}I_e}{1 + h_{fe}} + \frac{h_{oe}V_{ce}}{1 + h_{fe}}$$
$$\approx \frac{-h_{fe}I_e}{1 + h_{fe}} + \frac{h_{oe}V_{cb}}{1 + h_{fe}} \qquad (21.24)$$

since $V_{cb} = V_{ce} + V_{eb}$; and if the voltage gain is high, $V_{cb} \gg V_{eb}$, so that $V_{cb} \approx V_{ce}$.

Comparing eqns. (21.24) and (21.19), it will be seen that

$$h_{fb} = \frac{-h_{fe}}{1 + h_{fe}} \qquad (21.25)$$

$$h_{ob} = \frac{h_{oe}}{1 + h_{fe}} \qquad (21.25a)$$

It is sometimes convenient to remember that $h_{ob} \approx 1/r_c$ and $h_{fb} \approx -a$ in the common-base equivalent-T circuit.

Also

$$V_{eb} = -h_{ie}I_b = h_{ie}I_e + h_{ie}I_c$$

$$= h_{ie}I_e - \frac{h_{ie}h_{fe}}{1 + h_{fe}}I_e + \frac{h_{ie}h_{oe}}{1 + h_{fe}}V_{cb} \quad \text{(from eqn. (21.24))}$$

$$= \frac{h_{ie}I_e}{1 + h_{fe}} + \frac{h_{ie}h_{oe}}{1 + h_{fe}}V_{cb}$$

Then, since $h_{ie}h_{oe}V_{cb} \ll h_{ie}I_e$,

$$V_{eb} \approx \frac{h_{ie}}{1 + h_{fe}}I_e \tag{21.26}$$

Comparing eqns. (21.26) and (21.18),

$$h_{ib} \approx \frac{h_{ie}}{1 + h_{fe}} \tag{21.27}$$

It will thus be seen that the input resistance, output *admittance* and current gain of the C.B. amplifier are less by a factor of $(1 + h_{fe})$ than the corresponding quantities for the C.E. amplifier. Owing to the 'ow input and high output resistances, however, there is both voltage and power gain in a common-base stage. It is also evident that, since $h_{fb} = -h_{fe}/(1 + h_{fe})$, the value of h_{fb} will always be slightly less than unity. Note that h_{fe} corresponds to a term which is frequently found in manufacturers specifications under the symbols α' or β, and that h_{fb} corresponds to a.

21.4 Low-frequency Response of C.E. Amplifier

EFFECT OF COUPLING CAPACITORS

Consider the common-emitter stage shown in Fig. 21.6(a), which is fed from an a.c. source, I, of internal impedance R_g. If the reactance of C_3 is small enough, the impedance, R_{in}, seen looking into the transistor from the coupling capacitor C_1, consists of R_1, R_2 and h_{ie} in parallel (i.e. the impedance to the signal current), so that the equivalent a.c. input circuit is as shown in Fig. 21.6(b). At mid-frequencies, the reactance of C_1 is negligibly small, and the current, I_{in}, is given by

$$I_{in} = \frac{IR_g}{R_g + R_{in}}$$

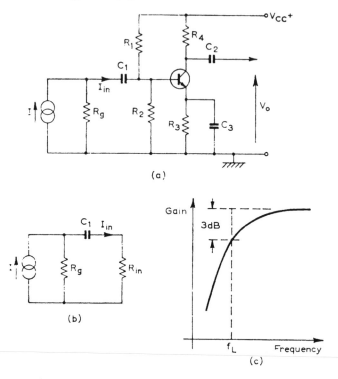

(b)

(c)

Fig. 21.6   COMMON-EMITTER AMPLIFIER STAGE

The low-frequency response is determined by the coupling and emitter bypass capacitors, C_1-C_3

As the frequency falls, however, the reactance of C_1 must be taken into account, and the current I_{in}' is then

$$I_{in}' = \frac{IR_g}{R_g + R_{in} + \dfrac{1}{j\omega C_1}}$$

As the frequency falls the term $1/j\omega C_1$ becomes comparable in size with $(R_g + R_{in})$ so that the current I'_{in} becomes smaller than I_{in}. The effective current gain I_0/I will therefore fall. At mid-frequencies I_0/I has the value

$$A_{io} = \frac{I_0}{I} = \frac{I_0}{I_{in}} \times \frac{I_{in}}{I} = \frac{-h_{fe}}{1 + h_{oe}R_L} \times \frac{R_g}{R_g + R_{in}}$$

at low frequencies this ratio becomes

$$A_i = \frac{I_o}{I} = \frac{I_o}{I_{in}} \times \frac{I_{in}'}{I_o} = \frac{-h_{fe}}{1 + h_{oe}R_L} \times \frac{R_g}{R_g + R_{in} + 1/j\omega C_1}$$

The difference between the expressions for A_{io} and A_i is the inclusion of the term $1/j\omega C_1$ in the denominator. This term causes a reduction in the size of I_o/I and also an increase in its phase angle. A convenient way of describing this effect is to determine the frequency for which

$$R_g + R_{in} = \frac{1}{\omega C_1}$$

At this frequency (called the lower cut-off frequency, f_L) $R_g + R_{in} + 1/j\omega C_1$ may be written as

$$R_g + R_{in} + \frac{1}{j2\pi f_L C_1} = (R_g + R_{in}) + \frac{1}{j}(R_g + R_{in})$$
$$= (R_g + R_{in})(1 - j1)$$
$$= \sqrt{2}\,(R_g + R_{in})\,\underline{/-45°}$$

i.e. the current gain will be $1/\sqrt{2}$ of its mid-frequency value and the phase angle of the current gain will be *increased* by 45°. The frequency at which this occurs is given by

$$2\pi f_L C_1 = \frac{1}{R_g + R_{in}}$$

i.e.

$$f_L = \frac{1}{2\pi C_1(R_g + R_{in})} \tag{21.28}$$

The current gain at any frequency, f, can now be written as

$$A_i = \frac{-h_{fe}}{1 + h_{oe}R_L} \frac{R_g}{R_g + R_{in} + f_L/j2\pi f f_L C_1}$$
$$= \frac{-h_{fe}}{1 + h_{oe}R_L} \frac{R_g}{R_g + R_{in} + (R_g + R_{in})f_L/jf}$$

Hence

$$A_i = \frac{A_{io}}{1 + f_L/jf} = \frac{A_{io}}{1 - jf_L/f} \tag{21.29}$$

When $f = f_L$, $A_i = A_{io}/(1 - j1) = A_{io}/\sqrt{2}\,\underline{/-45°}$ as derived previously. Note that *for this condition*

$$20 \log_{10}(A_i/A_{io}) = -20 \log_{10}\sqrt{2} = -10 \log_{10}2$$
$$= -3\,\text{dB}$$

The frequency f_L is often called the lower 3 dB cut-off frequency. The bandwidth of an amplifier is arbitrarily taken to extend between those frequencies for which the gain falls by 3 dB from the mid-frequency value (the upper 3 dB cut-off frequency is dealt with in Section 21.5).

Similar conditions apply to the voltage gain of the stage.

Very often it is sufficiently accurate when calculating the value of C_1 required for a given low-frequency response to assume that R_1 and R_2 are large enough to be neglected compared with h_{ie} so that R_{in} in eqn. (21.28) becomes simply h_{ie}. It is further obvious from eqn. (21.28) that for a given coupling capacitor, C_1, the higher the value of R_g the lower will be the cut-off frequency f_L. Hence a voltage signal source (which will normally be of low impedance) gives a poorer low-frequency performance than a current signal source (which will be of high impedance—e.g. a preceding common-base or common-emitter amplifier stage).

It is evident that the second coupling capacitor, C_2, will produce a similar effect, which can be calculated in the same way. The approximate shape of the gain/frequency characteristic for low frequencies is shown in Fig. 21.6(c).

EFFECT OF EMITTER BYPASS CAPACITOR

The emitter bypass capacitor, C_3, is required in order to connect the emitter to earth with respect to the signal. If it were removed there would be a signal feedback effect and a consequent fall in gain, which

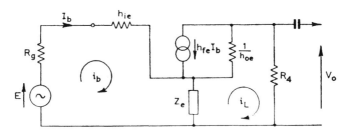

Fig. 21.7   A.C. EQUIVALENT CIRCUIT OF FIG 21.6, INCLUDING
EMITTER-CIRCUIT RESISTANCE

will be discussed further in the next chapter. At low frequencies the reactance of C_3 increases, and hence the emitter may no longer be considered as an earth point for the signal. This reduces the gain of the stage by an amount which may be determined as follows.

The equivalent a.c. circuit for the stage shown in Fig. 21.6 is shown in Fig. 21.7, including a finite emitter impedance, Z_e, and neglecting the effect of the coupling capacitors C_1 and C_2. The signal source is represented by the equivalent constant-voltage generator of e.m.f. E and internal impedance R_g. For simplicity the bias resistors, R_1 and R_2, are assumed high enough in value to carry negligible signal current, and additional loading on the stage is assumed incorporated in R_4. The mesh equations can then be written (using small letters for mesh currents) as

$$\text{Mesh 1} \quad E = (R_g + h_{ie} + Z_e)i_b - Z_ei_L$$
$$\approx (R_g + h_{ie})i_b - Z_ei_L \tag{21.30}$$

$$\text{Mesh 2} \quad 0 = -Z_ei_b + \left(\frac{1}{h_{oe}} + Z_e + R_4\right)i_L + \frac{h_{fe}}{h_{oe}}I_b$$
$$\approx \frac{h_{fe}}{h_{oe}}I_b + \left(\frac{1}{h_{oe}} + R_4\right)i_L \tag{21.31}$$

and $i_L \gg I_b$; $i_b = I_b$; hence $Z_ei_b \ll Z_ei_L$. Also $Z_e \ll 1/h_{oe}$.

From eqns. (21.30) and (21.31),

$$E = \frac{-(R_g + h_{ie})\left(\dfrac{1}{h_{oe}} + R_4\right)i_L}{h_{fe}/h_{oe}} - Z_ei_L$$
$$= -\frac{(R_g + h_{ie})(1 + h_{oe}R_4) + Z_eh_{fe}}{h_{fe}}i_L$$

so that

$$\frac{V_o}{E} = \frac{i_LR_4}{E} = \frac{-R_4h_{fe}}{(R_g + h_{ie})(1 + h_{oe}R_4) + Z_eh_{fe}} \tag{21.32}$$

As $Z_e \to 0$ this ratio gives the mid-frequency value of gain:

$$\frac{V_o}{E}\bigg|_{\text{mid-}f} = \frac{-R_4h_{fe}}{(R_g + h_{ie})(1 + h_{oe}R_4)}$$

If it is now assumed that Z_e is the reactance of C_3 alone (i.e. $Z_e = 1/j\omega C_3$), the ratio of V_o/E, or the voltage gain, will be 3 dB down on its mid-frequency value when the reactive component of the denominator in eqn. (21.32) is equal to the resistive component, i.e. when

$$\frac{h_{fe}}{\omega C_3} = (R_g + h_{ie})(1 + h_{oe}R_4)$$

or

$$\frac{1}{\omega C_3} = \frac{(R_g + h_{ie})(1 + h_{oe}R_4)}{h_{fe}}$$

Hence the frequency for a 3 dB fall due to C_3 alone is

$$f_L \approx \frac{h_{fe}}{2\pi(R_g + h_{ie})(1 + h_{oe}R_4)C_3} \qquad (21.33)$$

This again shows that the higher the source impedance, R_g, the lower will be the 3 dB cut-off frequency. Also note that, since $h_{oe}R_4$ is generally less than 1, the value of C_3 is generally h_{fe} times the value of the coupling capacitor for the same low-frequency cut-off.

In the actual circuit both the coupling-capacitor and emitter-capacitor effects will occur together. Generally the leakage through C_3 is less critical than that through C_1 or C_2, so that it may economically be chosen to have a high enough capacitance to make the emitter effect occur at a much lower frequency than the cut-off frequency due to C_1 or C_2.

The emitter circuit resistance, R_3, can affect the result, but if $1/\omega C_3 < 0.1 R_3$ at the low-frequency cut-off the effect is small.

EXAMPLE 21.2   The transistor used in the circuit of Fig. 21.6(a) has $h_{oe} = 40\,\mu\text{S}$; $h_{ie} = 1.5\,\text{k}\Omega$; $h_{fe} = 100$; R_1 and R_2 are large enough to be neglected compared with h_{ie} at signal frequencies; $R_3 = 1\,\text{k}\Omega$; and $R_4 = 5\,\text{k}\Omega$.

Find suitable values for C_1 and C_3 to give a low-frequency cut-off at 15 Hz for a source resistance of 5 kΩ. If the source resistance is reduced to 500 Ω what will be the low-frequency cut-off?

(a) From eqn. (21.28), the coupling capacitance is

$$C_1 = \frac{1}{2\pi f_L(R_g + R_{in})} \quad \text{where } R_{in} \text{ is taken as equal to } h_{ie}$$

$$= \frac{10^6}{2\pi \times 15(5 + 1.5) \times 10^3} = 1.63, \text{ or say } \underline{\underline{2\,\mu\text{F}}}$$

If the emitter decoupling capacitor is to have only a small effect at this frequency then, from eqn. (21.32),

$$C_3 \gg \frac{h_{fe}}{\omega(R_g + h_{ie})(1 + h_{oe}R_4)} \gg \frac{100}{2\pi \times 15 \times 6.5 \times 10^3 \times 1.2} \gg 150\,\mu\text{F}$$

A suitable value for C_3 would be $\underline{\underline{500\,\mu\text{F}}}$.

(b) For a source resistance of 500 Ω, eqn. (21.28) gives

$$f_L = \frac{1}{2\pi C_1(R_g + R_{in})} = \frac{10^6}{2\pi \times 2 \times 2 \times 10^3} = \underline{\underline{40\,\text{Hz}}}$$

21.5 Transistor Characteristics at High Frequencies

Whereas the low-frequency response of transistor amplifiers is determined mainly by the associated circuit components, the high-frequency response is largely a function of the transistor itself. By far the most important high-frequency effect is due to the time taken for charge carriers to cross from the emitter junction to the collector junction. This is called *base transit time.* In addition, the junction capacitances and stray circuit-wiring capacitances will affect the high-frequency performance of transistor amplifiers, but these effects are of less importance in general.* The following treatment represents a much simplified approach. In practical transistors, different physical constructions give different high-frequency characteristics, which are only approximated by the following derivation.

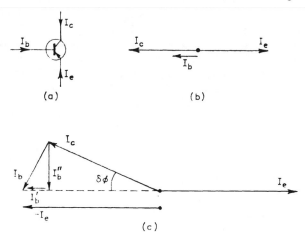

Fig. 21.8 EFFECT OF TRANSIT TIME ON THE HIGH-FREQUENCY
RESPONSE OF A TRANSISTOR

In order to understand the physical basis for the loss of gain at high frequencies in a transistor, consider the device as a "black box" in which there is an input current, I_e. This current divides between the collector and the base; i.e. if I_e enters the device, I_c and I_b must leave it—this corresponds to the physical operation. In the purely mathematical convention adopted earlier, all alternating currents are assumed to flow *into* the device as shown in Fig. 21.8(a). It follows that the currents are related both instantaneously and in complexor representation by the equation

$$I_e = -(I_b + I_c)$$

At low frequencies the transit time is only a small fraction of the period of the signal, and hence there will be negligible delay through the transistor and the complexor diagram will be as shown in Fig. 21.8(b), the 180° phase reversal being due to the conventional current directions chosen.

* The situation, however, is very dependent on current advances in production technology, particularly with silicon transistors.

Under short-circuit conditions at the output, eqn. (21.9) gives

$$I_c = h_{feo}I_b$$

where h_{feo} is the low-frequency value of h_{fe}.

In this equation, I_b is the normal recombination base current of the transistor.

As the frequency increases the time taken for the charge carriers to cross the base becomes a significant fraction of the signal period. The collector current, I_c, will therefore lag behind its low-frequency phase position by a small angle $\delta\phi = 2\pi f \tau_B$, where τ_B is the base transit time. This follows since for a signal period $T = 1/f$ second, a delay of T corresponds to a phase-angle lag of 2π radians, and hence a delay of τ_B corresponds to a lag of $2\pi\tau_B/T = 2\pi f\tau_B$ radians. But the base current, I_b, is equal to $-(I_e + I_c)$, so that I_b must be the complexor joining the end points of I_c and $-I_e$, as shown in Fig. 21.8(c). This current can be resolved into two components, I_b' in phase with $-I_e$, and I_b'' leading $-I_e$ by 90°. The current I_b' represents the normal low-frequency base-recombination current I_c/h_{feo}.

Assuming that $\delta\phi$ is small,

$$I_b'' \approx I_c\delta\phi = I_c 2\pi f\tau_B$$

and leads $-I_e$ by 90°. Hence

$$I_b = I_b' + I_b'' \approx \frac{I_c}{h_{feo}} + j2\pi f\tau_B I_c$$

The complex short-circuit current gain is defined as $h_{fe} = I_c/I_b$, so that

$$h_{fe} = \frac{1}{\dfrac{1}{h_{feo}} + j2\pi f\tau_B} = \frac{h_{feo}}{1 + j2\pi h_{feo}f\tau_B} \tag{21.34}$$

Obviously if τ_B is very small so that $2\pi h_{feo}f\tau_B \ll 1$, then $h_{fe} = h_{feo}$, and further, the value of h_{fe} will fall by 3 dB and will introduce a phase *lag* of 45° when $2\pi f h_{feo}\tau_B = 1$. The frequency at which this occurs is often called the β *cut-off frequency*, f_β or alternatively f_{hfe}, and is given by

$$f_{hfe} = f_\beta = \frac{1}{2\pi h_{feo}\tau_B} \tag{21.35}$$

It is now more usual to quote the frequency, f_1, at which h_{fe} falls to unity. This is obtained by letting f in eqn. (21.34) become so large that $2\pi f h_{feo}\tau_B \gg 1$. Then

$$h_{fe} \approx \frac{h_{feo}}{2\pi f h_{feo}\tau_B} = \frac{1}{2\pi f\tau_B}$$

This will have a magnitude of unity when

$$f = f_1 = \frac{1}{2\pi\tau_K} \tag{21.36}$$

Combining eqns. (21.35) and (21.36), the following important relation is obtained:

$$f_{hfe}h_{feo} = \frac{1}{2\pi\tau_B} = f_1$$

Also, for frequencies at which $2\pi f h_{feo}\tau_B \gg 1$, eqn. (21.34) gives

$$h_{fe}f \approx \frac{h_{feo}f}{2\pi f h_{feo}\tau_B} = \frac{1}{2\pi\tau_B} = f_1 \tag{21.37}$$

This is called the *gain-bandwidth product* of the transistor, and is an important design parameter.

Sometimes a frequency, f_T, is quoted in relation to transistors. This is the frequency at which the magnitude of h_{fe} falls to unity if the fall in gain is constant at 6 dB per octave from the break frequency f_{hfe}. Normally f_T is only slightly different from f_1.

From eqns. (21.34) and (21.35), a general expression for h_{fe} can be written as

$$h_{fe} = \frac{h_{feo}}{1 + j\dfrac{f}{f_\beta}} \tag{21.38}$$

Although this applies only so long as the frequency is below or not much above f_β, it gives some approximation to the actual characteristic up to the frequency f_1.

21.6 High-frequency Response of C.E. Amplifier

The characteristics of a common-emitter amplifier at high frequencies can be obtained by substituting the expression for h_{fe} given by eqn. (21.38) in eqns. (21.10)–(21.17). Thus the internal current gain is

$$A_i = \frac{-h_{fe}}{1 + h_{oe}R_L}$$

$$= \frac{-h_{feo}}{\left(1 + j\dfrac{f}{f_\beta}\right)(1 + h_{oe}R_L)} \tag{21.39}$$

This is 3 dB down on the mid-frequency gain when $f = f_\beta \ (= f_{hfe})$

and falls at a rate of 6 dB per octave for frequencies above f_β (i.e. for each doubling in frequency the gain falls by a further 6 dB).* Also when the frequency is f_β there will be an added 45° phase lag. A similar situation exists for voltage gain. In both cases there is an increase of phase lag up to an additional 90° as the frequency approaches infinity.

It is obvious from Fig. 21.8(c) that at sufficiently high frequencies there is a component of base (input) current which leads the normal low-frequency base current by 90°—i.e. there is an added capacitive component of input impedance. This is also demonstrated by substituting the expression for h_{fe} of eqn. (21.38) in eqn. (21.15), which gives the exact expression for input impedance.

21.7 High-frequency Response of C.B. Amplifier

The response of the common-base amplifier to high-frequency signals is obtained by substituting for h_{fe} from eqn. (21.38) in the expression representing the operation of this type of amplifier. Thus from eqn. (21.22) the current gain of the common-base amplifier is

$$
\begin{aligned}
A_i &= \frac{-h_{fb}}{1 + h_{ob}R_L} = \frac{h_{fe}}{(1 + h_{fe})(1 + h_{ob}R_L)} \\[2mm]
&= \frac{h_{feo}/(1 + jf/f_\beta)}{\left\{1 + \dfrac{h_{feo}}{1 + jf/f_\beta}\right\}(1 + h_{ob}R_L)} \\[2mm]
&= \frac{h_{feo}}{(1 + h_{feo} + jf/f_\beta)(1 + h_{ob}R_L)}
\end{aligned}
$$

Thus as the frequency increases the current gain falls, owing to the term jf/f_β, and is 3 dB down on its mid-frequency value of. $h_{feo}/(1 + h_{feo})(1 + h_{ob}R_L)$, when

$$
1 + h_{feo} = \frac{f}{f_\beta}
$$

* The ratio $\dfrac{\text{Gain at a high frequency,} f}{\text{Gain at mid-frequencies}} = \dfrac{1}{1 + j\dfrac{f}{f_\beta}}$

so that

$$
\text{Gain in decibels} = 20 \log_{10}\left|\frac{1}{(1 + jf/f_\beta)}\right| = -20 \log_{10}\sqrt{(1 + f^2/f_\beta^2)}
$$

$$
= -20 \log_{10}(f/f_\beta) \quad \text{for } f \gg f_\beta
$$

Hence for a frequency of $2f$ (an octave higher than f),

$$
\text{Gain} = -20 \log_{10} 2f/f_\beta = -20 \log_{10} 2 - 20 \log_{10} f/f_\beta
$$

$$
= (\text{Gain at frequency } f - 6) \text{ decibels}
$$

The frequency at which this occurs is called the α *cut-off frequency* f_α, given by

$$f_\alpha = (1 + h_{feo})f_\beta \tag{21.40}$$

The expression for current gain can therefore be written

$$A_i = \frac{-h_{fbo}}{\left(1 + j\dfrac{f}{f_\alpha}\right)(1 + h_{ob}R_L)} \quad \text{where} \quad h_{fbo} = \frac{h_{feo}}{1 + h_{feo}}$$

At the frequency f_α there will be a phase lag of 45°.

The response characteristics of a transistor in common-emitter and common-base connexions are shown in Fig. 21.9, where gain

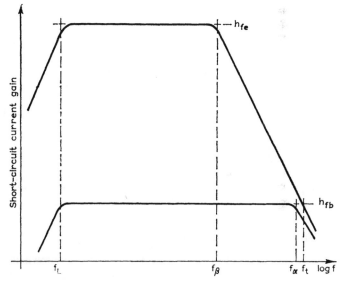

Fig. 21.9 FREQUENCY RESPONSE CHARACTERISTICS OF TRANSISTORS IN C.E. AND C.B. CONNEXIONS

in decibels is plotted against the logarithm of frequency. The bandwidth of the amplifier is the frequency range between the lower and upper cut-off frequencies. For a good high-frequency transistor the base transit time may be as short as 0·125 ns (i.e. 0·125 × 10⁻⁹ second), giving

$$f_1 = \frac{1}{2\pi \times 0.125 \times 10^{-9}} \approx 1\,\text{GHz}$$

Table 21.2 summarizes the characteristics of common-base and common-emitter amplifiers at mid-frequencies. The numerical values given are to be considered as giving orders of magnitude, and depend on the values of load and generator resistances. Thus the input resistance in common-base connexion rises significantly as the load resistance, R_L, increases (above about $10\,\text{k}\Omega$) while that in

TABLE 21.2

Connexion	Input resistance	Output resistance	Short-circuit current gain	Voltage gain	Mid-frequency phase characteristic
Common emitter	medium $1\cdot5\,\text{k}\Omega$ (Ge) $2\cdot5\,\text{k}\Omega$ (Si)	medium $25\,\text{k}\Omega$ (Ge) $70\,\text{k}\Omega$ (Si)	high $(20 -$ over 200)	high	$180°$
Common base	low $25\,\Omega$ (Ge) $40\,\Omega$ (Si)	high $1\,\text{M}\Omega$ (Ge) $1\,\text{M}\Omega$ (Si)	low <1	high	$0°$

common-emitter connexion falls slightly as R_L increases. Similarly the output resistance in common-base connexion falls appreciably as the source resistance, R_g, falls (below about $1\,\text{k}\Omega$) while that in common-emitter connexion rises slightly as R_g falls.

If the collector resistance $R_L \ll 1/h_{oe}$, then C.E. and C.B. amplifiers have approximately the same voltage gain. If R_L is very high the C.B. connexion will give the higher voltage gain.

21.8 Tuned Transistor Amplifiers

A simple tuned amplifier with transformer input is shown in Fig. 21.10(a), with the simplified equivalent a.c. circuit at (b). At resonance (angular frequency, $\omega_0 = 1/\sqrt{(LC)}$) the impedance of the tuned circuit is $R_0 = L/Cr$ and at an angular frequency ω near resonance it is

$$Z = \frac{R_0}{1 + j2Q\Delta} \tag{21.41}$$

where $Q = \omega L/r = R_0/\omega L$ and $\Delta \approx (\omega - \omega_0)/\omega_0$ is the per-unit frequency deviation.

Capacitors C_2 and C_3 provide a low-impedance path between input and emitter for signal currents. Their capacitances may be

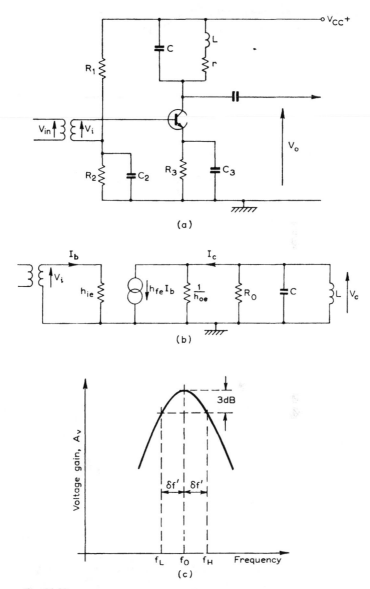

Fig. 21.10 **TUNED TRANSISTOR AMPLIFIER**

relatively small, since they are only required to bypass signal-frequency currents, and not low-frequency currents. Hence the input current is given by

$$I_b = \frac{V_i}{h_{ie}} \tag{21.42}$$

and the output current is

$$I_c = h_{fe}I_b + V_o h_{oe}$$

Hence the output voltage, V_o, is

$$V_o = -I_c Z = -\frac{(h_{fe}I_b + V_o h_{oe})R_0}{1 + j2Q\Delta}$$

or, from eqns. (21.41) and (21.42),

$$V_o + \frac{V_o R_0 h_{oe}}{1 + j2Q\Delta} = \frac{-h_{fe}V_i R_0}{h_{ie}(1 + j2Q\Delta)}$$

The voltage gain is thus

$$A_v = \frac{V_o}{V_i} = \frac{-h_{fe}R_0}{(1 + h_{oe}R_0 + j2Q\Delta)h_{ie}} \tag{21.43}$$

The relationship between gain and frequency around resonance is shown in Fig. 21.10(c). At resonance, $\Delta = 0$, and the gain has a maximum value of

$$A_{vm} = \frac{-h_{fe}R_0}{(1 + h_{oe}R_0)h_{ie}} \tag{21.44}$$

It follows that the gain falls 3 dB below its resonance value when (from eqn. (21.43))

$$2Q\Delta = 1 + h_{oe}R_0$$

i.e. when

$$\Delta = \frac{\delta f'}{f_0} = \frac{1 + h_{oe}R_0}{2Q} = \frac{1}{2Q'} \tag{21.45}$$

where $\delta f'$ is the half bandwidth at the 3 dB points, and Q' is the effective Q-factor of the circuit when it is loaded by the output resistance $(1/h_{oe})$ of the transistor; i.e.

$$Q' = \frac{R_{eq}}{\omega L}$$

where R_{eq} is R_0 in parallel with $1/h_{oe}$:

$$R_{eq} = \frac{R_0(1/h_{oe})}{R_0 + 1/h_{oe}} = \frac{R_0}{1 + h_{oe}R_0}$$

Hence

$$Q' = \frac{R_0}{(1 + h_{oe}R_0)\omega L} = \frac{Q}{1 + h_{oe}R_0}$$

The voltage gain (eqn. (21.43)) can thus be expressed in terms of R_{eq} and Q' as

$$A_v = \frac{-h_{fe}R_{eq}}{(1 + j2Q'\Delta h_{ie})} \tag{21.46}$$

If the lower and upper frequencies at which the gain falls 3 dB below the resonance value are f_L and f_H, then

$$f_L = f_0 - \delta f' = f_0 - \frac{f_0}{2Q'} \tag{21.47}$$

and

$$f_H = f_0 + \delta f' = f_0 + \frac{f_0}{2Q'} \tag{21.48}$$

while the overall bandwidth in hertz is

$$B = f_H - f_L = \frac{f_0}{Q'} \tag{21.49}$$

Note that the bandwidth between half-power frequencies of the unloaded tuned circuit is given by $B = f_0/Q$.

It is thus apparent that the active device lowers the effective Q-factor and increases the bandwidth between half-power points. In a C.E. transistor amplifier $1/h_{oe}$ may be appreciably smaller than R_0, and the bandwidth may thus be undesirably wide (the whole object of a tuned amplifier usually being to give selective amplification over a narrow bandwidth around the resonant frequency). In order to overcome this the Q-factor of the circuit must be increased. This may be done by increasing the C/L ratio while keeping the product LC constant (for a given resonant frequency).* However this can lead to unrealistically large values for C (if it is a tuning capacitor).

* $$Q = \frac{\omega L}{r} = \frac{\omega L}{L/CR_0}$$

since $R_0 = L/Cr$, where r is the series resistance of L: i.e.

$$Q = \omega CR_0 = R_0\sqrt{\frac{C}{L}}$$

since at resonance $\omega = 1/\sqrt{(LC)}$.

Hence for a high Q, C/L must be large. Note also that if L is reduced, then r will also fall; hence, since L is proportional to the square of the number of turns, and r, to the number of turns, Q will increase.

An alternative approach is to connect the transistor across a tapping of the tuning inductor as shown in Fig. 21.11(*a*). This gives the simplified equivalent circuit shown at (*b*), which may be further reduced to the circuit shown at (*c*), assuming that there is perfect coupling between the primary and secondary windings. The tapping is taken at a fraction, k_t, of the total number of turns. It is left as

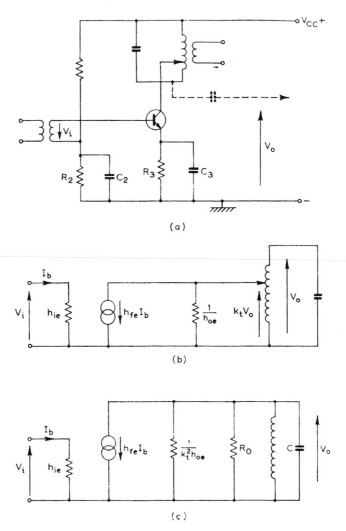

(*a*)

(*b*)

(*c*)

Fig. 21.11   USE OF A TAPPED COIL TO INCREASE THE EFFECTIVE Q-FACTOR OF A TUNED TRANSISTOR AMPLIFIER

an exercise for the reader to verify that circuits (*b*) and (*c*) are in fact equivalent.

The resultant Q-factor is then $Q' = R_{eq}'/\omega L$ where R_{eq}' is the resistance of R_0 in parallel with $1/(k_t^2 h_{oe})$. Without the tapping the value of R_{eq} has already been shown to be R_0 in parallel with $(1/h_{oe})$, so that the tapping has increased R_{eq} ($k_t < 1$). The voltage gain becomes

$$A_v = \frac{-k_t h_{fe} R_{eq}'}{(1 + j2Q'\Delta)h_{ie}} \tag{21.50}$$

where

$$R_{eq}' = \frac{R_0}{(1 + k_t^2 h_{oe} R_0)} \tag{21.51}$$

A further complication arises in transistor tuned amplifiers, since the input resistance, R_{in}, of the following stage may be low, and its input capacitance, C_{in}, high. This can be partially compensated by taking the output from a secondary winding as shown in Fig. 21.11. Then if the ratio of primary to secondary turns is k_{t2}, the input will reflect a parallel resistance $k_{t2}^2 R_{in}$ into the tuned circuit (where $k_{t2} > 1$). It will also reflect a parallel capacitance C_{in}/k_{t2}^2 into the tuned circuit, so that the effects of R_{in} and C_{in} on bandwidth and tuning are very much reduced.

Tuned amplifiers are designed on a *power gain* basis. Hence $1/h_{oe}$ is matched to the load by suitable transformer tapping. The tuned circuit acts as a negligible power-loss shunt at resonance. Off resonance it absorbs more signal power, and so reduces the gain.

21.9 Neutralization and Unilaterization

The capacitance, C_{bc}, which exists between the base and collector of a transistor can result in a part of the output voltage of a tuned amplifier being fed back to the input base connexion, as shown in Fig. 21.12(*a*) where only the a.c. components of the circuit have been shown. As will be seen in Chapter 22, this feedback can have undesirable effects, and may result in the amplifier becoming an oscillator. It can be eliminated by extending the tuning inductance, L_1, by a few turns, and connecting a *neutralizing capacitor*, C_n, between this extension and the base input. Then L_1, L_2, C_{bc} and C_n form a simple bridge circuit, in which at balance

$$\frac{V_{o1}}{V_{on}} = \frac{1/C_{bc}}{1/C_n} \quad \text{or} \quad C_n = \frac{V_{o1}}{V_{on}} C_{bc}$$

(i.e. the neutralizing current is equal to the feedback current, so that the alternating voltage between B and E is zero, and no base current can flow). Assuming perfect coupling the ratio of voltages is the same as the turns ratio, N_1/N_2, so that

$$C_n = \frac{N_1}{N_2} C_{bc} \qquad (21.52)$$

A practical alternative is to use the secondary winding to provide a connexion for the neutralizing capacitor as shown at (b).

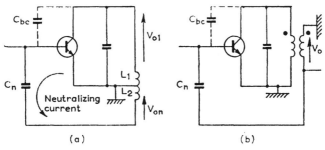

(a)   (b)

Fig. 21.12   NEUTRALIZATION OF A TUNED AMPLIFIER TO PREVENT
OSCILLATION

In some instances the internal feedback inherent in the transistor can provide the equivalent of a resistance connected across C_{bc}. In this case a resistance must also be connected in parallel with C_n, and the process is then called *unilaterialization*. In practice the parallel combination of C_n with a resistor would give a d.c. path, and it is usual to employ the equivalent series combination. This means that unilateralization is achieved at one particular frequency only.

Both these processes are nowadays important only at the highest frequencies, due to the cheapness and improved quality of high-frequency transistors, which means incidentally that they can be used for low-frequency amplification.

21.10 Transistor Amplifiers in Cascade

The gain of a single-stage small-signal transistor amplifier is normally between 30 and 50 dB. To increase the gain, amplifiers are connected in cascade and a typical two-stage amplifier is shown in Fig. 21.13.

Equivalent circuits may be used to determine the overall gain of such amplifiers. The load on the first stage is then the input resistance of the second stage, comprising the parallel combination of the bias resistors and the input resistance of the second transistor. The effective input current to the second transistor is only that fraction

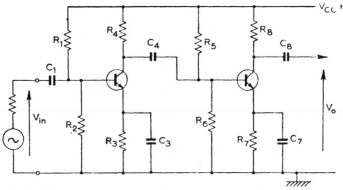

Fig. 21.13 TYPICAL TWO-STAGE *RC*-COUPLED TRANSISTOR
AMPLIFIER

of the output current of the first transistor which flows into the
second transistor base.

Similar considerations regarding overall bandwidth apply as in
cascaded valve amplifiers.

Sometimes the complementary properties of transistors are used
to give a cascaded circuit with direct coupling. This improves the

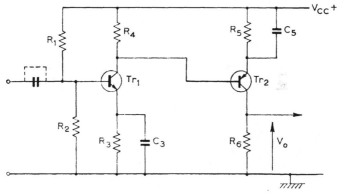

Fig. 21.14 USE OF COMPLEMENTARY SYMMETRY IN TWO-STAGE
DIRECT-COUPLED AMPLIFIER

frequency characteristics, since it eliminates the high-pass *CR*
coupling circuit. A typical circuit is shown in Fig. 21.14, where the
first transistor is of the *n–p–n* type and the second is a *p–n–p*. The
circuit is designed from the fact that the quiescent voltage at the
emitter of Tr_2 combined with the emitter resistance, R_5, gives the
operating current of Tr_2. The collector potential of Tr_1 must then
be the emitter voltage of Tr_2 plus the base–emitter voltage of Tr_2.

For a chosen value of quiescent current in Tr_1 this then gives the required value of R_4, remembering that the quiescent base current of Tr_2 flows towards Tr_1.

PROBLEMS

21.1 A common-emitter amplifier uses a transistor which has the following common-base T-parameters: $r_b = 700\,\Omega$; $r_e = 17\,\Omega$; $r_c = 900\,k\Omega$; $\alpha = 0.97$. The internal resistance of the signal source is $30\,k\Omega$, the collector load resistance is $2.2\,k\Omega$ and the input resistance of the next stage is $1\,k\Omega$. Calculate, from first principles and at mid-frequencies (1) the current gain, (2) the amplifier input resistance, (3) the voltage gain, (4) the power output for an input of 0.5 V r.m.s. Neglect the effect of bias circuit resistors on a.c. operation. (*Hint.* Draw the C.E. circuit, then replace the transistor by its equivalent-T circuit (common base) and solve). (H.N.D.)

Ans. -31.5; $1250\,\Omega$; -17.4; 0.18 mW.

21.2 From common-base tests on a transistor the following results were obtained: (i) with I_E constant, an increase of 1.0 V in collector–base voltage caused an increase in emitter–base voltage of 1 mV and an increase in collector current of $2\,\mu A$; (ii) with collector–base voltage constant an increase of 75 mV in emitter–base voltage produced increases in base and collector currents of $50\,\mu A$ and 5 mA respectively.

Determine the parameters of an equivalent T circuit for this transistor. Give two limitations in the use of this circuit.

Ans. $r_c = 0.5\,M\Omega$; $r_b = 500\,\Omega$; $r_e = 9.9\,\Omega$; $\alpha = 0.99$.

21.3 A small-signal single-stage C.E. amplifier employs a transistor for which $h_{ie} = 1.5\,k\Omega$, $h_{fe} = 80$, $h_{oe} = 20\,\mu S$. The collector load resistance is $4.7\,k\Omega$. Determine the mid-frequency current gain of the stage. If the input source has an e.m.f. of 50 mV r.m.s. determine the output signal voltage if the internal source resistance is (*a*) $600\,\Omega$, (*b*) $70\,\Omega$. Assume that the base-bias circuit is equivalent to a resistance of $5\,k\Omega$ at the input terminals, and neglect h_{re}.

Ans. -56; 7.57 V; 10.8 V.

21.4 Repeat the calculations of Problem 21.3 for the $600\,\Omega$ source resistance, but include the feedback factor $h_{re} = 4 \times 10^{-4}$.

Ans. -49; 8.1 V.

21.5 A single-stage transistor C.E. amplifier employs a transistor for which $h_{feo} = 100$ and $f_1 = 0.5$ GHz. Determine the frequency at which the current gain falls 3 dB below the mid-frequency value.

What will be the fall in current gain relative to the mid-frequency value at a frequency of (*a*) 10 MHz, (*b*) 20 MHz, (*c*) 40 MHz?

Ans. 5 MHz; 7.0 dB; 12.3 dB; 18.1 dB.

21.6 A single-stage transistor C.E. amplifier has a current gain which is 3 dB down on its mid-frequency value of 80 when the frequency is 12 MHz. Determine the approximate value of the base transit time and the gain-bandwidth product of the transistor.

Ans. 0.17 ns; 960 MHz.

21.7 The Q-factor of the tuned circuit of a tuned transistor C.E. amplifier is 80, the resonant frequency is 470 kHz, and the resonant impedance (dynamic resistance) is $100\,k\Omega$. Determine (*a*) the bandwidth of the simple amplifier, (*b*) the coil tapping required for a bandwidth of 10 kHz, and (*c*) the tuning capacitance required. The transistor has a value of h_{oe} of 40 μS.

Ans. 29·5 kHz; 0·42; 271 pF.

21.8 Derive an expression for the lower cut-off frequency of a transistor *RC*-coupled C.E. amplifier, stating any approximations made.

A transistor for which $h_{fe} = 60$ and $h_{oe} = 50$ μS is used in the first stage of an *RC* coupled C.E. amplifier. The input resistance of the second stage is 1·5 kΩ and the coupling capacitance is 2 μF. The gain falls by 3 dB at 16·8 Hz. Determine (*a*) the collector load resistance, (*b*) the mid-frequency gain, (*c*) the gain reduction and phase shift at 10 Hz. Neglect the effect of input coupling to the first stage.   (H.N.D.)

Ans. 3·9 kΩ; −41; −5·9 dB, 239°.

21.9 Derive the common-base *h*-parameters of a transistor from its common-emitter *h*-parameters assuming h_{re} to be negligible and that $h_{oe}h_{ie} \ll (1 + h_{fe})$.

Two identical transistors are used in the cascode arrangement of Fig. 21.15. The transistor parameters are $h_{ie} = 1$ kΩ; $h_{fe} = 80$; $h_{oe} = 10^{-4}$ S and h_{re} negligible. Calculate the current and power gains for the load resistor R_L at the

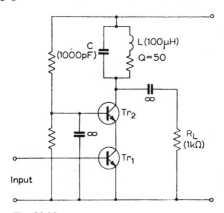

Fig. 21.15

resonant frequency of the tuned collector circuit. (*Hint.* Find the common-base *h*-parameters; then treat the circuit as having a C.E. input transistor followed by a C.B. output transistor).

Ans. −65·7; 4,310.

21.10 Sketch and explain biasing and coupling arrangements for a two-stage common-emitter *RC*-coupled transistor amplifier.

Such an amplifier has a second stage which employs a transistor with $h_{fe} = 60$, $h_{ie} = 2$ kΩ and $h_{oe} = 15 \mu$S. If the collector load resistance is 12 kΩ, determine the voltage gain of the second stage, neglecting bias and coupling components.

Ans. −305.

21.11 If the reverse voltage-feedback ratio in Problem 21.10 is $h_{re} = 2 \times 10^{-4}$, draw the exact equivalent circuit for the second stage and solve for the voltage gain.

Ans. −288.

21.12 Design a single-stage transistor amplifier using an *n–p–n* transistor for which $h_{ie} = 2\,\text{k}\Omega$; $h_{fe} = 60$; $h_{oe} = 25\,\mu\text{S}$. The collector circuit is to take 1 mA from a 10 V supply. Find the voltage gain. If the stage feeds a second, similar stage, to what value does the voltage gain fall? Assume a base–emitter voltage of 0·7 V, and a collector–emitter voltage equal to the voltage drop across the collector load resistance. (Start by choosing a suitable quiescent emitter voltage.)

Ans. Typical gains, -120; -40, with emitter resistance of 1 kΩ.

21.13 A common-emitter amplifier stage is fed through a *CR* coupling from a similar previous stage. The transistors used have $h_{ie} = 1·5\,\text{k}\Omega\mu\text{S}$, $h_{fe} = 50$ and $h_{oe} = 40\,\mu\text{S}$. The collector-circuit load resistance is 5 kΩ. Determine the coupling capacitance required for a low-frequency cut-off at 20 Hz. Neglect bias circuits and emitter decoupling.

Ans. Minimum value, 1·84 μF.

21.14 A common-emitter amplifier has a collector load resistance of 10 kΩ. Assuming typical values of $h_{feo} = 50$, $h_{oe} = 25\,\mu\text{S}$, determine the mid-frequency current gain. If the amplifier is to have an upper cut-off frequency of 3 MHz, determine a minimum value for f_1 and hence find the current gain at 4·5 MHz.

Ans. -40; 153 MHz; 22/124°.

21.15 Two common-emitter stages with *CR* coupling use transistors for which $h_{ie} = 2\,\text{k}\Omega$, $h_{re} = 5 \times 10^{-4}$, $h_{fe} = 60$ and $h_{oe} = 25\,\mu\text{S}$. The collector resistances are each 5 kΩ, the source impedance is 2·5 kΩ and the load is 2 kΩ. Determine the voltage gain (V_oE) and the current gain (i) neglecting h_{re}, (ii) including h_{re}, where E is the source e.m.f.

Ans. (i) 760; 1,710, (i) 756; 1,700.

21.16 A common-base amplifier is used as a current buffer stage to feed a load of 20 kΩ from a current source of 0·15 mA which has an internal resistance 10 kΩ. The transistor used has $h_{ib} = 35\,\Omega$, $h_{ob} = 1\,\mu\text{S}$ and $h_{fb} = -0·98$. Determine the current in the 20 kΩ load if it is (*a*) connected directly to the current source, (*b*) used as the collector resistor of the common-base amplifier.

Ans. (*a*) 0·05 mA, (*b*) 0·147 mA.

21.17 Show that the two-transistor circuit of Fig. 21.16 is equivalent to a single transistor of equivalent overall gain $h_{fe'} = h_{fe1} + h_{fe2} + h_{fe1}h_{fe2}$ and input

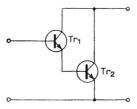

Fig. 21.16

impedance $h_{ie'} = h_{ie1} + (1 + h_{fe1})\,h_{ie2}$, where the subscript 1 refers to the first and the subscript 2 refers to the second transistor. (This is known as the *Darlington connexion*).

Chapter 22

FEEDBACK AMPLIFIERS

In the simple amplifiers discussed so far, the input signal current or voltage controls the output, without itself being affected by the output. It has been seen, however, that in the exact h-parameter equivalent circuit of a transistor, one of the terms (which has been neglected in the present treatment) represents an effect which the output voltage has on the input circuit. Also the emitter circuit impedance was shown to couple output and input circuits. This effect is called *feedback*, and by the intentional use of feedback the characteristics of amplifiers can be considerably modified.

If the feedback effect reduces the value of the overall gain, it is called *negative feedback*. The advantage of negative feedback is that it enables some of the output characteristics to be stabilized, i.e. to become less dependent on supply voltage and active device parameter changes. Thus negative voltage feedback (where the feedback signal is proportional to the output voltage) will reduce the voltage gain, but make it less liable to vary, and will also increase the amplifier voltage-gain bandwidth. Negative current feedback (where the feedback signal is proportional to the output current) will reduce the current gain, but will stabilize it and will increase the bandwidth. In both cases the distortion and noise will be reduced in the same ratio as the gain, so that the signal-to-noise ratio is unaltered, but it is, in fact, possible to increase this ratio because the active device may not be driven into a non-linear part of its characteristic when negative feedback is applied. In general, the selected gain characteristics become more dependent on circuit components than on active device parameters. Negative feedback will also affect amplifier input and output impedances, as will be seen. If the feedback effect causes an increase in the overall gain it is called *positive feedback*. In this case the amplifier stability suffers (i.e. the gain becomes more dependent on active device variations and supply voltage changes). If there is enough feedback the amplifier may become completely unstable and act as an oscillator. Often stray coupling (especially at high frequencies) can give unwanted positive feedback and result in undesired oscillation or instability of gain.

22.1 Series Voltage Feedback

Fig. 22.1 represents an amplifier having a constant voltage amplification A_v between the input and the output terminals, i.e.

$$A_v = \frac{V_o}{V'} \tag{22.1}$$

A fraction, β say, of the output voltage V_o is fed back and connected in series with the input voltage (e.g. by using a simple resistance

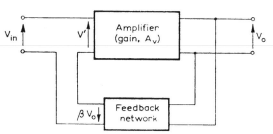

Fig. 22.1   SERIES VOLTAGE FEEDBACK IN GENERAL

voltage-divider). The effective input voltage to the amplifier, V', is then the complexor sum of the input voltage to the circuit, V_{in}, and the voltage which is fed back, βV_o:

$$V' = V_{in} + \beta V_o \tag{22.2}$$

Note that β can be a complex number if the feedback path contains reactive elements.

Substituting for V' in eqn. (22.1) gives

$$V_o = A_v(V_{in} + \beta V_o)$$

so that

$$V_o(1 - \beta A_v) = A_v V_{in}$$

and the overall gain, A_{vf}, with feedback is

$$A_{vf} = \frac{V_o}{V_{in}} = \frac{A_v}{1 - \beta A_v} \qquad (22.3)$$

If the magnitude of the denominator on the right-hand side of this equation is greater than unity the amplification is reduced and the feedback effect is negative, or degenerative. If the magnitude of the denominator is less than unity the amplification is increased and the feedback is positive, or regenerative.

Positive feedback apparently gives an attractive method of increasing the gain of an amplifier. It is seldom used in such a role because it leads to instability of amplification.

Note that if the feedback is taken over any odd number of common-emitter stages and is fed back in series aiding with the input then the overall gain at mid-frequencies, A_v, will be negative and β will be positive. The denominator of eqn. (22.3) will then be greater than unity, and the feedback will be negative. A similar result is obtained by taking the feedback over an even number of stages but feeding it in series opposition to the input. In this case A_v is positive but β is negative, again giving overall negative feedback.

Also if A_v is large so that $\beta A_v \gg 1$ then eqn. (22.3) reduces to

$$A_{vf} \approx -\frac{1}{\beta} \qquad (22.3a)$$

EXAMPLE 22.1 A 3-stage C.E. amplifier has a mid-frequency voltage gain, A_v of -10^5. If 2 per cent negative series voltage feedback is employed, determine the gain with feedback, A_{vf}. Also find the maximum percentage change in A_v so that A_{vf} will not decrease by more than 2 per cent.

From eqn. (22.3),

$$A_{vf} = \frac{-10^5}{1 + 0.02 \times 10^5} \approx \frac{-1}{0.02} = \underline{\underline{-50}}$$

Let A_v' be the value of A_v that results in a 2 per cent change in A_{vf}. Then

$$-0.98 \times 50 = \frac{A_v'}{1 - \beta A_v'} \qquad \text{or} \qquad -49 + 0.98A_v' = A_v'$$

whence

$$A_v' = -\frac{49}{0.02} = -2,450$$

so that the change in A_v is

$$\frac{10^5 - 2,450}{10^5} \times 100 = \underline{\underline{97.5 \text{ per cent}}}$$

22.2 Shunt Voltage Feedback

In shunt voltage feedback the feedback signal is applied in parallel with the input signal, and is proportional to the output voltage. The general circuit fed from a constant-voltage source is shown in

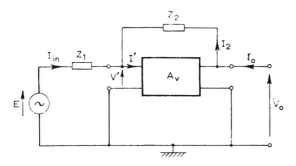

Fig. 22.2   SHUNT VOLTAGE FEEDBACK IN GENERAL

Fig. 22.2. In this circuit, feedback is achieved by adding the impedances Z_1 and Z_2. Then by Millman's theorem,

$$V' = \frac{E/Z_1 + V_0/Z_2}{1/Z_1 + 1/Z_2 + 1/R_{in}} = \frac{V_0}{A_v}$$

where R_{in} is the input resistance of the amplifier without feedback. Hence

$$\frac{E}{Z_1} + \frac{V_0}{Z_2} = \frac{V_0}{A_v}\frac{1}{Z_p}$$

where

$$Z_p = \frac{1}{\dfrac{1}{R_{in}} + \dfrac{1}{Z_1} + \dfrac{1}{Z_2}}$$

It follows that

$$\frac{V_0}{E} = \frac{A_v Z_p}{Z_1\left(1 - \dfrac{A_v Z_p}{Z_2}\right)}$$

Hence the overall voltage gain is

$$A_{vf} = \frac{V_0}{E} = \frac{\alpha A_v}{1 - \beta A_v} \tag{22.4}$$

where $\alpha = Z_p/Z_1$ and $\beta = Z_p/Z_2$.

This equation has the same form as eqn. (22.3). In the above Z_1 includes the source impedance. The derivation for a source of high impedance will be delayed until Section 22.7.

The current gain without feedback is I_o/I', and with feedback is I_o/I_{in}. It is shown in Section 22.7 that $I_{in} \gg I'$, so that shunt voltage feedback reduces the current gain.

Negative shunt voltage feedback is achieved automatically in a single-stage C.E. transistor amplifier when Z_1 and Z_2 are added, owing to the inherent 180° phase shift through the amplifier. It is also obtained if feedback is taken over any *odd* number of stages.

This type of amplifier is often referred to as a *virtual earth amplifier*, since if the gain of the amplifier is high, then for any normal output the input will be very small. The base of the first transistor must therefore be only a fraction of a volt above or below earth potential, and constitutes a *virtual earth* point (see Section 22.10).

22.3 Effect of Negative Feedback on Stability

If A_v (the gain of an amplifier without feedback) changes as the result of supply voltage changes or variation in the active device parameters then A_{vf} will also change, but by a much smaller fraction, i.e. the amplifier will have become more stable (see Example 22.1).

For example, consider an amplifier with shunt voltage feedback in which Z_2 and R_{in} are both very much greater than Z_1; then if $A_v = -100$, $\beta = 0.1$, $\alpha = 1$, gives eqn. (22.4).

$$A_{vf} = \frac{-100}{1 + 10} = -9.09$$

If A_v now falls to -90, the new value of overall gain is

$$A_{vf}' = \frac{-90}{1 + 9} = -9.00$$

This shows that in this case a 10 per cent change in A_v gives less than 1 per cent change in A_{vf}. The increase in stability is achieved at the expense of the corresponding reduction in gain. If the term βA_v is large, eqn. (22.4) becomes

$$A_{vf} \approx \frac{\alpha A_v}{-\beta A_v} = -\frac{\alpha}{\beta} \tag{22.5}$$

i.e. the voltage gain is independent of the active device parameters and the supply voltage provided only that the gain is large enough.

EXAMPLE 22.2 For the circuit shown in Fig. 22.3, determine the voltage gain V_o/V_1 and the quiescent d.c. potentials of the base, emitter and collector terminals. The transistor is a silicon *n–p–n* type for which $h_{ie} = 5\,\text{k}\Omega$, $h_{oe} = 40\,\mu\text{S}$

and $h_{fe} = 90$. The base-emitter voltage is $0.7\,V$ and $R_g = 50\,\Omega$; $R_1 = 5\,k\Omega$; $R_f = 250\,k\Omega$; $R_L = 7\,k\Omega$; and $R_E = 1.5\,k\Omega$. If the production spread of h_{fe} is $90 - 100$ determine the expected variation in gain.

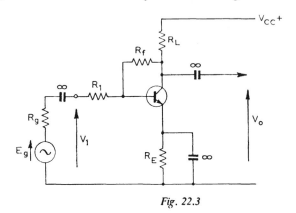

Fig. 22.3

In order to determine the d.c. operating potentials it may be assumed that the d.c. current gain, h_{FE}, is the same as the small signal value, h_{fe}, i.e. 90. Note that the feedback resistor, R_f, provides stabilization of the d.c. operating voltages, since its shunt feedback effect is operative on direct as well as alternating current. The direct currents through R_L and R_E are practically equal, so that, if this current is I,

$$V_{CE} = V_{CC} - IR_L - IR_E$$
$$V_{CB} = V_{CE} - 0.7$$

and

$$I_B = \frac{I}{1 + h_{fe}} = \frac{V_{CB}}{R_f}$$

Hence

$$\frac{IR_f}{1 + h_{fe}} = V_{CC} - IR_L - IR_E - 0.7$$

so that

$$I\left(\frac{R_f}{1 + h_{fe}} + R_L + R_E\right) = 9.3$$

and

$$I = 0.825\,mA$$

The various potentials can now be found.

Emitter potential $= 0.825 \times 1.5 \approx \underline{\underline{1.2\,V}}$

Base potential $= 1.2 + 0.7 \approx \underline{\underline{1.9\,V}}$

Collector potential $= 10 - (0.825 \times 7.0) \approx \underline{\underline{4.2\,V}}$

The small-signal voltage gain may be found by finding the gain without feedback from eqn. (21.13) and substituting this in eqn. (22.4). This is relatively straightforward when the signal source is of low impedance as in this case. Thus, neglecting R_g, the gain without feedback is

$$A_v = \frac{-h_{fe}R_L}{(1 + h_{oe}R_L)h_{ie}} = -1 \cdot 1 h_{fe}$$

Also $Z_p = h_{ie} \| R_f \| R_1 = 2 \cdot 5 \Omega$; $\alpha = Z_p/R_1 = 0 \cdot 5$; and $\beta = Z_p/Z_f = 0 \cdot 01$. The gain with feedback is therefore

$$A_{vf} = \frac{-0 \cdot 5 \times 1 \cdot 1 h_{fe}}{1 + (0 \cdot 01 \times 1 \cdot 1 h_{fe})}$$

For $h_{fe} = 90$, $A_{vf} = -25$, and for $h_{fe} = 100$, $A_{vf} = -26$.

In this case the gain of the single stage without feedback is fairly low, and the variation of the transistor parameters still has a significant (though smaller) effect on gain with feedback. Frequently the required characteristic is obtained by using a three-stage amplifier with overall feedback from the last to the first stage.

22.4 Effect of Negative Feedback on Bandwidth

The expression for the gain of a single-stage C.E. transistor amplifier is obtained from eqn. (21.39) as

$$A_v = \frac{A_i R_L}{R_{in}} = \frac{-h_{feo}R_L}{(1 + h_{oe}R_L)(1 + jf/f_\beta)R_{in}} = \frac{A_{vo}}{1 + jf/f_\beta}$$

If negative feedback is applied, with feedback factor β, then the general expression for the gain with feedback is

$$A_{vf} = \frac{A_v}{1 - \beta A_v}$$

which at mid-frequencies becomes

$$A_{vf} = \frac{A_{vo}}{1 - \beta A_{vo}}$$

At a high frequency, f, the overall gain is

$$A_{vf}' = \frac{A_{vo}/(1 + jf/f_\beta)}{1 - \beta A_{vo}/(1 + jf/f_\beta)}$$

where f_β is the upper 3 dB frequency of the amplifier without feedback. Simplifying,

$$A_{vf}' = \frac{A_{vo}}{1 - \beta A_{vo} + jf/f_\beta} = \frac{A_{vo}}{(1 - \beta A_{vo})\left(1 + j\dfrac{f}{f_\beta(1 - \beta A_{vo})}\right)}$$

$$(22.6)$$

This is $3\,dB$ down on the mid-frequency value when

$$\frac{f}{f_\beta(1 - \beta A_{vo})} = 1$$

i.e. when

$$f = f_\beta(1 - \beta A_{vo}) \tag{22.7}$$

where the product βA_{vo} is negative with negative feedback, so that $f > f_\beta$.

This shows that the "voltage-gain" bandwidth of the amplifier (i.e. the bandwidth between $3\,dB$ points on the voltage-gain/frequency characteristic) has been increased by a factor $(1 + \beta A_{vo})$, again at the expense of a corresponding reduction in voltage gain. If the feedback reduces the current gain, similar considerations show that "current-gain" bandwidth is increased in the same proportion. Note that the gain-bandwidth product remains constant for a single-stage amplifier since bandwidth is increased by the same factor as the gain is reduced by the feedback. This simple relationship does not apply to multi-stage amplifiers.

EXAMPLE 22.3 A single-stage amplifier has a voltage gain of $-100 + j0$ at $1\,kHz$ and $-10 + j30$ at $50\,kHz$. Find the upper $3\,dB$ cut-off frequency. If 20 per cent negative series voltage feedback is now introduced determine the gain at (i) $1\,kHz$, and (ii) $60\,kHz$.

Without feedback the gain at frequency f is

$$A_v = \frac{-100}{1 + jf/f_\beta}$$

Hence at $50\,kHz$,

$$-10 + j30 = \frac{-100}{(1 + j50 \times 10^3/f_\beta)}$$

so that

$$(-10 + j30)(1 + j50 \times 10^3/f_\beta) = -100$$

i.e.

$$-10 - 15 \times 10^5/f_\beta + j30 - j50 \times 10^4/f_\beta = -100$$

Equating reference terms,

$$\frac{15 \times 10^5}{f_\beta} = 90$$

so that $f_\beta = 16 \cdot 7\,kHz$.

As a check, the quadrate terms give $f_\beta = 50 \times 10^4/30 = 16 \cdot 7\,kHz$.
With 20 per cent negative feedback, $\beta = 0 \cdot 2$ and at $1\,kHz$ the gain is

$$A_{vf} = \frac{-100}{1 + 20} = -4 \cdot 8$$

At 60 kHz the gain is, from eqn. (22.6),

$$A_{vf}' = \frac{-4\cdot8}{1 + j\dfrac{60}{16\cdot7} \times \dfrac{1}{21}} = \frac{-4\cdot8}{1 + j0\cdot17} = \underline{\underline{4\cdot7/\underline{170^\circ}}}$$

The upper 3 dB frequency limit with feedback is, of course, $f_\beta(1 + 20) = \underline{\underline{350 \, \text{kHz}}}$.

22.5 Effect of Negative Feedback on Distortion and Noise

Distortion occurs in amplifiers when the signal is so large that the active device operates outside the linear range of its characteristics. The output then contains components which are not present at the input. These components are harmonics of the signal frequencies. Generally the last stage of an amplifier is the one in which most harmonic distortion occurs.

Noise also will occur in amplifiers, where "noise" is taken to mean unwanted voltages appearing at the output. These noise voltages may be due to (*a*) poor smoothing in the rectifier supply, (*b*) 50 Hz or 100 Hz hum pick-up, (*c*) mechanical vibrations of valve electrodes (*microphony*), (*d*) drift in transistors due to changes in temperature, (*e*) thermal noise in resistors due to molecular vibrations, (*f*) *shot noise* in the active device (due to the fact that the charge carriers are emitted randomly). Generally the first stage of an amplifier determines its noise performance, since it is in this stage that the signal is smallest, and hence the signal/noise ratio is also smallest. The size of the input signal which can be usefully amplified is determined largely by the noise in the first stage, since for any useful gain it must be possible to detect the signal in the presence of the noise. Hence generally the minimum input signal must be greater than the noise. When low signal levels are to be amplified special low-noise input circuits must be employed.

An important amplifier criterion is the ratio of signal/noise power at the input to that at the output. This is called the *noise factor*, F (in decibels it is $10 \log_{10} F$). Thus for a signal power S_i at the input and a power gain G in the amplifier the output signal power is GS_i. The input noise power N_i is also amplified and added to by the internal noise so that the output noise power, N_o, will be greater than GN_i. Then

$$F = \frac{S_i/N_i}{GS_i/N_o} = \frac{N_o}{GN_i} \tag{22.8}$$

Noise and distortion generated within an amplifier can be represented by a constant-voltage generator having an e.m.f. e_n, in series with the output as shown in the series-feedback circuit of Fig. 22.4. This e.m.f. can be assumed constant if the output signal is maintained at the level which it had without feedback (i.e. by increasing the input voltage).

Without feedback the harmonic or noise voltage output is e_n. Let this become $m e_n$ when feedback is present, where m is a fraction.

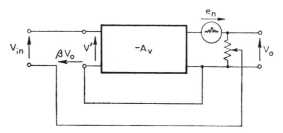

Fig. 22.4   FEEDBACK OF DISTORTION AND NOISE

Then the fraction of harmonic or noise voltage which is fed back is $\beta m e_n$. This is amplified in the amplifier (of gain $-A_v$) to give an output harmonic or noise voltage of $-A_v \beta m e_n$, so that

$$m e_n = e_n - A_v \beta m e_n$$

Hence

$$m = \frac{1}{1 + \beta A_v} \qquad (22.9)$$

where negative feedback has been assumed. This shows that the noise or distortion is reduced in the same ratio as the gain. Note that adding feedback alone does not improve the signal/noise ratio at the amplifier output. Indeed, since noise and harmonic distortion voltages may well cover a wider bandwidth than the signal, they may be present at the lower and upper frequency ranges where the phase shift of a multi-stage amplifier is such that the feedback becomes positive. In this case the noise may be increased by the feedback, and careful design of the frequency characteristics of the feedback path is essential to prevent this. Note that the same arguments apply to drift in d.c. amplifiers—i.e. the percentage drift is not reduced by negative feedback.

Generally feedback in the output stage of an amplifier can be used to reduce distortion, because extra stages are inserted before the feedback stage in order to restore the overall gain. Since these are low-signal-level stages they will hardly contribute any distortion themselves, and hence the overall distortion will be reduced, for the same output power.

Consideration of the shunt voltage feedback circuit gives the same results as those derived above.

22.6 Current Feedback

SERIES CURRENT FEEDBACK

The simplest way to obtain series current feedback (in which the feedback signal is proportional to the load current) is to have an unbypassed resistor in the emitter or cathode circuit. Then the

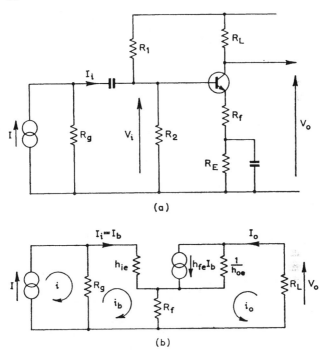

(a)

(b)

Fig. 22.5 SERIES CURRENT FEEDBACK

output current flowing through this resistor gives rise to a voltage (proportional to current) in series with the input. A typical transistor circuit with series current feedback is shown in Fig. 22.5(a).

It will be seen that the voltage drop across the unbypassed portion of the emitter-circuit resistor is common to both the input and output circuits, and hence causes feedback which an inspection of relative current directions shows is negative. Neglecting the effect of $R_p \, (=R_1 \| R_2)$ for simplicity, the small-signal equivalent circuit becomes that shown at (b). Using the mesh equations (with small letters to represent the complexor mesh currents),

Input mesh $0 = (R_g + h_{te} + R_f)i_b - R_g i + R_f i_o$ (22.10)

Output mesh $0 = \left(R_L + \dfrac{1}{h_{oe}} + R_f\right) i_o + R_f i_b - \dfrac{h_{fe}i_b}{h_{oe}}$ (22.11)

Since $i_b \ll i_o$, eqn. (22.11) can be simplified by neglecting $R_f i_b$ which is small compared to $R_f i_o$, so that

$$\frac{h_{fe}i_b}{h_{oe}} \approx \left(R_f + R_L + \frac{1}{h_{oe}}\right) i_o$$

For the mesh currents shown, the conductor currents I_b and I_o are equal to the corresponding mesh currents i_b and i_o, so that the internal current gain with feedback is

$$A_{if} = \frac{V_o/R_L}{I_b} = \frac{-I_o}{I_b} = \frac{-h_{fe}}{1 + h_{oe}(R_L + R_f)}$$ (22.12)

If $R_f \ll R_L$, the current gain is thus hardly affected by the feedback (without feedback $A_i = -h_{fe}/(1 + h_{oe}R_L)$).

The voltage gain with feedback is given by

$$A_{vf} = \frac{V_o}{V_i} = \frac{-I_o R_L}{I_b(h_{ie} + R_f) + I_o R_f} \approx \frac{-I_o R_L}{I_b h_{ie} + I_o R_f} \quad (\text{since } I_b \ll I_o)$$

$$= \frac{-I_o R_L}{I_b h_{ie} + \dfrac{I_b h_{fe} R_f}{1 + h_{oe}(R_L + R_f)}}$$

$$= \frac{-I_o R_L}{I_b h_{ie}\left\{1 + \dfrac{h_{fe} R_f}{h_{ie}(1 + h_{oe}(R_L + R_f))}\right\}} = \frac{A_v}{1 - \beta A_v}$$ (22.13)

where

$$\beta = \frac{R_f}{R_L} \frac{1 + h_{oe}R_L}{1 + h_{oe}(R_L + R_f)}$$

$$\approx \frac{R_f}{R_L} \quad \text{and} \quad A_v = \frac{-h_{fe}R_L}{h_{ie}(1 + h_{oe}R_L)}$$

is the voltage gain without feedback. Since eqn. (22.13) has the same form as eqn. (22.5), the same considerations regarding stability and bandwidth will apply.

SHUNT CURRENT FEEDBACK

With negative shunt current feedback the signal that is fed back to the input is proportional to the output current, and is fed back in parallel with the input. One possible circuit is shown in Fig. 22.6(*a*), where the effective a.c. circuit elements only are shown.

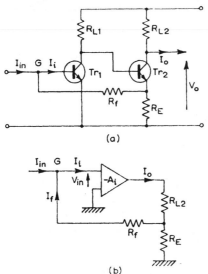

(a)

(b)

Fig. 22.6 SHUNT CURRENT FEEDBACK

In this case it may be shown that the current gain I_o/I_{in} is reduced by a factor of $(1 - \beta A_1)$, where $\beta = R_E/R_f$ and A_1 is the magnitude of the current gain over the two stages without feedback. The input impedance is reduced by the same factor, and the output impedance is increased.

For the general circuit shown at (*b*) the current gain without feedback is the ratio of output current I_o to input current I_i, i.e.

$$A_i = \frac{I_o}{I_i} = \frac{I_o}{I_{in}} \qquad \text{where } I_i = I_{in} = \text{source current}$$

When shunt current feedback is applied the current gain (assuming the same source current, I_{in}) becomes

$$A_{if} = \frac{I_{of}}{I_{in}} \qquad \text{where } I_{of} \text{ is the output current with feedback}$$

In this case the input current to the first transistor is

$$I_{if} = I_{in} + I_f = \frac{I_{of}}{A}$$

Since G is a virtual earth,

$$I_f \approx \frac{I_{of}R_E}{R_E + R_f} = \frac{I_{of}R_E}{R_f} \approx \beta I_{of}$$

provided that $R_f \gg R_E$ and where $\beta = R_E/R_f$. Hence

$$I_{in} + \beta I_{of} = \frac{I_{of}}{A_i}$$

so that

$$A_{if} = \frac{I_{of}}{I_{in}} = \frac{A_i}{1 - \beta A_i} \tag{22.14}$$

In the absence of feedback the input impedance of the circuit is

$$Z_{in} = \frac{V_{in}}{I_{in}} = \frac{V_{in}}{I_i} = \frac{A_i V_{in}}{I_o}$$

Assume now that feedback is applied and that the input current is increased until the output current has the same value as without feedback. The input current to the first transistor will be $I_i = I_o/A_i$ and the input voltage will be V_{in}—the same values as without feedback. The source current I_{in} must now be changed to

$$I_{in\,f} = I_i - I_f = I_i - \beta I_o = I_i(1 - \beta A_i)$$

so that the input impedance with feedback becomes

$$Z_{in\,f} = \frac{V_{in}}{I_{in\,f}} = \frac{V_{in}}{I_i(1 - \beta A_i)} = \frac{Z_{in}}{(1 - \beta A_i)} \tag{22.15}$$

The input impedance is thus changed in the same ratio as the gain. For negative feedback the input impedance is *reduced* in the same ratio as the gain. Note that for positive feedback the input impedance will be *increased* and may become negative after passing through the value infinity.

22.7 Effect of Feedback on Impedance Levels

It has been seen in the last section that negative shunt current feedback reduces the input impedance to an amplifier circuit. In general, feedback can affect both the input and output impedances of amplifiers. The way in which the feedback is introduced into the input circuit will affect the input impedance (i.e. the ratio of input voltage to input current). Thus, for series negative feedback, the input voltage required for a given input current must be increased—i.e. the input impedance is increased. For shunt negative feedback the input current can divide between the amplifier and the feedback path, and hence input impedance is reduced. Both of these effects are reversed if the feedback is positive, and indeed in this case the input resistance can become negative.

The effect of feedback on output impedance depends upon whether the feedback signal is proportional to output voltage or output current. In general voltage feedback reduces the output impedance while current feedback increases it.

Output impedance can be obtained by inspection of the voltage gain expression for an amplifier. Thus if the voltage gain is written in the form

$$A_v = \frac{PZ_L}{Q + Z_L} \tag{22.16}$$

where Z_L is the load, then P is the intrinsic voltage gain as $Z_L \to \infty$, and Q is the output impedance. The Thévenin equivalent output circuit is as shown in Fig. 22.7(a). The Norton equivalent circuit

$$(a) \qquad\qquad\qquad (b)$$

Fig. 22.7 EQUIVALENT OUTPUT CIRCUITS OF AN AMPLIFIER

is shown at (b). The output impedance seen by any other load which may be connected across Z_L is then simply Q in parallel with Z_L.

An alternative method of finding the output impedance is to apply a constant-voltage or a constant-current generator to the output terminals with the input signal source represented by its internal impedance. The output impedance is then the ratio of voltage at the output to the current flowing into the output terminals.

SERIES VOLTAGE FEEDBACK

In the series voltage feedback circuit of Fig. 22.8(a), let the amplifier (which will usually be multistage) have an intrinsic gain A_v and an output impedance of Z_o without feedback. A constant-voltage source V is applied across the output terminals, and the input

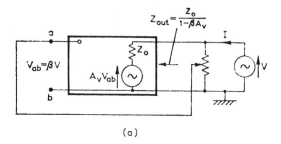

(a)

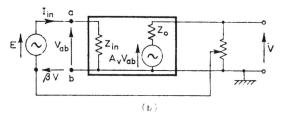

(b)

Fig. 22.8   IMPEDANCE LEVELS FOR SERIES VOLTAGE FEEDBACK

terminals are short-circuited. A fraction, β, of the voltage V at the output is fed back to the input. Then the voltage V_{ab} at the input is βV, and neglecting the impedance of the feedback network (which must be purely resistive for pure voltage feedback), the current, I, from the source is

$$I = \frac{V - A_v V_{ab}}{Z_0} = \frac{V - \beta V A_v}{Z_0}$$

Hence the output impedance is

$$Z_{out} = \frac{V}{I} = \frac{Z_0}{1 - \beta A_v} \tag{22.17}$$

If the feedback is negative the output impedance is smaller when feedback is applied than without feedback. Note that if A_v is positive the feedback is still negative if β is negative—this normally requires a transformer coupling.

The input impedance with feedback is obtained by considering the circuit at (b). A signal is applied to the input, and the output is open-circuited. Then neglecting the impedance of the feedback network, the feedback voltage is $\beta V = \beta V_{ab}A_v$ and the new input impedance is

$$Z_{inf} = \frac{E}{I_{in}} = \frac{V_{ab} - \beta V}{I_{in}} = \frac{V_{ab}(1 - \beta A_v)}{I_{in}}$$

$$= Z_{in}(1 - \beta A_v) \tag{22.18}$$

where Z_{in} is the input impedance without feedback. For negative feedback βA_v is negative, and the input impedance is therefore increased.

EXAMPLE 22.4 A transistor amplifier uses a transistor for which $h_{ie} = 1\,k\Omega$ in the first stage, which is in common-emitter connexion. The overall current gain without feedback is $-15,000$. If there is $0.05\,p.u.$ negative series voltage feedback, determine the input impedance of the feedback amplifier if the load resistance is $75\,\Omega$. If the output impedance without feedback is $20\,k\Omega$, determine the ouput impedance when feedback is applied.

The input resistance, R_{in}, without feedback will be almost equal to h_{ie} (neglecting the d.c. bias resistors).

The overall voltage gain without feedback is

$$A_v = \frac{A_i \times R_L}{R_{in}} = \frac{-15,000 \times 75}{1,000} = -1,125 \quad \text{(since } R_{in} \approx h_{ie}\text{)}$$

Hence

$$Z_{inf} = R_{in}(1 - \beta A_v) \quad \text{(eqn. (22.18))}$$
$$= 1,000(1 + 56\cdot3) = \underline{57\cdot3\,k\Omega}$$

From eqn. (22.17) the output impedance with feedback applied is

$$Z_{out\,f} = \frac{Z_o}{1 - \beta A_v} = \frac{20,000}{57\cdot3} = \underline{\underline{350\,\Omega}}$$

SHUNT VOLTAGE FEEDBACK

The input impedance of the shunt voltage feedback circuit shown in Fig. 22.9 is

$$Z_{inf} = \frac{E}{I_{in}} = Z_1 + Z_{GE}$$

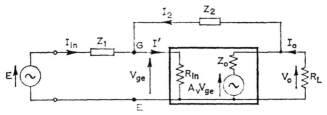

Fig. 22.9 MULTISTAGE AMPLIFIER WITH SHUNT VOLTAGE FEEDBACK

where $Z_{GE} = V_{GE}/(I' - I_2)$. The admittance looking into the amplifier at G is

$$
\begin{aligned}
Y_{GE} &= \frac{I' - I_2}{V_{GE}} = \frac{1}{R_{in}} - \frac{I_2}{V_{GE}} \\
&= \frac{1}{R_{in}} - \frac{(A_v V_{GE} - V_{GE})}{Z_2 V_{GE}} \\
&= \frac{1}{R_{in}} + \frac{1 - A_v}{Z_2}
\end{aligned}
\tag{22.19}
$$

where $A_v = V_o/V_{GE}$.

This represents the admittance of a circuit consisting of a resistance R_{in} in parallel with an impedance $Z_2/(1 - A_v)$. Hence the impedance at G is decreased by the negative feedback. The input impedance is thus $Z_1 + 1/Y_{GE}$. Note that if A_v is large, the impedance of the feedback path through Z_2 as seen looking in at point G is approximately $Z_2/(-A_v)$, which will generally be small.

The gain of the amplifier fed from a high-impedance source can now be found. The circuit is that of Fig. 22.9, but Z_1 is now the source impedance. The voltage gain $A_v (= V_o/V_{GE})$ is almost unaltered by the feedback, provided that $I_2 \ll I_o$, since the amplifier input terminals are now G and E.

With feedback applied the input current, I_{in}, divides between R_{in} and the equivalent impedance of Z_2 as seen looking into point G (i.e. $Z_2/(1 - A_v)$, from eqn. 22.19). The current gain is now

$$
\begin{aligned}
A_{if} &= \frac{I_o}{I_{in}} = \frac{I_o}{I'} \frac{I'}{I_{in}} \\
&= A_i \frac{Z_2}{(1 - A_v)} \frac{1}{R_{in} + Z_2/(1 - A_v)} \\
&= \frac{A_i}{\dfrac{R_{in}}{Z_2}(1 - A_v) + 1} \\
&\approx \frac{A_i}{\beta A_v} \quad \text{(assuming } A_v \gg 1\text{)}
\end{aligned}
\tag{22.20}
$$

and where $\beta = R_{in}/Z_2$.

The current gain thus is reduced by the negative feedback.

MILLER EFFECT

It should be noted that shunt voltage feedback may occur unintentionally in single-stage amplifiers owing to the stray capacitance between input and output circuits providing the feedback impedance, Z_2. This stray capacitance is called the *Miller capacitance*. In this case the impedance Z_2 is $1/j\omega C$, where C is the Miller capacitance. The admittance of the feedback circuit seen at the input to the amplifier is, from eqn. (22.19),

$$Y_f = \frac{1 - A_v}{Z_2} = j\omega C(1 - A_v) \qquad (22.21)$$

Three special cases are of interest. First, if A_v is a negative reference quantity, then Y_f is a capacitive susceptance of magnitude $\omega C(1 + A_v)$. Second, if A_v has a phase angle between 90° and 180° so that it can be represented as $A_v = -P + jQ$, then

$$Y_f = j\omega C(1 + P - jQ) = \omega C Q + j\omega C(1 + P) \qquad (22.22)$$

i.e. it consists of a conductance (frequency dependent), $\omega C Q$, in parallel with a capacitive susceptance, $\omega C(1 + P)$. Third, if A_v has a phase angle between 180° and 270° (e.g. in a single stage with an inductive load such as a tuned circuit operating below resonance), so that it can be represented as $A_v = -P - jQ$, then

$$Y_f = j\omega C(1 + P + jQ) = -\omega C Q + j\omega C(1 + P) \qquad (22.23)$$

i.e. there is a negative component of input conductance. If this is larger than the conductance of the source the amplifier will be unstable.

EXAMPLE 22.5 In the transistor amplifier shown in Fig. 22.10(a), find expressions for the input admittance and the output impedance. There are stray capacitances C_{cb} between collector and base and C_{be} between base and earth, and the collector load is an inductance $j\omega L$.

The small-signal equivalent circuit is shown at (b) neglecting R_4 and R_2 and assuming that the emitter resistor, R_3, is adequately decoupled. From this, the current gain without feedback is

$$A_i = \frac{-h_{fe}}{1 + h_{oe}Z_L} \quad \text{where } Z_L = j\omega L$$

Hence the voltage gain without feedback is

$$A_v = \frac{-h_{fe}Z_L}{h_{ie}(1 + h_{oe}Z_L)} \approx \frac{-h_{fe}j\omega L}{h_{ie}}$$

so that the input admittance with feedback is (from eqn. (22.21))

$$Y_{inf} = Y_{in} + Y_f = j\omega C_{be} + \frac{1}{h_{ie}} + j\omega C_{cb}\left(1 + \frac{jh_{fe}\omega L}{h_{ie}}\right)$$

$$= \frac{-\omega^2 L C_{cb} h_{fe}}{h_{ie}} + \frac{1}{h_{ie}} + j\omega(C_{be} + C_{cb})$$

where $Y_{in} = j\omega C_{be} + 1/h_{ie} = $ input admittance without feedback.

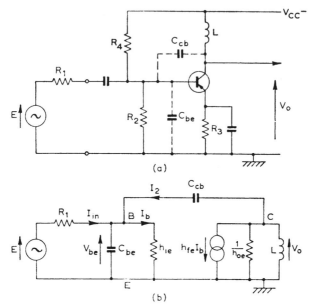

Fig. 22.10   MILLER CAPACITANCE

For a low-impedance source ($R_1 \ll h_{ie}$), the output impedance, obtained by replacing the load L by a constant-voltage source, V, and representing the input generator by its internal impedance, R_1, is very nearly $1/h_{oe}$ in parallel with $1/j\omega C_{cb}$.

For a high-impedance source ($R_1 \gg h_{ie}$) and neglecting C_{be}, the same procedure yields an alternating input current to the transistor of approximately $Vj\omega C_{cb}$ (assuming that $j\omega C_{cb} \gg 1/h_{ie}$). Hence at point C in Fig. 22.10(*b*) the currents are as follows:

$I_2 \approx Vj\omega C_{cb}$
Current in $h_{oe} = Vh_{oe}$
Current in constant-current generator, $h_{fe}I_b \approx h_{fe}I_2$

The sum of these gives the current, I, from the constant-voltage source which replaces L. Hence

$I = Vj\omega C_{cb} + Vh_{oe} + h_{fe}Vj\omega C_{cb}$,

so that the output impedance with feedback is

$$Z_{out} = \frac{V}{I} \approx \frac{1}{h_{oe} + j\omega C_{cb}(1 + h_{fe})}$$

which is even smaller than in the case of a low value of R_1.

SERIES CURRENT FEEDBACK

With this form of feedback the calculation of output impedance follows closely that used for series voltage feedback, but here a constant-current generator delivering a current I is connected across the output as shown in Fig. 22.11.

The voltage across R_f is IR_f and this is fed back to the input to give an input voltage of $-IR_f$. If the intrinsic voltage gain is A_v, then from the diagram,

$$V = I(Z_0 + R_f) + A_vV_1 = I(Z_0 + R_f) - IR_fA_v$$

Hence the output impedance with feedback is

$$Z_{out} = \frac{V}{I} = Z_0 + (1 - A_v)R_f \qquad (22.24)$$

For negative feedback A_v is negative, and hence the output impedance is greater with feedback than it is without it.

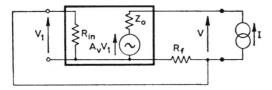

Fig. 22.11 TO FIND OUTPUT IMPEDANCE WITH SERIES CURRENT FEEDBACK

The input impedance of the amplifier can be found as before and will be higher with feedback than without it.

EXAMPLE 22.6 In the transistor series current-feedback amplifier of Fig. 22.5, $R_f = 100\,\Omega$; $R_L = 2,000\,\Omega$; $h_{ie} = 1\cdot5\,\mathrm{k}\Omega$; $h_{fe} = 80$; and $h_{oe} = 10^{-4}\,\mathrm{S}$. Determine approximately the input impedance, the voltage gain, and the output impedance.

Referring to Fig. 22.5(*b*), the input impedance with feedback is

$$Z_{inf} = \frac{V_i}{I_b} \approx \frac{I_b h_{ie} + I_0 R_f}{I_b} \quad \text{(assuming } I_b \ll I_0)$$

$$= \frac{I_b h_{ie} - A_i I_b R_f}{I_b}$$

$$= h_{ie} - A_i R_f$$

where A_i = current gain without feedback = $\dfrac{-h_{fe}}{1 + h_{oe}R_L}$.

Hence

$$Z_{inf} = h_{ie}(1 - \beta A_v)$$

where $\beta = R_f/R_L$ and $A_v = A_iR_L/R_{in} = A_iR_L/h_{ie}$. Note that this expression for impedance is in the same form as that obtained in eqn. (22.18) for series voltage feedback.

Inserting numerical values,

$$Z_{inf} = 1,500 + \frac{80 \times 100}{1 + 10^{-4} \times 2 \times 10^3} = \underline{\underline{8,200\,\Omega}}$$

From eqn. (22.13) the voltage gain with feedback is

$$A_{vf} = \frac{A_v}{1 - \beta A_v} = \frac{-h_{fe}R_L/h_{ie}(1 + h_{oe}R_L)}{1 + R_f h_{fe}/h_{ie}(1 + h_{oe}R_L)}$$

$$= \frac{-h_{fe}R_L}{h_{ie}\left(1 + h_{oe}R_L + \dfrac{R_f h_{fe}}{h_{ie}}\right)}$$

$$= \frac{-(h_{fe}/h_{ie}h_{oe})R_L}{\dfrac{1}{h_{oe}} + \dfrac{R_f h_{fe}}{h_{oe}h_{ie}} + R_L} \tag{i}$$

Comparing this with eqn. (22.16), it will be seen that the output impedance is

$$Z_{out} = \frac{1}{h_{oe}} + \frac{R_f h_{fe}}{h_{oe}h_{ie}}$$

$$= 10^4 + \frac{10^4 \times 100 \times 80}{1,500} = \underline{\underline{63\,k\Omega}}$$

This is the same result as would be obtained by the direct application of eqn. (22.24).

Inserting numerical values in (i), the voltage gain with feedback is

$$A_{vf} = \underline{\underline{-16 \cdot 5}}$$

22.8 Compound Feedback

In some instances it may be desired to achieve the increased stability that negative feedback gives together with definite levels of input

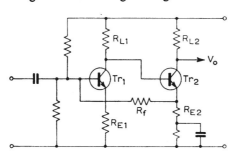

Fig. 22.12   TYPICAL COMPOUND FEEDBACK CIRCUIT

and output impedance. In such cases it may be necessary to use more than one feedback path. This is called *compound feedback*.

A typical circuit is shown in Fig. 22.12. The resistor R_{E1} gives negative series current feedback for the first stage (giving an increase in input impedance of the stage), and R_{E2} and R_f provide negative shunt current feedback over both stages (giving a decrease in input impedance). By suitable choice of components the input and output impedance levels may be made to have designed values.

An alternative arrangement is to remove R_{E2} and connect R_f between the collector of Tr_1 and the emitter of Tr_2. Note, however, that such a connexion no longer provides shunt-applied feedback.

EXAMPLE 22.7 In the circuit of Fig. 22.12 the value of R_{E1} is chosen to give 0·02 p.u. negative feedback in the first stage, and the values of R_f and R_{E2} give 0.01 p.u. negative feedback over the two stages. If h_{ie} for the first transistor is 1·2 kΩ determine the new value of input impedance. The loaded voltage gain of the first stage is $A_{v1} = -100$, and that of the second stage is $A_{v2} = -120$, both without feedback.

The resistor R_{E1} in the first stage gives negative series current feedback. This increases the input impedance of the first stage from $R_{in} \approx h_{ie}$ to

$$Z_{in} \approx h_{ie}(1 - \beta A_{v1})$$

The gain of the first stage with the series current feedback falls to

$$A_{v1f} = \frac{A_{v1}}{1 - \beta_1 A_{v1}} = \frac{-100}{1 + 2} = -33·3$$

Hence the overall gain of both stages with feedback in the first stage only is

$$A_v' = A_{v2} \times A_{v1f} = (-120)(-33·3) = 4,000$$

The new overall input impedance is reduced by the second form of feedback from $h_{ie}(1 - \beta A_{v1})$ to

$$Z_{inf} = \frac{h_{ie}(1 - \beta_1 A_{v1})}{1 - \beta_2 A_v'} = \frac{1,200\,(1 + 2)}{1 + 40} = \underline{\underline{88\,\Omega}}$$

22.9 The Emitter Follower (Common Collector)

The *emitter follower* circuit employs 100 per cent negative series voltage feedback, and therefore from Section 22.7(*a*) it will have a high input and a low output impedance. The voltage gain will be shown to be slightly less than unity, with no phase reversal, and the circuit is frequently used as a buffer stage between a high-impedance source and a low-impedance load. The circuit is shown in Fig. 22.13(*a*) and the small-signal a.c. equivalent circuit at (*b*). The circuit is often called the *common-collector circuit*, since the collector terminal is common to input and output circuits (through the low impedance of the d.c. supply).

The d.c. conditions are set by the resistors R_1 and R. Thus for a quiescent collector current I_C ($\approx I_E$) the emitter voltage is $I_C R$ and

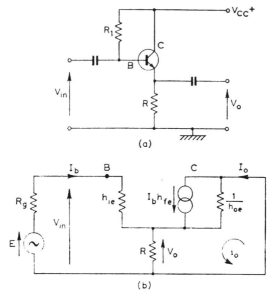

Fig. 22.13 EMITTER FOLLOWER

the base voltage is $I_C R + V_{BE}$, so that for a forward d.c. short-circuit current gain h_{FE} (to give $I_B = I_C/h_{FE}$), the value of R_1 is

$$R_1 = \frac{(V_{CC} - I_C R - V_{BE})h_{FE}}{I_C}$$

For the signal, the output mesh equation can be written

$$0 = \left(\frac{1}{h_{oe}} + R\right) I_o - \frac{h_{fe}}{h_{oe}} I_b \quad \text{(assuming } I_o R \gg I_b R)$$

so that

$$\frac{I_o}{I_b} = \frac{h_{fe}}{1 + h_{oe}R} = A_i \tag{22.25}$$

The current gain is thus the same as that derived for the common-emitter amplifier using the simplified equivalent circuit.

In the input mesh, the input signal voltage is

$$V_{in} = I_b h_{ie} + I_o R$$
$$= I_o \left\{ \frac{(1 + h_{oe}R)h_{ie}}{h_{fe}} + R \right\}$$
$$= \frac{V_o}{R} \left\{ \frac{(1 + h_{oe}R)h_{ie}}{h_{fe}} + R \right\}$$

Hence the voltage gain is

$$A_v = \frac{V_o}{V_{in}} = \frac{h_{fe}R}{h_{ie} + (h_{fe} + h_{oe}h_{ie})R} \qquad (22.26)$$

Note that

$$\frac{V_o}{E} = \frac{h_{fe}R}{h_{fe}R + (1 + h_{oe}R)(h_{ie} + R_g)} \qquad (22.27)$$

There is no phase reversal, and A_v is obviously smaller than unity.
The input impedance is readily obtained from the relation

$$V_{in} = I_b h_{ie} + I_o R$$

$$= I_b \left(h_{ie} + \frac{h_{fe}R}{1 + h_{oe}R} \right)$$

so that the input impedance is

$$Z_{in} = h_{ie} + \frac{h_{fe}R}{1 + h_{oe}R} \qquad (22.28)$$

The simplest way of determining the output impedance is by
inspection of eqn. (22.27) and comparison with eqn. (22.16). Thus
the ratio of V_o to E can be written

$$\frac{V_o}{E} = \frac{h_{fe}R}{\{h_{fe} + h_{oe}(h_{ie} + R_g)\} \left\{ \dfrac{h_{ie} + R_g}{h_{fe} + h_{oe}(h_{ie} + R_g)} + R \right\}}$$

It follows that the output impedance with feedback is

$$Z_{out} = \frac{h_{ie} + R_g}{h_{fe} + h_{oe}(h_{ie} + R_g)} \qquad (22.29)$$

and the output admittance is

$$Y_{out} = h_{oe} + \frac{h_{fe}}{h_{ie} + R_g} \qquad (22.29a)$$

For a constant-voltage source ($R_g \to 0$) Y_{out} becomes h_{fe}/h_{ie},
while for a constant-current source ($R_g \to \infty$), $Y_{out} \to h_{oe}$.
The frequency characteristic of the emitter-follower amplifier is
obtained by substituting $h_{fe} = h_{feo}/(1 + jf/f_\beta)$ (i.e. from eqn. 21.38)
in a simplified form of eqn. (22.26). Thus

$$A_v \approx \frac{h_{fe}R}{h_{ie} + h_{fe}R} \quad \text{(assuming } h_{oe}h_{ie} \ll h_{fe})$$

i.e.

$$A_v = \frac{h_{feo}R}{h_{ie}(1 + jf/f_\beta) + h_{feo}R}$$

This is 3 dB down on the mid-frequency gain when

$$\frac{h_{ief}}{f_\beta} = h_{ie} + h_{feo}R$$

i.e. when

$$f = f_\beta\left(1 + \frac{h_{feo}R}{h_{ie}}\right) \approx h_{feo}f_\beta \approx f_1 \qquad (22.30)$$

assuming R is of the same order as h_{ie}.

22.10 The Operational Amplifier

The shunt voltage feedback amplifier can be used to perform certain linear and non-linear "mathematical" operations (e.g. adding,

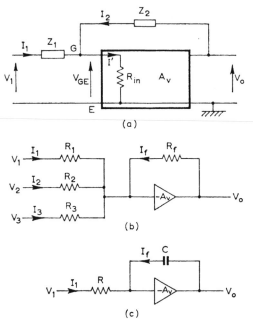

Fig. 22.14 OPERATIONAL AMPLIFIER

integrating and inverting) provided that the gain, A_v, of the amplifier without feedback is negative and very high (generally at least -10^5 and usually higher). The amplifier is then called a *virtual-earth* or *operational* amplifier. The basic circuit is shown in Fig. 22.14(a).

Several important features follow the assumption of high gain.

(a) If $A_v (= V_o/V_{GE})$ is very large it follows that, since V_o is limited by the supply voltage to the order of 100 V, V_{GE} must be very small and G is virtually at earth potential.

(b) The impedance of the feedback path seen looking into G is $Z_2/(1 - A_v)$ (from eqn. (22.19)), and as $A_v \rightarrow -\infty$ this tends to zero.

(c) Since $Z_2/(1 - A_v)$ will in general be very small compared to the input impedance, R_{in}, of the amplifier without feedback, it follows that $I' \ll -I_2$, so that all the input current can be assumed to flow in Z_2, or in other words $I_2 = -I_1$.

(d) Since $V_{GE} \rightarrow 0$, the input voltage, V_{in}, must all appear across Z_1 and the output voltage V_o across Z_2. Hence

$$V_{in} = I_1 Z_1 \quad \text{and} \quad V_o = I_2 Z_2$$

or

$$\frac{V_o}{V_{in}} = \frac{I_2 Z_2}{I_1 Z_1} = \frac{-Z_2}{Z_1} \tag{22.31}$$

At (b) the amplifier is shown arranged to add several input signals. For this circuit

$$I_f = -I_1 - I_2 - I_3$$

Hence

$$\frac{V_o}{R_f} = -\frac{V_1}{R_1} - \frac{V_2}{R_2} - \frac{V_3}{R_3}$$

or

$$V_o = -\frac{R_f}{R_1} V_1 - \frac{R_f}{R_2} V_2 - \frac{R_f}{R_3} V_3 \tag{22.32}$$

The output voltage is thus the inverse of the sum of the input voltages, taken in ratios determined by the input and feedback resistances. Obviously, if these resistances are all of equal magnitude,

$$V_o = -(V_1 + V_2 + V_3)$$

If only one input is used the amplifier becomes a simple invertor ($V_o = -V_1$) when the input and feedback resistances are equal.

At (c) the feedback element is a capacitor. In this case, using instantaneous values,

$$v_1 = i_1 R \quad \text{and} \quad v_o = \frac{1}{C} \int_0^t i_f \, dt$$

Since $i_1 = -i_f$, it follows that

$$v_o = -\frac{1}{CR} \int_0^t v_1 \, dt \tag{22.33}$$

The output voltage in this case is the time integral of the input voltage multiplied by the constant $1/CR$ (which is usually chosen as unity). If the input consists of a d.c. step, the output will be a ramp, which will rise linearly until the amplifier becomes saturated.

When a capacitor is used to provide the input impedance and a resistor to provide the feedback impedance, an analysis similar to the above shows that the amplifier will differentiate the input signal. Differentiators are not commonly used since they tend to introduce instability and noise because the gain rises with frequency.

It should be noted that it has been assumed that the gain of the amplifier without feedback is very much greater than unity, and is negative. For low-gain amplifiers, an analysis must use eqn. (22.4), and the addition and integrating properties are adversely affected.

22.11   D.C. Amplifiers—the Long-tailed Pair

The amplifiers described in Section 22.10 must have a high gain down to zero frequency—i.e. they are *d.c. amplifiers*. Such amplifiers find applications in other areas also, such as instrumentation and power-supply stabilizers. In general d.c. amplifiers are inherently more difficult to design and stabilize than the corresponding a.c. amplifiers.

In capacitively-coupled or transformer-coupled a.c. amplifiers the coupling capacitors or transformers act as d.c. blocks, so that small, slow changes in the quiescent point of one stage (due to slow change of temperature or supply voltage, for example) are not passed on to the next stage; sudden changes are not passed on permanently. Such couplings cannot be used in d.c. amplifiers which must be direct coupled, and hence any change in the quiescent point will affect the output and will not be distinguishable from a change in the d.c. input signal. Any change in the output of a d.c. amplifier that is not due to a change in the input is called *drift*. Amplifier drift is usually expressed in terms of the equivalent input signal that will produce the same change of output. Drift is one of the main problems in d.c. amplifier design.

The chief sources of drift are:

(a) Changes in temperature, which (i) can change transistor leakage currents and base–emitter voltages, (ii) can give rise to changes in contact potentials, and (iii) can alter the ohmic value of resistors (hence silicon transistors and low-temperature-coefficient resistors should be used, together with temperature-compensating techniques).

(b) Changes in the supply voltage which alter the quiescent point (hence the supplies to d.c. amplifiers should be very well stabilized.

(c) Random changes and ageing of components (hence stable, long-life components should be used, together with circuits balanced to give zero output with zero input.

In multi-stage d.c. amplifiers any drift in the first stage is amplified in succeeding stages, so that the overall drift is largely determined by that of the first stage.

Drift cannot be reduced by straightforward feedback since this reduces the gain in proportion; the technique of compensation may

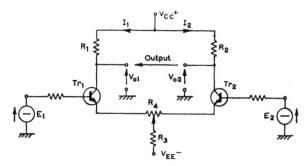

Fig. 22.15 BASIC LONG-TAILED PAIR CIRCUIT

however be employed. In this method drift in one part of a circuit is balanced against corresponding drift in another part. The long-tailed or emitter-coupled pair circuit shown in Fig. 22.15 is an example.

In this circuit, ideally, R_1 and R_2 are equal and the two transistors are matched and mounted together so that they are both subject to the same temperature variations. The "tail" resistor, R_3, should be large (ideally the tail should be a constant-current source). The circuit operates either as a *difference amplifier* (giving a differential output, $(V_{o1} - V_{o2})$), which is proportional to the difference $(E_1 - E_2)$, between the isolated inputs), or (when one of the inputs is set at zero) as a *single-ended stage*.

Drift is minimized by this configuration. Thus if the supply voltage changes, both halves of the circuit are equally affected, and the differential output is (ideally) unchanged. Similarly, temperature changes will (ideally) give the same change in leakage current and base–emitter voltage in both transistors, so that again the current changes in both halves of the circuit will be the same and the differential output will not alter. Drift due to changes in the current gain, h_{fe}, will also tend to cancel, and may be further reduced by operating at low bias currents, using low source impedances and having transistors with high values of matched current gain. In a practical

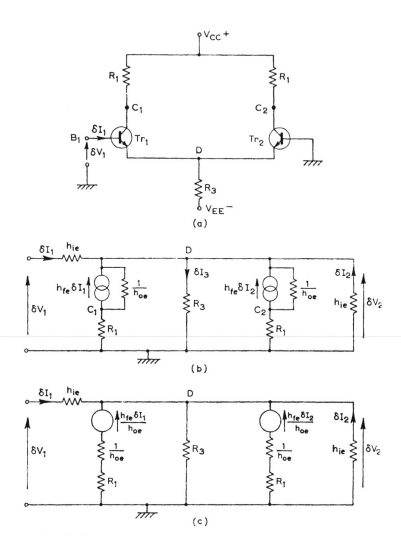

Fig. 22.16 EQUIVALENT CIRCUITS FOR LONG-TAILED PAIR

circuit there will, of course, be some residual drift, but the above compensating actions will operate to keep this small.

The initial balance of the circuit is achieved by setting both inputs to zero (or e_1 to zero in the single-ended case) and adjusting R_4 until the differential output is zero.

Consider a single-ended input applied to Tr_1 with the base of Tr_2 held at a fixed potential. Tr_1 acts as an emitter-follower stage with the emitter load consisting of the tail resistor R_3 in parallel with the input impedance of the Tr_2 circuit, which acts as a common-base circuit. If R_3 is large compared to this input impedance, then any change in current in Tr_1 gives rise to an almost equal and opposite change of current in Tr_2. Hence if V_{o1} rises, V_{o2} falls by an almost equal amount, and the circuit will act as a phase splitter with antiphase outputs δV_{o1} and δV_{o2}, where $\delta V_{o1} \approx -\delta V_{o2}$.

Assuming linear operation, suppose that the input to Tr_2 now varies by an equal but opposite amount to the input to Tr_1. The total change in output voltage will be twice that for the single-ended input. In general, therefore, the differential output will be proportional to the difference between the input signals, i.e. to the differential input.

If the two inputs change together by the same amount and in the same sense there will ideally be no change in the differential output voltage. The quality of a practical differential amplifier may be measured in terms of its ability to amplify only the difference between the input signals and to reject the signal common to both inputs. The property is called *common-mode rejection*, and numerically it is expressed as a ratio:

$$\text{Common-mode rejection ratio} = \frac{\text{Gain for antiphase inputs}}{\text{Gain for in-phase inputs}}$$
(22.34)

The higher this ratio (which is normally expressed in decibels) the better is the amplifier.

An approximate analysis of the long-tailed pair circuit of Fig. 22.16(a) follows from the approximate equivalent circuit at (b), where the transistors are represented by their common-emitter equivalents, neglecting h_{re}. The two collector resistors are assumed to be equal. Then for a single-ended input change of δV_1, the current changes will be

$$\delta I_1 = \frac{\delta V_1 - \delta V_2}{h_{ie}} \quad \text{and} \quad \delta I_2 = -\frac{\delta V_2}{h_{ie}}$$

The constant-current sources, $h_{fe}\delta I_1$ and $h_{fe}\delta I_2$, and their shunt resistors, $1/h_{oe}$, may be replaced (Norton/Thévenin conversion) by the constant-voltage sources of e.m.f. (h_{fe}/h_{oe}), $(h_{fe}/h_{oe})\delta I_2 I_1$ and series resistors $1/h_{oe}$ as shown at (c). Then, using Millman's theorem,

$$\delta V_2 = \frac{\delta V_1/h_{ie} + h_{fe}\delta I_1/h_{oe}(R_1 + 1/h_{oe}) + h_{fe}\delta I_2/h_{oe}(R_1 + 1/h_{oe})}{2/h_{ie} + 1/R_3 + 2/(R_1 + 1/h_{oe})}$$

Substituting for δI_1 and δI_2 and collecting like terms,

$$\delta V_2 \left\{ \frac{2}{h_{ie}} + \frac{1}{R_3} + \frac{2}{(R_1 + 1/h_{oe})} + \frac{2h_{fe}}{h_{oe}h_{ie}(R_1 + 1/h_{oe})} \right\}$$

$$= \delta V_1 \left\{ \frac{1}{h_{ie}} + \frac{h_{fe}}{h_{oe}h_{ie}(R_1 + 1/h_{oe})} \right\} \tag{22.35}$$

Since $2/h_{ie}$, $1/R_3$ and $2/(R_1 + 1/h_{oe})$ are all very much smaller than $h_{fe}/h_{oe}h_{ie}(R_1 + 1/h_{oe})$, it follows that

$$\delta V_2 \approx \frac{\delta V_1}{2} \tag{22.36}$$

To find the differential gain, the changes in the collector voltages at C_1 and at C_2 must be determined, and this can be done by referring back to Fig. 22.16(b). The change in output voltage at collector C_1 is then, since $1/h_{oe} \gg R_1$,

$$\delta V_{C1} = -h_{fe}R_1\delta I_1 = -h_{fe}R_1(\delta V_1 - \delta V_2)/h_{ie} \approx \frac{-h_{fe}R_1}{2h_{ie}}\delta V_1$$

and at collector C_2 is

$$\delta V_{C2} = -h_{fe}R_1\delta I_2 = +\frac{h_{fe}R_1\delta V_2}{h_{ie}} \approx \frac{h_{fe}R_1}{2h_{ie}}\delta V_1$$

The differential output voltage is therefore

$$\delta V_{C1} - \delta V_{C2} = -\frac{h_{fe}R_1\delta V_1}{h_{ie}} \tag{22.37}$$

By superposition, the gain with equal antiphase inputs, δV_1, will be

$$\delta V_{C1} - \delta V_{C2} = -\frac{2h_{fe}R_1\delta V_1}{h_{ie}} \tag{22.38}$$

The input impedance of the amplifier for single-ended operation will be

$$R_{in} = \frac{\delta V_1}{\delta I_1} = \frac{\delta V_1 h_{ie}}{(\delta V_1 - \delta V_2)} \approx 2h_{ie} \qquad (22.39)$$

since for differential input $(\delta V_1 - \delta V_2) = \delta V_1 - (-\delta V_1) = 2\delta V_1$.
Differential gains of several hundred can be readily achieved, with input impedances of well over a kilohm. The long-tailed pair forms the basis of all integrated-circuit operational amplifiers.

An alternative approach to the problem of d.c. amplifiers is to convert the d.c. input signal to a.c. by a mechanical or electronic chopper. The a.c. signal is then amplified in a conventional amplifier, and is reconstructed to d.c. by a phase-sensitive rectifier at the output. The main disadvantage of this approach is the reduction in high-frequency response which results. D.C. amplifiers can also be designed in which the chopper and a.c. amplifier are placed in the feedback loop. These are known as *chopper-stabilized amplifiers*, and have a considerably greater bandwidth than the simple chopper amplifiers, though still less than that of the long-tailed pair.

PROBLEMS

22.1 In an amplifier having an overall amplification of 50,000 without feedback, 0·02 per cent of the output voltage is fed back in antiphase to the input. Calculate the percentage reduction in amplification if the overall amplification, without feedback, falls to 40,000.

Ans. 2·2 per cent.

22.2 A resistance-capacitance-coupled amplifier has a voltage gain of −80. The coupling capacitor has negligible reactance at the frequency of operation. A very-high-resistance voltage divider across the output taps off one-tenth of the output voltage, and this is fed back and connected in series with the input voltage to give negative feedback. If there is an output ripple voltage of 1·8 V r.m.s. due to the supply to the amplifier without feedback calculate the ripple voltage appearing in the output of the amplifier when feedback is applied.

Ans. 200 mV.

22.3 A single-stage *RC* amplifier has a mid-frequency voltage gain of 46 dB and an upper 3 dB cut-off frequency of 1·5 MHz. Determine the mid-frequency gain and the bandwidth if 0·01 p.u. negative shunt voltage feedback is applied. Also evaluate the overall gain and bandwidth for two such stages in cascade.

Ans. −66·7; 4·5 MHz; 4,600; 2·9 MHz.

22.4 An amplifier without feedback has a voltage gain A_v of −1,500. Determine the percentage negative feedback that must be applied so that the sensitivity of the gain to changes in A_v is reduced to 0·01 of the value without feedback. What is the new overall gain?

Ans. 6·6 per cent; −15.

22.5 In the circuit of Fig. 22.17 the component values are $R_1 = 33$ kΩ, $R_2 = 10$ kΩ; $R_L = 4.7$ kΩ, $R_E = 330$ Ω. The *h*-parameters of the transistor are $h_{ie} = 2.5$ kΩ; $h_{fe} = 60$; $h_{oe} = 50$ μS. Determine the input impedance, current gain, voltage gain and output impedance. Do not neglect the effect of R_1 and R_2 on the equivalent a.c. circuit. Assume a constant-current source.

Ans. 5.4 kΩ; −14.3; −12.3; 50.4 kΩ.

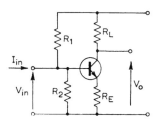

Fig. 22.17

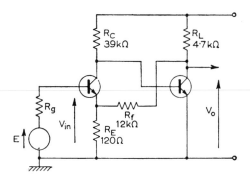

Fig. 22.18

22.6 Calculate the mid-frequency voltage gain and the input impedance of the simplified amplifier circuit of Fig. 22.18. For the transistors it may be assumed that $h_{fe} = 70$, $h_{ie} = 1.2$ kΩ, $h_{oe} = h_{re} = 0$. The d.c. bias circuit may be assumed to give an equivalent resistance of 30 kΩ at the input terminals.

Ans. 98.8; 25.2 kΩ.

22.7 The circuit shown in Fig. 22.19 has a two-stage amplifier with an overall gain of +100. The input and output impedances are very large. If $R_1 = 0.1$ MΩ and $C = 2.5$ μF, determine the value of R_2 which is required if 100e is to be equal to $K \int V_{in} dt$ and find the value of K.

The amplifier saturates when $e = 1.0$ V. Find the maximum time over which an input step of 5 V can be integrated. (Note that the feedback is positive.)

Ans. 9.9 MΩ; $K = 400$; 50 ms.

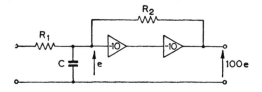

Fig. 22.19

22.8 A two-stage d.c. amplifier has a gain $A_{vo} = 1{,}550/(1 - f^2 10^{-8} + j2f10^{-4})$ where f is in hertz. Negative voltage feedback using a network of pure resistances is applied to reduce the d.c. gain to 50. Determine the fraction of the output voltage that is fed back, and calculate the gain with and without feedback at 10, 40 and 50 kHz. Sketch the modulus of the gain/frequency response with and without feedback, and comment on the shapes. (*L.U.*)

[Note that the feedback becomes positive as the frequency increases.]

Ans. 0·019; no feedback, $775\underline{/-90°}$; $92\underline{/-152°}$: $60\underline{/-157°}$; with feedback, $50\underline{/1°}$; $83\underline{/-25°}$; $121\underline{/-51°}$.

22.9 The load on a C.E. transistor amplifier is a tuned circuit of dynamic resistance 100 kΩ. The amplifier has series current feedback through an un-bypassed 100 Ω resistor in the emitter lead. The resonant circuit is tapped in the ratio 1 : 6 for matching to the transistor. If $1/h_{oe} = 20$ kΩ, $h_{ie} = 1$ kΩ, $h_{fe} = 60$, and the source generator has an internal resistance of 5 kΩ, determine the effective additional damping due to the transistor circuit, and the dynamic efficiency. (Dynamic efficiency = (power in tuned circuit)/(transistor output power).)

Ans. 710 kΩ; 0·87 p.u.

22.10 A transistor amplifier uses shunt voltage feedback. It is fed from a source of internal resistance 2 kΩ. The feedback resistance is 10 kΩ and the collector load resistance is 1 kΩ. If $h_{ie} = 1·2$ kΩ, $1/h_{oe} = 40$ kΩ and $h_{fe} = 70$, determine the output voltage if the signal generator e.m.f. is 200 mV.

Ans. 0·81 V.

22.11 An emitter-follower circuit is fed from a C.E. amplifier which has a collector circuit resistance of 5 kΩ. The emitter-circuit load of the emitter follower is 500 Ω. For a given current input to the first transistor, determine the ratio of the output voltage of the emitter follower to the voltage across a 500 Ω resistor used as the load on the first stage with the emitter follower disconnected. Also find the output impedance of the emitter follower. Both transistors have $h_{ie} = 1$ kΩ, $h_{oe} = 50$ μS and $h_{fe} = 90$.

Ans. 8·1; 60 Ω.

22.12 A transistor amplifier has a $2 \cdot 5\,k\Omega$ load resistor. Bias is obtained through a $60\,k\Omega$ resistor connected between collector and base. This resistor is divided into two $30\,k\Omega$ sections, with the mid-point connected to the supply rail by a very large capacitor C. Calculate the current gain and input impedance when (a) C is connected (hence no negative signal feedback), (b) C is removed. The transistor has $h_{ie} = 800\,\Omega$; $h_{oe} = 60\,\mu S$; $h_{fe} = 100$.

Ans. 87, 780 Ω; 19, 17 Ω.

Chapter 23

OSCILLATIONS AND PULSES

In the previous chapter it was seen that negative feedback in general improves the stability but decreases the gain of an amplifier. If the feedback is positive the stability will decrease but the gain will increase. This increase in gain may be such that an a.c. output is obta˙ned without an input, and the amplifier has become an oscillator, giving an a.c. output whose energy is obtained from the d.c. supply.

Oscillators may be divided into two types according to the output waveform:

(*a*) *Sinusoidal* oscillators, giving pure sine-wave outputs.

(*b*) *Relaxation* oscillators, giving pulse or square-wave outputs.

All oscillators work on a negative-resistance principle, and each of the above types may be further divided into (i) *retroaction* oscillators, where the negative resistance is obtained by use of the feedback principle, and (ii) *dynatron* oscillators, where the device itself introduces the negative resistance. Basically, if a tuned circuit is connected in parallel with a negative conductance (i.e. a system in which an *increase* in current gives rise to a *reduction* in voltage), then when the negative conductance is equal to the equivalent parallel conductance of the tuned circuit, the total parallel conductance is zero. The circuit damping will then be zero and oscillations will be sustained in the tuned circuit at its self-resonant frequency.

Only retroactive oscillators will be considered in this chapter.

23.1 General Condition for Oscillation

The block diagram of a voltage amplifier of gain A_v without feedback is shown in Fig. 23.1(a). As in the previous chapter, the expression for gain with feedback is

$$A_{vf} = \frac{V_o}{V_i} = \frac{A_v}{1 - \beta A_v}$$

If the feedback is positive the term βA_v is positive, and A_{vf} is greater than A_v, and indeed rises towards infinity when $1 - \beta A_v = 0$, or

$$\beta A_v = 1 \tag{23.1}$$

To investigate this condition further, consider the feedback loop broken as shown in Fig. 23.1(b). With a given a.c. signal, V',

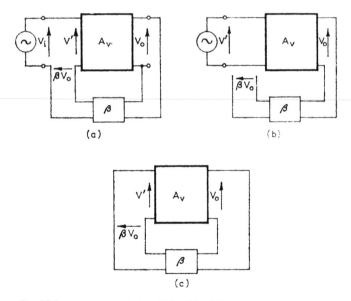

(a)   (b)

(c)

Fig. 23.1   DEVELOPMENT OF A FEEDBACK OSCILLATOR

applied, the loop gain is defined as $\beta V_o/V'$. If this is unity, then the input source can be removed and replaced by the feedback voltage as shown at (c). The circuit will continue to give a signal output, and has become an oscillator. In this condition the source of the signal is in fact the noise in the circuit. This noise is selectively amplified to give the output. For this to occur,

$$\frac{\beta V_o}{V'} = 1 \quad \text{or} \quad \beta A_v = 1 \quad \text{(as above)}$$

Since in general both β and A_v may be complex, eqn. (23.1) implies two conditions:

(*a*) In magnitude, $\beta A_v = 1$.
(*b*) The phase of βA_v is zero or any multiple of 2π.

For sinusoidal oscillators the open loop is designed to be frequency dependent, so that only at one frequency is the condition $\beta A_v = 1$ fulfilled, and this is the frequency of oscillation.

If the magnitude of βA_v is less than unity when the phase angle is zero or an integral multiple of 2π, then insufficient voltage will be fed back to maintain the output voltage in the absence of an input. If the external input voltage, V_i, shown at (*a*) is removed, the output will fall to zero.

If $\beta A_v > 1$ when the phase angle of βA_v is zero or $2\pi n$ then the feedback voltage is more than enough to maintain the output voltage even when the external input is removed, and any oscillation will increase in magnitude. With the large voltage swings which will become established, the amplifier will operate in a non-linear mode, so that A_v will fall. Eventually this fall in A_v will bring βA_v to unity, and the amplitude of the output voltage will become stabilized at the level at which $\beta A_v = 1$. Any further increase in the output voltage will cause A_v to fall further, and the output voltage will also tend to fall to its stable level. Similarly, any random decrease in the output voltage will cause A_v to rise, so that the output voltage will also tend to rise to its stable level. Thus, if the condition $\beta A_v > 1$ when arg $\beta A_v = 0$ or $2\pi n$ is satisfied for small voltage variations (i.e. for operation in the linear region of the amplifier), oscillations will build up to a stable value.

In addition, if $\beta A_v > 1$, not only will oscillations be maintained but they will be self-starting. This is due to the noise voltages which must exist in any circuit, and which can be shown to cover a wide range of frequencies. Selective amplification causes the build-up of oscillations at the frequency for which the phase condition arg $\beta A_v = 0$ or $2\pi n$ is satisfied.

EXAMPLE 23.1 A three-stage transistor amplifier uses transistors for which, at frequency f, $h_{fe} = 30/(1 + jf/10^6)$; $h_{ie} = 1\,\mathrm{k}\Omega$; and $h_{oe} = 50\,\mu\mathrm{S}$. The final stage feeds a load of $1\,\mathrm{k}\Omega$. Find the maximum series voltage feedback factor, β, that can be used if instability is to be avoided at high frequencies.

The collector load resistances are $5\,\text{k}\Omega$ in each case, and base-bias resistors can be neglected.

The effective load, R_L', of each stage is $5\,\text{k}\Omega$ in parallel with $1\,\text{k}\Omega$, i.e. $840\,\Omega$. The voltage gain per stage without feedback is thus, from eqn. (21.13),

$$A_v = \frac{-h_{fe}R_L'}{(1 + h_{oe}R_L')h_{ie}} = \frac{-30 \times 840}{(1 + jf/10^6)(1 + 0\cdot04) \times 1,000}$$

$$= \frac{24/180^\circ}{(1 + jf/10^6)}$$

At frequencies for which $f/10^6 \ll 1$, each stage introduces 180° phase shift, so that the three stages introduce a total phase shift of $360^\circ + 180^\circ$, i.e. a resultant phase shift of 180°. Any feedback from the last stage to the first will then be negative. If, however, each stage introduces a further phase shift of $\pm60^\circ$, the resultant phase shift will be zero, and the feedback will be positive, so that the amplifier may be unstable. In this case the phase angle of $(1 + jf/10^6)$ must be 60°. This means that

$$\arg(1 + jf/10^6) = 60^\circ$$

or

$$f/10^6 = \tan^{-1}60^\circ = \sqrt{3}$$

or

$$f = \sqrt{3} \times 10^6\,\text{Hz}$$

At this frequency the value of A_v per stage is $24/\sqrt{(1 + 3)} = 12$, so that for the three stages it is $12^3 = \underline{1,730}$.

Hence, if the feedback factor, β, is greater than $1/1,730$, eqn. (23.1) will be satisfied and oscillations will build up at a frequency of $\sqrt{3}\,\text{MHz}$.

It should be noted that, in an RC-coupled multistage amplifier, the coupling capacitors introduce phase shift at low frequencies. Oscillation may occur if this amounts to 60° per stage, and if the gain at such frequencies is large enough.

23.2 General Theory of Transistor *LC* Oscillators

In all oscillators the basic criterion is that the critical loop gain is unity (*Barkhausen criterion*); i.e. power, voltage and current gains are all unity. With bipolar junction transistors the input impedance is not infinite and it is most convenient to take the condition for oscillation that the loop current gain is unity, where this gain is calculated by opening the feedback loop and terminating it in an impedance, h_{ie}, equal to the input impedance of the transistor. The current in this terminating impedance with the open feedback loop will then be the input current to the first transistor when the loop is closed.

Consider the general circuit shown in Fig. 23.2(a), where only the a.c. components are shown. The a.c. equivalent circuit is shown at (b), where the resistance $1/h_{oe}$ has been assumed large compared

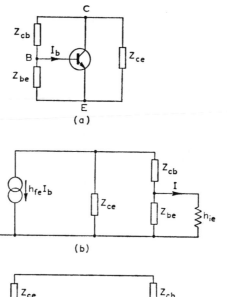

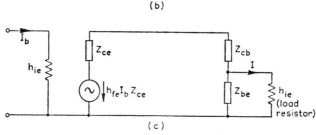

Fig. 23.2 GENERALIZED TRANSISTOR *LC* OSCILLATOR

with the load impedance, Z_{ce}. Thévenin's theorem is now used to transform the constant-current source $h_{fe}I_b$ and impedance Z_{ce} to the constant-voltage equivalent as shown at (c). The circuit is loaded by a resistor h_{ie} and the feedback connexion is not made. In this case the current I in the load resistor h_{ie} is

$$I = \frac{-h_{fe}I_bZ_{ce}}{Z_{ce} + Z_{cb} + Z_{be}h_{ie}/(Z_{be} + h_{ie})} \frac{Z_{be}}{(Z_{be} + h_{ie})}$$

$$= \frac{-h_{fe}Z_{ce}Z_{be}I_b}{(Z_{be} + h_{ie})(Z_{ce} + Z_{cb}) + Z_{be}h_{ie}}$$

Now, for oscillation to take place, $I = I_b$. Hence

$$I_b = \frac{-h_{fe}Z_{ce}Z_{be}I_b}{Z_{be}Z_{ce} + Z_{be}Z_{cb} + h_{ie}(Z_{ce} + Z_{cb} + Z_{be})}$$

so that

$$Z_{be}Z_{ce} + Z_{be}Z_{cb} + h_{ie}(Z_{ce} + Z_{cb} + Z_{be}) = -Z_{ce}Z_{be}h_{fe}$$

If it is now assumed that Z_{be}, Z_{ce} and Z_{cb} are pure reactances, it follows that the terms $Z_{bc}Z_{ce}$, $Z_{be}Z_{cb}$ and $Z_{ce}Z_{be}h_{fe}$ are positive or negative reference terms, while $h_{ie}(Z_{be} + Z_{cc} + Z_{eb})$ is quadrate. Equating reference terms and substituting reactances for impedances where appropriate*,

$$X_{be}X_{ce} + X_{be}X_{cb} = -X_{ce}X_{be}h_{fe}$$

or

$$X_{cb} + (1 + h_{fe})X_{ce} = 0 \tag{23.2}$$

so that X_{cb} and X_{ce} must be opposite types of reactance, i.e. if X_{cb} is inductive then X_{ce} must be capacitive and vice versa.

Equating the quadrate term to zero,

$$h_{ie}(X_{ce} + X_{cb} + X_{be}) = 0$$

and since $h_{ie} \neq 0$ it follows that

$$X_{ce} + X_{cb} + X_{be} = 0 \tag{23.3}$$

Substituting for X_{cb} from eqn. (23.2) in eqn. (23.3),

$$X_{ce} - (1 + h_{fe})X_{ce} + X_{be} = 0$$

or

$$X_{be} = h_{fe}X_{ce}$$

so that X_{be} and X_{ce} must be reactances of the same type.

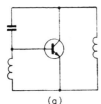 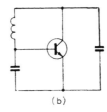

Fig. 23.3   THE TWO BASIC LC OSCILLATOR CIRCUITS

Note that eqn. (23.3) gives the frequency condition for oscillation. while eqn. (23.2) gives the necessary gain condition for the maintenance of oscillation. Also, the relation between the reactances is

$$X_{ce}:X_{be}:X_{cb} = 1:h_{fe}:-(1 + h_{fe}) \tag{23.4}$$

There are therefore basically two types of LC oscillator as shown in Fig. 23.3. These lead to several variants, depending on the position of the d.c. supply.

* X_{be} will represent $\pm jX_{be}$. and so on.

23.3 Colpitts and Hartley Oscillators

These oscillators illustrate the application of the general theory very simply. The Colpitts oscillator is basically the two-capacitor type shown in Fig. 23.4. The bias circuit is required to ensure that the transistor will be biased "on" when the supply is connected, so that

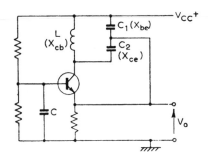

Fig. 23.4 PRACTICAL COLPITTS OSCILLATOR

the oscillator will be self-starting. Capacitor C acts as a base-decoupling capacitor for the signal. For this circuit, eqn. (23.3) gives the frequency condition as

$$\frac{1}{j\omega C_2} + j\omega L + \frac{1}{j\omega C_1} = 0$$

or

$$\omega^2 = \frac{1}{L}\left(\frac{1}{C_1} + \frac{1}{C_2}\right) \qquad (23.5)$$

while eqn. (23.2) gives the gain condition as

$$i\omega L + \frac{(1 + h_{fe})}{j\omega C_2} = 0$$

i.e.

$$1 + h_{fe} = \omega^2 L C_2 = \frac{C_2}{C_1} + 1$$

Hence

$$h_{fe} > \frac{C_2}{C_1} \qquad (23.6)$$

Frequencies over 5 GHz are possible with this oscillator. Similar expressions can readily be obtained for the Hartley oscillator shown in Fig. 23.5. It is left as an exercise for the reader to show that, if

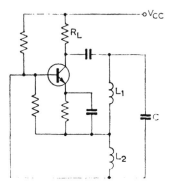

Fig. 23.5 PRACTICAL HARTLEY OSCILLATOR

there is no mutual coupling, M, between the coils L_1 and L_2, then the frequency of oscillation is given by

$$\omega^2 = \frac{1}{C(L_1 + L_2)} \tag{23.7}$$

and the gain condition by

$$h_{fe} \geqslant \frac{L_2}{L_1} \tag{23.8}$$

If $M \neq 0$, these expressions are modified to $\omega^2 = 1/C(L_1 + L_2 + 2M)$ and $h_{fe} = (L_1 + M)/(L_2 + M)$.

23.4 Frequency Stability

In the foregoing derivations, the frequency at which an oscillator operates has been found as a function of circuit components only. In practical oscillators, external loading, non-linearity of active device and component losses all introduce effects which make the frequency dependent on supply voltage, active device parameter changes and temperature. The frequency stability depends upon how rapidly the phase criterion expressed by eqn. (23.13) changes with frequency near the critical frequency. If the change is not rapid, small changes in parameters may cause a large frequency change as the closed loop adjusts itself to maintain the phase and gain response required for oscillation. Note that, if $\beta A_{vo} > 1$ when arg $\beta A_{vo} = 0$ or $2\pi n$, then oscillations build up in amplitude until the amplifier operates in a non-linear region so that A_{vo} falls until $\beta A_{vo} = 1$. To achieve frequency stability one of the frequency-determining elements is often chosen as a tuned circuit, operated very slightly off resonance, so that it behaves either as an inductance (at frequencies below resonance) or a capacitance (when operated slightly above resonance). The higher the Q-factor of the tuned circuit, the greater will be the change in phase with frequency near the critical frequency, and hence the greater will be the frequency stability. Expressed mathematically the condition for frequency stability

$$\frac{d}{d\omega}(X_{ce} + X_{cb} + X_{be})$$

is as high as possible when

$$X_{ce} + X_{cb} + X_{be} \approx 0$$

Since a piezo-electric crystal behaves as a tuned circuit of very high Q-factor, oscillators in which one of the elements is a crystal have very high frequency stability.

The piezo-electrical effect is the mechanical contraction or elongation along one axis of a crystal when a voltage is applied across another axis (and vice versa). If a voltage pulse is applied then on its removal the mechanical strain relaxes and the strain energy is converted to stored electrical energy—a voltage appears on the voltage axis. This in turn leads to further mechanical stress. The crystal will continue to oscillate between mechanical and electrical stored energy states until internal losses use up all the original stored energy. It is thus equivalent to a tuned circuit as shown in Fig 23.6(a). Q-factors of the order of 50,000 are readily obtained from quartz crystals. Because of their larger electrical output crystals of Rochelle salt are commonly used in piezo-electric pick-ups, microphones and transducers. The self-resonant frequency of crystals is determined largely by their thickness—the thicker the crystal the lower the self-resonant frequency.

A simple piezo-electric oscillator is shown at (b). This is effectively a Colpitts oscillator—with the coil replaced by the crystal, which must therefore operate slightly below its self-resonant frequency. This is often called the Pierce oscillator. A similar circuit in which the capacitors C_1 and C_2 are replaced by coils gives the crystal equivalent of the Hartley oscillator.

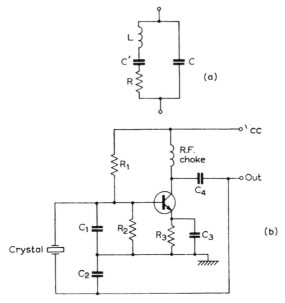

Fig. 23.6   PIERCE OSCILLATOR

23.5 Oscillators with Mutual Coupling

Mutual coupling is often used in oscillator circuits. When generalized theory is applied in these cases the mutual inductance should be replaced, as far as the calculations are concerned, by the equivalent π-circuit shown in Fig 23.7.

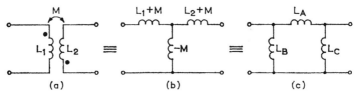

Fig. 23.7   EQUIVALENT CIRCUITS OF A MUTUAL COUPLING

By considering mesh equations it is easy to verify that the circuit at (*b*) with no mutual coupling is equivalent to that at (*a*), and the star–delta transformation then gives the equivalent π-circuit, where

$$L_A = \frac{L_1 L_2 - M^2}{-M} \qquad L_B = \frac{L_1 L_2 - M^2}{L_2 + M} \qquad L_C = \frac{L_1 L_2 - M^2}{L_1 + M}$$

$$(23.9)$$

Note that L_A is a negative inductance. This means that it has a negative reactance which increases with frequency. A negative inductance in parallel with a positive inductance does not, however, form a tuned circuit with a resonant frequency, but simply gives a *larger* positive inductance (if the negative reactance is larger than the positive reactance).

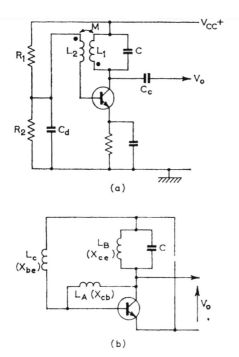

Fig. 23.8 TUNED-COLLECTOR OSCILLATOR
(*a*) Actual circuit   (*b*) Equivalent a.c. circuit

Using the above relations it is seen that the tuned collector oscillator of Fig. 23.8(*a*) reverts to the general form as shown at (*b*). The inductance L_C is positive and L_A is negative (i.e. X_{be} is positive and X_{cb} is negative). Hence X_{ce} must be positive, i.e. L_B in parallel with C must operate *below* the resonant frequency, $f = 1/2\pi\sqrt{(L_B C)}$. The conditions for oscillation are then fulfilled provided that the gain is high enough. In a practical transistor oscillator of this type the tuned circuit coil will normally be tapped in order to reduce the loading of the circuit due to the transistor. A third winding is often used for the output.

The frequency of the oscillator will be slightly below $1/2\pi\sqrt{(L_BC)}$, i.e. slightly above $1/2\pi\sqrt{(L_1C)}$. The frequency stability of the circuit can be high, but because the rate of change of reactance with frequency of the negative inductance, L_A, is negative, the stability is not so high as it would be if a capacitor were to replace L_A (the rate of change of reactance of the capacitor with frequency would be positive). Mathematically,

$$\frac{d}{d\omega}(X_{ce} + X_{cb} + X_{be})$$

is smaller if X_{cb} is the reactance of a negative inductance than if it is the reactance of a capacitor, and so the frequency stability is lower.

23.6 Transistor Wien Bridge Oscillator

For the circuit shown in Fig. 23.9(a) the voltage transfer function is

$$\frac{V_f}{V'} = \frac{1}{3 + j\omega CR + 1/j\omega CR}$$

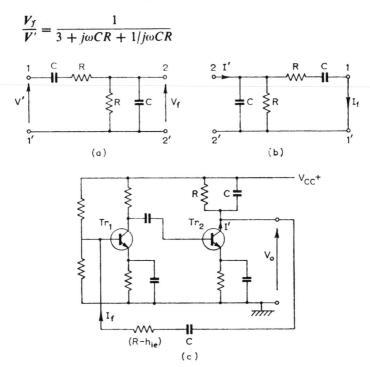

Fig. 23.9 SIMPLE TRANSISTOR WIEN BRIDGE OSCILLATOR

The current dual of this circuit is obtained from a simple application of the *principle of duality*. It follows from this principle that the ratio of output to input voltage for an *open-circuited* two-port passive network is the same as the ratio of the current through a *short-circuit* across the input terminals to the current fed into the output terminals. Thus, if the output terminals 2, 2' in Fig. 23.9(a) are made the current input terminals at (b) of the current dual, it is easy to verify that

$$\frac{I_f}{I'} = \frac{1}{3 + j\omega CR + 1/j\omega CR}$$

In the oscillator circuit shown at (c) the feedback resistance is chosen to be $(R - h_{te})$, so that when the transistor input resistance, h_{te}, is taken into account, the circuit is equivalent to a capacitor C and resistor R in series feeding into a short-circuit. The input base current of transistor Tr_1 is

$$I_f = \frac{I'}{3 + j\omega CR + 1/j\omega CR} \tag{23.10}$$

neglecting the effect of the base bias resistors. Hence, if

$$\frac{I'}{I_f} > 3 \tag{23.11}$$

the circuit will oscillate at the frequency which makes the quadrate terms in eqn. (23.10) vanish; i.e. at a frequency given by

$$\omega CR = \frac{1}{\omega CR}$$

or

$$\omega = \frac{1}{CR} \tag{23.12}$$

It is, of course, relatively simple to achieve a current gain of 3 in a two-stage amplifier, but the frequency stability of this circuit is poor. For this reason the full Wien bridge oscillator shown in Fig. 23.10 is to be preferred.

This requires a differential amplifier whose output is a constant, A_v times the *difference* between the inputs; i.e.

$$V_o = A_v(V_1 - V_2) \tag{23.13}$$

The gain A_v must be positive reference number—i.e. the differential amplifier must have an even number of stages, as in the half Wien bridge. The bridge is *almost* balanced, so that the voltage across AB is small.

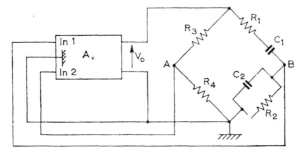

Fig. 23.10   THE FULL WIEN BRIDGE OSCILLATOR

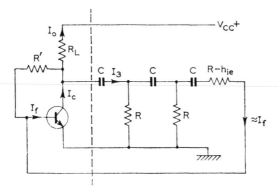

Fig. 23.11   SINGLE-STAGE *RC* OSCILLATOR

Analysis of the circuit shows that the frequency of oscillation is given by

$$f = \frac{1}{2\pi\sqrt{(C_1 R_1 C_2 R_2)}}$$

and, since the voltage across AB is small, the condition of oscillation will be that A_v is large enough to satisfy eqn. (23.10) (high gains are however, easily obtained). The frequency stability of the circuit is high; i.e. the effective Q-factor of the selective network is much greater than that of the half Wien bridge.

23.7 Single-stage Transistor *RC* Oscillator

The circuit of a single-stage *RC* phase-shift oscillator is shown in Fig. 23.11. The circuit to the left of the broken line is a simple amplifier, in which the bias is obtained by means of the collector–base resistor R'. Since the amplifier is in the common emitter connexion, the collector current, I_c, will be approximately in antiphase with the base input current, I_f. Hence, for oscillation, the phase-shift network of C's and R's must introduce a further 180° phase shift, and the transistor current gain must compensate for the network attenuation. Assuming a transistor input resistance of h_{te}, the following equation can be deduced (neglecting any current through the bias resistor R') by considering the current division between R_L and the feedback chain, and the current division in the feedback chain itself.

$$I_c = I_o + I_3$$
$$= I_f \left\{ 3 + \frac{R}{R_L} - \frac{1}{\omega^2 C^2 R} \left(\frac{1}{R} + \frac{5}{R_L} \right) + \frac{1}{j\omega C} \left(\frac{4}{R} + \frac{6}{R_L} \right) \right.$$
$$\left. - \frac{1}{j\omega^3 C^3 R^2 R_L} \right\} \quad (23.14)$$

Oscillation will occur at the frequency which makes the quadrate term zero, provided that the current gain makes I_c equal to I_f times the reference term. For zero quadrate term,

$$\frac{1}{\omega C} \left(\frac{4}{R} + \frac{6}{R_L} \right) = \frac{1}{\omega^3 C^3 R^2 R_L}$$

or

$$\omega^2 = \frac{R R_L}{C^2 R^2 R_L (4R_L + 6R)} = \frac{1}{C^2 R (4R_L + 6R)} \quad (23.15)$$

so that

$$\omega = \frac{1}{C \sqrt{[R(4R_L + 6R)]}} \quad (23.16)$$

When this condition is satisfied,

$$\frac{I_c}{I_f} = 3 + \frac{R}{R_L} - \frac{(4R_L + 6R)(R_L + 5R)}{R R_L} \quad (23.17)$$

(substituting for $\omega^2 C^2 R$ from eqn. (23.15) in the reference part of eqn. (23.14)).

The current gain, A_i, must be greater than this in order to sustain oscillations; i.e.

$$A_i > 3 + \frac{R}{R_L} - \frac{4R_L}{R} - 26 - \frac{30R}{R_L}$$

$$> - \left(23 + \frac{29R}{R_L} + \frac{4R_L}{R}\right)$$

Neglecting $1/h_{oe}$, since $1/h_{oe} \ll R_L$, and assuming $I_c \approx -h_{fe}I_b \approx -h_{fe}I_f$,

$$h_{fe} > 23 + \frac{29R}{R_L} + \frac{4R_L}{R} \qquad (23.18)$$

It has been assumed that $1/h_{oe} > 10R_L$, and that $I_c \approx -h_{fe}I_b$.

For example, if $C = 0.05\mu\text{F}$, $R_L = 3.3\,\text{k}\Omega$ and $R = 5.6\,\text{k}\Omega$, then the oscillation frequency is approximately

$$f = \frac{10^6}{2\pi \times 0.05\sqrt{[3.3(13.2 + 33.6) \times 10^6]}} = 250\,\text{Hz}$$

and $h_{fe} > 23 + 49 + 2 > 74$.

It is left as an exercise for the reader to show that if the resistor $(R - h_{ie})$ is omitted, the frequency of oscillation is given by

$$\omega^2 = \frac{1}{C^2(h_{ie}R_L + 3RR_L + 3R^2 + 3h_{ie}R)}$$

and the gain condition is that

$$h_{fe} > 1 + \frac{2h_{ie}}{R} + \frac{h_{ie}}{R_L}$$
$$- \frac{(R_L + h_{ie} + 4R)(h_{ie}R_L + 3RR_L + 3R^2 + 3h_{ie}R)}{R^2R_L}$$

If $h_{ie} \to 0$ and $R_L = R$, these expressions reduce to

$$\omega^2 = \frac{1}{6C^2R^2} \qquad \text{and} \qquad h_{fe} > 29$$

In general, RC oscillators are convenient at low frequencies, since no large tuning inductance is required. Further, since the frequency of oscillation is inversely proportional to the capacitance (while in LC oscillators it is inversely proportional to the square root of the capacitance), the same capacitance change gives a larger frequency variation in RC than in LC oscillators.

23.8 Astable Multivibrator (Relaxation Oscillator)

In the *astable multivibrator*, two cascaded RC coupled stages are given 100 per cent positive feedback, and this gives rise to a conduction process whereby one stage is fully conducting while the other is switched off, conduction being transferred from one stage to the other by the discharge of a capacitor through a resistor. The output consists of square waves or pulses whose repetition frequency depends upon the CR time constants involved.

A simple circuit is shown in Fig. 23.12. The basic design criterion is that when either transistor is conducting it will operate in the

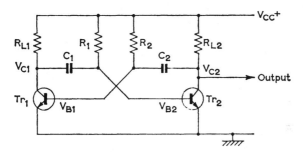

Fig. 23.12 CROSS-COUPLED ASTABLE MULTIVIBRATOR

saturation region. Up to the saturation value of collector current the relation $I_C \approx h_{FE}I_B$ applies (where h_{FE} is the large-signal d.c. current gain before saturation). If I_B is further increased there will be no corresponding increase in I_C since it has reached its saturation value (all that will happen is that I_E will increase slightly). Hence the criterion for saturation is $h_{FE}I_B > I_C$.

For transistor Tr_1, the base current in the saturated "on" state is approximately V_{CC}/R_2 (neglecting the transistor base–emitter voltage), and the collector current is V_{CC}/R_{L1}. Hence, for this state,

$$\frac{h_{FE}}{R_2} \gg \frac{1}{R_{L1}} \quad \text{or} \quad R_{L1} \gg \frac{R_2}{h_{FE}}$$

Similarly $R_{L2} \gg R_1/h_{FE}$ (normally R_{L1} and R_{L2} are chosen as equal).

The initial setting of the circuit can be seen by assuming that, when the d.c. supply is switched on, both transistors carry equal rising currents: some random variation will be bound to slightly increase or reduce one current with respect to the other—suppose that the current in Tr_2 is increased above that in Tr_1 by a small amount. The following reactions will occur:

(a) The voltage drop across R_{L2} will increase, so that V_{C2} will fall, and since the voltage across C_2 cannot change instantaneously, this fall in voltage is also effective at the base of Tr_1.

(b) Since V_{B1} has fallen, the base current in Tr₁ falls and the collector current is reduced, so that the collector voltage V_{C1} rises, this rise being transferred across C_1 to the base of Tr₂.

(c) Since V_{B2} rises it causes the current in Tr₂ to increase further, and the process repeats until Tr₂ is saturated and Tr₁ is cut off (provided that $R_{L2} \gg R_1/h_{FE}$).

The system is thus inherently unstable. It may be taken that the changes take place instantaneously since there are no reactive

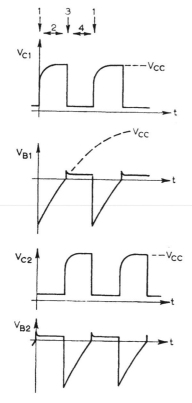

Fig. 23.13   WAVEFORMS OF ASTABLE MULTIVIBRATOR

elements except C_1 and C_2, whose instantaneous charges are constant. It is, of course, purely random which transistor becomes saturated and which is cut off initially.

The wave diagrams shown in Fig. 23.13 commence at the instant when Tr₁ is cut off, so that, to a fair approximation, $V_{B2} = 0$;

$V_{C2} = 0$; V_{B1} is rapidly going negative with respect to earth; and V_{C1} is rapidly rising to the supply voltage, V_{CC}. The reactions from this instant are as follows:

1. The positive-going voltage at the collector of Tr_1 is applied across C_1 and gives a base current flow in Tr_2, so that C_1 can rapidly charge through the forward-biased base–emitter diode of Tr_2. V_{B2} rises sharply above earth but rapidly falls to just above earth potential, so giving the small positive spike shown. It is convenient to assume that, just prior to the present voltage changes, V_{C2} has been at the supply voltage, V_{CC}, and that V_{B1} has been at zero voltage; this may not be so initially but becomes so after only a very few oscillations. Then the change in the collector voltage of Tr_2 is from V_{CC} (when Tr_2 is non-conducting) to zero (when Tr_2 is saturated). Since the charge on C_2 cannot change instantaneously, the base potential of Tr_1 must fall by the same amount, i.e. from about zero initially to $-V_{CC}$, so that the voltage across R_2 is $2V_{CC}$.

2. In the interval marked "2" in Fig. 23.13, relatively static conditions prevail. V_{C2} remains at about zero voltage, V_{C1} at $+V_{CC}$, and V_{B2} at about zero. However the voltage across R_2 is now $2V_{CC}$, and so a current flows in it which charges C_2 from $-V_{CC}$ towards $+V_{CC}$, the time constant of the exponential charging current being C_2R_2. The equation of this "relaxation" is

$$V_{B1} = -V_{CC} + 2V_{CC}(1 - e^{-t/C_2R_2})$$
$$= V_{CC} - 2V_{CC}\,e^{-t/C_2R_2} \qquad (23.19)$$

However, as soon as its base comes to zero voltage, Tr_2 starts conducting and a second stage of precipitate amplification (labelled "3") occurs. The variation of base voltage during phase 2 is shown, the broken curve indicating the "prospective" rise, which is cut short at instant 3.

3. As soon as V_{B1} reaches zero voltage, Tr_1 starts conducting. Hence V_{C1} falls, and the fall is transferred across C_1 (whose voltage cannot change instantaneously) to the base of Tr_2, which therefore becomes cut off. V_{C2} rises (short time constant, C_2R_{L2}) to V_{CC} so that (since the voltage across C_2 cannot change instantaneously) the base voltage of Tr_1 also rises, to saturate Tr_1 and cause its collector voltage to fall to about zero. The base voltage of Tr_2 falls to $-V_{CC}$ volts.

4. As soon as this precipitate amplification period is over, a further quiescent period follows, with $V_{C1} \approx 0$, $V_{C2} \approx V_{CC}$ and $V_{B1} \approx 0$. V_{B2} now rises (time constant, C_1R_1) from $-V_{CC}$ towards $+V_{CC}$. The equation for the rise is

$$V_{B2} = V_{CC} - 2V_{CC}\,e^{-t/C_1R_1} \qquad (23.20)$$

The whole process repeats as soon as V_{B2} reaches zero, when a further stage of precipitate amplification takes place.

The output from either collector is thus a series of rectangular pulses, whose repetition frequency can now be found. Thus for the first relaxation period, the time, τ_1, required for the base voltage to rise from $-V_{CC}$ to zero is given, from eqn. (23.19), by

$$0 = V_{CC} - 2V_{CC}\, e^{-\tau_1/C_2R_2}$$

Hence

$$\exp(\tau_1/C_2R_2) = 2 \quad \text{or} \quad \tau_1 = C_2R_2 \log_e 2 \qquad (23.21)$$

Similarly the time for the second relaxation period is

$$\tau_2 = C_1R_1 \log_e 2$$

and the *pulse repetition frequency* (p.r.f.) is

$$\text{P.R.F.} = \frac{1}{\tau_1 + \tau_2} = \frac{1}{(C_1R_1 + C_2R_2) \log_e 2} \qquad (23.22)$$

For the symmetrical multivibrator, $C_1R_1 = C_2R_2 = CR$, and hence

$$\text{P.R.F.} = \frac{1}{2CR \log_e 2} = \frac{1}{1\cdot4CR} \quad \text{pulses/second} \qquad (23.23)$$

The p.r.f. can be altered by (a) changing the CR constant or (b) returning the base resistors R_1 and R_2 to a different potential than that of the supply rail (increasing the voltage to which they are returned increases the frequency).

The frequency stability of transistor multivibrators is not high, and in particular the temperature sensitivity of the reverse leakage current I_{CEO} (especially in germanium transistors) causes frequency drift. This is because the leakage current at the base of the "off" transistor affects the charging rate of the "relaxing" capacitor. The basic circuit is not suitable for use with silicon planar transistors if $V_{CC} > 6\text{V}$ since the base–emitter junction may break down when it becomes reverse biased.

23.9 Synchronizing

It is often desirable to lock the frequency of a multivibrator to that of a more frequency-stable oscillator that can produce short pulses (*clock pulses*). This can be done in the circuit of Fig. 23.12 by feeding positive clock pulses through diodes to the base connexions. The transition between states can then be made to correspond to clock impulses as shown in Fig. 23.14. If the clock pulse frequency is slightly higher than that of the multivibrator, the multivibrator will be brought into exact synchronism. If the clock pulse frequency is several times the multivibrator basic frequency, the multivibrator will act as a dividing circuit (as in Fig. 23.14) giving one output pulse

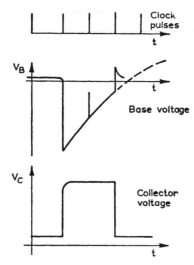

Fig. 23.14 SYNCHRONIZING WAVEFORMS FOR A MULTIVIBRATOR

for every 2, 3, 4 . . . etc. clock pulses. This division is not usually carried beyond 10, since it is then difficult to make synchronizing determinate and stable.

23.10 Monostable Multivibrator

The astable multivibrator has been seen to have no stable state, conduction being transferred continuously from one transistor to the other and back. The *monostable multivibrator* (Fig. 23.15(*a*)) has one stable state to which it will always revert. When an external trigger pulse is applied the circuit switches over, and then, after some delay, returns to the stable state until a further trigger pulse is applied. The circuit is used to give a time delay action, or to produce a known pulse width.

In the circuit shown, the voltage of the base of Tr_1 is set by the resistor chain R_{L2}, R_2 and R_3 to be below the emitter voltage. Thus Tr_1 is normally cut off. The base of Tr_2 is connected through R_1 to the positive supply, so that Tr_2 is normally conducting (i.e. in the "on" state). This is the stable state, in which $V_{C1} = V_{CC}$ and $V_{C2} \approx 0$.

When a negative trigger pulse is applied to the base of Tr$_2$, the current in this transistor falls, V_{C2} rises, and since the voltage across C_2 cannot change instantaneously, the voltage V_{B1} also rises. This starts Tr$_1$ conducting; its collector voltage falls, and since the voltage across C_1 cannot change instantaneously, the base of Tr$_2$ also falls

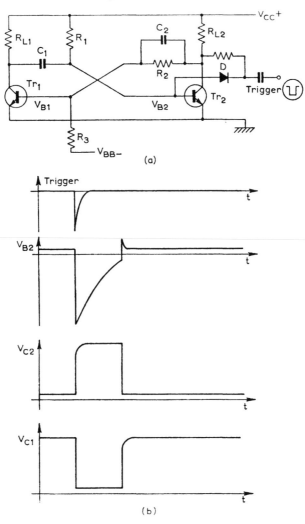

Fig. 23.15   MONOSTABLE MULTIVIBRATOR

to maintain Tr_2 in the cut-off condition. This precipitate amplification ends with Tr_1 on and Tr_2 off, so that $V_{C1} \approx 0$; $V_{C2} \approx V_{CC}$ (neglecting the voltage drop across R_{L2} due to the resistor chain current); $V_{B1} \approx 0$; and $V_{B2} = -V_{CC}$.

The quasi-stable state follows, as C_1 charges towards V_{CC} (time constant, $C_1 R_1$), with $V_{C1} \approx 0$ and $V_{C2} \approx V_{CC}$. The duration of this state is the time required for V_{B2} to rise from $-V_{CC}$ to zero, when aiming at $+V_{CC}$; i.e. it is given by eqn. (23.21) as

$$\tau = C_1 R_1 \log_e 2$$

As soon as V_{B2} reaches zero, Tr_2 starts conducting, and a further stage of precipitate amplification occurs to change Tr_1 to "off" and Tr_2 to "on." The circuit is then back in its stable state and awaits a further trigger pulse before again switching over. The waveforms are shown in Fig. 23.15(*b*).

Base leakage current in Tr_2 causes the same effect as in the astable circuit, and results in some temperature dependence of the output pulse-width.

EXAMPLE 23.2 In the monostable circuit of Fig. 23.15, $R_{L1} = R_{L2} = 1 \cdot 8 \, k\Omega$; $C_1 = C_2 = 10,000 \, pF$; $R_1 = 22 \, k\Omega$; $R_2 = 15 \, k\Omega$; $R_3 = 120 \, k\Omega$. The supply voltage, V_{CC}, is 6 V and the bias voltage is -6 V. Verify the stable-state currents and voltages, and show that Tr_1 is bottomed in the quasi-stable state. The transistors used have $h_{FE} = 70$.

In the quiescent state Tr_1 is off and Tr_2 is on. Assuming that Tr_2 is saturated (bottomed) and that the voltage drop across it is then negligible, the collector current in Tr_2 will be

$$I_{C2} = \frac{V_{CC}}{R_{L2}} = \frac{6}{1,800} A = 3 \cdot 3 \, mA$$

The base current is then

$$I_{B2} \approx \frac{V_{CC}}{R_1} = \frac{6}{22,000} A = 0 \cdot 27 \, mA$$

(neglecting the base–emitter voltage).

Since $I_{B2} > I_{C2}/h_{FE}$, the transistor is saturated.

The base voltage of Tr_1 is determined by the voltage-dividing action of R_2 and R_3 across V_{C2} and $-V_{BB}$, i.e.

$$V_{B1} = \frac{-V_{BB} R_2}{R_2 + R_3} = -\frac{6 \times 15}{135} = -0 \cdot 7 \, V \quad \text{(since } V_{C2} \approx 0\text{)}$$

Hence the base of Tr_1 is at a lower potential than the emitter, and so Tr_1 must be cut off. Note also that the drain of R_2 and R_3 is approximately

$$I = \frac{V_{BB}}{R_2 + R_3} \approx 45 \, \mu A$$

which gives a negligible voltage drop across R_{L2}.

The width of the output pulse when triggering occurs is

$$\tau = C_1 R_1 \log_e 2 = 10^{-8} \times 22 \times 10^3 \times 0.69 \, s = 152 \, \mu s$$

In the quasi-stable state Tr_2 is off. *If* there were no base current in Tr_1 the voltage at the mid-point of R_2 and R_3 would be

$$-V_{BB} + \frac{V_{CC} R_3}{R_{L2} + R_2 + R_3} = -6 + \frac{12 \times 120 \times 10^3}{137 \times 10^3} = +4.5 \, V$$

Hence the base must be forward biased, and neglecting the base–emitter voltage drop, the actual base voltage is zero. The actual current through R_2 is therefore

$$I_{R2} = \frac{6}{1,800 + 15,000} A = 0.357 \, mA$$

The current through R_3 is, similarly, $6/120,000 \, A = 0.05 \, mA$, so that the base current is $0.375 - 0.05 = 0.325 \, mA$. Since this is greater than I_{C1}/h_{FE} the transistor is bottomed.

23.11 The Bistable Multivibrator

The *bistable multivibrator* has two stable states, since either transistor may conduct and in doing so cuts the other one off. The circuit is shown in Fig. 23.16(*a*). Each transistor is d.c. coupled to the other. The capacitors C_1 and C_2 are called "speed-up capacitors" and act simply to ensure precipitate amplification during the transition from one stable state to the other.

Suppose that Tr_1 is on. The voltage-divider resistor chain is designed so that the low collector voltage of the "on" transistor is sufficient to keep the other transistor off under steady-state conditions. Hence if Tr_1 is on, Tr_2 is held off. If a negative trigger pulse is now routed through diode D_1 to the base of Tr_1, the base potential falls and Tr_1 takes less current. The collector potential of Tr_1 rises, and hence the base potential of Tr_2 rises. Tr_2 starts conducting and precipitate amplification then follows to turn Tr_2 fully on and to cut Tr_1 off. The circuit remains in this state until a negative reset pulse is applied through D_2 to the base of Tr_2. This will turn Tr_2 off and Tr_1 on. The relevant waveforms are shown in Fig. 23.16(*b*).

By interconnecting the trigger and reset pulse circuits, the input negative pulses may be routed by the diodes to whichever of the two bases is the more positive, so that successive input pulses will cause the bistable multivibrator to change state. The output at either collector will then be at half the pulse repetition frequency of the input pulse train. The circuit is then a *scale-of-two* counter or a *divide-by-two* circuit, which has many important applications in computers and static switching.

Leakage currents do not affect the operation of the bistable multivibrator, which is inherently stable and largely independent of temperature variations.

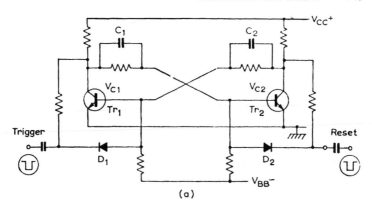

(a)

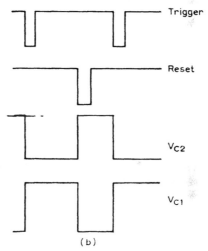

(b)

Fig. 23.16 BISTABLE MULTIVIBRATOR

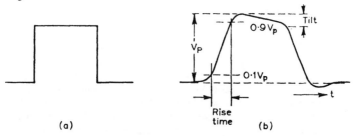

(a)   (b)

Fig. 23.17 IDEAL AND PRACTICAL PULSES
(a) Ideal   (b) Practical

23.12   Rise Time and Tilt of Pulse Waveforms

An ideal pulse is rectangular in shape, as shown in Fig. 23.17(*a*). In practice it is not possible for a pulse to rise or fall instantaneously, and often the top of the pulse is not level. These effects are exaggerated at (*b*) for clarity.

When the leading edge of any practical pulse is observed on a cathode-ray oscilloscope with a fast enough time base, it is impossible to determine the exact instant at which the rise starts, since (as shown at (*b*)) the rise starts on an asymptotic curve. For the same reason it is also impossible to determine the time at which the peak is first reached. It has therefore become usual to specify the rise time of a pulse as the time taken for the pulse to grow from 0·1 to 0·9 of its peak value.

The tilt of a pulse waveform is a measure of the slope of the top of the pulse. The per-unit tilt is the difference between the peak value and the value at the start of the trailing edge expressed as a fraction of the peak value.

23.13   Pulse Response of a Low-pass *CR* Network

The voltage transfer function of the low-pass circuit shown in Fig. 23.18 is

$$\frac{V_o}{V_i} = \frac{1/j\omega C}{(R + 1/j\omega C)} = \frac{1}{1 + j\omega CR}$$

If ω is small so that $\omega \ll 1/CR$, the voltage transfer function is unity. V_o/V_i will fall by 3 dB from this value when $\omega CR = 1$. The frequency, f_H, at which this occurs may be taken as a measure of the upper limit of frequency of the circuit, where

$$f_H = \frac{1}{2\pi CR} \tag{23.24}$$

Hence, for any frequency f,

$$\frac{V_o}{V_i} = \frac{1}{1 + j\dfrac{f}{f_H}} \tag{23.25}$$

When a pulse waveform is applied to the circuit, transient analysis shows that the output will be in the form of an exponential growth and decay. If the time constant is less than 0·2 of the pulse-width, the output pulse will eventually reach about 99 per cent of the peak value of the input as at (*b*) (a *CR* transient may be considered to reach

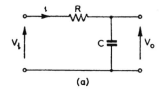

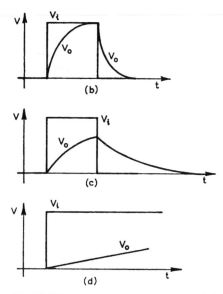

Fig. 23.18   PULSE RESPONSE OF LOW-PASS *RC* NETWORK

a steady value in five time constants). The expression for the leading edge of the output voltage waveform for a pulse of height V_i is

$$v_o = V_i(1 - e^{-t/CR})$$

The rise time is $(t_2 - t_1)$, where t_1 is the time for v_o to reach $0 \cdot 1 V_i$, and t_2 is the time to reach $0 \cdot 9 V_i$. Thus

$$0 \cdot 1 V_i = V_i(1 - e^{-t_1/CR}) \quad \text{or} \quad t_1 = 0 \cdot 1 CR$$

and

$$0 \cdot 9 V_i = V_i(1 - e^{-t_2/CR}) \quad \text{or} \quad t_2 = 2 \cdot 3 CR$$

so that

$$\text{Rise time} = t_1 - t_2 = 2 \cdot 2 CR = \frac{2 \cdot 2}{2\pi f_H} \quad \text{seconds} \quad (23.26)$$

On the trailing edge the output decays according to

$$v_0 = V_t \, e^{-t/CR}$$

where t is reckoned from the instant when the trailing edge of the input pulse commences.

If the pulse width is less than five time constants, the output pulse will never reach the peak of the input. This condition is shown in Fig. 23.18(c). The decay will, however, last at least five time constants, unless the next pulse arrives before this.

If the time constant, CR, is very large, the output of the circuit will be approximately equal to the time integral of the input for integrating times which are small compared to CR. Thus, for the circuit of Fig. 23.18(a), the instantaneous output voltage, v_0, is given by

$$v_0 = \frac{q}{C}$$

where q is the instantaneous charge on C. Therefore

$$\frac{dv_0}{dt} = \frac{i}{C}$$

and

$$V_i = iR + v_0 = CR \, \frac{dv_0}{dt} + v_0$$

If CR is large enough, so that $v_0 \ll V_i$, then

$$v_0 = \frac{1}{CR} \int_0^t V_i \, dt \tag{23.27}$$

The disadvantage of this form of integrating circuit is that the output voltage is very small. The output of such a circuit with a step input is shown at (d), and is simply the initial part of an exponential growth curve of long time constant.

23.14 Pulse Response of a High-pass CR Network

In the high-pass CR circuit shown in Fig. 23.19(a), the a.c. voltage transfer function is

$$\frac{V_o}{V_i} = \frac{R}{R + 1/j\omega C} = \frac{1}{1 + 1/j\omega CR}$$

For very high frequencies this ratio is unity. It falls 3 dB below unity at a frequency, f_L, for which

$$2\pi f_L = \frac{1}{CR} \tag{23.28}$$

The voltage transfer function can hence be written for any frequency, f, as

$$\frac{V_0}{V_i} = \frac{1}{1 - j\dfrac{f_L}{f}}$$

(23.29)

When a pulse waveform is applied to this circuit, on the rising edge the change of input causes the output voltage to rise instantaneously by the same amount, since the voltage across C cannot change

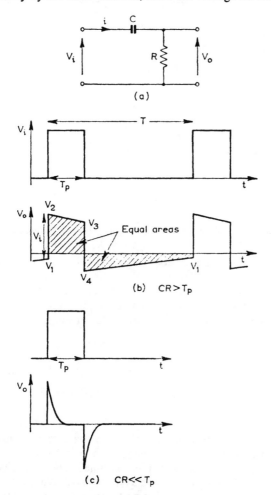

Fig. 23.19 PULSE RESPONSE OF HIGH-PASS *RC* NETWORK

instantaneously. C charges through R, and hence the output voltage falls exponentially towards zero. The fall in voltage stops when the negative edge of the input pulse arrives. A similar process then occurs at this falling edge as shown in Fig. 23.19(b). Since C acts as a d.c. block, the mean output voltage must be zero, so that the shaded areas above and below the time axis in the diagram must be equal.

The per-unit tilt of the output pulse can be found readily if CR is very much greater than the pulse-width, T_p. In this case $T_p/CR \ll 1$ and the exponent in $e^{-T_p/CR}$ is approximately equal to $1 - T_p/CR$. Also, if the output voltage rises from V_1 to V_2, then

$$V_2 = V_1 + V_t$$

where V_1 will be negative, and

$$V_3 = V_2\,e^{-T_p/CR} \approx V_2\left(1 - \frac{T_p}{CR}\right)$$

so that the per-unit tilt is

$$\frac{V_2 - V_3}{V_t} \approx \frac{V_2 - V_2(1 - T_p/CR)}{V_t} = \frac{V_2 T_p}{V_t CR} \qquad (23.30)$$

If V_1 is small, $V_t \approx V_2$ so that

$$\text{Per-unit tilt} = \frac{T_p}{CR} = 2\pi f_L T_p \qquad (23.30a)$$

For a CR time constant which is less than 0·2 of the pulse-width, the capacitor C can completely charge before the negative-going edge of the pulse arrives, and the output voltage will consist of positive- and negative-going spikes as shown at (c).

If the output voltage, v_o, is a very small fraction of the input voltage, v_i, then to a good approximation all the input voltage appears across the capacitor C so that the alternating current in C is

$$i \approx C\frac{dv_i}{dt}$$

Thus, if there is no external loading,

$$v_o = iR \approx CR\frac{dv_i}{dt} \qquad (23.31)$$

i.e. the circuit acts as approximate differentiating circuit. Note, however, that if the input is an ideal pulse with zero rise time, the derivative is infinite, and this cannot be obtained from the given circuit. In other words, the circuit acts as an approximate differentiator if (a) the CR time constant is small compared to the input rise time, (*b*) the output voltage is very small compared to the input, and (*c*) the rise time of the input is finite.

EXAMPLE 23.3 A pulse waveform of peak value 10 V is applied to the high-pass *CR* network of Fig. 23.19(*a*). The pulse-width is 100 μs and the *pulse repetition frequency* (p.r.f.) is 2,000 pulses/second. If $C = 0.1$ μF and $R = 100$ kΩ, determine the maximum positive and negative output voltage and the per-unit tilt.

Since the p.r.f. is 2,000 the pulse period is

$$T = \frac{1}{\text{p.r.f.}} = \frac{1}{2,000}\text{s} = 500\,\mu\text{s}$$

The output waveform is that shown in Fig. 23.19(*b*), where

$$V_2 = V_1 + V_i \tag{i}$$
$$V_3 = V_2\,e^{-T_p/CR} \tag{ii}$$
$$V_4 = V_3 - V_i$$

and

$$V_1 = V_4\,e^{-(T-T_p)/CR} \tag{iii}$$

For equal areas above and below the axis,

$$\int_0^{T_p} V_2\,e^{-t/CR}\,dt + \int_0^{T-T_p} V_4\,e^{-t/CR}\,dt = 0$$

Hence

$$-CRV_2(e^{-T_p/CR} - 1) = CRV_4(e^{-(T-T_p)/CR} - 1)$$

Substituting the given numerical values,

$$-V_2(e^{-0.01} - 1) = V_4(e^{-0.04} - 1)$$

and, to a close approximation,

$$-V_2(1 - 0.01 - 1) = V_4(1 - 0.04 - 1) \quad \text{or} \quad V_2 = -4V_4$$

From eqn. (iii),

$$V_1 \approx V_4(1 - 0.04) = 0.96V_4$$

Solving these equations for V_1 and V_2 gives $V_1 = -1.9$ V and $V_2 = 8.1$ V.

The per-unit tilt is found by substituting in eqn. (23.30):

$$\text{Per-unit tilt} \approx \frac{V_2 T_p}{V_i CR} = \frac{8.1 \times 10^{-4}}{10 \times 10^{-7} \times 10^5} = \underline{\underline{0.008}}$$

23.15 Preservation of Pulse Waveform

When a pulse waveform is applied to an *RC*-coupled amplifier the output will in general have a longer rise time than the input and will also have tilt. The Fourier analysis of a pulse waveform shows that it contains a large number of harmonics, and hence it is essential that the amplifier should have a wide bandwidth if the output is not to be distorted.

The high-frequency response of a single-stage amplifier determines the rise time of the pulse output. This is equivalent to the response of a low-pass *RC* network, and eqn. (23.25) gives the relation between rise time and the upper cut-off frequency, f_H, as

$$\text{Rise time} = \frac{2\cdot 2}{2\pi f_H}$$

The low-frequency response of a single-stage amplifier determines the per-unit tilt of the output, given by eqn. (23.30*a*) for a pulse-width, T_p, as

$$\text{Per-unit tilt} \approx 2\pi f_L T_p$$

provided that the pulse period, *T*, is long enough to ensure that $(T - T_p) > 5CR$.

Hence for best response f_H must be as high as possible and f_L as low as possible.

If a pulse waveform is applied to a circuit which has appreciable input capacitance, an increase in the rise time will occur unless the source is of zero impedance.

This is of particular importance if pulses are to be observed on a cathode-ray oscilloscope, since the trace obtained may not then correspond to the actual input, owing to the input capacitance of the oscilloscope. This effect may be partly overcome by the use of a *compensated attenuator*.

Consider the simple attenuator shown in Fig. 23.20(*a*). Using the normal transient analysis for a step input, it may be readily shown that the output for a step input is given by

$$V_o = \frac{V_i R_2}{(R_1 + R_2)} (1 - e^{-t/CR_p})$$

where $R_p = R_1 R_2/(R_1 + R_2)$. Hence the output rise time is $2\cdot2CR_p$ seconds.

If the attenuator is compensated by the addition of capacitor C_1, as shown at (*b*), then for r.m.s. voltages,

$$\frac{V_i}{V_o} = 1 + \frac{Y_p}{Y_s} = 1 + \frac{Z_s}{Z_p}$$

where $Y_s = 1/R_1 + j\omega C_1$ and $Y_p = 1/R_2 + j\omega C_2$. Hence

$$\frac{V_i}{V_o} = 1 + \frac{R_1(1 + j\omega C_2 R_2)}{R_2(1 + j\omega C_1 R_1)}$$

If $C_1 R_1 = C_2 R_2$ this expression becomes independent of frequency, and will therefore give no distortion of the pulse waveform. For this case,

$$\frac{V_o}{V_i} = \frac{1}{1 + \dfrac{R_1}{R_2}} = \frac{R_2}{R_1 + R_2} \tag{23.32}$$

If the attenuator is undercompensated ($C_1 R_1 < C_2 R_2$), there will be degradation of the rise time. If it is overcompensated ($C_1 R_1 > C_2 R_2$), there will be an overshoot on the output waveform. These effects are shown at (b).

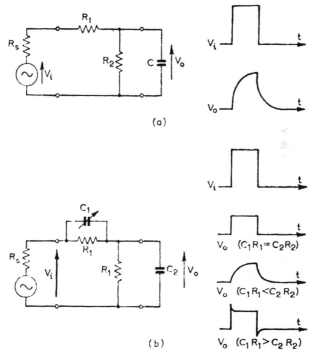

Fig. 23.20 PULSE RESPONSE OF ATTENUATOR
(a) Uncompensated   (b) Compensated

23.16 Sawtooth Generator

In many applications a voltage that increases linearly with time is required (e.g. for the deflexion of the spot on a cathode-ray tube; in television scanning circuits; in radar displays; etc.). Normally a slow *sweep* followed by a very rapid *flyback* is wanted, such as the sawtooth waveform shown in Fig. 23.21(*a*). This waveform may be

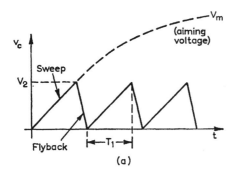

(a)

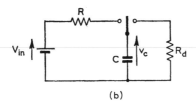

(b)

Fig. 23.21 SAWTOOTH WAVEFORM

generated by alternately charging a capacitor through a high resistance and discharging it through a low resistance. This is shown schematically at (*b*). The switching mechanism will be an electronic circuit. In order to achieve reasonable linearity, the flyback must be initiated when the capacitor voltage has reached a value V_2 which is a small fraction of the *aiming voltage*, V_m.

From transient theory, the instantaneous capacitor voltage, v_c, is

$$v_c = V_m(1 - e^{-t/CR})$$

$$= V_m\left(1 - 1 + \frac{t}{CR} - \tfrac{1}{2}\left(\frac{t}{CR}\right)^2 + \ldots\right)$$

$$= \frac{V_m t}{CR}\left(1 - \tfrac{1}{2}\frac{t}{CR} + \ldots\right)$$

For true linearity the capacitor voltage would be $V_m t/CR$, so that the term $\frac{1}{2}t/CR$ represents a fractional error. Since the neglected terms in the exponential series alternate in sign, and (provided that $t/CR \ll 1$) become smaller, the actual fractional error will always be less than $\frac{1}{2}t/CR$. If flyback occurs at time T_1 after the commencement of the sweep, when the capacitor voltage is V_2, then

$$V_2 \approx \frac{V_m T_1}{CR}\left(1 - \frac{1}{2}\frac{T_1}{CR}\right) \tag{23.33}$$

Hence the error in linearity is small provided that T_1 is a small fraction of the CR time constant of the charging circuit. If a constant-current source is used to charge C, there will be no error in linearity, since effectively the aiming voltage will then be infinite. The rate of growth of voltage can be varied by altering V_m, C or R.

An extremely simple sawtooth generator using a neon discharge tube is shown in Fig. 23.22(a). The capacitor, C, charges through

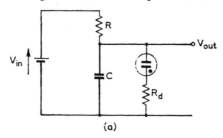

(a)

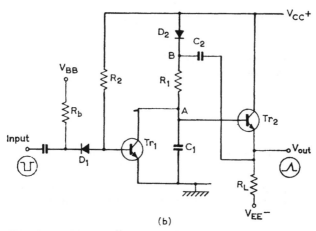

(b)

Fig. 23.22 SIMPLE SAWTOOTH GENERATORS
(a) Neon   (b) Bootstrap integrator

R until the breakdown voltage of the neon is reached, when it discharges rapidly through the protective resistance in the neon circuit. The neon is extinguished and the process repeats. The voltage rise will, of course, start from the extinction voltage of the neon and not from zero. For good linearity, V_m requires to be very large, and the sweep time, T_1, must be small compared to CR.

A more sophisticated circuit is the bootstrap integrator shown in Fig. 23.22(*b*). This relies on the fact that the integral of a square wave is a linearly rising curve. In principle, capacitor C_1 charges slowly through R_1 when transistor Tr_1 is non-conducting, and discharges rapidly through Tr_1 when this transistor is switched into conduction by the rising edge of the input square wave. The function of transistor Tr_2 is to give a constant voltage across the charging resistor R_1 (and hence to give a constant charging current for C_1). Since Tr_2 is in the emitter-follower connexion, it has a low-impedance output, so that the output waveform is very little affected by any load at the output terminals.

With no input, Tr_1 conducts, so that the voltage at A is almost zero (and hence C_1 is discharged), while the voltage across R_1 is almost V_{CC}. The output voltage is also almost zero, owing to emitter-follower action.

When the input voltage goes negative, Tr_1 becomes cut off, and C_1 starts to charge through R_1. The voltage at A rises, and owing to emitter-follower action, so does the output voltage. The output voltage rise is fed back through the large capacitor C_2 to point B, so that, if the gain of the emitter follower is unity, the voltages at A and B will both increase by the same amount and the charging current will remain constant. The rising voltage at B will cut off diode D_2. This linear rise continues as long as Tr_1 is cut off—the output of Tr_2 may be thought of as lifting itself during this time by its own bootstraps. The linearity depends on how closely the gain of the emitter follower approaches unity, and on the fact that $C_2 \gg C_1$.

PROBLEMS

23.1   The open loop gain of a feedback amplifier is given by

$$\beta A_{vo} = -10/(1 + jf(10^{-7}))^3$$

Determine whether the amplifier will be unstable. Determine the frequency at which the amplifier will become unstable if the numerator becomes -5.

Ans. 17·3 MHz.

23.2   For the Colpitts oscillator shown in Fig. 23.4 the value of L is 20 μH and the two tuned circuit capacitors each have a value of 500 pF. Determine from the generalized conditions for oscillation the frequency of oscillation and the minimum value of h_{fe} required for the transistor.

Ans. 2·25 MHz; 1.

23.3 Repeat Problem 23.2 for the Hartley oscillator of Fig. 23.5 given that $L_1 = 500\,\mu$H, $L_2 = 200\mu$H and $C = 1,000$ pF, and assuming no mutual coupling between the coils.

Ans. 190 kHz; 2·5.

23.4 Determine the figure of merit

$$\frac{d}{d\omega}(X_{ce} + X_{cb} + X_{be})$$

for frequency stability of the oscillator of Problem 23.3.

Ans. 40 × 10⁻⁶.

23.5 Repeat Problem 23.4 for the Hartley oscillator of Problem 23.3.

Ans. 1·40 × 10⁻³.

23.6 In the tuned-collector oscillator of Fig. 23.6 the tuned circuit has a dynamic resistance of R_0. Show that the circuit will oscillate at a frequency given by $\omega_0 \approx 1/\sqrt{(L_1C)}$ when

$$n(h_{ie} + R)(1/R_0 + h_{oe}) + 1/n = h_{fe}$$

where R is the resistance of R_1 and R_2 in parallel. The transformer has a turns ratio of $n:1$ and a coupling coefficient of unity.

23.7 Sketch the circuit and describe the operation of a free-running (astable) multivibrator.

In a symmetrical multivibrator the collector resistors are 1·8 kΩ and the coupling capacitors are 0·01 μF. The transistors are *p–n–p* and the base resistors of 27 kΩ are connected to the negative supply rail. Calculate from first principles the pulse repetition frequency. Also determine the rise time of the output pulses.

Ans. 2,650 pulses/second; 39·5 μs.

23.8 Explain the operation of a monostable transistor multivibrator.

For the circuit shown in Fig. 23.30 determine the output pulse-width and the

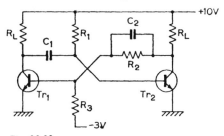

Fig. 23.23

base and collector voltages and currents in the quiescent state. $R_L = 4\cdot7$ kΩ, $C_1 = C_2 = 0\cdot001\,\mu$F; $R_1 = R_2 = 100$ kΩ; $R_3 = 47$ kΩ. Assume a base–emitter voltage of 0·7 V and negligible collector–emitter voltage at saturation.

Ans. 69 μs; $V_{C1} = 10$ V; $V_{B1} = -0\cdot96$ V; $V_{C2} = 0$; $V_{B2} = 0\cdot7$ V; $I_{C2} = 2\cdot13$ mA; $I_{B2} = 93\,\mu$A.

23.9   A train of positive pulses, of amplitude 20V, pulse-width 0·2ms and repetition frequency 1 kHz, is applied to a high-pass CR network in which $C = 0·01 \mu F$ and $R = 60k\Omega$. Sketch the output waveform and determine the maximum positive and negative values of the output voltage.   (*H.N.C.*)
   Ans. 18·2V;   −7V.

23.10   A square-wave pulse train (1:1 mark-to-space ratio) is applied to a low-pass RC network. The input pulses have a peak-to-peak amplitude of 10V. If $R = 100k\Omega$, $C = 0·1 \mu F$ and the pulse repetition frequency is 150 pulses per second, determine the peak-to-peak output voltage. Sketch the input and output waves.
   Ans. 0·28V.

23.11   Explain the action of a compensated attenuator.
   In a compensated attenuator as shown in Fig. 23.20(*b*), $C_2 = 250pF$, $R_1 = 10k\Omega$, $R_2 = 15k\Omega$. Determine the correct value of C_1. If C_1 is 10 per cent high, find the fractional overshoot when a rectangular step is applied at the input.
   Ans. 375pF; 0·054.

23.12   The symmetrical astable multivibrator shown in Fig. 23.24 uses *n–p–n*

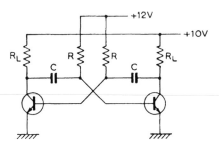

Fig. 23.24

transistors.   Determine   the   pulse   repetition   frequency   it $R_L = 4·7k\Omega$, $R = 100k\Omega$ and $C = 500pF$.
   Ans. 16,500 pulses/second.

Chapter 24

POWER AMPLIFIERS

Power amplifiers can be generally classed into two groups:

(*a*) Those intended to deliver power over a wide frequency range.

(*b*) Those intended to deliver power at one particular frequency only.

The first type is principally used for audio-frequency power (30–30,000 Hz) and may be either transformer coupled or direct coupled; the second type is principally used for radio-frequency power (100 kHz upwards) and is generally a tuned amplifier.

The power amplifier stage in an equipment is generally the final stage which must drive a given load, e.g. a mechanical indicator, a loudspeaker, cathode-ray-tube magnetic deflexion coils, a transmitting aerial of a communication network. Power amplifiers are also used in regulated power supplies, chopper amplifiers, motor controllers, etc. In general the load will demand a certain power from the amplifier and the output stage must be capable of dealing with this power. The design of an equipment often commences with the output stage and with its required output power. The preceding stage is then designed to give the input required to drive the output stage. Amplifiers are then added until there is the required amplification between the initial input signal and the output power amplifier.

In a power amplifier a large portion of the active-device characteristic is used (i.e. there is large alternating input voltage or current) to make efficient use of the device. This gives rise to non-linear distortion in the device due to the curvature of the characteristics.

759

It is important to keep the non-linear distortion within tolerable limits.

The power transistor is rapidly replacing the valve as the active device in most power amplifier applications. Triodes and tetrodes, however, are still used as output valves in broadcast transmitters, and special thermionic devices are still required at frequencies in the microwave range, where valves (Klystrons) having continuous output powers of over 50 kW at 10 GHz are used. It is not proposed to discuss these special devices in this book.

Power transistors are now available for operation at voltages in excess of 1 kV, or at currents in excess of 250 A (at low voltage). They can be obtained in complementary n-p-n and p-n-p configurations, which in many applications, gives them further advantages over valves, besides the great advantage of requiring no heater supply. The value of f_T for power transistors has been continuously increased over the past few years due to improvements in manufacturing technology, and for many nominally "low-frequency" power devices f_T exceeds 15 MHz.

The construction of the power transistor may differ mechanically from that of the small-signal transistor, in that the collector is usually directly connected to the metal outside case. This facilitates direct connexion (thermally and electrically) to a metal heat sink and enables the highest possible power rating to be achieved. At elevated temperatures care must be taken to observe the requirements of the derating curves supplied by the manufacturers.

24.1 Class A Power Amplifier

A Class A amplifier is one in which the current flows in the output valve or transistor at all instants throughout the cycle—other types will be described in following sections. A Class A valve amplifier is shown in Fig. 24.1(a). A triode, a pentode or a beam tetrode may be used—the triode has less distortion, but requires a larger alternating input grid voltage. For small powers (less than, say, 25 W) a pentode is more common, but for higher powers a triode or tetrode becomes essential as it is difficult to dissipate any appreciable power from the screen grid of a pentode.

A transistor power output stage is shown at (c). In transistor stages particular attention must be paid to heat dissipation (to prevent thermal runaway of the transistor), and to voltage breakdown.

Since power amplifiers may operate over wide ranges of the active-device characteristics, it is no longer permissible to use the small-signal linear equivalent circuit—actual device characteristic curves must be used. The load resistor, R, is usually connected through a step-down transformer in order to achieve maximum power output. This has the additional advantage that it prevents the flow of direct current through the load.

Typical output characteristics are shown at (*b*) and (*d*). In both cases the resistance of the transformer primary has negligible effect on the static load line, which is therefore the vertical ordinate from the supply voltage V_S. The quiescent (no signal) operating point is determined by the bias conditions, and is chosen to give equal

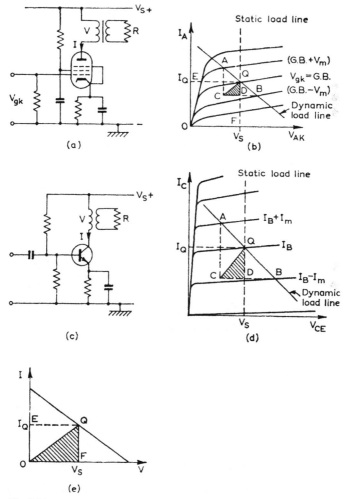

(a)

(b)

(c)

(d)

(e)

Fig. 24.1 CLASS A AMPLIFICATION

excursions along the dynamic load line above and below Q. For signal currents the load resistance referred to the primary of the output transformer is $k_t^2 R$, where k_t ($=N_1/N_2$) is the turns ratio. The dynamic load line is drawn through Q with a slope of $-1/k_t^2 R$, assuming that the transformer winding resistance is negligible (this may not always be true, however, particularly for small transformers).

For an alternating input the operating point moves along the dynamic load line between A and B, and the variation of the primary current of the output transformer is then proportional to AC, and the corresponding voltage variation, to CB.

Assuming sinusoidal variations, the r.m.s. alternating component of the transformer primary current, I, is proportional to $AC/2\sqrt{2}$. Also the r.m.s. alternating component of the transformer primary voltage, V, is proportional to $BC/2\sqrt{2}$. Therefore

$$\text{Alternating power developed} = VI \propto \frac{AC \cdot BC}{8} \qquad (24.1)$$

By geometry, this alternating power is represented by one-quarter of the area ABC, or, if distortion is neglected, by the shaded area CQD.

The power drawn from the supply is the supply voltage times the mean direct current, i.e. VI_Q. This is represented by the rectangle OEQF shown at (b).

$$\text{Efficiency of amplifier} = \frac{\text{Signal power output}}{\text{Power from supply}} = \frac{VI}{VI_Q}$$

$$= \frac{\frac{1}{4} \times \text{area ABC}}{\text{area OEQF}} \approx \frac{\text{area CQD}}{\text{area OEQF}} \qquad (24.2)$$

For the cases depicted the efficiency is less than $12\frac{1}{2}$ per cent.

$$\text{Power dissipated in active device} = V_S I_Q - VI \qquad (24.3)$$

If there is no signal input, the entire supply power, $V_S I_Q$, will be dissipated in the valve or transistor, and the quiescent operating point must be chosen so that this power does not exceed the permissible limit for the device.

Due to the low efficiency and the possibility of thermal runaway in the absence of an input signal, transistors are not commonly used in Class A power operation.

24.2 Maximum Efficiency in Class A

As the signal input increases, the output power and efficiency will increase. Neglecting distortion, and assuming that the output characteristics extend to the current and voltage axes in Figs. 24.1(*b*) and (*d*), maximum power output is obtained if the entire dynamic load line is used as at (*e*). Then

$$\text{Efficiency} = \frac{\text{area OQF}}{\text{area OEQF}} = 0.5 \qquad (24.4)$$

This is the theoretical maximum efficiency of any Class A amplifier. In valve amplifiers it is unusual for efficiencies to exceed 0·25 owing to the large amount of distortion produced with large grid-voltage swings. Some transistor amplifiers achieve a value much nearer the theoretical maximum.

Note that, if the input signal is a square wave or a pulse waveform, the maximum theoretical efficiency is increased to 100 per cent. Pulse operation of a power amplifier has been called Class D operation.*

EXAMPLE 24.1 The anode characteristics of a triode are shown in Fig. 24.2. The triode is to be used as a power amplifier valve feeding a load resistance of 50Ω through a 10:1 step-down transformer. The supply voltage, V_B, is 530V. The maximum power dissipation at the anode of the triode is 20W.

(i) Draw a line across the triode characteristics representing the maximum anode dissipation.

(ii) Choose a suitable grid-bias voltage and determine the corresponding cathode bias resistor.

(iii) Draw the dynamic load line.

(iv) Determine the power output and efficiency for peak grid voltages of (*a*) 20V, and (*b*) 40V. The transformer may be assumed perfect.

(i) For a power of 20W at the anode, $V_A I_A = 20$, where I_A is the anode current in amperes. The line representing 20W on the anode characteristics is the rectangular hyperbola given by the above equation and detailed in the following table:

V_A (volts)	200	300	400	500	600	800
I_A (mA)	100	66·7	50	40	33·3	25

* An oscillator is said to operate in Class D when the active device is switched on and off at successive angles of current flow of 180°. The maximum theoretical efficiency, like that of Class C, is 100 per cent. The power is limited only by the permissible peak current and voltage, and by the switching speed obtainable, as distinct from Class C, where the power tends to zero as the efficiency approaches 100 per cent.

Class D operation may be applied to amplifiers. The angle of current flow in the on (or off) condition is modulated by the signal to give a pulse-width-modulated output from which the signal is recovered by means of a low-pass filter.

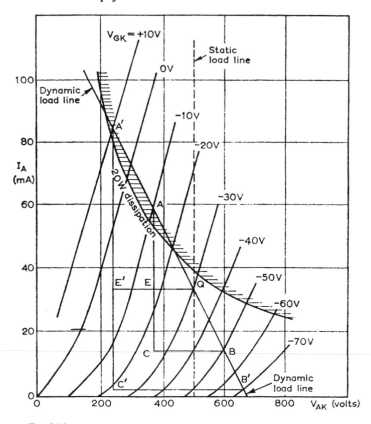

Fig. 24.2

(ii) To give operation in the linear region without exceeding the permissible anode power, a grid bias of $-30\,$V will be seen from Fig. 24.2 to be suitable with a supply voltage of $530\,$V. The anode voltage will thus be $530 - 30 = 500\,$V. Q represents the quiescent operating point and quiescent current. Therefore $I_Q = 33\,$mA, and

$$\text{Cathode bias resistance} = \frac{V_{GK}}{I_Q} = \frac{30}{33} \times 1{,}000 = \underline{\underline{910\,\Omega}}$$

(iii) The impedance in the anode circuit to alternating currents is $k_t^2 R$, i.e. $10^2 \times 50 = 5{,}000\,\Omega$. (This is about twice r_a for the valve.) Thus the anode load line is a line through Q with a slope of $-1/5{,}000$. (Although at some points the load line is above the 20 W permissible limit, the mean and quiescent power will still be less than 20 W.)

(iv) (a) For a peak grid voltage of 20 V the operation is between points A and B on the dynamic load line.

Peak-to-peak current swing = AC = 46 mA

Peak-to-peak voltage swing = BC = 222 V

Power output (assuming sine waves) = $\dfrac{46 \times 222}{8 \times 1,000}$ = $\underline{\underline{1\cdot3\,\text{W}}}$

Power input $\approx V_B I_Q = \dfrac{530 \times 33}{1,000}$ = 17·5 W

Efficiency = $\dfrac{1\cdot3}{17\cdot5} \times 100$ = 7·5 per cent

(*b*) For a peak grid voltage of 40 V the operation is between A' and B' on the dynamic load line. Thus

Power output (assuming sine waves) = $\dfrac{84 \times 420}{8,000}$ = $\underline{\underline{4\cdot4\,\text{W}}}$

and

Efficiency = $\dfrac{4\cdot4}{17\cdot5} \times 100$ = $\underline{\underline{25}}$ per cent

The following point may be noted: when the power is calculated from the peak-to-peak current and voltage swings, the assumption of sine waves gives negligible error. Indeed it gives the fundamental output power without error if only even-order curvature is present.

24.3 Choice of Transformer Ratio in Class A

When a power amplifier is designed the following will normally be specified (*a*) load resistance; (*b*) load power (which, after assuming a reasonable anode or collector efficiency, will determine the rating of the valve or transistor used); (*c*) permissible distortion. Sometimes the supply voltage is also specified, but more usually this is chosen to suit the active devices. A suitable output transformer ratio must then be selected. From the output characteristics of the active device the required bias and signal input are determined, and power output and distortion may then be checked graphically.

If the output characteristics are reasonably linear, and the excursion of the input is not too great, then the maximum power transfer theorem shows that the maximum power transfer occurs when the reflected load resistance, $k_t^2 R$, is equal to the slope of the output characteristics (i.e. the output impedance of the active device).

In practice, in order to obtain the best utilization of the active device, it is usual to operate into the non-linear region of the characteristics as shown in Fig. 24.3. The output characteristics of transistors, tetrodes and pentodes are generally closer together at both ends of the dynamic load line, and hence for a large output signal which extends into these regions there will be some flattening of both its positive and negative peaks. Such a flattening has been seen in Chapter 5 to introduce third-harmonic distortion. If only the negative or only the positive peaks are flattened the resulting distortion is mainly second harmonic—this is the most usual form of distortion in triode power amplifiers.

If the dynamic load line represents too high a reflected load impedance (e.g. the line marked (1) in Fig. 24.3), intolerable distortion occurs at the positive peak of the signal. On the other hand, if the reflected impedance is too low (line (2)), the maximum power output is small. By choosing the transformer ratio so that the dynamic load line passes through the knee of the output characteristic

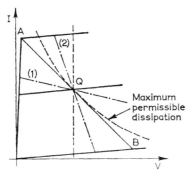

Fig. 24.3   PERTAINING TO THE CHOICE OF A TURNS RATIO TO GIVE A DYNAMIC LOAD LINE THROUGH THE KNEE OF THE OUTPUT CHARACTERISTIC

as shown by line AB on the diagram, a high maximum power output with minimum distortion is achieved. With transistors this knee voltage may be quite small (less than 1 V), but in valve circuits it is generally several tens of volts. This is a further important reason why transistors are so much more efficient power amplifiers than valves, since they enable a much larger range of the characteristic to be used. In all cases care must be taken not to exceed the maximum power dissipation of the transistor or valve.

It can be shown that in triode amplifiers, limited to a negative grid voltage, the optimum ratio is that which makes $k_t^2 R = 2r_a$.

24.4 Square-law Distortion

In many cases the relation between input and output signals (particularly in triode amplifiers) may be simply approximated by a square law. Thus let the relation between anode current, i_a, and grid voltage, v_g, for an output valve be given by

$$i_a = a_0 + a_1 v_g + a_2 v_g^2 \qquad (24.5)$$

Let the grid voltage be given by

$$v_g = V_{gb} + V_m \cos \omega t$$

i.e. a steady bias of V_{gb} in series with a sinusoidal signal. Then

$$i_a = a_0 + a_1V_{gb} + a_2V_{gb}^2 \tag{1}$$
$$+ a_1V_m \cos \omega t + a_2 . 2V_{gb} . V_m \cos \omega t \tag{2}$$
$$+ \tfrac{1}{2}a_2V_m^2 \tag{3}$$
$$+ \tfrac{1}{2}a_2V_m^2 \cos 2\omega t \tag{4}$$

since $\cos^2 \omega t = \tfrac{1}{2}(1 + \cos 2\omega t)$.

Part 1 of the anode current is the normal quiescent current I_g for the grid bias V_{gb}.

Part 2 is the anode current variation at the frequency of the input, i.e. the desired anode current variation.

Part 3 is an extra direct current introduced owing to the curvature of the characteristics and proportional to the square of the alternating input voltage. It also represents the rectified signal which is characteristic of a square-law transfer relation.

Part 4 is an unwanted double-frequency, or second harmonic, component in the output waveform which is not in the input waveform. This is the result of non-linear distortion. It should be noted that the amplitude of the second-harmonic term is the same as the magnitude of the direct current due to the curvature.

The component and total current waveforms are shown in Fig. 24.4. It will be seen that from quiescent current to peak positive current corresponds to peak fundamental current plus twice peak

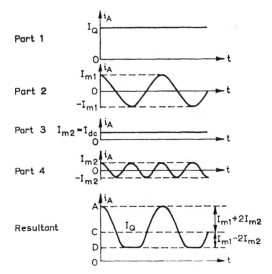

Fig. 24.4 COMPONENTS OF A WAVE WITH SQUARE-LAW DISTORTION

second-harmonic current, and that from quiescent current to peak negative current corresponds to peak fundamental current minus twice peak second-harmonic current.

Referring to the resultant waveform,

$$AC = I_{m1} + 2I_{m2}$$
$$CD = I_{m1} - 2I_{m2}$$

so that

$$I_{m1} = \frac{AC + CD}{2}$$

and

$$I_{m2} = \frac{AC - CD}{4}$$

Therefore

Percentage second-harmonic distortion

$$= \frac{I_{m2}}{I_{m1}} \times 100 = \frac{1}{2} \frac{AC - CD}{AC + CD} \times 100 \qquad (24.6)$$

Thus the second-harmonic distortion may be simply estimated from the dynamic characteristic. This is actually the distortion of the current waveform—the percentage distortion of the voltage waveform will be the same since the voltage and current are related by the dynamic resistance of the anode load. Though it is actually the second-harmonic power which is undesirable, it is always the percentage second-harmonic current or voltage which is referred to in specifications. "A limiting distortion of 5 per cent" means that the harmonic current should not exceed 5 per cent.

With triodes only the second harmonic is of importance. The distortion in pentodes and transistors has already been mentioned, in Section 24.3.

24.5 Class B and Class C Amplifiers

Operation in Class B is obtained if the active device is biased in the quiescent condition to the cut-off point. Current can therefore flow only during one half-cycle of the input signal. In a "single-ended" wide-band amplifier as shown in Fig. 24.1, this would, of course, give an intolerably distorted output. It may, however, be used in tuned power output stages, where the output frequency is determined by the resonant frequency of the tuned circuit, and hence distortion is negligible if the Q-factor of the tuned circuit is reasonable.

The advantage of this class of operation is a higher theoretical efficiency. This may be derived as follows.

Assume that the output current of the active device is a series of half sine-waves of peak value I_m. Then

$$\text{Mean output current} = \frac{1}{\pi} I_m$$

Since the device is cut off for one half-cycle, the power taken from the supply is

$$\text{Supply power} = \frac{1}{\pi} I_m V_{CC}$$

when V_{CC} is the supply voltage. There is an a.c. power output only during the conducting half-cycle, and this has a value given by

$$\text{A.C. power} = \frac{1}{2} \frac{V_m}{\sqrt{2}} \frac{I_m}{\sqrt{2}}$$

where V_m is the peak alternating output voltage. From this,

$$\text{Conversion efficiency} = \frac{V_m I_m}{4} \frac{\pi}{I_m V_{CC}} = \frac{\pi}{4} \frac{V_m}{V_{CC}} \qquad (24.7)$$

Since the maximum theoretical value of the peak alternating output voltage must be equal to the supply voltage, V_{CC}, if the full output characteristic is used, it follows that

$$\text{Maximum possible Class B efficiency} = \frac{\pi}{4} \times 100$$

$$= 78 \cdot 5 \text{ per cent}$$

Actual efficiencies of around 50 per cent may be achieved in tuned Class B amplifiers.

In Class C operation, the active device is biased well beyond the cut-off point, so that conduction occurs only during a fraction of a half-cycle of the signal. The output efficiency is higher than for Class B, but applications are limited, owing to the obvious disadvantage that there will be no output whatever if the input voltage falls below a critical value. Class C operation is, however, much used in tuned high-power valve amplifiers where the high efficiency is a great advantage.

Sometimes, particularly in push-pull circuits as described in the next section, an intermediate class of amplification designated as Class AB is employed. In this case the bias is set so that the active device conducts for more than half a cycle of the input. It is intermediate between Class A and Class B. Very frequently transistor amplifiers which are designated Class B are in fact Class AB.

The various classes of power amplification are illustrated for valve circuits on the mutual dynamic characteristics shown in Fig. 24.5,

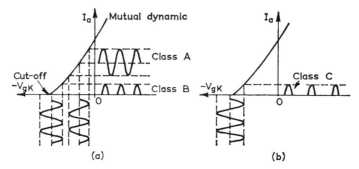

Fig. 24.5   CLASS A, B AND C OPERATION

where output current waveforms are shown for a sinusoidal input and resistive load superimposed on the appropriate grid bias. Notice that it is more usual to employ fixed (external) bias or grid current bias for Class B or C amplifiers, since the devices are cut-off in the quiescent state.

24.6 Class B Push-pull Operation

In Class B push-pull circuits the valves or transistors are arranged in such a way that each conducts for alternate half-cycles of the input. The outputs are combined to give a composite output which is the sum of those due to each active device alone.

A simple transistor output stage is shown in Fig. 24.6(*a*) using centre-tapped input and output transformers. The bases are driven in antiphase by the input transformer. During the positive half-cycles of the input, Tr_1 conducts and Tr_2 is cut off, so that the dotted end of the output transformer secondary goes negative. During the negative half-cycles of the input, Tr_1 is off and Tr_2 conducts, so that the dotted end of the output transformer secondary goes positive. Hence, theoretically, the output is an amplified version of the input.

In practice the base–emitter voltage, V_{BE}, required for conduction means that a transistor will not conduct until the input voltage exceeds V_{BE}. The output wave may therefore be considerably flattened near the zero current values. This flattening, which is shown in the output waveform at (*b*), gives rise to a type of distortion known as *crossover distortion*, which has a large third-harmonic component and is normally quite unacceptable. There is, however, almost no second-harmonic distortion, since positive and negative half-cycles are identical. If the transistors are not well matched, however, the positive and negative half-cycles of the output will not be identical so that second-harmonic distortion will appear. Cross-over distortion is minimized by forward biasing of the base by an amount which is sufficient to overcome V_{BE}. This is shown schematically at (*a*).

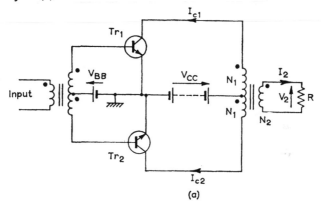

(a)

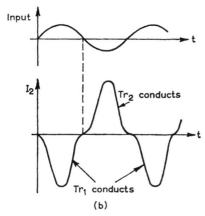

(b)

Fig. 24.6   CLASS B PUSH-PULL OUTPUT STAGE

The advantages of push-pull operation include the following:

1. Class B or Class AB operation is possible without the introduction of second-harmonic distortion (which generally has the most unpleasant effects in audio amplifiers). Hence high output efficiencies can be achieved, since larger inputs can be tolerated.

2. There is no resultant d.c. magnetization in the output transformer, since the mean currents in each half of the primary winding flow in opposite directions. It follows that no specially large iron cross-section need be used to reduce the second-harmonic distortion which would arise due to asymmetrical magnetization (and hence asymmetrical saturation) of the core. Any saturation effects will be symmetrical, and will therefore introduce odd harmonics only.
3. Since the fundamental signal currents are in anti-phase in each transistor, the signal current which flows through the d.c. supply (i.e. the sum of I_{c1} and I_{c2}) can contain no fundamental component.
4. Conversely, hum or noise in the d.c. supply will affect each transistor equally, and since the output is proportional to the *difference* between I_{c1} and I_{c2}, any such variations will cancel out.

24.7 Power Output—Class B Push Pull

For the circuit of Fig. 24.6(*a*), the load resistance reflected into each half of the primary of the output transformer is

$$R' = \left(\frac{N_1}{N_2}\right)^2 R \tag{24.8}$$

For a supply voltage of V_{CC}, the maximum possible collector current occurs if the whole output characteristic is used, i.e. if the transistor is driven up to saturation. The collector–emitter voltage will then be almost zero, so that the peak collector current is instantaneously

$$I_{cm} = \frac{V_{CC}}{R'}$$

and the voltage across the relevant half of the output transformer is V_{CC}.

Note that at this point the collector–emitter voltage of the cut-off transistor will be the supply voltage plus the voltage induced in its own half of the primary (which will have a peak value of V_{CC} also), i.e. it will have a possible maximum value of $2V_{CC}$. This helps to determine the output transistor to be used in a design (or the supply voltage for a given transistor), since it means that the transistor should have a peak voltage rating of at least twice the proposed supply voltage.

The maximum fundamental-frequency power output to the load R, assuming 100 per cent efficiency of the output transformer, is given by

$$P_m = \left(\frac{I_{cm}}{\sqrt{2}}\right)^2 R' = \left(\frac{V_{CC}}{\sqrt{2R'}}\right)^2 R' = \frac{V_{CC}^2}{2R}\left(\frac{N_2}{N_1}\right)^2 \qquad (24.9)$$

The collector dissipation can now be found. It is simply the power, P_S, taken from the d.c. supply less the a.c. load power, P_m. Thus assuming half sine-waves of collector current through each transistor, the mean supply current is $(2/\pi)I_{cm}$. Hence

$$P_S = \frac{2}{\pi} I_{cm} V_{CC}$$

so that

$$\text{Collector dissipation per transistor} = \frac{P_S - P_m}{2} \qquad (24.10)$$

The output current waveform for one half-cycle can readily be obtained from a load line drawn on the output characteristic of one transistor, the load line having a slope of $-R'$ and passing through the point $I_C = 0$, $V_{CE} = V_{CC}$. For identical transistors the other half-cycle of output current will be the same.

EXAMPLE 24.2   Two transistors whose characteristics are shown in Fig. 24.7(a) are used in Class B push-pull, to supply a $15\,\Omega$ load through an output transformer. The supply voltage is $15\,\text{V}$. Determine (a) a suitable transformer ratio, (b) the power output, and (c) the conversion efficiency. Sketch one half-cycle of the output waveform, assuming a sinusoidal base-current drive.

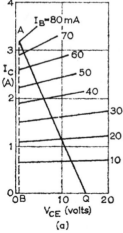

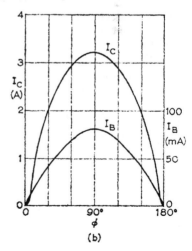

Fig. 24.7

(*a*) For maximum efficiency the dynamic load line (which starts at point Q in Fig. 24.7(*a*)) passes through the knee of the characteristic. From the diagram the effective primary resistance, R', is

$$R' = \frac{BQ}{AB} = \frac{14}{3\cdot2} = 4\cdot4\,\Omega$$

Hence, from eqn. (24.8), the turns ratio of the transformer is

$$\frac{N_1}{N_2} = \sqrt{\frac{R'}{R_L}} = \sqrt{\frac{4\cdot4}{15}} = \frac{1}{1\cdot86}$$

(*b*) With this turns ratio, the maximum power output to the load is

$$P_{cm} = \frac{I_{cm}{}^2}{2}\,R' = \frac{3\cdot2^2}{2} \times 4\cdot4 = 22\cdot5\,\text{W}$$

(*c*) Assuming that the output current consists of a sine wave of peak value 3·2 A, the mean current from the supply is $(2/\pi) \times 3\cdot2 = 2\cdot03\,\text{A}$, so that the supply power is $2\cdot03 \times 15 = 30\cdot5\,\text{W}$, and

$$\text{Conversion efficiency} = \frac{22\cdot5}{30\cdot5} = 0\cdot74\,\text{p.u.}$$

The collector-current waveform is obtained by sketching the sinusoidal base-current waveform over one half-cycle (i.e. to a peak of 80 mA) against phase angle as at (*b*). At any phase angle the collector current corresponding to the instantaneous base current is obtained from the intersection of the dynamic load line with the given base current.

Note that practical Class B amplifiers are normally arranged to have a small quiescent current in order to reduce crossover distortion.

24.8 Class AB Push-pull Amplifiers

In order to reduce crossover distortion, and at the same time to maintain a reasonably high conversion efficiency, it is usual to bias the output transistors (or valves) so that under quiescent conditions they are not quite cut off. This is Class AB operation. A typical circuit is shown in Fig. 24.8(*a*). Resistors R_1 and R_2 provide the base bias, and resistors R_3 and R_4 in the emitter leads help to stabilize the circuit against changes in temperature.

The graphical solution for the output may be obtained by drawing the composite characteristics of the push-pull pair. First the reflected load resistance is found from the following relations:

Primary voltage per turn = Secondary voltage per turn

i.e.

$$\frac{V_1}{N_1} = \frac{V_2}{N_2} \tag{24.11}$$

and

Total primary ampere-turns = Secondary ampere-turns

i.e.

$$(I_{c1} - I_{c2})N_1 = I_2 N_2 \tag{24.12}$$

where I_{c1} and I_{c2} are the collector signal currents.

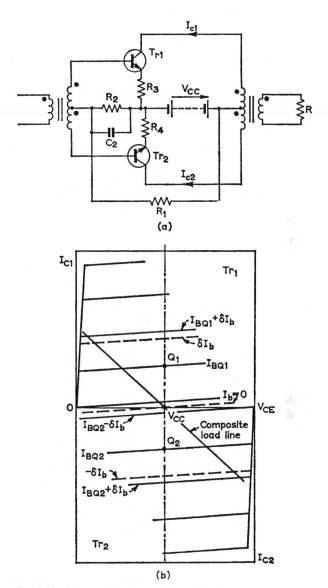

Fig. 24.8 CLASS AB PUSH-PULL OUTPUT STAGE

Dividing eqn. (24.11) by eqn. (24.12),

$$\frac{V_1}{(I_{c1} - I_{c2})N_1{}^2} = \frac{V_2}{I_2N_2{}^2}$$

or

$$\frac{V_1}{(I_{c1} - I_{c2})} = \frac{V_2N_1{}^2}{I_2N_2{}^2} = \left(\frac{N_1}{N_2}\right)^2 R = R' \tag{24.13}$$

This equation shows that the reflected load resistance, R', may be represented by a line of slope $-1/R'$ on a characteristic of the difference current $(I_{c1} - I_{c2})$ plotted to a base of collector voltage, V_{CE}. To obtain this characteristic it is convenient to plot two sets of collector characteristics, one of which has both current and voltage axes reversed. The voltage axes are arranged so that the points corresponding to the supply voltage, V_{CC}, coincide as shown in Fig. 24.8(b).

The composite characteristic for the quiescent base current (the same in each transistor) is then obtained by adding the currents shown by the static curves I_{BQ1} and I_{BQ2} at each value of voltage to give the broken line labelled $I_b = 0$ (i.e. no signal). Similarly for a signal current of δI_b the composite curve is obtained by adding the currents shown by the static curves $I_{BQ1} + \delta I_b$ and $I_{BQ2} - \delta I_b$ (corresponding to a signal current of $+\delta I_b$ in one transistor and $-\delta I_b$ in the other), to give the broken line labelled $+\delta I_b$; and for a signal current of $-\delta I_b$ by adding the currents shown by the static curves $I_{BQ1} - \delta I_b$ and $I_{BQ2} + \delta I_b$ to give the broken line labelled $-\delta I_b$. This process may be repeated for other values of δI_b as required.

The composite load line then has a slope of $-1/R'$ $(=N_2{}^2/N_1{}^2R)$ and passes through the point $I_C = 0$, $V_{CE} = V_{CC}$.

A similar construction will apply in the case of valve push-pull amplifiers.

Class A push-pull operation is sometimes used in valve amplifiers where a large output with low distortion is required. This class of operation is unusual in transistor circuits because of the large quiescent dissipation and the resulting difficulty in achieving adequate thermal stability. Also in Class A the efficiency is below 50 per cent, and for a given power output the cost of transistors is higher than that of valves.

EXAMPLE 24.3   The valve push-pull amplifier shown in Fig. 24.9(a) is operated with a supply voltage of 400 V and a grid bias of -12 V to give Class AB operation. The valves used are identical Mullard EL31 pentodes whose characteristics are shown at (b). The output transformer has an overall ratio of 20:1 and the load resistance is 10 Ω. At an anode voltage of 400 V, the valve is cut off if the grid voltage is less than -20 V.

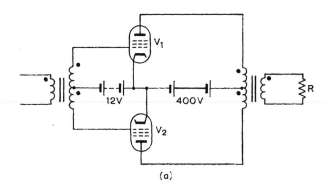

(a)

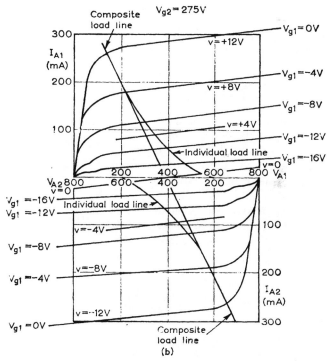

(b)

Fig. 24.9

Draw the composite characteristics, and determine the power output, and estimate the conversion efficiency if the sinusoidal signal input voltage is 12 V peak at each grid.

The composite characteristic is drawn by inverting one set of individual characteristics, and setting the supply voltage points (400 V) together. For a zero signal ($v = 0$) the composite characteristic is obtained by adding the currents shown by the static curves for $V_{g1} = -12$ V at each value of anode voltage. This is the line marked $v = 0$ at (*b*), passing through the 400 V zero-current point. For a signal of $+4$ V on valve V_1 (and hence -4 V on valve V_2) the composite curve is obtained by adding the currents of the static characteristic for $V_{g1} = -12 + 4 = -8$ V for V_1 and that for $V_{g1} = -12 - 4 = -16$ V for V_2. This is the line marked $v = +4$ V at (*b*).

When the signal exceeds ± 8 V one or other valve becomes cut off and the composite characteristics coincide with the static characteristics for each valve alone.

The composite load line has a slope of $-1/R'$ (where $R' = R(N_1/N_2)^2 = 10 \times (10)^2 = 1,000$, since the ratio for each half of the output transformer is 10:1). This load line passes through the 400 V zero-current point as shown. Notice that the individual load lines are curved—these represent the current/voltage relations for the individual valves. Thus, for zero resultant output current, the anode voltages are each 400 V and the grid voltages each -12 V, giving quiescent anode currents of 60 mA, so that (400 V, 60 mA) is a point on the individual load characteristics. When the signal voltage is $v = +8$ V, the anode voltage of the conducting valve is 200 V, and the current is 180 mA, while the anode voltage of the valve which is cut off is 600 V and its anode current is zero. Hence points (600 V, 0 mA) and (200 V, 180 mA) are points on each individual load line.

The d.c. power input is found by estimating the mean anode current when the ± 12 V signal is applied. This is done by drawing the anode-current waveform for one valve, using the individual load lines. In this problem, it is estimated that the mean anode current is 100 mA. Hence

D.C. power input $= 2 \times 400 \times 0.1 = 80$ W

The a.c. power output is obtained from the composite characteristics. Thus

Peak difference current $= 263$ mA (at $v = \pm 12$ V)

Hence

R.M.S. difference current $= 186$ mA

Peak alternating voltage on one side $= 260$ V

Hence

R.M.S. alternating voltage on one side $= 184$ V

It follows that

A.C. power output $= 184 \times 0.186 = \underline{\underline{34\,\text{W}}}$

Therefore

Conversion efficiency $= \frac{34}{80} = \underline{\underline{0.43\,\text{p.u.}}}$

24.9 Push-pull Complementary-symmetry Output Stages

The fact that both *p–n–p* and *n–p–n* transistors are available enables push-pull circuits to be designed without transformers—i.e. with single-ended inputs and outputs. Thus if a *p–n–p* and an *n–p–n* transistor are fed from the same drive, a given input swing will cause one transistor to conduct more while the other conducts less, giving a push-pull operation.

Consider the circuit of Fig. 24.10. Tr_1 acts as the driver, the bias being designed so that the voltages at A and B are just sufficient to cause both Tr_2 and Tr_3 to conduct slightly. The voltage at D is set by adjustment of R_1 to be approximately $V_{CC}/2$. Diode D_1 will be a silicon diode if Tr_2 and Tr_3 are silicon power transistors, in order to

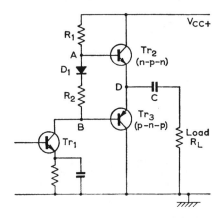

Fig. 24.10 TRANSFORMERLESS OUTPUT STAGE

improve the thermal stability of the circuit; i.e. the change of voltage between A and B with temperature will match the corresponding changes at the emitters of Tr_2 and Tr_3. R_2 is of low value so that A and B are approximately at the same voltage so far as the signal is concerned.

When the current in Tr_1 falls, the voltages at A and B both rise, causing Tr_3 to cut off and Tr_2 to take more current—the current flow being through the load resistor R_L. Under maximum power conditions the voltage at D approaches V_{CC}. When the current in Tr_1 rises above its quiescent value, the voltages at A and B fall, and Tr_2 becomes cut off while Tr_3 takes more current (by discharging capacitor C). The load current through R_L therefore reverses as C discharges, and for maximum power the voltage at D falls to almost zero. Positive and negative half-cycles will be identical if the transistors Tr_2 and Tr_3 have identical characteristics.

For high power outputs the complementary pair of transistors is used to drive two high-power transistors of the same type—added to a circuit such as that of Fig. 24.10, these would be two *n–p–n* power transistors as in Fig. 24.11.

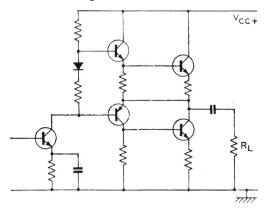

Fig. 24.11   HIGH-POWER COMPLEMENTARY-SYMMETRY OUTPUT STAGE

EXAMPLE 24.4   For the circuit of Fig. 24.10, the complementary transistors have the characteristics shown in Fig. 24.12. The supply voltage is 40 V. Assuming practical Class B operation (i.e. in the quiescent state both output transistors are almost cut off), determine the optimum load resistance, the approximate power output and the conversion efficiency.

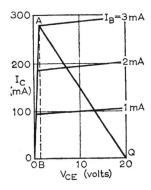

Fig. 24.12

In the quiescent state assume that $V_{CE} = 20$, $I_C = 0$ (point Q). The optimum load will give a load line from Q to the knee of the characteristic as shown. Then the optimum load resistance is

$$R_L = \frac{QB}{AB} = \frac{18\cdot6}{0\cdot28} = \underline{\underline{66\Omega}}$$

The peak voltage swings across the load will be $\pm 18 \cdot 6$ V, so that the maximum power output will be approximately

$$P_m = \frac{18 \cdot 6^2}{2R_L} = \frac{18 \cdot 6^2}{2 \times 66} = 2 \cdot 6\,\text{W}$$

Assuming sine waves, the peak output current is 280 mA, and hence the current from the supply is $(1/\pi) \times 0 \cdot 28 = 0 \cdot 09$ A. The conversion efficiency is thus

$$\eta = \frac{2 \cdot 6}{0 \cdot 09 \times 40} = 0 \cdot 72\,\text{p.u.}$$

Note that, since Tr$_2$ is cut off for half a cycle of the signal, current is drawn from the d.c. supply for only one half-cycle and hence the mean current is $(1/\pi)$ of the peak signal current.

PROBLEMS

24.1 A certain pentode has the following characteristics when operated at a screen voltage of 400 V:

V_g	V_A	200	300	400	800 volts
-8 V		100	110	116	120 mA
-16 V	I_A	51	59	63	67 mA
-24 V		19	21	22	23 mA

Find the amplification factor, r_a, and g_m, when the grid bias is -16 V and the anode voltage is 400 V. If the peak grid signal is 8 V, calculate the output power if the load is 4,000 Ω. *(L.U. part question)*

Ans. 170; 30 kΩ; 5·8 mA/V; 3·3 W.

24.2 Draw the anode characteristics for the triode from which the following test figures were obtained:

$V_g = 0$ V	$\begin{cases} V_A \\ I_A \end{cases}$	0 0	50 18	100 53	150 110	200 183	250 V 260 mA
$V_g = -40$ V	$\begin{cases} V_A \\ I_A \end{cases}$	100 3	150 20	200 50	250 100	300 170	350 V 240 mA
$V_g = -80$ V	$\begin{cases} V_A \\ I_A \end{cases}$	200 5	250 22	300 50	350 100	400 V 160 mA	
$V_g = -120$ V	$\begin{cases} V_A \\ I_A \end{cases}$	300 7	350 25	400 52	450 100	500 V 150 mA	
$V_g = -160$ V	$\begin{cases} V_A \\ I_A \end{cases}$	350 1	400 10	450 30	500 58	550 V 95 mA	
$V_g = -200$ V	$\begin{cases} V_A \\ I_A \end{cases}$	450 2	500 10	550 22	600 58	650 V 100 mA	
$V_g = -240$ V	$\begin{cases} V_A \\ I_A \end{cases}$	550 3	600 12	650 30	700 62	750 V 105 mA	
$V_g = -280$ V	$\begin{cases} V_A \\ I_A \end{cases}$	650 3	700 15	750 30	800 V 60 mA		

The triode is required to deliver power to a purely resistive load of 300 Ω. An output transformer of ratio 3:1 is available. A d.c. supply voltage of 500 V is to be used. Considering a mean anode current of 100 mA and a peak grid signal of 120 V, calculate (a) the power delivered to the load, (b) the anode efficiency, (c) the percentage second-harmonic distortion.

Ans. 7·5 W; 17 per cent; 6 per cent approx.

24.3 Draw the anode characteristics for the tetrode for which the following test figures were obtained when the screen voltage was 250 V.

		V_A 40	80	120	200	300	400 V
$V_g = 0$	I_A	62	72	76	79	81	83 mA
$V_g = -5$ V	I_A	50	57	61	64	67	70 mA
$V_g = -10$ V	I_A	37	43	46	48	50	52 mA
$V_g = -15$ V	I_A	27	32	34	37	37·5	37·5 mA
$V_g = -20$ V	I_A	17·5	22	24	26	27	27 mA
$V_g = -25$ V	I_A	10	13	15	17·5	18	18 mA
$V_g = -30$ V	I_A	4	7	8	10	11	11 mA

The tetrode is required to deliver power to a purely resistive load of 70 Ω. An output transformer of ratio 10:1 is available A d.c. supply voltage of 250 V is to be used. Considering a mean anode current of 30 mA and a peak grid signal of 10 V, calculate (a) the power delivered to the load, (b) the anode efficiency, (c) the percentage second-harmonic distortion.

Ans. 1·2 W; 15 per cent; 4 per cent.

24.4 A push-pull power amplifier utilizes two identical triodes, each of which has the anode characteristics given in Problem 24.2. If the anode supply is 500 V and the grid bias is −160 V for both valves, determine the power output and efficiency when the peak grid voltage is 160 V and the valves are coupled to a load of 15 Ω by a transformer of total primary turns to secondary turns ratio of 16·7:1.

Ans. 32 W; 30 per cent approx.

24.5 A push-pull power amplifier utilizes two identical tetrodes, each of which has the anode characteristics given in Problem 24.3. If the anode supply is 300 V and the grid bias is −25 V for both valves, determine the power output and efficiency when the peak grid voltage is 25 V and the valves are coupled to a load of 15 Ω by a transformer of total primary turns to secondary turns ratio of 31·6:1.

Ans. 8 W; 45 per cent approx.

24.6 Discuss, with the aid of sketches, the causes and effects of distortion in amplifiers.

The anode current flowing in an amplifier is given by

$$I_A = 0·05(15 + V_g)^2 \quad \text{milliamperes}$$

where V_g is the grid voltage. Calculate (a) the steady anode current with a grid bias of 8 V, (b) the mean anode current when an alternating signal of 6 sin ωt is superimposed on this grid bias, (c) the amplitude of the fundamental anode current in (b), and (d) the amplitude of the second-harmonic current in (b).

(*H.N.C.*)

Ans. 2·4 mA; 3·3 mA; 4·2 mA; 0·9 mA.

24.7 A power transistor is used in the circuit of Fig. 24.1(*c*). The operating region on the collector characteristic is bounded by (i) $I_{Cmax} = 2$ A; (ii) maximum collector dissipation = 10 W; (iii) maximum collector voltage = 30 V. The emitter resistor has a value of 1 Ω. Sketch the limits of the permissible operating region, and determine suitable values for the effective collector load resistance, supply voltage, quiescent collector current and maximum power output for a sinusoidal signal.

Ans. 10 Ω; 11 V; 1 A; 5 W.

24.8 Two transistors with the ratings given in Problem 24.7 are to be used in Class B push-pull. The supply voltage is 15 V. Determine a suitable value for the effective resistance presented to each collector, the corresponding maximum power output for a sinusoidal signal, and the maximum collector dissipation.

Ans. 7·5 Ω; 15 W; 2·1 W.

24.9 Fig. 24.13 shows the circuit for a popular linear power amplifier using a complementary pair of transistors Tr₁ and Tr₂. Describe the operation of this

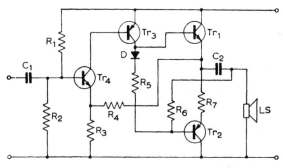

Fig. 24.13

circuit, and comment on the factors that affect the linearity. Discuss the use of the diode D.

Chapter 25

POWER RECTIFICATION

Since electrical distribution almost universally uses alternating currents, applications which require direct current necessitate the provision of either a d.c. generator or a rectifier equipment. For fixed voltage d.c. supplies, half-wave, full-wave, bridge or polyphase connected rectifiers may be used, employing vacuum or gas-filled valves, or semiconductor diodes. In these cases the required output voltage is obtained by a suitable choice of ratio for the transformer which supplies the rectifier circuit. Diode rectification is used for power and communication circuits over the range of frequencies extending from power frequencies up to microwave frequencies,* and including signal-detection circuits. Semiconductor devices can conveniently be used for most applications, ranging from high-power mains-frequency installations right through to applications at frequencies in the microwave region.

Powers involved range from microwatts in signal-detection circuits to megawatts in, for example, the rectifiers employed in d.c. power system interconnectors. Only power rectification will be considered in this chapter.

In many applications a variable d.c. supply is required (e.g. control of d.c. machines). This may involve the use of devices which can be controlled so that conduction takes place for a predetermined fraction of the a.c. cycle. Such devices include the thyratron, mercury-arc rectifier and silicon controlled rectifier (or thyristor).

---

* Electromagnetic waves whose frequencies are higher than 1 gigahertz are called *microwaves* (1 GHz = 10^9 Hz).

25.1 Conduction in a Gas

Gas-filled valves contain a gas at a pressure between 10^{-4} and 10 metres of mercury depending on the design and application of the valve. In these valves there are frequent collisions between gas atoms and electrons during the passage of electron current from cathode to anode. The most common gas filling is mercury vapour, but argon, neon, hydrogen and other gases are also used.

There are three types of collision which may occur between an electron and a gas atom depending on the speed or kinetic energy of the bombarding electron. If the speed is low (corresponding to an energy of 2 or $3\,eV$), the electrostatic fields of the bombarding electron and the outer electrons of the atom interact with one another and the electron is repelled. The "collision" follows the ordinary laws of mechanics, and the electron rebounds with only a small transference of energy to the very much more massive atom. This is called an *elastic collision*. If the bombarding electron has a somewhat greater energy, it may, on collision, cause a temporary disturbance of the electron orbits within the atom. In the disturbance the atom will, internally, absorb some of the energy of the bombarding electron so that the electron rebounds with reduced energy. The disturbance within the atom consists of the temporary raising of the energy associated with an internal electron followed by the return of that electron from its abnormal energy state to its normal energy state. During the return the excess energy, which has been absorbed from the bombarding electron, is emitted from the atom as an electromagnetic wave with a characteristic wavelength depending on the atom. For some gases the wavelength is in the visible band (e.g. sodium vapour gives a characteristic yellow light), and this is the basis of operation of some electric discharge lamps. During the disturbance the atom is said to be excited and the collision is known as an *excitation collision*.

The third type of collision is known as an *ionizing collision*. The energy of the bombarding electron is then from 10 to $20\,eV$, depending on the gas. With this energy an internal electron may be completely freed from the atom so that there are two free electrons and a positively charged atom, or *positive ion*. The atom will have absorbed from the bombarding electron an amount of energy equal to the amount required to free an electron from the atom.

The ionizing collision is by far the most important from an electrical circuit point of view, and this type of collision will be implied hereafter unless otherwise stated. The potential difference through which an electron must pass to gain sufficient energy to give an ionizing collision with a particular gas atom is called the *ionization potential* for the gas.

The above are the fundamental concepts of gaseous conduction upon which the operation of gas valves depend. It will be evident that a gas will be a good insulator unless ionizing collisions (ionization) occur. It will then be a fairly good conductor.

25.2 Gas Diodes

The gas diode has an indirectly heated cathode which provides a copious supply of primary electrons. The gas is almost invariably mercury vapour, since (i) this gas has a low ionization potential (10·6 V), which leads to efficient operation, and (ii) the mercury

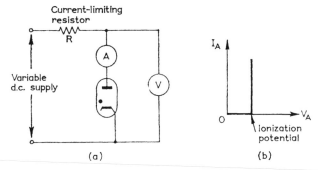

Fig. 25.1   CHARACTERISTIC OF A GAS DIODE

vapour does not react chemically with the oxide coating on the cathode. The principal disadvantage of mercury vapour is the large pressure variation which occurs with a moderate temperature variation (0·0001 mm Hg at 0°C to 0·1 mm Hg at 80°C). This makes the operation satisfactory over only a narrow temperature range (40–80°C). The anode is either solid carbon or carbon coated to give a good thermal emissivity.

The circuit of Fig. 25.1(*a*) may be used to determine the valve characteristics. If the supply voltage is raised from zero, the current will be found to be negligible until a critical voltage is reached. At this voltage ionization occurs and the valve becomes highly conducting. The current-limiting resistor *R* must be included in the circuit to limit the current when ionization occurs. For all values of current up to the permissible limit, the voltage drop across the valve remains almost constant at approximately the ionization potential. The current conduction continues until the applied potential falls to too low a value to maintain the ionization. The anode characteristic is shown in Fig. 25.1(*b*).

The action of the gas diode depends on the greater mobility of the electron compared with the positive ions (due to the much smaller mass of the electron). The electrons move much more quickly towards the anode than do the positive ions towards the cathode, so that a resultant positive space charge develops; this tends to increase the electric stress towards the cathode, and to decrease the electric stress towards the anode. This is depicted in Fig. 25.2(*a*), where it

(a)   (b)

Fig. 25.2   CONDITIONS IN CATHODE–ANODE SPACE OF A GAS DIODE

has been assumed that the positive space charge first builds up in the middle of the interspace. Owing to the high stress, electrons will move quickly from the negative space charge around the thermionic cathode to the positive space charge, and at the same time electrons will move slowly from the positive space charge to the anode, owing to the decreased electric field strength in this region. With the increased flow of electrons from the cathode space charge more positive ions will be formed and the positive space charge will continue to increase, the limit being the value of the anode potential, for if the space charge potential exceeded the anode potential, electrons would move back into the positive space charge and neutralize the effect of positive ions until the space charge potential again fell below the anode potential.

Between the positive space charge and the anode there is a region in which the electric stress becomes very low; this region is filled with about equal numbers of positive ions and negative electrons; it is highly conducting and is called the *plasma*. This is illustrated in Fig. 25.2(*b*). It will be seen that, in effect, the positive space charge has become the effective anode joined to the actual anode by the highly conducting plasma. Since the plasma contains about equal numbers of electrons and positive ions at any given instant, and since, even in the plasma, the electrons will be moving with a much greater velocity than the positive ions, there must be many more electrons passing through a given section in any given time than there are positive ions. Therefore by far the larger portion of the current through the valve is due to electrons thermionically emitted from the cathode. The proportions have been estimated to be of the order of 100 to 1. It should be clearly realized that the function of the positive ions is to assist the thermionically emitted electron current rather than to contribute to the current themselves.

The constant-voltage nature of the conduction results from the tendency of the positive space charge to approach the cathode surface. The closer the positive space charge is to the cathode the higher will be the electric stress. However, provided that saturation current for the given cathode temperature is not reached, the electric stress at the cathode will determine the electron current from the cathode space charge. For a given external current the positive space charge will only approach the cathode surface to a distance that will correspond with the required current. If the space charge approached more closely than this, there would be an excess electron emission which would cancel the positive space charge. If the external current is increased the space charge will move correspondingly nearer to the cathode and so give the additional current through the valve with no additional voltage drop across the valve. In most cases it is sufficiently accurate to assume that the voltage drop is constant for all normal currents and is equal to the voltage at which conduction will first start (*ignition voltage*) or the voltage at which conduction ceases (collapse or *extinction voltage*).

The most severe limitation on the use of gas valves is the peak-current limitation—this, usually, must not exceed two or three times the mean current even momentarily (bombardment of the coated cathode by ions may permanently destroy the surface if excess currents are permitted). Reservoir capacitors should not be used in gas-valve rectifier circuits as these tend to produce high peak currents.

EXAMPLE 25.1 A gas diode is connected in series with the output of a single-phase transformer to supply the following circuits in turn:

    (a) a pure resistance of $50\,\Omega$ in series with a choke of $0.2\,H$ inductance and negligible resistance,

    (b) a battery of constant e.m.f. $80\,V$ (connected for charging) in series with a limiting resistance of $10\,\Omega$.

If the gas diode has an effective voltage drop of $10\,V$ and the transformer gives a sinusoidal output of $100\,V$ (r.m.s.) at $50\,Hz$, calculate the mean and peak currents in each case.

    (a) The circuit is represented in Fig. 25.3(a) and the wave diagram is shown in Fig. 25.3(b). V_A is the potential of the anode with respect to earth potential. The instant at which conduction starts (i.e. when $V_A = 10\,V$) is taken as the zero reference for the time base. Therefore

$$V_A = 141 \sin(\omega t + \phi)$$

where $141 \sin \phi = 10$, i.e. $\phi = \sin^{-1}(10/141) = 4°$

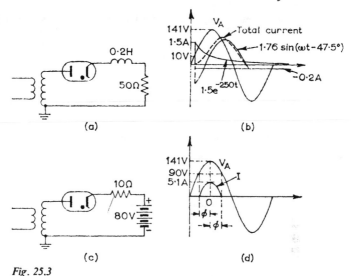

Fig. 25.3

Thus the valve may be taken as conducting between the instant $t = 0$ and the instant at which the cathode–anode voltage is again 10 V, i.e. the instant at which the current becomes zero. For this range,

$$L \frac{di}{dt} + Ri = 141 \sin (\omega t + \phi) - 10$$

This is a transient equation which will have a solution of the form $i = $ steady-state current (i_s) + transient current (i_t) (by Section 6.1), where

$$i_s = \left\{ \frac{141}{\sqrt{R^2 + \omega^2 L^2}} \sin \left(\omega t + \phi - \tan^{-1} \frac{\omega L}{R} \right) - \frac{10}{R} \right\}$$

$$= \{1 \cdot 76 \sin (\omega t - 47 \cdot 5^\circ) - 0 \cdot 2\} A$$

and

$$i_t = A e^{-\frac{R}{L} t}$$

where A is a constant.

At $t = 0$, $i = 0$, due to the inductor effect. Thus

$$i = i_t + i_s = 0 = A + 1 \cdot 76 \sin (-47 \cdot 5^\circ) - 0 \cdot 2$$

so that $A = 1 \cdot 5$, and the total current is

$$i = 1 \cdot 5 e^{-250t} + 1 \cdot 76 \sin (\omega t - 47 \cdot 5^\circ) - 0 \cdot 2$$

The three components of this current wave are drawn and added graphically in Fig. 25.3(*b*). It will be seen that the conduction is spread over more than half a cycle. From the graphs

$$\frac{\text{Peak current}}{\text{Mean current}} = \frac{0 \cdot 8}{0 \cdot 33} = \underline{\underline{2 \cdot 4}}$$

Inductive smoothing is very useful with gas valves for it helps to overcome the peak current limitations. Capacitor smoothing has the opposite effect.

(*b*) The circuit is shown in Fig. 25.3(*c*). It will be seen that the potential of the cathode is 80 V above earth when no current flows. Thus the anode potential must exceed 90 V above earth to start conduction. If the peak positive voltage is taken as the zero reference of time, then conduction will commence when

$$\phi = -\cos^{-1} \frac{90}{141} = -50\cdot4°$$

and conduction will cease when

$$\phi = \cos^{-1} \frac{90}{141} = +50\cdot4°$$

Between $-\phi$ and $+\phi$ the voltage across the current-limiting resistor is

$$v_R = 141 \cos \omega t - 90$$

Therefore

$$\text{Charging current} = \frac{v_R}{10} = \frac{141}{10} \cos \omega t - \frac{90}{10}$$

and

$$\text{Peak current} = 14\cdot1 - 9 = \underline{\underline{5\cdot1\,\text{A}}}$$

$$\text{Mean current} = \frac{1}{10}\frac{1}{2\pi} \int_{-\phi}^{+\phi} [141 \cos \omega t - 90]d(\omega t)$$

$$= \frac{141 \sin \phi}{10\pi} - \frac{90\phi}{10\pi} = \underline{\underline{0\cdot93\,\text{A}}}$$

Note that with a load consisting of a pure resistance the conduction will commence at the striking voltage of the valve and will cease at the extinction voltage.

25.3 Gas Triodes (Thyratrons)

A cross-section of the electrode assembly for a gas triode is shown in Fig. 25.4(*a*). The anode and cathode are screened from each other by a shield or grid electrode. The shield is designed to prevent any electron reaching the anode from the cathode except by way of the hole in the grid electrode. The circuit for testing a gas triode is shown in Fig. 25.4(*b*).

If the grid potential is held negative with respect to the cathode, then until the anode voltage reaches a critical potential (the *striking* voltage) no electrons will leave the cathode space charge and there will be neither ionization nor conduction. This corresponds to the cut-off conditions in a vacuum valve. If the striking voltage is exceeded, electrons will pass from the cathode space charge and give ionizing collisions with gas molecules.

When ionization commences, positive ions immediately collect around the negative grid and completely neutralize its effect. The conditions in the triode are then identical with those in the gas diode, i.e. the anode voltage falls to an almost constant value approximately equal to the ionization potential and nearly independent of the current through the valve. This current is limited only by the external impedances and will cease only when the supply voltage is reduced below the ionization or *maintaining voltage*. Any change in grid voltage after ionization has occurred will not affect the conduction through the valve, and the grid cannot be used to stop

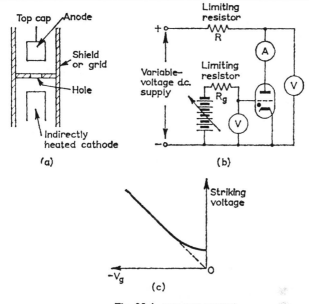

Fig. 25.4 THE GAS TRIODE

conduction. The grid of a gas triode only controls the value of anode voltage at which the valve will start to conduct.

Fig. 25.4(c) shows the grid-control characteristic for a gas triode (i.e. the anode striking voltage for a range of negative grid potentials). Above about 3 V the ratio of striking voltage to control voltage is constant at a value called the *control ratio*. This ratio is usually about 30.

Gas triodes are not of great use in circuits with direct current supplies, since once conduction has commenced it is difficult to interrupt the anode current except by interrupting the supply. The principal use of gas triodes is as grid-controlled rectifiers, in which case the grid regains control of the valve every negative half-cycle.

25.4 Grid-controlled Single-phase Rectification

Simple negative bias control of a thyratron is not normally employed, since it permits control only over the first quarter of the a.c. waveform applied to the anode.

PULSE CONTROL

The circuit is shown in Fig. 25.5(*a*). The grid has a large negative bias so that even the peak supply voltage will not be sufficient to

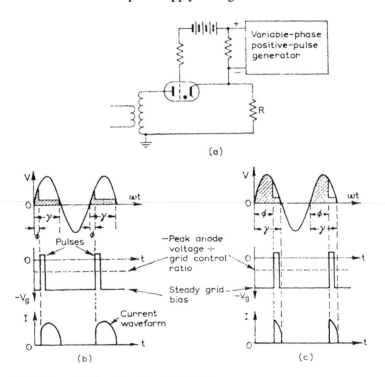

Fig. 25.5   PULSE CONTROL OF A GAS TRIODE

cause conduction through the valve in the absence of the positive pulses injected into the grid circuit. The amplitude of the positive grid pulses is at least equal to the magnitude of the grid bias voltage so that conduction will always commence at the instant in the cycle at which the positive pulse is applied, provided only that the supply voltage is positive and greater than the ionization potential at the instant at which the pulse occurs. Fig. 25.5(*b*) shows the anode voltage and current waveforms when the pulse occurs in the first positive quarter-cycle. Fig. 25.5(*c*) shows the same waveforms when the pulse occurs in the second positive quarter-cycle. It will be clear that this method has the advantages of (i) controlling the current from maximum mean value to zero mean value, and (ii) starting the conduction at a definite instant in the cycle independent of any supply voltage variations. The mean current for a resistive load may be determined from the following equation:

$$\text{Mean output current} = \frac{1}{2\pi R} \int_{\phi}^{\gamma} [V_m \sin \omega t - \text{volt drop}] d(\omega t)$$

(25.1)

where R = Load resistance
ϕ = Ignition angle
γ = Extinction angle

The pulse unit may be a mechanical contact driven by a synchronous motor, a peaking transformer supplied from a phase-shifting circuit, or an electronic pulse generator with a phase-shifting circuit giving pulses synchronized with the supply frequency.

PHASE-SHIFT CONTROL

This method has been developed to preserve the advantages of the pulse control circuit without the necessity of having a pulse generator. A sinusoidal voltage with or without a negative grid bias is applied to the grid of the gas triode. The sinusoidal voltage has the same frequency as the supply voltage and usually a constant magnitude. It will be found that, by altering the phase of the grid voltage relative to that of the supply voltage, the instant at which conduction commences may be controlled.

Fig. 25.6(*a*) shows a phase-controlled gas-triode rectifier circuit incorporating one possible method of deriving a constant-magnitude variable-phase voltage. By Section 1.7 an *RC* series circuit with variable resistance has a semicircular complexor locus diagram as shown in Fig. 25.6(*b*). It will be seen that the potential difference between the centre point of the winding an the junction of the resistor and capacitor has a constant magnitude and a phase which is variable over almost 180° as R is varied from zero to infinity. An additional 180° phase shift may be obtained by reversing the connexions to the auxiliary transformer winding.

The value of the mean output current may be obtained from eqn. (25.1), where the ignition angle ϕ is found in the following manner. In Fig. 25.6(c) the supply voltage waveform is drawn and also the waveform of critical grid voltage corresponding to the supply voltage, i.e. the negative grid voltage which would be just sufficient to prevent the valve from striking at the instantaneous value of supply voltage. The external grid-voltage wave is now superimposed

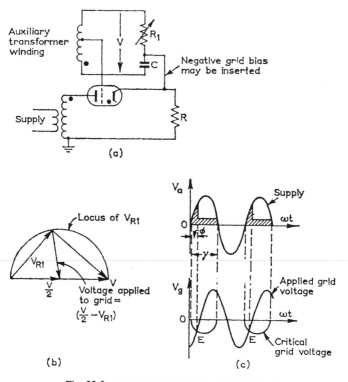

Fig. 25.6   PHASE-SHIFT CONTROL OF A GAS DIODE

on the critical grid-voltage wave in the correct phase. The point, E, of intersection of the external grid-voltage waveform and the critical grid-voltage waveform where the external voltage is going positive gives the striking point in the cycle. It is evident that by altering the phase of the external voltage the striking point and hence the mean current may be controlled.

EXAMPLE 25.2   A gas-filled triode, used as a phase-controlled rectifier on a 250 V supply, has a control ratio of 20 and a negligible tube drop. If the voltage applied to the grid is 100 V (r.m.s.) lagging behind the anode supply voltage by 60°, find the mean anode current in a load of 1,500 Ω.   (*H.N.C.*)

The supply voltage may be represented as 354 sin ωt volts. Therefore

$$\text{Critical grid voltage} = -\frac{354}{20} \sin \omega t = -17 \cdot 7 \sin \omega t$$

Voltage applied to grid = 141 sin ($\omega t - 60°$)

These waves may be drawn as in Fig. 25.6(c) or, since in this case the control ratio is assumed to be constant, the ignition angle ϕ may be found by solving the equation

$$
\begin{aligned}
-17 \cdot 7 \sin \phi &= 141 \sin (\phi - 60°) \\
&= 141 \sin \phi \cos 60° - 141 \cos \phi \sin 60° \\
&= 70 \cdot 7 \sin \phi - 122 \cos \phi
\end{aligned}
$$

whence tan $\phi = 122/88 \cdot 4 = 1 \cdot 38$; therefore the ignition angle, ϕ, is 54°.

Since the voltage drop is to be neglected, the extinction angle, γ, may be taken to be π.

$$\text{Mean load voltage} = \frac{1}{2\pi} \int_{54°}^{180°} 354 \sin \omega t \, d(\omega t) = 89 \cdot 5 \, \text{V}$$

and

$$\text{Mean anode current} = \frac{89 \cdot 5}{1,500} = \underline{\underline{59 \cdot 5 \, \text{mA}}}$$

25.5 Mercury-arc Rectifiers

In its single-phase form the mercury-arc rectifier consists of a graphite or carbon-coated iron anode and a mercury-pool cathode enclosed in an envelope from which all air has been removed. Mercury vapour fills the space between anode and cathode, the pressure of the vapour varying widely with temperature. Unlike the gas diode and triode, there is not an external power source to provide a supply of primary electrons. Thus conduction will not commence by merely raising the anode–cathode voltage above the ionizing potential for mercury vapour (10·6 V).

In one type of single-phase rectifier (called an *ignitron*) an auxiliary pointed electrode (*ignitor*) dips into the mercury pool. At the instant during the positive voltage cycle at which conduction to the anode should commence, a heavy short-duration current is passed through the ignitor to the mercury pool. This current raises the temperature at the point to a white heat so that some primary electrons are emitted. The primary electrons ionize the mercury vapour, whereupon positive ions travel back to the cathode. The action of the positive ions at the cathode is such that copious primary-electron emission develops at a particular spot, which becomes white hot. The mechanism by which electrons are emitted at the cathode spot is not clearly understood. It would seem that thermionic, secondary and field emission effects are all present. The conduction through the mercury vapour is of the same nature as in the gas diode, i.e. a positive space charge forms in front of the cathode and is joined to the anode by the plasma. The voltage drop between anode and cathode when the rectifier is conducting is somewhat greater than the ionization potential alone, since with the greater currents and greater anode–cathode separation there is an appreciable voltage drop across the plasma, and a voltage drop is also found to exist at the anode surface. The total voltage drop is usually between 20 and 30 V. Ignitrons capable of carrying currents of several thousand amperes have been constructed.

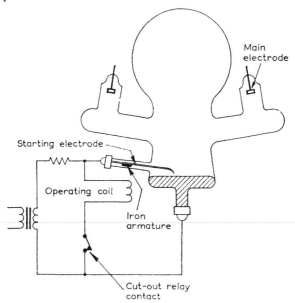

Fig. 25.7   BASIC MERCURY-ARC RECTIFIER

The mercury-arc rectifier has the advantage over the thermionic gas diode in that, though the mean current will naturally be limited by temperature rise as a whole, the peak current is almost unlimited and very high overloads may be tolerated for short periods. There is also the considerable advantage that, since the mercury cathode is almost indestructible, the life of the rectifier is exceedingly great.

The single-phase form of the rectifier is particularly useful for the control of welding currents. It will be noted that the mean current may be controlled by varying the instant at which conduction starts.

A glass-bulb permanently-evacuated mercury-arc rectifier is illustrated in Fig. 25.7. Two anodes only are shown, although rectifiers of this kind usually operate from a 3-phase supply and have

3, 6 or 12 main anodes with two auxiliary anodes. In this case the arc is continuous and does not require to be restruck each cycle. After conduction starts a relay cuts out the ignition system, which consists of a flexible electrode which is pulled into contact with the mercury pool by an electromagnet. When the flexible electrode contacts the mercury pool the electromagnet is short-circuited so that the flexible electrode springs back giving a spark as it breaks contact with the pool. This spark gives the initial electrons which are attracted by an anode at a positive potential and give rise to ionization of the mercury vapour.

In the glass-bulb rectifier the anodes are housed in arms which leave the bulb at acute angles. This is to prevent mercury globules forming on the anode surfaces. If these form, then *backfire* (reverse conduction) or *crossfire* (anode-to-anode conduction) may occur. Steel-tank rectifiers are also manufactured; in these the anodes are protected by insulating barriers. Another method of protection against crossfire and backfire is the use of grids controlling the instant at which an anode may start to conduct. The grids are in the form of a perforated metal sheet or a metal grill cutting off the anode from the rest of the rectifier. The grids are held at a negative potential except at the instant when the anode should start to conduct. The grids may also be used to reduce the conduction angle and hence the output voltage.

In addition to the main anodes which carry the load current there are usually two auxiliary anodes arranged to form a separate full-wave rectifier supplying a small dummy load. The purpose of this is to maintain the cathode hot spot should the main load current temporarily fall to zero. When the load is reimposed the main anodes will again conduct without the ignition system requiring to be separately operated.

Glass-bulb rectifiers are generally used for loads up to 500 A at 500 V. Permanently-evacuated steel-tank rectifiers are used for loads above this and up to 750 A at 750 V. Continuously-pumped steel-tank rectifiers are used for loads up to 3 MW at voltages exceeding 20 kV.

The internal efficiency of a rectifier largely depends on the voltage at which it operates since the voltage drop along the arc is constant.

Let V_{dc} and I_{dc} be the direct voltage (including arc drop) and current generated when the arc drop is V_{arc}. Then

$$\text{Efficiency, } \eta = \frac{V_{dc}I_{dc} - V_{arc}I_{dc}}{V_{dc}I_{dc}} \times \frac{100}{1} \text{ per cent}$$

$$= 1 - \frac{V_{arc}}{V_{dc}} \text{ p.u.} \qquad (25.2)$$

e.g. if the arc drop is 25 V the efficiency at 100 V output is only 75 per cent, but at 1,000 V it is 97·5 per cent.

A rectifier will, however, always require a transformer and various auxiliary circuits and the losses in these must also be taken into account.

25.6 Polyphase Rectification

The semiconductor diode rectifier is now replacing the mercury-arc rectifier for polyphase rectification in all applications except those involving the highest voltages. The forward voltage drop across a

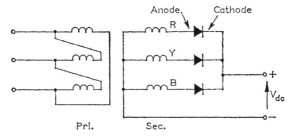

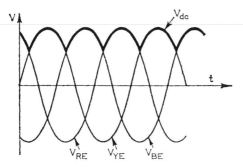

Fig. 25.8(a)   SIMPLE 3-PHASE RECTIFIER CONNEXIONS

semiconductor diode is only a fraction of a volt, and hence the efficiency as defined by eqn. (25.2) will be almost unity (replacing V_{arc} by the diode forward voltage). Germanium diodes are used for high-current and silicon diodes for high-voltage applications.

By using polyphase rectification, the ripple voltage in the d.c. output may be made small without the necessity of using smoothing capacitors—this is a great advantage in power rectification.

The secondary of the transformer supplying a polyphase rectifier must have a neutral point when a mercury-arc rectifier (which has only one cathode) is used in order to give the negative output connexion. The primary winding is generally delta-connected to avoid the difficulties which arise with star-star transformers. The primary is fed from a 3-phase supply, and the secondary is connected to give 3, 6 or 12 phase operation as required. A delta-connected primary is shown in Fig. 25.8(*a*) and is assumed in Figs. 25.8(*b*) and (*c*). In these diagrams diodes are shown, but the connexions apply

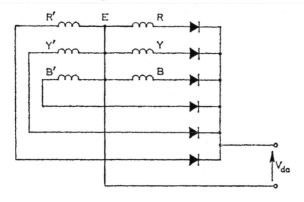

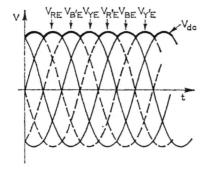

Fig. 25.8(*b*)   SIMPLE 6-PHASE RECTIFIER CONNEXIONS

equally to 3- and 6-anode mercury-arc rectifiers in which case all the "cathodes" are common and are in fact the mercury pool.

A simple 3-phase circuit is shown in Fig. 25.8(*a*), together with the corresponding wave diagrams. Conduction takes place through whichever diode (or to whichever anode in a mercury-arc rectifier) has its anode at the highest positive potential. The common cathode potential will then be equal to the potential of the most positive anode less any internal voltage drop. (Remember that this voltage drop may be 20 to 30 V in a mercury-arc rectifier). The d.c. output is shown by the heavy line in the wave diagram. The ripple is seen to be considerably less than that in a full-wave rectifier.

A simple 6-phase connexion is shown in Fig. 25.8(*c*). The supply transformer has three centre-tapped secondary windings, the centre taps forming the neutral. The e.m.f.s of each side of one phase winding will be in antiphase, and the wave diagram shows that the

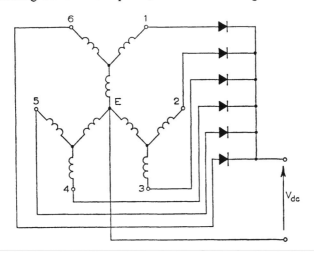

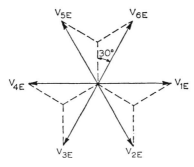

Fig. 25.8(c)   ZIGZAG RECTIFIER CONNEXIONS

e.m.f.s on the diodes will reach their positive maxima in the sequence R, B', Y, R', B, Y'. Each diode conducts for one-sixth of a cycle. The primary currents will not be sinusoidal, since each diode passes a block of current for only one-sixth of a cycle. The output d.c. ripple is less than for the 3-phase connexion.

An alternative connexion, shown in Fig. 25.8(*c*), is the zigzag connexion. This gives a more sinusoidal form of transformer primary current than the simple 6-phase circuit, but maintains 6-phase smoothing. The complexor diagram shows how the symmetrical 6-phase system is developed. Note that the transformer primary current is distributed over two phases no matter which diode is conducting.

In Fig. 25.9 is shown a 3-phase bridge circuit which is commonly used with diode rectifiers. It cannot be used with a mercury-arc

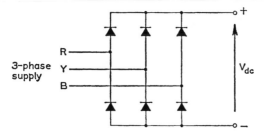

Fig. 25.9 THREE-PHASE BRIDGE RECTIFIER

rectifier since the cathodes are not common. This circuit does not require a star-connected secondary. The peak output direct voltage is the peak line voltage of the supply. Smoothing is 6-phase.

25.7 Voltage and Current Ratios

In polyphase rectifiers, conduction always occurs to the most positive anode. The "cathode" potential will therefore be the potential of the most positive anode minus the voltage drop V_a, across the rectifier

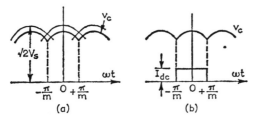

Fig. 25.10 WAVEFORMS IN POLYPHASE-RECTIFIER OPERATION

at any instant. Thus, as conduction commutates from anode to anode, the output voltage waveform for a 3-phase connexion will be as shown in Fig. 25.8(*a*). A 6-phase system could be dealt with similarly, and it will be seen that increasing the number of anodes reduces the ripple in the output voltage waveform.

Fig. 25.10 shows the general waveform for an *m*-phase rectifier, with the origin of the angular base at a peak value. Conduction to one anode will commence at $-\pi/m$ and be complete at $+\pi/m$.

Let V_s be the r.m.s. secondary output voltage per phase; then

$$\text{Output voltage} = \sqrt{2}V_s \cos \omega t - V_d \quad \text{for} \quad -\frac{\pi}{m} < \omega t < +\frac{\pi}{m}$$

Therefore

$$\text{Mean output voltage} = \frac{1}{2\pi/m} \int_{-\pi/m}^{+\pi/m} (\sqrt{2}V_s \cos \omega t - V_d)\, d(\omega t)$$

so that

$$\text{Mean output voltage,} \quad V_{dc} = \sqrt{2}V_s \frac{\sin \pi/m}{\pi/m} - V_d \qquad (25.3)$$

$$= E_{dc} - V_d$$

where

$$E_{dc} = \frac{\sqrt{2}V_s \sin \pi/m}{\pi/m} \qquad (25.4)$$

From eqn. (25.4) it will be seen that E_{dc} increases with the number of anodes and tends towards $\sqrt{2}V_s$ as $m \to \infty$.

The current ratio is usually calculated on the assumption that the load circuit has a smoothing choke or at least sufficient inherent inductance to eliminate any ripple in the current waveform as shown at (b). It is also assumed that the full current instantaneously commutates from one diode to the next (see Section 25.8).

Let I_{dc} be the output current and I_s be the r.m.s. current at a transformer terminal; then

$$I_s = \sqrt{(\text{mean square of diode current})} = \sqrt{\left(\frac{1}{2\pi} I_{dc}^2 \frac{2\pi}{m}\right)}$$

whence

$$I_s = \frac{I_{dc}}{\sqrt{m}} \qquad (25.5)$$

There is only one pulse of current through each diode per cycle. The rating of the transformer secondary winding is then given by

$$\text{Rating of secondary} = mV_s I_s \quad \text{volt-amperes} \qquad (25.6)$$

The primary winding may have a rating somewhat less than this owing to the improved current waveforms, but this depends on the method of connexion.

Substituting for V_s and I_s from eqns. (25.4) and (25.5),

$$\text{Secondary rating} = m \frac{E_{dc}}{\sqrt{2} \sin \pi/m} \frac{\pi}{m} \frac{I_{dc}}{\sqrt{m}}$$

$$= \frac{\pi/m}{\sin \pi/m} E_{dc} I_{dc} \sqrt{\frac{m}{2}} \qquad (25.7)$$

$E_{dc} I_{dc}$ is the actual power output (neglecting rectifier losses). Therefore

$$\frac{\text{Actual power}}{\text{Full-load rating}} = \frac{\sin \pi/m}{\pi/m} \sqrt{\frac{2}{m}} \qquad (25.8)$$

This is termed the *utilization coefficient* of the transformer secondary. It may be shown that the utilization coefficient has its maximum value of 0·675 for a 3-anode rectifier and is only 0·4 for a 12-anode rectifier. A low utilization coefficient will increase the size and cost of the transformer. Thus, though a large number of diodes gives a smooth output, it also leads to an expensive transformer.

25.8 Overlap Effects

In deriving the previous voltage and current relationships it has been assumed that the commutation of the current from one anode to the next is accomplished in zero time, i.e. the full current instantaneously stops flowing to one anode and starts flowing to the next. This is, in fact, impossible since the transformer windings in the anode circuits must have some leakage inductance, through which it is not possible to have an instantaneous change of current.

Fig. 25.11(*a*) shows two successive phases of a secondary winding with the equivalent reactance represented as external to the windings. The resistance may usually be neglected. Suppose that diode 1 is first conducting and that the full current I_{dc} flows to it. When the potential of diode 2 reaches that of diode 1 the current to diode 1 cannot immediately cease but must decay over a finite time while the current to diode 2 increases from zero to I_{dc}. Thus for a short period the current must split between diodes 1 and 2, and for this period the anodes of these diodes must have the same potential. This is called *overlap*. During the overlap period the output e.m.f. is the mean of the e.m.f.s of the two secondary phases alone as shown at (*b*). The length of time during which overlap takes place and the reduction of the mean output voltage due to it are proportional to the load current.

Since the current through the transformer secondary windings has no longer the rectangular waveform previously assumed, the relationship between the r.m.s. anode current and the direct load current will be modified from that given by eqn. (25.5). For 3-phase rectification the modification is negligible, and may often be neglected

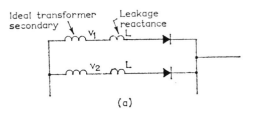

(a)

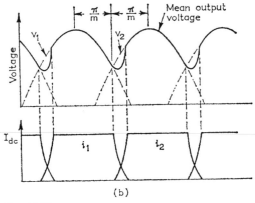

(b)

Fig. 25.11   EFFECT OF OVERLAP

for 6-phase rectification. For 12-phase rectification the modification is considerable.

25.9 Double Three-phase Connexion

By the addition of an *interphase reactor** to the simple 6-phase connexion it is possible to obtain rectifier action with the smoothness of normal 6-phase rectification and the utilization factor of 3-phase rectification. The principal characteristic of the double 3-phase connexion is that conduction is made to split into two parts at all instants of the cycle. The basic principle is similar to that of overlap where the transformer leakage inductance causes the current to split into two parts at each commutating point.

---

* Also called *interphase transformer*.

A centre-tapped iron-cored choke is used as shown in the circuit diagram of Fig. 25.12. Since the sides of the interphase winding have equal numbers of turns and are wound round the same core in the same direction, the e.m.f.s induced in the two halves of the winding will be equal in magnitude and in direction at all instants. Neglecting resistance and leakage reactance, the potential differences across the two halves will then be equal in magnitude and direction at all instants. It is to be particularly noted that successive anodes are connected on opposite sides of the interphase reactor, so that.

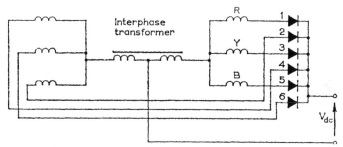

Fig. 25.12   INTERPHASE REACTOR OPERATION

if ordinary 6-phase operation occurred, the direct current would alternate between the sides of the interphase reactor.

Suppose that the whole current flows to anode 1 and attempts to commutate naturally to anode 2. This means that the direct current should cease to flow in the right-hand side of the interphase reactor and should suddenly start to flow in the left-hand side (Fig. 25.12). The inductance of the interphase reactor will react against these changes of current: the right-hand terminal of the reactor will be driven positive with respect to earth potential while the left-hand terminal becomes negative. This raises the potential of anodes 1, 3 and 5 while lowering that of anodes 2, 4 and 6. Provided that the inductance of the interphase reactor is sufficient, the current will not wholly transfer from anode 1 to anode 2, for the raising of the potential of the right-hand side and the lowering of the potential of the left-hand side will maintain anodes 1 and 2 at the same instantaneous potential with respect to earth. The instantaneous potentials are illustrated in Fig. 25.13(*a*). Eventually the potential of anode 3 will exceed that of anode 1 and the current which flows to anode 1 will transfer to anode 3, since this does not involve a change of current in the interphase reactor (there may be some delay due to simple overlap). The current is now shared between anodes 2 and 3. The potential of anode 3 will eventually tend to exceed that of anode 2 and this will cause the interphase reactor e.m.f.s to reverse, so that the potential of the left-hand side is raised and that of the right-hand side is lowered.

The current continues to be shared between anodes 2 and 3 until the current flowing to anode 2 transfers to anode 4. This is illustrated in Figs. 25.13(b)–(d). The process will continue in a similar manner round all the anodes.

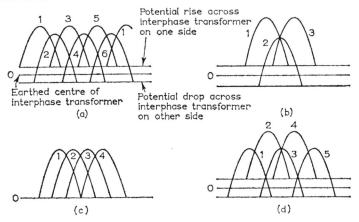

Fig. 25.13   WAVEFORMS WITH INTERPHASE REACTOR

It will be seen that there are always two successive anodes conducting and that the mean current in each half of the interphase reactor will be half the output current, i.e. $I_{dc}/2$. It should also be noted that the mean currents in each part of the interphase reactor are in opposite directions and will produce no net magnetizing effect in the iron core.

The output voltage at any instant is the mean of the e.m.f.s of the conducting secondary windings at the same instant. Fig. 25.14

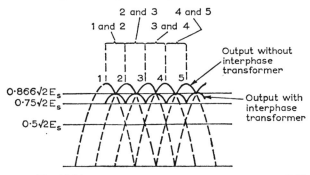

Fig. 25.14   OUTPUT VOLTAGE WITH INTERPHASE CONNEXION

shows the e.m.f.s of the secondary windings, and also the waveform of the average voltage of successive pairs of anodes as these pairs share the conduction. This latter waveform is also the waveform of the effective output voltage; the output voltage ripple obviously corresponds to 6-phase rectification.

If the load current is less than a critical value which depends on the inductance of the interphase reactor, then the operation will be almost normal 6-phase operation. As the load is increased the transition between normal 6-phase and double 3-phase operation is marked by a sudden sharp decrease in the output voltage from the 6- to the 3-phase value. Often a small permanent load is connected across the output of the rectifier so that I_{dc} does not fall below the critical value.

Exact analysis shows that the mean output voltage (neglecting overlap) is $0 \cdot 826\sqrt{2}E_s$, which happens to be exactly the output expected for a 3-phase rectifier where the r.m.s. secondary e.m.f. is E_s. Therefore

$$\text{Mean output voltage} = \sqrt{2}E_s \frac{\sin \pi/3}{\pi/3} - V_d \qquad (25.9)$$

It will also be seen that each anode and each secondary winding carries a current $I_{dc}/2$ for one-third of a cycle. Thus

$$\text{R.M.S. current per secondary phase} = \frac{I_{dc}}{2}\frac{1}{\sqrt{3}} \qquad (25.10)$$

This gives the arrangement the same utilization coefficient as is normally obtained for 3-phase rectification.

25.10 Primary Current Waveforms

In the following examination of waveforms, the effect of overlap on the current waveform will be neglected for the sake of clarity. In general, the overlap effect will give rise to smoother primary current waveforms than are obtained when the effect of overlap is neglected.

DELTA PRIMARY, SIX-PHASE SIMPLE SECONDARY (Fig. 25.15(a))

Fig. 25.15(b) shows the current waveforms for the primary phases and lines. It will be realized that current flowing in the secondary to anode 1 will have a corresponding current in the primary phase 1, and current flowing in the secondary to anode 4 will have a

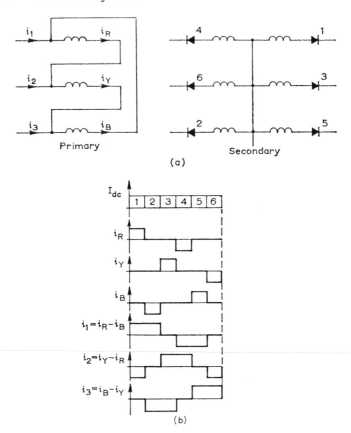

Fig. 25.15   PRIMARY CURRENTS FOR A DELTA-CONNECTED PRIMARY
AND SIMPLE 6-PHASE OPERATION

corresponding current in the same primary phase, but in the reverse direction. The currents in the other phases will have similar effects.

The primary phases each carry current for one-third of a cycle, and hence the primary utilization factor will be better than the secondary utilization factor.

DELTA PRIMARY, DOUBLE THREE-PHASE SECONDARY

Each secondary phase carries current over one-third of a cycle (see Section 25.9). The primary waveforms are shown in Fig. 25.16. It will be seen that each primary phase now carries current over two-thirds of a cycle so that the utilization factor is still further improved.

The primary line currents are also more nearly sinusoidal so that there will be less harmonic current in the lines.

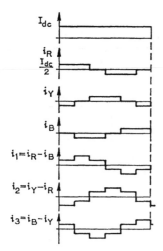

Fig. 25.16 PRIMARY CURRENTS WITH INTERPHASE CONNEXION

25.11 The Thyristor

The thyristor is a four-layer silicon semiconductor junction device, which has characteristics similar to those of a gas-filled triode. It can be triggered into the conducting state, and conducts unidirectionally until the voltage across it is reduced to almost zero. The voltage drop across it in the conducting state is small. This device is also known by the term silicon controlled rectifier (s.c.r.). Important characteristics of the thyristor are the extreme rapidity with which the device can be turned on and the very fast switch-off time.

Fig. 25.17(*a*) is a schematic drawing of the thyristor together with its circuit symbol. The operation can be deduced by considering that the four layers form two transistors and a diode as shown at (*b*). If reverse bias is applied (i.e. if the "anode" is connected to the negative of the supply), then junctions 1 and 3 are reverse biased, and only the small reverse-bias saturation current will flow until reverse breakdown occurs. This is almost independent of any voltage on the *gate* terminal.

Consider the gate connexion open-circuited, and a positive bias applied (i.e. the anode positive). Then junctions 1 and 3 will be forward biased and junction 2 will be reverse biased. The first three layers form a p–n–p transistor, with junction 1 forming an emitter-base junction. Hence holes injected across junction 1 will be collected across the reverse-biased junction 2 by normal transistor action to give a hole current $\alpha_1 I_h$ as shown at (b). In the same way, the last three layers form an n–p–n transistor, in which junction 3 forms the emitter–base junction. Electrons injected from right to left across junction 3 are collected across the reverse-biased junction 2. At the same time junction 2 acts as a reverse-biased diode junction, and

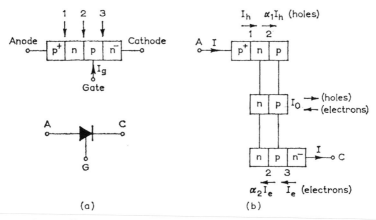

Fig. 25.17   THE TWO-TRANSISTOR REPRESENTATION OF A THYRISTOR

the reverse bias saturation current I_0 (consisting of holes moving from n to p and electrons from p to n) flows.

If I_h is the injected hole current across junction 1 (assuming that the p-layer is heavily doped), then $\alpha_1 I_h$ is the fraction of this flowing across junction 2. Also, if I_e is the electron current flowing from right to left across junction 3, then a fraction $\alpha_2 I_e$ is collected across junction 2. The total current across 2 is therefore

$$I_0 + \alpha_1 I_h + \alpha_2 I_e$$

flowing conventionally left to right. But by Kirchhoff's law, there is the same current across each junction, and hence across 1 and 3, $I_h = I_e = I$ and across 2, $I = I_0 + \alpha_1 I + \alpha_2 I$, or

$$I = \frac{I_0}{1 - (\alpha_1 + \alpha_2)} \qquad (25.11)$$

By suitable choice of doping, the current gain, α, of silicon transistors can be made to have a low value at very small emitter currents, the gain rising towards unity as the current increases.

For small values of applied voltage α_1 and α_2 can both be well below 0·5 so that the resultant current is small. As the applied voltage increases $(\alpha_1 + \alpha_2)$ can become unity, and eqn. (25.11) shows that in this condition I increases without limit (hence a current-limiting resistor must be included in the circuit). When this happens junction 2 breaks down, and the voltage across the thyristor falls to the small constant breakdown voltage across junction 2 plus the even smaller voltage drops across the bulk resistance of the layers and the forward-biased junctions 1 and 3.

The characteristic is shown in Fig. 25.18(a). Between 0 and A the term $(\alpha_1 + \alpha_2)$ is less than unity, but is increasing so that I

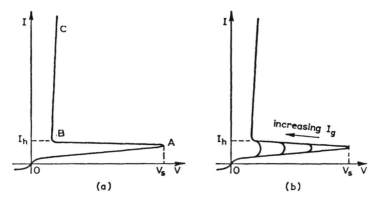

Fig. 25.18 CURRENT/VOLTAGE CHARACTERISTIC OF A THYRISTOR

(a) Gate current zero
(b) Effect of increasing gate current

increases. At A $(\alpha_1 + \alpha_2) = 1$ and breakdown occurs—the voltage falls abruptly (AB). Breakdown conditions will persist so long as the anode current is above the holding value I_h, giving an almost constant voltage across the device, irrespective of the current (BC on the characteristic).

If a current is now fed into the gate connexion, this increases the current across junction 3, so that α_2 increases and $(\alpha_1 + \alpha_2)$ now becomes equal to unity at a lower value of applied anode–cathode voltage, as shown at (b). If the applied voltage is less than the switching voltage, V_s, at which breakdown occurs when there is no gate current, then breakdown will not occur until a gate current flows. In this way the thyristor can be made to act as a switch.

In order to avoid undue heat dissipation at the gate it is usual to feed the gate with a pulse waveform. When an a.c. supply is connected across the thyristor, the instant in the positive half-cycle at

which the device "fires" is controlled by the time at which the gate
pulse is applied. A typical thyristor bridge rectifier circuit and waveforms are shown in Fig. 25.19.

Care must be taken in thyristor circuits when the load is inductive
since high reverse voltages may then appear across the rectifier when
the load is switched off. Sometimes an avalanche diode is connected
with reverse polarity across the load, as shown by the broken line in
Fig. 25.19, in order to minimize this effect.

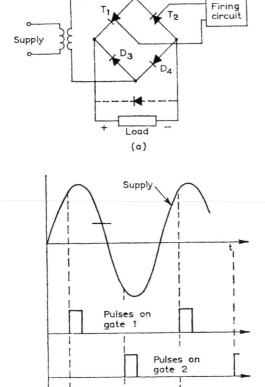

Fig. 25.19   THYRISTOR-CONTROLLED BRIDGE RECTIFIER

25.12   Grid-controlled Mercury-arc Rectification

Grid control may be applied to polyphase mercury-arc rectification in the same way as it was applied to single-phase rectifier circuits (Section 25.3). The electron path to an anode may be blocked by a negatively biased grid so that conduction to a given anode may not

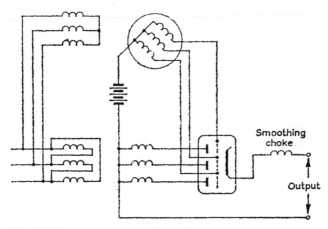

Fig. 25.20   GRID-CONTROLLED MERCURY-ARC RECTIFIER

commence until the negative grid potential is reduced. Usually phase shift control with an additional steady negative bias is used. A variable-phase voltage for grid control may be obtained from a phase-shifting transformer (which is constructed like an induction motor), with a 3-phase stator winding and a stationary rotor wound for the number of phases corresponding to the number of anodes. By varying the angular position of the rotor the grid voltage may be given any desired phase relationship to the supply voltage.

A connexion diagram for 3-phase grid-controlled rectification is shown in Fig. 25.20, a battery being shown as providing the steady negative grid bias where normally a single-phase metal rectifier circuit would be employed.

25.13   Controlled Polyphase Rectification–Output Voltage

The output voltage waveform for an m-phase rectifier in which grid or gate control has delayed the commutation by a phase angle α is shown in Fig. 25.21. Taking a voltage maximum as the reference zero, the angle of conduction per phase will be seen to be from $(-\pi/m + \alpha)$ to $(+\pi/m + \alpha)$. Therefore

$$\text{Mean output voltage} = \frac{1}{2\pi/m} \int_{-\frac{\pi}{m} + \alpha}^{\frac{\pi}{m} + \alpha} (\sqrt{2}E_s \cos \omega t - V_d)d(\omega t)$$

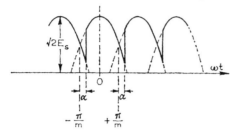

Fig. 25.21 CONTROLLED POLYPHASE RECTIFICATION WAVEFORMS

where V_d is the drop across the rectifier.

$$V_{out} = \frac{m}{2\pi}\sqrt{2}E_s \left[\sin\left(\frac{\pi}{m} + \alpha\right) - \sin\left(-\frac{\pi}{m} + \alpha\right)\right] - V_d$$

$$= \sqrt{2}E_s \frac{m}{\pi} \sin \frac{\pi}{m} \cos \alpha - V_d$$

$$= \sqrt{2}E_s \frac{\sin \pi/m}{\pi/m} \cos \alpha - V_d \qquad (25.12)$$

This equation, however, ceases to apply if the delay angle α is greater than $(\pi/2 - \pi/m)$, for in this case the current conduction becomes discontinuous. The waveform is shown in Fig. 25.22.

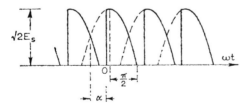

Fig. 25.22 DISCONTINUOUS CONDUCTION WITH GRID OR GATE CONTROL

For this case,

$$\text{Mean output voltage} = \frac{1}{2\pi/m} \int_{-\frac{\pi}{m} + \alpha}^{\frac{\pi}{2}} (\sqrt{2}E_s \cos \omega t - V_d)d(\omega t)$$

$$(25.13)$$

25.14 Invertor Operation

A controlled polyphase rectifier may be used to link a d.c. system
to an a.c. system so that energy flows from the d.c. system to the a.c.
system. In this case the rectifier is called an *invertor*. This operation
will only be considered briefly.

The a.c. system must have at least one synchronous generator
connected to it to determine the frequency of operation. The d.c.

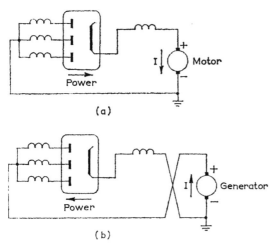

Fig. 25.23 INVERTED OPERATION OF A MERCURY-ARC RECTIFIER
(a) Rectifier   (b) Inversion

system must, naturally, include a d.c. source of energy. It should
be first noted that, when a rotating machine changes from motoring
to generating operation, the current direction rather than the e.m.f.
direction changes. With a rectifier, however, it is not possible for
the current direction to change and thus the e.m.f. direction should
be changed. (This illustrated in Fig. 25.23.) With respect to the a.c.
system, it will be realized that, if current conduction during a
positive voltage half-cycle led to the transmission of power from the
a.c. system to the d.c. system, then for power transmission in the
opposite sense and with the same direction of current conduction it
will be necessary to have the current conduction during a negative
voltage half-cycle. This is arranged by the connexion of the generator
and by using grid control to delay conduction to a negative half-
cycle.

As shown in Fig. 25.23 the generator voltage raises the potential
of the transformer neutral to a positive value with respect to earth

potential. The anode potentials are then the d.c. potential plus the
e.m.f. of the corresponding secondary phase. The anode potentials
are shown in Fig. 25.24. The direct voltage of the generator is
assumed to be approximately equal to the output voltage of the
same rectifier if used to deliver power from the same a.c. system.
This is usually approximately the case but is not necessarily so.
Except for the p.d. across the choke the cathode of the rectifier is at

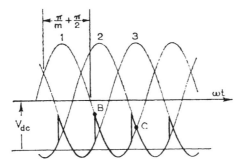

Fig. 25.24   WAVEFORMS FOR INVERTED OPERATION

earth potential, and the anodes are positive with respect to the
cathode over most of the cycle.

It is essential that commutation be delayed over at least
$(\pi/m + \pi/2)$ radians so that conduction to a given anode will only
occur when the secondary e.m.f. is negative. If conduction occurs
when the secondary e.m.f. is positive, there will be rectifier action
and short-circuit conditions will ensue. Conduction may be permitted to start at a point such as B in the alternating voltage cycle
of each anode. It is essential also that commutation occurs before
the point C is reached; otherwise it will not occur at all as the anode
"carrying" the arc will become more positive than the succeeding
anode after the point C is reached. If commutation to the succeeding
anode does not occur, the arc will remain with an anode whose
alternating potential will become positive, giving rectification and
short-circuit conditions. Thus the phase angle of the grid may only
be varied over a limited range.

Neglecting the arc drop, the cathode potential becomes the potential of the conducting anode at each instant. The difference between
the cathode potential and earth potential is the p.d. across the
smoothing choke.

It should be remembered that with a constant d.c. system voltage
and a constant a.c. system voltage the grid potential will control
the current flow between the two systems.

Exactly similar considerations apply in the case of thyristors.

D.C. links (with a rectifier at one end and an invertor at the other end of the link) are now commonly used to connect two a.c. power distribution systems together. By doing this the following advantages are obtained:

(*a*) There is no need to synchronize the two a.c. systems.

(*b*) Only *power* flow takes place in the d.c. link (i.e. there is no reactive flow through the link).

Such links are used to connect the British grid to the European power network, and are also used to interconnect parts of the British grid system.

PROBLEMS

25.1 A single gas diode is to be used to charge a 100 V battery of negligible internal resistance from a 250 V 50 Hz supply. The voltage drop and ignition voltage across the gas diode may be taken as 20 V. Find the series resistance so that the average charging current is 2 A.

Ans. 28 Ω.

25.2 An a.c. supply at 20 V is connected in series with a full-wave rectifier to a 12 V battery. The total resistance of the circuit is 5 Ω in the conducting direction and infinity in the reverse direction. Plot the current waveform and determine the mean value of the current in the circuit. (*L.U.*)

Ans. 0·7 A.

25.3 A thyratron with a control ratio of 20 is to be used as a half-wave rectifier from a 250 V 50 Hz supply. Plot a curve of mean output voltage to a base of grid bias voltage for a variation of grid voltage between 0 and −20 V. The ignition voltage and voltage drop may be taken as 20 V (assume a pure resistance load).

25.4 A thyratron used in a grid-controlled rectifier has a voltage drop of 25 V when conducting, this being also the extinction voltage. Its control ratio is 30. An alternating p.d. of 212 V (r.m.s.) is applied, via a load resistance between anode and cathode, and a p.d. of 10·6 V (r.m.s.) lagging a quarter of a cycle behind the anode p.d. is applied between the grid and cathode. During what fraction of the alternating voltage cycle does the valve conduct? What is the mean anode current if the resistance is 100 Ω?

Ans. 0·34; 0·69 A.

25.5 A gas-filled triode, having a control ratio of 15 and negligible tube drop, is used as a controlled rectifier on a 180 V 500 Hz supply. Find the mean current in a load of 1,000 Ω resistance when a bias voltage of −10 V is applied to the valve grid. (*H.N.C.*)

Ans. 73 mA.

25.6 A gas-filled triode has a control ratio of 14·1 and operates from a 250 V a.c. supply. The load resistance is 150 Ω. Calculate the mean current in the load if the grid voltage consists of a steady voltage of −20 V superimposed on an alternating voltage of r.m.s. value 70·7 V which lags behind the anode voltage by 90°. Neglect valve voltage drop. (*H.N.C.*)

Ans. 0·395 A.

25.7 A mercury-arc rectifier has an arc drop of 25 V. What will be its internal efficiency when giving (*a*) 250 V, (*b*) 2,000 V on the d.c. side?

Ans. 91 per cent; 98·8 per cent.

25.8 A 250 V 400 A 6-anode mercury-arc rectifier operates from a transformer star-connected on the secondary side. Ignoring arc drop and impedance, calculate (*a*) the transformer secondary voltage, (*b*) the r.m.s. anode current, and (*c*) the rating of the secondary winding.

Ans. 185 V; 163 A; 181 kVA.

25.9 A 6-anode mercury-arc rectifier is to be supplied from 3-phase mains. Discuss the possible transformer arrangements and compare their relative advantages.

In a particular case the 800-turn primary windings are connected in delta and supplied at 6·6 kV; the direct voltage is 480 V and the arc drop is 25 V. Neglecting other voltage drops, determine the requisite number of secondary turns for six-phase star connexion. (*L.U.*)

Ans. 45.

25.10 Explain the operation of a six-anode mercury-arc rectifier with an interphase reactor. If the output is 300 V and 100 A, determine the secondary current and voltage for the supply transformer. Neglect overlap and assume an arc voltage drop of 20 V. If the transformer is delta connected, sketch and explain the theoretical primary current waveform. (*H.N.C.*)

Ans. 28·9; 278 V.

25.11 In the 3-phase bridge rectifier of Fig. 25.8, the three top diodes are replaced by thyristors. The 3-phase line voltage is 400 V, and pulses are applied to the gates to delay conduction by an angle of 25°. Determine the mean output voltage (d.c.), and sketch the line current waveforms. The thyristor voltage drop may be neglected.

Ans. 425 V.

25.12 Repeat Problem 25.11 for a delay angle, α, of 35°.

Ans. 384 V.

25.13 A thyristor bridge rectifier, as shown in Fig. 25.18 is used to supply a load of 100 Ω resistance. The a.c. supply voltage is 40 V r.m.s., and the holding current of the thyristor is 35 mA. Determine (*a*) the extinction angle, and (*b*) the mean load voltage when the gate firing pulses are adjusted to give a firing delay of 90°. The thyristor voltage drop may be assumed constant at 1·2 V.

Ans. 175°; 17·3 V.

Hint. Remember that conduction is discontinuous.

Chapter 26

FIELD-EFFECT
TRANSISTORS

Field-effect transistors, or FETs, are a group of semiconductor devices that complement, and in some cases may replace, bipolar transistors. Although they have been known in principle for some time, production on a commercial scale has become possible only because of recent advances in metallurgical technology. The FET is so called because it uses a transverse electric field to modulate the longitudinal current.

In some ways FETs have characteristics that closely resemble those of thermionic valves—for example, both valves and FETs have very high input resistances. However, in common with bipolar transistors and unlike valves, FETs require no heater current and can be obtained in complementary forms. They may have a very high frequency response. In addition, at zero bias, they can be either cut off or conducting depending on type. They have very high current and power gains, and because they are majority carrier devices (bipolar transistors are, of course, basically minority carrier devices) they have a high resistance to radiation damage and are less temperature dependent. In particular FETs are very suited to integrated circuit technology.

This chapter will briefly review the two main types of FET and will give some circuit applications.

819

26.1 The Junction-gate Field-effect Transistor

The junction-gate field-effect transistor, or JUGFET, was the first commercially available type of FET. It can be fabricated by either alloy junction or planar techniques. An *n*-channel alloy device is shown schematically in Fig. 26.1(*a*). It consists of a thin wafer of *n*-type

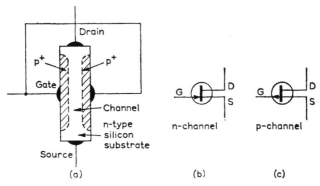

Fig. 26.1   THE JUGFET, WITH GRAPHICAL SYMBOLS

silicon into the opposite faces of which are diffused two very heavily doped *p*-type regions which are normally connected together to form the *gate* or control electrode. Ohmic connexions called the *drain* and *source* are made at the ends of the wafer for the load current. The gate, source and drain correspond to the base, emitter and collector of a bipolar transistor, or to the grid, cathode and anode of a thermionic triode.

With no external voltages the usual potential barrier builds up across the two *p–n* junctions and depletion layers are formed. Because the doping of the *p*-regions is much greater than that of the *n*-type wafer the depletion layers are mainly in the *n*-region. The drain and source are connected by the relatively narrow *n*-type *channel* and are separated from the gate connexions by the depletion layers. A similar device is possible using a *p*-type wafer and *n*-type gate connexions. giving a *p*-channel type. The graphical symbols are shown at (*b*) and (*c*) for an *n*-channel and a *p*-channel FET respectively.

In operation the gates are reverse biased with respect to source and drain, so that the gate current is only the very small reverse-bias saturation current. The reverse bias voltage extends the width of the depletion layers and hence narrows the conducting channel and so controls the drain-to-source current. This current is, of course, mainly a majority carrier current for the basic silicon wafer (electrons in the *n*-channel and holes in the *p*-channel type). The shape of the depletion layer is shown in Fig. 26.2 for a biased *n*-channel JUGFET.

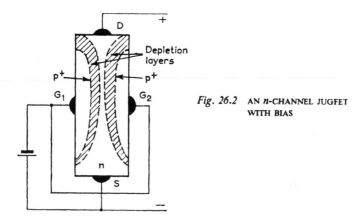

Fig. 26.2 AN *n*-CHANNEL JUGFET
WITH BIAS

Since D is more positive than S the reverse bias across the gate–channel junctions is greater towards the drain end than at the source end. This explains why the channel is narrower at the drain end than at the source end, as shown.

The drain characteristics of an *n*-channel JUGFET are shown in Fig. 26.3(*a*) where the drain current I_D is drawn as a function of

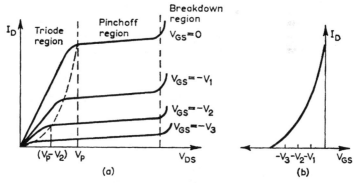

Fig. 26.3 CHARACTERISTICS OF AN *n*-CHANNEL JUGFET
(*a*) Drain characteristic
(*b*) Transfer characteristic

drain–source voltage, V_{DS}, for various fixed values of gate–source voltage, V_{GS}. The characteristics can be divided into three regions—the triode region, the pinchoff region and the breakdown region.

Consider $V_{GS} = 0$. As V_{DS} is increased the channel current, I_D, increases almost linearly. However, since the drain is now positive, while the gate remains at zero potential, a reverse bias builds up between the gate and the drain and the depletion layer takes the form shown in Fig. 26.2. As V_{DS} is further increased a point is reached when the two depletion layers almost meet. The voltage at which this occurs is called the *pinchoff voltage*, V_p. Further increase in V_{DS} causes practically no further increase in I_D all that happens is that the depletion layers almost meet over a longer distance and this gives rise to an increase in the channel resistance in proportion to the increase in voltage. If V_{DS} continues to increase a point is reached where breakdown of the channel occurs and the current may rise to destructively high values.

At any negative value of V_{GS}, the depletion layers will already be closer together than for $V_{GS} = 0$, so that the initial slope of the curve in the triode region is less and pinchoff will occur at a value of V_{DS} lower than V_p by about the value of gate bias applied.

The size of the pinchoff voltage depends on the thickness of the initial channel. For relatively thick channels V_p will be large and drain voltages will be of the same order as in pentode valve circuits. For thin channels V_p will be low, and operating voltages will correspond to those of bipolar transistors. Minimum values of V_p are around 0·5 V, but normally JUGFETs have much higher values of V_p. The transfer characteristic of a JUGFET at a fixed value of V_{DS} in the pinchoff region is shown in Fig. 26.3(b).

In the triode region of the JUGFET characteristics the curves of I_D against V_{DS} are reasonably linear, with a slope that depends on the applied gate–source voltage. The characteristics apply for negative as well as for positive values of V_{DS} (Fig. 26.4) as long as the negative

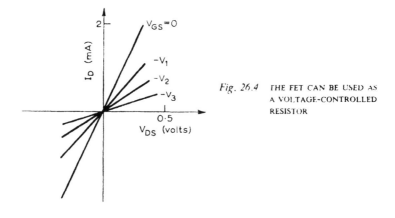

Fig. 26.4 THE FET CAN BE USED AS A VOLTAGE-CONTROLLED RESISTOR

value is not large enough to cause appreciable conduction of the
gate–drain diode. In silicon JUGFETS this negative voltage will be
approximately 0·6 V. Over the range of voltages of ±0·6 V, therefore, a JUGFET may be used as a linear resistor, and this is often
done in integrated circuits. In addition, the resistance will vary
with the bias voltage. This latter effect can be used in automatic
gain control circuits in amplifiers, and amplitude control circuits
in oscillators.

26.2 The Insulated-gate Field-effect Transistor

As with the JUGFET, the insulated-gate field-effect transistor, or
IGFET, can be constructed with either an *n*-type or a *p*-type channel.
In addition, it can be designed so that it conducts only when forward
gate–source bias is applied (*enhancement mode*) or conducts with
zero bias (*depletion mode*). Because of its construction, this type of
FET is often called a MOST (metal-oxide-semiconductor transistor).

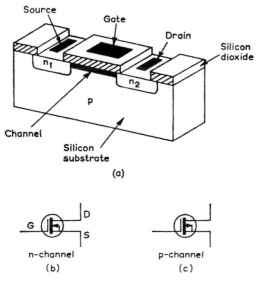

Fig. 26.5 TYPICAL CONSTRUCTION AND GRAPHICAL SYMBOLS
FOR AN IGFET

An enhancement-mode *n*-channel IGFET is shown in Fig. 26.5(*a*).
Two heavily doped *n*-regions, n_1 and n_2, are diffused into a lightly
doped *p*-type substrate. A thin ($\sim 0\cdot2\ \mu$m) *insulating* layer of silicon
dioxide is grown over the surface, with gaps to allow the ohmic

connexions to be made to n_1 and n_2. These source and drain "windows" are etched by photolithographic techniques. The gate connexion is a metallic layer (often aluminium) on the silicon dioxide that lies between the two n-regions. Similar metallic layers on n_1 and n_2 form the source and drain connexions.

With no gate voltage ($V_{GS} = 0$) the two n-regions are separated by a p-type region, typically 5 μm long. Hence for any voltage applied between drain and source there will always be two opposing p–n junctions in series and negligible current will flow no matter what polarity is applied between D and S. If the gate is made positive with

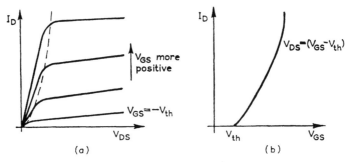

Fig. 26.6   CHARACTERISTICS OF AN n-CHANNEL ENHANCEMENT-
TYPE IGFET
(a) Drain characteristics
(b) Transfer characteristic

respect to the source, then an electric field is set up beneath the gate electrode. Since the silicon-dioxide layer is thin and since the silicon substrate is not a very good conductor this field will extend into the p-type silicon and will cause an attraction of electrons to just beneath the dioxide layer. Eventually as the gate–source voltage, V_{GS}, is increased it reaches a sufficiently positive value (called the *threshold voltage, V_{th}*) for the movement of electrons to the surface to be large enough for a predominantly n-type region to form near the surface. This then constitutes a conducting channel between n_1 and n_2 as shown in Fig. 26.5(a).

With V_{GS} more positive than the threshold value any voltage on the drain will cause conduction between drain and source. This drain current will be almost linearly related to both V_{DS} and V_{GS}. However, as the drain voltage is made more positive, the voltage between gate and drain falls, until eventually the pinchoff voltage is reached and the conducting channel becomes pinched off as in the JUGFET. The drain current cannot increase further and becomes almost constant until breakdown occurs, at which point the current rises very rapidly. The family of drain characteristics are shown in Fig. 26.6(*a*)—the corresponding transfer characteristic is shown at (*b*). The transfer characteristic approximates very closely to a square law, giving, in the pinchoff region,

$$I_D = -\frac{\beta}{2}(V_{GS} - V_{th})^2 \qquad (26.1)$$

where β is a constant that depends on the geometry of the device. It is because the application of forward bias increases the channel conductivity that this device is said to operate in the enhancement mode.

In the depletion-mode *n*-channel IGFET, an *n*-type surface layer is diffused between the drain and source, giving an initial channel which allows conduction when $V_{GS} = 0$. If the gate is made negative, electrons are repelled from the channel, which becomes a poorer conductor. If V_{GS} is made positive the channel becomes more definitely *n*-type and the drain current increases. For any values of V_{GS} the pinchoff effect described in considering the enhancement-mode IGFET will apply, so that the drain characteristics will be as shown in Fig. 26.7.

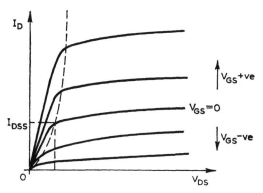

Fig. 26.7 DRAIN CHARACTERISTICS OF AN *n*-CHANNEL
DEPLETION-MODE IGFET

One parameter that is often quoted by manufacturers is the saturation drain current that flows when gate and source are connected together. This is given the symbol I_{DSS}.

Exactly similar relations will be obtained using an *n*-type substrate with a *p*-type channel and heavily doped *p*-type drain and source areas, i.e. the IGFET can be manufactured in complementary forms.

Transfer characteristics for *p*-channel and *n*-channel depletion-
and enhancement-mode IGFETs are shown in Fig. 26.8.

A depletion-mode IGFET is conducting when $V_{GS} = 0$. This can
sometimes be a useful property, e.g. in oscillator circuits. An
enhancement-mode IGFET must be biased for conduction to take
place. As with bipolar transistors, this has the advantages of (i)
allowing circuits to be designed for direct interstage coupling without
requiring an auxiliary d.c. supply of opposite polarity, and (ii)
giving automatic cut-off when $V_{GS} = 0$. Note that depletion-mode
devices operate by enhancement when the gate voltage increases.
Enhancement-mode devices, on the other hand, can never act in the
depletion mode—there is nothing to deplete!

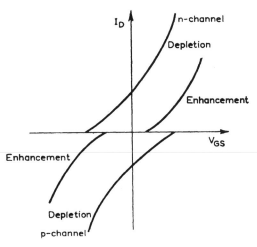

Fig. 26.8   IGFET TRANSFER CHARACTERISTICS

The substrate of an IGFET is normally connected to the source
but may be biased to alter the pinchoff voltage. In the pinchoff
region the equivalent circuit may be deduced by considering the
load line on drain/source characteristics that are assumed to be
parallel straight lines of slope $1/r_{DS}$ (r_{DS} is the *drain/source slope
resistance*). A simple enhancement-mode common source amplifier
circuit is shown in Fig. 26.9(*a*). Here the gate bias is obtained from
resistors R_1, R_2 and R_g. With this arrangement R_g can be of a high
value to give a high impedance bias voltage source as seen from the
gate. The drain/source characteristics with a load line AB are shown
at (*b*). This load line is drawn from B (where V_{DS} is equal to the
supply voltage V_{DD}) to A (where I_D is equal to V_{DD}/R_L). For any

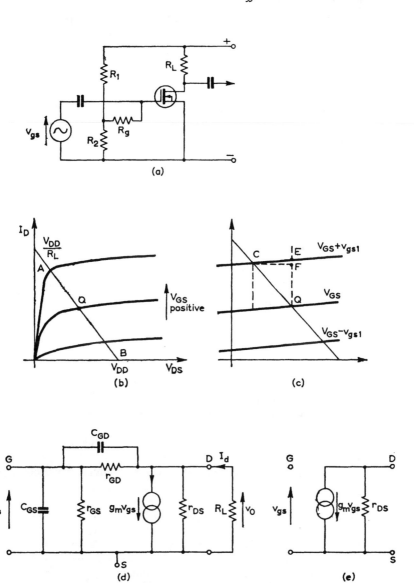

Fig. 26.9 LOAD LINE AND SMALL-SIGNAL EQUIVALENT CIRCUITS

variation in gate voltage about the quiescent value, the operating point moves up or down AB to the corresponding value of V_{GS}.

Consider the drain/source characteristics as at (c). With $R_L = 0$ any change in gate–source voltage would cause a proportional change in drain current I_D, given by

$$\delta I_D' = g_m \delta V_{GS} \quad \text{where} \quad g_m = \frac{\delta I_D}{\delta V_{GS}}\bigg|_{V_{DS} = \text{constant}}$$

g_m is known as the *mutual conductance* of the FET. The change in current, $\delta I_D'$, is represented on the characteristics by QE in Fig. 26.9(c). With R_L connected the actual drain current change, δI_D, will be QF (the operating point will actually move to C). Hence

$$\delta I_D = g_m \delta V_{GS} - \text{EF} = g_m \delta V_{GS} - \frac{\text{CF}}{r_{DS}}$$

But CF is the change in output voltage δV_0, where $\delta V_0 = -\delta I_D R_L$ (an increase in I_D causes a decrease in V_0). Hence

$$\delta I_D = g_m \delta V_{GS} - \frac{\delta I_D R_L}{r_{DS}}$$

or

$$\delta I_D = g_m \delta V_{GS} \frac{r_{DS}}{r_{DS} + R_L} \tag{26.2}$$

This is the current that would flow in a load resistance R_L connected across a constant-current source of value $g_m V_{GS}$ with internal impedance r_{DS} as shown at (d).

The above analysis applies to both JUGFETs and IGFETs, and to both depletion and enhancement modes.

The other components of the equivalent circuit represent the leakage resistances, r_{GS} and r_{GD}, between gate and source, and gate and drain, respectively, and in JUGFETs the depletion-layer capacitances, or in IGFETs the insulation-layer capacitances, C_{GS} and C_{GD}. Typical values for these quantities are $100\,\text{M}\Omega$ and $10\,\text{pF}$ for JUGFETs, and $10^8\,\text{M}\Omega$ and $2\,\text{pF}$ for IGFETs. Typical values of mutual conductance, g_m, are $2\,\text{mA/V}$ for JUGFETs and $1\cdot5\,\text{mA/V}$ for IGFETs, though in some devices much higher values are obtained. These values are considerably lower than the $40\,\text{mA/V}$ common in bipolar transistors. Note that the gate input resistances are generally so high that they may often be neglected, giving the low-frequency equivalent circuit shown in Fig. 26.9(e). At high frequencies, however, the gate input capacitance cannot be neglected.

The phase relationships in the common-source circuit which has been described (and which is equivalent to the bipolar common-emitter circuit) are obtained from the fact that any increase in gate–source voltage causes an increase in drain current. This in turn increases the voltage drop across the load resistance and causes a decrease in output voltage—i.e. there is an inherent 180° phase shift. Hence in complexor r.m.s. signal values

$$I_d = g_m V_{gs} \frac{r_{DS}}{r_{DS} + R_L} \qquad (26.3)$$

and

$$V_o = -I_d R_L = -\frac{g_m r_{DS} R_L}{r_{DS} + R_L} V_{gs} \qquad (26.4)$$

The voltage gain is thus

$$A_v = \frac{V_0}{V_{gs}} = -\frac{g_m r_{DS} R_L}{r_{DS} + R_L} \qquad (26.5)$$

The product $g_m r_{DS}$ is sometimes called the FET *amplification factor* μ. Note that subscripts in small letters are used to denote r.m.s. quantities.

EXAMPLE 26.1 An *n*-channel IGFET is connected as a simple amplifier as in Fig. 26.9(a). If $g_m = 1.5$ mA/V, $r_{DS} = 45$ kΩ and $R_L = 47$ kΩ, determine the overall mid-frequency voltage gain. If $R_1 = 100$ kΩ, $R_2 = 10$ kΩ and $R_g = 10$ MΩ find the current and power gains.

From eqn. (26.4),

$$\frac{V_o}{V_{gs}} = A_v = -\frac{g_m r_{DS} R_L}{r_{DS} + R_L}$$

$$= -\frac{1.5 \times 10^{-3} \times 45 \times 10^3 \times 47 \times 10^3}{92 \times 10^3}$$

$$= -34.5$$

The input resistance of the IGFET can be considered to be very much greater than the 10 MΩ of R_g. Hence the input current I_i is essentially that which flows through R_g and the parallel combination of R_1 and R_2—i.e. it is almost equal to V_{gs}/R_g, since $R_1 R_2/(R_1 + R_2) \ll R_g$. Since the output current $I_o = V_o/R_L$ the current gain is

$$\frac{I_o}{I_i} = A_i = \frac{V_o}{V_{gs}} \frac{R_g}{R_L} = -34.5 \times \frac{10^7}{47 \times 10^3} = -7.34 \times 10^3$$

and the power gain is $A_v A_i = 253 \times 10^3$

These results should be compared with those for bipolar transistors.

26.4 Temperature Effects

In JUGFETs the reverse biased gate diode exhibits the increase in leakage current with temperature that is common to all junction diodes; hence the input resistance falls as the temperature rises. This effect is much less pronounced in IGFETs because the input resistance is determined by the temperature coefficient of the insulating silicon-dioxide layer. In both types the channel resistance increases with temperature (due to a decrease in carrier mobility) giving a negative temperature coefficient of drain current. In JUGFETs the gate diode contact voltage drops with temperature (by about 2 mV/K at around 300 K). This causes a positive temperature coefficient of drain current which will be more pronounced the lower the value of pinchoff voltage, V_p. Thus JUGFETs with low values of V_p (~ 0.5 V) have normally an overall positive temperature coefficient, while those with $V_p > 1.5$ V have normally an overall negative temperature coefficient of drain current.

Since the carriers in the conducting channel do not tend to "freeze out", IGFETs can be operated down to liquid helium temperatures (~ 4 K), and temperature effects are generally small even up to 400 K ($\sim 130°$C). (In bipolar transistors the impurity centres tend to fix or "freeze" the excess holes or electrons at low temperatures so that they are no longer mobile. In FETs the channel field prevents this effect).

Because of the small changes that occur in drain current with increase in temperature, FETs are not subject to the thermal runaway that can occur in bipolar transistors.

26.5 High-frequency Performance

As with bipolar transistors, the high-frequency performance of FETs is basically limited by the source-to-drain transit time of the majority carriers. In addition, parasitic channel impedances and gate capacitance can limit high-frequency performance. In general the h.f. performance can be specified in the same manner as was developed in Chapter 21 for bipolar transistors. Operating frequencies extend well into the 100 MHz region, and are generally limited by the external circuit strays rather than the device itself. This is where the higher mutual conductance of the bipolar transistor can be of advantage.

One advantage of the FET over the bipolar transistor arises because it is a majority-carrier device. For this reason, on step inputs the turn-off time will be small due to the absence of minority-carrier storage effects. Turn-on and turn-off times in the nano-second region are possible.

26.6 Bias Arrangements

Some typical small-signal common-source circuits are shown in Fig. 26.10. At (a) the bias resistors, R_1 and R_2, can be of high value to give the required bias for operation in the enhancement mode

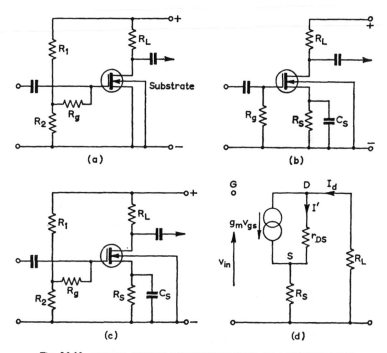

Fig. 26.10 TYPICAL BIASING ARRANGEMENTS FOR AN *n*-CHANNEL IGFET

    (a) Enhancement type
    (b) Depletion type
    (c) Either type
    (d) Equivalent circuit

but taking very little power from the supply. A possible bias arrangement for the depletion-mode IGFET is shown at (b). The circuits at (a) and (b) may be combined to give that at (c), which can be applied to both modes. The source resistance R_s gives negative feedback on direct current, and this increases the temperature stability. The voltage gain for these circuits is given by eqn. (26.5):

$$A_v = - \frac{g_m r_{DS} R_L}{r_{DS} + R_L}$$

If the bypass capacitor C_S is omitted from (b) or (c) then signal negative current feedback takes place. The equivalent circuit for both connexions is then as shown at (d). For this

$$V_{gs} = V_{in} - I_d R_S \tag{i}$$
$$I_d = I' + g_m V_{gs}$$

and

$$I_d(R_L + R_S) + I' r_{DS} = 0 = I_d(R_L + R_S + r_{DS}) - g_m r_{DS} V_{gs} \tag{ii}$$

Hence, substituting in (ii) from (i) to eliminate V_{gs},

$$I_d(R_L + R_S + r_{DS}) = g_m r_{DS}(V_{in} - I_d R_s)$$

so that

$$I_d = \frac{g_m r_{DS} V_{in}}{R_L + (1 + g_m r_{DS})R_S + r_{DS}}$$

and the voltage gain A_{vf} with feedback is

$$A_{vf} = \frac{V_o}{V_{in}} = \frac{I_d R_L}{V_{in}} = -\frac{g_m r_{DS} R_L}{r_{DS} + R_L + (1 + g_m r_{DS})R_S} \tag{26.6}$$

This expression should be compared with the corresponding result in Chapter 22, i.e.

$$A_{vf} \approx \frac{A_v}{1 - \beta A_v} \tag{26.7}$$

which can be derived from the expression for A_v and eqn. (26.6) if $\beta = R_S / R_L$.

Note that for all of these circuits the input resistance of the FET itself is very high (hundreds of megohms). The circuit input impedance depends on the bias circuit arrangements. The output impedance is r_{DS} without feedback. This impedance is increased by negative current feedback and reduced by negative voltage feedback.

26.7 Common-drain Connexion (Source Follower)

This connexion is shown in Fig. 26.11. The circuit is one with 100 per cent negative voltage feedback. From the equivalent circuit at (b),

$$V_{gs} = V_{in} - V_o \tag{i}$$
$$I_d = g_m V_{gs} + I' \tag{ii}$$

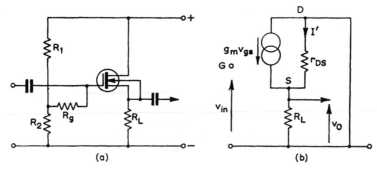

Fig. 26.11 SOURCE FOLLOWER AND EQUIVALENT CIRCUIT

and

$$I'r_{DS} = -V_o = -(I_d - g_m V_{gs})r_{DS} \qquad \text{(iii)}$$

Hence

$$-\frac{V_o}{r_{DS}} = I_d - g_m(V_{in} - V_o)$$

so that

$$V_o = I_d R_L \qquad \text{(Note that there is } no \text{ phase reversal with this connexion)}$$

$$= -\frac{V_o}{r_{DS}} R_L + g_m V_{in} R_L - g_m V_o R_L \qquad \text{(26.8)}$$

and the voltage gain is

$$A_{vf} = \frac{V_o}{V_{in}} = \frac{g_m R_L}{1 + \dfrac{R_L}{r_{DS}} + g_m R_L} \approx 1 \qquad \text{(26.9)}$$

provided that $g_m R_L \gg 1 + R_L/r_{DS}$.
    Rewriting eqn. (26.8),

$$A_{vf} = \frac{g_m R_L}{\left(\dfrac{1}{r_{DS}} + g_m\right)\left(\dfrac{r_{DS}}{1 + g_m r_{DS}} + R_L\right)}$$

so that (comparing with eqn. (22.16)) the output impedance is

$$Z_{out} = \frac{r_{DS}}{1 + g_m r_{DS}} \qquad \text{(26.10)}$$

$$\approx \frac{1}{g_m} \qquad \text{if } g_m r_{DS} \gg 1 \qquad \text{(26.11)}$$

i.e. the output impedance is very low.

The input impedance of the circuit will depend on the bias arrangement rather than the FET.

The source follower may be used as a buffer stage in the same way as the emitter follower.

26.8   Common-gate Circuit

The common-gate circuit corresponds to the common-base circuit of the bipolar transistor. It may be used as an impedance convertor from low input to high output values.

26.9   Cascaded Stages

FETS may be cascaded in the same way as bipolar transistors. Couplings may be *CR* or direct. A typical circuit is shown in Fig. 26.12.

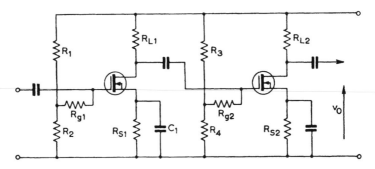

Fig. 26.12   CASCADED FET AMPLIFIER

Frequently FET circuits are cascaded with bipolar power transistors using the high-impedance FET to feed the low-impedance bipolar transistor. High power gains can be achieved, but the low input impedance of the bipolar transistor acting as the load on the FET reduces the voltage gain to a low value.

FETS are particularly suited to incorporation in integrated circuits. They may also be used in digital circuits and logic circuits. In integrated circuits the ability of the FET to act as a resistor is of great value. Normally in integrated circuits *p*-channel operation is used. In general FETS have been used only in small-signal or low-power applications due to difficulties with power dissipation. However, they have been employed from audio to ultra-high frequencies ($>$400 MHz).

PROBLEMS

The *n*-channel depletion-mode IGFET used in these problems has the following characteristics. I_D is in milliamperes, and all voltages are in volts.

$V_{GS} = 3$	V_{DS}	10	20	30		
	I_D	1·2	1·7	1·85		
$V_{GS} = 2$	V_{DS}	10	20	30		
	I_D	0·8	1·2	1·28		
$V_{GS} = 1$	V_{DS}	5	10	20	30	40
	I_D	0·38	0·65	0·72	0·77	0·82
$V_{GS} = 0$	V_{DS}	5	10	20		40
	I_D	0·20	0·31	0·35		0·38
$V_{GS} = -1$	V_{DS}	5	10			50
	I_D	0·06	0·08			0·09

26.1 Draw the FET characteristics. From these determine for a quiescent operating point at $V_{DS} = 25$ V and $V_{GS} = 1$ V approximate values for r_{DS}, g_m and μ.

Ans. 200 kΩ; 0·45 mA/V; 90.

26.2 A common-source amplifier uses the above FET. The supply voltage is 50 V and the load resistance is 25 kΩ. Draw the load line. If the quiescent gate voltage is 1 V determine from the characteristics the approximate small-signal voltage gain. Sketch the output voltage waveform for a sinusoidal signal input voltage of peak value 1 V.

Ans. 10.

26.3 The FET is used in the amplifier shown in Fig. 26.10(*b*). If the supply voltage is 40 V, $R_L = 24$ kΩ and $R_S = 1$ kΩ, determine the quiescent operating point.

Ans. $V_{DS} = 23$ V; $I_D = 0·66$ mA.

Chapter 27

LOGIC

Logic circuits are used in digital computers, data processors, and many forms of control and sequencing systems. The individual elements (or *gates*) of a logic system are normally obtained direct from the manufacturers as "black boxes", and it is the interconnexion of these elements that gives rise to a particular logic system. Hence it is very important to understand the interconnexions, and such overall systems will be dealt with before the individual electronic elements are considered. This "systems approach" applies whether the logic elements are pneumatic (fluidic) or electrical, electronic, etc. The chapter will close with a brief outline of some electronic means of achieving logic functions. It is emphasized, however, that the chapter as a whole attempts only to introduce the subject.

Essentially logic circuits operate on a digital or two-state basis. In positive electronic logic circuits the presence of a positive voltage at a point is conventionally designated the logic "1" state at that point (or ON state). The absence of a voltage is described as the logic "0" state (or OFF state). Note that the symbols "1" and "0" do not have their normal mathematical meaning, but indicate simply the presence or absence of voltage. Various voltage levels may be chosen for the "1" state, e.g. $+1$ V, $+5$ V, $+20$ V, etc. In negative logic the "1" state is represented by the more negative voltage of the two voltage levels chosen. Generally the output of any logic element or gate will be either almost the circuit supply voltage or almost zero. This output will depend on the states ("0" or "1") of the various inputs to the gate. (Note that in some instances the "1" state may be

836

represented by zero volts—e.g. in a positive logic system with reference voltages of -3 V and 0 V, or a negative logic system with reference voltages $+5$ V and 0 V—this however is not common.)

Logic circuits are so called because the output quantity depends on the logic inputs on a "yes-no" (or "true-false") basis, just as the results of classic logic depend on the initial statements.

27.1 The Logic OR Gate

This gate gives a logic "1" output if any one of its various inputs is in the "1" state. The number of inputs possible for the gate depends on the design and output loading of the element and is referred to as the *fan in*. The OR gate may be considered symbolically as resulting from the parallel connexion of switch contacts as shown for a 3-input gate in Fig. 27.1(a), where the switches are taken to be in the

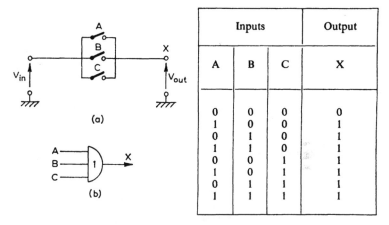

	Inputs			Output
A	B	C	X
0	0	0	0
1	0	0	1
0	1	0	1
1	1	0	1
0	0	1	1
1	0	1	1
0	1	1	1
1	1	1	1

Fig. 27.1 THREE-INPUT "OR" GATE

"1" state when closed. The logic symbol is shown at (b); the figure 1 inside the semicircle represents the minimum number of inputs that must be in the logic "1" state in order to give a logic "1" output—in this case one. The *truth table* on the right of the diagram relates the state of the output, X, to that of the three inputs, A, B and C. Symbolically, using the notation of the algebra developed by George Boole, and called *Boolean algebra*,

$$X = A + B + C \qquad (27.1)$$

where the + sign is taken to represent the logic function OR. This equation means that X takes the logic value "1" if, and only if, *at least one of* the inputs A OR B OR C has the logic value "1". The truth table shows how X varies with A, B and C.

27.2 The Logic AND Gate

The AND gate gives a logic "1" output only if *all* its inputs are in the "1" state. It may be considered to be the result of connecting a group of switch contacts in series as shown in Fig. 27.2(a). There

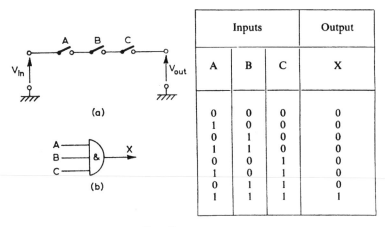

	Inputs			Output
	A	B	C	X
	0	0	0	0
	1	0	0	0
	0	1	0	0
	1	1	0	0
	0	0	1	0
	1	0	1	0
	0	1	1	0
	1	1	1	1

Fig. 27.2   THREE-INPUT "AND" GATE

will be an output only if contacts A AND B AND C are closed. The logic symbol is shown at (*b*), and the truth table on the right represents the Boolean expression

$$X = A \cdot B \cdot C \qquad (27.2)$$

where the points represent the logic function AND. The equation means that the output X takes the value "1" only if A AND B AND C are *all* in state "1".

Notice that in both the OR and the AND truth tables the conditions of the inputs have been arranged for convenience in a logical form that derives from binary arithmetic. Thus the first input (A) alternates between "0" and "1", the second input (B) is in pairs of zeros and of ones, and the third input (C) is in groups of four zeros and four ones. If this scheme is followed, no possible combinations of inputs will be missed from the truth table. Note that the total number of different combinations of input for an *n*-input gate will be 2^n.

27.3 The NOT Gate—NOR and NAND

The NOT gate is essentially an invertor with one input. The output, *X*, is "1" if the input is "0", and is "0" if the input is "1". The logic and circuit symbols are shown in Fig. 27.3. Notice that the

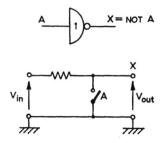

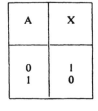

Fig. 27.3 THE "NOT" GATE

small circle in the logic symbol represents the negation or inversion of the output (in the same way a small circle at the input represents negation of that input).

In Boolean notation the output *X* is NOT *A*, and the negation is represented by a bar above the symbol, i.e.

$$X = \bar{A} \tag{27.3}$$

The truth table is shown on the right. Obviously two NOT gates in cascade (i.e. with the output of the first NOT gate providing the input to the second) cancel each other, i.e.

$$\bar{\bar{A}} = A$$

Usually the operations of OR and NOT and of AND and NOT are combined, since this normally results in some simplification of the electronic circuitry involved due to the inherent inversion of a single stage. These combined operations are termed NOR and NAND respectively. Truth tables for two-input NOR and NAND gates are shown in Figs. 27.4(*a*) and (*b*) respectively, the logic symbols being shown on the left.

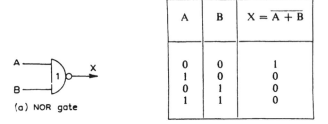

(a) NOR gate

A	B	$X = \overline{A + B}$
0	0	1
1	0	0
0	1	0
1	1	0

(b) NAND gate

A	B	$X = \overline{A . B}$
0	0	1
1	0	1
0	1	1
1	1	0

Fig. 27.4   TWO-INPUT "NOR" AND "NAND" GATES

The Boolean expressions for the NOR and NAND functions for multiple-input gates are

NOR $X = \overline{A + B + C +} \ldots$ (27.4)

NAND $X = \overline{A . B . C} \ldots$ (27.5)

Note that multiple-input NOR or NAND gates in which only one input is used simply give the NOT function. (See Section 27.5 and later).

It is left as an exercise for the reader to draw the truth tables for 3-input NOR and NAND gates.

27.4 Simple Boolean Relations

It is obvious that, for more than three of four inputs, truth tables become cumbersome. In such cases it is usually more convenient to simplify the logic system by using some of the following relationships that apply to Boolean algebra. The logic system required in any given instance is, of course, a function of the action required.

Boolean algebra obeys the three basic laws of ordinary algebra, namely commutation ($A + B = B + A$), association (($A + B) + C) = (A + (B + C)$) and distribution ($A . (B + C)) = (A . B + A . C$). It differs, however, from ordinary algebra in two significant respects, namely:

(a) "add", "negate" and "multiply" are the only allowed operations,

(b) the "answer" must be either "1" or "0".

The following equations illustrate this:

$0 + 0 = 0$	(27.6)	$0.0 = 0$	(27.9)
$0 + 1 = 1$	(27.7)	$0.1 = 0$	(27.10)
$1 + 1 = 1$	(27.8)	$1.1 = 1$	(27.11)

In particular, eqn. (27.8) expresses the fact that two closed switches in parallel are simply equivalent to a single closed switch. If A is taken as a logic gate input which can be either in the "1" or the "0" state, then the following Boolean expressions apply:

$$A + A + A + \ldots = A \tag{27.12}$$

$$A + \bar{A} = 1 \tag{27.13}$$

$$1 + A = 1 \tag{27.14}$$

$$0 + A = A \tag{27.15}$$

$$\overline{A + \bar{A}} = 0 = A . \bar{A} \tag{27.16}$$

$$A . A . A \ldots = A \tag{27.17}$$

$$A . \bar{A} = 0 \tag{27.18}$$

$$1 . A = A \tag{27.19}$$

$$0 . A = 0 \tag{27.20}$$

$$\overline{A . \bar{A}} = 1 = A + \bar{A} \tag{27.21}$$

For two inputs A and B (both of which can take the logic value "1" or "0") the following can readily be verified by drawing up the appropriate truth tables, or applying the relationships stated above:

$$A + A . B = A(1 + B) = A . 1 = A \tag{27.22}$$

$$A . (A + B) = A . A + A . B = A + A . B = A \tag{27.23}$$

$$A + \bar{A} . B = A + B \tag{27.24}$$

$$A . (\bar{A} + B) = A . \bar{A} + A . B = A . B \tag{27.25}$$

The following important relations, which are known as *de Morgan's theorem*, may also be verified by a truth table:

NOR $\qquad \overline{A + B + C} = \bar{A} \cdot \bar{B} \cdot \bar{C}$ (27.26)

NAND $\qquad \overline{A \cdot B \cdot C} = \bar{A} + \bar{B} + \bar{C}$ (27.27)

Extending these equations,

$$A + B + C = \overline{\bar{A} \cdot \bar{B} \cdot \bar{C}}$$ (27.28)

and

$$A \cdot B \cdot C = \overline{\bar{A} + \bar{B} + \bar{C}}$$ (27.29)

Equation (27.28) shows that a NAND gate may be used to perform the OR function provided that each input is inverted before being applied to the gate. In the same way, by inverting the inputs, a NOR gate will perform the AND function. Such inverted inputs are normally available from the input transducers. Thus, if the input is derived from a changeover microswitch as shown in Fig. 27.5, then both A and $\bar{A}$ inputs are available.

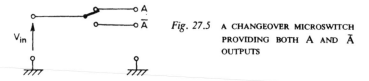

Fig. 27.5 A CHANGEOVER MICROSWITCH PROVIDING BOTH A AND $\bar{A}$ OUTPUTS

EXAMPLE 27.1 A coin-operated hot-drink dispenser will provide a paper cup of tea or coffee under the following conditions:

$\qquad$ (*a*) the correct coin is inserted (*I*)
AND $\quad$ (*b*) a paper cup is in position (*P*)
AND $\quad$ (*c*) hot water is available (*W*)
AND $\quad$ (*d*) the selector is set at "tea" (*T*) OR "coffee" (*C*)
$\qquad\qquad$ AND milk (*M*) OR no milk ($\bar{M}$) $\Big)$
$\qquad\qquad$ AND sugar (*S*) OR no sugar ($\bar{S}$) $\Big)$ are set

Obtain the logic expression for the conditions under which a drink *D*, may be obtained. Show how this may be implemented using (*a*) NOR gates only, (*b*) NAND gates only.

The Boolean expression for *D* is derived from the conditions under which action is required, i.e.

$$D = I \cdot P \cdot W \cdot ((T + C) \cdot (M + \bar{M}) \cdot (S + \bar{S}))$$
$$= I \cdot P \cdot W \cdot (T + C)$$

since $(M + \bar{M}) = (S + \bar{S}) = 1$ where a drink is obtained whenever $D = 1$. Note that a drink will be obtained whichever setting of the "milk" and "sugar" selectors is used.

(*a*) From eqn. (27.29),

$$D = \overline{\bar{I} + \bar{P} + \bar{W} + \overline{(T + C)}}$$

i.e. D is the output of a NOR gate whose inputs are $\bar{I}$, $\bar{P}$, $\bar{W}$ and $\overline{(T + C)}$. Note that $\overline{(T + C)}$ is itself the output of a NOR gate whose inputs are T and C. The resultant logic circuit is shown in Fig. 27.6(a).

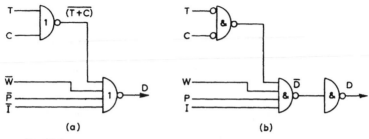

(a)   (b)

Fig. 27.6

(b) In order to be able to realize the expression for D using NAND gates only, the equation $D = I . P . W . (T + C)$ must be manipulated so that only NAND expressions appear on the right-hand side. The part in brackets is readily expressed in terms of a NAND gate using eqn. (27.28). Thus

$$T + C = \overline{\overline{T} . \overline{C}}$$

i.e. $T + C$ can be obtained using a NAND gate with inputs $\bar{T}$ and $\bar{C}$. It is not, however, possible to express D directly in terms of NAND gates. We must first obtain $\bar{D}$ as

$$\bar{D} = \overline{I . P . W . (\bar{T} . \bar{C})}$$

i.e. $\bar{D}$ is obtained as the output of a NAND gate whose inputs are I, P, W and $\overline{T} . \overline{C}$. Then D itself is readily obtained by adding one further NAND gate as shown in Fig. 27.6(b) and recalling that $D = \bar{\bar{D}}$. Note that the inverted inputs $\bar{T}$ and $\bar{C}$ are represented at (b) by the small circles at the input of the first NAND gate.

27.5 Electronic Logic Families

Electronic logic circuits are commonly constructed in integrated circuit form, where a complete gate (or indeed several complete gate circuits and their interconnexions) are formed on a single monocrystalline silicon chip. They may also be constructed from discrete components. Such circuits may be obtained in AND, OR, NAND, NOR etc., configurations over a range of positive or negative logic voltages. The various types of circuit may be considered as families that are distinguished from one another on the basis of the following properties.

(a) *Cost per gate.* Very frequently this is the most important criterion.

(b) *Propagation delay time per gate.* It always takes a finite time for an electronic circuit to change over from one steady state to another.

(c) *Threshold voltage.* This is the input voltage level that is required to make the circuit change from one logic state to another. It is important since logic inputs are often the outputs from other logic elements.

(d) *Noise margin.* It is undesirable that an unwanted signal (or "noise") should cause malfunction of any logic gate. The *noise margin* of a logic circuit is the difference between the operating voltage and the threshold voltage. If any unwanted disturbance exceeds this noise margin the gate may change state without any change in the true logic input.

(e) *Maximum fan-in.* This is the total maximum number of logic inputs with which any particular logic circuit is designed to operate. If the design number is exceeded, faulty logic operation may result.

(f) *Maximum fan-out.* This is the total maximum number of logic circuits that any one gate is capable of driving. If it is exceeded, operating voltages may fall below the threshold value.

(g) *Power dissipation per gate.* Where there are a great many logic elements, it becomes important to keep the overall power dissipation to as low a value as possible.

Four families of logic circuit will be considered in the next four sections. There are:

RTL. Resistor-transistor logic
DTL. Diode-transistor logic
TTL. Transistor-transistor logic
ECL. Emitter-coupled logic

Generally the cost, power dissipation per gate and fan-out all increase as we go from RTL to ECL, while the propagation delay time decreases. The noise margin is typically highest for the TTL family of circuits, while the fan-in is highest for DTL and TTL. It is left as an exercise for the reader to account for these facts from the following circuit descriptions.

27.6 Resistor-Transistor Logic (RTL)

A typical circuit using RTL to produce the NOR logic function with positive logic is shown in Fig. 27.7. A NAND form of circuit is also possible.

In the circuit shown, the voltage $-V_{BB}$, the input resistors, and the resistor R_4 are so chosen that with input voltages of near zero at A, B and C the base–emitter junction of transistor Tr_1 is reverse biased. The transistor is therefore cut off, and the output at X will be very nearly V_{CC} or logic "1", depending on the external loading of the gate; obviously this loading must be restricted or the voltage at X will be considerably below V_{CC}). If any one of the inputs has a sufficient positive voltage applied to it then the transistor will be turned "on" and will saturate, with the result that the output voltage

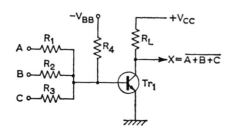

Fig. 27.7 THREE-INPUT RESISTOR-TRANSISTOR-LOGIC "NOR" CIRCUIT

falls to almost zero. Thus, if A or B or C is "up" in voltage (i.e. logic "1") the output at X is "down" (i.e. logic "0") and the NOR function has been produced.

Note that, if only one input is used and the others are left unconnected, then the circuit acts as an invertor or NOT circuit, i.e. the output is "up" if the input is "down" and vice versa. A NOR followed by a NOT gives an OR circuit.

In some RTL circuits each input is applied through a resistor direct to the base of a separate transistor (hence there are as many transistors as inputs). The transistors all have one common-collector load resistor from which the output is taken. This form of circuit has the advantage that it does not require a separate negative supply rail. The added complexity of additional transistors does not increase the relative cost of integrated circuits in the proportion that it would do in discrete circuits.

The relatively high propagation delay time of RTL circuits arises because of (a) the combination of input resistors and the transistor base–emitter capacitance, giving rise to a CR delay, and (b) the fact that the transistor saturates (giving high turn-off times).

RTL circuits, which were the first commercially available, are not now common in new equipments.

27.7 Diode-Transistor Logic (DTL)

A typical 3-input DTL NAND gate is shown in Fig. 27.8. Again positive logic is used. Only if all the inputs, A, B and C are positive together will none of the diodes D_1, D_2, or D_3 conduct. In this case point K will be at a positive potential, D_4 and the base–emitter junction of transistor Tr_1 will conduct, Tr_1 will saturate and the output at X will be almost zero (typically 0.2 V). The voltage between K and

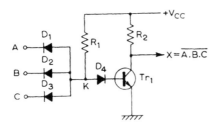

Fig. 27.8   THREE-INPUT DIODE-TRANSISTOR-LOGIC "NAND" GATE

earth will be 1.4 V (twice the mean forward conducting voltage of a silicon junction diode, since two junctions are in series between K and earth). This may be increased in 0.7 V steps by adding diodes in series with D_4 in order to increase the noise margin.

If any input voltage goes "down", the corresponding input diode conducts and the voltage at K becomes approximately 0.7 V. D_4 and the transistor are then both cut off and the output voltage rises to almost $+V_{CC}$ depending on the output loading. This represents the logic NAND operation. If only one input is connected the circuit acts as a simple invertor.

The propagation delay time is less than for RTL on account of the low forward resistance of the input diodes.

In integrated-circuit form, DTL may be cheaper than RTL since diodes are more readily produced than resistors.

21.8 Transistor–Transistor Logic (TTL)

The circuit of a multi-emitter TTL NAND gate is shown in Fig. 27.9. This form of emitter has been developed with integrated circuits and provides considerable flexibility in circuit design. The second transistor, Tr_2, provides a power output stage and permits a high fan-out.

With each input "up" (i.e. at a positive voltage, or logic "1"), the emitter–base junctions of Tr_1 are reverse biased, but the collector junction is forward biased, so permitting the flow of base current to transistor Tr$_2$, which saturates. The output voltage is therefore nearly zero (typically 0·2 V).

If any of the inputs A, B or C goes "down" (logic "0") to almost zero, then the corresponding emitter–base junction of Tr$_1$ is forward biased. The base voltage of Tr$_1$ becomes almost zero, and hence so does the base voltage of Tr$_2$, which cuts off, so that the output rises (logic "1").

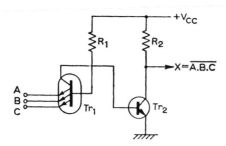

Fig. 27.9   THREE-INPUT TRANSISTOR-TRANSISTOR-LOGIC "NAND" GATE

Thus only if A AND B AND C are all in the logic "1" state together will the output be at logic "0"—this is the NAND gate condition, i.e.

$$X = \overline{A \cdot B \cdot C}$$

Alternative connexions allow for TTL NOR gates, while the inclusion of invertors allows AND and OR operations to be performed.

27.9 Emitter-Coupled Logic (ECL)

Because it operates in the non-saturated condition the ECL family of logic circuits provides the lowest propagation delay times achieved so far commercially, but at the expense of higher cost and increased power dissipation per gate. When designing a logic system, the usual engineering compromises are thus seen to be necessary, balancing speed of operation and overall performance, etc., against cost per gate.

A basic ECL NOR gate is shown in Fig. 27.10. The emitter resistor, R_3, is assumed to be large enough to give almost constant-current operation for transistors Tr$_1$–Tr$_4$. With inputs A, B and C all at about zero volts (logic "0"), Tr$_1$, Tr$_2$ and Tr$_3$ are all cut off so that their common collector leads rise towards V_{CC} volts. This turns transistor Tr$_5$ on and gives an output voltage.

Transistor Tr$_6$ and diodes D$_1$ and D$_2$ provide a constant reference voltage, V_{ref}, at the base of Tr$_4$. If, now, any one of the inputs rises above V_{ref}, that input transistor is turned on (current is still limited to below the saturation value by R_3) and the collector voltage falls so that Tr_5 is cut off and the output falls to nearly zero (logic "0"). This describes the logic NOR function. The reference voltage defines the circuit threshold voltage, but since non-saturated operation is used the on and off voltage levels are less precisely defined. The emitter-follower output gives the possibility of a very large maximum fan-out.

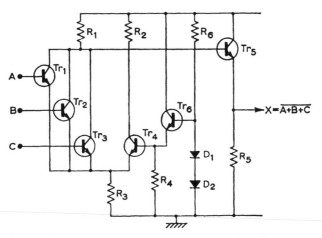

Fig. 27.10   THREE-INPUT EMITTER-COUPLED-LOGIC "NOR" GATE

ECL circuits are sometimes supplied with OR outputs but are not suitable for generating NAND or AND functions.

PROBLEMS

A knowledge of binary arithmetic is NOT required in solving these problems.

27.1   The logic equations of a binary half-adder, which can be used in computer circuits as part of the process of binary addition for two inputs A and B, and a "carry" function C, are as follows.

Sum $= S = \bar{A}.B + A.\bar{B}$ Carry $= C = A.B$

Show that S can be represented by $(A + B).\overline{AB}$. Hence or otherwise construct the logic circuits required to realize S and C using AND, OR and NOT logic elements.

27.2   Repeat Problem 27.1 using NOR logic elements only.

27.3   The truth table for an "exclusive OR" or "modulo 2 adder" is given below. The inputs are A and B, and a logic "1" output is obtained only when A and B

have different logic states. Using any logic gates you wish, construct a suitable logic circuit to realize this function.

A	B	$X = A.\bar{B} + \bar{A}.B$
0	0	0
1	0	1
0	1	1
1	1	0

27.4   Repeat Problem 27.3 using NOR elements only.

27.5   In a 4-bit coincidence detector a logic "1" output is required if and only if the inputs A, B, C and D are equal to four reference inputs A_r, B_r, C_r and D_r respectively. Obtain a suitable logic circuit using (*a*) NAND gates, (*b*) NOR gates.

Chapter **28**

THE RELIABILITY OF ELECTRICAL AND ELECTRONIC EQUIPMENT

The concept of reliability has always been associated, in a qualitative way, with good design, endurance, consistent quality and dependability. In recent years, however, the much greater complexity of electrical and electronic equipment and the seriousness of a failure in the system have made it necessary to attempt not only to improve the reliability of equipment but also to assess it in quantitative terms.

In order to appreciate some of the difficulties which are involved in the quantitative assessment of reliability, imagine a discussion concerning the relative merits of two types of television receiver. In the first place the specifications are compared objectively and this is followed by, say, comparisons of the picture quality and styling. The discussion may then turn to the likelihood of faults developing in the sets. This is important not only because of the annoyance caused to the viewer by a failure but also because of the cost of repair. The customer should be prepared to pay a higher initial cost for a receiver in return for an assurance that the extra cost will mean smaller maintenance costs.

If the reliability of each type of set is to be compared the number of faults occurring in their operation will have to be measured. For the comparison to be meaningful the measurements will have to be made on a reasonably large sample of each type of set, operated

850

under the same environmental conditions, for the same length of time. The type of fault would also need to be considered since faults vary in their seriousness and in the maintenance costs they cause.

From these considerations alone it will be seen that the assessment of reliability is not a simple matter and it will be appreciated that the achievement of high reliability is an aim in which many people must be involved. The component manufacturer, the designers, the production team, the test and quality control engineers, the installation engineer and the customer must all contribute to this aim, and it is the object of this chapter to indicate some of the considerations which are involved.

28.1   Quantitative Measurement of Reliability

The important factors which must be included in any statement of reliability have already been mentioned, and while it is difficult to evolve a definition satisfactory in all circumstances, the following meets most requirements:

> Reliability is the characteristic of a component or of a system which may be expressed by the *probability* that it will perform a *required function* under *stated conditions* for a *specified period of time.*

There are a number of difficulties which arise when this definition is applied to the assessment of an equipment. For example, it is not always easy to specify precisely its required function, or to determine the environmental conditions in which the equipment must operate reliably.

The reliability characteristic is unlike the other equipment characteristics in that it is based on statistical concepts. Whereas, for example, the gain of an amplifier can be specified either as, say, $55 \pm 2 \, \text{dB}$ or as, say, $\ll 37 \, \text{dB}$, the reliability cannot be expressed as "this equipment will function for $1{,}000 \pm 15$ hours", or even as "this equipment will function for not less than 700 hours".

Since a probability of zero means that the event cannot happen and a probability of unity means that the event certainly will happen, a practical reliability figure will be between 0 and 1, and values which are very close to unity mean that it is very improbable that the equipment will fail.

EXAMPLE 28.1 Explain what is meant by the statement: "the probability that a certain capacitor will not fail within the next 50 hours is 0·9995".

Imagine that a very large number of similar capacitors are tested under the stated conditions for 50 hours each and that the number which have failed in that time is recorded. Then, *on average*, 5 capacitors in every 10,000 tested will have failed.

Although it is very unlikely that a selected capacitor will fail in the specified time, it is important to note that in complex equipment which uses many components it is quite likely that an equipment fault will occur within the 50 hour period. This is because it is assumed that the equipment fault will be produced by a failure in any one capacitor.

28.2 Reliability and Unreliability

If the number of components tested is N_O, the number of components which fail in time t is N_F and the number which survive is N_S, then

$$\text{Reliability}, \ R(t) = \frac{N_S}{N_O} \tag{28.1}$$

and

$$\text{Unreliability}, \ Q(t) = \frac{N_F}{N_O} \tag{28.2}$$

provided that N_O is very large.

Since $N_O = N_S + N_F$, then

$$R(t) + Q(t) = 1 \tag{28.3}$$

Thus reliability is a function which varies with time from unity at the beginning of the test ($t = 0$) to zero at a time when all the components have failed. Note that the term "failure" includes a change in a parameter to a value outside the permitted tolerance, as well as complete failures, such as short- and open-circuits. A component or system failure is *any* inability of the item to carry out its specified function.

Failures may be either *partial* or *complete*, *gradual* or *sudden*, and may be caused by an *inherent weakness* or by *misuse*.

Catastrophic failures are both sudden and complete, whereas *degradation* failures are both gradual and partial.

Primary failures are failures in components which are not caused by a failure or failures in another part of the system. *Secondary failures* are those which are caused by the failure of another part of the system.

28.3 Mean Time to Failure (MTTF)

This is a term which is applied to non-repairable parts and is a measure of the average time to failure of a large number of similar parts which operate under specified conditions. In general, the MTTF of a part will be altered by a change in the stress conditions.

For example, an increase in the operating temperature of a capacitor will reduce its MTTF.

MTTF may be calculated from the equation

$$\text{MTTF} = \frac{\text{Sum of time to failure of each component}}{\text{Number of components under test}}$$

In practice, however, the MTTF is often calculated from data taken over a period of time in which not all the components fail. In this case,

$$\text{MTTF} = \frac{\text{Total operating time for all components}}{\text{Number of failures in that time}} \qquad (28.4)$$

EXAMPLE 28.2 Five hundred parts are operated under specified stress conditions for a period of 312 h, and the following fault data were recorded.

Time from start of test, t (hours)	No. of failures during time interval, n_f	Cumulative failures, n_c	No. of survivors
0			500
	19		
24		19	481
	15		
48		34	466
	13		
72		47	453
	17		
96		64	436
	12		
120		76	424
	16		
144		92	408
	12		
168		104	396
	14		
192		118	382
	11		
216		129	371
	14		
240		143	357
	12		
264		155	345
	9		
288		164	336
	13		
312		177	323

Estimate the mean time to failure of the part when operated under the specified conditions.

The total number of operating hours for all components may be determined by using the mid-ordinate rule to find the area under the curve of component survivors plotted to a base of time, as set out in the table below.

Time from start of test (hours) (1)	Number of survivors (2)	Average no. of survivors in each 24 hour period (mid-ordinate) (3)	Total operating time for all components in each 24 hour period (4)
0	500		
		490·5	490·5 × 24
24	481		
		473·5	473·5 × 24
48	466		
		459·5	459·5 × 24
72	453		
		444·5	444·5 × 24
96	436		
		430	430 × 24
120	424		
		416	416 × 24
144	408		
		402	402 × 24
168	396		
		389	389 × 24
192	382		
		376·5	376·5 × 24
216	371		
		364	364 × 24
240	357		
		351	351 × 24
264	345		
		340·5	340·5 × 24
288	336		
		329·5	329·5 × 24
312	323		

Total operating hours for all components = Σ Col. (4) = 5266·5 × 24

From eqn. (28.4),

$$\text{Estimated mean time to failure} = \frac{5266 \cdot 5 \times 24}{177} = \underline{714 \text{ h}}$$

28.4 Mean Time Between Failures (MTBF)

This is a term which is applied to repairable items, and is a measure of the average time that a particular equipment will remain in service. The MTBF of an equipment depends on the operating stresses, including the environmental conditions, but it may be reduced by potential defects introduced by poor maintenance procedures.

If the time between failures is long compared with the repair time, and if faults occur at times $t_1, t_2 \ldots t_n$, then

$$\text{MTBF} \approx \frac{1}{n} \sum_{k=1}^{k=n} (t_k - t_{k-1}) = \frac{t_n - t_0}{n} = \frac{t_n}{n}$$

since $t_0 = 0$. Hence

$$\text{MTBF} = \frac{\text{Total operating time}}{\text{Number of failures in that time}} \qquad (28.5)$$

Comparing eqns (28.4) and (28.5) it can be seen that

$$\text{MTTF} = \text{MTBF} = \bar{m} = \frac{\text{Total operating time}}{\text{Number of failures in that time}} \qquad (28.6)$$

It should be clear that a significant number of faults must be recorded in order to obtain a high confidence that the measured value of the MTBF is close to the true value. If, for example, an equipment which has an MTBF of 4,000 h is tested for only 1,000 h, there is a high probability that no failure will occur.

Instead of testing one equipment for a very long time, which would be impracticable in most cases, it is usual to test a number of equipments simultaneously for a shorter period each, and to determine the total number of faults in the total operating time of all the equipments. It must be understood that this method assumes that failures occur by chance at any time during the test and that wearout failure has not occurred.

28.5 Failure Rate

Although *failure rate* is related to the number of failures per unit time, it is not defined simply as that number, because the number of items which fail in a given time depends, not only on the quality of the item and on the stresses applied to it, but also on the number of the components which are in operation. If the number of components in operation at the time of a failure is N_s the failure rate $\lambda(t)$ is defined by

$$\lambda(t) = \lim_{\Delta t \to 0} \frac{1}{N_s} \frac{\Delta N_F}{\Delta t} = \frac{1}{N_s} \frac{dN_F}{dt} \qquad (28.7)$$

where ΔN_F is the number of failures which occur in the time Δt.

The way in which $\lambda(t)$ varies with time depends on weaknesses in the part or equipment and on the operating stresses. In many cases the variation of $\lambda(t)$ takes the form shown in Fig. 28.1, which is often called the *bath-tub* curve. The curve shows three distinct phases in the life of the equipment. The first phase, the *early failure period*, is the time when very weak components fail. The weakness of these components is due to inevitable minor defects in the materials and in the manufacturing processes. These components are commonly removed by testing for a time t_1, with the result that the operational reliability of the equipment is improved. The second phase, the *constant failure rate period*, is the time during which

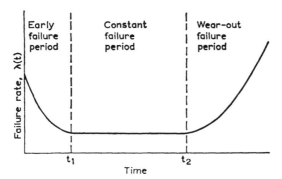

Fig. 28.1   THE BATH-TUB CURVE

the component or equipment is most usefully employed since the failure rate is lowest and the failures which occur are due only to fortuitous combinations of high stresses. This period is terminated by the *wear-out failure period* beginning at time t_2.

Preventative maintenance should be carried out before the time t_2 and the appropriate part replaced.

28.6 Constant Failure Rate Period

It will be shown that, if the failure rate is constant, then there is a simple relationship between the reliability at time t and the failure rate. For the purpose of this analysis, failures occurring in the early failure period have been eliminated and it has been assumed that the constant failure rate régime prevails, so that operational time is measured from t_1; i.e. at time t_1 in Fig. 28.1, $t = 0$. Then

$$\lambda(t) = \text{constant} = \lambda \text{ for } t < t_2 - t_1.$$

From eqn. (28.7), $\lambda = \dfrac{1}{N_s} \dfrac{dN_F}{dt}$

where N_S is the number which have survived up to time t, N_F is the number which have failed up to time t, and $N_0 = N_S + N_F$, where N_0 is the total number in operation at $t = 0$. Hence

$$\lambda = \left(\frac{1}{N_O - N_F}\right) \frac{dN_F}{dt}$$

or

$$\int_0^t \lambda \, dt = \int_0^{N_F} \frac{dN_F}{N_O - N_F}$$

Therefore

$$-\lambda t = \left[\log_e (N_O - N_F)\right]_0^{N_F}$$

and

$$e^{-\lambda t} = 1 - \frac{N_F}{N_O} \tag{28.8}$$

From eqn. (28.1),

$$R(t) = \frac{N_S}{N_O} = \frac{N_O - N_F}{N_O} \tag{29.9}$$

$$= 1 - \frac{N_F}{N_O} \tag{28.10}$$

Comparing eqns. (28.8) and (28.10), evidently

$$R(t) = e^{-\lambda t} \tag{28.11}$$

This equation does not apply to the early failure period nor to the wear-out period; a discussion of the reliability during these periods is outside the scope of this book.

28.7 Relation between MTTF, MTBF and Failure Rate

If failures are due to chance and if the failure rate is constant, then it is immaterial whether one equipment is tested for T hours or N equipments are each tested for T/N hours, since the probability of failure in a specified time will be the same in each case.

Also from eqn. (28.7),

$$\lambda = \frac{1}{N_S} \frac{dN_F}{dt} \qquad \text{so that} \qquad \int_0^t N_S \, dt = \frac{1}{\lambda} \int_0^{N_F} dN_F = \frac{N_F}{\lambda}$$

where N_F failures occur in time t. From eqn. (28.6),

$$\bar{m} = \frac{\text{total operating time}}{\text{number of failures}}$$

Hence

$$\bar{m} = \frac{1}{N_F} \int_0^t N_S \, dt = \frac{1}{\lambda} \tag{28.12}$$

It follows that the constant failure rate λ is

$$\lambda = \frac{1}{\text{MTTF}} \quad \text{for non-repairable parts}$$

or

$$\lambda = \frac{1}{\text{MTBF}} \quad \text{for repairable parts or equipments}$$

The basic unit of mean time used in reliability calculations is the hour, and the unit of failure rate is therefore the per-unit failures per hour. Because this is a very large unit, failure rate is also expressed as percentage failures in 1,000 h, as per-unit failures in 1,000,000 h, or as parts failing per 1,000,000 parts in 1 h.

EXAMPLE 28.3 One thousand similar equipments which are known to have constant failure rates of 5 per cent per 1,000 h are put into operation at the same time. Calculate the predicted times which will elapse before (a) 50 and (b) 500 equipments have failed in service.

The reliability is given by eqns. (28.10) and (28.11) as

$$R(t) = e^{-\lambda t} = 1 - \frac{N_F}{N_O}$$

$$\lambda = \frac{5}{100} \times \frac{1}{1,000} = 5 \times 10^{-5}$$

(a) Let t_1 hours be the time required for 50 failures to occur. Then

$$R(t_1) = \exp(-5 \times 10^{-5} t_1) = 1 - \frac{50}{1,000}$$

$$\exp(-5 \times 10^{-5} t_1) = 0.95$$

$$5 \times 10^{-5} t_1 = \log_e \left(\frac{1}{0.95} \right) \qquad t_1 = 1,020 \, \text{h}$$

Since $R(t_1) \approx 1$, in this case an approximate solution may be obtained by writing $e^{-\lambda t} = 1 - \lambda t$.

(b) Let t_2 hours be the time required for 500 failures to occur. Then

$$\exp(-5 \times 10^{-5} t_2) = 1 - \frac{500}{1,000}$$

so that

$$5 \times 10^{-5} t_2 = \log_e 2 \qquad \text{and} \qquad t_2 = \underline{\underline{13,900 \, \text{h}}}$$

Note that since 50 out of 1,000 equipments fail during the first 1,000 h, we may say that the probability of an equipment failing in that time is 5 per cent. Also, the probability is 50 per cent that an equipment will fail during the first 13,900 h.

While it is impossible to eliminate failures entirely the chances of a failure may be made very small by designing the equipment to have a very large MTBF compared with its operating time. In the example above, if the operating time of the equipment were only 1 hour then the reliability $R(t) \approx 1 - 5 \times 10^{-5}$.

This means that, on average, there will be only 50 failures in every million equipments, and it follows that most groups of 1,000 equipments will not have any failures during the first hour.

The acceptable value for the reliability of an equipment at a given time depends on many factors, such as human safety, cost of repair and operational usefulness.

There are many examples of the relationship between reliability and human safety, but the most dramatic is that of astronauts. In this case the unreliability of the component parts of the system must be made so low that failure becomes only a remote possibility.

Repeaters which are used in underwater telecommunication systems must operate without failure for a considerable time compared with the duration of a space mission. In this case it is the very high cost of repair and the cost of the *down time* (the period during which the equipment is non-operational) that dictate the need for high reliability.

A third example, that of a digital computer used for scientific or business applications, is also built with a high reliability specification. However, neither safety nor very high repair costs are important in this case, but the cost of down time in large computers is considerable, and thousands of calculations must be made without error; thus operational usefulness is the important characteristic in this case.

28.8 Series Reliability

If a number of parts of a system are operated in such a way that the failure of any one part causes a failure of the system, then those parts are considered functionally to be in series. If a failure of any part is independent of the operation of the other parts then the reliability of the system is given by the product of the reliabilities of the parts. Thus,

$$R(t) = R_1(t) \times R_2(t) \times R_3(t) \ldots R_k(t) \ldots R_n(t) \qquad (28.13)$$

where $R(t)$ is the system reliability and $R_k(t)$ is the reliability of the kth part at a time t.

If, in particular, λ is constant then

$$R_k(t) = e^{-\lambda_k t}$$

From eqn. (28.13),

$$
\begin{aligned}
R(t) &= e^{-\lambda_1 t} \times e^{-\lambda_2 t} \ldots e^{-\lambda_k t} \ldots e^{-\lambda_n t} \\
&= e^{-(\lambda_1 + \lambda_2 + \cdots + \lambda_k \cdots + \lambda_n)t} \qquad\qquad (28.14) \\
&= e^{-\lambda t}
\end{aligned}
$$

where

$$\lambda = \lambda_1 + \lambda_2 \ldots + \lambda_k \ldots + \lambda_n \qquad\qquad (28.15)$$

is the system failure rate.

Note that the condition of independence is not always fulfilled. A secondary failure is an example of one failure occurring because of another.

28.9 Parallel Reliability

An important method of improving the reliability of a system is the use of redundancy, i.e. the inclusion of additional equipment in such a way that a failure of one part does not cause a failure of the whole system. Redundancy may be used in the parts of an equipment or in the complete system. For example, if two components are connected in parallel the equipment can be designed to function with an open-circuit in one of the components. However, in this case, the stresses in the good component may be increased, the equipment performance may be degraded, and the faulty component must be repaired or replaced quickly if the advantages of redundancy are to be restored. As a second example, consider the use of computers in an air traffic control system. For reasons of safety it is necessary that the computing system should function continuously, and however high the reliability of one computer, these requirements can only be met by duplication of the computer.

If the reliability of each part is assumed to be independent of the other parts then the probability of a system failure is

$$Q(t) = Q_1(t) \times Q_2(t) \times Q_3(t) \ldots Q_k(t) \ldots Q_n(t) \qquad (28.16)$$

where $Q(t)$ is the system unreliability and $Q_k(t)$ is the part unreliability. Since $R(t) = 1 - Q(t)$ the system reliability is easily calculated.

EXAMPLE 28.4   After 100 h an equipment has a reliability of 0·75. Calculate the overall reliabilities which would be obtained if the equipment were (a) duplicated and (b) triplicated.

Since $R_1(100) = 0·75$, $Q_1(100) = 0·25$. Assume that

$$Q_1(t) = Q_2(t) = Q_3(t)$$

(a) $$Q_D(100) = [Q_1(100)]^2 = 0·0625$$

Therefore

$$R_D(100) = 1 - Q_D(100) = \underline{\underline{0·9375}}$$

(b) $$Q_T(100) = [Q_1(100)]^3 = 0·0156$$

Therefore

$$R_T(100) = \underline{\underline{0·9844}}$$

Hence, if an equipment has a reliability of 75 per cent, two such equipments used in such a way that the system fails only if both fail, will have a reliability of 93·8 per cent, and three equipments a reliability of 98·4 per cent.

28.10   Environmental Conditions

Most components or equipments having a high reliability when operated under well-controlled laboratory conditions are found to have a greater failure rate when they are subjected to increased stress of one kind or another. The following environmental conditions for both storage and operation of components and equipments must always be taken into account.

Extremes of *temperature* invariably reduce reliability because of (a) the effects of expansion and contraction of materials with change of temperature, (b) changes of component values, (c) melting, softening or freezing of some component materials, and (d) the effect of temperature on chemical action.

High humidity is another cause of increased failure rate, particularly when it is associated with high temperatures. Under these conditions a thin film of water can form on a component, and printed-circuit board surfaces can become ionized and thus form a conducting path. This path, together with the capacitive path provided by the high relative permittivity of water, is likely to cause a component defect. Moisture can also enter equipment by diffusing through a material of which it is made or by entry through a hole in the sealing of the equipment.

Other effects which may be associated with the temperature and humidity effects are *dust, air pressure, salt spray* and *mould growth.* Dust can reduce surface insulation resistance and spoil the performance of lubricants; low pressure, by causing the release of water vapour or other trapped gases, can change the electrical properties of the component; it can also affect cooling by convection. Salt spray may produce rapid corrosion.

Vibration is a serious cause of unreliability unless the equipment is very carefully designed to cope with it. The important characteristics of the vibration are the amplitude and frequency of its periodic components. The mechanical resonance of parts in the equipment must be designed to be at frequencies outside the range of the expected vibrations and they must be suitably damped.

28.11 Reliability Assessment

From what has already been written it should be clear that no single value of failure rate can be given for a component because the value changes under different stress conditions. Although a simple relationship between failure rates at two stress levels should not be taken for granted, it is found that in many cases the failure rate at one stress condition may be calculated from the failure rate at another by the use of weighting factors. Three important weighting factors are (1) *environment weighting factor* (EWF), (2) *rating weighting factor* (RWF) and (3) *temperature weighting factor* (TWF). Hence

Failure rate = Basic failure rate $\times$ EWF $\times$ RWF $\times$ TWF   (28.17)

Figures for basic failure rates and for various weighting factors have not been included in this book because it is important always to use the latest available information. This should be obtained from recently published papers, and from manufacturers' technical information services.

The temperature at which a component operates will depend on (i) the ambient temperature, (ii) self-heating, (iii) heating from other components in the equipment, and (iv) the method of cooling used. Devices such as resistors and transistors, which are given a specified power rating at 25°C ambient temperature, must be derated as the ambient temperature is increased.

The method of calculating the power dissipated in a device must take into account the permitted tolerances of all the components in the circuit and of the supply. The calculation is usually based on either *worst-case design* or *statistical design*.

In the worst-case method, all the component values are assumed to be at an extreme acceptable value and are so chosen that the circuit is most likely to fail. This idea will be illustrated by a very simple example.

EXAMPLE 28.5 A d.c. power supply which has an output of 24 V $\pm$ 10 per cent is connected to a resistor $R_1 = 100\ \Omega \pm 20$ per cent in series with a resistor $R_2 = 200\ \Omega \pm 10$ per cent. Determine the worst-case dissipation of power in R_2.

If the power dissipated in R_2 is P_2, then

$$P_2 = \frac{V^2 R_2}{(R_1 + R_2)^2}$$

where $V = V_0(1 \pm a) = 24(1 \pm 0{\cdot}1)$ V

$$R_1 = R_{10}(1 \pm a_1) = 100(1 \pm 0{\cdot}2)\ \Omega$$
$$R_2 = R_{20}(1 \pm a_2) = 200(1 \pm 0{\cdot}1)\ \Omega$$

Hence

$$P_2 = \frac{V_0^2 R_{20}(1 \pm a)^2(1 \pm a_2)}{(R_{10} + R_{20} \pm \Delta R_1 \pm \Delta R_2)^2}$$

$$= P_{20}\frac{(1 \pm a)^2(1 \pm a_2)}{\left[1 \pm \left(\dfrac{R_{10}}{R_{10} + R_{20}}\right)a_1 \pm \left(\dfrac{R_{20}}{R_{10} + R_{20}}\right)a_2\right]^2}$$

where $P_{20} = V_0^2 R_{20}/(R_{10} + R_{20})^2$.

The power in R_2 will be a maximum when V is greatest and when R_1 is least, but the correct limit for R_2 is not so obvious. However, from the maximum power transfer theorem (Section 2.4) it can be deduced that the power in R_2 is greatest when the value of R_2 is closest to the value of R_1, and in this case the lower limit of R_2 should be chosen. Hence

$$P_2 = \left\{\begin{array}{l}\text{Power dissipated by } R_2 \\ \text{when all components} \\ \text{have nominal values}\end{array}\right\} \times \text{Tolerance factor}$$

In this example,

$$P_{20} = \frac{24^2 \times 200}{300^2} = 1{\cdot}28 \text{ W}$$

and

$$\text{Tolerance factor} = \frac{1{\cdot}1^2 \times 0{\cdot}9}{(1 - 0{\cdot}067 - 0{\cdot}067)^2} = 1{\cdot}45$$

Hence

$$(P_2)_{worst\ case} = 1{\cdot}28 \times 1{\cdot}45 = \underline{\underline{1{\cdot}85 \text{ W}}}$$

Thus the worst-case power dissipation in R_2 is 45 per cent higher than the power dissipated if all the components and the supply had actual values equal to the nominal design values.

The solution given above illustrates a general method. In this particular case a more direct method could have been used.

The disadvantage of using worst-case design is that components with high ratings must be used to take into account the unlikely event that all the tolerances are at their extreme values. Some reduction in the ratings is obtained by using components with closer tolerances, and where this is not possible, as with transistors, by using special techniques which make the circuit operation largely independent of characteristics of the device.

However, the cost of a component and the probability of a degradation failure are both increased by a reduction in the permitted tolerance. It is for these reasons that statistical design should be considered. In this case it is assumed that extreme values will not occur in the same circuit, and the design is based on a calculated probability that certain ratings will not be exceeded. A study of this method of design is beyond the scope of this book.

28.12 Maintainability

The number of times that an equipment becomes faulty is a direct function of its unreliability, but the length of time for which an equipment is not available is a function both of unreliability and of the time it takes to repair it and return it to service. *Maintainability* is a characteristic of an equipment related to the ease with which it can be repaired. There are four parts to the problem of achieving good maintainability:

(i) The method of determining that a defect exists. It may be obvious to a viewer that a defect exists in his television receiver, but it may not be so obvious that the arithmetic unit in a computer is producing some wrong answers.

(ii) The method of quickly identifying a defective component or assembly. The solution to this problem includes the training of servicemen, provision of test apparatus and test procedures, clear identification of assemblies and of parts, provision of test points and also of purpose-built test equipment. It is very important that designers should consider maintainability from the beginning of the project and not add service extras to the equipment as an after-thought.

(iii) Rectifying the fault, which may include replacement of a defective part. All components should be made as accessible as possible, particularly those known to have a high failure rate. Wiring looms should not be positioned so that they prevent the easy removal of an assembly, and if a large number of soldered connections have to be removed then this operation may well produce subsequent faults in the equipment. Preset controls should be sited where adjustments may be made without undue difficulty, but on the other hand they should not be directly accessible because they would then be open to misuse.

(iv) Finally it will be necessary to verify that the system is functioning correctly after the completion of the repair.

A useful definition, of maintainability is:

"Maintainability is the probability that a device will be restored to operational effectiveness within a given period of time when the maintenance action is performed in accordance with prescribed procedures".*

28.13 Reliability Costs

It should be clear that, in general, designing an equipment for high reliability and building a system with a guaranteed reliability will add to the initial cost. However, this higher initial cost will be offset, to a greater or lesser extent, by reduced maintenance costs for any given degree of maintainability. Reduced maintenance costs arise not only because of a reduction in man-hours devoted to maintenance but also due to a reduction in the provision of test equipment and stock of spares required.

The graph of costs against reliability takes the form shown in Fig. 28.2.

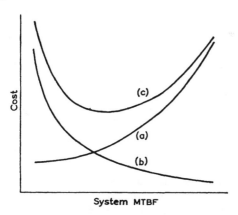

Fig. 28.2   THE COST OF UNRELIABILITY
(a) Initial cost
(b) Maintenance cost
(c) Total cost

* CALABRO, S. R., *Reliability Principles and Practice*, Chap. 9 (McGraw-Hill, 1962).

It is obvious that minimum total cost is not always the most important consideration and that safety must be a first priority.

A part of the cost of achieving high reliability is bound up with the problems of testing components and equipments and analysing these data. If only those components which have been adequately tested and approved are used in a design, then one part of this cost will be kept to a minimum and it will be possible to obtain increased confidence in the reliability of the components.

High reliability is possible only if everyone connected with the equipment plays his part. A reliability statement must be regarded as a part of the specification as important as, say, the function details, and must be designed into the system from the outset. A realization of the relation between the part and its probable total environment is very important. This includes, not only the environmental factors considered above, but also the important part that *human engineering* should play. The position of the equipment in relation to other equipments, the design and position of the controls, the type of indicators and the methods of labelling are all obvious considerations but they are all too often overlooked.* All production personnel should be trained to understand fully the importance of reliability. Components and assemblies may be damaged by poor handling, and bad joints are more easily corrected during manufacture than later, during test.

A programme of *tests* and *quality control* must be carefully considered and carried out by trained technicians of high professional integrity. Also, care must be exercised to see that stresses set up during the tests do not make the components potentially unreliable.

The MTBF of an equipment will be reduced seriously unless sufficient attention is paid to the hazards of *transport* and *installation*. Consideration must also be given to various environmental conditions during transport and storage as well as to the need to protect the equipment from mechanical damage. Rules for installation should be worked out with the known variations in human characteristics kept in mind. *Installation engineers* should not have to be weight lifters, contortionists or giants in order to do their job, and as with other members of the team, they should be made aware of the importance of their part in ensuring the reliability of the system. In particular, they should pay attention to problems of dust, cooling and the protection of interconnecting cables. The *customer* is an important link in the reliability chain because incorrect operation of the

---

* DUMMER, G. W. A, and WINTON, R. C., *An Elementary Guide to Reliability* (Pergamon Press, 1967).

equipment will usually reduce its reliability. This again is a matter of training; but a good instruction manual and sufficient attention to the ergonomic problems involving the use of the equipment are also of importance. This means that attention should be given to operator fatigue and error due, for example, to the operator having to stand rather than sit, to controls and dials located in awkward positions, to excessive noise, to too little or too much light or heat, to draughts and vibration. While it is not always possible to achieve a perfect operator environment, a considerable improvement can be made by giving the problem careful thought. The term *operability* is used to describe that characteristic of an equipment which is measured by the probability of an operator not making an error in the use of the equipment.

Finally the importance of accurate recording of *fault data*, and the speedy *feedback of information* to the designer, cannot be over-stated. It is by these means that the designer can investigate apparent weaknesses in his equipment, and then take remedial action. The report form should be designed with care. It should be as simple as possible, but it should include questions which will give the essential information about the type of defect found, the environmetal conditions, and whether the failure is a primary or secondary one.

PROBLEMS

28.1 A radio receiver has an MTBF of 1,500 h. What is the probability of a failure occurring during the period of the Wimbledon tennis championships?

Assume that the receiver is functioning at the beginning of the period and that it is switched on for a total of 40 h. State any other assumptions which must be made.

Ans. 0·0263.

28.2 A reliability test was made on a batch of 2,000 similar paper-dielectric capacitors at a temperature of 70°C and at the rated voltage. The number of survivors at the end of the 100th hour was 1,960, and at the end of the 200th hour was 1,930. What was the average failure rate over the second 100 h period? Why is the measurement of low failure rate a costly procedure?

Ans. 15·5 per cent per 1,000 h.

28.3 If the failure rate of the capacitors in Problem 28.2 is assumed to be constant, calculate the reliability of the capacitors at the end of the first 500 h period of test. Why is this assumption of constant failure rate probably wrong? Is the true reliability during the first 500 h period likely to be greater or less than that calculated in the first part of the problem?

Ans. 0·92.

28.4 A particular circuit which consists of two transistors and six resistors is operated under conditions for which the MTTF of the transistors is 2×10^6 h and that of the resistors is 10^6 h. Calculate the MTBF of the circuit. What is the reliability of the circuit after 9,000 h?

   Ans. 143,000 h; 0·939.

28.5 An equipment uses 20 of the circuits described in Problem 28.4. Calculate the reliability of the equipment after 9,000 h assuming that a defect in any one circuit causes a failure in the equipment.

   Ans. 0·284.

Index